DATE DUE

PRINTED IN U.S.A.

THIRD EDITION

Electric Power Distribution Engineering

THIRD EDITION

Electric Power Distribution Engineering

Turan Gönen

CRC Press
Taylor & Francis Group
Boca Raton   London   New York

CRC Press is an imprint of the
Taylor & Francis Group, an **informa** business

MATLAB® is a trademark of The MathWorks, Inc. and is used with permission. The MathWorks does not warrant the accuracy of the text or exercises in this book. This book's use or discussion of MATLAB® software or related products does not constitute endorsement or sponsorship by The MathWorks of a particular pedagogical approach or particular use of the MATLAB® software.

CRC Press
Taylor & Francis Group
6000 Broken Sound Parkway NW, Suite 300
Boca Raton, FL 33487-2742

© 2014 by Taylor & Francis Group, LLC
CRC Press is an imprint of Taylor & Francis Group, an Informa business

No claim to original U.S. Government works

Printed on acid-free paper
Version Date: 20131023

International Standard Book Number-13: 978-1-4822-0700-2 (Hardback)

Visit the Taylor & Francis Web site at
http://www.taylorandfrancis.com

and the CRC Press Web site at
http://www.crcpress.com

To an excellent engineer,
a great teacher, and a dear friend,
Dr. David D. Robb

and

In the memory of another
great teacher, my father
Hajruddin Muezinovich

There is a Turkish proverb to the effect that
"the world belongs to the dissatisfied."
I believe in this saying absolutely.
For me the one great underlying principle
of all human progress is that "divine discontent"
makes men strive for better conditions
and improved methods.

Charles P. Steinmetz

A man knocked at the heavenly gate
His face was scared and old.
He stood before the man of fate
For admission to the fold.
"What have you done," St. Peter asked
"To gain admission here?"
"I've been a distribution engineer, Sir," he said
"For many and many a year."
The pearly gates swung open wide;
St. Peter touched the bell.
"Come in and choose your harp," he said,
"You've had your share of hell."

Author Unknown

Life is the summation of confusions.
The more confused you are, the more alive you are.
When you are not confused any longer,
You are dead!

Turan Gönen

Contents

Preface

Today, there are many excellent textbooks that deal with topics in power systems. Some of them are considered to be classics. However, they do not particularly address, nor concentrate on, topics dealing with electric power distribution engineering. Presently, to the best of this author's knowledge, the only book available in the electric power systems literature that is totally devoted to power distribution engineering is the one by the Westinghouse Electric Corporation entitled *Electric Utility Engineering Reference Book—Distribution Systems*. However, as the title suggests, it is an excellent reference book but unfortunately not a textbook. Therefore, the intention here is to fill the gap, at least partially, that has existed so long in the power system engineering literature.

This book has evolved from the content of courses that have conducted by the author at the University of Missouri at Columbia, the University of Oklahoma, and Florida International University. It has been written for senior-level undergraduate and beginning-level graduate students, as well as practicing engineers in the electric power utility industry. It can serve as a text for a two-semester course, or by a judicious selection, the material in the text can also be condensed to suit a single-semester course.

Most of the material presented in this book was included in my book entitled *Electric Power Distribution System Engineering*, which was published by McGraw-Hill in 1986 previously. The book includes topics on distribution system planning, load characteristics, application of distribution transformers, design of subtransmission lines, distribution substations, primary systems, and secondary systems, voltage drop and power-loss calculations, application of capacitors, harmonics on distribution systems, voltage regulation, and distribution system protection, and reliability and electric power quality. It includes numerous new topics, examples, problems, as well as MATLAB® applications.

This book has been particularly written for students or practicing engineers who may want to teach themselves and/or enhance their learning. Each new term is clearly defined when it is first introduced; a glossary has also been provided. Basic material has been explained carefully and in detail with numerous examples. Special features of the book include ample numerical examples and problems designed to use the information presented in each chapter. A special effort has been made to familiarize the reader with the vocabulary and symbols used by the industry. The addition of the appendixes and other back matter makes the text self-sufficient.

MATLAB® is a registered trademark of The Mathworks, Inc. For product information, please contact:

The MathWorks, Inc.
3 Apple Hill Drive
Natick, MA 01760-2098 USA
Tel: 508-647-7000
Fax: 508-647-7001
E-mail: info@mathworks.com
Web: www.mathworks.com

Acknowledgments

This book could not have been written without the unique contribution of Dr. David D. Robb, of D. D. Robb and Associates, in terms of numerous problems and his kind encouragement and friendship over the years. I would also like to express my sincere appreciation to Dr. Paul M. Anderson of Power Math Associates and Arizona State University for his continuous encouragement and suggestions.

I am very grateful to numerous colleagues, particularly Dr. John Thompson, who provided moral support for this project, and Dr. James Hilliard of Iowa State University; Dr. James R. Tudor of the University of Missouri at Columbia; Dr. Earl M. Council of Louisiana Tech University; Dr. Don O. Koval of the University of Alberta; Late Dr. Olle I. Elgerd of the University of Florida; and Dr. James Story of Florida International University for their interest, encouragement, and invaluable suggestions.

A special thank you is extended to John Freed, chief distribution engineer of the Oklahoma Gas & Electric Company; C. J. Baldwin, Advanced Systems Technology, Westinghouse Electric Corporation; W. O. Carlson, S&C Electric Company; L. D. Simpson, Siemens-Allis, Inc.; E. J. Moreau, Balteau Standard, Inc.; and T. Lopp, General Electric Company, for their kind help and encouragement.

I would also like to express my thanks for the many useful comments and suggestions provided by colleagues who reviewed this text during the course of its development, especially to John. J. Grainger, North Carolina State University; James P. Hilliard, Iowa State University; Syed Nasar, University of Kentucky; John Pavlat, Iowa State University; Lee Rosenthal, Fairleigh Dickinson University; Peter Sauer, University of Illinois; and R. L. Sullivan, University of Florida.

A special thank you is also extended to my students Margaret Sheridan, for her contribution to the MATLAB work, and Joel Irvine for his kind help with the production of this book.

Finally, my deepest appreciation to my wife, Joan Gönen, for her limitless patience and understanding.

Author

Turan Gönen is a professor of electrical engineering at California State University, Sacramento (CSUS). He received his BS and MS in electrical engineering from Istanbul Technical College in 1964 and 1966, respectively, and his PhD in electrical engineering from Iowa State University in 1975. Professor Gönen also received an MS in industrial engineering in 1973 and a PhD comajor in industrial engineering in 1978 from Iowa State University, as well as an MBA from the University of Oklahoma in 1980.

Professor Gönen is the director of the Electrical Power Educational Institute at CSUS. Prior to this, he was professor of electrical engineering and director of the Energy Systems and Resources Program at the University of Missouri–Columbia. He also held teaching positions at the University of Missouri-Rolla, the University of Oklahoma, Iowa State University, Florida International University, and Ankara Technical College. He has taught electrical electric power engineering for over 40 years.

Professor Gönen has a strong background in the power industry; for eight years, he worked as a design engineer in numerous companies both in the United States and abroad. He has served as a consultant for the United Nations Industrial Development Organization, Aramco, Black & Veatch Consultant Engineers, San Diego Gas & Electric Corporation, Aero Jet Corporation, and the public utility industry. He has written over 100 technical papers as well as four other books: *Modern Power System Analysis*, *Electric Power Transmission System Engineering: Analysis and Design*, *Electrical Machines*, and *Engineering Economy for Engineering Managers*.

Professor Gönen is a fellow of the Institute of Electrical and Electronics Engineers and a senior member of the Institute of Industrial Engineers. He served on several committees and working groups of the IEEE Power Engineering Society and is a member of numerous honor societies, including Sigma Xi, Phi Kappa Phi, Eta Kappa Nu, and Tau Alpha Pi. He is also the recipient of the Outstanding Teacher Award twice at CSUS in 1997 and 2009.

1 Distribution System Planning and Automation

To fail to plan is to plan to fail.

A.E. Gasgoigne, 1985

Those who know how can always get a job, but those who know why, may be your boss!

Author Unknown

To make an end is to make a beginning. The end is where we start from.

T.S. Eliot

1.1 INTRODUCTION

The electric utility industry was born in 1882 when the first electric power station, Pearl Street Electric Station in New York City, went into operation. The electric utility industry grew very rapidly, and generation stations and transmission and distribution networks have spread across the entire country. Considering the energy needs and available fuels that are forecasted for the next century, energy is expected to be increasingly converted to electricity.

In general, the definition of an electric power system includes a generating, a transmission, and a distribution system. In the past, the distribution system, on a national average, was estimated to be roughly equal in capital investment to the generation facilities, and together they represented over 80% of the total system investment [1]. In recent years, however, these figures have somewhat changed. For example, Figure 1.1 shows the investment trends in electric utility plants in service. The data represent the privately owned class A and class B utilities, which include 80% of all the electric utility in the United States. The percentage of electric plants represented by the production (i.e., generation), transmission, distribution, and general plant sector is shown in Figure 1.2. The major investment has been in the production sector, with distribution a close second. Where expenditures for individual generation facilities are visible and receive attention due to their magnitude, the data indicate the significant investment in the distribution sector.

Production expense is the major factor in the total electrical operation and maintenance (O&M) expenses, which typically represents two-thirds of total O&M expenses. The main reason for the increase has been rapidly escalating fuel costs. Figure 1.3 shows trends in the ratio of maintenance expenses to the value of plant in service for each utility sector, namely, generation, transmission, and distribution. Again, the major O&M expense has been in the production sector, followed by the one for the distribution sector.

Succinctly put, the economic importance of the distribution system is very high, and the amount of investment involved dictates careful planning, design, construction, and operation.

1.2 DISTRIBUTION SYSTEM PLANNING

System planning is essential to assure that the growing demand for electricity can be satisfied by distribution system additions that are both technically adequate and reasonably economical. Even though considerable work has been done in the past on the application of some types of systematic

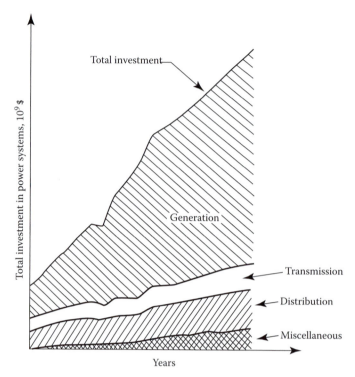

FIGURE 1.1 Typical investment trends in electric utility plants in service.

approach to generation and transmission system planning, its application to distribution system planning has unfortunately been somewhat neglected. In the future, more than in the past, electric utilities will need a fast and economical planning tool to evaluate the consequences of different proposed alternatives and their impact on the rest of the system to provide the necessary economical, reliable, and safe electric energy to consumers.

The objective of *distribution system planning* is to assure that the growing demand for electricity, in terms of increasing growth rates and high load densities, can be satisfied in an optimum way by additional distribution systems, from the secondary conductors through the bulk power substations, which are both technically adequate and reasonably economical. All these factors and others, for example, the scarcity of available land in urban areas and ecological considerations, can put the problem of optimal distribution system planning beyond the resolving power of the unaided human mind.

Distribution system planners must determine the load magnitude and its geographic location. Then the distribution substations must be placed and sized in such a way as to serve the load at maximum cost effectiveness by minimizing feeder losses and construction costs, while considering the constraints of service reliability.

In the past, the planning for other portions of the electric power supply system and distribution system frequently has been authorized at the company division level without the review of or coordination with long-range plans. As a result of the increasing cost of energy, equipment, and labor, improved system planning through use of efficient planning methods and techniques is inevitable and necessary.

The distribution system is particularly important to an electrical utility for two reasons: (1) its close proximity to the ultimate customer and (2) its high investment cost. Since the distribution system of a power supply system is the closest one to the customer, its failures affect customer service more directly than, for example, failures on the transmission and generating systems, which usually do not cause customer service interruptions.

Therefore, distribution system planning starts at the customer level. The demand, type, load factor, and other customer load characteristics dictate the type of distribution system required.

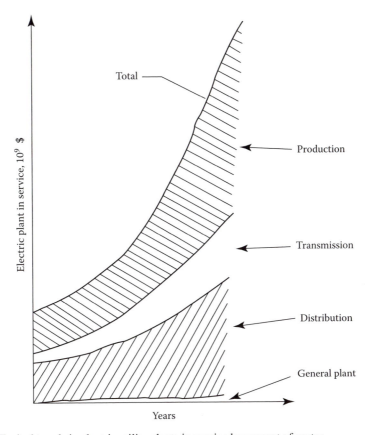

FIGURE 1.2　Typical trends in electric utility plants in service by percent of sector.

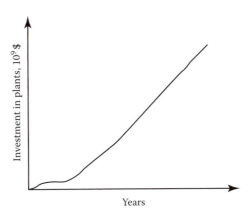

FIGURE 1.3　Typical ratio of maintenance expenses to plant in service for each utility sector. The data are for privately owned class A and class B electric utilities.

Once the customer loads are determined, they are grouped for service from secondary lines connected to distribution transformers that step down from primary voltage.

The distribution transformer loads are then combined to determine the demands on the primary distribution system. The primary distribution system loads are then assigned to substations that step down from transmission voltage. The distribution system loads, in turn, determine the size and location, or siting, of the substations as well as the routing and capacity of the associated transmission lines. In other words, each step in the process provides input for the step that follows.

The distribution system planner partitions the total distribution system planning problem into a set of subproblems that can be handled by using available, usually ad hoc, methods and techniques. The planner, in the absence of accepted planning techniques, may restate the problem as an attempt to minimize the cost of subtransmission, substations, feeders, laterals, etc., and the cost of losses. In this process, however, the planner is usually restricted by permissible voltage values, voltage dips, flicker, etc., as well as service continuity and reliability. In pursuing these objectives, the planner ultimately has a significant influence on additions to and/or modifications of the subtransmission network, locations and sizes of substations, service areas of substations, location of breakers and switches, sizes of feeders and laterals, voltage levels and voltage drops in the system, the location of capacitors and voltage regulators, and the loading of transformers and feeders.

There are, of course, some other factors that need to be considered such as transformer imped-ance, insulation levels, availability of spare transformers and mobile substations, dispatch of genera-tion, and the rates that are charged to the customers.

Furthermore, there are factors over which the distribution system planner has no influence but which, nevertheless, have to be considered in good long-range distribution system planning, for example, the timing and location of energy demands; the duration and frequency of outages; the cost of equipment, labor, and money; increasing fuel costs; increasing or decreasing prices of alternative energy sources; changing socioeconomic conditions and trends such as the growing demand for goods and services; unexpected local population growth or decline; changing public behavior as a result of technological changes; energy conservation; changing environmental concerns of the pub-lic; changing economic conditions such as a decrease or increase in gross national product (GNP) projections, inflation, and/or recession; and regulations of federal, state, and local governments.

1.3 FACTORS AFFECTING SYSTEM PLANNING

The number and complexity of the considerations affecting system planning appear initially to be staggering. Demands for ever-increasing power capacity, higher distribution voltages, more auto-mation, and greater control sophistication constitute only the beginning of a list of such factors. The constraints that circumscribe the designer have also become more onerous. These include a scarcity of available land in urban areas, ecological considerations, limitations on fuel choices, the undesirability of rate increases, and the necessity to minimize investments, carrying charges, and production charges.

Succinctly put, the planning problem is an attempt to minimize the cost of subtransmission, sub-stations, feeders, laterals, etc., as well as the cost of losses. Indeed, this collection of requirements and constraints has put the problem of optimal distribution system planning beyond the resolving power of the unaided human mind.

1.3.1 Load Forecasting

The load growth of the geographic area served by a utility company is the most important factor influencing the expansion of the distribution system. Therefore, forecasting of load increases and system reaction to these increases is essential to the planning process. There are two common time scales of importance to load forecasting: long range, with time horizons on the order of 15 or 20 years away, and short range, with time horizons of up to 5 years distant. Ideally, these forecasts would predict future loads in detail, extending even to the individual customer level, but in practice, much less resolution is sought or required.

Figure 1.4 indicates some of the factors that influence the load forecast. As one would expect, load growth is very much dependent on the community and its development. Economic indica-tors, demographic data, and official land use plans all serve as raw input to the forecast procedure. Output from the forecast is in the form of load densities (kilovoltamperes per unit area) for long-range forecasts. Short-range forecasts may require greater detail.

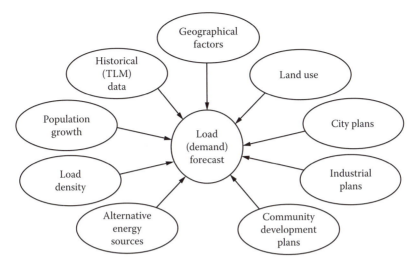

FIGURE 1.4 Factors affecting load forecast.

Densities are associated with a coordinate grid for the area of interest. The grid data are then available to aid configuration design. The master grid presents the load forecasting data, and it provides a useful planning tool for checking all geographic locations and taking the necessary actions to accommodate the system expansion patterns.

1.3.2 SUBSTATION EXPANSION

Figure 1.5 presents some of the factors affecting the substation expansion. The planner makes a decision based on tangible or intangible information. For example, the forecasted load, load density, and load growth may require a substation expansion or a new substation construction. In the system expansion plan, the present system configuration, capacity, and the forecasted loads can play major roles.

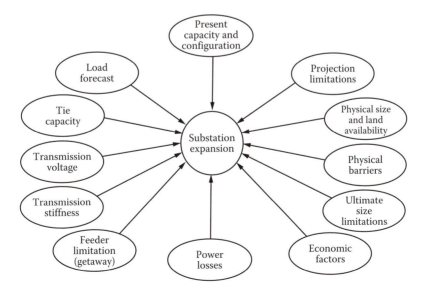

FIGURE 1.5 Factors affecting substation expansion.

1.3.3 SUBSTATION SITE SELECTION

Figure 1.6 shows the factors that affect substation site selection. The distance from the load centers and from the existing subtransmission lines as well as other limitations, such as availability of land, its cost, and land use regulations, is important.

The substation siting process can be described as a screening procedure through which all possible locations for a site are passed, as indicated in Figure 1.7. The service region is the area under evaluation. It may be defined as the service territory of the utility. An initial screening is applied

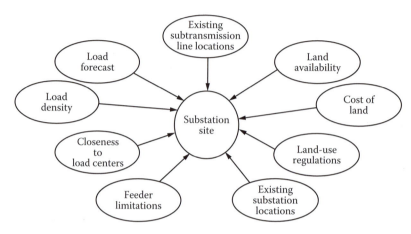

FIGURE 1.6   Factors affecting substation siting.

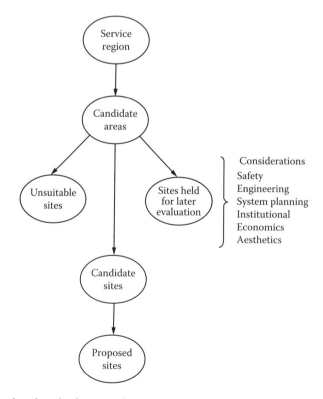

FIGURE 1.7   Substation site selection procedure.

by using a set of considerations, for example, safety, engineering, system planning, institutional, economics, and aesthetics. This stage of the site selection mainly indicates the areas that are unsuitable for site development.

Thus the service region is screened down to a set of candidate sites for substation construction. Further, the candidate sites are categorized into three basic groups: (1) sites that are unsuitable for development in the foreseeable future, (2) sites that have some promise but are not selected for detailed evaluation during the planning cycle, and (3) candidate sites that are to be studied in more detail.

The emphasis put on each consideration changes from level to level and from utility to utility. Three basic alternative uses of the considerations are (1) quantitative vs. qualitative evaluation, (2) adverse vs. beneficial effects evaluation, and (3) absolute vs. relative scaling of effects. A complete site assessment should use a mix of all alternatives and attempt to treat the evaluation from a variety of perspectives.

1.3.4 OTHER FACTORS

Once the load assignments to the substations are determined, then the remaining factors affecting primary voltage selection, feeder route selection, number of feeders, conductor size selection, and total cost, as shown in Figure 1.8, need to be considered.

In general, the subtransmission and distribution system voltage levels are determined by company policies, and they are unlikely to be subject to change at the whim of the planning engineer unless the planner's argument can be supported by running test cases to show substantial benefits that can be achieved by selecting different voltage levels.

Further, because of the standardization and economy that are involved, the designer may not have much freedom in choosing the necessary sizes and types of capacity equipment. For example, the designer may have to choose a distribution transformer out of a fixed list of transformers that are presently stocked by the company for the voltage levels that are already established by the company. Any decision regarding the addition of a feeder or adding on to an existing feeder will, within limits, depend on the adequacy of the existing system and the size, location, and timing of the additional loads that need to be served.

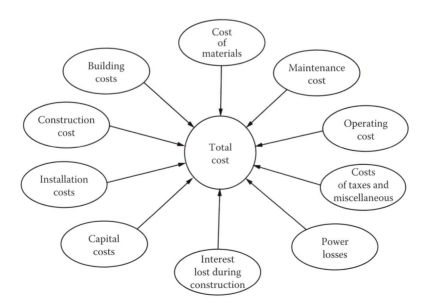

FIGURE 1.8 Factors affecting total cost of the distribution system expansion.

1.4 PRESENT DISTRIBUTION SYSTEM PLANNING TECHNIQUES

Today, many electric distribution system planners in the industry utilize computer programs, usually based on ad hoc techniques, such as load flow programs, radial or loop load flow programs, short-circuit and fault-current calculation programs, voltage drop calculation programs, and total system impedance calculation programs, as well as other tools such as load forecasting, voltage regulation, regulator setting, capacitor planning, reliability, and optimal siting and sizing algorithms.

However, in general, the overall concept of using the output of each program as input for the next program is not in use. Of course, the computers do perform calculations more expeditiously than other methods and free the distribution engineer from detailed work. The engineer can then spend time reviewing results of the calculations, rather than actually making them.

Nevertheless, there is no substitute for engineering judgment based on adequate planning at every stage of the development of power systems, regardless of how calculations are made. In general, the use of the aforementioned tools and their bearing on the system design is based purely on the discretion of the planner and overall company operating policy.

Figure 1.9 shows a functional block diagram of the distribution system planning process currently followed by most of the utilities. This process is repeated for each year of a long-range (15–20 years) planning period. In the development of this diagram, no attempt was made to represent the planning procedure of any specific company but rather to provide an outline of a typical planning process. As the diagram shows, the planning procedure consists of four major activities: load forecasting, distribution system configuration design, substation expansion, and substation site selection.

Configuration design starts at the customer level. The demand type, load factor, and other customer load characteristics dictate the type of distribution system required. Once customer loads are determined, secondary lines are defined, which connect to distribution transformers. The latter provides the reduction from primary voltage to customer-level voltage.

The distribution transformer loads are then combined to determine the demands on the primary distribution system. The primary distribution system loads are then assigned to substations that step down from subtransmission voltage. The distribution system loads, in turn, determine the size and location (siting) of the substations as well as the route and capacity of the associated subtransmission lines. It is clear that each step in this planning process provides input for the steps that follow.

Perhaps what is not clear is that in practice, such a straightforward procedure may be impossible to follow. A much more common procedure is the following. Upon receiving the relevant load projection data, a system performance analysis is done to determine whether the present system is capable of handling the new load increase with respect to the company's criteria. This analysis, constituting the second stage of the process, requires the use of tools such as a distribution load flow program, a voltage profile, and a regulation program. The acceptability criteria, representing the company's policies, obligations to the consumers, and additional constraints, can include

1. Service continuity
2. The maximum allowable peak-load voltage drop to the most remote customer on the secondary
3. The maximum allowable voltage dip occasioned by the starting of a motor of specified starting current characteristics at the most remote point on the secondary
4. The maximum allowable peak load
5. Service reliability
6. Power losses

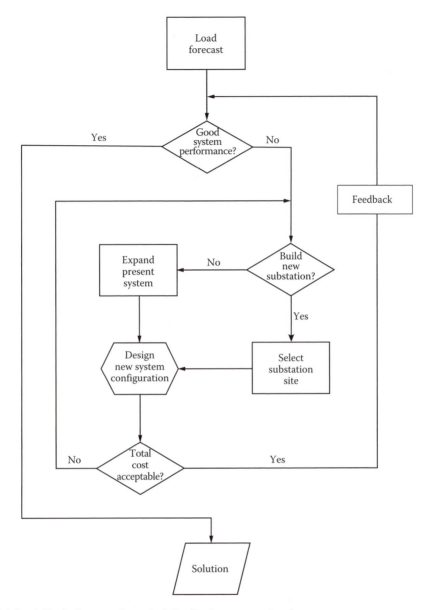

FIGURE 1.9 A block diagram of a typical distribution system planning process.

As illustrated in Figure 1.9, if the results of the performance analysis indicate that the present system is not adequate to meet future demand, then either the present system needs to be expanded by new, relatively minor, system additions, or a new substation may need to be built to meet the future demand. If the decision is to expand the present system with minor additions, then a new additional network configuration is designed and analyzed for adequacy.

If the new configuration is found to be inadequate, another is tried, and so on, until a satisfactory one is found. The cost of each configuration is calculated. If the cost is found to be too high, or adequate performance cannot be achieved, then the original expand-or-build decision is reevaluated.

If the resulting decision is to build a new substation, a new placement site must be selected. Further, if the purchase price of the selected site is too high, the expand-or-build decision may need further reevaluation. This process terminates when a satisfactory configuration is attained, which provides a solution to existing or future problems at a reasonable cost. Many of the steps in the earlier procedures can feasibly be done only with the aid of computer programs.

1.5 DISTRIBUTION SYSTEM PLANNING MODELS

In general, distribution system planning dictates a complex procedure due to a large number of variables involved and the difficult task of the mathematical presentation of numerous requirements and limitations specified by system configuration.

Therefore, mathematical models are developed to represent the system and can be employed by distribution system planners to investigate and determine optimum expansion patterns or alternatives, for example, by selecting

1. Optimum substation locations
2. Optimum substation expansions
3. Optimum substation transformer sizes
4. Optimum load transfers between substations and demand centers
5. Optimum feeder routes and sizes to supply the given loads subject to numerous constraints to minimize the present worth of the total costs involved

Some of the operations research techniques used in performing this task include

1. The alternative-policy method, by which a few alternative policies are compared and the best one is selected
2. The decomposition method, in which a large problem is subdivided into several small problems and each one is solved separately
3. The linear-programming, integer-programming, and mixed-integer programming methods that linearize constraint conditions
4. The quadratic programming method
5. The dynamic-programming method
6. Genetic algorithms method

Each of these techniques has its own advantages and disadvantages. Especially in long-range planning, a great number of variables are involved, and thus there can be a number of feasible alternative plans that make the selection of the optimum alternative a very difficult one [7].

The distribution system costs of an electric utility company can account for up to 60% of investment budget and 20% of operating costs, making it a significant expense [10]. Minimizing the cost of distribution system can be a considerable challenge, as the feeder system associated with only a single substation may present a distribution engineer with thousands of feasible design options from which to choose. For example, the actual number of possible plans for a 40-node distribution system is over 15 million, with the number of feasible designs being in about 20,000 variations.

Finding the overall least cost plan for the distribution system associated with several neighboring substations can be a truly intimidating task. The use of computer-aided tools that help identify the lowest cost distribution configuration has been a focus of much R&D work in the last three decades. As a result, today a number of computerized optimization programs can be used as tools to find the best design from among those many possibilities. Such programs never consider all aspects of the problem, and most include approximations that slightly limit accuracy. However, they can help to deduce distribution costs even with the most conservative estimate by 5%–10%, which is more than enough reason to use them [10].

Expansion studies of a distribution system have been done in practice by planning engineers. The studies were based on the existing system, forecasts of power demands, extensive economic and electrical calculations, and planner's past experience and engineering judgment. However, the development of more involved studies with a large number of alternating projects using mathematical models and computational optimization techniques can improve the traditional solutions that were achieved by the planners. As expansion costs are usually very large, such improvements of solutions represent valuable savings.

For a given distribution system, the present level of electric power demand is known and the future levels can be forecasted by one stage, for example, 1 year, or several stages. Therefore, the problem is to plan the expansion of the distribution system (in one or several stages, depending on data availability and company policy) to meet the demand at minimum expansion cost. In the early applications, the overall distribution system planning problem has been dealt with by dividing it into the following two subproblems that are solved successfully:

1. The subproblem of the optimal sizing and/or location of distribution substations. In some approaches, the corresponding mathematical formulation has taken into account the present feeder network either in terms of load transfer capability between service areas or in terms of load times distance. What is needed is the full representation of individual feeder segments, that is, the network itself.
2. The subproblem of the optimal sizing and/or locating feeders. Such models take into account the full representation of the feeder network but without taking into account the former subproblem.

However, there are more complex mathematical models that take into account the distribution planning problem as a global problem and solving it by considering minimization of feeder and substation costs simultaneously. Such models may provide the optimal solutions for a single planning stage. The complexity of the mathematical problems and the process of resolution become more difficult because the decisions for building substations and feeders in one of the planning stages have an influence on such decisions in the remaining stages.

1.5.1 Computer Applications

Today, there are various innovative algorithms based on optimization programs that have been developed based on the earlier fundamental operations research techniques. For example, one such distribution design optimization program now in use at over 25 utilities in the United States. It works within an integrated Unix or Windows NT graphical user interface (GUI) environment with a single open SQL circuit database that supports circuit analysis, various equipment selection optimization routes such as capacitor-regulator sizing and locating, and a constrained linear optimization algorithm for the determination of multifeeder configurations.

The key features include a database, editor, display, and GUI structure specifically designed to support optimization applications in augmentation planning and switching studies. This program uses a linear trans-shipment algorithm in addition to a postoptimization radialization. For the program, a linear algorithm methodology was selected over nonlinear methods even though it is not the best in applications involving augmentation planning and switching studies.

The reasons for this section include its stability in use in terms of consistently converging performance, its large problem capacity, and reasonable computational requirements. Using this package, a system of 10,000 segments/potential segments, which at a typical 200 segments per feeder means roughly 8 substation service areas, can be optimized in one analysis on a DEC 3000/600 with 64 Mbyte RAM in about 1 min [10]. From the application point of view, distribution system planning can be categorized as (1) new system expansion, (2) augmentation of existing system, and (3) operational planning.

1.5.2 New Expansion Planning

It is the easiest of the earlier-provided three categories to optimize. It has received the most attention in the technical literature partially due to its large capital and land requirements. It can be envisioned as the distribution expansion planning for the growing periphery of a thriving city. Willis et al. [10] names such planning *greenfield planning* due to the fact that the planner starts with essentially nothing, or greenfield, and plans a completely new system based on the development of a region. In such planning problem, obviously there are a vast range of possibilities for the new design.

Luckily, optimization algorithms can apply a clever linearization that shortens computational times and allows large problem size, at the same time introducing only a slight approximation error. In such linearization, each segment in the potential system is represented with only two values, namely, a linear cost vs. kVA slope based on segment length, and a capacity limit that constrains its maximum loading. This approach has provided very satisfactory results since 1070s. According to Willis et al. [10], more than 60 utilities in this country alone use this method routinely in the layout of major new distribution additions today. Economic savings as large as 19% in comparison to good manual design practices have been reported in IEEE and Electric Power Research Institute (EPRI) publications.

1.5.3 Augmentation and Upgrades

Much more often than a greenfield planning, a distribution planner faces the problem of economical upgrade of a distribution system that is already in existence. For example, in a well-established neighborhood where a slowly growing load indicates that the existing system will be overloaded pretty soon.

Even though such planning may be seen as much easier than the greenfield planning, in reality, this perception is not true for two reasons. First of all, new routes, equipment sites, and permitted upgrades of existing equipment are very limited due to practical, operational, aesthetic, environmental, or community reasons. Here, the challenge is the balancing of the numerous unique constraints and local variations in options. Second, when an existing system is in place, the options for upgrading existing lines generally cannot be linearized. Nevertheless, optimization programs have long been applied to augmentation planning partially due to the absence of better tools. Such applications may reduce costs in augmentation planning approximately by 5% [10].

As discussed in Section 7.5, fixed and variable costs of each circuit element should be included in such studies. For example, the cost for each feeder size should include (1) investment cost of each of the installed feeder and (2) cost of energy lost due to I^2R losses in the feeder conductors. It is also possible to include the cost of demand lost, that is, the cost of useful system capacity lost (i.e., the demand cost incurred to maintain adequate and additional system capacity to supply I^2R losses in feeder conductors) into such calculations.

1.5.4 Operational Planning

It determines the actual switching pattern for operation of an already-built system, usually for the purpose of meeting the voltage drop criterion and loading while having minimum losses. Here, contrary to the other two planning approaches, the only choice is switching. The optimization involved is the minimization of I^2R losses while meeting properly the loading and operational restrictions.

In the last two decades, a piecewise linearization-type approximation has been effectively used in a number of optimization applications, providing good results.

However, operational planning in terms of determining switching patterns has very little effect if any on the initial investment decisions on ether feeder routes and/or substation locations. Once the investment decisions are made, then the costs involved become fixed investment costs. Any switching activities that take place later on in the operational phase only affect the minimization of losses.

1.5.5 BENEFITS OF OPTIMIZATION APPLICATIONS

Furthermore, according to Ramirez-Rosado and Gönen [11], the optimal solution is the same when the problem is resolved considering only the costs of investment and energy loses, as expected having a lower total costs. In addition, they have shown that the problem can successfully be resolved considering only investment costs. For example, one of their studies involving multi-stage planning has shown that the optimal network structure is almost the same as before, with the exception of building a particular feeder until the fourth year. Only a slight influence of not including the cost of energy losses is observed in the optimal network structure evolved in terms of delay in building a feeder.

It can easily be said that cost reduction is the primary justification for application of optimization. According to Willis et al. [10], a nonlinear optimization algorithm would improve average savings in augmentation planning to about the same level as those of greenfield results. However, this is definitely not the case with switching. For example, tests using a nonlinear optimization have shown that potential savings in augmentation planning are generally only a fourth to a third as much as in greenfield studies.

Also, a linear optimization delivers on the order of 85% of savings achievable using nonlinear analysis. An additional benefit of optimization efforts is that it greatly enhances the understanding of the system in terms of the interdependence between costs, performance, and tradeoffs. Willis et al. [10] report that in a single analysis that lasted less than a minute, the optimization program results have identified the key problems to savings and quantified how it interacts with other aspects of the problems and indicated further cost reduction possibilities.

1.6   DISTRIBUTION SYSTEM PLANNING IN THE FUTURE

In the previous sections, some of the past and present techniques used by the planning engineers of the utility industry in performing the distribution system planning have been discussed. Also, the factors affecting the distribution system planning decisions have been reviewed. Furthermore, the need for a systematic approach to distribution planning has been emphasized.

The following sections examine what today's trends are likely to portend for the future of the planning process.

1.6.1   ECONOMIC FACTORS

There are several economic factors that will have significant effects on distribution planning in the 1980s. The first of these is inflation. Fueled by energy shortages, energy source conversion cost, environmental concerns, and government deficits, inflation will continue to be a major factor.

The second important economic factor will be the increasing expense of acquiring capital. As long as inflation continues to decrease the real value of the dollar, attempts will be made by government to reduce the money supply. This in turn will increase the competition for attracting the capital necessary for expansions in distribution systems.

The third factor that must be considered is increasing difficulty in raising customer rates. This rate increase "inertia" also stems in part from inflation as well as from the results of customers being made more sensitive to rate increases by consumer activist groups.

1.6.2 DEMOGRAPHIC FACTORS

Important demographic developments will affect distribution system planning in the near future. The first of these is a trend that has been dominant over the last 50 years: the movement of the population from the rural areas to the metropolitan areas.

The forces that initially drove this migration—economic in nature—are still at work. The number of single-family farms has continuously declined during this century, and there are no visible trends that would reverse this population flow into the larger urban areas. As population leaves the countrysides, population must also leave the smaller towns that depend on the countrysides for economic life. This trend has been a consideration of distribution planners for years and represents no new effect for which account must be taken.

However, the migration from the suburbs to the urban and near-urban areas is a new trend attributable to the energy crisis. This trend is just beginning to be visible, and it will result in an increase in multifamily dwellings in areas that already have high population densities.

1.6.3 TECHNOLOGICAL FACTORS

The final class of factors, which will be important to the distribution system planner, has arisen from technological advances that have been encouraged by the energy crisis. The first of these is the improvement in fuel cell technology. The output power of such devices has risen to the point where in the areas with high population density, large banks of fuel cells could supply significant amounts of the total power requirements.

Other nonconventional energy sources that might be a part of the total energy grid could appear at the customer level. Among the possible candidates would be solar and wind-driven generators. There is some pressure from consumer groups to force utilities to accept any surplus energy from these sources for use in the total distribution network. If this trend becomes important, it would change drastically the entire nature of the distribution system as it is known today.

1.7 FUTURE NATURE OF DISTRIBUTION PLANNING

Predictions about the future methods for distribution planning must necessarily be extrapolations of present methods. Basic algorithms for network analysis have been known for years and are not likely to be improved upon in the near future.

However, the superstructure that supports these algorithms and the problem-solving environment used by the system designer is expected to change significantly to take advantage of new methods that technology has made possible. Before giving a detailed discussion of these expected changes, the changing role of distribution planning needs to be examined.

1.7.1 INCREASING IMPORTANCE OF GOOD PLANNING

For the economic reasons listed earlier, distribution systems will become more expensive to build, expand, and modify. Thus, it is particularly important that each distribution system design be as cost effective as possible. This means that the system must be optimal from many points of view over the time period from the 1st day of operation to the planning-time horizon.

In addition to the accurate load growth estimates, components must be phased in and out of the system so as to minimize capital expenditure, meet performance goals, and minimize losses. These requirements need to be met at a time when demographic trends are veering away from what have been their norms for many years in the past and when distribution systems are becoming more complex in design due to the appearance of more active components (e.g., fuel cells) instead of the conventional passive ones.

1.7.2 Impacts of Load Management (or Demand-Side Management)

In the past, the power utility companies of this nation supplied electric energy to meet all customer demands when demands occurred. Recently, however, because of the financial constraints (i.e., high cost of labor, materials, and interest rates), environmental concerns, and the recent shortage (or high cost) of fuels, this basic philosophy has been reexamined and customer load management investigated as an alternative to capacity expansion.

Load management's benefits are systemwide. Alteration of the electric energy use patterns will not only affect the demands on system generating equipment but also alter the loading of distribution equipment. The load management (or demand-side management) may be used to reduce or balance loads on marginal substations and circuits, thus even extending their lives. Therefore, in the future, the implementation of load management policies may drastically affect the distribution of load, in time and in location, on the distribution system, subtransmission system, and the bulk power system. Since distribution systems have been designed to interface with controlled load patterns, the systems of the future will necessarily be designed somewhat differently to benefit from the altered conditions. However, the benefits of load management (or demand-side management) cannot be fully realized unless the system planners have the tools required to adequately plan incorporation into the evolving electric energy system. The evolution of the system in response to changing requirements and under changing constraints is a process involving considerable uncertainty.

The requirements of a successful load management program are specified by Delgado [19] as follows:

1. It must be able to reduce demand during critical system load periods.
2. It must result in a reduction in new generation requirements, purchased power, and/or fuel costs.
3. It must have an acceptable cost/benefit ratio.
4. Its operation must be compatible with system design and operation.
5. It must operate at an acceptable reliability level.
6. It must have an acceptable level of customer convenience.
7. It must provide a benefit to the customer in the form of reduced rates or other incentives.

1.7.3 Cost/Benefit Ratio for Innovation

In the utility industry, the most powerful force shaping the future is that of economics. Therefore, any new innovations are not likely to be adopted for their own sake but will be adopted only if they reduce the cost of some activity or provide something of economic value, which previously had been unavailable for comparable costs. In predicting that certain practices or tools will replace current ones, it is necessary that one judge their acceptance on this basis.

The expected innovations that satisfy these criteria are planning tools implemented on a digital computer that deals with distribution systems in network terms. One might be tempted to conclude that these planning tools would be adequate for industry use throughout the 1980s. That this is not likely to be the case may be seen by considering the trends judged to be dominant during this period with those that held sway over the period in which the tools were developed.

1.7.4 New Planning Tools

Tools to be considered fall into two categories: network design tools and network analysis tools. The analysis tools may become more efficient but are not expected to undergo any major changes, although the environment in which they are used will change significantly. This environment will be discussed in the next section.

The design tools, however, are expected to show the greatest development since better planning could have a significant impact on the utility industry. The results of this development will show the following characteristics:

1. Network design will be optimized with respect to many criteria by using programming methods of operations research.
2. Network design will be only one facet of distribution system management directed by human engineers using a computer system designed for such management functions.
3. So-called *network editors* [20] will be available for designing trial networks; these designs in digital form will be passed to extensive simulation programs, which will determine if the proposed network satisfies performance and load growth criteria.

1.8 CENTRAL ROLE OF THE COMPUTER IN DISTRIBUTION PLANNING

As is well known, distribution system planners have used computers for many years to perform the tedious calculations necessary for system analysis. However, it has only been in the past few years that technology has provided the means for planners to truly take a system approach to the total design and analysis. It is the central thesis of this book that the development of such an approach will occupy planners in the future and will significantly contribute to their meeting the challenges previously discussed.

1.8.1 SYSTEM APPROACH

A collection of computer programs to solve the analysis problems of a designer necessarily constitutes neither an efficient problem-solving system nor such a collection even when the output of one can be used as the input of another. The system approach to the design of a useful tool for the designer begins by examining the types of information required and its sources. The view taken is that this information generates decisions and additional information that pass from one stage of the design process to another. At certain points, it is noted that the human engineer must evaluate the information generated and add his or her input. Finally, the results must be displayed for use and stored for later reference.

With this conception of the planning process, the system approach seeks to automate as much of the process as possible, ensuring in the process that the various transformations of information are made as efficiently as possible. One representation of this information flow is shown in Figure 1.10, where the outer circle represents the interface between the engineer and the system. Analysis programs forming part of the system are supported by a database management system (DBMS) that stores, retrieves, and modifies various data on distribution systems [21].

1.8.2 DATABASE CONCEPT

As suggested in Figure 1.10, the database plays a central role in the operation of such a system. It is in this area that technology has made some significant strides in the past 5 years so that not only is it possible to store vast quantities of data economically, but it is also possible to retrieve desired data with access times on the order of seconds.

The DBMS provides the interface between the process that requires access to the data and the data themselves. The particular organization that is likely to emerge as the dominant one in the near future is based on the idea of a relation. Operations on the database are performed by the *DBMS*.

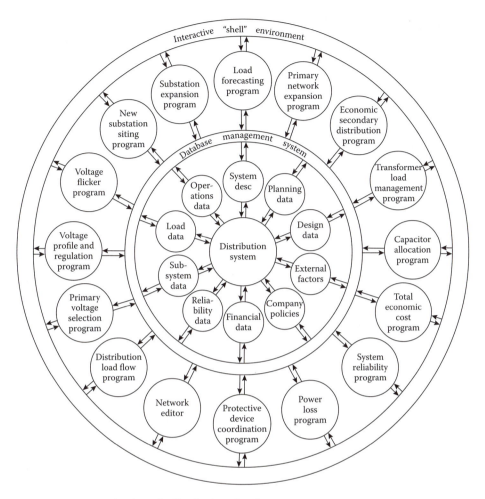

FIGURE 1.10   A schematic view of a distribution planning system.

1.8.3   New Automated Tools

In addition to the database management program and the network analysis programs, it is expected that some new tools will emerge to assist the designer in arriving at the optimal design. One such new tool that has appeared in the literature is known as a network editor [20]. The network consists of a graph whose vertices are network components, such as transformers and loads, and edges that represent connections among the components.

The features of the network editor may include network objects, for example, feeder line sections, secondary line sections, distribution transformers, or variable or fixed capacitors, control mechanisms, and command functions. A primitive network object comprises a name, an object class description, and a connection list. The control mechanisms may provide the planner with natural tools for correct network construction and modification [21].

1.9   IMPACT OF DISPERSED STORAGE AND GENERATION

Following the oil embargo and the rising prices of oil, the efforts toward the development of alternative energy sources (preferably renewable resources) for generating electric energy have been increased. Furthermore, opportunities for small power producers and cogenerators have been enhanced by recent legislative initiatives, for example, the *Public Utility Regulatory Policies Act* (PURPA) of 1978 and by the subsequent interpretations by the *Federal Energy Regulatory Commission* in 1980 [22,25].

The following definitions of the criteria affecting facilities under PURPA are given in Section 201 of PURPA:

A small power production facility is one which produces electric energy solely by the use of primary fuels of biomass, waste, renewable resources, or any combination thereof. Furthermore, the capacity of such production sources together with other facilities located at the same site must not exceed 80 MW.

A cogeneration facility is one which produces electricity and steam or forms of useful energy for industrial, commercial, heating, or cooling applications.

A qualified facility is any small power production or cogeneration facility which conforms to the previous definitions and is owned by an entity not primarily engaged in generation or sale of electric power.

In general, these generators are small (typically ranging in size from 100 kW to 10 MW and connectable to either side of the meter) and can be economically connected only to the distribution system. They are defined as *dispersed storage and generation* (DSG) devices. If properly planned and operated, DSG may provide benefits to distribution systems by reducing capacity requirements, improving reliability, and reducing losses.

Examples of DSG technologies include hydroelectric, diesel generators, wind electric systems, solar electric systems, batteries, storage space and water heaters, storage air conditioners, hydroelectric pumped storage, photovoltaics, and fuel cells. Table 1.1 gives the results of a comparison of DSG devices with respect to the factors affecting the energy management system (EMS) of a utility system [26]. Table 1.2 gives the interactions between the DSG factors and the functions of the EMS or energy control center.

As mentioned before, it has been estimated that the installed generation capacity will be about 1200 GW in the United States by the year 2000 (Table 1.3). The contribution of the DSG systems to this figure has been estimated to be in a range of 4%–10%. For example, if 5% of installed capacity is DSG in the year 2000, it represents a contribution of 60 GW.

According to Chen [27], as power distribution systems become increasingly complex due to the fact that they have more DSG systems, as shown in Figure 1.11, distribution automation will be indispensable for maintaining a reliable electric supply and for cutting down operating costs.

In distribution systems with DSG, the feeder or feeders will no longer be radial. Consequently, a more complex set of operating conditions will prevail for both steady state and fault conditions. If the dispersed generator capacity is large relative to the feeder supply capacity, then it might be considered as backup for normal supply. If so, this could improve service security in instances of loss of supply.

In a given fault, a more complex distribution of higher-magnitude fault currents will occur due to multiple supply sources. Such systems require more sophisticated detection and isolation techniques than those adequate for radial feeders. Therefore, distribution automation, with its multiple point monitoring and control capability, is well suited to the complexities of a distribution system with DSG.

1.10   DISTRIBUTION SYSTEM AUTOMATION

The main purpose of an electric power system is to efficiently generate, transmit, and distribute electric energy. The operations involved dictate geographically dispersed and functionally complex monitoring and control systems, as shown in Figure 1.12. As noted in the figure, the *EMS* exercises overall control over the total system.

The *supervisory control and data acquisition* (SCADA) system involves generation and transmission systems. The *distribution automation and control* (DAC) system oversees the distribution system, including connected load. Automatic monitoring and control features have long been a part of the SCADA system.

TABLE 1.1
Comparison of DSG Devices

DSG Devices	Size	Factors							
		Power Source Availability	Power Source Stability	DSG Energy Limitation	Voltage Control	Response Speed	Harmonic Generation	Special Automatic Start	DSG Factors
Biomass	Variable	Good	Good	No	Yes	Fast	No	Yes	Yes
Geothermal	Medium	Good	Good	No	Yes	Medium	No	Yes	No
Pumped hydro	Large	Good	Good	Yes	Yes	Fast	No	Yes	No
Compressed air storage	Large	Good	Good	Yes	Yes	Fast	No	Yes	No
Solar thermal	Variable	Uncertain	Poor	No	Uncertain	Variable	Uncertain	Uncertain	Yes
Photovoltaics	Variable	Uncertain	Poor	No	Uncertain	Fast	Yes	Yes	Yes
Wind	Small	Uncertain	Poor	No	Uncertain	Fast	Uncertain	Yes	Yes
Fuel cells	Variable	Good	Good	No	Yes	Fast	Yes	Yes	No
Storage battery	Variable	Good	Good	Yes	Yes	Fast	Yes	Yes	No
Low-head hydro	Small	Variable	Good	No	Yes	Fast	No	Yes	No
Cogeneration:									
Gas turbine	Medium	Good	Good	No	Yes	Fast	No	Yes	No
Burning refuse	Medium	Good	Good	No	Yes	Fast	No	Yes	No
Landfill gas	Small	Good	Good	No	Yes	Fast	No	Yes	No

Source: Kirkham, H. and Klein, J., *IEEE Trans. Power Appar. Syst.*, PAS-102(2), 339, 1983.

TABLE 1.2

Interaction between DSG Factors and Energy Management System Functions

		Factors							
Functions	Size	Power Source Availability	Power Source Stability	Energy Limitation	DSG Voltage Control	Response Speed	Harmonic Generation	Automatic Start	Special DSG Factors
Automatic generation control	1	1	1	1	0	1	0	0	0
Economic dispatch	1	1	1	1	?	0	0	1	0
Voltage control	1	0	1	0	1	1	?	0	0
Protection	1	0	1	0	1	1	1	1	1
State estimation	1	0	0	0	0	0	0	?	0
On-line load flow	1	0	0	0	0	0	0	0	0
Security monitoring	1	0	0	0	0	0	0	0	0

Source: Kirkham, H. and Klein, J., *IEEE Trans. Power Appar. Syst.*, PAS-102(2), 339, 1983.
1, Interaction probable; 0, interaction unlikely; ?, interaction possible.

TABLE 1.3

Profile of the Electric Utility Industry in the United States in the Year 2000

Total US population	250×10^6
Number of electric meters	110×10^6
Number of residence	
With central air conditioners	33×10^6
With electric water heaters	25×10^6
With electric space heating	7×10^6
Number of electric utilities	3100

Source: Vaisnys, A., *A Study of a Space Communication System for the Control and Monitoring of the Electric Distribution System*, JPL Publication 80-48, Jet Propulsion Laboratory, California Institute of Technology, Pasadena, CA, May 1980. With permission.

More recently, automation has become a part of the overall energy management, including the distribution system. The motivating objectives of the DAC system are [28]

1. Improved overall system efficiency in the use of both capital and energy
2. Increased market penetration of coal, nuclear, and renewable domestic energy sources Reduced reserve requirements in both transmission and generation
3. Increased reliability of service to essential loads

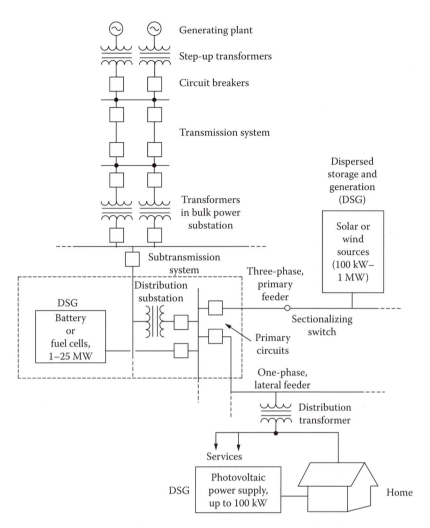

FIGURE 1.11 In the future, small dispersed-energy-storage-and-generation units attached to a customer's home, a power distribution feeder, or a substation would require an increasing amount of automation and control. (From Chen, A.C.M., Automated power distribution, *IEEE Spectrum*, pp. 55–60, April 1982. Used by permission © 1982 IEEE.)

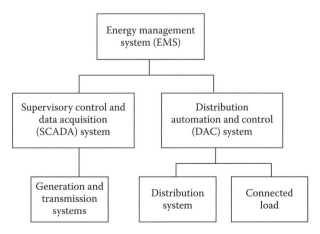

FIGURE 1.12 Monitoring and controlling of an electric power system.

Advances in digital technology are making true distribution automation a reality. Recently, inexpensive minicomputers and powerful microprocessors (computer on a chip) have provided distribution system engineers with new tools that are making many distribution automation concepts achievable. It is clear that future distribution systems will be more complex than those of today. If the systems being developed are to be optimal with respect to construction cost, capitalization, performance reliability, and operating efficiency, better automation and control tools are required.

The term *distribution automation* has a very broad meaning, and additional applications are added every day. To some people, it may mean a communication system at the distribution level that can control customer load and can reduce peak-load generation through load management. To others, the distribution automation may mean an unattended distribution substation that could be considered attended through the use of an on-site microprocessor.

The microprocessor, located at a distribution substation, can continuously monitor the system, make operating decisions, issue commands, and report any change in status to the *distribution dispatch center*, store it on-site for later use, or forget it, depending on the need of the utility.

1.10.1 DISTRIBUTION AUTOMATION AND CONTROL FUNCTIONS

There is no universal consensus among the utilities as to the types of functions that should be handled by a DAC system. Table 1.4 gives some of the automated distribution functions that can be categorized as the load management (or demand-side management) functions, real-time operational management functions, and remote meter reading functions. Some of these functions will be discussed in further detail.

Discretionary load switching: This function is also called the customer load management. It involves direct control of loads at individual customer sites from a remote central location. Control may be exercised for the purpose of overall system peak-load reduction or to reduce the load on a particular substation or feeder that is becoming overloaded.

Customer loads that are appropriate for control are water heating, air-conditioning, space heating, thermal storage heating, etc., and industrial loads supplied under interruptible service contracts. While this function is similar to peak-load pricing, the dispatching center controls the individual customer loads rather than only the meters.

Peak-load pricing: This function allows the implementation of peak-load pricing programs by remote switching of meter registers automatically for the purpose of time-of-day metering.

Load shedding: This function permits the rapid dropping of large blocks of load, under certain conditions, according to an established priority basis.

Cold load pickup: This function is a corollary to the load-shedding function. It entails the controlled pickup of dropped load. Here, cold load pickup describes the load that causes a high magnitude, short duration inrush current, followed by the undiversified demand experienced when reenergizing a circuit following an extended, that is, 20 min or more, interruption.

Fast completion of a fault isolation and service restoration operation will reduce the undiversified component of cold load pickup considerably. Significant service interruption will be limited to those customers supplied from the faulted and isolated line section. An extended system interruption may be due to upstream events beyond the control of the *distribution automation* system. When this occurs, the undiversified demand cold load pickup can be suppressed.

This is achieved by designing the system to disconnect loads controlled by the *load management* systems that customer loads are reduced when energy is restored. Reconnection of loads can be timed to match the return of diversity to prevent exceeding feeder loading limits.

Load reconfiguration: This function involves remote control of switches and breakers to permit routine daily, weekly, or seasonal reconfiguration of feeders or feeder segments for the purpose of

TABLE 1.4
Automated Distribution Functions Correlated with Locations

	Customer Sites					Power System Elements		
	Residential	Commercial and Industrial	Agricultural	Distribution Circuits	Industrial Substation	Distribution Substation	Power Substation	Bulk DSG Facilities
Load management								
Discretionary load switching	x	x	x					
Peak-load pricing	x	x	x					
Load shedding	x	x	x					
Cold load pickup	x	x	x					
Operational management								
Load reconfiguration				x	x	x	x	
Voltage regulation				x	x	x	x	
Transformer load management							x	x
Feeder load management						x		
Capacitor control				x	x			
Dispersed storage and generation					x			x
Fault detection, location, and isolation				x	x	x		
Load studies	x	x	x	x		x	x	
Condition and state monitoring		x	x	x	x	x	x	x
Remote meter reading								
Automatic customer meter reading	x	x	x					
DSG, dispersed storage and generation								

Source: Vaisnys, A., *A Study of a Space Communication System for the Control and Monitoring of the Electric Distribution System*, JPL Publication 80-48, Jet Propulsion Laboratory, California Institute of Technology, Pasadena, CA, May 1980.

taking advantage of load diversity among feeders. It enables the system to effectively serve larger loads without requiring feeder reinforcement or new construction. It also enables routine maintenance on feeders without any customer load interruptions.

Voltage regulation: This function allows the remote control of selected voltage regulators within the distribution network, together with network capacitor switching, to effect coordinated systemwide voltage control from a central facility.

Transformer load management (TLM): This function enables the monitoring and continuous reporting of transformer loading data and core temperature to prevent overloads, burnouts, or abnormal operation by timely reinforcement, replacement, or reconfiguration.

Feeder load management (FLM): This function is similar to TLM, but the loads are monitored and measured on feeders and feeder segments (known as the line sections) instead. This function permits loads to be equalized over several feeders.

Capacitor control: This function permits selective and remote-controlled switching of distribution capacitors.

Dispersed storage and generation: Storage and generation equipment may be located at strategic places throughout the distribution system, and they may be used for peak shaving. This function enables the coordinated remote control of these sites.

Fault detection, location, and isolation: Sensors located throughout the distribution network can be used to detect and report abnormal conditions. This information, in turn, can be used to automatically locate faults, isolate the faulted segment, and initiate proper sectionalization and circuit reconfiguration. This function enables the dispatcher to send repair crews faster to the fault location and results in lesser customer interruption time.

Load studies: This function involves the automatic online gathering and recording of load data for special off-line analysis. The data may be stored at the collection point, at the substation, or transmitted to a dispatch center. This function provides accurate and timely information for the planning and engineering of the power system.

Condition and state monitoring: This function involves real-time data gathering and status reporting from which the minute-by-minute status of the power system can be determined.

Automatic customer meter reading: This function allows the remote reading of customer meters for total consumption, peak demand, or time-of-day consumption, and saves the otherwise necessary man-hours involved in meter reading.

Remote service connect or disconnect: This function permits remote control of switches to connect or disconnect an individual customer's electric service from a central control location.

1.10.2 LEVEL OF PENETRATION OF DISTRIBUTION AUTOMATION

The level of penetration of distribution automation refers to how deeply into the distribution system the automation will go. Table 1.5 gives the present and near-future functional scope of power distribution automation systems.

Recently, the need for gathering substation and power plant data has increased. According to Gaushell et al. [29], this is due to

1. Increased reporting requirements of reliability councils and government agencies
2. Operation of the electric system closer to design limits

TABLE 1.5

Functional Scope of Power Distribution Automation System

Present	Within up to 5 Years	After 5 Years
Protection		
Excessive current over long time	Breaker failure protection	Dispersed storage and generation (DSG) protection
Instantaneous overcurrent	Synchronism check	
Under frequency		Personnel safety
Transformer protection		
Bus protection		
Operational control and monitoring		
Automatic bus sectionalizing	Integrated voltage and var control:	DSG command and control: power, voltage, synchronization
Alarm annunciation	Capacitor bank control	
Transformer tap-change control	Transformer tap-change control	DSG scheduling
Instrumentation	Feeder deployment switching and automatic sectionalizing	Automatic generation control
Load control	Load shedding	Security assessment
	Data acquisition, logging, and display	
	Sequence-of-events recording	
	Transformer monitoring	
	Instrumentation and diagnostics	
Data collection and system planning		
Remote supervisory control and data acquisition (SCADA) at a substation	Distribution SCADA	Distribution dispatching center
	Automatic meter reading	Distribution system database
		Automatic billing
		Service connecting and disconnecting
Communications		
One-way load control	Two-way communication, using one medium	Two-way communication, using many media

Source: Chen, A.C.M., *IEEE Spectrum*, pp. 55–60, April 1982. With permission.

3. Increased efficiency requirements because of much higher fuel prices
4. The tendency of utilities to monitor lower voltages than previously

These needs have occurred simultaneously with the relative decline of the prices of computer and other electronic equipment. The result has been a quantum jump in the amount of data being gathered by an SCADA system or EMS.

A large portion of these data consists of analog measurements of electrical quantities, such as watts, vars, and volts, which are periodically sampled at a remote location, transmitted to a control center, and processed by computer for output on CRT displays, alarm logs, etc. However, as the amount of information to be reported grows, so do the number of communication channels and the amount of control center computer resources required.

Therefore, as equipment are controlled or monitored further down the feeder, the utility obtains more information, can have greater control, and has greater flexibility. However, costs increase as well as benefits. As succinctly put by Markel and Layfield [30],

1. The number of devices to be monitored or controlled increases drastically.
2. The communication system must cover longer distances, connect more points, and transmit greater amounts of information.
3. The computational requirements increase to handle the larger amounts of data or to examine the increasing number of available switching options.
4. The time and equipment needed to identify and communicate with each individually controlled device increases as the addressing system becomes more finely grained.

Today, microprocessors use control algorithms, which permit real-time control of distribution system configurations. For example, it has become a reality that normal loadings of substation transformers and of looped (via a normally open tie recloser) sectionalized feeders can be economically increased through software-controlled load-interrupting switches. SCADA remotes, often computer directed, are being installed in increasing numbers in distribution substations.

They provide advantages such as continuous scanning, higher speed of operation, and greater security. Furthermore, thanks to the falling prices of microprocessors, certain control practices (e.g., protecting power systems against circuit-breaker failures by energizing backup equipment, which is presently done only in transmission systems) are expected to become cost-effective in distribution systems.

The EPRI and the US Department of Energy singled out power-line, telephone, and radio carriers as the most promising systems for their research; other communication techniques are certainly possible. However, at the present time, these other techniques involve greater uncertainties [31].

In summary, the choice of a specific communication system or combination of systems depends upon the specific control or monitoring functions required, amount and speed of data transmission required, existing system configuration, density of control points, whether one-way or two-way communication is required, and, of course, equipment costs.

It is possible to use hybrid systems, that is, two or more different communication systems, between utility and customer. For example, a radio carrier might be used between the control station and the distribution transformer, a power-line carrier (PLC) between the transformer and the customer's meter. Furthermore, the command (forward) link might be one communication system, for example, broadcast radio, and the return (data) link might be another system, such as VHF radio.

An example of such a system is shown in Figure 1.13. The forward (control) link of this system uses commercial broadcast radio. Utility phase-modulated digital signals are added to amplitude-modulated (AM) broadcast information. Standard AM receivers cannot detect the utility signals, and vice versa. The return data link uses VHF receivers that are synchronized by the broadcast station to significantly increase data rate and coverage range [32].

Figure 1.14 shows an experimental system for automating power distribution at the LaGrange Park Substation of Commonwealth Edison Company of Chicago. The system includes two mini-computers, a commercial VHF radio transmitter and receiver, and other equipment installed at a special facility called *Probe*. Microprocessors atop utility poles can automatically connect or disconnect two sections of a distribution feeder upon instructions from the base station.

Figure 1.15 shows a substation control and protection system that has also been developed by EPRI. It features a common signal bus to control recording, comparison, and follow-up actions. It includes line protection and transformer protection. The project is directed toward developing

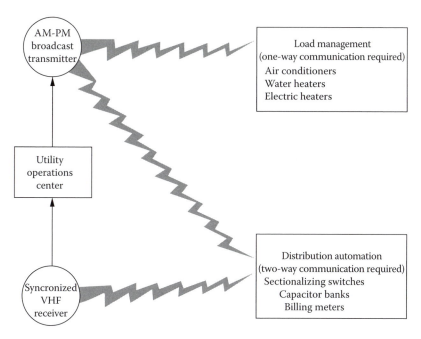

FIGURE 1.13   Applications of two-way radio communications. (From *EPRI J.*, 46, September 1982.)

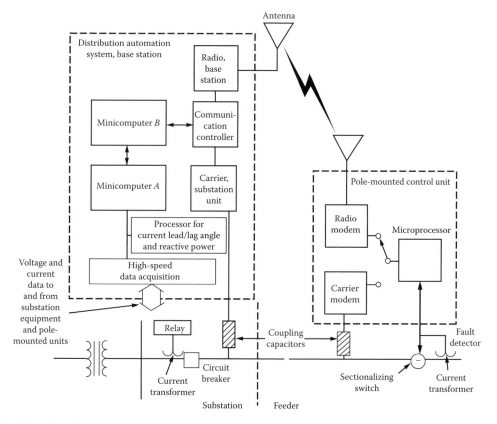

FIGURE 1.14   The research system consisted of two minicomputers with distributed high-speed data acquisition processing units at the La Grange Park Substation. (From Chen, A.C.M., Automated power distribution, *IEEE Spectrum*, pp. 55–60, April 1982. Used by permission © 1982 IEEE.)

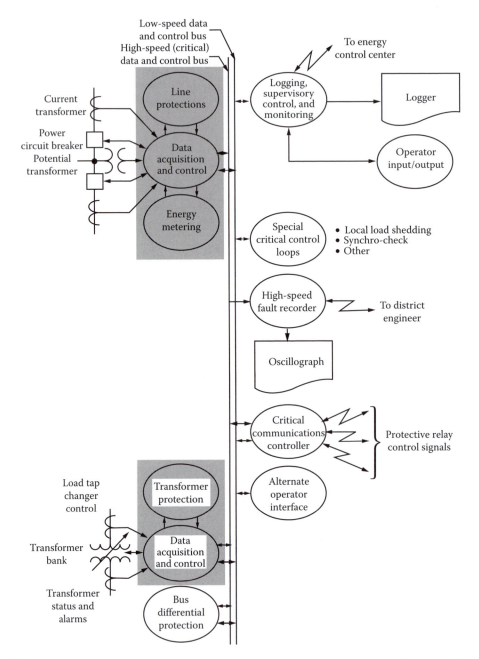

FIGURE 1.15 Substation control and protection system that features a common signal bus (center lines) to control recording, comparison, and follow-up actions (right). Critical processes are shaded. (From *EPRI J.*, 53, June 1978.)

microprocessor-based digital relays capable of interfacing with conventional current and potential transformers and of accepting digital data from the substation yard.

These protective devices can also communicate with substation microcomputer controls capable of providing sequence of events, fault recording, and operator control display. They are also able to interface upward to the dispatcher's control and downward to the distribution system control [44].

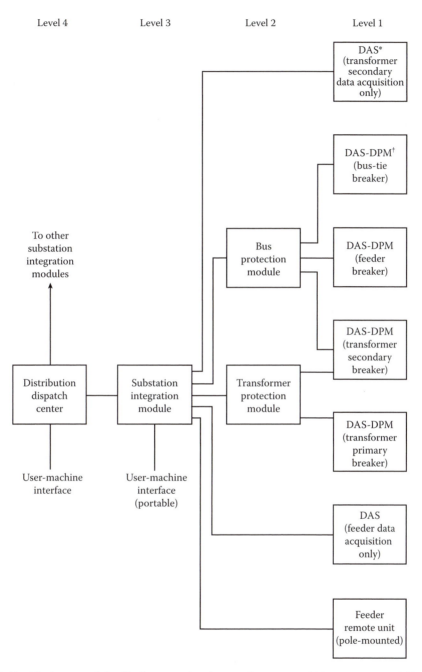

FIGURE 1.16   The integrated distribution control and protection system of EPRI. *Data acquisition system; †digital protection module. (From *EPRI J.*, 43, May 1983.)

Figure 1.16 shows an integrated distribution control and protection system developed by EPRI. The integrated system includes four subsystems: a *substation integration module* (SIM), a *data acquisition system* (DAS), a *digital protection module* (DPM), and a *feeder remote unit* (FRU).

The SIM coordinates the functions of the data acquisition and control system, the DPM, and FRUs by collecting data from them and forming the real-time database required for substation and feeder control. The digital protection module operates in coordination with the DAS and is also a stand-alone device.

1.10.3 ALTERNATIVES OF COMMUNICATION SYSTEMS

There are various types of communication systems available for distribution automation:

1. PLC
2. Radio carrier
3. Telephone (lines) carrier
4. Microwave
5. Private cables, including optical fibers

Power-line carrier (PLC) systems use electric distribution lines for the transmission of communication signals. The advantages of the PLC system include complete coverage of the entire electric system and complete control by the utility. Its disadvantages include the fact that under mass failure or damage to the distribution system, the communication system could also fail and that additional equipment must be added to the distribution system.

In radio carrier systems, communication signals are transmitted point to point via radio waves. Such systems would be owned and operated by electric utilities. It is a communication system that is separate and independent of the status of the distribution system. It can also be operated at a very high data rate. However, the basic disadvantage of the radio system is that the signal path can be blocked, either accidentally or intentionally.

Telephone carrier systems use existing telephone lines for signal communication, and therefore they are the least expensive. However, existing telephone tariffs probably make the telephone system one of the more expensive concepts at this time. Other disadvantages include the fact that the utility does not have complete control over the telephone system and that not all meters have telephone service at or near them. Table 1.6 summarizes the advantages and disadvantages of the aforementioned communication systems.

TABLE 1.6
Summary of Advantages and Disadvantages of the Power-Line, Radio, and Telephone Carriers

Advantages	Disadvantages
Power-line carrier	
Owned and controlled by utility	Utility system must be conditioned
	Considerable auxiliary equipment
	Communication system fails if poles go down
Radio carrier	
Owned and controlled by utility	Subject to interference by buildings and trees
Point-to-point communication	
Terminal equipment only	
Telephone carrier	
Terminal equipment only	Utility lacks control
Carrier maintained by phone company	Ongoing tariff costs
	New telephone drops must be added
	Installation requires house wiring
	Communication system fails if poles go down

Source: *Proceedings Distribution Automation and Control Working Group*, JPL Publication 79-35, Jet Propulsion Laboratory, California Institute of Technology, Pasadena, CA, March 1979. With permission.

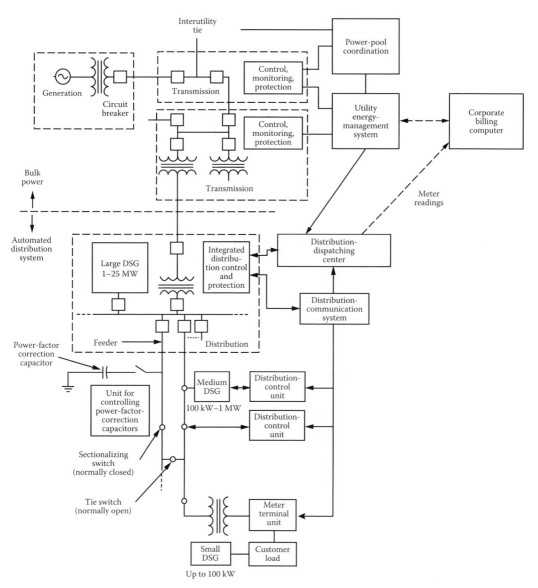

FIGURE 1.17 A control hierarchy envisaged for future utilities. (From Chen, A.C.M., Automated power distribution, *IEEE Spectrum*, pp. 55–60, April 1982. Used by permission © 1982 IEEE.)

Furthermore, according to Chen [27], utilities would have to change their control hierarchies substantially in the future to accommodate the DSG systems in today's power distribution systems, as shown in Figure 1.17.

1.11 SUMMARY AND CONCLUSIONS

In summary, future distribution systems will be more complex than those of today, which means that the distribution system planner's task will be more complex. If the systems being planned are to be optimal with respect to construction cost, capitalization, performance reliability, and operating efficiency, better planning and operation tools are required.

While it is impossible to foresee all the effects that technology will have on the way in which distribution planning and engineering will be done, it is possible to identify the major forces that are beginning to institute a change in the methodology and extrapolate.

REFERENCES

1. Westinghouse Electric Corporation: *Electric Utility Engineering Reference Book-Distribution Systems*, Vol. 3, Westinghouse Electric Corporation, East Pittsburgh, PA, 1965.
2. Energy Information Administration: *Energy Data Reports-Statistics of Privately-Owned Electric Utilities in the United States*, US Department of Energy, Washington, DC, 1975–1978.
3. US Department of Energy: *The National Electric Reliability Study: Technical Study Reports*, US Department of Energy, DOE/EP-0005, Office of Emergency Operations, Washington, DC, April 1981.
4. Economic Regulatory Administration: *The National Power Grid Study*, Vol. 2, US Department of Energy, DOE/ERA-0056–2, Economic Regulatory Administration, Office of Utility Systems, Washington, DC, September 1979.
5. Gönen, T. et al.: Toward automated distribution systems planning, *Proceedings of the IEEE Control of Power Systems Conference*, Texas A&M University, College Station, TX, March 19–21, 1979, pp. 23–30.
6. Munasinghe, M.: *The Economics of Power System Reliability and Planning*, Johns Hopkins, Baltimore, MD, 1979.
7. Gönen, T. and J. C. Thompson: Distribution system planning: The state-of-the-art and the future trends, *Proceedings, Southwest Electrical Exposition and the IEEE Conference*, Houston, TX, January 22–24, 1980, pp. 13–18.
8. Gönen, T. and J. C. Thompson: An interactive distribution system planning model, *Proceedings of the 1979 Modeling and Simulation Conference*, University of Pittsburgh, Pittsburgh, PA, April 25–27, 1979, Vol. 10, pt. 3, pp. 1123–1131.
9. Sullivan, R. L.: *Power System Planning*, McGraw-Hill, New York, 1977.
10. Willis, H. L. et al.: Optimization applications to power distribution, *IEEE Comp. Appl. In Power*, 2(10), October 1995, 12–17.
11. Ramirez-Rosado, I. J. and T. Gönen: Optimal multi-stage planning of power distribution systems, *IEEE Trans. Power Delivery*, 2(2), April 1987, 512–519.
12. Gönen, T. and B. L. Foote: Distribution system planning using mixed-integer programming, *IEEE Proc.*, 128(pt. C, no. 2), March 1981, 70–79.
13. Knight, U. G.: *Power Systems Engineering and Mathematics*, Pergamon, Oxford, England, 1972.
14. Gönen, T., B. L. Foote, and J. C. Thompson: *Development of Advanced Methods for Planning Electric Energy Distribution Systems*, US Department of Energy, Washington, DC, October 1979.
15. Gönen, T. and D. C. Yu: A distribution system planning model, *Proceedings of the IEEE Control of Power Systems Conference (COPS)*, Oklahoma City, OK, March 17–18, 1980, pp. 28–34.
16. Gönen, T. and B. L. Foote: Mathematical dynamic optimization model for electrical distribution system planning, *Electr. Power Energy Syst.*, 4(2), April 1982, 129–136.
17. Ludot, J. P. and M. C. Rubinstein: Méthodes pour la Planification á Court Terme des Réseaux de Distribution, *Proceedings of the Fourth PSCC*, Paper 1.1/12, Grenoble, France, 1972.
18. Launay, M.: Use of computer graphics in data management systems for distribution network planning in Electricite De France (E.D.F.), *IEEE Trans. Power Appar. Syst.*, PAS-101(2), 1982, 276–283.
19. Delgado, R.: Load management—A planner's view, *IEEE Trans. Power Appar. Syst.*, PAS-102(6), 1983, 1812–1813.
20. Gönen, T. and B. L. Foote: Application of mixed-integer programming to reduce suboptimization in distribution systems planning, *Proceedings of the 1979 Modeling and Simulation Conference*, University of Pittsburgh, Pittsburgh, PA, April 25–27, 1979, Vol. 10, pt. 3, pp. 1133–1139.
21. Gönen, T. and D. C. Yu: Bibliography of distribution system planning, *Proceedings of the IEEE Control of Power Systems Conference (COPS)*, Oklahoma City, OK, March 17–18, 1980, pp. 23–34.
22. Public Utility Regulatory Policies Act (PURPA), House of Representatives, Congressional Report No. 95–1750, *Conference Report*, Library of Congress, October 10, 1980.
23. Ma, F., L. Isaksen, and R. Patton: *Impacts of Dispersing Storage and Generation in Electric Distribution Systems*, Final report, US Department of Energy, Washington, DC, July 1979.
24. Vaisnys, A.: *A Study of a Space Communication System for the Control and Monitoring of the Electric Distribution System*, JPL Publication 80-48, Jet Propulsion Laboratory, California Institute of Technology, Pasadena, CA, May 1980.

25. Federal Energy Regulatory Commission Regulations under Sections 201 and 210 of PURPA, Sections 292.101, 292.301–292.308, and 292.401–292.403. Congressional Report No. 95–181, Library of Congress, 1981.

26. Kirkham, H. and J. Klein: Dispersed storage and generation impacts on energy management systems, *IEEE Trans. Power Appar. Syst.*, PAS-102(2), 1983, 339–345.

27. Chen, A. C. M.: Automated power distribution, *IEEE Spectrum*, April 1982, pp. 55–60.

28. Distribution automation and control on the electric power system, *Proceedings of the Distribution Automation and Control Working Group*, JPL Publication 79-35, Jet Propulsion Laboratory, California Institute of Technology, Pasadena, CA, March 1979.

29. Gaushell, D. J., W. L. Frisbie, and M. H. Kuchefski: Analysis of analog data dynamics for supervisory control and data acquisition systems, *IEEE Trans. Power Appar. Syst.*, PAS-102(2), February 1983, 275–281.

30. Markel, L. C. and P. B. Layfield: Economic feasibility of distribution automation, *Proceedings of Control of Power Systems Conference*, Texas A&M University, College Station, TX, March 14–16, 1977, pp. 58–62.

31. Two-way data communication between utility and customer, *EPRI J.*, May 1980, 17–19.

32. Distribution, communication and load management, in R&D Status Report-Electrical Systems Division, *EPRI J.*, September 1982, 46–47.

33. Kaplan, G.: Two-way communication for load management, *IEEE Spectrum*, August 1977, 47–50.

34. Bunch, J. B. et al.: Probe and its implications for automated distribution systems, *Proceedings of the American Power Conference*, Chicago, IL, April 1981, Vol. 43, pp. 683–688.

35. Castro, C. H., J. B. Bunch, and T. M. Topka: Generalized algorithms for distribution feeder deployment and sectionalizing, *IEEE Trans. Power Appar. Syst.*, PAS-99(2), March/April 1980, pp. 549–557.

36. Morgan, M. G. and S. N. Talukdar: Electric power load management: Some technical, economic, regulatory and social issues, *Proc. IEEE*, 67(2), February 1979, 241–313.

37. Bunch, J. B., R. D. Miller, and J. E. Wheeler: Distribution system integrated voltage and reactive power control, *Proceedings of the PICA Conference*, Philadelphia, PA, May 5–8, 1981, pp. 183–188.

38. Redmon, J. R. and C. H. Gentz: Effect of distribution automation and control on future system configuration, *IEEE Trans. Power Appar. Syst.*, PAS-100(4), April 1981, 1923–1931.

39. Chesnut, H. et al.: Monitoring and control requirements for dispersed storage and generation, *IEEE Trans. Power Appar. Syst.*, PAS-101(7), July 1982, 2355–2363.

40. Inglis, D. J., D. L. Hawkins, and S. D. Whelan: Linking distribution facilities and customer information system data bases, *IEEE Trans. Power Appar, Syst.*, PAS-101(2), February 1982, 371–375.

41. Gönen, T. and J. C. Thompson: Computerized interactive model approach to electrical distribution system planning, *Electr. Power Ener. Syst.*, 6(1), January 1984, 55–61.

42. Gönen, T., A. A. Mahmoud, and H. W. Colburn: Bibliography of power distribution system planning, *IEEE Trans. Power Appar. Syst.*, 102(6), June 1983, 1778–1187.

43. Gönen, T. and I. J. Ramirez-Rosado: Review of distribution system planning models: A model for optimal multistage planning, *IEE Proc.*, 133(2, part.C), March 1981, 397–408.

44. Control and protection systems, in R&D Status Report-Electrical Systems Division, *EPRI J.*, June 1978, 53–55.

45. Ramirez-Rosado, I. J., R. N. Adams, and T. Gönen: Computer-aided design of power distribution systems: Multi-objective mathematical simulations, *Int. J. Power Ener. Syst.*, 14(1), 1994, 9–12.

46. Distribution automation, in R&D Status Report-Electrical Systems Division, *EPRI J.*, May 1983, 43–45.

47. Ramirez-Rosado, I. J. and T. Gönen: Review of distribution system planning models: A model for optimal multistage planning, *IEE Proc.*, 133(part C, no. 7), November 1986, 397–408.

48. Ramirez-Rosado, I. J. and T. Gönen: Pseudo-dynamic planning for expansion of power distribution systems, *IEEE Trans. Power Syst.*, 6(1), February 1991, 245–254.

2 Load Characteristics

Only two things are infinite, the universe and human stupidity.
And I am not so sure about the former.

Albert Einstein

2.1 BASIC DEFINITIONS

Demand: "The demand of an installation or system is the load at the receiving terminals averaged over a specified interval of time" [1]. Here, the load may be given in kilowatts, kilovars, kilovoltamperes, kiloamperes, or amperes.

Demand interval: It is the period over which the load is averaged. This selected Δt period may be 15 min, 30 min, 1 h, or even longer. Of course, there may be situations where the 15 and 30 min demands are identical.

The demand statement should express the demand interval Δt used to measure it. Figure 2.1 shows a daily demand variation curve, or load curve, as a function of demand intervals. Note that the selection of both Δt and total time t is arbitrary. The load is expressed in per unit (pu) of peak load of the system. For example, the maximum of 15-min demands is 0.940 pu, and the maximum of 1-h demands is 0.884, whereas the average daily demand of the system is 0.254. The data given by the curve of Figure 2.1 can also be expressed as shown in Figure 2.2. Here, the time is given in per unit of the total time. The curve is constructed by selecting the maximum peak points and connecting them by a curve. This curve is called the *load duration curve*. The load duration curves can be daily, weekly, monthly, or annual. For example, if the curve is a plot of all the 8760 hourly loads during the year, it is called an *annual load duration curve*. In that case, the curve shows the individual hourly loads during the year, but not in the order that they occurred, and the number of hours in the year that load exceeded the value is also shown.

The hour-to-hour load on a system changes over a wide range. For example, the daytime peak load is typically double the minimum load during the night. Usually, the annual peak load is, due to seasonal variations, about three times the annual minimum.

To calculate the average demand, the area under the curve has to be determined. This can easily be achieved by a computer program.

Maximum demand: "The maximum demand of an installation or system is the greatest of all demands which have occurred during the specified period of time" [1]. The maximum demand statement should also express the demand interval used to measure it. For example, the specific demand might be the maximum of all demands such as daily, weekly, monthly, or annual.

Example 2.1

Assume that the loading data given in Table 2.1 belongs to one of the primary feeders of the No Light & No Power (NL&NP) Company and that they are for a typical winter day. Develop the idealized daily load curve for the given hypothetical primary feeder.

Solution

The solution is self-explanatory, as shown in Figure 2.3.

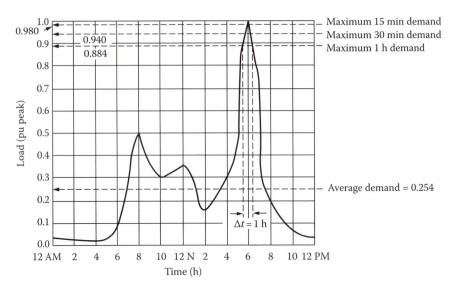

FIGURE 2.1   A daily demand variation curve.

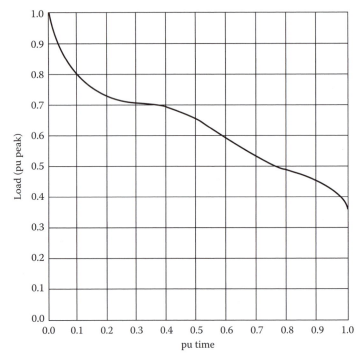

FIGURE 2.2   A load duration curve.

Diversified demand (or coincident demand): It is the demand of the composite group, as a whole, of somewhat unrelated loads over a specified period of time. Here, the maximum diversified demand has an importance. It is the maximum sum of the contributions of the individual demands to the diversified demand over a specific time interval.

For example, "if the test locations can, in the aggregate, be considered statistically representative of the residential customers as a whole, a load curve for the entire residential class of customers can be prepared. If this same technique is used for other classes of customers, similar load

TABLE 2.1
Idealized Load Data for the NL&NP's Primary Feeder

| Time | Load, kW | | |
	Street Lighting	Residential	Commercial
12 AM	100	200	200
1	100	200	200
2	100	200	200
3	100	200	200
4	100	200	200
5	100	200	200
6	100	200	200
7	100	300	200
8		400	300
9		500	500
10		500	1000
11		500	1000
12 noon		500	1000
1		500	1000
2		500	1200
3		500	1200
4		500	1200
5		600	1200
6	100	700	800
7	100	800	400
8	100	1000	400
9	100	1000	400
10	100	800	200
11	100	600	200
12 PM	100	300	200

curves can be prepared" [3]. As shown in Figure 2.4, if these load curves are aggregated, the system load curve can be developed. The interclass coincidence relationships can be observed by comparing the curves.

Noncoincident demand: Manning [2] defines it as "the sum of the demands of a group of loads with no restrictions on the interval to which each demand is applicable." Here, again the maximum of the noncoincident demand is the value of some importance.

Demand factor: It is the "ratio of the maximum demand of a system to the total connected load of the system" [1]. Therefore, the demand factor (DF) is

$$DF \triangleq \frac{\text{Maximum demand}}{\text{Total connected demand}} \qquad (2.1)$$

The DF can also be found for a part of the system, for example, an industrial or commercial customer, instead of for the whole system. In either case, the DF is usually less than 1.0. It is an indicator of the simultaneous operation of the total connected load.

Connected load: It is "the sum of the continuous ratings of the load-consuming apparatus connected to the system or any part thereof" [1]. When the maximum demand and total connected demand have the same units, the DF is dimensionless.

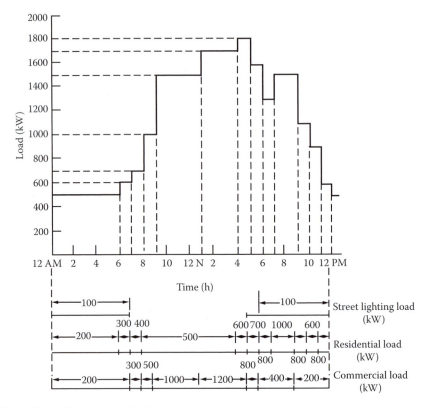

FIGURE 2.3   The daily load curve for Example 2.1.

Utilization factor: It is "the ratio of the maximum demand of a system to the rated capacity of the system" [1]. Therefore, the utilization factor (F_u) is

$$F_u \triangleq \frac{\text{Maximum demand}}{\text{Rated system capacity}} \tag{2.2}$$

The utilization factor can also be found for a part of the system. The rated system capacity may be selected to be the smaller of thermal- or voltage-drop capacity [2].

Plant factor: It is the ratio of the total actual energy produced or served over a designated period of time to the energy that would have been produced or served if the plant (or unit) had operated continuously at maximum rating. It is also known as the *capacity factor* or the *use factor*. Therefore,

$$\text{Plant factor} = \frac{\text{Actual energy produced or served} \times T}{\text{Maximum plant rating} \times T} \tag{2.3}$$

It is mostly used in generation studies. For example,

$$\text{Annual plant factor} = \frac{\text{Actual annual energy generation}}{\text{Maximum plant rating}} \tag{2.4}$$

or

$$\text{Annual plant factor} = \frac{\text{Actual annual energy generation}}{\text{Maximum plant rating} \times 8760} \tag{2.5}$$

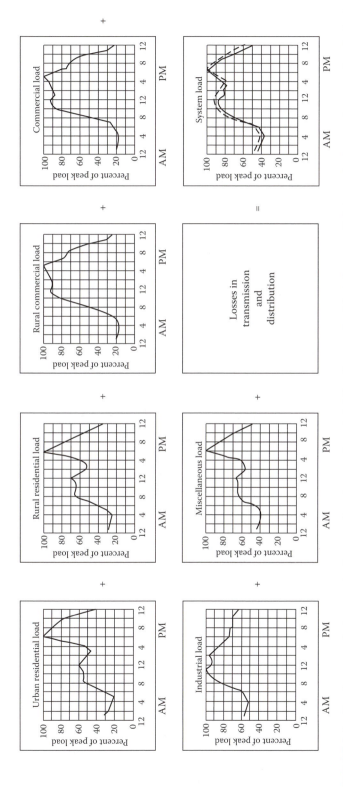

FIGURE 2.4 Development of aggregate load curves for winter peak period. Miscellaneous load includes street lighting and sales to other agencies. Dashed curve shown on system load diagram is actual system generation sent out. Solid curve is based on group load study data. (From Sarikas, R.H. and Thacker, H.B., *AIEE Trans.*, 31(pt. III), 564, August 1957. Used by permission.)

Load factor: It is "the ratio of the average load over a designated period of time to the peak load occurring on that period" [1]. Therefore, the load factor F_{LD} is o average load:

$$F_{LD} \triangleq \frac{\text{Average load}}{\text{Peak load}} \tag{2.6}$$

or

$$F_{LD} \triangleq \frac{\text{Average load} \times T}{\text{Peak load} \times T}$$

$$= \frac{\text{Units served}}{\text{Peak load} \times T} \tag{2.7}$$

where T is the time, in days, weeks, months, or years. The longer the period T, the smaller the resultant factor. The reason for this is that for the same maximum demand, the energy consumption covers a larger time period and results in a smaller average load. Here, when time T is selected to be in days, weeks, months, or years, use it in 24, 168, 730, or 8760 h, respectively. It is less than or equal to 1.0.

Therefore,

$$\text{Annual load factor} = \frac{\text{Total annual energy}}{\text{Annual peak load} \times 8760} \tag{2.8}$$

Diversity factor: It is "the ratio of the sum of the individual maximum demands of the various subdivisions of a system to the maximum demand of the whole system" [1]. Therefore, the diversity factor (F_D) is

$$F_D \triangleq \frac{\text{Sum of individual maximum demands}}{\text{Coincident maximum demand}} \tag{2.9}$$

or

$$F_D = \frac{D_1 + D_2 + D_3 + \cdots + D_n}{D_g} \tag{2.10}$$

or

$$F_D = \frac{\sum_{i=1}^{n} D_i}{D_g} \tag{2.11}$$

where

D_i is the maximum demand of load i, disregarding time of occurrence

$D_g = D_{1+2+3+\cdots+n}$

= coincident maximum demand of group of n loads

The diversity factor can be equal to or greater than 1.0.

From Equation 2.1,

$$DF = \frac{\text{Maximum demand}}{\text{Total connected demand}}$$

or

$$\text{Maximum demand} = \text{Total connected demand} \times DF \tag{2.12}$$

Substituting Equation 2.12 into 2.11, the diversity factor can also be given as

$$F_D = \frac{\sum_{i=1}^{n} TCD_i \times DF_i}{D_g} \tag{2.13}$$

where

TCD$_i$ is the total connected demand of group, or class, i load

DF$_i$ is the demand factor of group, or class, i load

Coincidence factor: It is "the ratio of the maximum coincident total demand of a group of consumers to the sum of the maximum power demands of individual consumers comprising the group both taken at the same point of supply for the same time" [1]. Therefore, the coincidence factor (F_c) is

$$F_c = \frac{\text{Coincident maximum demand}}{\text{Sum of individual maximum demands}} \tag{2.14}$$

or

$$F_c = \frac{D_g}{\sum_{i=1}^{n} D_i} \tag{2.15}$$

Thus, the coincidence factor is the reciprocal of diversity factor, that is,

$$F_c = \frac{1}{F_D} \tag{2.16}$$

These ideas on the diversity and coincidence are the basis for the theory and practice of north-to-south and east-to-west interconnections among the power pools in this country. For example, in the United States during winter, energy comes from south to north, and during summer, just the opposite occurs. Also, east-to-west interconnections help to improve the energy dispatch by means of sunset or sunrise adjustments, that is, the setting of clocks 1 h late or early.

Load diversity: It is "the difference between the sum of the peaks of two or more individual loads and the peak of the combined load" [1]. Therefore, the load diversity (LD) is

$$LD \triangleq \left(\sum_{i=1}^{n} D_i \right) - D_g \tag{2.17}$$

Contribution factor: Manning [2] defines c_i as "the contribution factor of the ith load to the group maximum demand." It is given in per unit of the individual maximum demand of the ith load. Therefore,

$$D_g \triangleq c_1 \times D_1 + c_2 \times D_2 + c_3 \times D_3 + \cdots + c_n \times D_n \tag{2.18}$$

Substituting Equation 2.18 into 2.15,

$$F_c = \frac{c_1 \times D_1 + c_2 \times D_2 + c_3 \times D_3 + \cdots + c_n \times D_n}{\sum_{i=1}^{n} D_i} \tag{2.19}$$

or

$$F_c = \frac{\sum_{i=1}^{n} c_i \times D_i}{\sum_{i=1}^{n} D_i} \tag{2.20}$$

Special cases

Case 1: $D_1 = D_2 = D_3 = \cdots = D_n = D$. From Equation 2.20,

$$F_c = \frac{D \times \sum_{i=1}^{n} c_i}{n \times D} \tag{2.21}$$

or

$$F_c = \frac{\sum_{i=1}^{n} c_i}{n} \tag{2.22}$$

That is, the coincidence factor is equal to the average contribution factor.

Case 2: $c_1 = c_2 = c_3 = \cdots = c_n = c$. Hence, from Equation 2.20,

$$F_c = \frac{c \times \sum_{i=1}^{n} D_i}{\sum_{i=1}^{n} D_i} \tag{2.23}$$

or

$$F_c = c \tag{2.24}$$

That is, the coincidence factor is equal to the contribution factor.

Loss factor: It is "the ratio of the average power loss to the peak-load power loss during a specified period of time" [1]. Therefore, the loss factor (F_{LS}) is

$$F_{LS} \triangleq \frac{\text{Average power loss}}{\text{Power loss at peak load}} \tag{2.25}$$

Equation 2.25 is applicable for the copper losses of the system but not for the iron losses.

Example 2.2

Assume that the annual peak load of a primary feeder is 2000 kW, at which the power loss, that is, total copper, or $\sum I^2 R$ loss, is 80 kW per three phase. Assuming an annual loss factor of 0.15, determine

a. The average annual power loss
b. The total annual energy loss due to the copper losses of the feeder circuits

Solution

a. From Equation 2.25,

$$\text{Average power loss} = \text{power loss at peak load} \times F_{LS}$$

$$= 80 \text{ kW} \times 0.15$$

$$= 12 \text{ kW}$$

b. The total annual energy loss is

$$\text{TAEL}_{Cu} = \text{average power loss} \times 8760 \text{ h/year}$$

$$= 12 \times 8760 = 105{,}120 \text{ kWh}$$

Example 2.3

There are six residential customers connected to a distribution transformer (DT), as shown in Figure 2.5. Notice the code in the customer account number, for example, 4276. The first figure, 4, stands for feeder F4; the second figure, 2, indicates the lateral number connected to the F4 feeder; the third figure, 7, is for the DT on that lateral; and finally the last figure, 6, is for the house number connected to that DT.

Assume that the connected load is 9 kW per house and that the DF and diversity factor for the group of six houses, either from the NL&NP Company's records or from the relevant handbooks, have been decided as 0.65 and 1.10, respectively. Determine the diversified demand of the group of six houses on the DT DT427.

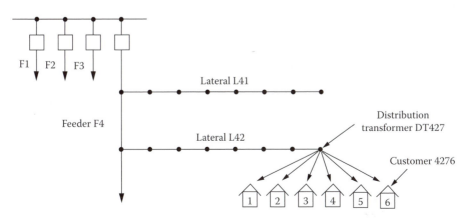

FIGURE 2.5 Illustration of load connected to a distribution transformer.

Solution

From Equation 2.13, the diversified demand of the group on the DT is

$$D_g = \frac{\left(\sum_{i=1}^{6} TCD_i\right) \times DF}{F_D}$$

$$= \frac{\left(\sum_{i=1}^{6} 9kW\right) \times 0.65}{1.1}$$

$$= \frac{6 \times 9kW \times 0.65}{1.1}$$

$$= 31.9kW$$

Example 2.4

Assume that feeder 4 of Example 2.3 has a system peak of 3000 kVA per phase and a copper loss of 0.5% at the system peak. Determine the following:

a. The copper loss of the feeder in kilowatts per phase
b. The total copper losses of the feeder in kilowatts per three phase

Solution

a. The copper loss of the feeder in kilowatts per phase is

$$I^2R \triangleq 0.5\% \times \text{system peak}$$

$$= 0.005 \times 3000 \text{ kVA}$$

$$= 15 \text{ kW per phase}$$

b. The total copper losses of the feeder in kilowatts per three phase is

$$3I^2R \triangleq 3 \times 15$$

$$= 45 \text{ kW per three phase}$$

Example 2.5

Assume that there are two primary feeders supplied by one of the three transformers located at the NL&NP's Riverside distribution substation, as shown in Figure 2.6. One of the feeders supplies an industrial load that occurs primarily between 8 AM and 11 PM, with a peak of 2000 kW at 5 PM. The other one feeds residential loads that occur mainly between 6 AM and 12 PM, with a peak of 2000 kW at 9 PM, as shown in Figure 2.7. Determine the following:

a. The diversity factor of the load connected to transformer T3
b. The load diversity of the load connected to transformer T3
c. The coincidence factor of the load connected to transformer T3

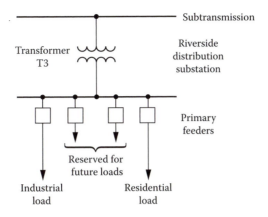

FIGURE 2.6   NL&NP's riverside distribution substation.

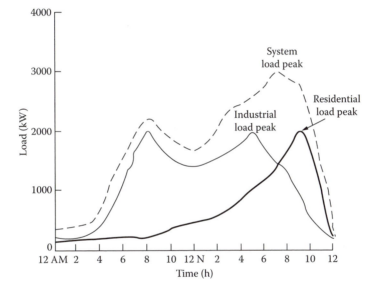

FIGURE 2.7   Daily load curves of a substation transformer.

Solution

a. From Equation 2.11, the diversity factor of the load is

$$F_D = \frac{\sum_{i=1}^{2} D_i}{D_g}$$

$$= \frac{2000 + 2000}{3000} = 1.33$$

b. From Equation 2.17, the load diversity of the load is

$$LD = \sum_{i=1}^{2} D_i - D_g$$

$$= 4000 - 3000 = 1000\,kW$$

c. From Equation 3.16, the coincidence factor of the load is

$$F_c = \frac{1}{F_D}$$

$$= \frac{1}{1.33}$$

$$\cong 0.752$$

Example 2.6

Use the data given in Example 2.1 for the NL&NP's load curve. Note that the peak occurs at 4 PM. Determine the following:

a. The class contribution factors for each of the three load classes
b. The diversity factor for the primary feeder
c. The diversified maximum demand of the load group
d. The coincidence factor of the load group

Solution
a. The class contribution factor is

$$c_i \cong \frac{\text{Class demand at time of system (i.e., group) peak}}{\text{Class noncoincident maximum demand}}$$

For street lighting, residential, and commercial loads,

$$C_{\text{street}} = \frac{0\ \text{kW}}{100\ \text{kW}} = 0$$

$$C_{\text{residential}} = \frac{600\ \text{kW}}{1000\ \text{kW}} = 0.6$$

$$C_{\text{commercial}} = \frac{1200\ \text{kW}}{1200\ \text{kW}} = 1.0$$

b. From Equation 2.11, the diversity factor is

$$F_D = \frac{\sum_{i=1}^{n} D_i}{D_g}$$

and from Equation 2.18,

$$D_g \triangleq c_1 \times D_1 + c_2 \times D_2 + c_3 \times D_3 + \cdots + c_n \times D_n$$

Substituting Equation 2.18 into 2.11,

$$F_D = \frac{\sum_{i=1}^{n} D_i}{\sum_{i=1}^{n} c_i \times D_i}$$

Therefore, the diversity factor for the primary feeder is

$$F_D = \frac{\sum_{i=1}^{3} D_i}{\sum_{i=1}^{3} c_i \times D_i}$$

$$= \frac{100 + 1000 + 1200}{0 \times 100 + 0.6 \times 1000 + 1.0 \times 1200}$$

$$= 1.278$$

c. The diversified maximum demand is the coincident maximum demand, that is, D_g. Therefore, from Equation 2.13, the diversity factor is

$$F_D = \frac{\sum_{i=1}^{n} \mathrm{TCD}_i \times \mathrm{DF}_i}{D_g}$$

where the maximum demand, from Equation 2.12, is

$$\text{Maximum demand} = \text{Total connected demand} \times \mathrm{DF}$$

Substituting Equation 2.12 into 2.13,

$$F_D = \frac{\sum_{i=1}^{n} D_i}{D_g}$$

or

$$D_g = \frac{\sum_{i=1}^{n} D_i}{F_D}$$

Therefore, the diversified maximum demand of the load group is

$$D_g = \frac{\sum_{i=1}^{3} D_i}{F_D}$$

$$= \frac{100 + 1000 + 1200}{1.278}$$

$$= 1800 \text{ kW}$$

d. The coincidence factor of the load group, from Equation 2.15, is

$$F_c = \frac{D_g}{\sum_{i=1}^{n} D_i}$$

or, from Equation 2.16,

$$F_c = \frac{1}{F_D}$$

$$= \frac{1}{1.278}$$

$$= 0.7825$$

2.2 RELATIONSHIP BETWEEN THE LOAD AND LOSS FACTORS

In general, the loss factor cannot be determined from the load factor. However, the limiting values of the relationship can be found [2]. Assume that the primary feeder shown in Figure 2.8 is connected to a variable load. Figure 2.9 shows an arbitrary and idealized load curve. However, it does

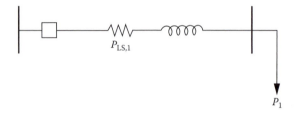

FIGURE 2.8 A feeder with a variable load.

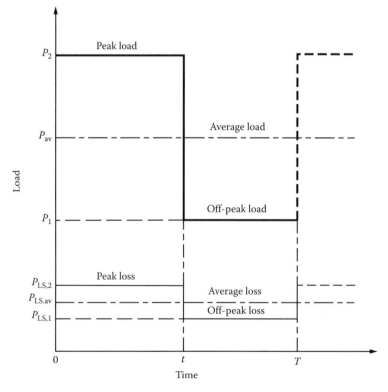

FIGURE 2.9 An arbitrary and ideal load curve.

not represent a daily load curve. Assume that the off-peak loss is $P_{LS,1}$ at some off-peak load P_1 and that the peak loss is $P_{LS,2}$ at the peak load P_2. The load factor is

$$F_{LD} = \frac{P_{av}}{P_{max}} = \frac{P_{av}}{P_2} \tag{2.26}$$

From Figure 2.9,

$$P_{av} = \frac{P_2 \times t + P_1 \times (T - t)}{T} \tag{2.27}$$

Substituting Equation 2.27 into 2.26,

$$F_{LD} = \frac{P_2 \times t + P_1 \times (T - t)}{P_2 \times T}$$

or

$$F_{LD} = \frac{t}{T} + \frac{P_1}{P_2} \times \frac{T - t}{T} \tag{2.28}$$

The loss factor is

$$F_{LS} = \frac{P_{LS,av}}{P_{LS,max}} = \frac{P_{LS,av}}{P_{LS,2}} \tag{2.29}$$

where
$P_{LS,av}$ is the average power loss
$P_{LS,max}$ is the maximum power loss
$P_{LS,2}$ is the peak loss at peak load

From Figure 2.9,

$$P_{LS,av} = \frac{P_{LS,2} \times t + P_{LS,1} \times (T - t)}{T} \tag{2.30}$$

Substituting Equation 2.30 into 2.29,

$$F_{LS} = \frac{P_{LS,2} \times t + P_{LS,1} \times (T - t)}{P_{LS,2} \times T} \tag{2.31}$$

where
$P_{LS,1}$ is the off-peak loss at off-peak load
t is the peak-load duration
$T - t$ is the off-peak-load duration

The copper losses are the function of the associated loads. Therefore, the off-peak and peak loads can be expressed, respectively, as

$$P_{LS,1} = k \times P_1^2 \tag{2.32}$$

and

$$P_{LS,2} = k \times P_2^2 \tag{2.33}$$

where k is a constant. Thus, substituting Equations 2.32 and 2.33 into 2.31, the loss factor can be expressed as

$$F_{LS} = \frac{\left(k \times P_2^2\right) \times t + \left(k \times P_1^2\right) \times (T-t)}{\left(k \times P_2^2\right) \times T} \tag{2.34}$$

or

$$F_{LS} = \frac{t}{T} + \left(\frac{P_1}{P_2}\right)^2 \times \frac{T-t}{T} \tag{2.35}$$

By using Equations 2.28 and 2.35, the load factor can be related to loss factor for three different cases.

Case 1: *Off-peak load is zero.* Here,

$$P_{LS,1} = 0$$

since $P_1 = 0$. Therefore, from Equations 2.28 through 2.35,

$$F_{LD} = F_{LS} = \frac{t}{T} \tag{2.36}$$

That is, the load factor is equal to the loss factor, and they are equal to the t/T constant.

Case 2: *Very short-lasting peak.* Here,

$$t \to 0$$

hence in Equations 2.28 and 2.35,

$$\frac{T-t}{T} \to 1.0$$

therefore,

$$F_{LS} \to (F_{LD})^2 \tag{2.37}$$

That is, the value of the loss factor approaches the value of the load factor squared.

Case 3: *Load is steady.* Here,

$$t \to T$$

That is, the difference between the peak load and the off-peak load is negligible. For example, if the customer's load is a petrochemical plant, this would be the case. Thus, from Equations 2.28 through 2.35,

$$F_{LS} \to F_{LD} \tag{2.38}$$

That is, the value of the loss factor approaches the value of the load factor.

Therefore, in general, the value of the loss factor is

$$F_{LD}^2 < F_{LS} < F_{LD} \tag{2.39}$$

Therefore, the loss factor cannot be determined directly from the load factor. The reason is that the loss factor is determined from losses as a function of time, which, in turn, are proportional to the time function of the square load [2–4].

However, Buller and Woodrow [5] developed an approximate formula to relate the loss factor to the load factor as

$$F_{LS} = 0.3F_{LD} + 0.7F_{LD}^2 \tag{2.40a}$$

where

F_{LS} is the loss factor, pu

F_{LD} is the load factor, pu

Equation 2.40a gives a reasonably close result. Figure 2.10 gives three different curves of loss factor as a function of load factor. Relatively recently, the formula given earlier has been modified for rural areas and expressed as

$$F_{LS} = 0.16F_{LD} + 0.84F_{LD}^2 \tag{2.40b}$$

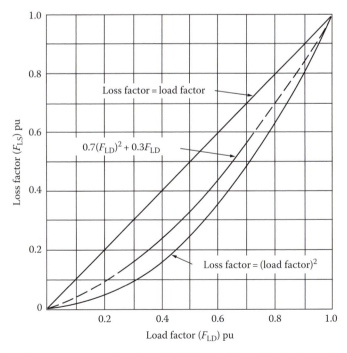

FIGURE 2.10 Loss factor curves as a function of load factor. (From Westinghouse Electric Corporation, *Electric Utility Engineering Reference Book-Distribution Systems*, Vol. 3, Westinghouse Electric Corporation, East Pittsburgh, PA, 1965.)

Example 2.7

The average load factor of a substation is 0.65. Determine the average loss factor of its feeders, if the substation services

    a. An urban area
    b. A rural area

Solution

    a. For the urban area,

$$F_{LS} = 0.3F_{LD} + 0.7(F_{LD})^2$$

$$= 0.3(0.65) + 0.7(0.65)^2$$

$$= 0.49$$

    b. For the rural area,

$$F_{LS} = 0.16F_{LD} + 0.84(F_{LD})^2$$

$$= 0.16(0.65) + 0.84(0.65)^2$$

$$= 0.53$$

Example 2.8

Assume that the Riverside distribution substation of the NL&NP Company supplying Ghost Town, which is a small city, experiences an annual peak load of 3500 kW. The total annual energy supplied to the primary feeder circuits is 10,000,000 kWh. The peak demand occurs in July or August and is due to air-conditioning load.

    a. Find the annual average power demand.
    b. Find the annual load factor.

Solution

Assume a monthly load curve as shown in Figure 2.11.

    a. The annual average power demand is

$$\text{Annual } P_{av} = \frac{\text{Total annual energy}}{\text{Year}}$$

$$= \frac{10^7 \text{ kWh/year}}{8760 \text{ h/year}}$$

$$= 1141 \text{ kW}$$

    b. From Equation 2.6, the annual load factor is

$$F_{LD} = \frac{\text{Annual average load}}{\text{Annual peak demand}}$$

$$= \frac{1141 \text{ kW}}{3500 \text{ kW}}$$

$$= 0.326$$

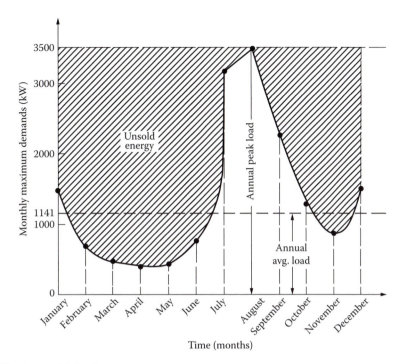

FIGURE 2.11 A monthly load curve.

or, from Equation 2.8,

$$\text{Annual load factor} = \frac{\text{Total annual energy}}{\text{Annual peak load} \times 8760}$$

$$= \frac{10^7 \text{ kWh/year}}{3500 \text{ kW} \times 8760}$$

$$= 0.326$$

The unsold energy, as shown in Figure 2.11, is a measure of capacity and investment cost. Ideally, it should be kept at a minimum.

Example 2.9

Use the data given in Example 2.8 and suppose that a new load of 100 kW with 100% annual load factor is to be supplied from the Riverside substation. The investment cost, or capacity cost, of the power system upstream, that is, toward the generator, from this substation is $18.00/kW per month. Assume that the energy delivered to these primary feeders costs the supplier, that is, NL&NP, $0.06/kWh.

    a. Find the new annual load factor on the substation.
    b. Find the total annual cost to NL&NP to serve this load.

Solution

Figure 2.12 shows the new load curve after the addition of the new load of 100 kW with 100% load.

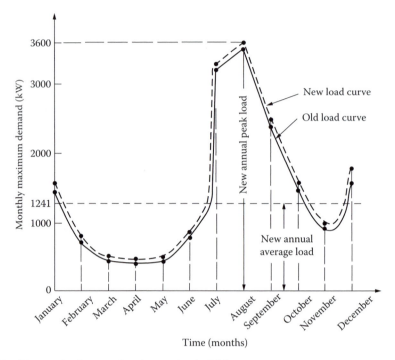

FIGURE 2.12   The new load curve after the new load addition.

a. The new annual load factor on the substation is

$$F_{LD} = \frac{\text{Annual average load}}{\text{Annual peak demand}}$$

$$= \frac{1141+100}{3500+100}$$

$$= 0.345$$

b. The total annual and additional cost to NL&NP to serve the additional 100 kW load has two cost components, namely, (1) annual capacity cost and (2) annual energy cost. Therefore,

$$\text{Annual additional capacity cost} = \$18/\text{kW/month} \times 12 \text{ month/year} \times 100 \text{ kW}$$

$$= \$21,600$$

and

$$\text{Annual energy cost} = 100 \text{ kW} \times 8,760 \text{ h/year} \times \$0.06/\text{kWh}$$

$$= \$52,560$$

Therefore,

$$\text{Total annual additional costs} = \text{Annual capacity cost} + \text{Annual energy cost}$$

$$= \$21,600 + \$52,560$$

$$= \$74,160$$

Example 2.10

Assume that the annual peak-load input to a primary feeder is 2000 kW. A computer program that calculates voltage drops and I^2R losses shows that the total copper loss at the time of peak load is $\sum I^2R = 100$ kW. The total annual energy supplied to the sending end of the feeder is 5.61×10^6 kWh.

a. By using Equation 2.40, determine the annual loss factor.
b. Calculate the total annual copper loss energy and its value at $0.06/kWh.

Solution

a. From Equation 2.40, the annual loss factor is

$$F_{LS} = 0.3F_{LD} + 0.7F_{LD}^2$$

where

$$F_{LD} = \frac{5.61 \times 10^6 \text{ kWh}}{2000 \text{ kW} \times 8760 \text{ h/year}}$$

$$= 0.32$$

Therefore,

$$F_{LS} = 0.3 \times 0.32 + 0.7 \times 0.32^2$$

$$\cong 01677$$

b. From Equation 2.25,

$$F_{LS} \triangleq \frac{\text{Average power loss}}{\text{Power loss at peak load}}$$

or

$$\text{Average power loss} = 0.1677 \times 100 \text{ kW}$$

$$= 16.77 \text{ kW}$$

Therefore,

$$\text{Total annual copper loss} = 16.77 \text{ kW} \times 8760 \text{ h/year}$$

$$= 146,905 \text{ kWh}$$

and

$$\text{Cost of total annual copper loss} = 146,905 \text{ kWh} \times \$0.06/\text{kWh}$$

$$= \$8,814$$

Example 2.11

Assume that one of the DTs of the Riverside substation supplies three primary feeders. The 30 min annual maximum demands per feeder are listed in the following table, together with the power factor (PF) at the time of annual peak load.

	Demand	
Feeder	kW	PF
1	1800	0.95
2	2000	0.85
3	2200	0.90

Assume a diversity factor of 1.15 among the three feeders for both real power (P) and reactive power (Q).

a. Calculate the 30 min annual maximum demand on the substation transformer in kilowatts and in kilovoltamperes.
b. Find the load diversity in kilowatts.
c. Select a suitable substation transformer size if zero load growth is expected and if company policy permits as much as 25% short-time overloads on the distribution substation transformers. Among the standard three-phase (3ϕ) transformer sizes available are the following:

   2500/3125 kVA self-cooled/forced-air-cooled
   3750/4687 kVA self-cooled/forced-air-cooled
   5000/6250 kVA self-cooled/forced-air-cooled
   7500/9375 kVA self-cooled/forced-air-cooled

d. Now assume that the substation load will increase at a constant percentage rate per year and will double in 10 years. If the 7500/9375 kVA-rated transformer is installed, in how many years will it be loaded to *its fans-on* rating?

Solution

a. From Equation 2.10,

$$F_D = \frac{1800 + 2000 + 2200}{D_g} = 1.15$$

Therefore,

$$D_g = \frac{6000}{1.15} = 5217\,kW = P$$

To find power in kilovoltamperes, find the PF angles. Therefore,

$$PF_1 = \cos\theta_1 = 0.95 \rightarrow \theta_1 = 18.2°$$

$$PF_2 = \cos\theta_2 = 0.85 \rightarrow \theta_2 = 31.79°$$

$$PF_3 = \cos\theta_3 = 0.90 \rightarrow \theta_3 = 25.84°$$

Thus, the diversified reactive power (Q) is

$$Q = \frac{\sum_{i=1}^{3} P_i \times \tan\theta}{F_D}$$

$$= \frac{1800 \times \tan 18.2° + 2000 \times \tan 31.79° + 2200 \times \tan 25.84°}{1.15}$$

$$= 2518.8 \text{ kvar}$$

Therefore,

$$D_g = (P^2 + Q^2)^{1/2} = S$$

$$= (5217^2 + 2518.8^2)^{1/2} = 5793.60 \text{ kVA}$$

b. From Equation 2.17, the load diversity is

$$LD = \sum_{i=1}^{3} D_i - D_g$$

$$= 6000 - 5217 = 783 \text{ kW}$$

c. From the given transformer list, it is appropriate to choose the transformer with the 3750/4687-kVA rating since with the 25% short-time overload, it has a capacity of

$$4687 \times 1.25 = 5858.8 \text{ kVA}$$

which is larger than the maximum demand of 5793.60 kVA as found in part (a).

d. Note that the term *fans-on* rating means the forced-air-cooled rating. To find the increase (g) per year,

$$(1+g)^{10} = 2$$

hence,

$$1 + g = 1.07175$$

or

$$g = 7.175\%/\text{year}$$

Thus,

$$(1.07175)^n \times 5793.60 = 9375 \text{ kVA}$$

or

$$(1.07175)^n = 1.6182$$

Therefore,

$$n = \frac{\ln 1.6182}{\ln 1.07175}$$

$$= \frac{0.48130}{0.06929} = 6.946, \quad \text{or} \quad 7 \text{ years}$$

Therefore, if the 7500/9375 kVA-rated transformer is installed, it will be loaded to its *fans-on* rating in about 7 years.

2.3 MAXIMUM DIVERSIFIED DEMAND

Arvidson [7] developed a method of estimating DT loads in residential areas by the diversified-demand method, which takes into account the diversity between similar loads and the noncoincidence of the peaks of different types of loads.

To take into account the noncoincidence of the peaks of different types of loads, Arvidson introduced the hourly variation factor. It is "the ratio of the demand of a particular type of load coincident with the group maximum demand to the maximum demand of that particular type of load [2]." Table 2.2 gives the hourly variation curves for various types of household appliances. Figure 2.13 shows a number of curves for various types of household appliances to determine the average maximum diversified demand per customer in kilowatts per load. In Figure 2.13, each curve represents a 100% saturation level for a specific demand.

To apply Arvidson's method to determine the maximum diversified demand for a given saturation level and appliance, the following steps are suggested [2]:

1. Determine the total number of appliances by multiplying the total number of customers by the per-unit saturation.
2. Read the corresponding diversified demand per customer from the curve, in Figure 2.13, for the given number of appliances.
3. Determine the maximum demand, multiplying the demand found in step 2 by the total number of appliances.
4. Finally, determine the contribution of that type load to the group maximum demand by multiplying the resultant value from step 3 by the corresponding hourly variation factor found from Table 2.2.

Example 2.12

Assume a typical DT that serves six residential loads, that is, houses, through six service drops (SDs) and two spans of secondary line (SL). Suppose that there are a total of 150 DTs and 900 residences supplied by this primary feeder. Use Figure 2.13 and Table 2.2. For the sake of illustration, assume that a typical residence contains a clothes dryer, a range, a refrigerator, and some lighting and miscellaneous appliances. Determine the following:

  a. The 30 min maximum diversified demand on the DT.
  b. The 30 min maximum diversified demand on the entire feeder.
  c. Use the typical hourly variation factors given in Table 2.2 and calculate the small portion of the daily demand curve on the DT, that is, the total hourly diversified demands at 4, 5, and 6 PM, on the DT, in kilowatts.

TABLE 2.2
Hourly Variation Factors

| Hour | Lighting and Miscellaneous | Refrigerator | Home Freezer | Range | Air-Conditioning[a] | Heat Pump[a] | | | Water Heater[b] OPWH[c] | | | Clothes[d] Dryer |
						Cooling Season	Heating Season	House[a] Heating	Both Elements Restricted	Only Bottom Elements Restricted	Uncontrolled	
12 AM	0.32	0.93	0.92	0.02	0.40	0.42	0.34	0.11	0.41	0.61	0.51	0.03
1	0.12	0.89	0.90	0.01	0.39	0.35	0.49	0.07	0.33	0.46	0.37	0.02
2	0.10	0.80	0.87	0.01	0.36	0.35	0.51	0.09	0.25	0.34	0.30	0
3	0.09	0.76	0.85	0.01	0.35	0.28	0.54	0.08	0.17	0.24	0.22	0
4	0.08	0.79	0.82	0.01	0.35	0.28	0.57	0.13	0.13	0.19	0.15	0
5	0.10	0.72	0.84	0.02	0.33	0.26	0.63	0.15	0.13	0.19	0.14	0
6	0.19	0.75	0.85	0.05	0.30	0.26	0.74	0.17	0.17	0.24	0.16	0
7	0.41	0.75	0.85	0.30	0.41	0.35	1.00	0.76	0.27	0.37	0.46	0
8	0.35	0.79	0.86	0.47	0.53	0.49	0.91	1.00	0.47	0.65	0.70	0.08
9	0.31	0.79	0.86	0.28	0.62	0.58	0.83	0.97	0.63	0.87	1.00	0.20
10	0.31	0.79	0.87	0.22	0.72	0.70	0.74	0.68	0.67	0.93	1.00	0.65
11	0.30	0.85	0.90	0.22	0.74	0.73	0.60	0.57	0.67	0.93	0.99	1.00
12 noon	0.28	0.85	0.92	0.33	0.80	0.84	0.57	0.55	0.67	0.93	0.98	0.98
1	0.26	0.87	0.96	0.25	0.86	0.88	0.49	0.51	0.61	0.85	0.86	0.70
2	0.29	0.90	0.98	0.16	0.89	0.95	0.46	0.49	0.55	0.76	0.82	0.65
3	0.30	0.90	0.99	0.17	0.96	1.00	0.40	0.48	0.49	0.68	0.81	0.63
4	0.32	0.90	1.00	0.24	0.97	1.00	0.43	0.44	0.33	0.46	0.79	0.38
5	0.70	0.90	1.00	0.80	0.99	1.00	0.43	0.79	0	0.09	0.75	0.30
6	0.92	0.90	0.99	1.00	1.00	1.00	0.49	0.88	0	0.13	0.75	0.22
7	1.00	0.95	0.98	0.30	0.91	0.88	0.51	0.76	0	0.19	0.80	0.26
8	0.95	1.00	0.98	0.12	0.79	0.73	0.60	0.54	1.00	1.00	0.81	0.20
9	0.85	0.95	0.97	0.09	0.71	0.72	0.54	0.42	0.84	0.98	0.73	0.18
10	0.72	0.88	0.96	0.05	0.64	0.53	0.51	0.27	0.67	0.77	0.67	0.10
11	0.50	0.88	0.95	0.04	0.55	0.49	0.34	0.23	0.54	0.69	0.59	0.04
12 PM	0.32	0.93	0.92	0.02	0.40	0.42	0.34	0.11	0.44	0.61	0.51	0.03

Source: From Sarikas, R.H. and Thacker, H.B., *AIEE Trans.*, 31(pt. III), 564, August 1957. With permission.

[a] Load cycle and maximum diversified demand are dependent on outside temperature, dwelling construction and insulation, among other factors.

[b] Load cycle and maximum diversified demands are dependent on tank size, and heater element rating; values shown apply to 52 gal tank, 1500 and 1000 W elements.

[c] Load cycle dependent on schedule of water heater restriction.

[d] Hourly variation factor is dependent on living habits of individuals; in a particular area, values may be different from those shown.

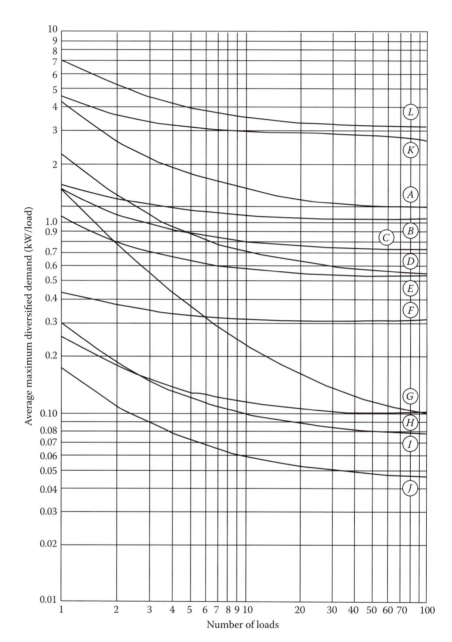

FIGURE 2.13 Maximum diversified 30 min demand characteristics of various residential loads: *A*, clothes dryer; *B*, off-peak water heater, "off-peak" load; *C*, water heater, uncontrolled, interlocked elements; *D*, range; *E*, lighting and miscellaneous appliances; *F*, 0.5-hp room coolers; *G*, off-peak water heater, "on-peak" load, upper element uncontrolled; *H*, oil burner; *I*, home freezer; *J*, refrigerator; *K*, central air-conditioning, including heat-pump cooling, 5-hp heat pump (4-ton air conditioner); *L*, house heating, including heat-pump-heating-connected load of 15 kW unit-type resistance heating or 5 hp heat pump. (From Westinghouse Electric Corporation, *Electric Utility Engineering Reference Book-Distribution Systems*, Vol. 3, Westinghouse Electric Corporation, East Pittsburgh, PA, 1965.)

Solution

a. To determine the 30 min maximum diversified demand on the DT, the average maximum diversified demand per customer is found from Figure 2.13. Therefore, *when the number of loads is six*, the average maximum diversified demands per customer are

$$P_{av,max} = \begin{cases} 1.6 \text{ kW/house} & \text{for dryer} \\ 0.8 \text{ kW/house} & \text{for range} \\ 0.066 \text{ kW/house} & \text{for refrigerator} \\ 0.61 \text{ kW/house} & \text{for lighting and misc. appliances} \end{cases}$$

Thus,

$$\sum_{i=1}^{4} \left(P_{av,max}\right)_i = 1.6 + 0.8 + 0.066 + 0.61$$

$$= 3.076 \text{ kW/house}$$

and for six houses

$$(3.076 \text{ kW/house})(6 \text{ houses}) = 18.5 \text{ kW}$$

Thus, the contributions of the appliances to the 30 min maximum diversified demand on the DT is approximately 18.5 kW.

b. As in part (a), the average maximum diversified demand per customer is found from Figure 2.13. Therefore, *when the number of loads is 900* (note that, due to the given curve characteristics, the answers would be the same as the ones for the number of loads of 100), then the average maximum diversified demands per customer are

$$P_{av,max} = \begin{cases} 1.2 \text{ kW/house} & \text{for dryer} \\ 0.53 \text{ kW/house} & \text{for range} \\ 0.044 \text{ kW/house} & \text{for refrigerator} \\ 0.52 \text{ kW/house} & \text{for lighting and miscellaneous appliances} \end{cases}$$

Hence,

$$\sum_{i=1}^{4} (P_{av,max})_i = 1.2 + 0.53 + 0.044 + 0.0.52$$

$$= 2.294 \text{ kW/house}$$

Therefore, the 30 min maximum diversified demand on the entire feeder is

$$\sum_{i=1}^{4} (P_{av,max})_i = 900 \times 2.294$$

$$= 2064.6 \text{ kW/feeder}$$

However, if the answer for the 30 min maximum diversified demand on one DT found in part (a) is multiplied by 150 to determine the 30 min maximum diversified demand on the entire feeder, the answer would be

$$150 \times 18.5 \cong 2775 \ \text{kW}$$

which is greater than the demand 2064.6 kW found previously. This discrepancy is due to the application of the appliance diversities.

c. From Table 2.2, the *hourly variation factors* can be found as 0.38, 0.24, 0.90, and 0.32 for dryer, range, refrigerator, and lighting and miscellaneous appliances. Therefore, the total hourly diversified demands on the DT can be calculated as given in the following table in which

(1.6 kW/house)(6 houses)	= 9.6 kW
(0.8 kW/house)(6 houses)	= 4.8 kW
(0.066 kW/house)(6 houses)	= 0.4 kW
(0.61 kW/house)(6 houses)	= 3.7 kW

Time	Dryers, kW	Ranges, kW	Refrigerators, kW	Lighting and Misc. Appliances, kW	Total Hourly Diversified Demand, kW
(1)	(2)	(3)	(4)	(5)	(6)
4 PM	9.6×0.38	4.8×0.24	0.4×0.90	3.7×0.32	6.344
5 PM	9.6×0.30	4.8×0.80	0.4×0.90	3.7×0.70	9.670
6 PM	9.6×0.22	4.8×1.00	0.4×0.90	3.7×0.92	10.674

Note: The results given in column (6) are the sum of the contributions to demand given in columns (2)–(5).

2.4 LOAD FORECASTING

The load growth of the geographical area served by a utility company is the most important factor influencing the expansion of the distribution system. Therefore, forecasting of load increases is essential to the planning process.

Fitting trends after transformation of data is a common practice in technical forecasting. An arithmetic straight line that will not fit the original data may fit, for example, the logarithms of the data as typified by the exponential trend

$$y_t = ab^x \tag{2.41}$$

This expression is sometimes called a *growth equation*, since it is often used to explain the phenomenon of growth through time. For example, if the load growth rate is known, the load at the end of the *n*th year is given by

$$P_n = P_0(1+g)^n \tag{2.42}$$

where
  P_n is the load at the end of the *n*th year
  P_0 is the initial load
  g is the annual growth rate
  n is the number of years

Now, if it is set so that $P_n = y_t$, $P_0 = a$, $1 + g = b$, and $n = x$, then Equation 2.42 is identical to the exponential trend equation (2.41). Table 2.3 gives a MATLAB® computer program to forecast the future demand values if the past demand values are known.

TABLE 2.3

MATLAB® Demand-Forecasting Computer Program

```
%RLXD = read past demand values in MW
%RLXC = predicted future demand values in MW
%NP = number of years in the past up to the present
%NF = number of years from the present to the future that will be predicted
NP = input('Enter the number of years in the past up to the present:');
NF = input('Enter the number of years from the present to the future that
  will be predicted:');
for I = 1:NP
fprintf('Enter the past demand values in MW:" I); RLXD(I) = input(");
end
SXIYI = 0; SXISQ = 0; SXI = 0; SYI = 0; SYISQ = 0;
for I = I:NP XI = I-I;
Y(I) = 10g(RLXD(I)); SXIYI = SXIYI+XI*Y(I); SXI = SXI+X1; SY1 = SY1+ Y (I);
  SXISQ = SXISQ+ XI/\2; SY1SQ = SY1SQ+ Y(1)A2;
end
A = (SXIY1-(SXI*SY1)/NP)/(SXISQ-(SXI/\2)/NP); B = SYI/NP-A *SXI/NP;
R = exp(A);
RLXC(1) = exp(B);
RG = R-1;
fprintf('\n\nRate of growth =%f\n\n', RG);
NN = NP+NF;
for I = 2:NN XI = I-I;
DY = A * XI + B; RLXC(I) = exp(DY);
end
fprintf('\tRLXD\t\tRLXC\n'); for I = I:NP
fprintf('\t%f\t%f\n', RLXD(I), RLXC(I)); end'
for I = I:NF
IP = I + NP; fprintf('\t\t\t\t%f\n', RLXC(IP)); end
```

In order to plan the resources required to supply the future loads in an area, it is necessary to forecast the magnitude and distribution of these loads as accurately as possible. Such forecasts are normally based upon projections of the historical growth trend for the area and the existing load distribution within the area. Adjustments must be made for load transfers into and out of the area and for the addition or removal of block loads that are too large to be considered part of normal growth.

Before the 1973–1974 oil embargo, an exponential projection of adjusted historical peak loads provided satisfactory load forecasts for most distribution study areas. The growth in customers was reasonably steady, and the demand per customer continued to increase. However, in recent years, the picture has drastically changed. Energy conservation, load management, increasing electric rates, and a slow economy have combined to slow the growth. As a result, an exponential growth rate, such as the one given in the first part of this section, is no longer valid in most study areas.

Methods that forecast future demand by location divide the utility service area into a set of small areas forecasting the load growth in each. Most modern small-area forecast methods work with a uniform grid of small areas that covers the utility service area, as explained in Section 1.3.1, but the more traditional approach was to forecast the growth on a substation-by-substation or feeder-by-feeder basis, letting equipment service areas implicitly define the small areas.

Regardless of how small areas are defined, most forecasting methods themselves invariably fall into one of two categories, trending or land use. Trending methods extrapolate past historical peak loads using curve fitting or some other methods.

Contrarily, the behavior of load growth, in any relatively small area (served by substation, or feeder), is not a smooth curve, but is more like a sharp Gompertz curve, commonly referred to as an "S" curve. The S curve exhibits the distinct phases, namely, dormant, growth, and saturation phases. In the dormant phase, the small area has no load growth. In the growth phase, the load growth happens at a relatively rapid rate, usually due to new construction. In the saturation phase, the small area is fully developed. Any increase in load growth is extremely small.

By contrast, land-use simulation involves mapping existing and likely additions to land coverage by customer class definitions like residential, commercial, and industrial, in order to forecast growth. Either way, the ultimate goal is to project changes in the density of peak demand on a locality basis.

In order to plan a T&D system, it is necessary not just to study overall load in a region, but to study and forecast load on a *spatial basis*, that is, *analyzing it in total and on a local area* basis throughout the system, determining the *where* aspect of the load growth as well as the *how much*. Both are essential for determining T&D expansion needs.

Trend (or *regression analysis*) is the study of the behavior of a time series or a process in the past and its mathematical modeling so that future behavior can be extrapolated from it. Two usual approaches followed for trend analysis are

1. The fitting of continuous mathematical functions through actual data to achieve the least overall error, known as *regression analysis*
2. The fitting of a sequence on discontinuous lines or curves to the data

The second approach is more widespread in short-term forecasting. A time-varying event such as distribution system load can be broken down into the following four major components:

a. Basic trend.
b. Seasonal variation, that is, monthly or yearly variation of load.
c. Cyclic variation that includes influences of periods longer than that provided earlier and causes the load pattern to be repeated for 2 or 3 years or even longer cycles.
d. Random variations that occur on account of the day-to-day changes and in the case of power systems are usually dependent on weather and the time of the week, for example, weekday and weekend.

The principle of regression theory is that any function $y = f(x)$ can be fitted to a set of points (x_1, y_1), (x_2, y_2) so as to minimize the sum of errors squared to each point, that is,

$$\varepsilon^2 = \sum_{i=1}^{n} [y_i - f(x)]^2 = \text{minimum}$$

Sum of squared errors is used as it gives a significant indication of *goodness of fit*. Typical regression curves used in power system forecasting are

Linear	$y = a + bx$
Exponential	$y = a(1 + b)^x$
Power	$y = ax^b$
Polynomial	$y = a + bx + cx^2$
Gompertz	$y = ae^{-be^{-cx}}$

The coefficients used in these equations are called *regression coefficients*. The following are some of the methodologies used in applying some of the regression curves provided earlier:

Linear regression: It is applied by using the *method of least squares*. Here, the line $y = a + bx$ is fitted to the sets of points (x_1, y_1), (x_2, y_2), ..., (x_n, y_n), that is,

$$\varepsilon^2 = \sum_{i=1}^{n} [y_i - (a + bx_i)]^2 = \text{minimum}$$

By taking partial differentiation with respect to the regression coefficients and setting the resultant equations to zero to achieve the minimum error criterion,

$$a = \frac{\left(\sum y\right)\left(\sum x^2\right) - \left[\sum (x)\right] \cdot \left(\sum xy\right)}{n \sum x^2 - \left(\sum x\right)^2} \tag{2.43}$$

and

$$b = \frac{n\left(\sum xy\right) - \left(\sum x\right) \cdot \left(\sum y\right)}{n \sum x^2 - \left(\sum x\right)^2} \tag{2.44}$$

The earlier process is also referred to as the *least square line* method.

Least square parabola: The parabola curve of $y = a + bx + cx^2$ is fitted to minimize the sum of squared errors, that is,

$$\varepsilon^2 = \sum_{i=1}^{n} [y_i - (a + bx + cx^2)]^2 = \text{minimum}$$

By taking partial differentiation with respect to the regression coefficients and setting the resultant equations to zero give simultaneous equations that can be solved for a, b, and c coefficients.

Least square exponential: Here, the same approach that has been used in linear regression can be used at first, but $\sum y$ is replaced by $\sum \ln y$ in Equations 2.43 and 2.44, and the regression coefficients are found. The resultant coefficients are then transformed back.

Multiple regression: Two or more variables can be treated by an extension of the same principle. For example, if an equation of $z = a + bx + cy$ is required to fit to a series of points (x_1, y_1, z_1), (x_2, y_2, z_2),... then this is a multiple linear regression. Multi-nonlinear regressions are also used. Just like before, set the sum of squared errors,

$$\varepsilon^2 = \sum_{i=1}^{n} \left[z_i - (a + bx + cy^2)\right]^2 = \text{minimum}$$

then differentiate it with respect to a, b, and c so that one can get the following three simultaneous equations:

$$an + b\sum x_i + c\sum y_i = \sum z_i$$

$$a\sum x_i + b\sum x_i^2 + c\sum x_i y_i = \sum x_i z_i$$

$$a\sum y_i + b\sum x_i y_i + c\sum y_i^2 = \sum y_i z_i$$

which can be solved for a, b, and c.

2.4.1 Box–Jenkins Methodology

This method uses a stochastic time series to forecast future load demands. It is a popular method for short-term (5 years or less) forecasting. Box and Jenkins [8] developed this method of forecasting by trying to account for repeated movements in the historical series (those movements comprising a trend), leaving a series made up of only random, that is, irregular movements. To model the systematic patterns inherent in this series, the method relies upon autoregressive and moving average processes to account for cyclical movements and upon differencing to account for seasonal and secular movements. The Box–Jenkins methodology is an iterative procedure by which a stochastic model is constructed. The process starts from the most simple structure with the least number of parameters and develops into as complex a structure as necessary to obtain an *adequate model*, in the sense of yielding white noise only [9].

2.4.2 Small-Area Load Forecasting

In this type of forecasting, the utility service area is divided into a set of small areas, and the future load growth in each area is forecasted. Most modern small-area forecast methods work with a uniform grid of small area that covers the utility service area, but the more traditional approach was to forecast growth on a substation-by-substation or feeder-by-feeder basis, letting equipment service areas implicitly define the small areas. Regardless of how small areas are defined, most forecasting methods are based on trending or land use.

Trending methods have been explained in Section 2.5; by contrast, land-use simulation involves mapping existing and likely additions to land coverage by customer class definitions like residential, commercial, and industrial, in order to forecast future growth. In either way, the final goal is to project changes in the density of peak demand on a locality basis.

According to Willis [10], small-area growth is not a smooth, continuous process from year to year. Instead, growth in a small area is intense for several years, then drops to very low levels while high growth suddenly begins in other areas. This led to the characterization of small-area growth with Gompertz or the S curve. Its use does not imply that small-area growth always follows an S-shaped load history, but only that there is seldom a middle ground between high and low growth rates.

Therefore, small-area forecasting is less a process of extrapolating trends; it is a determination of when small areas transition among zero, high, and low growth states. Land-use methods are much better at predicting such growth-state transitions. Furthermore, the forecaster gets better and more meaningful answers to "what-if?" type questions by using land-use-based simulation methods.

2.4.3 Spatial Load Forecasting

In general, small-area load growth is a spatial process. Also the majority of load growth effects in any small area are due to effects from other small areas, some very far away, and a function of the distances to those areas. Therefore, the forecast of any one area must be based upon an assessment data not only for that area, but also for many other neighboring areas.

The best available trending method in terms of tested accuracy is load-trend-coupled (LTC) extrapolation, using a modified form of Markov regression, in which the peak-load histories of up to several hundred small areas are extrapolated in a single computation, with the historical trend in each area influencing the extrapolation of others.

The influence of one area's trend on other's is found by using pattern recognition as another's is found by using pattern recognition as a function of past trends and locations, making LTC trending a true spatial method. Its main advantage is economy of use. Only the peak-load histories of substations and feeders and X–Y locations of substations are required as input [10].

Figure 2.14 illustrates this method. It works with land-use classes that correspond to utility rate classes, differentiating electric consumption within each by end-use category, for example, heating, lighting, using peak day load curves on a 15 min demand-period basis. It is applied on a grid basis, with a spatial resolution of 2.5 acres (square areas 1/16 miles across). Base spatial data include

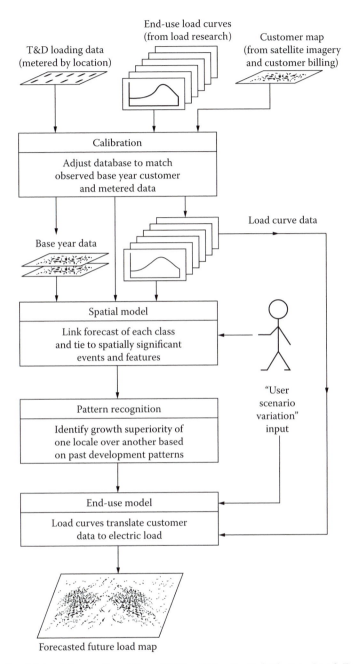

FIGURE 2.14 Spatial load forecasting. (From Willis, H.L., *Spatial Electric Load Forecasting*, Marcel Dekker, New York, 1996.)

multispectral satellite imagery of the region, used for land-use identification and mapping purposes, customer/billing/rate class data, end-use load curve and load research surveys, and metered load curve readings by substation throughout the system.

There are two inputs that control the forecast. The first one is the utility system-wide rate and marketing forecast. The second one is an optional set of scenario descriptors that allow the user to change future conditions to answer "what-if?" questions. It is very important that the base year model must provide accurately all known readings about customers, customer density, metered load curves, and their simultaneous variations in location and time.

Example 2.13

Write a simple MATLAB demand forecasting computer program based on the least-square exponential.

Solution

The MATLAB demand forecasting computer program is given in Table 2.4:

TABLE 2.4
Demand Forecasting MATLAB Program

```
%%%%%%%%%%%%%%%%%%%%%%%%%%%%%%%%%%%%%%%%%%%%%%%%%%%%%%%%%%%%%%%%%%%%
% demand forecasting matlab program
%%%%%%%%%%%%%%%%%%%%%%%%%%%%%%%%%%%%%%%%%%%%%%%%%%%%%%%%%%%%%%%%%%%%
fprintf('\nDemand Forecast\n');
fprintf('\nEnter an array of demand values in the form:\n');
fprintf('\t[yr1 ld1; yr2 ld2; yr3 ld3; yr4 ld4; yr5 ld5]\n');
past_dem = input('\nEnter year/demand values: ');
sizepd = size(past_dem);

% get the # of past years of data and the # of cols in the array
np = sizepd(1); cols = sizepd(2);

% get the number of years to predict
nf = input('\nEnter the number of year to predict: ');
ntotal = np + nf;

% obtain the least-square terms to estimate the ld growth value g
% y = ab^x must be transformed to ln(y) = ln(a) + x*ln(b)
Y = log(past_dem(:,2))'; X = 0:np - 1;
sumx2 = (X - mean(X))*(X - mean(X))';
sumxy = (Y - mean(Y))*(X - mean(X))';

% get the coeffs of the transformed data A = ln(b) and B = ln(a)
A = sumxy/sumx2; B = mean(Y) - A*mean(X);

% solve for the initial value, Po and g
Po = exp(B);
g = exp(A) - 1;

fprintf('\n\tRate of growth =%2.2f%%\n\n', g*100);
fprintf('\tYEAR\tACTUAL\t\tFORECAST\n');

% calculate the estimated values
est_dem = 0;
for i = 1:ntotal
 n = i - 1;
 % year = first year + n
 est_dem(i, 1) = past_dem(1, 1) + n;
 % load growth equation
 est_dem(i, 2) = Po*(1+g)^n;
 if i < = np
  fprintf('\t%4d\t%6.2f\t\t%6.2f\n', est_dem(i,1),past_dem(i,2),est_
    dem(i,2));
 else
  fprintf('\t%4d\t-\t\t\t%6.2f\n', est_dem(i,1),est_dem(i,2));
 end
end
plot(past_dem(:,1),past_dem(:,2), 'k-s', est_dem(:,1), est_dem(:,2), 'k-+');
xlabel('Year'); ylabel('Demand'); legend('Actual', 'Forecast');
```

Example 2.14

Assume the peak MW July demands for the last 8 years have been the following: 3094, 2938, 2714, 3567, 4027, 3591, 4579, and 4436. Use the MATLAB program given in Example 2.13 as a curve-fitting technique and determine the following:

a. The average rate of growth of the demand.
b. Find out the ideal data based on growth for the past 8 years to give the correct demand forecast.
c. The forecasted future demands for the next 10 years.
d. Plot the results found in parts (a) and (b).

Solution

Here is the program output showing the answers for the parts (a) through (c). The answer for part (d) is given in Figure 2.15.

```
Program Output
EDU» load_growth

Demand Forcast

Enter an array of demand values in the following form:
        [yr1 Id1; yr2 Id2; yr3 Id3; yr4 Id4; yr5 Id5; yr6 Id6; yr7 Id7]

An example is shown below:
        [1997 3094; 1998 2938; 1999 2714; 2000 3567; 2001 4027; 2002
        3591; 2003 4579]

Enter year/demand values: [1997 3094; 1998 2938; 1999 2714; 2000 3567;
  2001 4027; 2002 3591; 2003 4579; 2004 4436]
```

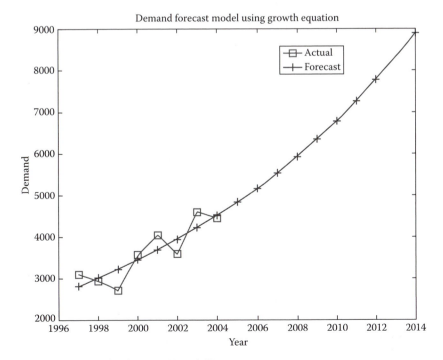

FIGURE 2.15 The answers for the parts (a) and (b).

```
Enter the number of year to predict: 10
```

Rate of growth = 5.55%		
YEAR	ACTUAL	FORECAST
1997	3094	3094
1998	2938	3266
1999	2714	3447
2000	3567	3639
2001	4027	3841
2002	3591	4054
2003	4579	4279
2004	4436	4516
2005	–	4767
2006	–	5032
2007	–	5311
2008	–	5606
2009	–	5918
2010	–	6246
2011	–	6593
2012	–	6959
2013	–	7345
2014	–	7753
2015	–	8184

EDU»

2.5 LOAD MANAGEMENT

The *load management* process involves controlling system loads by remote control of individual customer loads. Such control includes suppressing or biasing automatic control of cycling loads, as well as load switching. Load management can also be affected by inducing customers to suppress loads during utility-selected daily periods by means of time-of-day rate incentives. Such activities are called demand-side management.

Demand-side management (DSM): It includes all measures, programs, equipment, and activities that are directed toward improving efficiency and cost-effectiveness of energy usage on the *customer side* of the meter.

In general, such load control results in a load reduction at time t, that is, $\Delta S(t)$, that can be expressed as

$$\Delta S(t) = S_{av} \times [D_{uncont}(t) - D_{cont}(t)] \times N$$

where
  S_{av} is the average connected load of controlled devices
  $D_{uncont}(t)$ is the average duty cycle of uncontrolled units at time t
  $D_{cont}(t)$ is the duty cycle allowed by the load control at time t
  N is the number of units under control

Distribution automation provides the control and monitoring ability required for both load management scenarios. It provides for direct control of customer loads and the monitoring necessary to verify that programmed levels are achieved. It also provides for the appropriate selection of energy metering registers where time-of-use rates are in effect.

The use of load management provides various benefits to the utility and its customers. Maximizing utilization of existing distribution system can lead to deferrals of capital expenditures. This is achieved by shaping the daily (monthly, annual) load characteristic in the following manner:

1. By suppressing loads at peak times and/or encouraging energy consumption at off-peak times
2. By minimizing the requirement for more costly generation or power purchases by suppressing loads
3. By relieving the consequences of significant loss of generation or similar emergency situations by suppressing loads
4. By reducing cold load pickup during reenergization of circuits using devices with cold load pickup features

Load management monitoring and control functions include the following:

1. *Monitoring of substations and feeder loads*: To verify that the required magnitude of load suppression is accomplished for normal and emergency conditions as well as switch status
2. *Controlling individual customer loads*: To suppress total system, substation, or feeder loads for normal or emergency conditions, and switching meter registers in order to accommodate time of use, that is, time of day, rate structures, where these are in effect

The effectiveness of direct control of customer loads is increased by choosing the larger and more significant customer loads. These include electric space and water heating, air-conditioning, electric clothes dryers, and others.

Also customer-activated load management is achieved by incentives such as time-of-use rates or customer alert to warn customers so that they can alter their use. In response to the economic penalty in terms of higher energy rates, the customers will limit their energy consumption during peak-load periods. Distribution automation provides for remotely adjusting and reading the time-of-use meters.

Example 2.15

Assume that a 5 kW air conditioner would run 80% of the time (80% duty cycle) and, during the peak hour, might be limited by utility remote control to a duty cycle of 65%. Determine the following:

a. The number of minutes of operation denied at the end of 1 h of control of the unit
b. The amount of reduced energy consumption during the peak hour if such control is applied simultaneously to 100,000 air conditioners throughout the system
c. The total amount of reduced energy consumption during the peak
d. The total amount of additional reduction in energy consumption in part (c) if T&D losses of the T&D system at peak is 8%

Solution

a. The number of minutes of operation denied is

$$(0.80 - 0.65) \times (60 \text{ min/h}) = 9 \text{ min}$$

b. The amount of reduced energy consumption during the peak is

$$(0.80 - 0.65) \times (5 \text{ kW}) = 0.75 \text{ kW}$$

c. The total amount of energy reduction for 100,000 units is

$$\Delta S(t) = S_{av} \times [D_{uncont}(t) - D_{cont}(t)] \times N$$

$$= (5 \text{ kW}) \times [0.80 - 0.65] \times 100,000$$

$$= 75 \text{ MW}$$

  d. The total additional amount of energy reduction due to the reduction in the T&D losses is

$$(75 \text{ MW}) \times 0.08 = 6 \text{ MW}$$

Thus, the overall total reduction is

$$75 \text{ MW} + 6 \text{ MW} = 81 \text{ MW}$$

The earlier example shows attractiveness of controlling air conditioners to utility company.

2.6   RATE STRUCTURE

Even after the so-called *deregulation*, most public utilities are monopolies, that is, they have the exclusive right to sell their product in a given area. Their rates are subject to government regulation. The total revenue that a utility may be authorized to collect through the sales of its services should be equal to the company's total cost of service. Therefore,

$$(\text{Revenue requirement}) = (\text{operating expenses}) + (\text{depreciation expenses})$$

$$+ (\text{taxes}) + (\text{rate base or net valuation}) \times (\text{rate of return}) \quad (2.45)$$

The determination of the revenue requirement is a matter of regulatory commission decision. Therefore, designing schedules of rates that will produce the revenue requirement is a management responsibility subject to commission review. However, a regulatory commission cannot guarantee a specific rate of earnings; it can only declare that a public utility has been given the opportunity to try to earn it.

The rate of return is partly a function of local conditions and should correspond with the return being earned by comparable companies with similar risks. It should be sufficient to permit the utility to maintain its credit and attract the capital required to perform its tasks.

However, the rate schedules, by law, should avoid unjust and unreasonable discrimination, that is, customers using the utility's service under similar conditions should be billed at similar prices. It is a matter of necessity to categorize the customers into classes and subclasses, but all customers in a given class should be treated the same. There are several types of rate structures used by the utilities, and some of them are

1. Flat demand rate structure
2. Straight-line meter rate structure
3. Block meter rate structure
4. Demand rate structure
5. Season rate structure
6. Time-of-day (or peak-load pricing) structure

The flat rate structure provides a constant price per kilowatthour, which does not change with the time of use, season, or volume. The rate is negotiated knowing connected load; thus metering is not required. It is sometimes used for parking lot or street lighting service. The straight-line meter rate structure is similar to the flat structure. It provides a single price per kilowatthour without considering customer demand costs.

The block meter rate structure provides lower prices for greater usage, that is, it gives certain prices per kilowatthour for various kilowatthour blocks where the price per kilowatthour decreases for succeeding blocks. Theoretically, it does not encourage energy conservation and *off-peak* usage. Therefore, it causes a greater than necessary peak and, consequently, excess idle generation capacity during most of the time, resulting in higher rates to compensate the cost of a greater *peak-load* capacity.

The demand rate structure recognizes load factor and consequently provides separate charges for demand and energy. It gives either a constant price per kilowatthour consumed or a decreasing price per kilowatthour for succeeding blocks of energy used.

The seasonal rate structure specifies higher prices per kilowatthour used during the season of the year in which the system peak occurs (*on-peak season*) and lower prices during the season of the year in which the usage is the lowest (*off-peak season*).

The time-of-day rate structure (or *peak-load pricing*) is similar to the seasonal load rate structure. It specifies higher prices per kilowatthour used during the peak period of the day and lower prices during the off-peak period of the day.

The seasonal rate structure and the time-of-day rate structure are both designed to reduce the system's peak load and therefore reduce the system's idle standby capacity.

2.6.1 CUSTOMER BILLING

Customer billing is done by taking the difference in readings of the meter at two successive times, usually at an interval of 1 month. The difference in readings indicates the amount of electricity, in kilowatthours, consumed by the customer in that period. This amount is multiplied by the appropriate rate or the series of rates and the adjustment factors, and the bill is sent to the customer.

Figure 2.16 shows a typical monthly bill rendered to a residential customer. The monthly bill includes the following items in the indicated spaces:

1. The customer's account number.
2. A code showing which of the rate schedules was applied to the customer's bill.
3. A code showing whether the customer's bill was estimated or adjusted.

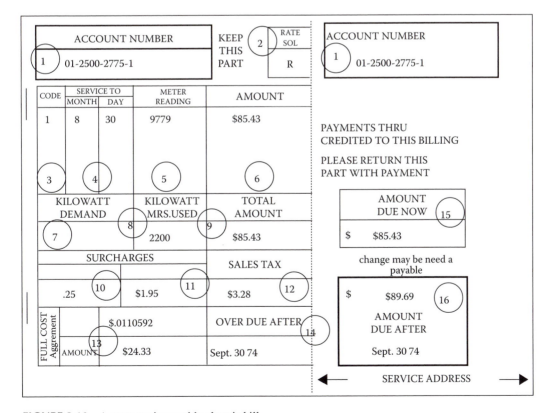

FIGURE 2.16 A customer's monthly electric bill.

4. Date on which the billing period ended.
5. Number of kilowatthours the customer's meter registered when the bill was tabulated.
6. Itemized list of charges. In this case, the only charge shown in box 6 of Figure 2.14 is a figure determined by adding the price of the electricity the customer has used to the routine taxes and surcharges. However, had the customer received some special service during this billing period, a service charge would appear in this space as a separate entry.
7. Information appears in this box only when the bill is sent to a nonresidential customer using more than 6000 kWh electricity a month.
8. The number of kilowatthours the customer used during the billing period.
9. Total amount that the customer owes.
10. Environmental surcharge.
11. County energy tax.
12. State sales tax.
13. Fuel cost adjustment. Both the total adjustment and the adjustment per kilowatthour are shown.
14. Date on which bill, if unpaid, becomes overdue.
15. Amount due now.
16. Amount that the customer must pay if the bill becomes overdue.

The sample electrical bill, shown in Figure 2.16, is based on the following rate schedule. Note that there is a minimum charge regardless of how little electricity the customer uses and that the first 20 kWh that the customer uses is covered by this flat rate. Included in the minimum, or service, charge is the cost of providing service to the customer, including metering, meter reading, billing, and various overhead expenses.

Rate schedule

Minimum Charge (Including First 20 kWh or Fraction Thereof)	$2.25/Month
Next 80 kWh	$0.0355/kWh
Next 100 kWh	$0.0321/kWh
Next 200 kWh	$0.0296/kWh
Next 400 kWh	$0.0265/kWh
Consumption in excess of 800 kWh	$0.0220/kWh

The sample bill shows a consumption of 2200 kWh, which has been billed according to the following schedule:

First 20 kWh @ $2.25 (flat rate)	= $2.25
Next 80 kWh × 0.0355	= $2.84
Next 100 kWh × 0.0321	= $3.21
Next 200 kWh × 0.096	= $5.92
Next 400 kWh × 0.0265	= $10.60
Additional 1400 kWh × 0.0220	= $30.80
2200 kWh	= $55.62
Environmental surcharge	= $0.25
County energy tax	= $1.95
Fuel cost adjustment	= $24.33
State sales tax	= $3.28
Total amount	= $85.43

TABLE 2.5

Typical Energy Rate Schedule for Commercial Users

On-peak season (June 1–October 31)

First 50 kWh or less/month for	$4.09
Next 50 kWh/month	@5.5090/kWh
Next 500 kWh/month	@4.8430/kWh
Next 1400 kWh/month	@4.0490/kWh
Next 3000 kWh/month	@3.8780/kWh
All additional kWh/month	@3.3390/kWh

Off-peak season (November 1–May 31)

First 50 kWh or less/month for	4.09
Next 50 kWh/month	@5.5090/kWh
Next 500 kWh/month	@4.2440/kWh
Next 1400 kWh/month	@3.1220/kWh
Next 3000 kWh/month	@2.7830/kWh
All additional kWh/month	@2.6490/kWh

The customer is billed according to the utility company's rate schedule. In general, the rates vary according to the season. In most areas, the demand for electricity increases in the warm months. Therefore, to meet the added burden, electric utilities are forced to use spare generators that are often less efficient and consequently more expensive to run. As an example, Table 2.5 gives a typical energy rate schedule for the on-peak and off-peak seasons for commercial users.

2.6.2 Fuel Cost Adjustment

The rates stated previously are based upon an average cost, in dollars per million Btu, for the cost of fuel burned at the NL&NP's thermal generating plants. The monthly bill as calculated under the previously stated rate is increased or decreased for each kilowatthour consumed by an amount calculated according to the following formula:

$$\text{FCAF} = A \times \frac{B}{10^6} \times C \times \frac{1}{1-D} \qquad (2.46)$$

where

FCAF is the fuel cost adjustment factor, $/kWh, to be applied per kilowatthour consumed

A is the weighted average Btu per kilowatthour for net generation from the NL&NP's thermal plants during the second calendar month preceding the end of the billing period for which the kilowatthour usage is billed

B is the amount by which average cost of fuel per million Btu during the second calendar month preceding the end of the billing period for which the kilowatthour usage is billed exceeds or is less than $1/million Btu

C is the ratio, given in decimal, of the total net generation from all the NL&NP's thermal plants during the second calendar month preceding the end of the billing period for which the kilowatthour usage is billed to the total net generation from all the NL&NP's plants including hydro generation owned by the NL&NP, or kilowatthours produced by hydro generation and purchased by the NL&NP, during the same period

D is the loss factor, which is the ratio, given in decimal, of kilowatthour losses (total kilowatthour losses less losses of 2.5% associated with off-system sales) to net system input, that is, total system input less total kilowatthours in off-system sales, for the year ending December 31 preceding; this ratio is updated every year and applied for 12 months

Example 2.16

Assume that the NL&NP Utility Company has the following, and typical, commercial rate schedule:

1. Monthly billing demand = 30 min monthly maximum kilowatt demand multiplied by the ratio of (0.85/monthly average PF). The PF penalty shall not be applied when the consumer's monthly average PF exceeds 0.85.
2. Monthly demand charge = $2.00/kW of monthly billing demand.
3. Monthly energy charges shall be as follows:
    2.50 cents/kWh for the first 1000 kWh
    2.00 cents/kWh for the next 3000 kWh
    1.50 cents/kWh for all kWh in excess of 4000
4. The total monthly charge shall be the sum of the monthly demand charge and the monthly energy charge.

    Assume that two consumers, as shown in Figure 2.17, each requiring a DT, are supplied from a primary line of the NL&NP.

    a. Assume that an average month is 730 h and find the monthly load factor of each consumer.
    b. Find a reasonable size, that is, continuous kilovoltampere rating, for each DT.
    c. Calculate the monthly bill for each consumer.
    d. It is not uncommon to measure the average monthly PF on a monthly energy basis, where both kilowatthours and kilovarhours are measured. On this basis, what size capacitor, in kilovars, would raise the PF of customer B to 0.85?
    e. Secondary-voltage shunt capacitors, in small sizes, may cost about $30/kvar installed with disconnects and short-circuit protection. Consumers sometimes install secondary capacitors to reduce their billings for utility service. Using the 30/kvar figure, find the number of months required for the PF correction capacitors found in part (d) to pay back for themselves with savings in demand charges.

Solution

a. From Equation 2.7, the monthly load factors for each consumer are the following. For customer A,

$$F_{LD} = \frac{Units\,served}{Peak\,load \times T}$$

$$= \frac{7000\,kWh}{22\,kW \times 730\,h}$$

$$= 0.435$$

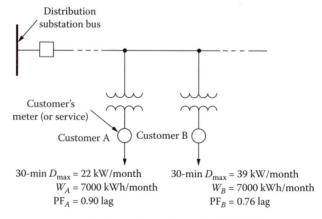

FIGURE 2.17   Two customers connected to a primary line of the NL&NP.

and for customer B,

$$F_{LD} = \frac{Units\,served}{Peak\,load \times T}$$

$$= \frac{7000\,kWh}{39\,kW \times 730\,h}$$

$$= 0.246$$

b. The continuous kilovoltamperes for each DT are the following:

$$S_A = \frac{P_A}{\cos\theta}$$

$$= \frac{22\,kW}{0.90}$$

$$= 24.4\,kVA$$

and

$$S_B = \frac{P_B}{\cos\theta}$$

$$= \frac{39\,kW}{0.76}$$

$$= 51.2\,kVA$$

Therefore, the continuous sizes suitable for the DTs A and B are 25 and 50 kVA ratings, respectively.

c. The monthly bills for each customer are the following:

For customer A

$$Monthly\ billing\ demand* = 22\,kW \times \frac{0.85}{0.90} \cong 22\,kW$$

$$Monthly\ demand\ charge = 22\ kW \times \$2.00/kW \cong \$44$$

Monthly energy charge

First 1000 kWh = $0.025/kWh × 1000 kWh	= $25
Next 3000 kWh = $0.02/kWh × 3000 kWh	= $60
Excess kWh = $0.015/kWh × 3000 kWh	= $45
Monthly energy charge	= $130

Therefore,

$$Total\ monthly\ bill = Monthly\ demand\ charge\ +\ Monthly\ energy\ charge$$

$$= \$44 + \$130 = \$174$$

---

* It is calculated from $P(0.85/PF)$. However, if the PF is greater than 0.85, then still the actual amount of P is used, rather than the resultant kW.

For customer B

$$\text{Monthly billing demand} = 39\,\text{kW} \times \frac{0.85}{0.76} = 43.6\,\text{kW}$$

$$\text{Monthly demand charge} = 43.6\text{ kW} \times \$2.00/\text{kW} = \$87.20$$

Monthly energy charge

First 1000 kWh = \$0.025/kWh × 1000 kWh	= \$25
Next 3000 kWh = \$0.02/kWh × 3000 kWh	= \$60
Excess kWh = \$0.015/kWh × 3000 kWh	= \$45
Monthly energy charge	= \$130

Therefore,

$$\text{Total monthly bill} = \$87.20 + \$130$$

$$= \$217.20$$

d. Currently, customer B at the lagging PF of 0.76 has

$$\frac{7000\,\text{kWh}}{0.76} \times \sin(\cos^{-1}0.76) = 5986.13\,\text{kvarh}$$

If its PF is raised to 0.85, customer B would have

$$\frac{7000\,\text{kWh}}{0.85} \times \sin(\cos^{-1}0.85) = 4338\,\text{kvarh}$$

Therefore, the capacitor size required is

$$\frac{5986.13\,\text{kvarh} - 4338\,\text{kvarh}}{730\,\text{h}} = 2.258\,\text{kvar} \cong 2.3\,\text{kvar}$$

e. The new monthly bill for customer B would be as follows:

$$\text{Monthly billing demand} = 39\text{ kW}$$

$$\text{Monthly demand charge} = 39\text{ kW} \times \$2.00 = \$78$$

$$\text{Monthly energy charge} = \$130\text{ as before}$$

Therefore,

$$\text{Total monthly bill} = \$78 + \$130$$

$$= \$208$$

Hence, the resultant savings due to the capacitor installation is the difference between the before-and-after total monthly bills. Thus,

$$\text{Savings} = \$217.20 - \$208$$

$$= \$9.20/\text{month}$$

or

$$\text{Savings} = \$87.20 - \$78$$

$$= \$9.20/\text{month}$$

The cost of the installed capacitor is

$$\$30/\text{kvar} \times 2.3 \text{ kvar} = \$69$$

Therefore, the number of months required for the capacitors to "payback" for themselves with savings in demand charges can be calculated as

$$\text{Payback period} = \frac{\text{Capacitor cost}}{\text{Savings}}$$

$$= \frac{\$69}{\$9.20/\text{month}}$$

$$= 7.5$$

$$\cong 8 \text{ months}$$

However, in practice, the available capacitor size is 3 kvar instead of 2.3 kvar. Therefore, the realistic cost of the installed capacitor is

$$\$30/\text{kvar} \times 3 \text{ kvar} = \$90$$

Therefore,

$$\text{Payback period} = \frac{\$90}{\$9.20/\text{month}}$$

$$\cong 10 \text{ months}$$

2.7 ELECTRIC METER TYPES

An electric meter is the device used to measure the electricity sold by the electric utility company. It deals with two basic quantities: *energy* and *power*. Energy is equivalent to work. Power is the rate of doing the work. Power applied (or consumed) for any length of time is energy. In other words, power integrated over time is energy. The basic unit of power is *watt*. The basic electrical unit of energy is *watthour*.

A watthour meter is used to measure the electric energy delivered to residential, commercial, and industrial customers and also used to measure the electric energy passing through various parts of generation, transmission, and distribution systems. A wattmeter measures the rate of energy, that is, power (in watthours per hour or simply watts). For a constant power level, power multiplied by time is energy. Electric meters could be of two types: the electromechanical meters and electronic (or also called digital) meters.

Figure 2.18 shows a single-phase (electromechanical) watthour meter; Figure 2.19 shows its basic parts; Figure 2.20 gives a diagram of a typical motor and magnetic retarding system for a single-phase watthour meter. The magnetic retarding system causes the rotor disk to establish, in combination with the stator, the speed at which the shaft will turn for a given load condition to determine the watthour constant. Figure 2.21a shows a typical socket-mounted two-stator polyphase watthour meter. It is a combination of single-phase watthour meter stators that drive a rotor at a speed proportional to the total power in the circuit.

FIGURE 2.18   Single-phase electromechanical watthour meter. (From General Electric Company, *Manual of Watthour Meters*, Bulletin GET-1840C.)

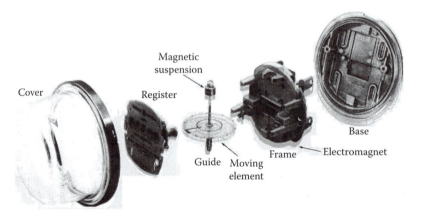

FIGURE 2.19   Basic parts of a single-phase electromechanical watthour meter. (From General Electric Company, *Manual of Watthour Meters*, Bulletin GET-1840C.)

The watthour meters used to measure the electric energy passing through various parts of generation, transmission, and distribution systems are required to measure large quantities of electric energy at relatively high voltages. For those applications, transformer-rated meters are developed. They are used in conjunction with standard instrument transformers, that is, current transformers (CTs) and potential transformers (PTs). These transformers reduce the voltage and the current to values that are suitable for low-voltage and low-current meters. Figure 2.21b shows a typical transformer-rated meter. Figure 2.22 shows a single-phase, two-wire watthour electromechanical meter connected to a high-voltage circuit through CTs and PTs.

A demand meter is basically a watthour meter with a timing element added. The meter functions as an integrator and adds up the kilowatthours of energy used in a certain time interval, for example, 15, 30, or 60 min. Therefore, the demand meter indicates energy per time interval, or average power, which is expressed in kilowatts. Figure 2.23 shows a demand register.

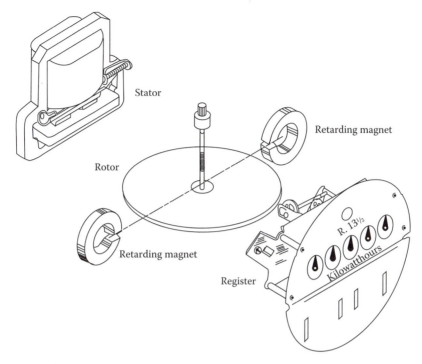

FIGURE 2.20 Diagram of a typical motor and magnetic retarding system for a single-phase electromechanical watthour meter. (From General Electric Company, *Manual of Watthour Meters*, Bulletin GET-1840C.)

    (a)   (b)

FIGURE 2.21 Typical polyphase (electromechanical) watthour meters: (a) self-contained meter (socket-connected cyclometer type). (b) transformer-rated meter (bottom-connected pointer type). (From General Electric Company, *Manual of Watthour Meters*, Bulletin GET-1840C.)

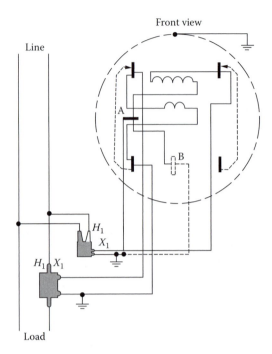

FIGURE 2.22 Single-phase, two-wire electromechanical watthour meter connected to a high-voltage circuit through current and potential transformers. (From General Electric Company, *Manual of Watthour Meters*, Bulletin GET-1840C.)

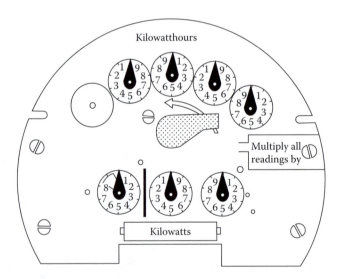

FIGURE 2.23 The register of an electromechanical demand meter for large customers. (From General Electric Company, *Manual of Watthour Meters*, Bulletin GET-1840C.)

2.7.1 ELECTRONIC (OR DIGITAL) METERS

Utility companies have started to use new electronic (or digital) meters with *programmable demand registers* (PDRs) since 1980s. At the first stage of the evolution, the same electromechanical devices used electronic registers. Such electronic register provided a digital display of energy and demand to an electromechanical meter. In the last stage of the evolution, the meters became totally electronic (or solid-state) designed meters.

Such electronic meters have no moving parts. They are built instead around large-scale integrated circuits, other solid-state components, and digital logic. The operation of an electronic meter is very different than the electromechanical meters. The electronic circuitry samples the voltage and current waveforms during each electrical cycle and converts them into digital quantities. Other circuitry then manipulates these values to determine numerous electrical parameters, such as kW, kWh, kvar, kvarh, PF, kVA, rms current, and rms voltage.

A PDR can also measure demand, whereas a traditional register measures only the amount of electricity used in a month. A demand profile shows how much electricity a customer used in a month. Industrial and commercial customers are billed according to their peak demand for the month, as well as their kWh consumption. Utilities have been using supplementary devices with the traditional meters to measure demand. But the PDR measures total kWh used, demand, and cumulative demand by itself.

Here, measuring cumulative demand is a security measure. If the cumulative demand doesn't equal to the sum of the monthly demands, then someone may have tampered with the meter. It will automatically add the demand reading to the cumulative each time it is reset, so a meter will know if someone reset it since he or she was there last.

The PDR may also be programmed to record the date each time it is reset. The PDR can also be programmed in many other ways. For example, it can alert a customer when he reaches a certain demand level, so that the customer could cut back if he or she wants it.

Today, electronic meters can also measure some or all of the following capabilities:

- *Time of use (TOU)*: The meter keeps up with energy and demand in multiple daily periods.
- *Bidirectional*: The meter measures (as separate quantities) energy delivered to and received from a customer. (It can be used by a customer who is able to generate electricity and sell to a utility company.)
- *Interval data recording*: The meter has solid-state memory in which it can record up to several months of interval-by-interval data.
- *Remote communications*: Its built-in communication capabilities allow the meter to be interrogated remotely via radio, telephone, or other communications media.
- *Diagnostics*: The meter checks for the proper voltage, currents, and phase angles on the meter conductors.
- *Loss compensation*: It can be programmed to automatically calculate watt and var losses in transformers and electrical conductors based on defined or tested loss characteristics of the transformers and conductors.

2.7.2 Reading Electric Meters

By reading the register, that is, the revolution counter, the customers' bills can be determined. There are primarily two different types of registers: (1) conventional dial and (2) cyclometer.

Figure 2.24 shows a conventional dial-type register. To interpret it, read the dials from left to right. (Note that numbers run clockwise on some dials and counterclockwise on others.) The figures above each dial show how many kilowatthours are recorded each time the pointer makes a complete revolution.

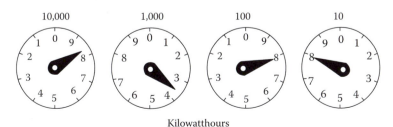

Kilowatthours

FIGURE 2.24   A conventional dial-type register of electromechanical meter.

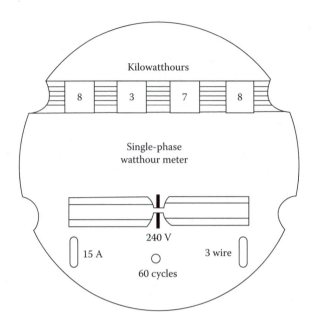

FIGURE 2.25 A cyclometer-type register.

As shown in Figure 2.24, if the pointer is between numbers, read the smaller one. The number 0 stands for 10. If the pointer is pointed directly at a number, look at the dial to the right. If that pointer has not yet passed 0, record the smaller number; if it has passed 0, record the number the pointer is on. For example, in Figure 2.24, the pointer on the first dial is between 8 and 9; therefore read 8. The pointer on the second dial is between 3 and 4; thus read 3. The pointer on the third dial is almost directly on 8, but the dial on the right has not reached 0 so the reading on the third dial is 7. The fourth dial reading is 8. Therefore, the total reading is 8378 kWh. The third dial would be read as 8 after the pointer on the 10-kWh dial reaches 0. This reading is based on a cumulative total, that is, since the meter was last set at 0, 8378 kWh of electricity has been used.

To find the customer's monthly use, take two readings 1 month apart and subtract the earlier one from the later one. Some electric meters have a constant, or multiplier, indicated on the meter. This type of meter is primarily for high-usage customers.

Figure 2.25 shows a cyclometer-type register. Here, even though the procedure is the same as in the conventional type, the wheels, which indicate numbers directly, replace the dials. Therefore, it makes possible the reading of the meter simply and directly, from left to right.

2.7.3 INSTANTANEOUS LOAD MEASUREMENTS USING ELECTROMECHANICAL WATTHOUR METERS

The instantaneous kilowatt demand of any customer may be determined by making field observations of the kilowatthour meter serving the customer. However, the instantaneous load measurement should not replace demand meters that record for longer time intervals. The instantaneous demand may be determined by using one of the following equations:

1. For a self-contained watthour meter,

$$D_i = \frac{3.6 \times K_r \times K_h}{T} \, \text{kW} \tag{2.47}$$

2. For a transformer-rated meter (where instrument transformers are used with a watthour meter),

$$D_i = \frac{3.6 \times K_r \times K_h \times CTR \times PTR}{T} \, kW \tag{2.48}$$

where
  D_i is the instantaneous demand, kW
  K_r is the number of meter disk revolutions for a given time period
  K_h is the watthour meter constant (given on the register), Wh/rev
  T is the time, s
  CTR is the current transformer ratio
  PTR is the potential transformer ratio

Since the kilowatt demand is based on a short-time interval, two or more demand intervals should be measured. The average value of these demands is a good estimate of the given customer's kilowatt demand during the intervals measured.

Example 2.17

Assume that the load is measured twice with a watthour meter that has a meter constant of 7.2 and the following data are obtained:

	First Reading	Second Reading
Revolutions of disk	32	27
Time interval for revolutions of disks	59	40

Determine the instantaneous demands and the average demand.

Solution
From Equation 2.47, for the first reading,

$$D_1 = \frac{3.6 \times K_r \times K_h}{T}$$

$$= \frac{3.6 \times 32 \times 7.2}{59}$$

$$= 14.058 \, kW$$

and for the second reading,

$$D_2 = \frac{3.6 \times K_r \times K_h}{T}$$

$$= \frac{3.6 \times 27 \times 7.2}{40}$$

$$= 17.496 \, kW$$

Therefore, the average demand is

$$D_{av} = \frac{D_1 + D_2}{2}$$

$$= \frac{14.058 + 17.496}{2}$$

$$= 15.777 \text{ kW}$$

Example 2.18

Assume that the data given in Example 2.17 are the results of load measurement with watthour meters and instrument transformers. Suppose that the new meter constant is 1.8 and that the ratios of the CTs and PTs used are 200 and 1, respectively. Determine the instantaneous demands for both readings and the average demand.

Solution

Therefore, from Equation 2.48,

$$D_1 = \frac{3.6 \times K_r \times K_h \times \text{CTR} \times \text{PTR}}{T}$$

$$= \frac{3.6 \times 32 \times 1.8 \times 200 \times 1}{59} = 702.9 \text{ kW}$$

and

$$D_2 = \frac{3.6 \times 27 \times 1.8 \times 200 \times 1}{40}$$

$$= 874.8 \text{ kW}$$

Thus the average demand is

$$D_{av} = \frac{D_1 + D_2}{2}$$

$$= \frac{702.9 + 874.8}{2}$$

$$\cong 788.9 \text{ kW}$$

Example 2.19

Assume that the load is measured with watthour and varhour meters and instrument transformers and that the following readings are obtained:

	Watthour Readings		Varhour Readings	
	First Set	Second Set	First Set	Second Set
Revolutions of disk	20	30	10	20
Time interval for revolutions of disks	50	60	50	60

Assume that the new meter constants are 1.2 and that the ratios of the current and potential transformers used are 80 and 20, respectively. Determine the following:

a. The instantaneous kilowatt demands
b. The average kilowatt demand
c. The instantaneous kilovar demands
d. The average kilovar demand
e. The average kilovoltampere demand

Solution

a. The instantaneous kilowatt demands are

$$D_1 = \frac{3.6 \times 20 \times 1.2 \times 80 \times 20}{50}$$

$$= 2764.8 \text{ kW}$$

and

$$D_2 = \frac{3.6 \times 30 \times 1.2 \times 80 \times 20}{60}$$

$$= 3456 \text{ kW}$$

b. The average kilowatt demand is

$$D_{av} = \frac{D_1 + D_2}{2}$$

$$= \frac{2764.8 + 3456}{2}$$

$$= 3110.4 \text{ kW}$$

c. The instantaneous kilovar demands are

$$D_1 = \frac{3.6 \times K_r \times K_h \times CTR \times PTR}{T}$$

$$= \frac{3.6 \times 10 \times 1.2 \times 80 \times 20}{50}$$

$$= 1382.4 \text{ kW}$$

and

$$D_2 = \frac{3.6 \times 20 \times 1.2 \times 80 \times 20}{60}$$

$$= 2304 \text{ kW}$$

d. The average kilovar demand is

$$D_{av} = \frac{D_1 + D_2}{2}$$

$$= \frac{1382.4 + 2304}{2}$$

$$= 1843.2 \text{ kW}$$

e. The average kilovoltampere demand is

$$D_{av} = \left[(D_{av,kW})^2 + (D_{av,kW})^2 \right]^{1/2}$$

$$= (3110.4^2 + 1843.2^2)^{1/2}$$

$$\cong 3615.5$$

PROBLEMS

2.1 Use the data given in Example 2.1 and assume that the feeder has the peak loss of 72 kW at peak load and an annual loss factor of 0.14. Determine the following:
  a. The daily average load of the feeder
  b. The average power loss of the feeder
  c. The total annual energy loss of the feeder

2.2 Use the data given in Example 2.1 and the equations given in Section 2.2 and determine the load factor of the feeder.

2.3 Use the data given in Example 2.1 and assume that the connected demands for the street lighting load, the residential load, and the commercial load are 100, 2000, and 2000 kW, respectively. Determine the following:
  a. The DF of the street lighting load
  b. The DF of the residential load
  c. The DF of the commercial load
  d. The DF of the feeder

2.4 Using the data given in Table P.2.1 for a typical summer day, repeat Example 2.1 and compare the results.

2.5 Use the data given in Problem 2.4 and repeat Problem 2.2.

2.6 Use the data given in Problem 2.4 and repeat Problem 2.3.

2.7 Use the result of Problem 2.2 and calculate the associated loss factor.

2.8 Assume that a load of 100 kW is connected at the Riverside substation of the NL&NP Company. The 15-min weekly maximum demand is given as 75 kW, and the weekly energy consumption is 4200 kWh. Assuming a week is 7 days, find the DF and the 15-min weekly load factor of the substation.

2.9 Assume that the total kilovoltampere rating of all DTs connected to a feeder is 3000 kVA. Determine the following:
  a. If the average core loss of the transformers is 0.50%, what is the total annual core loss energy on this feeder?
  b. Find the value of the total core loss energy calculated in part (a) at $0.025/kWh.

2.10 Use the data given in Example 2.6 and also consider the following added new load. Suppose that several buildings that have electric air-conditioning are converted from gas-fired heating to electric heating. Let the new electric heating load average 200 kW during 6 months of heating (and off-peak) season. Assume that off-peak energy delivered to these primary feeders costs the NL&NP Company 2 cents/kWh and that the capacity cost of the power system remains at $3.00/kW per month.
  a. Find the new annual load factor on the substation.
  b. Find the total annual cost to NL&NP to serve this new load.
  c. Why is it that the hypothetical but illustrative energy cost is smaller in this problem than the one in Example 2.8?

2.11 The input to a subtransmission system is 87,600,000 kWh annually. On the peak-load day of the year, the peak is 25,000 kW and the energy input that day is 300,000 kWh. Find the load factors for the year and for the peak-load day.

TABLE P.2.1
Typical Summer-Day Load, in kW

Time	Street Lighting	Residential	Commercial
12 AM	100	250	300
1	100	250	300
2	100	250	300
3	100	250	300
4	100	250	300
5	100	250	300
6	100	250	300
7		350	300
8		450	400
9		550	600
10		550	1100
11		550	1100
12 noon		600	1100
1		600	1100
2		600	1300
3		600	1300
4		600	1300
5		650	1300
6		750	900
7		900	500
8	100	1100	500
9	100	1100	500
10	100	900	300
11	100	700	300
12 PM	100	350	300

2.12 The electric energy consumption of a residential customer has averaged 1150 kWh/month as follows, starting in January: 1400, 900, 1300, 1200, 800, 700, 1000, 1500, 700, 1500, 1400, and 1400 kWh. The customer is considering purchasing equipment for a hobby shop that he has in his basement. The equipment will consume about 200 kWh each month. Estimate the additional annual electric energy cost for operation of the equipment. Use the electrical rate schedule given in the following table.

Residential

Rate: (net) per month per meter

Energy Charge	
For the first 25 kWh	6.00 ¢/kWh
For the next 125 kWh	3.2 ¢/kWh
For the next 850 kWh	2.00 ¢/kWh
All in excess of 1000 kWh	1.00 ¢/kWh
Minimum: $1.50 per month	

Commercial

A rate available for general, commercial, and miscellaneous power uses where consumption of energy does not exceed 10,000 kWh in any month during any calendar year.

Rate: (net) per month per meter

Energy Charge	
For the first 25 kWh	6.0 ¢/kWh
For the next 375 kWh	4.0 ¢/kWh
For the next 3600 kWh	3.0 ¢/kWh
All in excess of 4000 kWh	1.5 ¢/kWh
Minimum: $1.50 per month	

General power

A rate available for service supplied to any commercial or industrial customer whose consumption in any month during the calendar year exceeds 10,000 kWh. A customer who exceeds 10,000 kWh per month in any 1 month may elect to receive power under this rate.

A customer who exceeds 10,000 kWh in any 3 months or who exceeds 12,000 kWh in any 1 month during a calendar year shall be required to receive power under this rate at the option of the supplier.

A customer who elects at his own option to receive power under this rate may not return to the commercial service rate except at the option of the supplier.

Rate: (net) per month per meter kW is rate of flow. 1 kW for 1 h is 1 kWh.

Demand charge

For the first 30 kW of maximum demand per month	$2.50/kW
For all maximum demand per month in excess of 30 kW	$1.25/kW

Energy Charge	
For the first 100 kWh per kW of maximum demand per month	2.00 ¢/kWh
For the next 200 kWh per kW of maximum demand per month	1.2 ¢/kWh
All in excess of 300 kWh per kW of maximum demand per month	0.5 ¢/kWh

Minimum charge: The minimum monthly bill shall be the demand charge for the month. Determination of maximum demand: The maximum demand shall be either the highest integrated kW load during any 30 min period occurring during the billing month for which the determination is made or 75% of the highest maximum demand that has occurred in the preceding month, whichever is greater.

Water heating: 1.00/kWh with a minimum monthly charge of $1.00.

2.13 The Zubits International Company, located in Ghost Town, consumed 16,000 kWh of electric energy for Zubit production this month. The company's monthly average energy consumption is also 16,000 kWh due to some unknown reasons. It has a 30 min monthly maximum demand of 200 kW and a connected demand of 580 kW. Use the electrical rate schedule given in Problem 2.12.
   a. Find the Zubits International's total monthly electrical bill for this month.
   b. Find its 30 min monthly load factor.
   c. Find its DF.
   d. The company's newly hired plant engineer, who recently completed a load management course at Ghost University, suggested that, by shifting the hours of a certain production from the peak-load hours to off-peak hours, the maximum monthly demand can be reduced to 140 kW at a cost of $50/month. Do you agree that this will save money? How much?

2.14 Repeat Example 2.12, assuming that there are eight houses connected on each DT and that there are a total of 120 DTs and 960 residences supplied by the primary feeder.

2.15 Repeat Example 2.15, assuming that the 30 min monthly maximum demands of customers A and B are 27 and 42 kW, respectively. The new monthly energy consumptions by customers A and B are 8000 and 9000 kWh, respectively. The new lagging load PFs of A and B are 0.90 and 0.70, respectively.

2.16 A customer transformer has 12 residential customers connected to it. Connected load is 20 kW per house, DF is 0.6, and diversity factor is 1.15. Find the diversified demand of the group of 12 houses on the transformer.

2.17 A distribution substation is supplied by total annual energy of 100,000 MWh. If its annual average load factor is 0.6, determine the following:
   a. The annual average power demand
   b. The maximum monthly demand

2.18 Suppose that one of the transformers of a substation supplies four primary feeders. Among the four feeders, the diversity factor is 1.25 for both real power (P) and reactive power (Q). The 30 min annual demands of per feeder with their PFs at the time of annual peak load are shown as follows:

Feeder	Demand (kW)	PF
1	900	0.85
2	1000	0.9
3	2100	0.95
4	2000	0.9

   a. Find the 30 min annual maximum demand on the substation transformer in kW and in kVA.
   b. Find the load diversity in kW.
   c. Select a suitable substation transformer size if zero load growth is expected and company policy permits as much as 25% short-time overloads on the transformer. Among the standard three-phase transformer sizes available are the following:

   2500/3125 kVA self-cooled/forced-air-cooled
   3750/4687 kVA self-cooled/forced-air-cooled
   50006250 kVA self-cooled/forced-air-cooled
   7500/9375 kVA self-cooled/forced-air-cooled

   d. Now assume that the substation load will increase at a constant percentage rate per year and will double in 10 years. If 7500/9375 kVA-rated transformer is installed, in how many years will it be loaded to its *fans-on* rating?

2.19 Suppose that a primary feeder is supplying power to a variable load. Every day and all year long, the load has a daily constant peak value of 50 MW between 7 PM and 7 AM and a daily constant off-peak value of 5 MW between 7 AM and 7 PM. Derive the necessary equations and calculate the following:
   a. The load factor of the feeder
   b. The factor of the feeder

2.20 A typical DT serves four residential loads, that is, houses, through six SDs and two spans of SL. There are a total of 200 DTs and 800 residences supplied by this primary feeder. Use Figure 2.13 and Table 2.2 and assume that a typical residence has a clothes dryer, a range, a refrigerator, and some lighting and miscellaneous appliances. Determine the following:
   a. The 30 min maximum diversified demand on the transformer
   b. The 30 min maximum diversified demand on the entire feeder
   c. Use the typical hourly variation factors given in Table 2.2 and calculate the small portion of the daily demand curve on the DT, that is, the total hourly diversified demands at 6 AM, 12 noon, and 7 PM, on the DT, in kilowatts

2.21 Repeat Example 2.15, assuming that the monthly demand charge is $15/kW and that the monthly energy charges are 12 cents/kWh for the first 1000 kWh, 10 cents/kWh for the next 3000 kWh, and 8 cents/kWh for all kWh in excess of 4000. The 30 min maximum diversified demands for customers A and B are 40 kW each. The PFs for customers A and B are 0.95 lagging and 0.50 lagging, respectively.

2.22 Consider the MATLAB demand forecasting computer program given in Table 2.3. Assume that the peak MW July demands for the last 8 years have been the following: 3094, 2938, 2714, 3567, 4027, 3591, 4579, and 4436. Use the given MATLAB program as a curve-fitting technique and determine the following:
   a. The average rate of growth of the demand
   b. The ideal data based on rate of growth for the past 8 years to give the correct future demand forecast
   c. The forecasted future demands for the next 10 years

2.23 The annual peak load of the feeder is 3000 kWh. Total copper loss at peak load is 300 kW. If the total annual energy supplied to the sending end of the feeder is 9000 MWh, determine the following:
   a. The annual loss factor for an urban area
   b. The annual loss factor for a rural area
   c. The total amount of energy lost due to copper losses per year in part (a) and its value at $0.06/kWh
   d. The total amount of energy lost due to copper losses per year in part (b) and its value at $0.06/kWh.

REFERENCES

1. ASA: *American Standard Definitions of Electric Terms*, Group 35, Generation, Transmission and Distribution, ASA C42.35, Alexandra, VA, 1957.
2. Westinghouse Electric Corporation: *Electric Utility Engineering Reference Book-Distribution Systems*, Vol. 3, Westinghouse Electric Corporation, East Pittsburgh, PA, 1965.
3. Sarikas, R. H. and H. B. Thacker: Distribution system load characteristics and their use in planning and design, *AIEE Trans.*, 31(pt. III), August 1957, 564–573.
4. Seelye, H. P.: *Electrical Distribution Engineering*, McGraw-Hill, New York, 1930.
5. Buller, F. H. and C. A. Woodrow: Load factor-equivalent hour values compared, *Electr. World*, 92(2), July 14, 1928, 59–60.
6. General Electric Company: *Manual of Watthour Meters*, Bulletin GET-1840C, Schenectady, New York.
7. Arvidson, C. E.: Diversified demand method of estimating residential distribution transformer loads, *Edison Electr. Inst. Bull.*, 8, October 1940, 469–479.
8. Box, G. P. and G. M. Jenkins: *Time Series Analysis, Forecasting and Control*, Holden-Day, San Francisco, CA, 1976.
9. ABB Power T & D Company: *Introduction to Integrated Resource T & D Planning*, Cory, NC, 1994.
10. Willis, H. L.: *Spatial Electric Load Forecasting*, Marcel Dekker, New York, 1996.
11. Gönen, T. and J. C. Thompson: A new stochastic load forecasting model to predict load growth on radial feeders, *Int. J. Comput. Math. Electr. Electron. Eng. (COMPEL)*, 3(1), 1984, 35–46.
13. Thompson, J. C. and T. Gönen: A developmental system simulation of growing electrical energy demand, *IEEE Mediterranean Electrotechnical Conference (MELECON 83)*, Rome, Italy, May 24–26, 1983.
14. Thompson, J. C. and T. Gönen: Simulation of load growth developmental system models for comparison with field data on radial networks, *Proceedings of the 1982 Modeling and Simulation Conference*, University of Pittsburgh, Pittsburgh, PA, April 22–23, 1982, Vol. 13, pt. 4, pp. 1549–1554.
15. Gönen, T. and A. Saidian: Electrical load forecasting, *Proceedings of the 1981 Modeling and Simulation Conference*, University of Pittsburgh, Pittsburgh, PA, April 30–May 1, 1981.
16. Ramirez-Rosado, I. J. and T. Gönen: Economical and energetic benefits derived from selected demand-side management actions in the electric power distribution, *International Conference on Modeling, Identification, and Control*, Zurich, Switzerland, February 1998.
17. Gellings, C. W.: *Demand Forecasting for Electric Utilities*, The Fairmont Press, Lilburn, GA, 1992.

3 Application of Distribution Transformers

Now that I'm almost up the ladder,
I should, no doubt, be feeling gladder,
It is quite fine, the view and such,
If just it didn't shake so much.

Richard Armour

3.1 INTRODUCTION

In general, distribution transformers are used to reduce primary system voltages (2.4–34.5 kV) to utilization voltages (120–600 V). Table 3.1 gives standard transformer capacity and voltage ratings according to ANSI Standard C57.12.20-1964 for single-phase distribution transformers. Other voltages are also available, for example, 2400 × 7200, which is used on a 2400 V system that is to be changed later to 7200 V.

Secondary symbols used are the letter Y, which indicates that the winding is connected or may be connected in wye, and Gnd Y, which indicates that the winding has one end grounded to the tank or brought out through a reduced insulation bushing. Windings that are delta connected or may be connected delta are designated by the voltage of the winding only.

In Table 3.1, further information is given by the order in which the voltages are written for low-voltage (LV) windings. To designate a winding with a mid-tap that will provide half the full-winding kilovoltampere rating at half the full-winding voltage, the full-winding voltage is written first, followed by a slant, and then the mid-tap voltage. For example, 240/120 is used for a three-wire connection to designate a 120 V mid-tap voltage with a 240 V full-winding voltage. A winding that is appropriate for series, multiple, and three-wire connections will have the designation of multiple voltage rating followed by a slant and the series voltage rating, for example, the notation 120/240 means that the winding is appropriate either for 120 V multiple connection, for 240 V series connection, and for 240/120 three-wire connection. When two voltages are separated by a cross (×), a winding that is appropriate for both multiple and series connections but not for three-wire connection is indicated. The notation 120 × 240 is used to differentiate a winding that can be used for 120 V multiple connection and 240 V series connection, but not a three-wire connection. Examples of all symbols used are given in Table 3.2.

To reduce installation costs to a minimum, small distribution transformers are made for pole mounting in overhead distribution. To reduce size and weight, preferred oriented steel is commonly used in their construction. Transformers 100 kVA and below are attached directly to the pole, and transformers larger than 100 up to 500 kVA are hung on crossbeams or support lugs. If three or more transformers larger than 100 kVA are used, they are installed on a platform supported by two poles.

In underground distribution, transformers are installed in street vaults, in manholes direct-buried, on pads at ground level, or within buildings. The type of transformer may depend upon soil content, lot location, public acceptance, or cost.

The distribution transformers and any secondary service junction devices are installed within elements, usually placed on either the front or the rear lot lines of the customer's premises.

TABLE 3.1

Standard Transformer Kilovoltamperes and Voltages

Kilovoltamperes		High Voltages		Low Voltages	
Single Phase	**Three Phase**	**Single Phase**	**Three Phase**	**Single Phase**	**Three Phase**
5	30	2,400/4,160 Y	2,400	120/240	208 Y/120
10	45	4,800/8,320 Y	4,160 Y/2,400	240/480	240
15	75	4,800Y/8,320 YX	4,160 Y	2400	480
25	112½	7,200/12,470 Y	4,800	2520	480 Y/277
37½	150	12,470 Gnd Y/7,200	8,320 Y/4,800	4800	240 × 480
50	225	7,620/13,200 Y	8,320 Y	5040	2,400
75	300	13,200 Gnd Y/7,620	7,200	6900	4,160 Y/2,400
100	500	12,000	12,000	7200	4,800
167		13,200/22,860 Gnd Y	12,470 Y/7,200	7560	12,470 Y/7,200
250		13,200	12,470 Y	7980	13,200 Y/7,620
333		13,800 Gnd Y/7,970	13,200 Y/7,620		
500		13,800/23,900 Gnd Y	13,200 Y		
		13,800	13,200		
		14,400/24,940 Gnd Y	13,800		
		16,340	22,900		
		19,920/34,500 Gnd Y	34,400		
		22,900	43,800		
		34,400	67,000		
		43,800			
		67,000			

TABLE 3.2

Designation of Voltage Ratings for Single- and Three-Phase Distribution Transformers

Single Phase		Three Phase	
Designation	**Meaning**	**Designation**	**Meaning**
120/240	Series, multiple, or three-wire connection	2,400/4,160 Y	Suitable for delta or wye connection
240/120	Series or three-wire connection only	4,160 Y	Wye connection only (no neutral)
240 × 480	Series or multiple connection only	4,160 Y 2,400	Wye connection only (with neutral available)
120/208 Y	Suitable for delta or wye connection three phase	12,470 Gnd Y/7,200	Wye connection only (with reduced insulation neutral available)
12,470 Gnd Y/7,200	One end of winding grounded to tank or brought out through reduced insulation bushing	4,160	Delta connection only

The installation of the equipment to either front or rear locations may be limited by customer preference, local ordinances, landscape conditions, etc. The rule of thumb requires that a transformer be centrally located with respect to the load it supplies in order to provide proper cable economy, voltage drop, and aesthetic effect.

Secondary service junctions for an underground distribution system can be of the pedestal, hand-hole, or direct-buried splice types. No junction is required if the service cables are connected directly from the distribution transformer to the user's apparatus.

Secondary or service conductors can be either copper or aluminum. However, in general, aluminum conductors are mostly used for cost savings. The cables are either single conductor or triplex conductors. Neutrals may be either bare or covered, installed separately, or assembled with the power conductors. All secondary or service conductors are rated 600 V, and their sizes differ from # 6 AWG to 1000 kcmil.

3.2 TYPES OF DISTRIBUTION TRANSFORMERS

Heat is a limiting factor in transformer loading. Removing the coil heat is an important task. In liquid-filled types, the transformer coils are immersed in a smooth-surfaced, oil-filled tank. Oil absorbs the coil heat and transfers it to the tank surface, which, in turn, delivers it to the surrounding air. For transformers 25 kVA and larger, the size of the smooth tank surface required to dissipate heat becomes larger than that required to enclose the coils. Therefore, the transformer tank may be corrugated to add surface, or external tubes may be welded to the tank. To further increase the heat-disposal capacity, air may be blown over the tube surface. Such designs are known as forced-air-cooled, with respect to self-cooled types. Presently, however, all distribution transformers are built to be self-cooled.

Therefore, the distribution transformers can be classified as (1) dry type and (2) liquid-filled type. The dry-type distribution transformers are air-cooled and air-insulated. The liquid-filled-type distribution transformers can further be classified as (1) oil filled and (2) inerteen filled.

The distribution transformers employed in overhead distribution systems can be categorized as

1. Conventional transformers
2. Completely self-protecting (CSP) transformers
3. Completely self-protecting for secondary banking (CSPB) transformers

The conventional transformers have no integral lightning, fault, or overload protective devices provided as a part of the transformer. The CSP transformers are, as the name implies, self-protecting from lightning or line surges, overloads, and short circuits. Lightning arresters mounted directly on the transformer tank, as shown in Figure 3.1, protect the primary winding against the lightning and line surges. The overload protection is provided by circuit breakers inside the transformer tank. The transformer is protected against an internal fault by internal protective links located between the primary winding and the primary bushing. Single-phase CSP transformers (oil-immersed, pole-mounted, 65°C, 60 Hz, 10–500 kVA) are available for a range of primary voltages from 2,400 to 34,400 V. The secondary voltages are 120/240 or 240/480/277 V. The CSPB distribution transformers are designed for banked secondary service. They are built similar to the CSP transformers, but they are provided with two sets of circuit breakers. The second set is used to sectionalize the secondary when it is needed.

The distribution transformers employed in underground distribution systems can be categorized as

1. Subway transformers
2. Low-cost residential transformers
3. Network transformers

Subway transformers are used in underground vaults. They can be conventional type or current-protected type. Low-cost residential transformers are similar to those conventional transformers employed in overhead distribution. Network transformers are employed in secondary networks. They have the primary disconnecting and grounding switch and the network protector mounted integrally on the transformer. They can be either liquid filled, ventilated dry type, or sealed dry type.

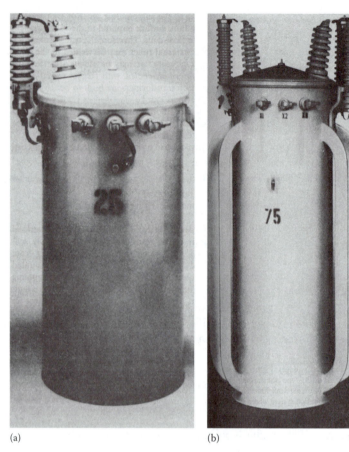

(a)   (b)

FIGURE 3.1 Overhead pole-mounted distribution transformers: (a) single-phase completely self-protecting (or conventional) and (b) three phase. (From Westinghouse Electric Corporation, East Pittsburgh, PA. With permission.)

The ABB Corporation developed *resibloc* dry-type distribution transformers from 112.5 through 25,000 kVA, and from 2,300 through 34.5 kV primary voltage level and 120 V through 15 kV secondary voltage level. Such transformers have windings that are hermetically cast in epoxy without the use of mold. The epoxy insulation system is reinforced by a special glass fiber roving technique that binds the coil together into virtual winding block. As a result, they have unsurpassed mechanical strength with design optimization through 25,000 kVA. Figure 3.2 shows such resibloc network transformer. Figure 3.3 shows a dry-type pole-mounted resibloc transformer. Figure 3.4 shows a dry-type resibloc network transformer. Figure 3.5 shows an outdoor three-phase dry-type resibloc transformer. Figure 3.6 shows a pad-mount-type single-phase resibloc transformer. Figure 3.7 shows a pad-mount three-phase resibloc transformer. Figure 3.8 shows an arch flash-resistant dry-type three-phase resibloc transformer. Figure 3.9 shows a TRIDRY dry-type resibloc transformer. Figure 3.10 shows a VPI dry resibloc transformer. Figure 3.11 shows a pad-mount installation of three-phase resibloc transformer.

These resibloc transformers provide the ultimate withstand to thermal and mechanical stresses from severe climates, cycling loads, and short circuit forces. The epoxy insulation system is highly resistant to moisture, freezing, and chemicals, and is used in most demanding applications. Such transformers are nonexplosive with resistance to flame and do not require vaults, containment dikes, or expensive fire suppression systems. Primary basic impulse insulation level (BIL) is up to 150 kV and secondary BIL is up to 75 kV.

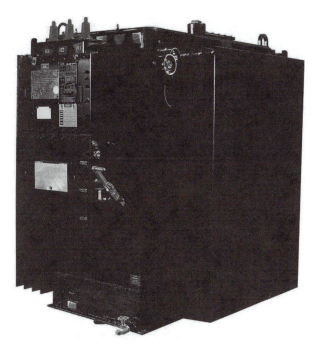

FIGURE 3.2   Network transformer. (Courtesy of ABB.)

FIGURE 3.3   Dry-type pole-mounted resibloc transformer. (Courtesy of ABB.)

They can withstand a temperature rise of 80°C. They have no danger of fire and explosion and have no liquids to leak. Thus, they require minimal maintenance. They can be used indoor and outdoor enclosures. The resibloc transformers will not ignite during an electrical arc of nominal duration. If ignited with a direct flame, resibloc will self-extinguish when the flame is removed.

Resibloc transformers are utilized in some of the harshest indoor and outdoor environments imaginable. However, while core and coil technologies have been enhanced to combat caustic and humid environments, resibloc transformers still require the protection of a properly designed enclosure.

FIGURE 3.4 Dry-type resibloc network transformer. (Courtesy of ABB.)

FIGURE 3.5 Outdoor three-phase dry-type resibloc transformer. (Courtesy of ABB.)

An enclosure that flexes or bends under high wind loading can compromise electrical clearances from the transformer to the enclosure, which can lead to transformer failures as well as electrical safety hazards.

For example, an enclosure that allows excess water entry into it also poses unique risk. Such enclosure designs have been used along coastal areas and frigid northern slopes where high winds and driving rain are common. They can be also supplied with forced air cooling. The temperature sensors are located in the LV windings and are connected to the three-phase winding temperature monitor that controls the forced air cooling automatically.

FIGURE 3.6   Pad-mount-type single-phase resibloc transformer. (Courtesy of ABB.)

FIGURE 3.7   Pad-mount three-phase resibloc transformer. (Courtesy of ABB.)

FIGURE 3.8 An arch flash-resistant dry-type three-phase resibloc transformer. (Courtesy of ABB.)

FIGURE 3.9 TRIDRY dry-type resibloc transformer. (Courtesy of ABB.)

FIGURE 3.10   VPI dry resibloc transformer. (Courtesy of ABB.)

FIGURE 3.11   Pad-mount installation of three-phase resibloc transformer. (Courtesy of ABB.)

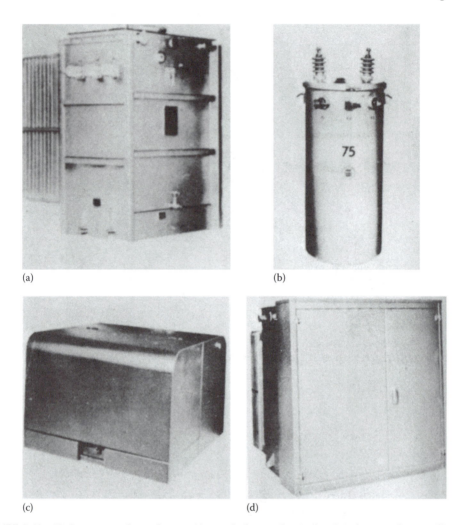

FIGURE 3.12 Various types of transformer: (a) a typical secondary unit substation transformer, (b) a typical single-phase pole-type tansformer, (c) a single-phase pad-mounted transformer, and (d) three-phase pad-mounted transformer. (From Baleau Standard Inc., with permission.)

Figure 3.12 shows various types of transformers. Figure 3.12a shows a typical secondary-unit substation with the high voltage (HV) and the LV on opposite ends and full-length flanges for close coupling to HV and LV switchgears. These units are normally made in sizes from 75 to 2500 kVA, three-phase, to 35 kV class. A typical single-phase pole-type transformer for a normal utility application is shown in Figure 3.12b. These are made from 10 to 500 kVA for delta and wye systems (one-bushing or two-bushing HV).

Figure 3.12c shows a typical single-phase pad-mounted (minipad) utility-type transformer. These are made from 10 to 167 kVA. They are designed to do the same function as the pole type except they are for the underground distribution system where all cables are below grade. A typical three-phase pad-mounted (stan-pad) transformer used by utilities as well as industrial and commercial applications is shown in Figure 3.12d. They are made from 45 to 2500 kVA normally, but have been made to 5000 kVA on special applications. They are also designed for underground service.

Figure 3.13 shows various types of transformers. Figure 3.13a shows a typical three-phase subsurface-vault-type transformer used in utility applications in vaults below grade where there is no room to place the transformer elsewhere. These units are made for 75–2500 kVA and are made of a heavier-gage steel, special heavy corrugated radiators for cooling, and a special coal-tar type of paint.

(a)

(b)

(c)

FIGURE 3.13 Various types of transformers: (a) three-phase sub-surface-vault type transformer, (b) a typical mobile transformer, and (c) a typical power transformer. (From Baleau Standard Inc., with permission.)

A typical mobile transformer is shown in Figure 3.13b. These units are made for emergency applications and to allow utilities to reduce inventory. They are made typically for 500–2500 kVA. They can be used on underground service as well as overhead service. Normally they can have two or three primary voltages and two or three secondary voltages, so they may be used on any system the utility may have. For an emergency outage, this unit is simply driven to the site, hooked up, and the power to the site is restored. This allows time to analyze and repair the failed unit. Figure 3.13c shows a typical power transformer. This class of unit is manufactured from 3700 kVA to 30 MVA up to about 138 kV class. The picture shows removable radiators to allow for a smaller size during shipment, and fans for increased capacity when required, including an automatic on-load tap changer that changes as the voltage varies.

Table 3.3 presents electrical characteristics of typical single-phase distribution transformers. Table 3.4 gives electrical characteristics of typical three-phase pad-mounted transformers. (For more accurate values, consult the individual manufacturer's catalogs.)

TABLE 3.3
Electrical Characteristics of Typical Single-Phase Distribution Transformers[a]

kVA	Percent of Av. Excit. Curr.	Watts Loss No Load	Watts Loss Total	Watts Loss 1.0 PF	Watts Loss 0.8 PF	120/240 V Low Voltage% Regulation %Z	%R	%X	Watts Loss No Load	Watts Loss Total	Watts Loss 1.0 PF	Watts Loss 0.8 PF	240/480 and 277/480 Y V Low Voltage% Regulation %Z	%R	%X
						2400/4160 Y V high voltage									
5	2.4	34	137	2.06	2.12	2.2	2.1	0.8							
10	1.6	68	197	1.30	1.68	1.7	1.3	1.1	68	202	1.35	1.69	1.7	1.3	1.0
15	1.4	84	272	1.27	1.59	1.6	1.3	1.0	84	277	1.30	1.60	1.6	1.3	1.1
25	1.3	118	385	1.10	1.65	1.7	1.1	1.1	118	390	1.11	1.65	1.7	1.1	1.3
38	1.1	166	540	1.00	1.55	1.6	1.0	1.3	166	550	1.04	1.54	1.6	1.0	1.2
50	1.0	185	615	0.88	1.58	1.7	0.9	1.5	185	625	0.90	1.58	1.7	0.9	1.5
75	1.3	285	910	0.85	1.41	1.5	0.8	1.2	285	925	0.86	1.33	1.4	0.9	1.1
100	1.2	355	1175	0.84	1.55	1.7	0.8	1.5	355	1190	0.85	1.49	1.6	0.8	1.4
167	1.0	500	2100	0.99	1.75	1.9	1.0	1.6	500	2000	0.90	1.57	1.7	0.9	1.4
250	1.0	610	3390	1.16	2.16	2.4	1.1	2.1	610	3280	1.11	2.02	2.2	1.1	1.9
333	1.0	840	4200	1.08	2.51	3.0	1.0	2.8	840	3690	0.88	1.90	2.2	0.9	1.9
500	1.0	1140	5740	0.97	2.50	3.1	0.9	3.0	1140	4810	0.95	2.00	2.3	0.7	2.2
						7,200/12,470 Y V high voltage									
5	2.4	41	144	2.07	2.11	2.2	2.1	0.8							
10	1.6	68	204	1.37	1.80	1.8	1.4	1.2	68	209	1.43	1.80	1.8	1.4	1.1
15	1.4	84	282	1.33	1.69	1.7	1.3	1.2	84	287	1.35	1.70	1.7	1.4	1.0
25	1.3	118	422	1.22	1.69	1.7	1.2	1.2	118	427	1.24	1.69	1.7	1.2	1.2
38	1.1	166	570	1.10	1.64	1.7	1.1	1.3	166	575	1.10	1.65	1.7	1.1	1.3

kVA															
50	1.0	185	720	1.10	1.71	1.8	1.1	1.4	185	725	1.10	1.71	1.8	1.1	1.4
75	1.3	285	985	0.95	1.60	1.7	0.9	1.4	285	1000	0.97	1.52	1.6	1.0	1.3
100	1.2	355	1275	0.95	1.72	1.9	1.9	1.7	355	1290	0.95	1.60	1.7	1.9	1.4
167	1.0	500	2100	0.98	1.90	2.1	1.0	1.9	500	2000	0.91	1.70	1.9	0.9	1.7
250	1.0	610	3490	1.22	2.45	2.8	1.2	2.6	610	3250	1.17	2.19	2.4	1.1	2.2
333	1.0	840	4255	1.07	2.50	3.0	1.0	2.8	840	3690	0.89	2.03	2.4	0.9	2.2
500	1.0	1140	5640	0.95	2.55	3.2	0.9	3.1	1140	4810	0.78	1.99	2.4	0.7	2.3

13,200/22,860 Gnd Y or 13,800/23,900 Gnd Y or 14,400/24,940 Gnd Y V high voltage

kVA															
5	2.4	42	154	2.25	2.30	2.4	2.3	0.9		220	1.49	1.89	1.9	1.5	1.2
10	1.6	73	215	1.45	1.89	1.9	1.4	1.3	73	310	1.52	1.80	1.8	1.5	1.0
15	1.4	84	305	1.48	1.80	1.8	1.5	1.0	84	442	1.30	1.78	1.8	1.3	1.2
25	1.3	118	437	1.29	1.79	1.8	1.3	1.3	118	590	1.16	1.72	1.8	1.1	1.4
38	1.1	166	585	1.15	1.72	1.8	1.1	1.4	166	740	1.15	1.81	1.9	1.1	1.5
50	1.0	185	735	1.14	1.81	1.9	1.1	1.4	185	1065	1.06	1.78	1.8	1.0	1.5
75	1.4	285	1050	1.05	1.78	1.8	1.0	1.5	285	1310	0.98	1.74	1.9	1.0	1.6
100	1.3	355	1300	0.97	1.81	2.0	0.9	1.8	355	2060	0.95	1.80	2.0	0.9	1.8
167	1.0	500	2160	0.98	1.96	2.2	1.0	2.0	500	3285	1.11	2.16	2.5	1.1	2.3
250	1.0	610	3490	1.22	2.52	2.9	1.2	2.7	610	3750	0.91	2.05	2.4	0.9	2.2
333	1.0	840	4300	1.09	2.60	3.1	1.0	2.9	840	4760	0.76	1.98	2.4	0.7	2.3
500	1.0	1140	5640	0.95	2.55	3.2	1.1	3.0	1140						

TABLE 3.4
Electrical Characteristics of Typical Three-Phase Pad-Mounted Transformers

		Watts Loss				208 Y/120 V Low Voltage % Regulation			Watts Loss				480 Y/277 V Low Voltage % Regulation		
KVA	Percent of Av. Excit. Curr.	No Load	Total	1.0 PF	0.8 PF	%Z	%R	%X	No load	Total	1.0 PF	0.8 PF	%Z	%R	%X
						4,160 Gnd Y/2,400 X12,470 Gnd Y/7,200 V high voltage									
75	1.5	360	1,350	1.35	2.1	2.1	1.3	1.6	360	1,350	1.35	2.1	2.1	1.3	1.6
112	1.0	530	1,800	1.15	1.7	1.7	1.1	1.3	530	1,800	1.15	1.7	1.7	1.1	1.3
150	1.0	560	2,250	1.15	1.9	1.9	1.1	1.6	560	2,250	1.15	1.9	1.9	1.1	1.6
225	1.0	880	3,300	1.10	1.9	1.9	1.1	1.6	800	3,300	1.10	1.9	1.9	1.1	1.6
300	1.0	1050	4,300	1.10	1.9	2.0	1.1	1.7	1050	4,100	1.05	1.8	1.8	1.0	1.5
500	1.0	1600	6,800	1.15	2.2	2.3	1.0	2.1	1600	6,500	1.10	2.0	2.0	1.0	1.7
750	1.0	1800	10,200	1.28	4.4	5.7	1.1	5.6	1800	9,400	1.18	4.3	5.7	1.0	5.6
1000	1.0	2100	12,500	1.20	4.3	5.7	1.0	5.6	2100	10,900	1.04	4.2	5.7	0.9	5.7
1500	1.0	2900	19,400	1.26	4.3	5.7	1.1	5.6	3300	16,500	1.04	4.2	5.7	0.9	5.7
2500	1.0								4800	26,600	1.03	4.2	5.7	0.9	5.7
3750	1.0								6500	35,500	0.95	4.1	5.7	0.8	5.7
						12,470 Gnd Y/7,200 V high voltage									
75	1.5	360	1,350	1.4	1.7	1.7	1.3	1.1	360	1,350	1.4	1.5	1.5	1.3	0.8
112	1.0	530	1,800	1.2	1.5	1.5	1.1	1.0	530	1,800	1.2	1.3	1.3	1.1	0.7
150	1.0	560	2,250	1.2	1.8	1.9	1.1	1.6	560	2,250	1.2	1.7	1.7	1.1	1.3
225	1.0	880	3,300	1.1	1.8	1.8	1.1	1.4	880	3,300	1.1	1.6	1.6	1.1	1.2
300	1.0	1050	4,300	1.1	1.6	1.6	1.1	1.2	1050	4,100	1.1	1.4	1.4	1.0	1.0

kVA									*2400/4160 Y/2400 V low voltage*						
500	1.0	1600	6,800	1.1	1.7	1.7	1.0	1.4	1600	6,500	1.1	1.4	1.4	1.0	1.0
750	1.0	1800	10,200	1.3	4.4	5.7	1.1	5.6	1800	9,400	1.2	4.3	5.7	1.0	5.6
1000	1.0	2100	12,500	1.2	4.3	5.7	1.0	5.6	2100	10,900	1.0	4.2	5.7	0.9	5.7
1500	1.0	2900	19,400	1.3	4.3	5.7	1.1	5.6	3300	16,500	1.0	4.2	5.7	0.9	5.7
2500	1.0								4800	26,600	1.0	4.2	5.7	0.9	5.7
3750	1.0								6500	35,500	0.9	4.1	5.7	0.8	5.7
12,470 Delta V high voltage															
1000	1.38	2443	11,480	1.06	4.09	5.56	0.89	5.49							
1500	1.33	3455	15,716	0.98	4.04	5.56	0.81	5.51							
2500	1.29	4956	23,193	0.92	3.97	5.56	0.73	5.52							
3750	1.37	6775	33,100	0.89	3.97	5.50	0.70	5.45							
5000	1.33	8800	42,125	0.86	3.94	5.50	0.67	5.45							
24,940 Delta V high voltage															
1000	1.42	2533	11,588	1.07	4.09	5.56	0.91	5.49							
1500	1.37	3625	15,213	0.96	4.03	5.56	0.80	5.50							
2500	1.31	5338	23,213	0.88	3.98	5.56	0.72	5.52							
3750	1.42	7075	33,700	0.90	3.97	5.50	0.71	5.44							
5000	1.33	8725	43,550	0.88	3.96	5.50	0.69	5.44							

To find the resistance (R') and the reactance (X') of a transformer of equal size and voltage, which has a different impedance value (Z') than the one shown in tables, multiply the tabulated percent values of R and X by the ratio of the new impedance value to the tabulated impedance value, that is, Z'/Z. Therefore, the resistance and the reactance of the new transformer can be found from

$$R' = R \times \frac{Z'}{Z} \quad \%\Omega \tag{3.1}$$

and

$$X' = X \times \frac{Z'}{Z} \quad \%\Omega \tag{3.2}$$

3.3 REGULATION

To calculate the transformer regulation for a kilovoltampere load of power factor $\cos\theta$, at rated voltage, any one of the following formulas can be used:

$$\%\text{regulation} = \frac{S_L}{S_T}\left[\%IR\,\cos\theta + \%IX\,\sin\theta + \frac{(\%IX\,\cos\theta - \%IR\,\sin\theta)^2}{200} \right] \tag{3.3}$$

or

$$\%\text{regulation} = \frac{I_{op}}{I_{ra}}\left[\%R\,\cos\theta + \%X\,\sin\theta + \frac{(\%X\,\cos\theta - \%R\,\sin\theta)^2}{200} \right] \tag{3.4}$$

or

$$\%\text{regulation} = V_R\,\cos\theta + V_X\,\sin\theta + \frac{(V_X\,\cos\theta - V_R\,\sin\theta)^2}{200} \tag{3.5}$$

where
θ is the power factor angle of load
V_R is the percent resistance voltage
$= \dfrac{\text{copper loss}}{\text{output}} \times 100$
S_L is the apparent load power
S_T is the rated apparent power of transformer
I_{op} is the operating current
I_{ra} is the rated current
V_X is the percent leakage reactance voltage
$= \left(V_Z^2 - V_R^2\right)^{1/2}$
V_Z is the percent impedance voltage

Note that the percent regulation at unity power factor is

$$\%\text{Regulation} = \frac{\text{Copper loss}}{\text{Output}} \times 100 + \frac{(\%\text{reactance})^2}{200} \tag{3.6}$$

3.4 TRANSFORMER EFFICIENCY

The efficiency of a transformer can be calculated from

$$\%\text{Efficiency} = \frac{\text{Output in watts}}{\text{Output in watts} + \text{Total losses in watts}} \times 100 \tag{3.7}$$

The total losses include the losses in the electric circuit, magnetic circuit, and dielectric circuit. Stigant and Franklin [3, p. 97] state that a transformer has its highest efficiency at a load at which the iron loss and copper loss are equal. Therefore, the load at which the efficiency is the highest can be found from

$$\%\text{Load} = \left(\frac{\text{Iron loss}}{\text{Copper loss}}\right)^{1/2} \times 100 \tag{3.8}$$

Figures 3.14 and 3.15 show nomograms for quick determination of the efficiency of a transformer. (For more accurate values, consult the individual manufacturer's catalogs.) With the cost of electric energy presently 5–6 cents/kWh and projected to double within the next 10–15 years, as shown in Figure 3.16, the cost efficiency of transformers now shifts to align itself with energy efficiency.

Note that the iron losses (or core losses) include (1) hysteresis loss and (2) eddy-current loss. The hysteresis loss is due to the power requirement of maintaining the continuous reversals of the elementary magnets (or individual molecules) of which the iron is composed as a result of the flux alternations in a transformer core. The eddy-current loss is the loss due to circulating currents in the core iron, caused by the magnetic flux in the iron cutting the iron, which is a conductor. The eddy-current loss is proportional to the square of the frequency and the square of the flux density. The core is built up of thin laminations insulated from each other by an insulating coating on the iron to reduce the eddy-current loss. Also, in order to reduce the hysteresis loss and the eddy-current loss, special grades of steel alloyed with silicon are used. The iron or core losses are practically independent of the load. On the other hand, the copper losses are due to the resistance of the primary and secondary windings.

In general, the distribution transformer costs can be classified as (1) the cost of investment, (2) the cost of lost energy due to the losses in the transformer, and (3) the cost of demand lost (i.e., the cost of lost capacity) due to the losses in the transformer. Of course, the cost of investment is the largest cost component, and it includes the cost of the transformer itself and the costs of material and labor involved in the transformer installation.

Figure 3.17 shows the annual cost per unit load vs. load level. At low-load levels, the relatively high costs result basically from the investment cost, whereas at high-load levels, they are due to the cost of additional loss of life of the transformer, the cost of lost energy, and the cost of demand loss in addition to the investment cost. Figure 3.15 indicates an operating range close to the bottom of the curve. Usually, it is economical to install a transformer at approximately 80% of its nameplate rating and to replace it later, at approximately 180%, by one with a larger capacity. However, presently, increasing costs of capital, plant and equipment, and energy tend to reduce these percentages.

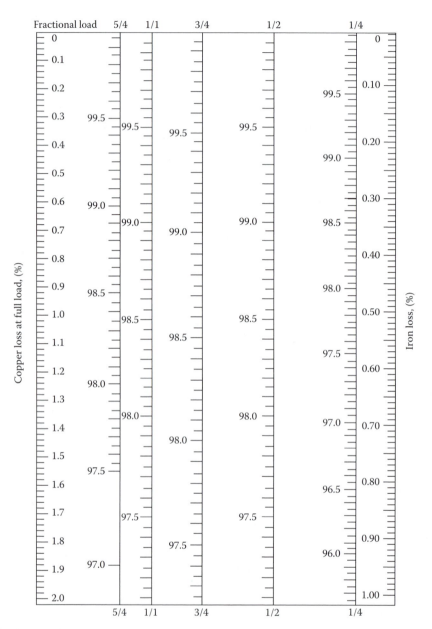

FIGURE 3.14 Transformer efficiency chart applicable only to the unity PF condition. To obtain the efficiency at a given load, lay a straightedge across the iron and copper loss values and read the efficiency at the point where the straightedge cuts the required load ordinate. (From Stigant, S.A. and Franklin, A.C., *The J&P Transformer Book*, Butterworth, London, U.K., 1973.)

3.5 TERMINAL OR LEAD MARKINGS

The terminals or leads of a transformer are the points to which external electric circuits are connected. According to NEMA and ASA standards, the higher-voltage winding is identified by HV or *H*, and the lower-voltage winding is identified by LV or *x*. Transformers with more than two windings have the windings identified as *H*, *x*, *y*, and *z*, in the order of decreasing voltage. The terminal H_1 is located on the right-hand side when facing the HV side of the transformer. On single-phase

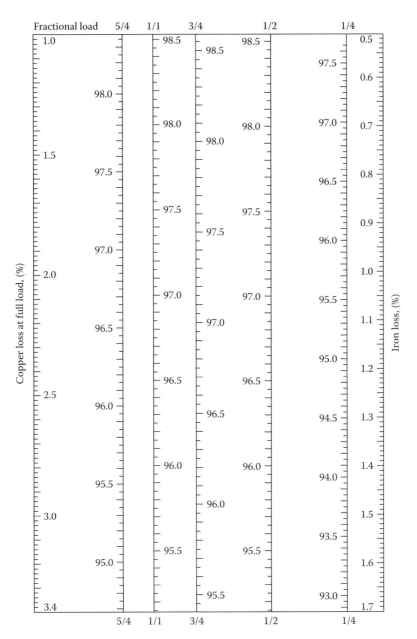

FIGURE 3.15 Transformer efficiency chart applicable only to the unity PF condition. To obtain the efficiency at a given load, lay a straightedge across the iron and copper loss values and read the efficiency at the point where the straightedge cuts the required load ordinate. (From Stigant, S.A. and Franklin, A.C., *The J&P Transformer Book*, Butterworth, London, U.K., 1973.)

transformers, the leads are numbered so that when H_1 is connected to x_1, the voltage between the highest-numbered H lead and the highest-numbered x lead is less than the voltage of the HV winding.

On three-phase transformers, the terminal H_1 is on the right-hand side when facing the HV winding, with the H_2 and H_3 terminals in numerical sequence from right to left. The terminal x_1 is on the left-hand side when facing the LV winding, with the x_2 and x_3 terminals in numerical sequence from left to right.

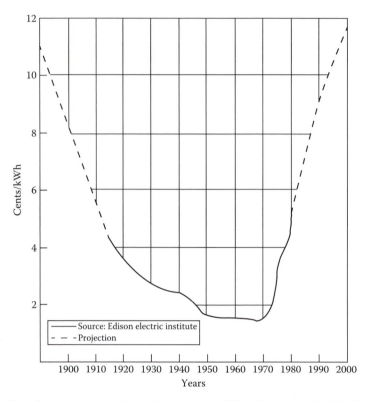

FIGURE 3.16   Cost of electric energy. (From Stigant, S.A. and Franklin, A.C., *The J&P Transformer Book*, Butterworth, London, U.K., 1973. With permission.)

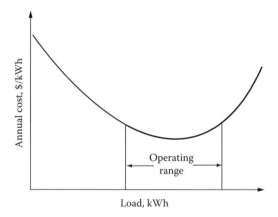

FIGURE 3.17   Annual cost per unit load vs. load level.

3.6   TRANSFORMER POLARITY

Transformer-winding terminals are marked to show polarity, to indicate the HV from the LV side. Primary and secondary are not identified as such because which is which depends on input and output connections.

Transformer polarity is an indication of the direction of current flowing through the HV leads with respect to the direction of current flow through the LV leads at any given instant. In other words, the transformer polarity simply refers to the relative direction of induced voltages between the HV leads and the LV terminals. The polarity of a single-phase distribution transformer may be

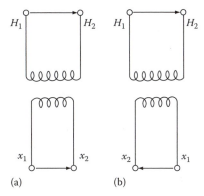

FIGURE 3.18 Additive and subtractive polarity connections: (a) subtractive polarity and (b) additive polarity.

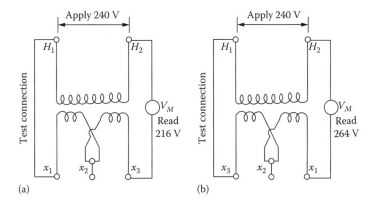

FIGURE 3.19 Polarity test: (a) subtractive polarity and (b) additive polarity.

additive or subtractive. With standard markings, the voltage from H_1 to H_2 is always in the same direction or in phase with the voltage from X_1 to X_2. In a transformer where H_1 and X_1 terminals are adjacent, as shown in Figure 3.18a, the transformer is said to have *subtractive* polarity. On the other hand, when terminals H_1 and X_1 are diagonally opposite, as shown in Figure 3.18b, the transformer is said to have *additive* polarity.

Transformer polarity can be determined by performing a simple test in which two adjacent terminals of the HV and LV windings are connected together and a moderate voltage is applied to the HV winding, as shown in Figure 3.19, and then the voltage between the HV and LV winding terminals that are not connected together are measured. The polarity is subtractive if the voltage read is less than the voltage applied to the HV winding, as shown in Figure 3.19a. The polarity is additive if the voltage read is greater than the applied voltage, as shown in Figure 3.19b.

By industry standards, all single-phase distribution transformers 200 kVA and smaller, having HVs 8660 V and below (winding voltages), have additive polarity. All other single-phase transformers have a subtractive polarity. Polarity markings are very useful when connecting transformers into three-phase banks.

3.7 DISTRIBUTION TRANSFORMER LOADING GUIDES

The rated kilovoltamperes of a given transformer is the output that can be obtained continuously at rated voltage and frequency without exceeding the specified temperature rise. Temperature rise is used for rating purposes rather than actual temperature, since the ambient temperature may vary considerably under operating conditions. The life of insulation commonly used in transformers depends upon the temperature the insulation reaches and the length of time that this temperature is sustained. Therefore, before the overload capabilities of the transformer can be determined, the ambient temperature, preload conditions, and the duration of peak loads must be known.

Based on Appendix C57.91 entitled *The Guide for Loading Mineral Oil-Immersed Overhead-Type Distribution Transformers with 55°C and 65°C Average Winding Rise* [4], which is an appendix to the ANSI Overhead Distribution Standard C57.12, 20 transformer insulation-life curves were developed. These curves indicate a minimum life expectancy of 20 years at 95°C and 110°C hotspot temperatures for 55°C and 65°C rise transformers. Previous transformer loading guides were based on the so-called 8°C insulation-life rule. For example, for transformers with class A insulation (usually oil filled), the rate of deterioration doubles approximately with each 8°C increase in temperature. In other words, if a class A insulation transformer were operated 8°C above its rated temperature, its life would be cut in half.

3.8 EQUIVALENT CIRCUITS OF A TRANSFORMER

It is possible to use several equivalent circuits to represent a given transformer. However, the general practice is to choose the simplest one, which would provide the desired accuracy in calculations.

Figure 3.20 shows an equivalent circuit of a single-phase two-winding transformer. It represents a practical transformer with an iron core and connected to a load (Z_L). When the primary winding is excited, a flux is produced through the iron core. The flux that links both primary and secondary is called the *mutual flux*, and its maximum value is denoted as ϕ_m. However, there are also leakage fluxes ϕ_{l1} and ϕ_{l2} that are produced at the primary and secondary windings, respectively. In turn, the ϕ_{l1} and ϕ_{l2} leakage fluxes produce x_{l1} and x_{l2}, that is, primary and secondary inductive reactances, respectively. The primary and secondary windings also have their internal resistances of r_1 and r_2.

Figure 3.21 shows an equivalent circuit of a loaded transformer. Note that I_2' current is a primary-current (or load) component that exactly corresponds to the secondary current I_2, as it does for an ideal transformer. Therefore,

$$I_2' = \frac{n_2}{n_1} \times I_2 \qquad (3.9)$$

or

$$I_2' = \frac{I_2}{n} \qquad (3.10)$$

where
  I_2 is the secondary current
  n_1 is the number of turns in primary winding
  n_2 is the number of turns in secondary winding

  n is the turns ratio $= \dfrac{n_1}{n_2}$

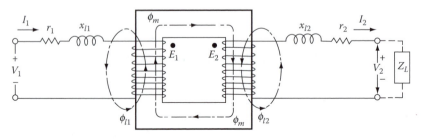

FIGURE 3.20   Basic circuit of a practical transformer.

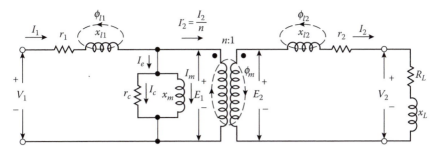

FIGURE 3.21 Equivalent circuit of a loaded transformer.

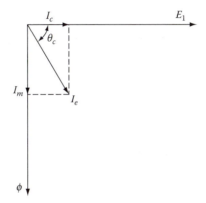

FIGURE 3.22 Phasor diagram corresponding to the excitation current components at no load.

The I_e current is the excitation current component of the primary current I_1 that is needed to produce the resultant mutual flux. As shown in Figure 3.22, the excitation current I_e also has two components, namely, (1) the magnetizing current component I_m and (2) the core-loss component I_c. The r_c represents the equivalent transformer power loss due to (hysteresis and eddy-current) iron losses in the transformer core as a result of the magnetizing current I_m. The x_m represents the inductive reactance components of the transformer with an open secondary.

Figure 3.23 shows an approximate equivalent circuit with combined primary and reflected secondary and load impedances. Note that the secondary current I_2 is seen by the primary side as I_2/n and that the secondary and load impedances are transferred (or *referred*) to the primary side as $n^2(r_2 + jx_{l2})$ and $n^2(R_L + jX_L)$, respectively. Also note that the secondary-side terminal voltage V_2 is transferred as nV_2.

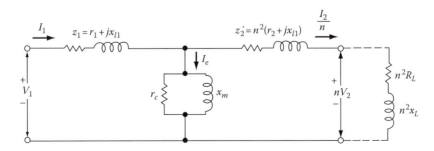

FIGURE 3.23 Equivalent circuit with the referred secondary values.

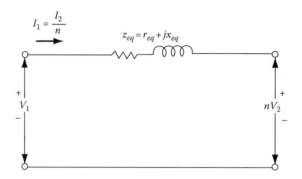

FIGURE 3.24 Simplified equivalent circuit assuming negligible excitation current.

Since the excitation current I_e is very small with respect to I_2/n for a loaded transformer, the former may be ignored, as shown in Figure 3.24. Therefore, the equivalent impedance of the transformer referred to the primary is

$$Z_{eq} = Z_1 + Z_2'$$

$$= Z_1 + n^2 Z_2$$

$$= r_{eq} + jx_{eq} \tag{3.11}$$

where

$$Z_1 = r_1 + jx_{l2} \tag{3.12}$$

$$Z_2 = r_2 + jx_{l2} \tag{3.13}$$

and therefore the equivalent resistance and the reactance of the transformer referred to the primary are

$$r_{eq} = r_1 + n^2 r_2 \tag{3.14}$$

and

$$x_{eq} = x_{l1} + n^2 x_{l2} \tag{3.15}$$

As before in Figure 3.25, for large-size power transformers,

$$r_{eq} \rightarrow 0$$

therefore the equivalent impedance of the transformer becomes

$$Z_{eq} = jx_{eq} \tag{3.16}$$

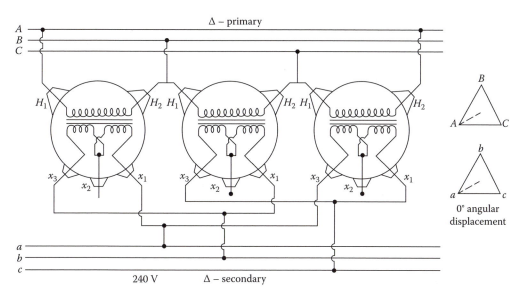

FIGURE 3.25 Simplified equivalent circuit for a large-sized power transformer.

3.9 SINGLE-PHASE TRANSFORMER CONNECTIONS

3.9.1 GENERAL

At present, the single-phase distribution transformers greatly outnumber the polyphase ones. This is partially due to the fact that lighting and the smaller power loads are supplied at single phase from single-phase secondary circuits. Also, most of the time, even polyphase secondary systems are supplied by single-phase transformers that are connected as polyphase banks.

Single-phase distribution transformers have one HV primary winding and two LV secondary windings that are rated at a nominal 120 V. Earlier transformers were built with four insulated secondary leads brought out of the transformer tank, the series or parallel connection being made outside the tank. Presently, in modern transformers, the connections are made inside the tank, with only three secondary terminals being brought out of the transformer.

Single-phase distribution transformers have one HV primary winding and two LV secondary windings. Figure 3.26 shows various connection diagrams for single-phase transformers supplying

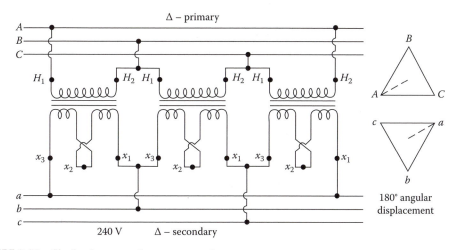

FIGURE 3.26 Single-phase transformer connections.

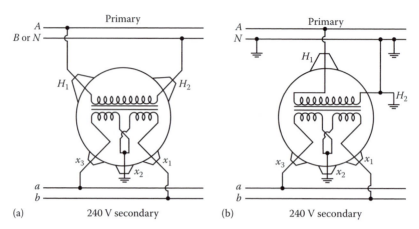

FIGURE 3.27   Single-phase transformer connections.

single-phase loads. Secondary coils each rated at a nominal 120 V may be connected in parallel to supply a two-wire 120 V circuit, as shown in Figure 3.26a and b, or they may be connected in series to supply a three-wire 120/240 V single circuit, as shown in Figure 3.26c and d. The connections shown in Figure 3.26a and b are used where the loads are comparatively small and the length of the secondary circuits is short. It is often used for a single customer who requires only 120 V single-phase power.

However, for modern homes, this connection usually is not considered adequate. If a mistake is made in polarity when connecting the two secondary coils in parallel (see Figure 3.26a) so that the LV terminal 1 is connected to terminal 4 and terminal 2 to terminal 3, the result will be a short-circuited secondary, which will blow the fuses that are installed on the HV side of the transformer (they are not shown in the figure). On the other hand, a mistake in polarity when connecting the coils in series (see Figure 3.26c) will result in the voltage across the outer conductors being zero instead of 240 V. Taps for voltage adjustment, if provided, are located on the HV winding of the transformer. Figure 3.26b and d shows single-bushing transformers connected to a multigrounded primary. They are used on 12,470 Gnd Y/7,200, 13,200 Gnd Y/7,620, and 24,940 Gnd Y/14,400 V multigrounded neutral systems. It is crucial that good and solid grounds are maintained on the transformer and on the system. Figure 3.27 shows single-phase transformer connections for single- and two-bushing transformers to provide customers who require only 240 V single-phase power. These connections are used for small industrial applications.

In general, however, the 120/240 V three-wire connection system is preferred since it has twice the load capacity of the 120 V system with only 12 times the amount of the conductor. Here, each 120 V winding has one-half the total kilovoltampere rating of the transformer. Therefore, if the connected 120 V loads are equal, the load is balanced, and no current flows in the neutral conductor. Thus the loads connected to the transformer must be held as nearly balanced as possible to provide the most economical usage of transformer capacity and to keep regulation to a minimum. Normally, one leg of the 120 V two-wire system and the middle leg of the 240 V two-wire or 120/240 V three-wire system are grounded to limit the voltage to ground on the secondary circuit to a minimum.

3.9.2   Single-Phase Transformer Paralleling

When greater capacity is required in emergency situations, two single-phase transformers of the same or different kilovoltampere ratings can be connected in parallel. The single-phase transformers

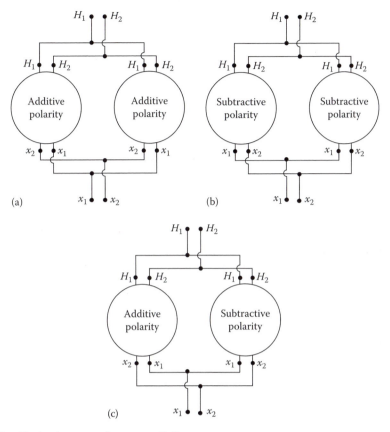

FIGURE 3.28   Single-phase transformer paralleling.

can be of either additive or subtractive polarity as long as the following conditions are observed and connected, as shown in Figure 3.28:

1. All transformers have the same turns ratio.
2. All transformers are connected to the same primary phase.
3. All transformers have identical frequency ratings.
4. All transformers have identical voltage ratings.
5. All transformers have identical tap settings.
6. Per unit impedance of one transformer is between 0.925 and 1.075 of the other in order to maximize capability.

However, paralleling two single-phase transformers is not economical since the total cost and losses of two small transformers are much larger than one large transformer with the same capacity. Therefore, it should be used only as a temporary remedy to provide for increased demands for single-phase power in emergency situations. Figure 3.29 shows two single-phase transformers, each with two bushings, connected to a two-conductor primary to supply 120/240 V single-phase power on a three-wire secondary.

To illustrate load division among the parallel-connected transformers, consider the two transformers connected in parallel and feeding a load, as shown in Figure 3.30. Assume that the aforementioned conditions for paralleling have already been met (Figure 3.31).

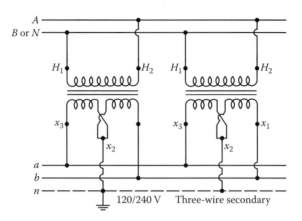

FIGURE 3.29   Parallel operation of two single-phase transformers.

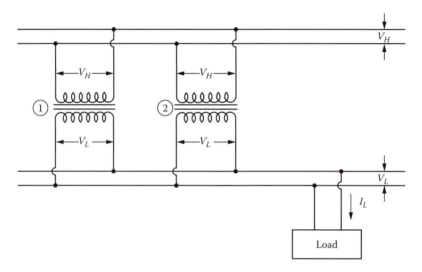

FIGURE 3.30   Two transformers connected in parallel and feeding a load.

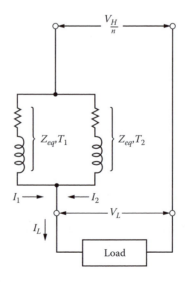

FIGURE 3.31   Equivalent circuit.

Figure 3.21 shows the corresponding equivalent circuit referred to as the LV side. Since the transformers are connected in parallel, the voltage drop through each transformer must be equal. Therefore,

$$I_1(Z_{eq,T1}) = I_2(Z_{eq,T2}) \tag{3.17}$$

from which

$$\frac{I_1}{I_2} = \frac{Z_{eq,T2}}{Z_{eq,T1}} \tag{3.18}$$

where
I_1 is the secondary current of transformer 1
I_2 is the secondary current of transformer 2
I_L is the load current
$Z_{eq,1}$ is the equivalent impedance of transformer 1
$Z_{eq,2}$ is the equivalent impedance of transformer 2

From Equation 3.18, it can be seen that the load division is determined only by the relative ohmic impedance of the transformers. If the ohmic impedances in Equation 3.18 are replaced by their equivalent in terms of percent impedance, the following equation can be found:

$$\frac{I_1}{I_2} = \frac{(\%Z)_{T2}}{(\%Z)_{T1}} \frac{S_{T1}}{S_{T2}} \tag{3.19}$$

where
$(\%Z)_{T1}$ is the percent impedance of transformer 1
$(\%Z)_{T2}$ is the percent impedance of transformer 2
S_{T1} is the kilovoltampere rating of transformer 1
S_{T2} is the kilovoltampere rating of transformer 2

Equation 3.19 can be expressed in terms of kilovoltamperes supplied by each transformer since the primary and secondary voltages for each transformer are the same, respectively. Therefore,

$$\frac{S_{L1}}{S_{L2}} = \frac{(\%Z)_{T2}}{(\%Z)_{T1}} \frac{S_{T1}}{S_{T2}} \tag{3.20}$$

where
S_{L1} is the kilovoltamperes supplied by transformer 1 to the load
S_{L2} is the kilovoltamperes supplied by transformer 2 to the load

Example 3.1

Figure 3.32 shows an equivalent circuit of a single-phase transformer with three-wire secondary for three-wire single-phase distribution. The typical distribution transformer is rated as 25 kVA, 7200-120/240 V, 60 Hz, and has the following *per unit** impedance based on the transformer ratings and based on the use of the entire LV winding with zero neutral current:

$$R_T = 0.014\,\text{pu}$$

---

* Per unit systems are explained in Appendix D.

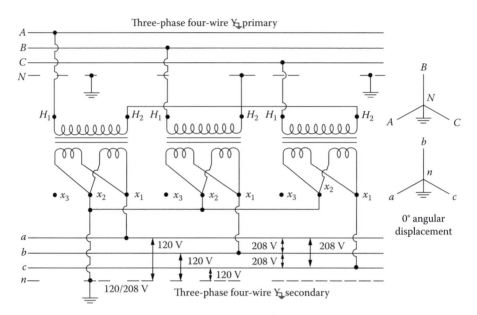

FIGURE 3.32 An equivalent circuit of a single-phase transformer with three-wire secondary. (From Lloyd, B., *Electric Utility Engineering Reference Book-Distribution Systems*, Vol. 3, Westinghouse Electric Corporation, East Pittsburgh, PA, 1965.)

and

$$X_T = 0.012\,\text{pu}$$

Here, the two halves of the LV may be independently loaded, and, in general, the three-wire secondary load will not be balanced. Therefore, in general, the equivalent circuit needed is that of a three-winding single-phase transformer as shown in Figure 3.32, when voltage drops and/or fault currents are to be computed. Thus, use the meager amount of data (it is all that is usually available) and evaluate numerically all the impedances shown in Figure 3.32.

Solution

Figure 3.32 is based on the reference by Lloyd [1]. To determine $\bar{Z}_{HX_{1-2}}$ approximately, Lloyd gives the following formula:

$$\bar{Z}_{HX_{1-2}} = 1.5R_T + j1.2X_T \tag{3.21}$$

where $\bar{Z}_{HX_{1-2}}$ is the transformer impedance referred to HV winding when the section of the LV winding between the terminals X_2 and X_3 is short-circuited.

From Figure 3.32, the turns ratio of the transformer is

$$n = \frac{V_H}{V_X} = \frac{7200\,\text{V}}{120\,\text{V}} = 60$$

Since the given per unit impedances of the transformer are based on the use of the entire LV winding,

$$\bar{Z}_{HX_{1-3}} = R_T + jX_T$$

$$= 0.014 + j0.012\,\text{pu}$$

Also, from Equation 3.21,

$$\overline{Z}_{HX_{1-2}} = 1.5R_T + j1.2X_T$$

$$= 1.5 \times 0.014 + j1.2 \times 0.012$$

$$= 0.021 + j0.0144\,\text{pu}$$

Therefore,

$$2\overline{Z}_{HX_{1-3}} - \overline{Z}_{HX_{1-2}} = 2(0.014 + j0.012) - (0.021 + j0.0144)$$

$$= 0.007 + j0.0096\,\text{pu}$$

$$= 14.515 + j19.906 = 24.637\angle 53.9°\,\Omega$$

and

$$\frac{2\overline{Z}_{HX_{1-2}} - 2\overline{Z}_{HX_{1-3}}}{n^2} = \frac{2(0.021 + j0.0144) - 2(0.014 + j0.012)}{60^2}$$

$$= 3.89 \times 10^{-6} + j1.334 \times 10^{-6}\,\text{pu}$$

$$= 0.008064 + j0.0028 = 8.525 \times 10^{-3}\,\angle 18.9°\,\Omega$$

Example 3.2

Using the transformer equivalent circuit found in Example 3.1, determine the line-to-neutral (120 V) and line-to-line (240 V) fault currents in three-wire single-phase 120/240 V secondaries shown in Figures 3.33 and 3.34, respectively. In the figures, R represents the resistance of service-drop cable per conductor. Usually R is much larger than X for such cable and therefore X may be neglected.
   Using the given data, determine the following:

a. Find the symmetrical rms fault currents in the HV and LV circuits for a 120 V fault if R of the service-drop cable is zero.
b. Find the symmetrical rms fault currents in the HV and LV circuits for a 240 V fault if R of the service-drop cable is zero.
c. If the transformer is a CSPB type, find the minimum allowable interrupting capacity (in symmetrical rms amperes) for a circuit breaker connected to the transformer's LV terminals.

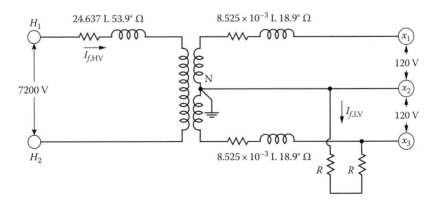

FIGURE 3.33 Secondary line-to-neutral fault.

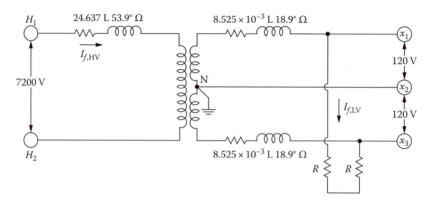

FIGURE 3.34   Secondary line-to-line fault.

Solution

a. When $R = 0$, from Figure 3.33, the line-to-neutral fault current in the secondary side of the transformer is

$$\bar{I}_{f,LV} = \frac{120}{8.525 \times 10^{-3} \angle 18.9° + (1/60)^2 (24.637 \angle 53.9°)}$$

$$= 8181.7 \angle -34.4° \text{ A}$$

Thus the fault current in the HV side is

$$\bar{I}_{f,HV} = \frac{\bar{I}_{f,LV}}{n}$$

$$= \frac{8181.7}{60} = 136.4 \text{ A}$$

Note that the turns ratio is found as

$$n = \frac{7200 \text{ V}}{120 \text{ V}} = 60$$

b. When $R = 0$, from Figure 3.24, the line-to-line fault current in the secondary side of the transformer is

$$\bar{I}_{f,LV} = \frac{240}{2(8.525 \times 10^{-3} \angle 18.9°) + (1/30)^2 (24.637 \angle 53.9°)}$$

$$= 5649 \angle -40.6° \text{ A}$$

Thus the fault current in the HV side is

$$\bar{I}_{f,HV} = \frac{\bar{I}_{f,LV}}{n}$$

$$= \frac{5649}{30} = 188.3 \text{ A}$$

Note that the turns ratio is found as

$$n = \frac{7200\,V}{240\,V} = 30$$

c. Therefore, the minimum allowable interrupting capacity for a circuit breaker connected to the transformer LV terminals is 8181.7 A.

Example 3.3

Using the data given in Example 3.2, determine the following:

a. Estimate approximately the value of R, that is, the service-drop cable's resistance, which will produce equal line-to-line and line-to-neutral fault currents.
b. If the conductors of the service-drop cable are aluminum, find the length of the service-drop cable that would correspond to the resistance R found in part (a) in case of (1) #4 AWG conductors with a resistance of 2.58 Ω/mi and (2) #1/0 AWG conductors with a resistance of 1.03 Ω/mi.

Solution

a. Since the line-to-line and the line-to-neutral fault currents are supposed to be equal to each other,

$$\frac{240}{2R + 0.032256 + j0.02765} = \frac{120}{2R + 0.012096 + j0.0083}$$

or

$$R \cong 0.0075\,\Omega$$

b. The length of the service-drop cable is
   1. If #4 AWG aluminum conductors with a resistance of 2.58 Ω/mi or 4.886×10^{-4} Ω/ft are used,

$$\text{Service-drop length} = \frac{R}{4.886 \times 10^{-4}}$$

$$= \frac{0.0075\,\Omega}{4.886 \times 10^{-4}\,\Omega/\text{ft}}$$

$$\cong 15.35\,\text{ft}$$

   2. If #1/0 AWG aluminum conductors with a resistance of 1.03 Ω/mi or 1.9508×10^{-4} Ω/ft are used,

$$\text{Service-drop length} = \frac{0.0075\,\Omega}{1.9508 \times 10^{-4}\,\Omega/\text{ft}}$$

$$\cong 38.45\,\text{ft}$$

Example 3.4

Assume that a 250 kVA transformer with 2.4% impedance is paralleled with a 500 kVA transformer with 3.1% impedance. Determine the maximum load that can be carried without overloading either transformer. Assume that the maximum allowable transformer loading is 100% of the rating.

Solution

Designating the 250 and 500 kVA transformers as transformers 1 and 2, respectively, and using Equation 3.20,

$$\frac{S_{L1}}{S_{L2}} = \frac{(\%Z)_{T2}}{(\%Z)_{T1}} \frac{S_{T1}}{S_{T2}}$$

$$= \frac{3.1}{2.4} \times \frac{250}{500} = 0.6458$$

Assume a load of 500 kVA on the 500 kVA transformer. The preceding result shows that the load on the 250 kVA transformer will be $(0.6458) \times (500 \text{ kVA}) = 322.9$ kVA when the load on the 500 kVA transformer is 500 kVA. Therefore, the 250 kVA transformer becomes overloaded before the 500 kVA transformer. The load on the 500 kVA transformer when the 250 kVA transformer is carrying rated load is

$$S_{L2} = \frac{S_{L1}}{0.6458}$$

$$= \frac{250}{0.6458}$$

$$= 387.1 \text{kVA}$$

Thus the total load is

$$\sum_{i=1}^{2} S_{Li} = S_{L1} + S_{L2}$$

$$= 250 + 387.1$$

$$= 637.1 \text{kVA}$$

3.10 THREE-PHASE CONNECTIONS

To raise or lower the voltages of three-phase distribution systems, either single-phase transformers can be connected to form three-phase transformer banks or three-phase transformers (having all windings in the same tank) are used. Figure 3.35 shows an eco-dry three-phase (resibloc) transformer. Figure 3.36 shows an air-to-water cooled resibloc three-phase transformer. Figure 3.37 shows an eco-dry (resibloc) three-phase transformer. Figure 3.38 shows a vacuum cast dry-type three-phase transformer.

Common methods of connecting three single-phase transformers for three-phase transformations are the delta–delta (Δ–Δ), wye–wye (Y–Y), wye–delta (Y–Δ), and delta–wye (Δ–Y) connections. Here, it is assumed that all transformers in the bank have the same kilovoltampere rating.

3.10.1 Δ–Δ Transformer Connection

Figures 3.39 and 3.40 show the Δ–Δ connection formed by tying together single-phase transformers to provide 240 V service at 0° and 180° angular displacements, respectively.

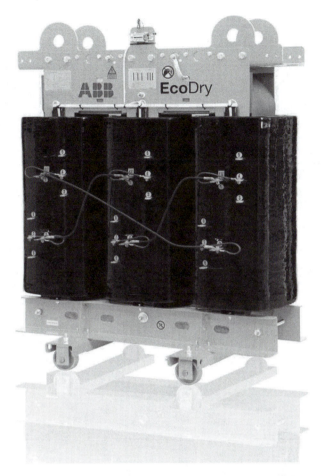

FIGURE 3.35 Eco-dry three-phase (RESIBLOC) transformer. (Courtesy of ABB, Munich, Germany.)

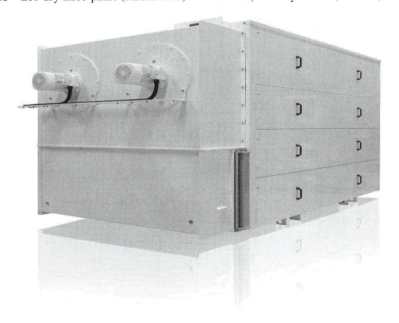

FIGURE 3.36 Air-to-water cooled RESIBLOC three-phase transformer. (Courtesy of ABB, Munich, Germany.)

FIGURE 3.37 Eco-dry (RESIBLOC) three-phase transformer. (Courtesy of ABB, Munich, Germany.)

FIGURE 3.38 Vacuum cast dry-type transformer. (Courtesy of ABB, Munich, Germany.)

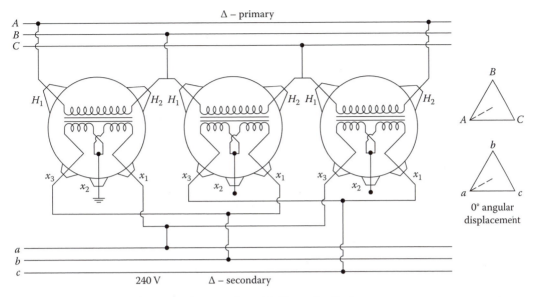

FIGURE 3.39   Δ–Δ transformer bank connection with 0° angular displacement.

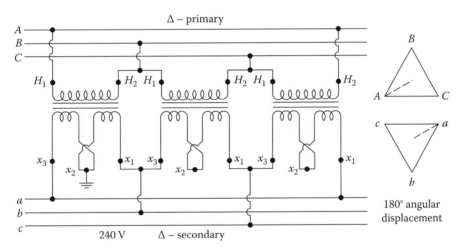

FIGURE 3.40   Δ–Δ transformer bank connection with 180° angular displacement.

This connection is often used to supply a small single-phase lighting load and three-phase power load simultaneously. To provide this type of service, the mid-tap of the secondary winding of one of the transformers is grounded and connected to the secondary neutral conductor, as shown in Figure 3.41. Therefore, the single-phase loads are connected between the phase and neutral conductors. Thus the transformer with the mid-tap carries two-thirds of the 120/240 V single-phase load and one-third of the 240 V three-phase load. The other two units each carry one-third of both the 120/240 and 240 V loads.

There is no problem from third-harmonic overvoltage or telephone interference. However, high circulating currents will result unless all three single-phase transformers are connected on the same regulating taps and have the same voltage ratios. The transformer bank rating is decreased unless all transformers have identical impedance values. The secondary neutral bushing can be grounded on only one of the three single-phase transformers, as shown in Figure 3.41.

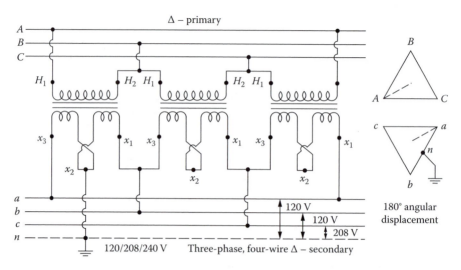

FIGURE 3.41 Δ–Δ transformer bank connection to provide 120/208/240 V three-phase four-wire service.

Therefore, to get balanced transformer loading, the conditions include the following:

1. All transformers have identical voltage ratios.
2. All transformers have identical impedance values.
3. All transformers are connected on identical taps.

However, if two of the units have the identical impedance values and the third unit has an impedance value that is within, plus or minus, 25% of the impedance value of the like transformers, it is possible to operate the Δ–Δ bank, with a small unbalanced transformer loading, at reduced bank output capacity. Table 3.5 gives the permissible amounts of load unbalanced on the odd and like transformers. Note that Z_1 is the impedance of the odd transformer unit and Z_2 is the impedance of the like transformer units. Therefore, with unbalanced transformer loading, the load values have to be checked against the values of the table so that no one transformer is overloaded.

Assume that Figure 3.42 shows the equivalent circuit of a Δ–Δ-connected transformer bank referred to the LV side. A voltage-drop equation can be written for the LV windings as

$$\bar{V}_{ba} + \bar{V}_{ac} + \bar{V}_{cb} = \bar{I}_{ba}\bar{Z}_{ab} + \bar{I}_{ac}\bar{Z}_{ca} + \bar{I}_{cb}\bar{Z}_{bc} \qquad (3.22)$$

TABLE 3.5

Permissible Percent Loading on Odd and Like Transformers as a Function of the Z_1/Z_2 Ratio

	% Load on	
Z_1/Z_2 Ratio	Odd Unit	Like Unit
0.75	109.0	96.0
0.80	107.0	96.5
0.85	105.2	97.3
0.90	103.3	98.3
1.10	96.7	102.0
1.15	95.2	102.2
1.20	93.8	103.1
1.25	92.3	103.9

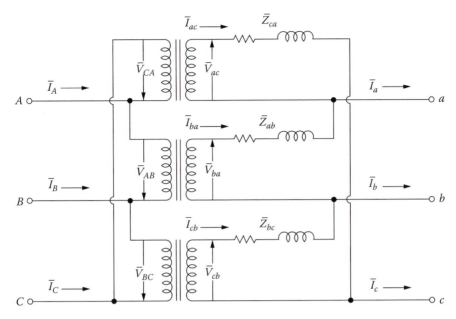

FIGURE 3.42 Equivalent circuit of a Δ–Δ-connected transformer bank.

where

$$\bar{V}_{ba} + \bar{V}_{ac} + \bar{V}_{cb} = 0 \tag{3.23}$$

Therefore, Equation 3.22 becomes

$$\bar{I}_{ba}\bar{Z}_{ab} + \bar{I}_{ac}\bar{Z}_{ca} + \bar{I}_{cb}\bar{Z}_{bc} = 0 \tag{3.24}$$

For the delta-connected secondary,

$$\bar{I}_a = \bar{I}_{ba} - \bar{I}_{ac} \tag{3.25}$$

$$\bar{I}_b = \bar{I}_{cb} - \bar{I}_{ba} \tag{3.26}$$

$$\bar{I}_c = \bar{I}_{ac} - \bar{I}_{cb} \tag{3.27}$$

From Equation 3.24,

$$\bar{I}_{ba}\bar{Z}_{ab} = -\bar{I}_{ac}\bar{Z}_{ca} - \bar{I}_{cb}\bar{Z}_{bc} \tag{3.28}$$

Adding the terms of $\bar{I}_{ba}\bar{Z}_{bc}$ and $\bar{I}_{ba}\bar{Z}_{ca}$ to either side of Equation 3.28 and substituting Equation 3.25 into the resultant equation,

$$\bar{I}_{ba} = \frac{\bar{I}_a\bar{Z}_{ca} - \bar{I}_b\bar{Z}_{bc}}{\bar{Z}_{ab} + \bar{Z}_{bc} + \bar{Z}_{ca}} \tag{3.29}$$

and similarly,

$$\bar{I}_{ac} = \frac{\bar{I}_c\bar{Z}_{bc} - \bar{I}_a\bar{Z}_{ab}}{\bar{Z}_{ab} + \bar{Z}_{bc} + \bar{Z}_{ca}} \tag{3.30}$$

and

$$\overline{I}_{cb} = \frac{\overline{I}_b \overline{Z}_{ab} - \overline{I}_c \overline{Z}_{ca}}{\overline{Z}_{ab} + \overline{Z}_{bc} + \overline{Z}_{ca}} \qquad (3.31)$$

If the three transformers shown in Figure 3.42 have equal percent impedance and equal ratios of percent reactance to percent resistance, then Equations 3.29 through 3.31 can be expressed as

$$\overline{I}_{ba} = \frac{(\overline{I}_a / S_{T,ca}) - (\overline{I}_b / S_{T,bc})}{(1/S_{T,ab}) + (1/S_{T,bc}) + (1/S_{T,ca})} \qquad (3.32)$$

$$\overline{I}_{ac} = \frac{(\overline{I}_c / S_{T,bc}) - (\overline{I}_a / S_{T,ab})}{(1/S_{T,ab}) + (1/S_{T,bc}) + (1/S_{T,ca})} \qquad (3.33)$$

$$\overline{I}_{cb} = \frac{(\overline{I}_b / S_{T,ab}) - (\overline{I}_c / S_{T,a})}{(1/S_{T,ab}) + (1/S_{T,bc}) + (1/S_{T,ca})} \qquad (3.34)$$

where
$S_{T,ab}$ is the kilovoltampere rating of the single phase between phases a and b
$S_{T,bc}$ is the kilovoltampere rating of the single phase between phases b and c
$S_{T,ca}$ is the kilovoltampere rating of the single phase between phases c and a

Example 3.5

Three single-phase transformers are connected in Δ–Δ to provide power for a three-phase wye-connected 200 kVA load with a 0.80 lagging power factor and an 80 kVA single-phase light load with a 0.90 lagging power factor, as shown in Figure 3.43.

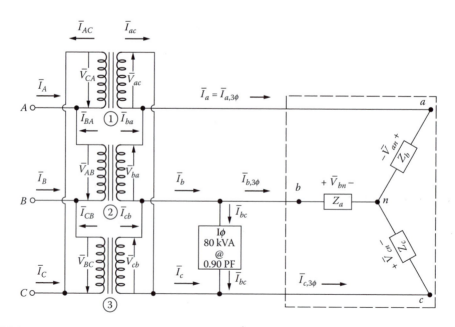

FIGURE 3.43 For Example 3.5.

Assume that the three single-phase transformers have equal percent impedance and equal ratios of percent reactance to percent resistance. The primary-side voltage of the bank is 7,620/13,200 V and the secondary-side voltage is 240 V. Assume that the single-phase transformer connected between phases b and c is rated at 100 kVA and the other two are rated at 75 kVA. Determine the following:

a. The line current flowing in each secondary-phase wire
b. The current flowing in the secondary winding of each transformer
c. The load on each transformer in kilovoltamperes
d. The current flowing in each primary winding of each transformer
e. The line current flowing in each primary-phase wire

Solution

(a) Using the voltage drop $\overline{V}_{an}$ as the reference, the three-phase components of the line currents can be found as

$$\left|\overline{I}_{a,3\phi}\right| = \left|\overline{I}_{b,3\phi}\right| = \left|\overline{I}_{c,3\phi}\right| = \frac{S_{L,3\phi}}{\sqrt{3} \times V_{L-L}}$$

$$= \frac{200}{\sqrt{3} \times 0.240}$$

$$= 481.7\,\text{A}$$

Since the three-phase load has a lagging power factor of 0.80,

$$\overline{I}_{a,3\phi} = \left|\overline{I}_{a,3\phi}\right|(\cos\theta - j\sin\theta)$$

$$= 481.7(0.80 - j0.60)$$

$$= 385.36 - j289.02$$

$$= 481.7\angle{-36.9°}\,\text{A}$$

$$\overline{I}_{b,3\phi} = a^2 \overline{I}_{a,3\phi}$$

$$= (1\angle240°)481.7\angle{-36.9°}$$

$$= -443.08 - j188.99$$

$$= 481.7\angle203.1°\,\text{A}$$

$$\overline{I}_{c,3\phi} = a\overline{I}_{a,3\phi}$$

$$= (1\angle120°)481.7\angle{-36.9°}$$

$$= 57.87 + j478.21$$

$$= 481.7\angle83.1°\,\text{A}$$

The single-phase component of the line currents can be found as

$$\left|\overline{I}_{bc}\right| = \frac{S_{L,1\phi}}{V_{L-L}} = \frac{80}{0.240}333.33\,\text{A}$$

Since the single-phase load has a lagging power factor of 0.90, the current phasor $\overline{I}_{bc}$ lags its voltage phasor $\overline{V}_{bc}$ by −25.8°. Also, since the voltage phasor $\overline{V}_{bc}$ lags the voltage reference

$\bar{V}_{an}$ by 90° (see Figure 3.26), then the current phasor $\bar{I}_{bc}$ will lag the voltage reference $\bar{V}_{an}$ by −115.8° (= −25.8° −90°). Therefore,

$$\bar{I}_{bc} = 333.33\angle -115.8°$$

$$= -145.3 - j300 \text{ A}$$

Hence, the line currents flowing in each secondary-phase wire can be found as

$$\bar{I}_a = \bar{I}_{a,3\phi}$$

$$= 481.7\angle -36.9° \text{ A}$$

$$\bar{I}_b = \bar{I}_{b,3\phi} + \bar{I}_{bc}$$

$$= 481.7\angle 203.1° + 333.33\angle -115.8°$$

$$= -588.38 - j488.99$$

$$= 765.05\angle 219.7° \text{ A}$$

$$\bar{I}_c = \bar{I}_{c,3\phi} - \bar{I}_{bc}$$

$$= 481.7\angle 83.1° - 333.33\angle -115.8°$$

$$= -87.43 + j178.21$$

$$= 198.5\angle -63.8° \text{ A}$$

(b) By using Equation 3.33, the current flowing in the secondary winding of transformer 1 can be found as

$$\bar{I}_{ac} = \frac{(\bar{I}_c/S_{T,bc}) - (\bar{I}_a/S_{T,ab})}{(1/S_{T,ab}) + (1/S_{T,bc}) + (1/S_{T,ca})}$$

$$= \frac{((198.5\angle -63.8°)/100) - ((481.7\angle -36.9°)/75)}{(1/75) + (1/100) + (1/75)}$$

$$= \frac{1.985\angle -63.8° - 6.4227\angle -36.9°}{0.0367}$$

$$= -116.07 + j56.55$$

$$= 129.11\angle -33.1° \text{ A}$$

Similarly, by using Equation 3.32,

$$\bar{I}_{ba} = \frac{(\bar{I}_a/S_{T,ca}) - (\bar{I}_b/S_{T,bc})}{(1/S_{T,ab}) + (1/S_{T,bc}) + (1/S_{T,ca})}$$

$$= \frac{((481.7\angle -36.9°)/75) - ((765.05\angle 219.7°)/100)}{(1/75) + (1/100) + (1/75)}$$

$$= \frac{6.4227\angle -36.9° - 7.6505\angle 219.7°}{0.0367}$$

$$= 300.34 + j28.08$$

$$= 301.65\angle 5.3° \text{ A}$$

and using Equation 3.34,

$$\bar{I}_{cb} = \frac{(\bar{I}_b/S_{T,ab}) - (\bar{I}_c/S_{T,ca})}{(1/S_{T,ab}) + (1/S_{T,bc}) + (1/S_{T,ca})}$$

$$= \frac{((765.05\angle 219.7°)/75) - ((198.5\angle -63.8°)/75)}{0.0367}$$

$$= -245.6 - j112.95$$

$$= 270.3\angle 204.7° \, A$$

c. The kilovoltampere load on each transformer can be found as

$$S_{L,ab} = V_{ba} \times \left| \bar{I}_{ba} \right|$$

$$= 0.240 \times 301.65$$

$$= 72.4 \, kVA$$

$$S_{L,bc} = V_{cb} \times \left| \bar{I}_{cb} \right|$$

$$= 0.240 \times 270.33$$

$$= 64.88 \, kVA$$

$$S_{L,ca} = V_{ac} \times \left| \bar{I}_{ac} \right|$$

$$= 0.240 \times 129.11$$

$$= 30.99 \, kVA$$

d. The current flowing in the primary winding of each transformer can be found by dividing the current flow in each secondary winding by the turns ratio. Therefore,

$$n = \frac{7620\,V}{240\,V} = 31.75$$

and hence

$$\bar{I}_{AC} = \frac{\bar{I}_{ac}}{n}$$

$$= \frac{129.11\angle -33.1°}{31.75}$$

$$= 4.07\angle -33.1° \, A$$

$$\bar{I}_{BA} = \frac{\bar{I}_{ba}}{n}$$

$$= \frac{301.65\angle 5.3°}{31.75}$$

$$= 9.5\angle 5.3° \, A$$

$$\overline{I}_{CB} = \frac{\overline{I}_{cb}}{n}$$

$$= \frac{270.3\angle 204.7°}{31.75}$$

$$= 8.51\angle 204.7° \, \text{A}$$

e. The line current flowing in each primary-phase wire can be found as

$$\overline{I}_A = \overline{I}_{AC} - \overline{I}_{BA}$$

$$= 4.07\angle -33.1° - 9.5\angle 5.3°$$

$$= -6.05 - j3.1$$

$$= 6.8\angle 270.1° \, \text{A}$$

$$\overline{I}_B = \overline{I}_{BA} - \overline{I}_{CB}$$

$$= 9.5\angle 5.3° - 8.51\angle 204.7°$$

$$= 17.19 + j4.44$$

$$= 17.76\angle 14.5° \, \text{A}$$

$$\overline{I}_C = \overline{I}_{CB} - \overline{I}_{AC}$$

$$= 8.51\angle 204.7° - 4.07\angle -33.1°$$

$$= -11.14 - j1.34$$

$$= 11.22\angle 186.8° \, \text{A}$$

3.10.2 Open-Δ Open-Δ Transformer Connection

The Δ–Δ connection is the most flexible of the various connection forms. One of the advantages of this connection is that if one transformer becomes damaged or is removed from service, the remaining two can be operated in what is known as the *open-Δ* or *V connection*, as shown in Figure 3.44.

Assume that a balanced three-phase load with unity power factor is served by all three transformers of a Δ–Δ bank. The taking out of one of the transformers from the service will result in having the currents in the other two transformers increase by a ratio of 1.73, even though the output of the transformer bank is the same with a unity power factor as before. However, the individual transformers now function at a power factor of 0.866. One of the transformers delivers a leading load and the other a lagging load. To operate the remaining portion of the Δ–Δ transformer bank (i.e., the open-Δ open-Δ bank) safely, the connected load has to be decreased by 57.7%, which can be found as follows:

$$S_{\Delta-\Delta} = \frac{\sqrt{3}V_{L-L}I_L}{1000} \, \text{kVA} \tag{3.35}$$

and

$$S_{\angle-\angle} = \frac{\sqrt{3}V_{L-L}I_L}{\sqrt{3} \times 1000} \, \text{kVA} \tag{3.36}$$

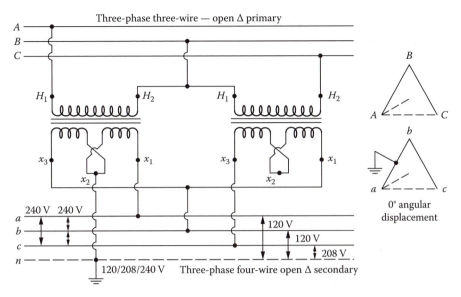

FIGURE 3.44 Three-phase four-wire open-delta connection. (Note that three-phase four wire means a three-phase system made up of four wires.)

Therefore, by dividing Equation 3.35 by 3.36, side by side,

$$\frac{S_{\angle-\angle}}{S_{\Delta-\Delta}} = \frac{1}{\sqrt{3}}$$

$$= 0.577, \text{ or } 57.7\% \qquad (3.37)$$

where
   $S_{\Delta-\Delta}$ is the kilovoltampere rating of the $\Delta-\Delta$ bank
   $S_{\angle-\angle}$ is the kilovoltampere rating of the open-Δ bank
   V_{L-L} is the line-to-line voltage, V
   I_L is the line (or full load) current, A

Note that the two transformers of the open-Δ bank make up 66.6% of the installed capacity of the three transformers of the $\Delta-\Delta$ bank, but they can supply only 57.7% of the three. Here, the ratio of 57.7/66.6 = 0.866 is the power factor at which the two transformers operate when the load is at unity power factor. By being operated in this way, the bank still delivers three-phase currents and voltages in their correct phase relationships, but the capacity of the bank is reduced to 57.7% of what it was with all three transformers in service since it has only 86.6% of the rating of the two units making up the three-phase bank. Open-Δ banks are quite often used where the load is expected to grow, and when the load does grow, the third transformer may be added to complete a $\Delta-\Delta$ bank.

Figure 3.45 shows an open-Δ connection for 240 V three-phase three-wire secondary service at 0° angular displacement. The neutral point n shown in the LV phasor diagram exists only on the paper.

For the sake of illustration, assume that a balanced three-phase load, for example, an induction motor as shown in the figure, with a lagging power factor is connected to the secondary. Therefore the a, b, c phase currents in the secondary can be found as

$$\bar{I}_a = \frac{S_{3\phi}}{\sqrt{3}V_{L-L}} \angle\theta_{\bar{I}_a} \qquad (3.38)$$

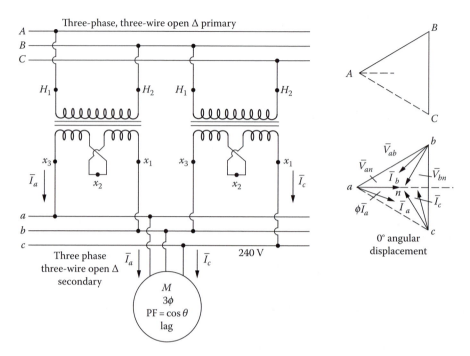

FIGURE 3.45   Three-phase three-wire open-delta connection.

$$\bar{I}_b = \frac{S_{3\phi}}{\sqrt{3}V_{L-L}} \angle\theta_{\bar{I}_b} \tag{3.39}$$

$$\bar{I}_c = \frac{S_{3\phi}}{\sqrt{3}V_{L-L}} \angle\theta_{\bar{I}_c} \tag{3.40}$$

The transformer kilovoltampere loads can be calculated as follows. The kilovoltampere load on the first transformer is

$$S_{T1} = V_{L-L} \times \left|\bar{I}_a\right|$$

$$= V_{L-L} \times \frac{S_{3\phi}}{\sqrt{3} \times V_{L-L}}$$

$$= \frac{S_{3\phi}}{\sqrt{3}} \text{ kVA} \tag{3.41}$$

and the kilovoltampere load on the second transformer is

$$S_{T2} = V_{L-L} \times \left|\bar{I}_b\right|$$

$$= V_{L-L} \times \frac{S_{3\phi}}{\sqrt{3} \times V_{L-L}}$$

$$= \frac{S_{3\phi}}{\sqrt{3}} \text{ kVA} \tag{3.42}$$

Therefore, the total load that the transformer bank can be loaded to (or the total "effective" transformer bank capacity) is

$$\sum_{i=1}^{2} S_{T_i} = \frac{2 \times S_{3\phi}}{\sqrt{3}} \qquad (3.43)$$

and hence,

$$S_{3\phi} = \frac{\sqrt{3}}{2} \sum_{i=1}^{2} S_{T_i} \text{ kVA} \qquad (3.44)$$

For example, if there are two 50 kVA transformers in the open-Δ bank, even though the total transformer bank capacity appears to be

$$\sum_{i=1}^{2} S_{T_i} = 100 \text{ kVA}$$

in reality the bank's "effective" maximum capacity is

$$S_{3\phi} = \frac{\sqrt{3}}{2} \times 100 = 86.6 \text{ kVA}$$

If there are three 50 kVA transformers in the Δ–Δ bank, the bank's maximum capacity is

$$S_{3\phi} = \sum_{i=1}^{3} S_{T_i} = 150 \text{ kVA}$$

which shows an increase of 73% over the 86.6 kVA load capacity.

Assume that the load power factor is $\cos \theta$ and its angle can be calculated as

$$\theta = \theta_{\bar{V}_{an}} - \theta_{\bar{I}_a} \qquad (3.45)$$

or using $\bar{V}_{an}$ as the reference,

$$\theta = 0° - \theta_{\bar{I}_a} \qquad (3.46)$$

If $\theta_{\bar{I}_a}$ is negative, then θ is positive, which means it is the angle of a lagging load power factor. Also, it can be shown that

$$\theta = \theta_{\bar{V}_{bn}} - \theta_{\bar{I}_b} \qquad (3.47)$$

or

$$\theta = -120° - \theta_{\bar{I}_b} \qquad (3.48)$$

and

$$\theta = \theta_{\bar{V}_{cn}} - \theta_{\bar{I}_c} \tag{3.49}$$

or

$$\theta = +120 - \theta_{\bar{I}_c} \tag{3.50}$$

The transformer power factors for transformers 1 and 2 can be calculated as

$$\cos\theta_{T_1} = \cos(\theta_{\bar{V}_{ab}} - \theta_{\bar{I}_a}) \tag{3.51}$$

or

$$\text{if } \theta_{\bar{I}_a} = -30°,$$

$$\cos\theta_{T_1} = \cos(\theta_{\bar{V}_{ab}} + 30°) \tag{3.52}$$

and

$$\cos\theta_{T_2} = \cos(\theta_{\bar{V}_{bc}} - \theta_{\bar{I}_c}) \tag{3.53}$$

or

$$\text{if } \theta_{\bar{I}_c} = +30°,$$

$$\cos\theta_{T_2} = \cos(\theta_{\bar{V}_{bc}} - 30°) \tag{3.54}$$

Therefore, the total real power output of the bank is

$$P_T = P_{T_1} + P_{T_2}$$
$$= V_{L-L}|\bar{I}_a|\cos(\theta + 30°) + V_{L-L}|\bar{I}_c|\cos(\theta - 30°)$$
$$= \sqrt{3}V_{L-L}I_L\cos\theta \text{ kW} \tag{3.55}$$

and, similarly, the total reactive power output of the bank is

$$Q_T = Q_{T_1} + Q_{T_2}$$
$$= V_{L-L}|\bar{I}_a|\sin(\theta + 30°) + V_{L-L}|\bar{I}_c|\sin(\theta - 30°)$$
$$= \sqrt{3}V_{L-L}I_L \sin\theta \text{ kvar} \tag{3.56}$$

As shown in Table 3.6, when the connected bank load has a lagging power factor of 0.866, it has a 30° power factor angle and, therefore, transformer 1, from Equation 3.52, has a 0.5 lagging power factor and transformer 2, from Equation 3.54, has a unity power factor. However, when the bank load has a unity power factor, of course, its angle is zero, and therefore transformer 1 has a 0.866 lagging power factor and transformer 2 has a 0.866 leading power factor.

TABLE 3.6

Effects of the Load Power Factor on the Transformer Power Factors

Load Power Factor		Transformer Power Factors	
$\cos\theta$	θ	$\cos\theta_{T_1} = \cos(\theta + 30°)$	$\cos\theta_{T_2} = \cos(\theta - 30°)$
0.866 lag	+30°	0.5 lag	1.0
1.0	0°	0.866 lag	0.866 lead

3.10.3 Y–Y TRANSFORMER CONNECTION

Figure 3.46 shows three transformers connected in Y–Y on a typical three-phase four-wire multigrounded system to provide for 120/208Y-V service at 0° angular displacement. This particular system provides a 208 V three-phase power supply for three-phase motors and a 120 V single-phase power supply for lamps and other small single-phase loads. An attempt should be made to distribute the single-phase loads reasonably equally among the three phases.

One of the advantages of the Y–Y connection is that when a system has changed from delta to a four-wire wye to increase system capacity, existing transformers can be used. For example, assume that the old distribution system was 2.4 kV delta and the new distribution system is 2.4/4.16Y kV. Here the existing 2.4/4.16Y kV transformers can be connected in wye and used.

In the Y–Y transformer bank connection, only 57.7% (or 1/1.73) of the line voltage affects each winding, but full line current flows in each transformer winding. Power-distribution circuits supplied from a Y–Y bank often create series disturbances in communication circuits (e.g., *telephone interference*) in their immediate vicinity.

Also, the primary neutral point should be solidly grounded and tied firmly to the system neutral; otherwise, excessive voltages may be developed on the secondary side. For example, if the

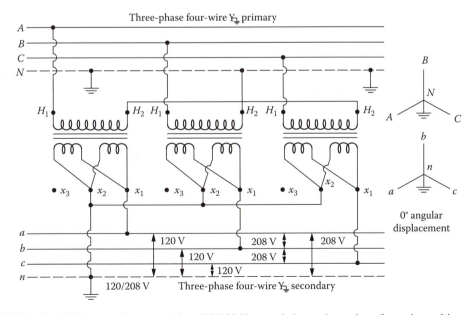

FIGURE 3.46 Y–Y connection to provide a 120/208 V grounded-wye three-phase four-wire multigrounded service.

neutral of the transformer is isolated from the system neutral, an unstable condition results at the transformer neutral, caused primarily by third-harmonic voltages. If the transformer neutral is connected to ground, the possibility of telephone interference is greatly enhanced, and there is also a possibility of resonance between the line capacitance to ground and the magnetizing impedance of the transformer.

3.10.4 Y–Δ Transformer Connection

Figure 3.47 shows three single-phase transformers connected in Y–Δ on a three-phase three-wire ungrounded-wye primary system to provide for 120/208/240 V three-phase four-wire delta secondary service at 30° angular displacement.

Figure 3.48 shows three transformers connected in Y–Δ on a typical three-phase four-wire grounded-wye primary system to provide for 240 V three-phase three-wire delta secondary service at 210° angular displacement.

The Y–Δ connection is advantageous in many cases because the voltage between the outside legs of the wye is 1.73 times the voltage to neutral, so that higher distribution voltage can be gained by using transformers with primary winding of only the voltage between any leg and the neutral. For example, 2.4 kV primary single-phase transformers can be connected in wye on the primary to a 4.16 kV three-phase wye circuit.

In the Y–Δ connection, the voltage/transformation ratio of the bank is 1.73 times the voltage/transformation ratio of the individual transformers. When transformers of different capacities are used, the maximum safe bank rating is three times the capacity of the smallest transformers.

The primary supply, usually a grounded wye circuit, may be either three wire or four wire including a neutral wire. The neutral wire, running from the neutral of the wye-connected substation transformer bank supplying the primary circuit, may be completely independent of the secondary or may be united with the neutral of the secondary system.

In the case of having the primary neutral independent of the secondary system, it is used as an isolated neutral and is grounded at the substation only.

In the case of having the same wire serving as both the primary neutral and the secondary neutral, it is grounded at many points, including each customer's service and is a multigrounded

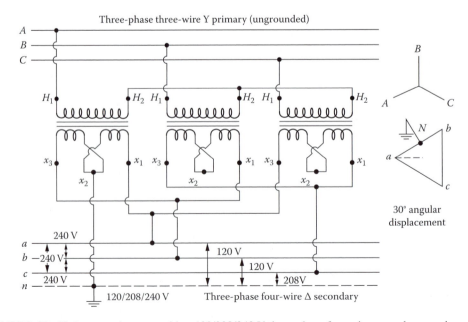

FIGURE 3.47 Y–Δ connection to provide a 120/208/240 V three-phase four-wire secondary service.

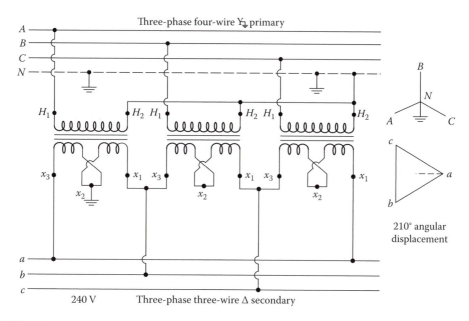

FIGURE 3.48 Y–Δ connection to provide a 240 V three-phase three-wire secondary service.

common neutral. However, in either case, the primary-bank neutral is usually not connected to the primary-circuit neutral since it is not necessary and prevents a burned-out transformer winding during phase-to-ground faults and extensive blowing of fuses throughout the system.

In the case of the Y–Y connection, neglecting the neutral on the primary side causes the voltages to be deformed from the sine-wave form. In the case of the Y–Δ connection, if the neutral is spared on the primary side, the voltage waveform tends to deform, but this deformation causes circulating currents in the delta, and these currents act as magnetizing currents to correct the deformation. Thus there is no objection to neglecting the neutral. However, if the transformer supplies a motor load, a damaging overcurrent is produced in each three-phase motor circuit, causing an equal amount of current to flow in two wires of the motor branch circuit and the total of the two currents to flow in the third. If the highest of the three currents occurs in the unprotected circuit, motor burnout will probably happen. This applies to ungrounded Y–Δ and Δ–Y banks.

If the transformer bank is used to supply three-phase and single-phase loads, and *if the bank neutral is solidly connected,* disconnection of the large transformer by fuse operation causes an even greater overload on the remaining two transformers. Here, the blowing of a single fuse is hard to detect since no decrease in service quality is noticeable right away, and one of the two remaining transformers may be burned out by the overload.

On the other hand, *if the bank neutral is not connected to the primary-circuit neutral, but left isolated,* disconnection of one transformer results in a partial service interruption without danger of a transformer burnout. The approximate rated capacity required in a Y–Δ-connected bank with an isolated bank neutral to serve a combined three-phase and single-phase loads, assuming unity power factor, can be found as

$$\frac{2S_{1\phi} + S_{3\phi}}{3}$$

which is equal to rated transformer capacity across lighting phase, where

$S_{1\phi}$ is the single-phase load, kVA
$S_{3\phi}$ is the three-phase load, kVA

In summary, when the primary-side neutral of the transformer bank is not isolated but connected to the primary-circuit neutral, the Y–Δ transformer bank may burn out due to the following reasons:

1. The transformer bank may act as a grounding transformer bank for unbalanced primary conditions and may supply fault current to any fault on the circuit to which it is connected, reducing its own capacity for connected load.
2. The transformer bank may be overloaded if one of the protective fuses opens on a line-to-ground fault, leaving the bank with only the capacity of an open-Y open-Δ bank.
3. The transformer bank causes circulating current in the delta in an attempt to balance any unbalanced load connected to the primary line.
4. The transformer bank provides a delta in which triple-harmonic currents circulate.

All the aforementioned effects can cause the transformer bank to carry current in addition to its normal load current, resulting in the burnout of the transformer bank.

3.10.5 OPEN-Y OPEN-Δ TRANSFORMER CONNECTION

As shown in Figure 3.49, in the case of having one phase of the primary supply opened, the transformer bank becomes open-Y open-Δ and continues to serve the three-phase load at a reduced capacity.

Example 3.6

Two single-phase transformers are connected in open-Y open-Δ to provide power for a three-phase wye-connected 100 kVA load with a 0.80 lagging power factor and a 50 kVA single-phase load with a 0.90 lagging power factor, as shown in Figure 3.50. Assume that the primary-side voltage of the bank is 7,620/13,200 V and the secondary-side voltage is 240 V. Using the given information, calculate the following:

a. The line current flowing in each secondary-phase wire
b. The current flowing in the secondary winding of each transformer
c. The kilovoltampere load on each transformer
d. The current flowing in each primary-phase wire and in the primary neutral

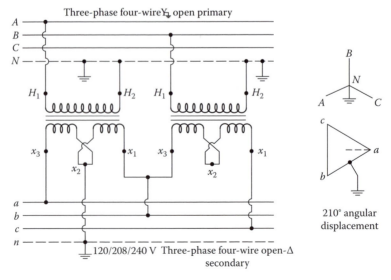

FIGURE 3.49 Open-wye open-delta connection.

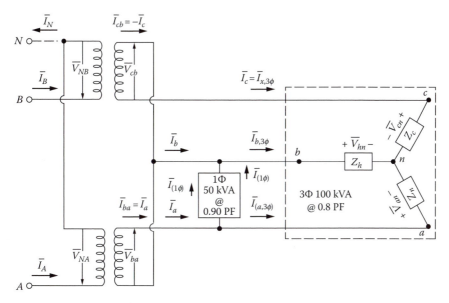

FIGURE 3.50   Open-wye open-delta connection for Example 3.6.

Solution

a. Using the voltage drop $\bar{V}_{an}$ as the reference, the three-phase components of the line currents can be found as

$$
\begin{aligned}
\left|\bar{I}_{a,3\phi}\right| &= \left|\bar{I}_{b,3\phi}\right| \\
&= \left|\bar{I}_{c,3\phi}\right| \\
&= \frac{S_{L,3\phi}}{\sqrt{3} \times V_{L-L}} \\
&= \frac{100}{\sqrt{3} \times 0.240} = 240.8\,\text{A}
\end{aligned}
\tag{3.57}
$$

Since the three-phase load has a lagging power factor of 0.80,

$$
\begin{aligned}
\bar{I}_{a,3\phi} &= \left|\bar{I}_{a,3\phi}\right|(\cos\theta - j\sin\theta) \\
&= 240.8(0.80 - j0.60) \\
&= 192.68 - j144.5 \\
&= 240.8\angle{-36.9°}\,\text{A}
\end{aligned}
\tag{3.58}
$$

$$
\begin{aligned}
\bar{I}_{b,3\phi} &= a^2\bar{I}_{a,3\phi} \\
&= (1\angle 240°)(240.8\angle{-36.9°}) \\
&= 240.8\angle 203.1° \\
&= -221.5 - j94.5\,\text{A}
\end{aligned}
\tag{3.59}
$$

$$\bar{I}_{c,3\phi} = a\bar{I}_{a,3\phi}$$

$$= (1\angle 120°)(240.8\angle -36.9°)$$

$$= 240.8\angle 83.1°$$

$$= 28.9 + j239.1\,\text{A} \tag{3.60}$$

The single-phase component of the line currents can be found as

$$\left|\bar{I}_{1\phi}\right| = \frac{S_{L,1\phi}}{V_{L-L}}$$

$$= \frac{50}{0.240} = 208.33\,\text{A} \tag{3.61}$$

therefore

$$\bar{I}_{1\phi} = \left|\bar{I}_{1\phi}\right|[\cos(30° - \theta_1) + j\sin(30° - \theta_1)]$$

$$= 208.33\,[\cos(30° - 25.8°) + j\sin(30° - 25.8°)]$$

$$= 207.78 + j15.26\,\text{A} \tag{3.62}$$

Hence, the line currents flowing in each secondary-phase wire can be found as

$$\bar{I}_a = \bar{I}_{a,3\phi} + \bar{I}_{1\phi}$$

$$= 192.68 - j144.5 + 207.78 + j15.26$$

$$= 400.46 - j129.24$$

$$= 420.8\angle -17.9°\,\text{A}$$

$$\bar{I}_b = \bar{I}_{b,3\phi} - \bar{I}_{1\phi}$$

$$= -221.5 - j94.5 - 207.78 - j15.26$$

$$= 429.28 - j109.76$$

$$= 442.8\angle -165.7°\,\text{A}$$

$$\bar{I}_c = \bar{I}_{c,3\phi}$$

$$= 240.8\angle 83.1°\,\text{A}$$

b. The current flowing in the secondary winding of each transformer is

$$\bar{I}_{ba} = \bar{I}_a$$

$$= 420.8\angle -17.9°\,\text{A}$$

$$\bar{I}_{cb} = -\bar{I}_c$$

$$= -240.8\angle 83.1°$$

$$= 240.8\angle 83.1° + 180°$$

$$= 240.8\angle 263.1°\,\text{A}$$

c. The kilovoltampere load on each transformer can be found as

$$S_{L,ba} = V_{ba} \times \left| \overline{I}_{ba} \right|$$

$$= 0.240 \times 420.8$$

$$= 101 \, \text{kVA} \tag{3.63a}$$

$$S_{L,cb} = V_{cb} \times \left| \overline{I}_{cb} \right|$$

$$= 0.240 \times 240.8$$

$$= 57.8 \, \text{kVA} \tag{3.63b}$$

d. The current flowing in each primary-phase wire can be found by dividing the current flow in each secondary winding by the turns ratio. Therefore,

$$n = \frac{7620 \, \text{V}}{240 \, \text{V}} = 31.7$$

and hence

$$\overline{I}_A = \frac{\overline{I}_{ba}}{n}$$

$$= \frac{420.8 \angle -17.9°}{31.75}$$

$$= 12.6 - j4.07$$

$$= 13.25 \angle -17.9° \, \text{A} \tag{3.64a}$$

$$\overline{I}_B = \frac{\overline{I}_{cb}}{n}$$

$$= \frac{240.8 \angle 263.1°}{31.75}$$

$$= -0.91 - j7.52$$

$$= 7.58 \angle 263.1° \, \text{A} \tag{3.64b}$$

Therefore, the current in the primary neutral is

$$\overline{I}_N = \overline{I}_A + \overline{I}_B$$

$$= 13.25 \angle -17.9° + 7.58 \angle 263.1°$$

$$= 11.69 - j11.6$$

$$= 16.47 \angle -44.8° \, \text{A} \tag{3.65}$$

3.10.6 Δ–Y Transformer Connection

Figures 3.51 and 3.52 show three single-phase transformers connected in Δ–Y to provide for 120/208 V three-phase four-wire grounded-wye service at 30° and 210° angular displacements, respectively.

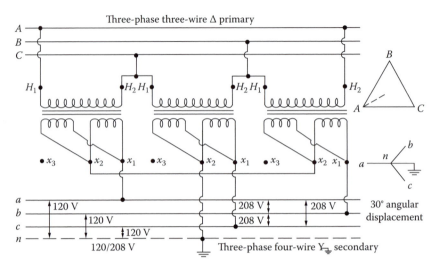

FIGURE 3.51   Δ–Y connection with 30° angular displacement.

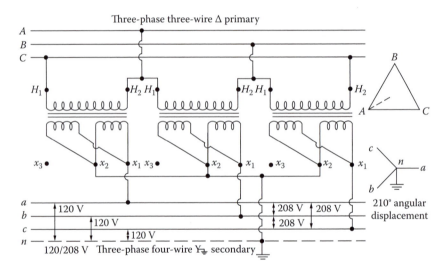

FIGURE 3.52   Δ–Y connection with 210° angular displacement.

In the previously mentioned transformer banks, the single-phase lighting load is all on one phase, resulting in unbalanced primary currents in any one bank. To eliminate this difficulty, the Δ–Y system finds many uses. Here the neutral of the secondary three-phase system is grounded and single-phase loads are connected between the different phase wires and the neutral while the three-phase loads are connected to the phase wires. Therefore, the single-phase loads can be balanced on three phases in each bank, and banks may be paralleled if desired.

When transformers of different capacities are used, maximum safe transformer bank rating is three times the capacity of the smallest transformer. If one transformer becomes damaged or is removed from service, the transformer bank becomes inoperative.

With both the Y–Y and Δ–Δ connections, the line voltages on the secondaries are in phase with the line voltages on the primaries, but with the Y–Δ or Δ–Y connections, the line voltages on the secondaries are at 30° to the line voltages on the primaries. Consequently, a Y–Δ or Δ–Y transformer bank cannot be operated in parallel with a Δ–Δ or Y–Y transformer bank. Having the identical angular displacements becomes especially important when three-phase transformers are interconnected into the same secondary system or paralleled with three-phase banks of single-phase transformers. The additional conditions to successfully parallel three-phase distribution transformers are the following:

1. All transformers have identical frequency ratings.
2. All transformers have identical voltage ratings.
3. All transformers have identical tap settings.
4. Per unit impedance of one transformer is between 0.925 and 1.075 of the other.

The Δ–Y step-up and Y–Δ step-down connections are especially suitable for HV transmission systems. They are economical in cost, and they supply a stable neutral point to be solidly grounded or grounded through resistance of such value as to damp the system critically and prevent the possibility of oscillation.

3.11   THREE-PHASE TRANSFORMERS

Three-phase voltages may be transformed by means of three-phase transformers. The core of a three-phase transformer is made with three legs, a primary and a secondary winding of one phase being placed on each leg. It is possible to construct the core with only three legs since the fluxes established by the three windings are 120° apart in time phase. Two core legs act as the return for the flux in the third leg. For example, if flux is at a maximum value in one leg at some instant, the flux is half that value and in the opposite direction through the other two legs at the same instant.

The three-phase transformer takes less space than do three single-phase transformers having the same total capacity rating since the three windings can be placed together on one core. Furthermore, three-phase transformers are usually more efficient and less expensive than the equivalent single-phase transformer banks. This is especially noticeable at the larger ratings. On the other hand, if one phase winding becomes damaged, the entire three-phase transformer has to be removed from the service. Three-phase transformers can be connected in any of the aforementioned connection types. The difference is that all connections are made inside the tank.

Figures 3.53 through 3.57 show various connection diagrams for three-phase transformers. Figure 3.53 shows a Δ–Δ connection for 120/208/240 V three-phase four-wire secondary service at 0° angular displacement. It is used to supply 240 V three-phase loads with small amounts of 120 V single-phase load. Usually, transformers with a capacity of 150 kVA or less are built in such a design

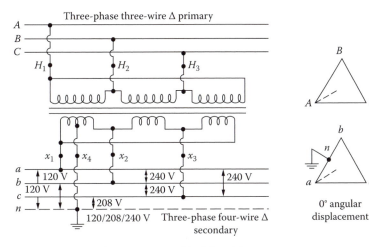

FIGURE 3.53   Three-phase transformer connected in delta–delta.

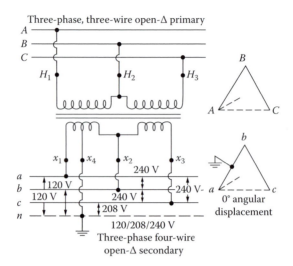

FIGURE 3.54 Three-phase transformer connected in open-delta.

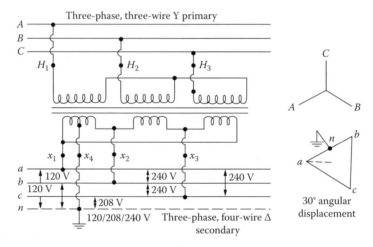

FIGURE 3.55 Three-phase transformer connected in Y–Δ.

that when 5% of the rated kilovoltamperes of the transformer is taken from the 120 V tap on the 240 V connection, the three-phase capacity is decreased by 25%.

Figure 3.54 shows a three-phase open-Δ connection for 120/240 V service. It is used to supply large 120 and 240 V single-phase loads simultaneously with small amounts of three-phase load. The two sets of windings in the transformer are of different capacity sizes in terms of kilovoltamperes. The transformer efficiency is low especially for three-phase loads. The transformer is rated only 86.6% of the rating of two sets of windings when they are equal in size, and less than this when they are unequal.

Figure 3.55 shows a three-phase Y–Δ connection for 120/240 V service at 30° angular displacement. It is used to supply three-phase 240 V loads and small amounts of 120 V single-phase loads.

Figure 3.56 shows a three-phase open-Y open-Δ connection for 120/240 V service at 30° angular displacement. The statements on efficiency and capacity for three-phase open-Δ connection are also applicable for this connection.

Figure 3.57 shows a three-phase transformer connected in Y–Y for 120/208Y-V service. The connection allows single-phase loads to balance among the three phases.

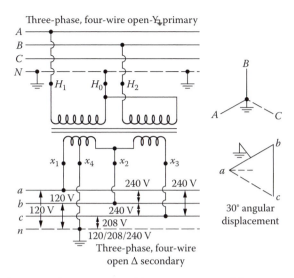

FIGURE 3.56 Three-phase transformer connected in open-wye open-delta.

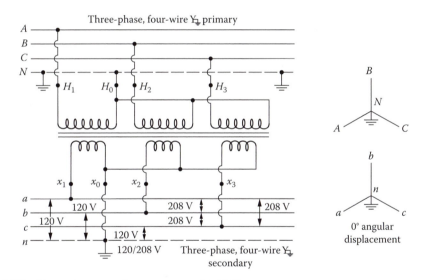

FIGURE 3.57 Three-phase transformer connected in Y–Y.

3.12 T OR SCOTT CONNECTION

In some localities, two phase is required from a three-phase system. The T or Scott connection, which employs two transformers, is the most frequently used connection for three-phase to two-phase (or even three-phase) transformations. In general, *the T connection is primarily used for getting a three-phase transformation, whereas the Scott connection is mainly used for getting a two-phase transformation.* In either connection type, the basic design is the same. Figures 3.58 through 3.60 show various types of the Scott connection. This connection type requires two single-phase transformers with Scott taps. The first transformer is called the main transformer and connected from line to line, and the second one is called the teaser transformer and connected from the midpoint of the first transformer to the third line. It dictates that the midpoints of both primary and secondary windings be available for connections. The secondary may be either three, four, or five wire, as shown in the figures.

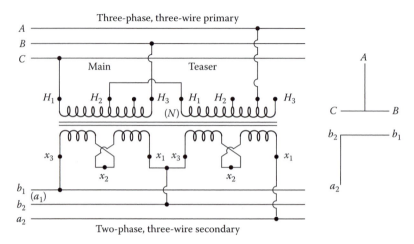

FIGURE 3.58   T or Scott connection for three-phase to two-phase, three-wire transformation.

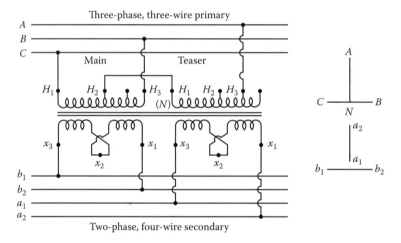

FIGURE 3.59   T or Scott connection for three-phase to two-phase, four-wire transformation.

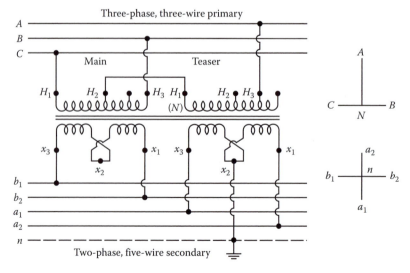

FIGURE 3.60   T or Scott connection for three-phase to two-phase, five-wire transformation.

In either case, the connection needs specially wound single-phase transformers. The main transformer has a 50% tap on the primary-side winding, whereas the teaser transformer has an 86.6% tap. (In usual design practice, both transformers are built to be identical so that both have a 50% and an 86.6% tap in order to be used interchangeably as main and teaser transformers.)

Although only two single-phase transformers are required, their total rated kilovoltampere capacity must be 15.5% greater if the transformers are interchangeable, or 7.75% greater if noninterchangeable, than the actual load supplied (or than the standard single-phase transformer of the same kilovoltampere and voltage). It is very important to keep the relative phase sequence of the windings the same so that the impedance between the two half windings is a minimum to prevent excessive voltage drop and the resultant voltage unbalance between phases.

The T or Scott connections change the number of phases but not the power factor, which means that a balanced load on the secondary will result in a balanced load on the primary. When the two-phase load at the secondary has a unity power factor, the main transformer operates at 86.6% power factor and the teaser transformer operates at unity power factor. These connections can transform power in either direction, that is, from three phase to two phase or from two phase to three phase.

Example 3.7

Two transformer banks are sometimes used in distribution systems, as shown in Figure 3.61, especially to supply customers having large single-phase lighting loads and small three-phase (motor) loads.

The LV connections are three-phase four-wire 120/240 V open-Δ. The HV connections are either open-Δ or open-Y. If it is open-Δ, the transformer-rated HV is the primary line-to-line voltage. If it is open-Y, the transformer-rated HV is the primary line-to-neutral voltage.

In preparing wiring diagrams and phasor diagrams, it is important to understand that all odd-numbered terminals of a given transformer, that is, H_1, x_1, x_3, etc., have the same instantaneous voltage polarity. For example, if all the odd-numbered terminals are positive (+) at a particular instant of time, then all the even-numbered terminals are negative (−) at the same instant. In other words, the no-load phasor voltages of a given transformer, for example, $\bar{V}_{H_1H_2}$, $\bar{V}_{x_1x_2}$, and $\bar{V}_{x_3x_4}$, are all in phase.

Assume that ABC phase sequence is used in the connections for both HV and LV and the phasor diagrams and

$$\bar{V}_{AC} = 13,200\angle 0°\,\text{V}$$

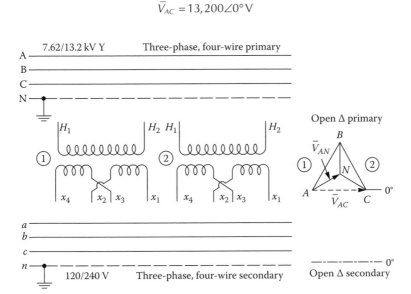

FIGURE 3.61 For Example 3.7.

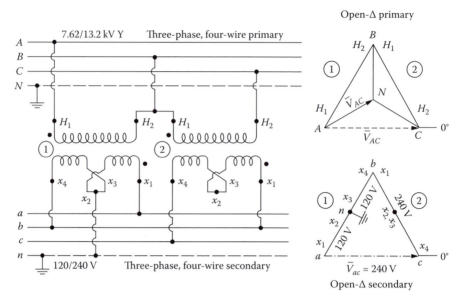

FIGURE 3.62   For Example 3.7.

and

$$\bar{V}_{AN} = 7620\angle 30° \, \text{V}$$

Also assume that the left-hand transformer is used for lighting. To establish the two-transformer bank with open-Δ primary and open-Δ secondary,

   a. Draw and/or label the voltage phasor diagram required for the open-Δ primary and open-Δ secondary on the 0° references given
   b. Show the connections required for the open-Δ primary and open delta secondary

Solution

Figure 3.62 illustrates the solution. Note that because of Kirchhoff's voltage law, there are $\bar{V}_{AC}$ and $\bar{V}_{ac}$ voltages between **A** and **C** and between **a** and **c**, respectively. Also note that the midpoint of the left-hand transformer is grounded to provide the 120 V for lighting loads.

Example 3.8

Figure 3.63 shows another two-transformer bank, which is known as the T–T connection. Today, some of the so-called three-phase distribution transformers now marketed contain two single-phase cores and coils mounted in one tank and connected in T–T. The performance is substantially like banks of three identical single-phase transformers or classical core- or shell-type three-phase transformers. However, perfectly balanced secondary voltages do not occur even though the load and the primary voltages are perfectly balanced. In spite of that, the unbalance in secondary voltages is small.

Figure 3.63 shows a particular T–T connection diagram and an arbitrary set of balanced three-phase primary voltages. Assume that the no-load line-to-line and line-to-neutral voltages are 480 and 277 V, respectively, exactly like wye circuitry, and abc sequence.

Based on the given information and Figure 3.63, determine the following:

   a. Draw the LV phasor diagram, correctly oriented on the 0° reference shown.
   b. Find the value of the $\bar{V}_{ab}$ phasor.

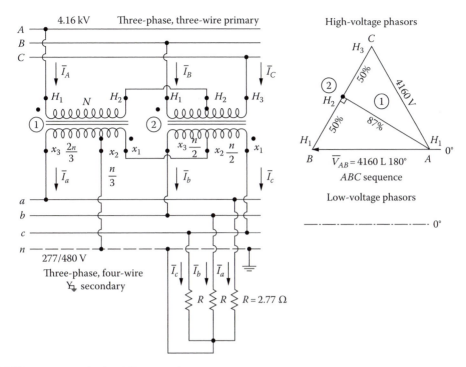

FIGURE 3.63   A particular T–T connection.

c. Find the magnitudes of the following rated winding voltages:
   i.   The voltage $V_{H_1H_2}$ on transformer 1
   ii. The voltage $V_{x_1x_2}$ on transformer 1
   iii. The voltage $V_{x_2x_3}$ on transformer 1
   iv. The voltage $V_{H_1H_2}$ on transformer 2
   v.   The voltage $V_{H_2H_3}$ on transformer 2
   vi. The voltage $V_{x_1x_2}$ on transformer 2
   vii. The voltage $V_{x_2x_3}$ on transformer 2
d. Would it be possible to parallel a T–T transformer bank with the following?
   i.   A Δ–Δ bank
   ii. A Y–Y bank
   iii. A Δ–Y bank

Solution

a. Figure 3.64 shows the required LV phasor diagram. Note the 180° phase shift among the corresponding phasors.
b. The value of the voltage phasor is

$$\bar{V}_{ab} = 480 \angle 0° \, \text{V}$$

c. The magnitudes of the rated winding voltages
   i.   From the HV phasor diagram shown in Figure 3.64,

$$\left|\bar{V}_{H_1H_2}\right| = (4160^2 - 2080^2)^{1/2}$$

$$= 3600 \, \text{V}$$

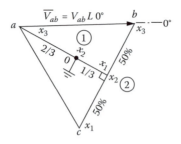

FIGURE 3.64 The required low-voltage phasor diagram.

  ii. From Figures 3.63 and 3.64,

$$\left|\overline{V}_{x_1x_2}\right| = \frac{1}{2}(480^2 - 240^2)^{1/2}$$

$$= 139 \text{ V}$$

$$\left|\overline{V}_{x_1x_2}\right| = \frac{1}{3}(480^2 - 240^2)^{1/2}$$

$$= 139 \text{ V}$$

  iii. From Figures 3.63 and 3.64,

$$\left|\overline{V}_{x_2x_3}\right| = \frac{2}{3}(480^2 - 240^2)^{1/2}$$

$$= 277 \text{ V}$$

  iv. From Figure 3.63,

$$\left|\overline{V}_{H_1H_2}\right| = 50\%(4160 \text{ V})$$

$$= 2080 \text{ V}$$

  v. From Figure 3.63,

$$\left|\overline{V}_{H_2H_3}\right| = 2080 \text{ V}$$

  vi. From Figure 3.64,

$$\left|\overline{V}_{x_2x_3}\right| = 240 \text{ V}$$

  d. i. No, (ii) no, (iii) yes.

Example 3.9

Assume that the T–T transformer bank of Example 3.8 is to be loaded with the balanced resistors ($R = 2.77 \; \Omega$) shown in Figure 3.63. Also assume that the secondary voltages are to be perfectly balanced and that the necessary HV applied voltages then are not perfectly balanced. Determine the following:

  a. The LV current phasors.
  b. The LV current-phasor diagram.

c. At what power factor does the transformer operate?
d. What power factor is seen by winding x_2x_3 of transformer 2?
e. What power factor is seen by winding x_1x_2 of transformer 2?

Solution

a. The LV phasor diagram of Figure 3.64 can be redrawn as shown in Figure 3.65a. Therefore, from Figure 3.65a, the LV current phasors are

$$\bar{I}_a = \frac{\bar{V}_{a0}}{R}$$

$$= \frac{277\angle -30°}{2.77}$$

$$= 100\angle -30° \text{ A}$$

$$\bar{I}_b = \frac{\bar{V}_{b0}}{R}$$

$$= \frac{-277\angle +30°}{2.77}$$

$$= -100\angle +30°$$

$$= 100\angle -150° \text{ A}$$

$$\bar{I}_c = \frac{\bar{V}_{c0}}{R}$$

$$= \frac{277\angle 90°}{2.77}$$

$$= 100\angle 90° \text{ A}$$

b. Figure 3.65b shows the LV current-phasor diagram.
c. From part (a), the power factor of transformer 1 can be found as

$$\cos\theta_{T1} = \cos(\theta_{\bar{V}_{a0}} - \theta_{\bar{I}_a})$$

$$= \cos[(-30°) - (-30°)]$$

$$= 1.0$$

d. The power factor seen by the winding x_3x_2 of transformer 2 is 0.866, lagging.
e. The power factor seen by the winding x_1x_2 of transformer 2 is 0.866, leading.

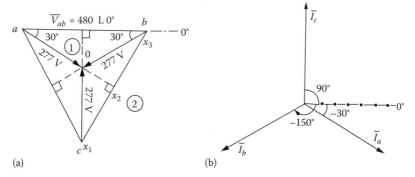

(a)　　　　　　　　　　　　　(b)

FIGURE 3.65 Phasor diagram for Example 3.9.

Example 3.10

Consider Example 3.9 and Figure 3.64, and determine the following:

a. The necessary voltampere rating of the x_2x_3 LV winding of transformer 1
b. The necessary voltampere rating of the x_2x_1 LV winding of transformer 1
c. Total voltampere output from transformer 1
d. The necessary voltampere rating of the x_1x_2 LV winding of transformer 2
e. The necessary voltampere rating of the x_2x_3 LV winding of transformer 2
f. Total voltampere output from transformer 2
g. The ratio of total voltampere rating of all LV windings in the transformer bank to maximum continuous voltampere output from the bank

Solution

a. From Figure 3.64, the necessary voltampere rating of the x_2x_3 LV winding of transformer 1 is

$$S_{x_2x_3} = \frac{2}{3}\left(\frac{\sqrt{3}}{2}V\right)I$$

$$= \frac{VI}{\sqrt{3}} \text{ VA}$$

b. Similarly,

$$S_{x_2x_1} = \frac{1}{3}\left(\frac{\sqrt{3}}{2}V\right)I$$

$$= \frac{VI}{2\sqrt{3}} \text{ VA}$$

c. Therefore, total voltampere output rating from transformer 1 is

$$\sum S_{T_1} = S_{x_2x_1} + S_{x_2x_3}$$

$$= \frac{\sqrt{3}}{2}VI \text{ VA}$$

d. From Figure 3.64, the necessary voltampere rating of the x_1x_2 LV winding of transformer 2 is

$$S_{x_1x_2} = \frac{V}{2}\times I \text{ VA}$$

e. Similarly,

$$S_{x_2x_3} = \frac{V}{2}\times I \text{ VA}$$

f. Therefore, total voltampere output rating from transformer 2 is

$$\sum S_{T_2} = S_{x_1x_2} + S_{x_2x_3}$$

$$= VI \text{ VA}$$

g. The ratio is

$$\frac{\sum \text{Installed core and coil capacity}}{\text{Max continuous output}} = \frac{(\sqrt{3}/2)+1}{\sqrt{3}} = 1.078$$

The same ratio for two-transformer banks connected in open-Δ HV open-Δ LV, or open-Y HV open-Δ LV is 1.15.

Example 3.11

In general, except for unique unbalanced loads, two-transformer banks do not deliver balanced three-phase LV terminal voltages even when the applied HV terminal voltages are perfectly balanced. Also, the two transformers do not, in general, operate at the same power factor or at the same percentages of their rated kilovoltamperes. Hence, the two transformers are likely to have unequal percentages of voltage regulation.

Figure 3.66 shows two single-phase transformers connected in open-Y HV and open-Δ LV. The two-transformer bank supplies a large amount of single-phase lighting and some small amount of three-phase power loads. Both transformers have 7200/120–240 V ratings and have equal transformer impedance of

$$\bar{Z}_T = 0.01 + j0.03 \, \text{pu}$$

based on their ratings. Here, neglect transformer magnetizing currents.

Figure 3.67 shows the LV phasor diagram. In this problem, the secondary voltages are to be assumed to be perfectly balanced and the primary voltages then unbalanced as required. Note that, in Figure 3.67, 0 indicates the three-phase neutral point. Based on the given information, determine the following:

a. Find the phasor currents $\bar{I}_a$, $\bar{I}_b$, and $\bar{I}_c$.
b. Select suitable standard kilovoltampere ratings for both the transformers. Overloads, as much as 10%, will be allowable as an arbitrary criterion.
c. Find the per unit kilovoltampere load on each transformer.

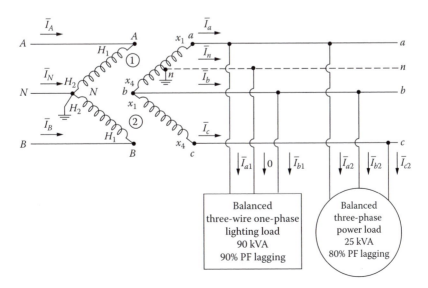

FIGURE 3.66 Two single-phase transformers connected in open-wye and open-delta.

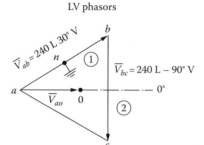

FIGURE 3.67　The low-voltage phasor diagram for Example 3.11.

    d. Find the power factor of the output of each transformer.

    e. Find the phasor currents $\bar{I}_A$, $\bar{I}_B$, and $\bar{I}_N$ in the HV leads.

    f. Find the HV terminal voltages V_{AN} and V_{BN}. Therefore, this part of the question can indicate the amount of voltage unbalance that may be encountered with typical equipment and typical loading conditions.

    g. Also write the necessary codes to solve the problem in MATLAB®.

Solution

    a. For the three-wire single-phase balanced lighting load,

$$\cos \theta = 0.90 \text{ lagging} \quad \text{or} \quad \theta = 25.8°$$

therefore, using the symmetrical-components theory,

$$\bar{I}_{a1} = \frac{90 \text{ kVA}}{0.240 \text{ kV}} \angle \theta_{\bar{V}_{ab}} - \theta$$

$$= 375 \angle 30° - 25.8°$$

$$= 375 \,(\cos 4.2° + j\sin 4.2°)$$

$$= 374 + j27.5$$

$$= 375 \angle 4.2° \text{ A}$$

Also

$$\bar{I}_{b1} = -\bar{I}_{a1}$$

$$= -374 - j27.5$$

$$= -375 \angle 4.2° \text{ A}$$

For the three-phase balanced power load,

$$\cos \theta = 0.80 \text{ lagging} \quad \text{or} \quad \theta = 36.8°$$

therefore,

$$\bar{I}_{a2} = \frac{25 \text{ kVA}}{\sqrt{3} \times 0.240 \text{ kV}} \angle \theta_{\bar{V}_{aO}} - \theta$$

$$= 60.2 \angle 0° - 36.8°$$

$$= 60.2 \angle -36.8° \text{ A}$$

Also,

$$\bar{I}_{b_2} = a^2 \bar{I}_{a_2}$$
$$= 1\angle 240° \times 60.2\angle -36.8°$$
$$= 60.2\angle 203.2° \text{ A}$$

and

$$\bar{I}_{c_2} = a \bar{I}_{a_2}$$
$$= 1\angle 120° \times 60.2\angle -36.8°$$
$$= 60.2\angle 83.2° \text{ A}$$

Therefore, the phasor currents in the transformer secondary are

$$\bar{I}_a = \bar{I}_{a1} + \bar{I}_{a2}$$
$$= 375\angle 4.2° + 60.2\angle -36.8°$$
$$= 422.04 - j8.44$$
$$= 422.12\angle -1.15° \text{ A}$$

$$\bar{I}_b = \bar{I}_{b1} + \bar{I}_{b2}$$
$$= -375\angle 4.2° + 60.2\angle 203.2°$$
$$= -429.33 - j51.22$$
$$= 432.37\angle -173.2° \text{ A}$$

$$\bar{I}_c = \bar{I}_{c1} + \bar{I}_{c2}$$
$$= 0 + 60.2\angle 83.2°$$
$$= 60.2\angle 83.2° \text{ A}$$

b. For transformer 1,

$$S_{T_1} = 0.240 \text{ kV} \times I_a$$
$$= 0.240 \times 422.12$$
$$= 101.3 \text{ kVA}$$

If a transformer with 100 kVA is selected,

$$S_{T_1} = 1.013 \text{ pu kVA}$$

with an overload of 1.3%. For transformer 2,

$$S_{T_2} = 0.240 \text{ kV} \times I_c$$
$$= 0.240 \times 60.2 = 14.4 \text{ kVA}$$

If a transformer with 15 kVA is selected,

$$S_{T_2} = 0.96\,\text{pu kVA}$$

with a 4% excess capacity.

c. From part (b),

$$S_{T_1} = 1.013\,\text{pukVA}$$

$$S_{T_2} = 0.96\,\text{pukVA}$$

d. Since the power factor that a transformer sees is not the power factor that the load sees, for transformer 1,

$$\cos\theta_{T_1} = \cos(\theta_{\bar{V}_{ab}} - \theta_{\bar{I}_a})$$

$$= \cos[30° - (-1.15°)]$$

$$= \cos 31.15°$$

$$= 0.856\ \text{lagging}$$

and for transformer 2,

$$\cos\theta_{T_2} = \cos(\theta_{\bar{V}_{cb}} - \theta_{\bar{I}_c})$$

$$= \cos[90° - 83.2°]$$

$$= \cos 6.8°$$

$$= 0.993\ \text{lagging}$$

e. The turns ratio is

$$n = \frac{7200\,\text{V}}{240\,\text{V}} = 30$$

therefore

$$\bar{I}_A = \frac{\bar{I}_a}{n}$$

$$= \frac{422.12\angle -1.15°}{30}$$

$$= 14.07\angle -1.15°\ \text{A}$$

and

$$\bar{I}_B = -\frac{\bar{I}_c}{n}$$

$$= -\frac{60.2\angle 83.2°}{30}$$

$$\cong -2\angle 83.2°\ \text{A}$$

Thus,

$$\bar{I}_N = -(\bar{I}_A + \bar{I}_B)$$
$$= -(14.07\angle -1.15° - 2\angle 83.2°)$$
$$= -14.02\angle -9.3° \text{ A}$$

f. In per units,

$$\bar{V}_{AN,\text{pu}} = \bar{V}_{ab,\text{pu}} + \bar{I}_{a,\text{pu}} \times \bar{Z}_{T,\text{pu}}$$

where

$$\bar{I}_{\text{base,LV}} = \frac{100 \text{kVA}}{0.240 \text{kV}} = 416.67 \text{ A}$$

$$\bar{I}_{a,\text{pu}} = \frac{\bar{I}_a}{I_{\text{base,LV}}}$$
$$= \frac{422.12\angle -1.15°}{416.67} = 1.013\angle -1.15° \text{ pu A}$$

$$\bar{V}_{ab,\text{pu}} = \frac{\bar{V}_{ab}}{V_{\text{base,LV}}}$$
$$= \frac{0.240\angle 30°}{0.240} = 1.0\angle 30° \text{ pu V}$$

$$\bar{Z}_{T,\text{pu}} = 0.01 + j0.03 \text{ pu}\Omega$$

Therefore,

$$\bar{V}_{AN,\text{pu}} = 1.0\angle 30° + (1.013\angle -1.15°)(0.01 + j0.03)$$
$$= 1.024\angle 31.15° \text{ pu V}$$

or

$$\bar{V}_{AN} = \bar{V}_{AN,\text{pu}} \times V_{\text{base,HV}}$$
$$= (1.024\angle 31.15°)(7200 \text{ V})$$
$$= 7372.8\angle 31.15° \text{ V}$$

Also,

$$\bar{V}_{BN,\text{pu}} = \bar{V}_{bc,\text{pu}} - \bar{I}_{c,\text{pu}} \times \bar{Z}_{T,\text{pu}}$$

where

$$\bar{I}_{c,pu} = \frac{\bar{I}_c}{I_{base,LV}}$$

$$= \frac{60.2\angle83.2°}{416.67} = 0.144\angle83.2° \text{ pu A}$$

$$\bar{V}_{bc,pu} = \frac{\bar{V}_{bc}}{V_{base,LV}}$$

$$= \frac{0.240\angle-90°}{0.240} = 1.0\angle-90° \text{ pu V}$$

Therefore,

$$\bar{V}_{BN,pu} = 1.0\angle-90° + (0.144\angle83.2°)(0.01 + j0.03)$$

$$= 1.00195\angle-89.76° \text{ pu V}$$

or

$$\bar{V}_{BN} = \bar{V}_{BN,pu} \times V_{base,HV}$$

$$= (1.00195\angle-89.76°)(7200\,\text{V})$$

$$= 7214.04\angle-89.76°\ \text{V}$$

Note that the difference between the phase angles of the $\bar{V}_{AN}$ and $\bar{V}_{BN}$ voltages is almost 120°, and the difference between their magnitudes is almost 80 V.

g. Here is the MATLAB script:

~~~~~~

```
clc
clear

% System parameters
ZT = 0.01 + j*0.03;
PFll = 0.9;
Smagll = 90;% kVA
PFpl = 0.8;
Smagpl = 25;% kVA
kVa = 0.24;
thetaVab = (pi*30)/180;
thetaVcb = (pi*90)/180;
thetaVa0 = 0;
a = -0.5 + j*0.866;
n = 7200/240;% turns ratio

% Solution for part a

% Phasor currents Ia, Ib and Ic
Ia1 = (Smagll/kVa)*(cos(thetaVab - acos(PFll)) + j*sin(thetaVab -
   acos(PFll)))
```

```
Ia2 = (Smagpl/(sqrt(3)*kVa))*(cos(thetaVa0 - acos(PFpl)) + j*sin(thetaVa0
  - acos(PFpl)))
Ib1 = -Ia1
Ib2 = a^2*Ia2
Ic2 = a*Ia2

Ia = Ia1 + Ia2
Ib = Ib1 + Ib2
Ic = Ic2

% Solution for part b and part c

% For transformer 1
ST1 = kVa*abs(Ia)
ST1pu100kVA = ST1/100

% For transformer 2
ST2 = kVa*abs(Ic)
ST2pu15kVA = ST2/15

% Solution for part d
PFT1 = cos(thetaVab - atan(imag(Ia)/real(Ia)))
PFT2 = cos(thetaVcb - atan(imag(Ic)/real(Ic)))

% Solution for part e
IA = Ia/n
IB = Ib/n
IN = -(IA + IB)

% Solution for part f
IbaseLV = 100/kVa
Iapu = Ia/IbaseLV
Vabpu = (kVa*(cos(thetaVab) + j*sin(thetaVab)))/kVa
VANpu = Vabpu + Iapu*ZT
VAN = VANpu*7200

Icpu = Ic/IbaseLV
Vbcpu = (kVa*(cos(-thetaVcb) + j*sin(-thetaVcb)))/kVa
VBNpu = Vbcpu - Icpu*ZT
VBN = VBNpu*7200
Vmagdiff = abs(VAN) - abs(VBN)
Thetadiff = 180*(atan(imag(VAN)/real(VAN)) - atan(imag(VBN)/real(VBN)))/pi
```

3.13   AUTOTRANSFORMER

The usual transformer has two windings (not including a tertiary, if there is any) that are not con-nected to each other, whereas an autotransformer is a transformer in which one winding is con-nected in series with the other as a single winding. In this sense, an autotransformer is a normal transformer connected in a special way. It is rated on the basis of output kilovoltamperes rather than the transformer's kilovoltamperes. It has lower leakage reactance, lower losses, smaller excitation current requirements, and, most of all, it is cheaper than the equivalent two-winding transformer (especially when the voltage ratio is 2:1 or less).

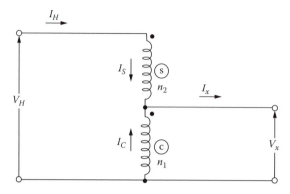

FIGURE 3.68 Wiring diagram of a single-phase autotransformer.

Figure 3.68 shows the wiring diagram of a single-phase autotransformer. Note that S and C denote the series and common portions of the winding. There are two voltage ratios, namely, circuit and winding ratios. The circuit ratio is

$$\frac{V_H}{V_x} = n$$

$$= \frac{n_1 + n_2}{n_1} = 1 + \frac{n_2}{n_1} \qquad (3.66)$$

where
  V_H is the HV-side voltage
  V_x is the LV-side voltage
  n is the turns ratio of the autotransformer
  n_1 is the number of turns in common winding
  n_2 is the number of turns in series winding

As can be observed from Equation 3.66, the circuit ratio is always larger than 1.
  On the other hand, the winding-voltage ratio is

$$\frac{V_S}{V_C} = \frac{n_2}{n_1}$$

$$= n - 1 \qquad (3.67)$$

where
  V_S is the voltage across the series winding
  V_C is the voltage across the common winding

Similarly, the current ratio is

$$\frac{I_C}{I_S} = \frac{I_C}{I_H}$$

$$= \frac{I_x - I_H}{I_H} = n - 1 \qquad (3.68)$$

where

I_C is the current in common winding
I_S is the current in series winding
I_x is the output current at the LV side
I_H is the input current at the HV side

Therefore, the circuits voltampere rating for an ideal autotransformer is

$$\text{Circuits VA rating} = V_H I_H$$

$$= V_x I_x \tag{3.69}$$

and the windings voltampere rating is

$$\text{Windings VA rating} = V_S I_S$$

$$= V_C I_C \tag{3.70}$$

which describes the capacity of the autotransformer in terms of core and coils.

Therefore, the capacity of an autotransformer can be compared to the capacity of an equivalent two-winding transformer (assuming the same core and coils are used) as

$$\frac{\text{Capacity as autotransformer}}{\text{Capacity as two-winding transformer}} = \frac{V_H I_H}{V_S I_S}$$

$$= \frac{V_H I_H}{(V_H - V_x) I_H}$$

$$= \frac{V_H / V_x}{(V_H - V_x)/V_x}$$

$$= \frac{n}{n-1} \tag{3.71}$$

For example, if n is given as 2, the ratio, given by Equation 3.71, is 2, which means that

$$\text{Capacity as autotransformer} = 2 \times \text{Capacity as two-winding transformer}$$

Therefore, one can use a 500 kVA autotransformer instead of using a 1000 kVA two-winding transformer. Note that as n approaches 1, which means that the voltage ratios approach 1, such as 7.2 kV/6.9 kV, then the savings, in terms of the core and coil sizes of autotransformer, increases. An interesting case happens when the voltage ratio (or the turns ratio) is unity: the maximum savings is achieved, but then there is no need for any transformer since the HV and LV are the same.

Figure 3.69 shows a single-phase autotransformer connection used in distribution systems to supply 120/240 V single-phase power from an existing 208Y/120 V three-phase system, the most economically.

Figure 3.70 shows a three-phase autotransformer Y–Y connection used in distribution systems to increase voltage at the ends of feeders or where extensions are being made to existing feeders. It is also the most economical way of stepping down the voltage. It is necessary that the neutral of the autotransformer bank be connected to the system neutral to prevent excessive voltage development on the secondary side. Also, the system impedance should be large enough to restrict the short-circuit current to about 20 times the rated current of the transformer to prevent any transformer burnouts.

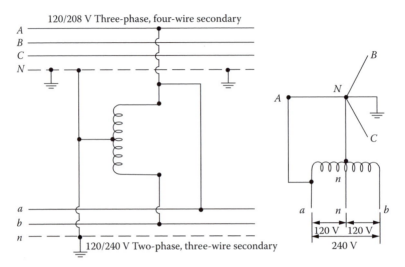

FIGURE 3.69 Single-phase autotransformer.

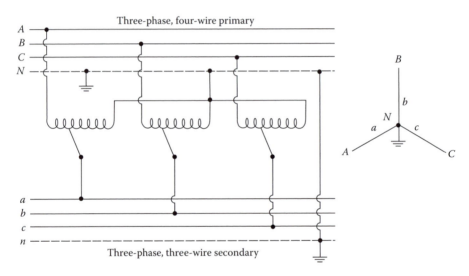

FIGURE 3.70 Three-phase autotransformer.

3.14 BOOSTER TRANSFORMERS

Booster transformers are also called the *buck-and-boost transformers* and provide a fixed buck or boost voltage to the primary of a distribution system when the line voltage drop is excessive. The transformer connection is made in such a way that the secondary is in series and in phase with the main line.

Figure 3.71 shows a single-phase booster transformer connection. The connections shown in Figure 3.71a and b boost the voltage 5% and 10%, respectively. In Figure 3.71a, if the lines to the LV bushings x_3 and x_1 are interchanged, a 5% buck in the voltage results. Figure 3.72 shows a three-phase three-wire booster transformer connection using two single-phase booster transformers.

Figure 3.73 shows a three-phase four-wire booster transformer connection using three single-phase booster transformers. Both LV and HV windings and bushings have the same level of insulation. To prevent harmful voltage induction by series winding, the transformer primary must never be open under any circumstances before opening or unloading the secondary. Also, the primary side of the transformer should not have any fuses or disconnecting devices. Boosters are often used in distribution feeders where the cost of tap-changing transformers is not justified.

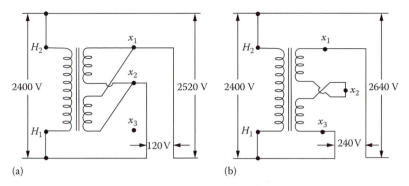

FIGURE 3.71   Single-phase booster transformer connection: (a) for 5% boost and (b) for 10% boost.

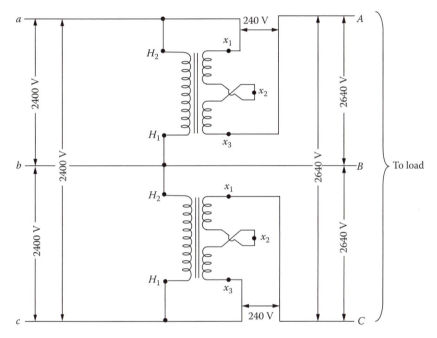

FIGURE 3.72   Three-phase three-wire booster transformer connection using two single-phase booster transformers.

3.15   AMORPHOUS METAL DISTRIBUTION TRANSFORMERS

The continuing importance of distribution system efficiency improvement and its economic evaluation has focused greater attention on designing equipment with exceptionally high efficiency levels. For example, because of extremely low magnetic losses, amorphous metal offers the opportunity to reduce the core loss of distribution transformers by approximately 60% and thereby reduce operating costs. For example, core loss of a 25 kVA, 7200/12,470Y-120/240V silicon steel transformer is 86 W, whereas it is only 28 W for an amorphous transformer.

Also, it is quieter (with 38 db) than its equivalent silicon steel transformer (with 48 db). There are more than 25 million distribution transformers installed in this country. Replacing them with amorphous units could result in an energy savings of nearly 15 billion kWh/year. Nationally, this could represent a savings of more than $700 million, which is annually equivalent to the energy consumed by a city of four million people. Each year, approximately one million distribution transformers are installed on US utility systems. Application of amorphous metal transformers is a substantial opportunity to reduce utility operating costs and defer generating capacity additions.

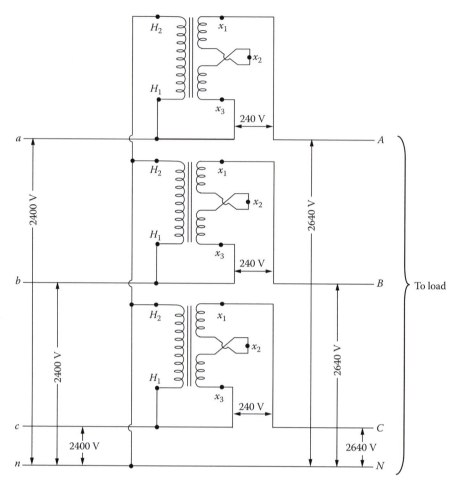

FIGURE 3.73 Three-phase three-wire booster transformer connection using three single-phase booster transformers.

3.16 NATURE OF ZERO-SEQUENCE CURRENTS

Consider the representation of the three-phase four-wire system shown in Figure 3.74a. If the system is balanced, there will not be any residual current in the neutral wire. However, if the system is unbalanced, then the current in the neutral wire is the nonzero residual current so that

$$\overline{I}_n = -(\overline{I}_a + \overline{I}_b + \overline{I}_c)$$

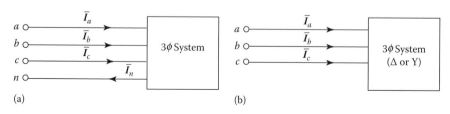

FIGURE 3.74 A three-phase system: (a) with four wires and wye-connected load and (b) three wires and wye- or delta-connected load.

But since the zero-sequence current is the average sum of the unbalanced currents

$$\overline{I}_{a0} = -\frac{\overline{I}_a + \overline{I}_b + \overline{I}_c}{3}$$

$$= -\frac{\overline{I}_n}{3}$$

Hence, the neutral current is

$$\overline{I}_n = -3\overline{I}_{a0}$$

which is also known as the *ground current* $\overline{I}_g$. Now consider the three-phase system of Figure 3.75b. The system could be delta or wye system. Note that there is no neutral wire this time. As a result,

$$\overline{I}_n = -3\overline{I}_{a0} = 0$$

Hence,

$$\overline{I}_{a0} = 0$$

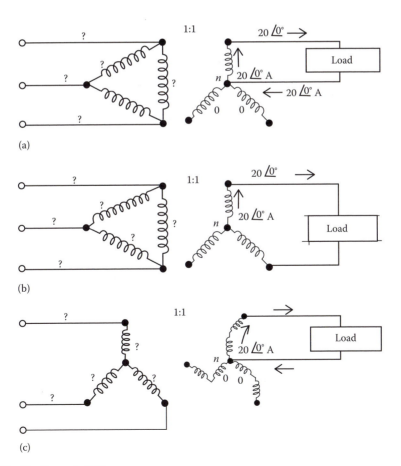

(a)

(b)

(c)

FIGURE 3.75 For Example 3.12.

Thus, it can be shown that

$$\bar{I}_a = 0 + \bar{I}_{a1} + \bar{I}_{a2}$$

$$\bar{I}_a = 0 + a^2\bar{I}_{a1} + a\bar{I}_{a2}$$

$$\bar{I}_a = 0 + a\bar{I}_{a1} + a^2\bar{I}_{a2}$$

Therefore, there is no zero-sequence current in the line currents. However, the positive- and negative-sequence currents will always exist in a three-wire or four-wire systems.

Example 3.12

Consider the three-phase transformers shown in Figure 3.75 and assume that the transformer ratios are given as 1:1. Given the load connections on the secondary side, determine the unknown currents that are indicated by the question marks in the figure:

a. Part (a)
b. Part (b)
c. Part (c)

Solution

Figure 3.76 shows the answers for all three parts.

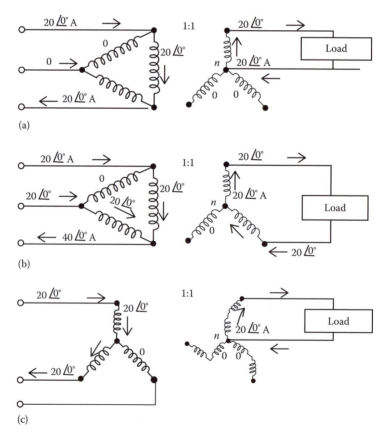

(a)

(b)

(c)

FIGURE 3.76   Solutions for Example 3.12.

Example 3.13

Consider the three-phase transformer connections shown in Figure 3.77. Determine the indicated unknown currents:

   a. For the wye-grounded delta system that is shown in Figure 3.77a
   b. For the ungrounded-wye ungrounded-wye system that is shown in Figure 3.77b
   c. For the three-wire grounded-wye grounded-wye system that is shown in Figure 3.77c
   d. For the four-wire grounded-wye grounded-wye system that is shown in Figure 3.77d

Solution

   a. It is shown in Figure 3.78a.
   b. It is shown in Figure 3.78b.
   c. It is shown in Figure 3.78c.
   d. It is shown in Figure 3.78d.

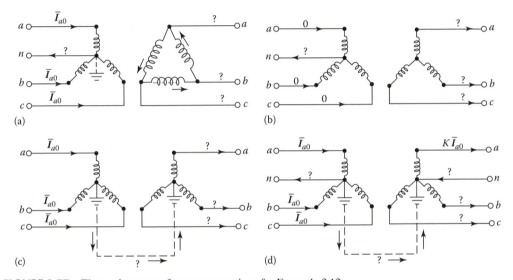

FIGURE 3.77 Three-phase transformer connections for Example 3.12.

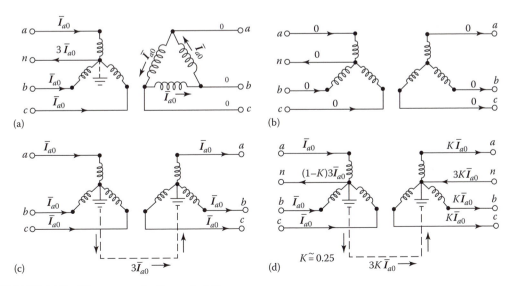

FIGURE 3.78 The answers for Example 3.12.

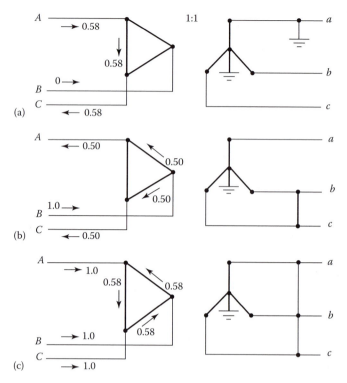

FIGURE 3.79 Δ–Y transformer bank connections for Example 3.13.

Example 3.14

Consider the system shown in Figure 3.79. Determine the indicated current flows.

Solution

   a. It is shown in Figure 3.80a.
   b. It is shown in Figure 3.80b.
   c. It is shown in Figure 3.80c.

Example 3.15

Consider the three-phase transformer connections shown in Figure 3.81. Determine the indicated unknown fault currents:

   a. For the wye-grounded delta system that is shown in Figure 3.81a when there is three-phase fault as indicated
   b. For the grounded-wye delta system that is shown in Figure 3.81b when there is line-to-line fault as indicated
   c. For the three-wire grounded-wye grounded-wye system that is shown in Figure 3.81c when there is three-phase fault as indicated
   d. For the three-wire grounded-wye grounded-wye system that is shown in Figure 3.81d when there is line-to-neutral fault as indicated
   e. For the three-wire grounded-wye grounded-wye system that is shown in Figure 3.81e when there is line-to-line fault as indicated

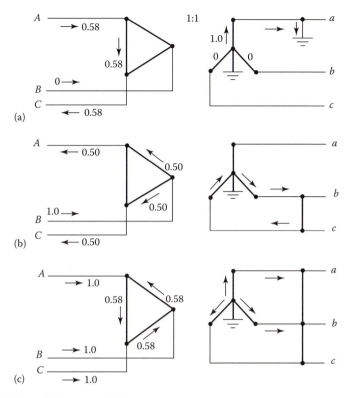

FIGURE 3.80   Solutions for Example 3.13.

Solution

   a. It is shown in Figure 3.82a.
   b. It is shown in Figure 3.82b.
   c. It is shown in Figure 3.82c.
   d. It is shown in Figure 3.82d.

Example 3.16

Consider the three-phase transformer connections shown in Figure 3.83. Determine the indicated unknown fault currents:

   a. For the Δ–Δ system that is shown in Figure 3.83a when there is three-phase fault as indicated
   b. For the Δ–Δ system that is shown in Figure 3.83b when there is line-to-line fault as indicated
   c. For the four-wire delta grounded-wye system that is shown in Figure 3.83c when there is three-phase fault as indicated
   d. For the four-wire delta grounded-wye system that is shown in Figure 3.83d when there is line-to-line fault as indicated
   e. For the four-wire delta grounded-wye system that is shown in Figure 3.83e when there is line-to-neutral fault as indicated

Solution

   a. It is shown in Figure 3.84a.
   b. It is shown in Figure 3.84b.
   c. It is shown in Figure 3.84c.
   d. It is shown in Figure 3.84d.

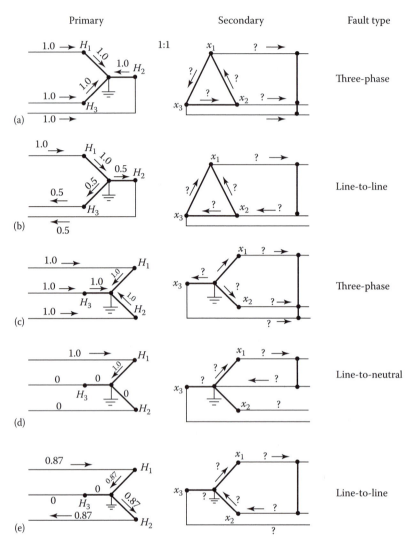

Primary   Secondary   Fault type

FIGURE 3.81 Three-phase transformer connections for Example 3.12.

3.17 ZIGZAG POWER TRANSFORMERS

As shown in Figure 3.85, the zigzag transformers are most often used as grounding transformers. It shows two different methods of connecting a grounding transformer to the system. Zigzag transformers are three-phase autotransformers with no secondary windings. They have 1:1 turn ratios among the autotransformer windings. Figure 3.85a shows a method of connecting a grounding transformer with individual line breaker directly to main bus of the system. On the other hand, Figure 3.85b shows a method of connecting a grounding transformer to main bus of the system without an individual line breaker. Note that in the last case, the grounding transformer is connected between the main transformer bank and its breaker.

As illustrated in Figure 3.86, a Y–Δ transformer can also be used as a grounding transformer. However, the delta must be closed to provide a path for zero-sequence current. The wye winding of the transformer has to be at the same level of voltage as the circuit that is to be grounded. Figure 3.86 also shows current distribution in such Y–Δ grounding transformer when a line-to-ground fault takes place on a three-phase system.

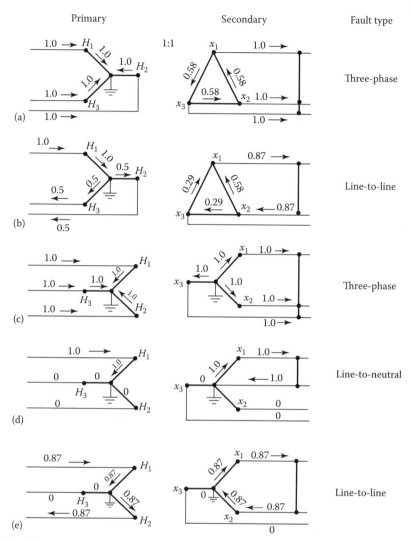

FIGURE 3.82   The answers for Example 3.12.

Figure 3.86a shows that primary and secondary windings are interconnected. Note that when positive- or negative-sequence voltages are applied to this connector, the impedance which is usually assumed to be infinite. Hence, the positive- and negative-sequence networks are unaffected by the grounding transformer. However, when zero-sequence currents are applied, the currents are shown to be all in phase and are connected in a manner that the mmf produced in each coil is opposed by an equal mmf from another phase winding.

Note that the impedance of the transformer to three-phase currents is so high that there is no fault current on the system, and only a small magnetizing current flows in transformer windings. But the impedance of the transformer to I_G ground current is so low (due to a special winding) that it permits high ground current I_G to flow.

Figure 3.87a shows the wiring diagram with the opposing sense of the coil connections in a zigzag transformer. This is valid for three single-phase units or one three-phase unit. Figure 3.87b shows the normal positive-sequence condition. It shows the normal voltage phasor diagram. Also note that the impedance to the zero-sequence currents is due to the leakage flux of the windings. As said before, for positive- and negative-sequence currents neglecting magnetizing current,

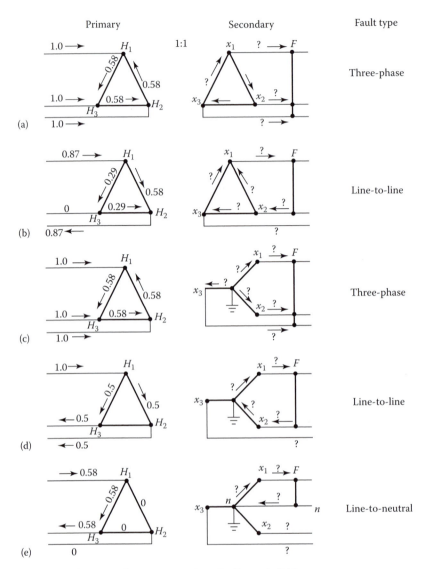

FIGURE 3.83 Three-phase transformer connections for Example 3.16.

the connection has an infinite impedance. In Figure 3.87c, the currents in the winding sections a_1 and c_2 must be equal as those are in series. By the same token, all currents must be equal, balancing the mmfs in each leg. As shown in Figure 3.87c, the windings a_1 and a_2 are located on the same leg and have the same number of turns but are wound in the opposite direction. Hence, the zero-sequence currents in the two windings on the same leg are canceling the ampere-turns. Thus, a zigzag transformer is often used to obtain a neutral for the grounding of a Δ–Δ connected system, as shown in Figure 3.87d.

Figure 3.87d shows the distribution of zero-sequence current and its return path for a single line-to-ground fault on one of the phases. The ground current is divided equally through the zigzag transformer; one-third of the current returns directly to the fault point and the remaining two-thirds must pass through two phases of the delta-connected windings to return to the fault point. Two phases and windings on the primary delta must carry current to balance the ampere-turns of the secondary winding currents, as shown in Figure 3.87c. An impedance is added between the artificially derived neutral and the ground to limit the ground fault current.

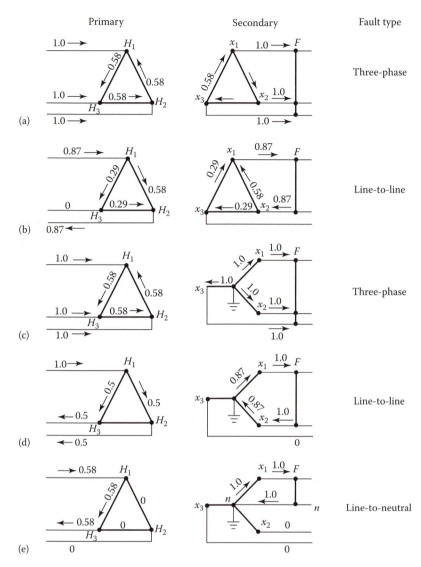

FIGURE 3.84 Solutions for Example 3.16.

3.18 GROUNDING TRANSFORMERS USED IN THE UTILITY SYSTEMS

The best way to get the system neutral is to use the source transformers or generators with wye-connected windings. Then, the neutral is readily available. Such transformers are available for all voltages. On new systems, 208Y/120 V or 480Y/277 V can be used to good advantage instead of 240 V. For 2400 and 4800 V systems, special 2400Y- or 4800Y-connected transformers may be used or grounding transformers may be employed.

In old systems (with 600 V or less, and many 2.4, 4.8, 6.9 kV), the system neutrals may not be available. When delta-connected systems require having neutrals, grounding transformers are used to provide solidly grounded neutral. Similarly, 2.4–15 kV systems with only delta-connected equipment can be grounded by adding grounding transformers and neutral resistors. In general, grounding transformers can be either of zigzag or of wye type.

The type of grounding transformer most commonly used is a three-phase zigzag transformer with no secondary winding. The impedance of the transformer to three-phase currents is so high that,

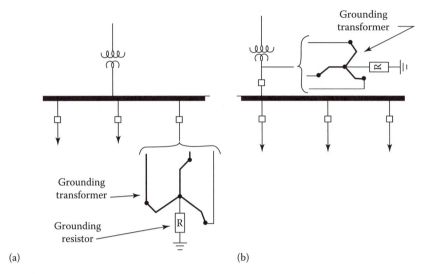

FIGURE 3.85 Two different ways of connecting a grounding transformer to the system.

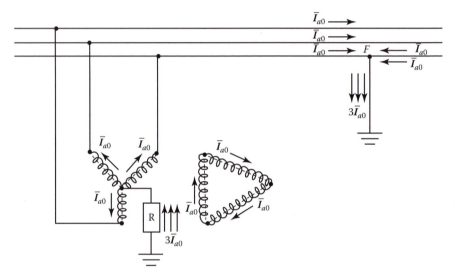

FIGURE 3.86 A Y–Δ transformer can also be used as a grounding transformer.

when there is no fault on the system, only a small magnetizing current flows in the transformer windings. The transformer impedance to ground current, however, is so low that it permits high ground currents to flow. The transformer divides the ground current into three equal components; these currents are in phase with each other and flow in the three windings of the grounding transformer.

As can be seen in Figure 3.87c, due to special winding of the zigzag transformer, when these three equal currents flow, the current in one section of the winding of each leg of the core is in a direct or opposite to that in the other section of the winding on that leg. Thus, the only magnetic flux that results from the zero-sequence ground currents is the leakage field of each winding section. This is the reason for the transformer having low impedance to the flow of ground current.

The short-time kVA rating of a grounding transformer is equal to the rated line-to-neutral voltage ties its rated neutral current. In general, a grounding transformer is designed to carry its rated current for a limited time only, such as 10 s or 1 min. Thus, it is normally about one-tenth as large, physically, as an ordinary three-phase transformer for the same rated kVA. Figure 3.80 shows two different types of zigzag connections to the LV bus of a utility system.

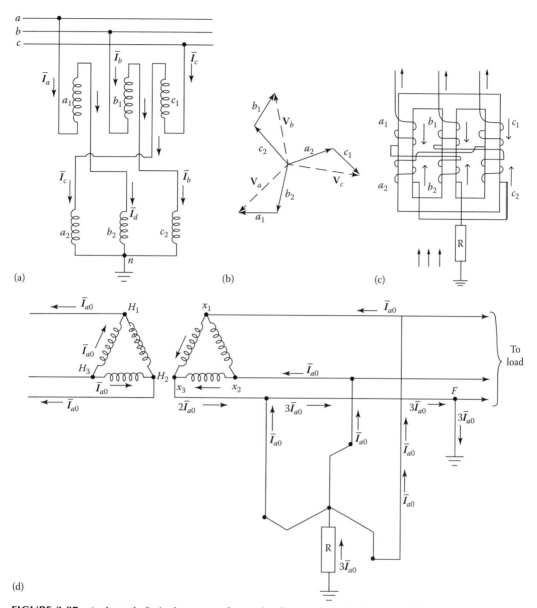

FIGURE 3.87 (a through d) A zigzag transformer is often used to obtain a neutral for the grounding of a Δ–Δ connected system.

As said before, a Y–Δ transformer can also be used as a grounding transformer, as shown in Figure 3.86. Hence, the delta must be closed to provide a path for zero-sequence current, but the delta can have any voltage desired. However, the wye winding must be of the same *voltage* rating as the circuit that is to be grounded.

3.19 PROTECTION SCHEME OF A DISTRIBUTION FEEDER CIRCUIT

Figure 3.88 shows the protection scheme of a distribution feeder circuit. As shown in the figure, each distribution transformer has a fuse that is located either externally, that is, in a fuse cutout next to the transformer, or internally, that is, inside the transformer tank as is the case for a CSP transformer.

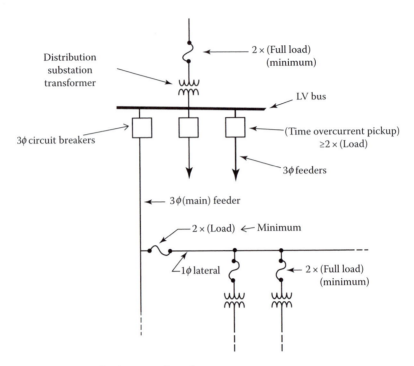

FIGURE 3.88   A distribution feeder protection scheme.

As shown in the figure, it is a common practice to install a fuse at the head of each lateral (or branch). The fuse must carry twice the expected load as a minimum, and it must coordinate with load-side transformer fuses or other devices. It is customary to select the rating of each lateral fuse adequately large so that it is protected from damage by the transformer fuses on the lateral. Furthermore, the lateral fuse is usually expected to clear faults occurring at the ends of the lateral. If the fuse does not clear the faults, then one or more additional fuses may be installed on the lateral.

As shown in the figure, a recloser, or circuit breaker with reclosing relays, is located at the substation to provide a backup protection. It clears the temporary faults in its protective zone. At the limit of the protective zone, the minimum available fault current, determined by calculation, is equal to the smallest value of current (called *minimum pickup current*), which will trigger the recloser, or circuit breaker, to operate. However, a fault beyond the limit of this protection zone may not trigger the recloser, or circuit breaker, to operate.

PROBLEMS

3.1   Repeat Example 3.7, assuming an open-Y primary and an open-Δ secondary and using the 0° references given in Figure P3.1.
Also determine
    a.   The value of the open-Δ HV phasor between A and B, that is, $\bar{V}_{AB}$
    b.   The value of the open-Y HV phasor between A and N, that is, $\bar{V}_{AN}$

3.2   Repeat Example 3.10, if the LV line current I is 100 A and the line-to-line LV is 480 V.

3.3   Consider the T–T connection given in Figure P3.3 and determine the following:
    a.   Draw the LV diagram, correctly oriented on the 0° reference shown.
    b.   Find the value of the $\bar{V}_{ab}$ and $\bar{V}_{an}$ phasors.
    c.   Find the magnitudes of the following rated winding voltages:
       i.   The voltage $V_{H_1H_2}$ on transformer 1
      ii.   The voltage $V_{x_1\phi}$ on transformer 1

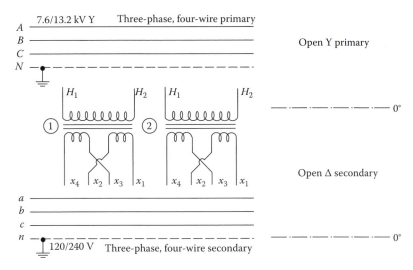

FIGURE P3.1   For Problem 3.1.

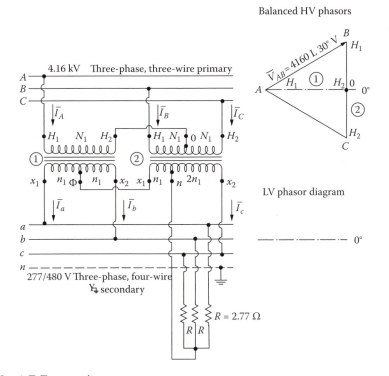

FIGURE P3.3   A T–T connection.

iii. The voltage $V_{\phi x_2}$ on transformer 1
iv. The voltage $V_{H_1 0}$ on transformer 2
v. The voltage V_{0H} on transformer 2
vi. The voltage $V_{x_1 n}$ on transformer 2
vii. The voltage V_{nx} on transformer 2

3.4 Assume that the T–T transformer bank given in Problem 3.3 is loaded with the balanced resistors given. Assume that the secondary voltages are perfectly balanced; the necessary HV applied voltages then are not perfectly balanced. Use secondary voltages of 480 V and neglect magnetizing currents. Determine the following:

a. The LV current phasors
b. The HV current phasors

3.5 Use the results of Problems 3.3 and 3.4 and apply the complex power formula $S = P + jQ = \overline{V}\overline{I}^*$ four times, once for each LV winding, for example, a part of the output of transformer 1 is $\overline{V}_{X_1X_2}\overline{I}_a^*$. Based on these results, find the following:

a. Total complex power output from the T–T bank. (Does your result agree with that which is easily computed as input to the resistors?)
b. The necessary kilovoltampere ratings of both LV windings of both the transformers.
c. The ratio of total kilovoltampere ratings of all LV windings in the transformer bank to the total kilovoltampere output from the bank.

3.6 Consider Figure P3.6 and assume that the motor is rated 25 hp and is mechanically loaded so that it draws 25.0 kVA three-phase input at $\cos\theta = 0.866$ lagging power factor:

a. Draw the necessary HV connections so that the LVs shall be as shown, that is, of *abc* phase sequence.
b. Find the power factors $\cos\theta_{T_1}$ and $\cos\theta_{T_2}$ at which each transformer operates.
c. Find the ratio of voltampere load on one transformer to total voltamperes delivered to the load.

3.7 Consider Figure P3.4 and assume that the two-transformer T–T bank delivers 120/208 V three-phase four-wire service from a three-phase three-wire 4160 V primary line. The problem is to determine if this bank can carry unbalanced loads even though the primary neutral terminal N is not connected to the source neutral. (If it can, the T–T performance is quite different from the three-transformer wye-grounded wye bank.) Use the ideal-transformer theory and pursue the question as follows:

a. Load phase a-n with $R = 1.20\ \Omega$ resistance and then find the following six complex currents numerically: $\overline{I}_a, \overline{I}_b, \overline{I}_c, \overline{I}_A, \overline{I}_B,$ and $\overline{I}_C$.
b. Find the following complex powers of windings by using the $S = P + jQ = \overline{V}\overline{I}^*$ equation numerically:

$$S_{T_{1(x_1-n)}} = \text{complex power of } x_1 - n \text{ portion of transformer 1}$$

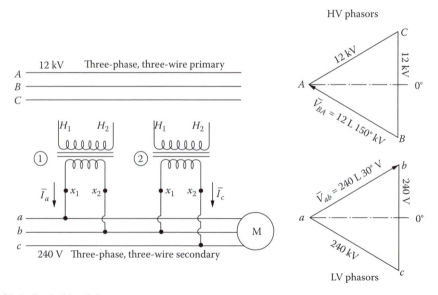

FIGURE P3.6 For Problem 3.6.

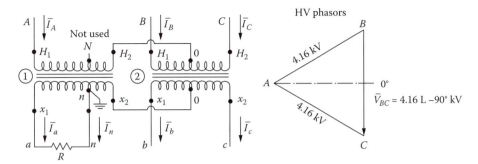

FIGURE P3.7   For Problem 3.7.

$$S_{T_{1(H_1-H_2)}} = \text{Complex power of } H_1 - H_2 \text{ portion of transformer 1}$$

$$S_{T_{2(H_1-0)}} = \text{Complex power of } H_1 - 0 \text{ portion of transformer 2}$$

$$S_{T_{2(H_2-0)}} = \text{Complex power of } H_2 - 0 \text{ portion of transformer 2}$$

    c. Do your results indicate that this bank will carry unbalanced loads successfully? Why?

3.8   Figure P3.8 shows two single-phase transformers, each with a 7620 V HV winding and two 120 V LV windings. The diagram shows the proposed connections for an open-Y to open-Δ bank and the HV applied phasor-voltage drops. Here, *abc* phase sequence at LV and HV sides and 120/240 V are required.

    a. Sketch the LV phasor diagram, correctly oriented on the 0° reference line. Label it adequately with *x*'s (1), and (2), etc., to identify.

    b. State whether or not the proposed connections will output the required three-phase four-wire 120/240 V delta LV.

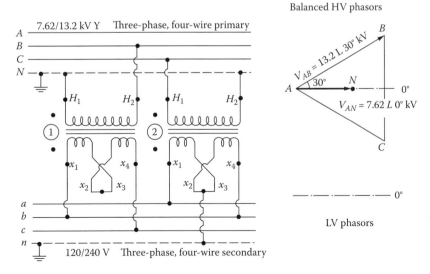

FIGURE P3.8   For Problem 3.8.

3.9 A large number of 25 kVA distribution transformers are to be purchased. Two competitive bids have been received. The bid data are tabulated as follows:

Transformer	Cost of Transformer Delivered to NL&NP's Warehouse	Core Loss at Rated Load (W)	Copper Loss at Rated Voltage and Frequency (W)	Per Unit Exciting Current
A	$355	360	130	0.015
B	$345	380	150	0.020

Evaluate the bids on the basis of total annual cost (TAC) and recommend the purchase of the one having the least TAC. The cost of installing a transformer is not to be included in this study. The following system data are given:

Annual peak load on transformer = 35 kVA
Annual loss factor = 0.15
Per unit annual fixed charge rate = 0.15
Installed cost of shunt capacitors = $10/kvar
Incremental cost of off-peak energy = $0.01/kWh
Incremental cost of on-peak energy = $0.012/kWh
Investment cost of power system upstream from distribution transformers = $300/kVA.

    Calculate the TAC of owning and operating one such transformer and state which transformer should be purchased. (*Hint*: Study the relevant equations in Chapter 6 before starting to calculate.)

3.10 Assume that a 250 kVA distribution transformer is used for single-phase pole mounting. The transformer is connected phase-to-neutral 7200 V on the primary and 2520 V phase-to-neutral on the secondary side. The leakage impedance of the transformer is 3.5%. Based on the given information, determine the following:

  a. Assume that the transformer has 0.7 pu A in the HV winding. Find the actual current values in the HV and LV windings. What is the value of the current in the LV winding in per units?

  b. Find the impedance of the transformer as referred to the HV and LV windings in ohms.

  c. Assume that the LV terminals of the transformer are short-circuited and 0.22 per unit voltage is applied to the HV winding. Find the HV and LV winding currents that exist as a result of the short circuit in per units and amps.

  d. Determine the internal voltage drop of the transformer, due to its leakage impedance, if a 1.2 per unit current flows in the HV winding. Give the result in per units and volts.

3.11 Resolve Example 3.11 by using MATLAB. Assume that all the quantities remain the same.

REFERENCES

1. Lloyd, B: *Electric Utility Engineering Reference Book-Distribution Systems*, Vol. 3, Westinghouse Electric Corporation, East Pittsburgh, PA, 1965.
2. Lloyd, B: *Electrical Transmission and Distribution Reference Book*, Westinghouse Electric Corporation, East Pittsburgh, PA, 1964.
3. Stigant, S. A. and A. C. Franklin: *The J&P Transformer Book*, Butterworth, London, U.K., 1973.
4. American National Standards Institute: *The Guide for Loading Mineral Oil-Immersed Overhead-Type Distribution Transformers with 55°C and 65°C Average Winding Rise*, Appendix C57.91, 1969.
5. Fink, D. G. and H. W. Beaty: *Standard Handbook for Electrical Engineers*, 11th edn., McGraw-Hill, New York, 1978.
6. Clarke, E.: *Circuit Analysis of AC Power Systems*, Vol. 1, General Electric Series, Schenectady, NY, 1943.
7. General Electric Company: *Distribution Transformer Manual*, Hickory, NC, 1975.

4 Design of Subtransmission Lines and Distribution Substations

A teacher affects eternity.

Author Unknown

Education is the best provision for old age.

Aristotle, 365 BC

Education is…hanging around until you've caught on.

Will Rogers

4.1 INTRODUCTION

In a broad definition, the distribution system is that part of the electric utility system between the bulk power source and the customers' service switches. This definition of the distribution system includes the following components:

1. Subtransmission system
2. Distribution substations
3. Distribution or primary feeders
4. Distribution transformers
5. Secondary circuits
6. Service drops

However, some distribution system engineers prefer to define the distribution system as that part of the electric utility system between the distribution substations and the consumers' service entrance. Figure 4.1 shows a one-line diagram of a typical distribution system. The subtransmission circuits deliver energy from bulk power sources to the distribution substations. The subtransmission voltage is somewhere between 12.47 and 245 kV. The distribution substation, which is made of power transformers together with the necessary voltage-regulating apparatus, buses, and switchgear, reduces the subtransmission voltage to a lower primary system voltage for local distribution. The three-phase primary feeder, which is usually operating in the range of 4.16–34.5 kV, distributes energy from the low-voltage bus of the substation to its load center where it branches into three-phase subfeeders and single laterals.

Distribution transformers, in ratings from 10 to 500 kVA, are usually connected to each primary feeder, subfeeders, and laterals. They reduce the distribution voltage to the utilization voltage. The secondaries facilitate the path to distribute energy from the distribution transformer to consumers through service drops. This chapter covers briefly the design of subtransmission and distribution substations.

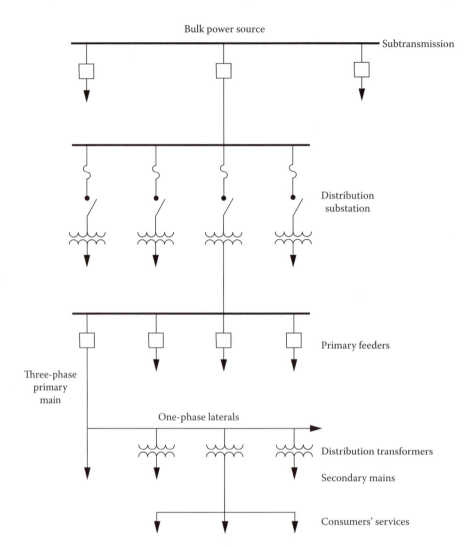

FIGURE 4.1   One-line diagram of a typical distribution system.

4.2 SUBTRANSMISSION

The subtransmission system is that part of the electric utility system that delivers power from bulk power sources, such as large transmission substations. The subtransmission circuits may be made of overhead open-wire construction on wood poles or of underground cables. The voltage of these circuits varies from 12.47 to 245 kV, with the majority at 69, 115, and 138 kV voltage levels. There is a continuous trend in the usage of the higher voltage as a result of the increasing use of higher primary voltages.

The subtransmission system designs vary from simple radial systems to a subtransmission network. The major considerations affecting the design are cost and reliability. Figure 4.2 shows a radial subtransmission system. In the radial system, as the name implies, the circuits radiate from the bulk power stations to the distribution substations. The radial system is simple and has a low first cost but it also has a low service continuity. Because of this reason, the radial system is not generally used. Instead, an improved form of radial-type subtransmission design is preferred, as shown in Figure 4.3. It allows relatively faster service restoration when a fault occurs on one of the subtransmission circuits.

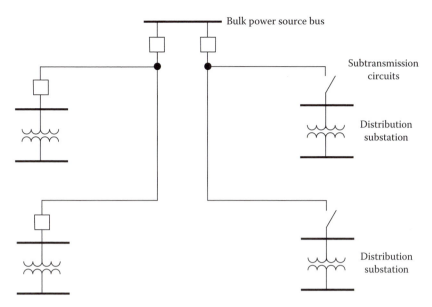

FIGURE 4.2 Radial-type subtransmission.

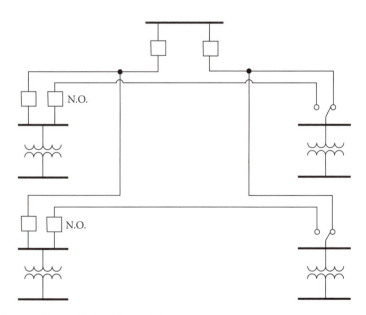

FIGURE 4.3 Improved form of radial-type subtransmission.

In general, due to higher service reliability, the subtransmission system is designed as loop circuits or multiple circuits forming a subtransmission grid or network. Figure 4.4 shows a loop-type subtransmission system. In this design, a single circuit originating from a bulk power bus runs through a number of substations and returns to the same bus.

Figure 4.5 shows a grid-type subtransmission that has multiple circuits. The distribution substations are interconnected, and the design may have more than one bulk power source. Therefore, it has the greatest service reliability, and it requires costly control of power flow and relaying. It is the most commonly used form of subtransmission.

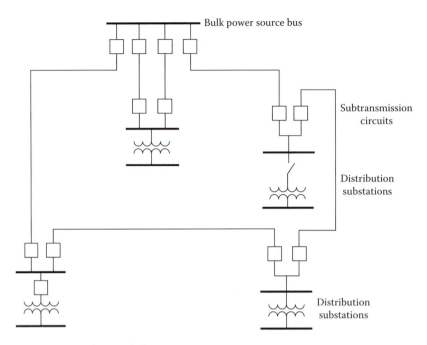

FIGURE 4.4   Loop-type subtransmission.

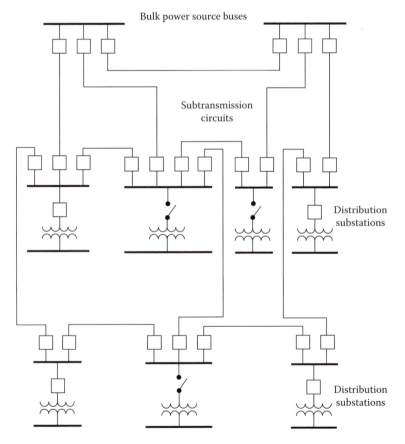

FIGURE 4.5   Grid- or network-type subtransmission.

4.2.1 SUBTRANSMISSION LINE COSTS

Subtransmission line costs are based on a per mile cost and a termination cost at the end of the line associated with the substation at which it is terminated. According to the ABB Guidebook [15], based on 1994 prices, costs can run from as low as $50,000 per mile for a 46 kV wooden pole subtransmission line with perhaps 50 MVA capacity ($1 per kVA-mile) to over $1,000,000 per mile for a 500 kV double circuit construction with 2,000 MVA capacity ($0.5 per kVA-mile).

4.3 DISTRIBUTION SUBSTATIONS

Distribution substation design has been somewhat standardized by the electric utility industry based upon past experiences. However, the process of standardization is a continuous one. Figures 4.6 and 4.7 show typical distribution substations. The attractive appearance of these substations is enhanced by the use of underground cable in and out of the station as well as between the transformer secondary and the low-voltage bus structure. Automatic switching is used for sectionalizing in some of these stations and for preferred-emergency automatic transfer in others.

FIGURE 4.6   A typical distribution substation. (Courtesy of S&C Electric Company, Chicago, IL.)

FIGURE 4.7   A typical small distribution substation. (Courtesy of S&C Electric Company, Chicago, IL.)

Figure 4.8 shows an overall view of a modern substation. The figure shows two 115 kV, 1200 A vertical-break-style circuit switchers to switch and protect two transformers supplying power to a large tire manufacturing plant. The transformer located in the foreground is rated 15/20/28 MVA, 115/4.16 kV, 8.8% impedance, and the second transformer is rated 15/20/28 MVA, 115/13.8 kV, 9.1% impedance. Figure 4.9 shows a close view of a typical modern distribution substation transformer. Figure 4.10 shows a primary unit substation transformer. Figure 4.11 shows a secondary unit substation transformer. Figure 4.12 shows a 630 kVA, 10/0.4 kV GEAFOL solid dielectric transformer. Figure 4.13 shows an oil distribution transformer: cutaway of a TUMERIC transformer with an oil expansion tank showing in the foreground and TUNORMA with an oil expansion tank shown in the background. Figure 4.14 shows a 630 kVA, 10/0.4 kV GEAFOL solid dielectric transformer.

FIGURE 4.8   Overview of a modern substation. (Courtesy of S&C Electric Company.)

FIGURE 4.9   Close view of typical modern distribution substation transformer. (Courtesy of ABB.)

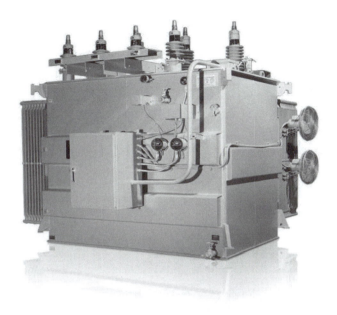

FIGURE 4.10   Primary unit substation transformer. (Courtesy of ABB.)

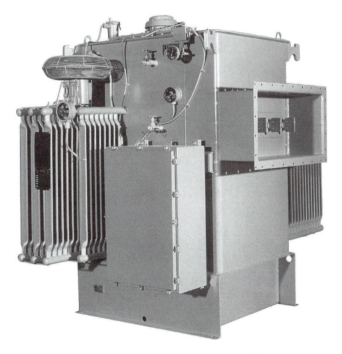

FIGURE 4.11   Secondary unit substation transformer. (Courtesy of ABB.)

Figure 4.15 shows a 40 MVA, 110 kV ± 16%/21 kV, three-phase, core-type transformer, 5.2 m high, 9.4 m long, 3 m wide, and weighing 80 tons. Figure 4.16 shows an 850/950/1100 MVA, 415 kV ± 11%/27 kV, three-phase, shell-type transformer, 11.3 m high, 14 m long, 5.7 m wide, and weighing (without cooling oil) 552 tons. Figure 4.17 shows a completely assembled 910 MVA, 20.5/500 kV, three-phase, step-up transformer of 39 ft high, 36 ft long, 29 ft wide, and weighing 562 tons. Figure 4.18 shows a typical core and coil assembly of a three-phase, core-type power transformer.

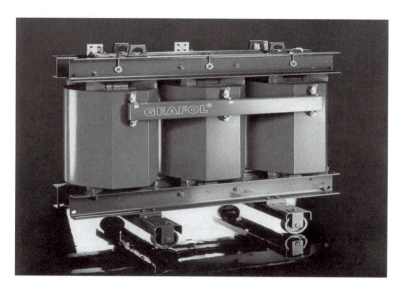

FIGURE 4.12   A 630 kVA, 10/0.4 kV GEAFOL solid dielectric transformer. (Courtesy of Siemens.)

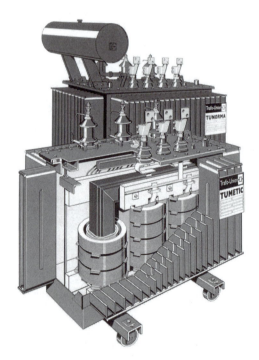

FIGURE 4.13   Oil distribution transformer: cutaway of a TUMERIC transformer with an oil expansion tank showing in the foreground and TUNORMA with an oil expansion tank shown in the background. (Courtesy of Siemens.)

A typical substation may include the following equipment: (1) power transformers, (2) circuit breakers, (3) disconnecting switches, (4) station buses and insulators, (5) current-limiting reactors, (6) shunt reactors, (7) current transformers, (8) potential transformers, (9) capacitor voltage transformers, (10) coupling capacitors, (11) series capacitors, (12) shunt capacitors, (13) grounding system, (14) lightning arresters and/or gaps, (15) line traps, (16) protective relays, (17) station batteries, and (18) other apparatus.

FIGURE 4.14 A 630 kVA, 10/0.4 kV GEAFOL solid dielectric transformer. (Courtesy of Siemens Press.)

FIGURE 4.15 A 40 MVA, 110 kV ± 16%/21 kV, three-phase, core-type transformer, 5.2 m high, 9.4 m long, 3 m wide, and weighing 80 tons. (Courtesy of Siemens Press.)

4.3.1 Substation Costs

Substation costs include all the equipment and labor required to build a substation, including the cost of land and easements (i.e., rights of way). For planning purposes, substation costs can be categorized into four groups:

1. *Site costs*: the cost of buying the site and preparing it for a substation
2. *Transmission cost*: the cost of terminating transmission at the site

FIGURE 4.16   A850/950/1100 MVA, 415 kV ± 11%/27 kV, three-phase, shell-type transformer, 11.3 m high, 14 m long, 5.7 m wide, and weighing (without cooling oil) 552 tons. (Courtesy of Siemens Press.)

FIGURE 4.17   A completely assembled 910 MVA, 20.5/500 kV, three-phase, step-up transformer, about 12 m high, 11 m long, 9 m wide, and weighing 562 tons. (Courtesy of ABB.)

3. *Transformer cost*: the transformer and all metering, control, oil spill containment, fire prevention, cooling, noise abatement, and other transformer-related equipment, along with typical bus work, switches, metering, relaying, and breakers associated with this type of transformer and their installations
4. *Feeder bus-work/getaway costs*: the cost of beginning distribution at the substation, which includes getting feeders out of the substation

The substation site costs depend on local land prices, that is, the real-estate market. It includes the cost of preparing the site in terms of grading, grounding mat, foundations, buried ductwork, control building, lighting, fence, landscaping, and access road. Often, estimated costs of feeder bus work and gateways are folded into the transformer costs. Substation costs vary greatly depending on type, capacity, local land prices, and other variable circumstances. According to the ABB Guidebook, substation

FIGURE 4.18   A typical core and coil assembly of a three-phase, core-type, power transformer. (Courtesy of Siemens Press.)

costs can vary from $1.8 million to $5.5 million, based on 1994 prices. It depends on land costs, labor costs, the utility equipment and installation standards, and other circumstances. Typical total substation cost could vary from between about $36 per kW and $110 per kW, depending on circumstances.

Example 4.1

Consider a typical substation that might be fed by two incoming 138 kV lines feeding two 32 MVA, 138/12.47 kV transformers, each with a low-voltage bus. Each bus has four outgoing distribution feeders of 9 MVA peak capacity each. The total site cost of the substation is $600,000. The total transmission cost including high-side bus circuit breakers is estimated to be $900,000. The total costs of the two transformers and associated equipment is $1,100,000. The feeder buswork/getaway cost is $400,000. Determine the following:

    a. The total cost of this substation
    b. The utilization factor of the substation, if it is going to be used to serve a peak load of about 50 MVA
    c. The total substation cost per kVA based on the previously mentioned utilization rate

Solution

    a. The total cost of this substation is

$$\$600,000 + \$900,000 + \$1,100,000 + \$400,000 = \$3,000,000$$

    b. The utilization factor of the substation is

$$F_u = \frac{\text{maximum demand}}{\text{rated system capacity}} = \frac{50\,\text{MVA}}{2(32\,\text{MVA})} \cong 0.78 \text{ or } 78\%$$

c. The total substation cost per kVA is

$$\frac{\$3,000,000}{50,000\,kVA} = \$60/kVA$$

4.4 SUBSTATION BUS SCHEMES

The electrical and physical arrangements of the switching and busing at the subtransmission voltage level are determined by the selected substation scheme (or diagram). On the other hand, the selection of a particular substation scheme is based upon safety, reliability, economy, simplicity, and other considerations.

The most commonly used substation bus schemes include (1) single bus scheme, (2) double bus–double breaker (or double main) scheme, (3) main-and-transfer bus scheme, (4) double bus–single breaker scheme, (5) ring bus scheme, and (6) breaker-and-a-half scheme.

Figure 4.19 shows a typical single bus scheme; Figure 4.20 shows a typical double bus–double breaker scheme; Figure 4.21 illustrates a typical main-and-transfer bus scheme; Figure 4.22 shows a typical double bus–single breaker scheme; Figure 4.23 shows a typical ring bus scheme; Figure 4.24 illustrates a typical breaker-and-a-half scheme.

Each scheme has some advantages and disadvantages depending on the economical justification of a specific degree of reliability. Table 4.1 gives a summary of switching schemes' advantages and disadvantages.

4.5 SUBSTATION LOCATION

The location of a substation is dictated by the voltage levels, voltage regulation considerations, subtransmission costs, substation costs, and the costs of primary feeders, mains, and distribution transformers. It is also restricted by other factors, as explained in Chapter 1, which may not be technical in nature.

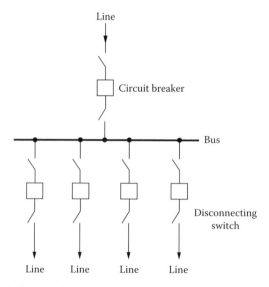

FIGURE 4.19 A typical single bus scheme.

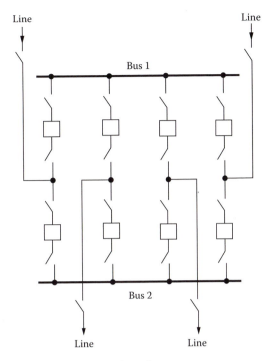

FIGURE 4.20 A typical double bus–double breaker scheme.

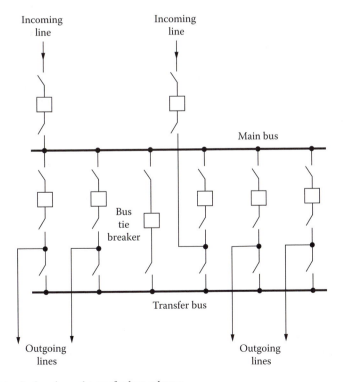

FIGURE 4.21 A typical main-and-transfer bus scheme.

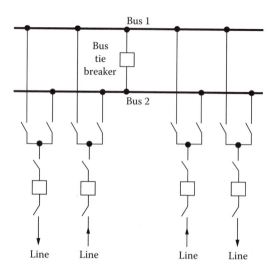

FIGURE 4.22   A typical double bus–single breaker scheme.

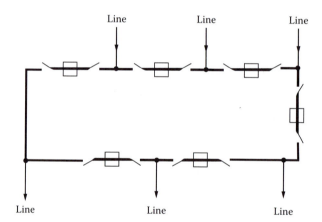

FIGURE 4.23   A typical ring bus scheme.

However, to select an ideal location for a substation, the following rules should be observed [2]:

1. Locate the substation as much as feasible close to the load center of its service area, so that the addition of load times distance from the substation is a minimum.
2. Locate the substation such that proper voltage regulation can be obtained without taking extensive measures.
3. Select the substation location such that it provides proper access for incoming subtransmission lines and outgoing primary feeders.
4. The selected substation location should provide enough space for the future substation expansion.
5. The selected substation location should not be opposed by land-use regulations, local ordinances, and neighbors.
6. The selected substation location should help minimize the number of customers affected by any service discontinuity.
7. Other considerations, such as adaptability and emergency.

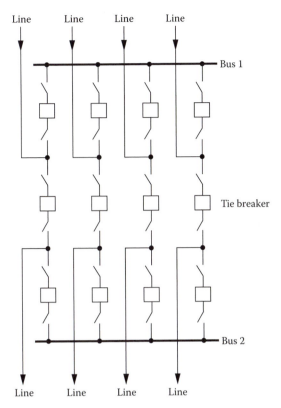

FIGURE 4.24   A typical breaker-and-a-half scheme.

4.6   RATING OF A DISTRIBUTION SUBSTATION

The additional capacity requirements of a system with increasing load density can be met by

1. Either holding the service area of a given substation constant and increasing its capacity
2. Or developing new substations and thereby holding the rating of the given substation constant

It is helpful to assume that the system changes (1) at constant load density for short-term distribution planning and (2) at increasing load density for long-term planning. Further, it is also customary and helpful to employ geometric figures to represent substation service areas, as suggested by Van Wormer [3], Denton and Reps [4], and Reps [5]. It simplifies greatly the comparison of alternative plans that may require different sizes of distribution substation, different numbers of primary feeders, and different primary-feeder voltages.

Reps [5] analyzed a square-shaped service area representing a part of, or the entire service area of, a distribution substation. It is assumed that the square area is served by four primary feeders from a central feed point, as shown in Figure 4.25. Each feeder and its laterals are of three phase. Dots represent balanced three-phase loads lumped at that location and fed by distribution transformers.

Here, the percent voltage drop from the feed point a to the end of the last lateral at c is

$$\%\mathrm{VD}_{ac} = \%\mathrm{VD}_{ab} + \%\mathrm{VD}_{bc}$$

TABLE 4.1

Summary of Comparison of Switching Schemes[a]

Switching Scheme	Advantages	Disadvantages
1. Single bus	1. Lowest cost	1. Failure of bus or any circuit breaker results in shutdown of entire substation 2. Difficult to do any maintenance 3. Bus cannot be extended without completely de-energizing substation 4. Can be used only where loads can be interrupted or have other supply arrangements
2. Double bus–double breaker	1. Each circuit has two dedicated breakers 2. Has flexibility in permitting feeder circuits to be connected to either bus 3. Any breaker can be taken out of service for maintenance 4. High reliability	1. Most expensive 2. Would lose half the circuits for breaker failure if circuits are not connected to both buses
3. Main-and-transfer	1. Low initial and ultimate cost 2. Any breaker can be taken out of service for maintenance 3. Potential devices may be used on the main bus for relaying	1. Requires one extra breaker for the bus tie 2. Switching is somewhat complicated when maintaining a breaker 3. Failure of bus or any circuit breaker results in shutdown of entire substation
4. Double bus–single breaker	1. Permits some flexibility with two operating buses 2. Either main bus may be isolated for maintenance 3. Circuit can be transferred readily from one bus to the other by use of bus-tie breaker and bus selector disconnect switches	1. One extra breaker is required for the bus tie 2. Four switches are required per circuit 3. Bus protection scheme may cause loss of substation when it operates if all circuits are connected to that bus 4. High exposure to bus faults 5. Line breaker failure takes all circuits connected to that bus out of service 6. Bus-tie breaker failure takes entire substation out of service
5. Ring bus	1. Low initial and ultimate cost 2. Flexible operation for breaker maintenance 3. Any breaker can be removed for maintenance without interrupting load 4. Requires only one breaker per circuit 5. Does not use main bus 6. Each circuit is fed by two breakers 7. All switching is done with breakers	1. If a fault occurs during a breaker maintenance period, the ring can be separated into two sections 2. Automatic reclosing and protective relaying circuitry rather complex 3. If a single set of relays is used, the circuit must be taken out of service to maintain the relays (common on all schemes) 4. Requires potential devices on all circuits since there is no definite potential reference point. These devices may be required in all cases for synchronizing, live line, or voltage indication 5. Breaker failure during a fault on one of the circuits causes loss of one additional circuit owing to operation of breaker-failure relaying

TABLE 4.1 (continued)
Summary of Comparison of Switching Schemes[a]

Switching Scheme	Advantages	Disadvantages
6. Breaker-and-a-half	1. Most flexible operation 2. High reliability 3. Breaker failure of bus side breakers removes only one circuit from service 4. All switching is done with breakers 5. Simple operation; no disconnect switching required for normal operation 6. Either main bus can be taken out of service at any time for maintenance 7. Bus failure does not remove any feeder circuits from service	1. 1½ breakers per circuit 2. Relaying and automatic reclosing are somewhat involved since the middle breaker must be responsive to either of its associated circuits

Source: Fink, D.G. and Beaty, H.W., *Standard Handbook for Electrical Engineers*, 11th edn., McGraw-Hill, New York, 1978.

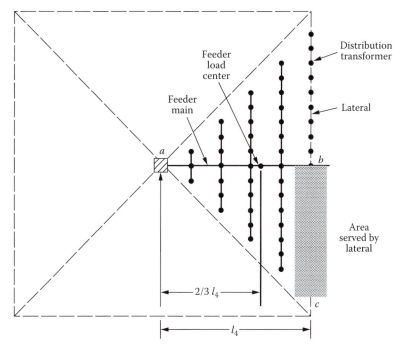

FIGURE 4.25 Square-shaped distribution substation service area. (Based on Westinghouse Electric Corporation, *Electric Utility Engineering Reference Book-Distribution Systems*, Vol. 3, East Pittsburgh, PA, 1965.)

Reps [5] simplified the previously mentioned voltage-drop calculation by introducing a constant *K* that can be defined as *percent voltage drop per kilovolt-ampere-mile*. Figure 4.26 gives the *K* constant for various voltages and copper conductor sizes. Figure 4.26 is developed for three-phase overhead lines with an equivalent spacing of 37 in. between phase conductors. The following analysis is based on the work done by Denton and Reps [4] and Reps [5].

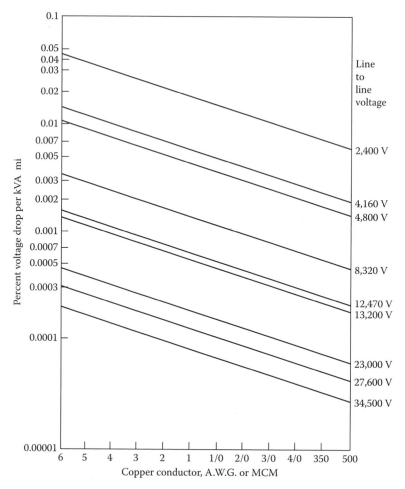

FIGURE 4.26 The K constant for copper conductors, assuming a lagging-load power factor of 0.9.

In Figure 4.25, each feeder serves a total load of

$$S_4 = A_4 \times D \text{ kVA} \tag{4.1}$$

where
S_4 is the kilovolt-ampere load served by one of four feeders emanating from a feed point
A_4 is the area served by one of four feeders emanating from a feed point, mi^2
D is the load density, kVA/mi^2

Equation 4.1 can be rewritten as

$$S_4 = l_4^2 \times D \text{ kVA} \tag{4.2}$$

since

$$A_4 = l_4^2 \tag{4.3}$$

where l_4 is the linear dimension of the primary-feeder service area, mi. Assuming uniformly distributed load, that is, equally loaded and spaced distribution transformers, the voltage drop in the primary-feeder main is

$$\%VD_{4,\text{main}} = \frac{2}{3} \times l_4 \times K \times S_4 \qquad (4.4)$$

or substituting Equation 4.2 into Equation 4.4,

$$\%VD_{4,\text{main}} = 0.667 \times K \times D \times l_4^3 \qquad (4.5)$$

In Equations 4.4 and 4.5, it is assumed that the total or lumped-sum load is located at a point on the main feeder at a distance of $(2/3) \times l_4$ from the feed point a.

Reps [5] extends the discussion to a hexagonally shaped service area supplied by six feeders from the feed point that is located at the center, as shown in Figure 4.27. Assume that each feeder service area is equal to one-sixth of the hexagonally shaped total area, or

$$A_6 = \frac{l_6}{\sqrt{3}} \times l_6$$

$$= 0.578 \times l_6^2 \qquad (4.6)$$

where

A_6 is the area served by one of six feeders emanating from a feed point, mi^2

l_6 is the linear dimension of a primary-feeder service area, mi

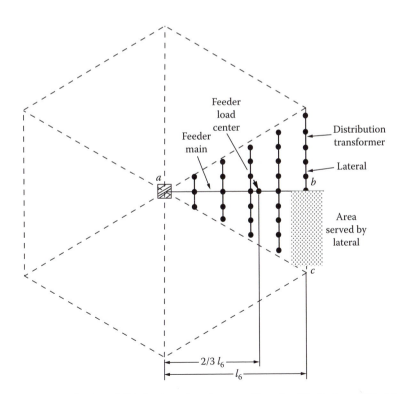

FIGURE 4.27 Hexagonally shaped distribution substation area. (Based on Westinghouse Electric Corporation, *Electric Utility Engineering Reference Book-Distribution Systems*, Vol. 3, East Pittsburgh, PA, 1965.)

Here, each feeder serves a total load of

$$S_6 = A_6 \times D \, \text{kVA} \tag{4.7}$$

or substituting Equation 4.6 into Equation 4.7,

$$S_6 = 0.578 \times D \times l_6^2 \tag{4.8}$$

As before, it is assumed that the total or lump sum is located at a point on the main feeder at a distance of $(2/3) \times l_6$ from the feed point. Hence, the percent voltage drop in the main feeder is

$$\%\text{VD}_{6,\text{main}} = \frac{2}{3} \times l_6 \times K \times S_6 \tag{4.9}$$

or substituting Equation 4.8 into Equation 4.9,

$$\%\text{VD}_{6,\text{main}} = 0.385 \times K \times D \times l_6^3 \tag{4.10}$$

4.7 GENERAL CASE: SUBSTATION SERVICE AREA WITH n PRIMARY FEEDERS

Denton and Reps [4] and Reps [5] extend the discussion to the general case in which the distribution substation service area is served by n primary feeders emanating from the point, as shown in Figure 4.28. Assume that the load in the service area is uniformly distributed and each feeder serves an area of triangular shape. The differential load served by the feeder in a differential area of dA is

$$dS = D \, dA \, \text{kVA} \tag{4.11}$$

where
  dS is the differential load served by the feeder in the differential area of dA, kVA
  D is the load density, kVA/mil
  dA is the differential service area of the feeder, mi^2

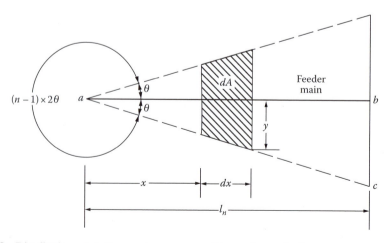

FIGURE 4.28 Distribution substation service area served by n primary feeders.

In Figure 4.28, the following relationship exists:

$$\tan\theta = \frac{y}{x+dx} \tag{4.12}$$

or

$$y = (x+dx)\tan\theta$$
$$\cong x \times \tan\theta \tag{4.13}$$

The total service area of the feeder can be calculated as

$$A_n = \int_{x=0}^{l_n} dA$$
$$= l_n^2 \times \tan\theta \tag{4.14}$$

The total kilovolt-ampere load served by one of n feeders can be calculated as

$$S_n = \int_{x=0}^{l_n} dS$$
$$= D \times l_n^2 \times \tan\theta \tag{4.15}$$

This total load is located, as a lump-sum load, at a point on the main feeder at a distance of $(2/3) \times l_4$ from the feed point a. Hence, the summation of the percent voltage contributions of all such areas is

$$\%VD_n = \frac{2}{3} \times l_n \times K \times S_n \tag{4.16}$$

or, substituting Equation 4.15 into Equation 4.16,

$$\%VD_n = \frac{2}{3} \times K \times D \times l_n^3 \times \tan\theta \tag{4.17}$$

or, since

$$n(2\theta) = 360° \tag{4.18}$$

Equation 4.17 can also be expressed as

$$\%VD_n = \frac{2}{3} \times K \times D \times l_n^3 \times \tan\frac{360°}{2n} \tag{4.19}$$

Equations 4.18 and 4.19 are only applicable when $n \geq 3$. Table 4.2 gives the results of the application of Equation 4.17 to square and hexagonal areas.

TABLE 4.2
Application Results of Equation 4.17

n	θ	$\tan\theta$	$\%\mathrm{VD}_n$
4	45°	1.0	$\dfrac{2}{3}\times K\times D\times l_4^3$
6	30°	$\dfrac{1}{\sqrt{3}}$	$\dfrac{2}{3\sqrt{3}}\times K\times D\times l_6^3$

For $n = 1$, the percent voltage drop in the feeder main is

$$\%\mathrm{VD}_1 = \frac{1}{2}\times K\times D\times l_1^3 \tag{4.20}$$

and for $n = 2$ it is

$$\%\mathrm{VD}_2 = \frac{1}{2}\times K\times D\times l_2^3 \tag{4.21}$$

To compute the percent voltage drop in uniformly loaded lateral, lump and locate its total load at a point halfway along its length and multiply the kilovolt-ampere-mile product for that line length and loading by the appropriate K constant [5].

4.8 COMPARISON OF THE FOUR- AND SIX-FEEDER PATTERNS

For a square-shaped distribution substation area served by four primary feeders, that is, $n = 4$, the area served by one of the four feeders is

$$A_4 = l_4^2 \ \mathrm{mi}^2 \tag{4.22}$$

The total area served by all four feeders is

$$TA_4 = 4A_4$$
$$= 4l_4^2 \ \mathrm{mi}^2 \tag{4.23}$$

The kilovolt-ampere load served by one of the feeders is

$$S_4 = D\times l_4^2 \ \mathrm{kVA} \tag{4.24}$$

Thus, the total kilovolt-ampere load served by all four feeders is

$$TS_4 = 4D\times l_4^2 \ \mathrm{kVA} \tag{4.25}$$

The percent voltage drop in the main feeder is

$$\%\mathrm{VD}_{4,\mathrm{main}} = \frac{2}{3}\times K\times D\times l_4^3 \tag{4.26}$$

The load current in the main feeder at the feed point a is

$$I_4 = \frac{S_4}{\sqrt{3} \times V_{L-L}}$$ (4.27)

or

$$I_4 = \frac{D \times l_4^2}{\sqrt{3} \times V_{L-L}}$$ (4.28)

The ampacity, that is, the current-carrying capacity, of a conductor selected for the main feeder should be larger than the current values that can be obtained from Equations 4.27 and 4.28.

On the other hand, for a hexagonally shaped distribution substation area served by six primary feeders, that is, $n = 6$, the area served by one of the six feeders is

$$A_6 = \frac{1}{\sqrt{3}} \times l_6^2 \text{ mi}^2$$ (4.29)

The total area served by all six feeders is

$$TA_6 = \frac{6}{\sqrt{3}} \times l_6^2 \text{ mi}^2$$ (4.30)

The kilovolt-ampere load served by one of the feeders is

$$S_6 = \frac{1}{\sqrt{3}} D \times l_6^2 \text{ kVA}$$ (4.31)

Therefore, the total kilovolt-ampere load served by all six feeders is

$$TS_6 = \frac{6}{\sqrt{3}} \times D \times l_6^2 \text{ kVA}$$ (4.32)

The percent voltage drop in the main feeder is

$$\%VD_{6,\text{main}} = \frac{2}{3\sqrt{3}} \times K \times D \times l_6^3$$ (4.33)

The load current in the main feeder at the feed point a is

$$I_6 = \frac{S_6}{\sqrt{3} \times V_{L-L}}$$ (4.34)

or

$$I_6 = \frac{D \times l_6^2}{3 \times V_{L-L}}$$ (4.35)

The relationship between the service areas of the four- and six-feeder patterns can be found under two assumptions: (1) feeder circuits are thermally limited (TL) and (2) feeder circuits are voltage-drop-limited (VDL).

1. *For TL feeder circuits*: For a given conductor size and neglecting voltage drop,

$$I_4 = I_6 \tag{4.36}$$

Substituting Equations 4.28 and 4.35 into Equation 4.36,

$$\frac{D \times l_4^2}{\sqrt{3} \times V_{L-L}} = \frac{D \times l_6^2}{3 \times V_{L-L}} \tag{4.37}$$

from Equation 4.37,

$$\left(\frac{l_6}{l_4}\right)^2 = \sqrt{3} \tag{4.38}$$

Also, by dividing Equation 4.30 by Equation 4.23,

$$\frac{TA_6}{TA_4} = \frac{6/\sqrt{3}l_6^2}{4l_4^2}$$

$$= \frac{\sqrt{3}}{2}\left(\frac{l_6}{l_4}\right)^2 \tag{4.39}$$

Substituting Equation 4.38 into Equation 4.39,

$$\frac{TA_6}{TA_4} = \frac{3}{2} \tag{4.40}$$

or

$$TA_6 = 1.50\, TA_4 \tag{4.41}$$

Therefore, the six feeders can carry 1.50 times as much load as the four feeders if they are thermally loaded.

2. *For VDL feeder circuits:* For a given conductor size and assuming equal percent voltage drop,

$$\%VD_4 = \%VD_6 \tag{4.42}$$

Substituting Equations 4.26 and 4.33 into Equation 4.42 and simplifying the result,

$$I_A = 0.833 \times I_6 \tag{4.43}$$

From Equation 4.30, the total area served by all six feeders is

$$TA_6 = \frac{6}{\sqrt{3}} \times l_6^2 \tag{4.44}$$

Substituting Equation 4.43 into Equation 4.23, the total area served by all four feeders is

$$TA_4 = 2.78 \times l_6^2 \qquad (4.45)$$

Dividing Equation 4.44 by Equation 4.45,

$$\frac{TA_6}{TA_4} = \frac{5}{4} \qquad (4.46)$$

or

$$TA_6 = 1.25\,TA_4 \qquad (4.47)$$

Therefore, the six feeders can carry only 1.25 times as much load as the four feeders if they are VDL.

4.9 DERIVATION OF THE *K* CONSTANT

Consider the primary-feeder main shown in Figure 4.29. Here, the effective impedance *Z* of the three-phase main depends upon the nature of the load. For example, when a lumped-sum load is connected at the end of the main, as shown in the figure, the effective impedance is

$$\bar{Z} = z \times l \ \Omega/\text{phase} \qquad (4.48)$$

where
  z is the impedance of three-phase main line, $\Omega/(\text{mi phase})$
  l is the length of the feeder main, mi

When the load is uniformly distributed, the effective impedance is

$$\bar{Z} = \frac{1}{2} \times z \times l \ \Omega/\text{phase} \qquad (4.49)$$

When the load has an increasing load density, the effective impedance is

$$\bar{Z} = \frac{2}{3} \times z \times l \ \Omega/\text{phase} \qquad (4.50)$$

Taking the receiving-end voltage as the reference phasor,

$$\bar{V}_r = V_r \angle 0° \qquad (4.51)$$

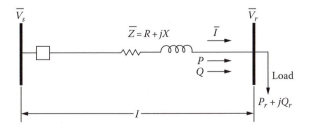

FIGURE 4.29 An illustration of a primary-feeder main.

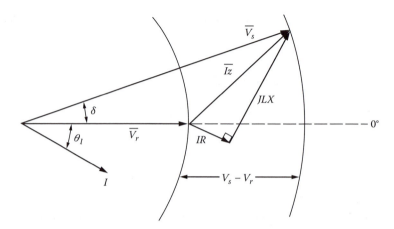

FIGURE 4.30 Phasor diagram.

and from the phasor diagram given in Figure 4.30, the sending-end voltage is

$$\overline{V}_s = V_s \angle \delta \tag{4.52}$$

The current is

$$\overline{I} = I \angle -\theta \tag{4.53a}$$

and the power-factor angle is

$$\theta = \theta_{\overline{V}_r} - \theta_{\overline{I}}$$

$$= 0° - \theta_{\overline{I}} = -\theta_{\overline{I}} \tag{4.53b}$$

and the power factor is a lagging one. When the real power P and the reactive power Q flow in opposite directions, the power factor is a *leading* one.

Here, the per unit voltage regulation is defined as

$$VR_{pu} \triangleq \frac{V_s - V_r}{V_r} \tag{4.54}$$

and the percent voltage regulation is

$$VR_{pu} = \frac{V_s - V_r}{V_r} \times 100 \tag{4.55}$$

or

$$\%VR = VR_{pu} \times 100 \tag{4.56}$$

whereas the per unit voltage drop is defined as

$$VR_{pu} \triangleq \frac{V_s - V_r}{V_B} \tag{4.57}$$

where V_B is normally selected to be V_r.

Hence, the percent voltage drop is

$$\%VD = \frac{V_s - V_r}{V_B} \times 100 \qquad (4.58)$$

or

$$\%VD = VD_{pu} \times 100 \qquad (4.59)$$

where V_B is the arbitrary base voltage. The base secondary voltage is usually selected as 120 V. The base primary voltage is usually selected with respect to the potential transformation (PT) ratio used.

Common PT Ratios	V_B (V)
20	2,400
60	7,200
100	12,000

From Figures 4.29 and 4.30, the sending-end voltage is

$$\overline{V_s} = V_r + \overline{IZ} \qquad (4.60)$$

or

$$V_s = (\cos\delta + j\sin\delta) = V_r \angle 0° + I(\cos\theta - j\sin\theta)(R + jX) \qquad (4.61)$$

The quantities in Equation 4.61 can be either all in per units or in the mks (or SI) system. Use line-to-neutral voltages for single-phase three-wire or three-phase three- or four-wire systems.

In typical *distribution* circuits,

$$R \cong X$$

and the voltage angle δ is closer to zero or typically

$$0° \leq \delta \leq 4°$$

whereas in typical *transmission* circuits,

$$\delta \cong 0°$$

since X is much larger than R.

Therefore, for a typical *distribution* circuit, the $\sin\delta$ can be neglected in Equation 4.61. Hence,

$$V_s \cong V_s \cos\delta$$

and Equation 4.61 becomes

$$V_s \cong V_r + IR\cos\theta + IX\sin\theta \qquad (4.62)$$

Therefore, the per unit voltage drop, for a lagging power factor, is

$$VD_{pu} = \frac{IR\cos\theta + IX\sin\theta}{V_B} \tag{4.63}$$

and it is a positive quantity. The VD_{pu} is negative when there is a leading power factor due to shunt capacitors or when there is a negative reactance X due to series capacitors installed in the circuits.

The complex power at the receiving end is

$$P_r + jQ_r = \bar{V}_r\bar{I}\,{}^* \tag{4.64}$$

Therefore,

$$\bar{I} = \frac{P_r - jQ_r}{\bar{V}_r} \tag{4.65}$$

since

$$\bar{V}_r = V_r\angle 0°$$

Substituting Equation 4.65 into Equation 4.61, which is the exact equation since the voltage angle δ is not neglected, the sending-end voltage can be written as

$$\bar{V}_s = V_r\angle 0° + \frac{RP_r + XQ_r}{V_r\angle 0°} - j\frac{RQ_r - XP_r}{V_r\angle 0°} \tag{4.66}$$

or approximately,

$$\bar{V}_s \cong V_r + \frac{RP_r + XQ_r}{V_r} \tag{4.67}$$

Substituting Equation 4.67 into Equation 4.57,

$$VD_{pu} \cong \frac{RP_r + XQ_r}{V_r V_B} \tag{4.68}$$

or

$$VD_{pu} \cong \frac{(S_r/V_r)R\cos\theta + (S_r/V_r)X\sin\theta}{V_B} \tag{4.69}$$

or

$$VD_{pu} \cong \frac{S_r \times R\cos\theta + S_r \times X\sin\theta}{V_r V_B} \tag{4.70}$$

since

$$P_r = S_r\cos\theta \ \text{W} \tag{4.71}$$

and

$$Q_r = S_r \sin\theta \text{ var} \tag{4.72}$$

Equations 4.69 and 4.70 can also be derived from Equation 7.63, since

$$S_r = V_r I \text{ VA} \tag{4.73}$$

The quantities in Equations 4.68 and 4.70 can be either all in per units or in the SI system. Use the line-to-neutral voltage values and per phase values for the P_r, Q_r, and S_r.

To determine the K constant, use Equation 4.68,

$$VD_{pu} \cong \frac{RP_r + XQ_r}{V_r V_B}$$

or

$$VD_{pu} \cong \frac{(S_{3\phi})(s)(r\cos\theta + x\sin\theta)((1/3)\times 1000)}{V_r V_B} \text{ puV} \tag{4.74}$$

or

$$VD_{pu} = s \times K \times S_{3\phi} \text{ puV} \tag{4.75}$$

or

$$VD_{pu} = s \times K \times S_n \text{ puV} \tag{4.76}$$

where

$$K \cong \frac{(r\cos\theta + x\sin\theta)((1/3)\times 1000)}{V_r V_B} \tag{4.77}$$

Therefore,

$$K = f (\text{conductor size, spacing, } \cos\theta, V_B)$$

and it has the unit of

$$\frac{VD_{pu}}{\text{Arbitrary no. of kVA} \cdot \text{mi}}$$

To get the percent voltage drop, multiply the right side of Equation 4.77 by 100, so that

$$K \cong \frac{(r\cos\theta + x\sin\theta)((1/3)\times 1000)}{V_r V_B} \times 100 \tag{4.78}$$

which has the unit of

$$\frac{\%VD}{\text{Arbitrary no. of kVA} \cdot \text{mi}}$$

In Equations 4.74 through 4.76, s is the effective length of the feeder main that depends upon the nature of the load. For example, *when the load is connected at the end of the main as lumped sum,* the effective feeder length is

$$s = 1 \text{ unit length}$$

when the load is uniformly distributed along the main,

$$s = \frac{1}{2} \times l \text{ unit length}$$

when the load has an increasing load density,

$$s = \frac{2}{3} \times l \text{ unit length}$$

Example 4.2

Assume that a three-phase 4.16 kV wye-grounded feeder main has #4 copper conductors with an equivalent spacing of 37 in. between phase conductors and a lagging-load power factor of 0.9.

    a. Determine the K constant of the main by employing Equation 4.77.
    b. Determine the K constant of the main by using the precalculated percent voltage drop per kilovolt-ampere-mile curves and compare it with the one found in part (a).

Solution
    a. From Equation 4.77,

$$K \cong \frac{(r\cos\theta + x\sin\theta)((1/3)\times 1000)}{V_r V_B}$$

    where
        $r = 1.503 \ \Omega/\text{mi}$ from Table A.1 for 50°C and 60 Hz
        $x_L = x_a + x_d = 0.7456 \ \Omega/\text{mi}$
        $x_a = 0.609 \ \Omega/\text{mi}$ from Table A.1 for 60 Hz
        $x_d = 0.1366 \ \Omega/\text{mi}$ from Table A.10 for 60 Hz and 37 –in. spacing $\cos\theta = 0.9$, lagging
        $V_r = V_B = 2400$ V, line-to-neutral voltage

    Therefore, the per unit voltage drop per kilovolt-ampere-mile is

$$K \cong \frac{(1.503\times 0.9 + 0.7456 \times 0.4359)((1/3)\times 1000)}{2400^2}$$

$$\cong 0.0001 \text{ VD}_{pu}/(\text{kVA}\cdot\text{mi})$$

or

$$K \cong 0.01\%VD/(kVA \cdot mi)$$

b. From Figure 4.26, the K constant for #4 copper conductors is

$$K \cong 0.01\%VD/(kVA \cdot mi)$$

which is the same as the one found in part (a).

Example 4.3

Assume that the feeder shown in Figure 4.31 has the same characteristics as the one in Example 4.2, and a lumped-sum load of 500 kVA with a lagging-load power factor of 0.9 is connected at the end of a 1 mi long feeder main. Calculate the percent voltage drop in the main.

Solution

The percent voltage drop in the main is

$$\%VD = s \times K \times S_n$$

$$= 1.0\,mi \times 0.01\%VD/(kVA \times mi) \times 500\,kVA$$

$$= 5.0\%$$

Example 4.4

Assume that the feeder shown in Figure 4.32 has the same characteristics as the one in Example 4.3, but the 500 kVA load is uniformly distributed along the feeder main. Calculate the percent voltage drop in the main.

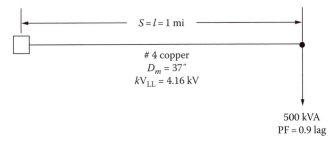

FIGURE 4.31   The feeder of Example 4.2.

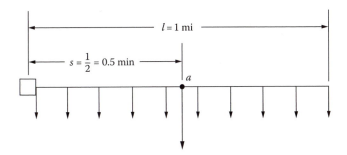

FIGURE 4.32   The feeder of Example 4.4.

Solution

The percent voltage drop in the main is

$$\%VD = s \times K \times S_n$$

where the effective feeder length s is

$$s = \frac{1}{2} = 0.5 \, \text{mi}$$

Therefore,

$$\%VD = s \times K \times S_n$$

$$= 0.5\,\text{mi} \times 0.01\,\%VD/(\text{kVA} \times \text{mi}) \times 500 \, \text{kVA}$$

$$= 2.5\%$$

Therefore, it can be seen that the negative effect of the lumped-sum load on the %VD is worse than the one for the uniformly distributed load. Figure 4.32 also shows the conversion of the uniformly distributed load to a lumped-sum load located at point a for the voltage-drop calculation.

Example 4.5

Assume that the feeder shown in Figure 4.33 has the same characteristics as the one in Example 4.3, but the 500 kVA load has an increasing load density. Calculate the percent voltage drop in the main.

Solution

The percent voltage drop in the main is

$$\%VD = s \times K \times S_n$$

where the effective feeder length s is

$$s = \frac{2}{3}\ell = 0.6667 \, \text{mi}$$

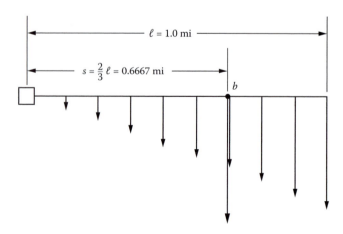

FIGURE 4.33   The feeder of Example 4.5.

Therefore,

$$\%VD = \frac{2}{3}\ell \times K \times S_n$$

$$= 0.6667\,\text{mi} \times 0.01\%VD/(kVA \times mi) \times 500\,kVA$$

$$= 3.33\%$$

Thus, it can be seen that the negative effect of the load with an increasing load density is worse than the one for the uniformly distributed load but is better than the one for the lumped-sum load. Figure 4.33 also shows the conversion of the load with an increasing load density to a lumped-sum load located at point b for the voltage-drop calculation.

Example 4.6

Use the results of the calculations of Examples 4.3 through 4.5 to calculate and compare the percent voltage drop ratios and reach conclusions.

Solution

a. The ratio of the percent voltage drop for the lumped-sum load to the one for the uniformly distributed load is

$$\frac{\%VD_{lumped}}{\%VD_{uniform}} = \frac{5.0}{2.5} = 2.0 \tag{4.79}$$

Therefore,

$$\%VD_{lumped} = 2.0(\%VD_{uniform}) \tag{4.80}$$

b. The ratio of the percent voltage drop for the lumped-sum load to the percent voltage drop for the load with increasing load density is

$$\frac{\%VD_{lumped}}{\%VD_{increasing}} = \frac{5.0}{3.33} = 1.5 \tag{4.81}$$

Therefore,

$$\%VD_{lumped} = 1.5(\%VD_{increasing}) \tag{4.82}$$

c. The ratio of the percent voltage drop for the load with increasing load density to the one for the uniformly distributed load is

$$\frac{\%VD_{increasing}}{\%VD_{uniform}} = \frac{3.33}{2.50} = 1.33 \tag{4.83}$$

Therefore,

$$\% VD_{inceasing} = 1.33(\%VD_{uniform}). \tag{4.84}$$

4.10 SUBSTATION APPLICATION CURVES

Reps [5] derived the following formula to relate the application of distribution substations to load areas:

$$\%VD_n = \frac{((2/3) \times \ell_n) K (n \times D \times A_n)}{n} \tag{4.85}$$

where

$\%VD_n$ is the percent voltage drop in primary-feeder circuit
$2/3 \times \ell_n$ is the effective length of primary feeder
K is the %VD/(kVA·mi) of the feeder
A_n is the area served by one feeder
n is the number of primary feeders
D is the load density

Reps [5] and Denton and Reps [4] developed an alternative form of Equation 4.85 as

$$\%VD_n = \frac{TS_n^{3/2}}{n^{3/2} \times D^{1/2}} \frac{(2/3) \times K}{(\tan \theta)^{1/2}} \tag{4.86}$$

where TS_n is the total kVA supplied from a substation ($= n \times D \times A_n$).

Based on Equation 4.86, they have developed the distribution substation application curves, as shown in Figures 4.34 and 4.35. These application curves relate the load density, substation load kilovolt-amperes, primary-feeder voltage, and permissible feeder loading.

The distribution substation application curves are based on the following assumptions [5]:

1. #4/0 AWG copper conductors are used for the three-phase primary-feeder mains.
2. #4 AWG copper conductors are used for the three-phase primary-feeder laterals.
3. The equivalent spacing between phase conductors is 37 in.
4. A lagging-load power factor of 0.9.

The curves are the plots of number of primary feeders n versus load density D for numerous values of TS_n, that is, total kilovolt-ampere loading of all n primary feeders including a pattern serving the load area of a substation or feed point. In Figures 4.34 and 4.35, the curves for n versus D are given for both constant TS_n and constant TA_n, that is, total area served by all n feeders emanating from the feed point or substation. The curves are drawn for five primary-feeder voltage levels and for two different percent voltage drops, that is, 3% and 6%. The percent voltage drop is [5] measured from the feed point or distribution substation bus to the last distribution transformer on the farthest lateral on a feeder.

The combination of distribution substations and primary feeders applied in a given system is generally designed to give specified percent voltage drop or a specified kVA loading in primary feeders. In areas where load density is light and primary feeders must cover long distances, the allowable maximum percent voltage in a primary feeder usually determines the kVA loading limit on that feeder. In areas where load density is relatively heavy and primary feeders are relatively short, the maximum allowable loading on a primary feeder is usually governed by its current-carrying capacity, which may be attained as a feeder becomes more heavily loaded and before voltage drop becomes a problem.

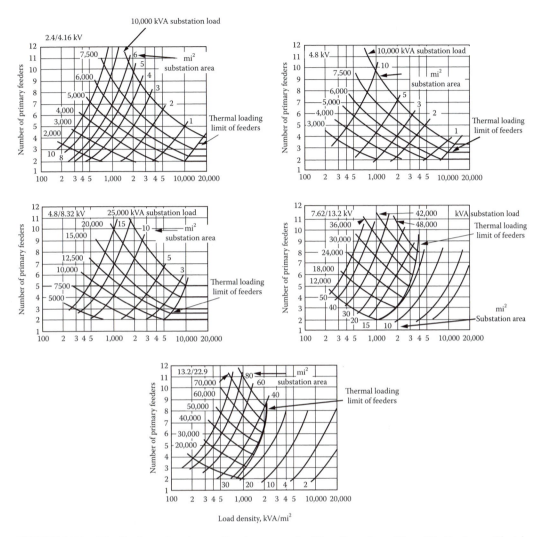

FIGURE 4.34 Distribution substation application curves for 3% voltage drop. (From Westinghouse Electric Corporation, *Electric Utility Engineering Reference Book-Distribution Systems*, Vol. 3, East Pittsburgh, PA, 1965.)

The application curves readily show whether the loading of primary feeders in a given substation area is limited by voltage drop or feeder current-carrying capacity. For each substation or feed point kVA loading, a curve of constant loading may be followed (from upper left toward lower right) as load density increases. As such a curve is followed, load density increases, and the number of primary feeders required to serve that load decreases. But eventually the number of primary feeders diminishes to the minimum number required to carry the given kVA load from the standpoint of feeder current-carrying, or kVA thermal, capacity. Further decrease in the number of primary feeders is not permissible, and the line of constant feed-point loading abruptly changes slope and becomes horizontal. For the horizontal portion of the curve, feeder loading is constant, but percent voltage drop decreases as load density increases. Hence, each set of planning curves may be divided into two general regions, one region in which voltage drop is constant and the other region within which primary-feeder loading is constant. In the region of constant primary-feeder loading, percent voltage drop decreases as load density increases.

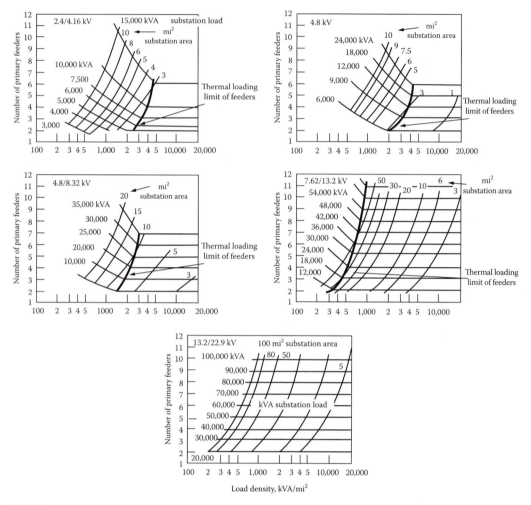

FIGURE 4.35 Distribution substation application curves for 6% voltage drop. (From Westinghouse Electric Corporation, *Electric Utility Engineering Reference Book-Distribution Systems*, Vol. 3, East Pittsburgh, PA, 1965.)

Example 4.7

Refer to previous text, and note that the distribution substation application curves, given in Figures 4.34 and 4.35, are valid only for the conductor sizes, spacing, and load power factor stated.

    a. Use the substation application curves and the data given in Table 4.3 for eight different cases and determine (1) the substation sizes, (2) the required number of feeders, and (3) whether the feeders are TL or VDL. Tabulate the results.

    b. In case thermally loaded or TL feeders are encountered, attempt to deduce if it is the #4/0 AWG copper main or the #4 AWG copper lateral that is TL. Show and explain your reasoning and calculations.

Solution

    a. For case # 1, the total substation kilovolt-ampere load is

$$TS_n = D \times TA_n$$

$$= 500 \times 6.0 = 3000 \text{ kVA}$$

TABLE 4.3
Data for Example 4.7

Case No.	Load Density D, kVA/mi^2	Substation Area Coverage TA_n, mi^2	Maximum Total Primary Feeder, %VD	Base Feeder Voltage, kV$_{L-L}$
1	500	6.0	3.0	4.16
2	500	6.0	6.0	4.16
3	2,000	3.0	3.0	4.16
4	2,000	3.0	6.0	4.16
5	10,000	1.0	3.0	4.16
6	10,000	1.0	6.0	4.16
7	2,000	15.0	3.0	13.2
8	2,000	15.0	6.0	13.2

From the appropriate figure (the one with 3.0% voltage drop and 4.16 kV line-to-line voltage base) among the figures given in Figure 4.34, for 3000 kVA substation load, 500 kVA/mi^2 load density, and 6.0 mi^2 substation area coverage, the number of required feeders can be found as 3.8 or 4. Since the corresponding point in the figure is located on the left-hand side of the curve for the thermal-loading limit of feeders (the one with darker line), the feeders are VDL. The remaining cases can be answered in a similar manner as given in Table 4.4. Note that cases # 6 and 8 are of TL feeders since their corresponding points are located on the right-hand side of the thermal-loading limit curves.

b. Cases # 6 and 8 have feeders that are thermally loaded. From Table A.1, the conductor ampacities for a #4/0 copper main and a #4 copper lateral can be found as 480 and 180 A, respectively.
For case 6, the kilovolt-ampere load of one feeder is

$$S_n = \frac{TS_n}{n}$$

$$= \frac{10,000 \text{ kVA}}{4} = 2500 \text{ kVA}$$

Therefore, the load current is

$$I = \frac{S_n}{\sqrt{3} \times V_{L-L}}$$

$$= \frac{2500 \text{ kVA}}{\sqrt{3} \times 4.16 \text{ kV}} = 347.4 \text{ A}$$

TABLE 4.4
Cases of Example 4.7

Case No.	Substation Size TS_n	Required No. of Feeders, n	VDL or TL Feeders
1	3,000	3.8 (or 4)	VDL
2	3,000	2	VDL
3	6,000	5	VDL
4	6,000	3	VDL
5	10,000	5	VDL
6	10,000	4	TL
7	30,000	5.85 (or 6)	VDL
8	30,000	5	TL

Since the conductor ampacity of the lateral is less than the load current, it is TL but not the main feeder.

For case # 8, the kilovolt-ampere load of one feeder is

$$S_n = \frac{30,000 \text{ kVA}}{5} = 6000 \text{ kVA}$$

The load current is

$$I = \frac{6000 \text{ kVA}}{\sqrt{3} \times 13.2 \text{ kV}} = 262.4 \text{ A}$$

Therefore, only the lateral is TL.

4.11   INTERPRETATION OF PERCENT VOLTAGE DROP FORMULA

Equation 4.85 can be rewritten in alternative forms to illustrate the interrelationship of several parameters guiding the application of distribution substations to load areas

$$\% \text{ VD}_n = \frac{((2/3) \times \ell_n) K (n \times D \times A_n)}{n}$$

$$= \frac{((2/3) \times \ell_n \times K) TS_n}{n}$$

$$= ((2/3) \times \ell_n \times K) S_n$$

where
   $\%VD_n$ is the percent voltage drop in primary-feeder circuit
   $2/3 \times \ell_n$ is the effective length of primary feeder
   $TS_n = n \times D \times A_n$ is the total kVA supplied from feed point
   K is the %VD/(kVA·mi) of the feeder
   A_n is the area served by one feeder
   n is the number of primary feeders
   D is the load density

To illustrate the use and interpretation of the equation, assume five different cases, as shown in Table 4.5.

Case # 1 represents an increasing service area as a result of geographic extensions of a city. If the length of the primary feeder is doubled (shown in the table by × 2), holding everything else constant, the service area A_n of the feeder increases four times, which in turn increases TS_n and S_n four times, causing the $\%VD_n$ in the feeder to increase eight times. Therefore, increasing the feeder length should be avoided as a remedy due to the severe penalty.

Case # 2 represents load growth due to load density growth. For example, if the load density is doubled, it causes TS_n and S_n to be doubled, which in turn increases the $\%VD_n$ in the feeder to be doubled. Therefore, increasing load density also has a negative effect on the voltage drop.

Case # 3 represents the addition of new feeders. For example, if the number of the feeders is doubled, it causes S_n to be reduced by half, which in turn causes the $\%VD_n$ to be reduced by half. Therefore, new feeder additions help to reduce the voltage drop.

TABLE 4.5

Illustration of the Use and Interpretation of Equation 4.85

Case	ℓ_n	K	Base, kV_{L-L}	N	D	A_n	TS_n	S_n	$\%VD_n$
1. Geographic extensions	×2 ↑	×1	×1	×1	×1	×4 ↑	×4 ↑	×4 ↑	×8 ↑
2. Load growth	×1	×1	×1	×1	×2 ↑	×1	×2 ↑	×2 ↑	×2 ↑
3. Add new feeders	×1	×1	×1	×2 ↑	×1	$\times\frac{1}{2}$ ↓	×1	$\times\frac{1}{2}$ ↓	$\times\frac{1}{2}$ ↓
4. Feeder reconductoring	×1	$\times\frac{1}{2}$ ↓	×1	×1	×1	×1	×1	×1	$\times\frac{1}{2}$ ↓
5. Δ-to-Y grounded conversion	×1	$\times\frac{1}{3}$ ↓	$\times\sqrt{3}$ ↑	×1	×1	×1	×1	×1	$\times\frac{1}{3}$ ↓

Case # 4 represents feeder reconductoring. For example, if the conductor size is doubled, it reduces the K constant by half, which in turn reduces the $\%VD_n$ by half.

Case # 5 represents the delta-to-grounded-wye conversion. It increases the line-to-line base kilovoltage by $\sqrt{3}$, which in turn decreases the K constant, causing the $\%VD_n$ to decrease to one-third its previous value.

Example 4.8

To illustrate distribution substation sizing and spacing, assume a square-shaped distribution substation service area as shown in Figure 4.25. Assume that the substation is served by four three-phase four-wire 2.4/4.16 kV grounded-wye primary feeders. The feeder mains are made of either #2 AWG copper or #1/0 ACSR conductors. The three-phase open-wire overhead lines have a geometric mean spacing of 37 in. between phase conductors. Assume a lagging-load power factor of 0.9 and a 1000 kVA/mi² uniformly distributed load density. Calculate the following:

a. Consider thermally loaded feeder mains and find the following:
    i.   Maximum load per feeder
    ii. Substation size
    iii. Substation spacing, both ways
    iv. Total percent voltage drop from the feed point to the end of the main
b. Consider VDL feeders that have 3% voltage drop and find the following:
    i.   Substation spacing, both ways
    ii. Maximum load per feeder
    iii. Substation size
    iv. Ampere loading of the main in per unit of conductor ampacity
c. Write the necessary codes to solve the problem in MATLAB®.

Solution

From Tables A.1 and A.5, the conductor ampacities for #2 AWG copper and #1/0 ACSR conductors can be found as 230 A.

a. *Thermally loaded mains*:
    i.   Maximum load per feeder is

$$S_n = \sqrt{3} \times V_{L-L} \times I_{max}$$

$$= \sqrt{3} \times 4.16 \times 230 = 1657.2 \text{ kVA}$$

   ii. Substation size is

$$TS_n = 4 \times S_n$$

$$= 4 \times 1657.2 = 6628.8\,kVA$$

   iii. Substation spacing, both ways, can be found from

$$S_n = A_n \times D$$

$$= I_4^2 \times D$$

or

$$I_4 = \left(\frac{S_n}{D}\right)^{1/2}$$

$$= \left(\frac{1657.2\,kVA}{1000\,kVA/mi^2}\right)^{1/2}$$

$$= 1.287\,mi$$

Therefore,

$$2I_4 = 2 \times 1.287$$

$$= 2.575\,mi$$

   iv. Total percent voltage drop in the main is

$$\%\,VD_n = \frac{2}{3} \times K \times D \times I_4^3$$

$$= \frac{2}{3} \times 0.007 \times 1000 \times (1.287)^3$$

$$= 9.95\%$$

   where K is 0.007 and found from Figure 4.26.
 b. VDL *feeders*:
   i. Substation spacing, both ways, can be found from

$$\%\,VD_n = \frac{2}{3} \times K \times D \times I_4^3$$

or

$$I_4 = \left(\frac{3 \times \%VD_n}{2 \times K \times D}\right)^{1/3}$$

$$= \left(\frac{3 \times 3}{2 \times 0.007 \times 1000}\right)^{1/3}$$

$$= 0.86\,mi$$

Therefore,

$$2l_4 = 2 \times 0.86$$

$$= 1.72 \, \text{mi}$$

ii.   Maximum load per feeder is

$$S_n = D \times l_4^2$$

$$= 1000 \times (0.86)^2 \cong 750 \, \text{kVA}$$

iii.   Substation size is

$$TS_n = 4 \times S_n$$

$$= 4 \times 750 = 3000 \, \text{kVA}$$

iv.   Ampere loading of the main is

$$I = \frac{S_n}{\sqrt{3} \times V_{L-L}}$$

$$= \frac{750 \, \text{kVA}}{\sqrt{3} \times 4.16 \, \text{kV}}$$

$$= 104.09 \, \text{A}$$

Therefore, the ampere loading of the main in per unit of conductor ampacity is

$$I_{pu} = \frac{104.09 \, \text{A}}{230 \, \text{A}}$$

$$= 0.4526 \, \text{pu}$$

c. *Here is the MATLAB script*:

~~~~~~

```
clc
clear

% System parameters
VLL = 4.16;% kV
Iamp = 230;% ampacity from Tables A.1 and A.5
D = 1000;% uniformly distributed load density in kVA/mi^2
K = 0.007;% from Figure 4.17
pVDn_b = 3;% voltage-drop-limited feeders

% Solution for part a (thermally loaded mains)

% (i) Maximum load per feeder
Sn_a = sqrt(3)*VLL*Iamp

% (ii) Substation size is
TSn_a = 4*Sn_a

% (iii) Substation spacing, both ways
l4 = sqrt(Sn_a/D)
lsp_a = 2*l4
```

```
% (iv) Substation spacing
pVDn_a = (2/3)*K*D*14^3

% Solution for part b (voltage-drop-limited feeders)

% (i) Substation spacing, both ways
14 = ((3*pVDn_b)/(2*K*D))^(1/3)
lsp_b = 2*14

% (ii) Maximum load per feeder
Sn_b = D*14^2

% (iii) Substation size is
TSn_b = 4*Sn_b

% (iv) Ampere loading of the mains
I = Sn_b/(sqrt(3)*VLL)
Ipu = I/Iamp
```

Example 4.9

Assume a square-shaped distribution substation service area as shown in Figure 4.36. The square area is 4 mi and has numerous three-phase laterals. The designing distribution engineer has the following design data that are assumed to be satisfactory estimates.

The load is uniformly distributed, and the connected load density is 2000 kVA/mi². The demand factor, which is an average value for all loads, is 0.60. The diversity factor among all loads in the area is 1.20. The load power factor is 0.90 lagging, which is an average value applicable for all loads.

For some unknown reasons (perhaps, due to the excessive distance from load centers or transmission lines or other limitations, such as availability of land, its cost, and land-use ordinances and regulations), the only available substation sites are at locations A and B. If the designer selects site A as the substation location, there will be a 2 mi long feeder main and 16 three-phase 2 mi long laterals.

On the other hand, if the designer selects site B as the substation location, there will be a 3 mi long feeder main (including a 1 mi long express feeder main) and 32 three-phase 1 mi long laterals.

The designer wishes to select the better one of the given two sites by investigating the total peak-load voltage drop at the end of the most remote lateral, that is, at point a.

Assume 7.62/13.2 kV three-phase four-wire grounded-wye primary-feeder mains that are made of #2/0 copper overhead conductors. The laterals are of #4 copper conductors, and they are all three-phase, four-wire, and grounded-wye.

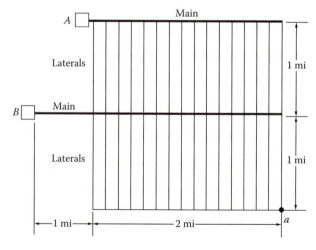

FIGURE 4.36 For Example 4.9.

Using the precalculated percent voltage drop per kilovolt-ampere-mile curves given in Figure 4.26, determine the better substation site by calculating the percent voltage drops at point a that correspond to each substation site and select the better one.

Solution

The maximum diversified demand is

$$\text{Diversified demand} = \frac{\sum_{i=1}^{n} \text{demand factor}_i \times \text{connected load}_i}{\text{diversity factor}}$$

$$= \frac{0.60 \times 2000 \text{ kVA/mi}^2}{1.20}$$

$$= 1000 \text{ kVA/mi}^2$$

The peak loads of the substations A and B are the same

$$TS_n = 1000 \text{ kVA/mi}^2 \times 4 \text{ mi}^2$$

$$= 4000 \text{ kVA}$$

From Figure 4.26, the K constants for #2/0 and #4 conductors are found as 0.0004 and 0.00095, respectively.

The maximum percent voltage drop for substation A occurs at point a, and it is the summation of the percent voltage drops in the main and the last lateral. Therefore,

$$\%VD_a = \frac{1}{2}K_m S_m + \frac{\ell}{2}K_l S_l$$

$$= \frac{2}{2} \times 0.0004 \times 4000 + \frac{2}{2} \times 0.00095 \times \frac{4000}{16}$$

$$\cong 1.84\%$$

The maximum percent voltage drop for substation B also occurs at point a. Therefore,

$$\%VD_a = 2 \times 0.0004 \times 4000 + \frac{\ell}{2} \times 0.00095 \times \frac{4000}{32}$$

$$\cong 3.26\%$$

Therefore, substation site A is better than substation site B from the voltage drop point of view.

Example 4.10

Assume a square-shaped distribution substation service area as shown in Figure 4.37. The four-feeder substation serves a square area of $2a \times 2a$ mi^2.

The load density distribution is D kVA/mi^2 and is uniformly distributed. Each feeder main is three-phase four-wire grounded wye with multigrounded common neutral open-wire line.

Since dimension d is much smaller than dimension a, assume that the length of each feeder main is approximately a mi, and the area served by the last lateral, which is indicated in the figure as the cross-hatched area, is approximately $a \times d$ mi^2. The power factor of all loads is $\cos \theta$ lagging. The impedance of the feeder main line per phase is

$$z_m = r_m + jx_m \; \Omega/\text{mi}$$

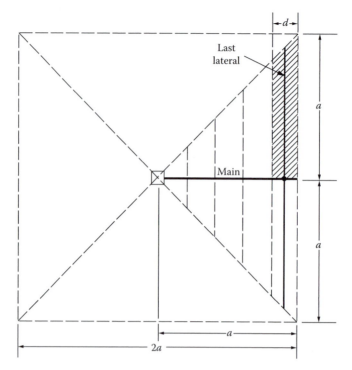

FIGURE 4.37   Service area for Example 4.10.

The impedance of the lateral line per phase is

$$z_l = r_l + jx_l \ \Omega/\text{mi}$$

The V_{L-L} is the base line-to-line voltage in kilovolts, which is also the nominal operating voltage.

   a. Assume that laterals are also three-phase four-wire grounded wye with multigrounded common neutral open-wire line. Show that the percent voltage drop at the end of the last lateral is

$$\%\text{VD} = \frac{2D \times a^3 (r_m \cos\theta + jx_m \sin\theta)}{30 \times V_{L-L}^2} + \frac{D \times a^2 \times d(r_l \cos\theta + jx_l \sin\theta)}{20 \times V_{L-L}^2} \tag{4.87}$$

   b. Assume that the laterals are single-phase two-wire with multigrounded common neutral open-wire line. Apply Morrison's approximation [6] and modify the equation given in part (a).

Solution

   a. The total kilovolt-ampere load served by one main is

$$S_m = D \times \frac{(2a)^2}{4}$$

$$= D \times a^2 \ \text{kVA} \tag{4.88}$$

The current in the main of the substation is

$$I_m = \frac{D \times a^2}{\sqrt{3} \times V_{L-L}}$$

(4.89)

Therefore, the percent voltage drop at the end of the main is

$$
\begin{aligned}
\%VD_m &= \frac{D \times a^2}{\sqrt{3} \times V_{L-L}} (r_m \cos\theta + x_m \sin\theta) \frac{\sqrt{3}}{1000 \times V_{L-L}} \left(\frac{2}{3} \times a\right) 100 \\
&= \frac{2D \times a^3}{30 \times V_{L-L}^2} (r_m \cos\theta + x_m \sin\theta)
\end{aligned}
$$

(4.90)

The kilovolt-ampere load served by the last lateral is

$$S_l = D \times a \times d \text{ kVA}$$

(4.91)

The current in the lateral is

$$I_l = \frac{D \times a \times d}{\sqrt{3} \times V_{L-L}}$$

(4.92)

Thus, the percent voltage drop at the end of the lateral is

$$
\begin{aligned}
\%VD_l &= \frac{D \times a \times d}{\sqrt{3} \times V_{L-L}} (r_l \cos\theta + x_l \sin\theta) \frac{\sqrt{3}}{1000 \times V_{L-L}} \left(\frac{1}{2} \times a\right) 100 \\
&= \frac{D \times a^2 \times d}{20 \times V_{L-L}^2} (r_l \cos\theta + x_l \sin\theta)
\end{aligned}
$$

(4.93)

Therefore, the addition of Equations 4.90 and 4.93 gives Equation 4.87.

b. According to Morrison [6], the percent voltage drop of a single-phase circuit is approximately four times that for a three-phase circuit, assuming the usage of the same-size conductors. Therefore,

$$\%VD_{1\phi} = 4 \times (\%VD_{3\phi})$$

(4.94)

Hence, the percent voltage drop in the main is the same as given in part a, but the percent voltage drop for the lateral is not the same and is

$$
\begin{aligned}
\%VD_{l,1\phi} &= 4 \times \frac{D \times a^2 \times d}{20 \times V_{L-L}^2} (r_l \cos\theta + x_l \sin\theta) \\
&= \frac{D \times a^2 \times d}{5 \times V_{L-L}^2} (r_l \cos\theta + x_l \sin\theta)
\end{aligned}
$$

(4.95)

Thus, the total percent voltage drop will be the sum of the percent voltage drop in the three-phase main, given by Equation 4.90, and the percent voltage drop in the single-phase lateral, given by Equation 4.95. Therefore, the total voltage drop is

$$\%VD = \frac{2D \times a^3}{30 \times V_{L-L}^2} (r_m \cos\theta + x_m \sin\theta) + \frac{D \times a^2 \times d}{5 \times V_{L-L}^2} (r_l \cos\theta + x_l \sin\theta)$$

(4.96)

Example 4.11

Figure 4.38 shows a pattern of service area coverage (not necessarily a good pattern) with primary-feeder mains and laterals. There are five substations shown in the figure, each with two feeder mains. For example, substation A has two mains like A, and each main has many closely spaced laterals such as a–a.

If the laterals are not three phase, the load in the main is assumed to be well balanced among the three phases. The load tapped off the main decreases linearly with the distance s, as shown in Figure 4.39.

Using the following notation and the notation given in the figures, analyze a feeder main:

D = uniformly distributed load density, kVA/mi²
V_{L-L} = base voltage and nominal operating voltage, line-to-line kV
A_2 i = area supplied by one feeder main
TA_2 = area supplied by one substation
S_2 = kVA input at the substation to one feeder main
TS_2 = total kVA load supplied by one substation
K_2 = %VD/(kVA · mi) for conductors and load power factor being considered
z_2 = impedance of three-phase main line, Ω/(mi · phase)
VD_2 = voltage drop at end of main, for example, A_1

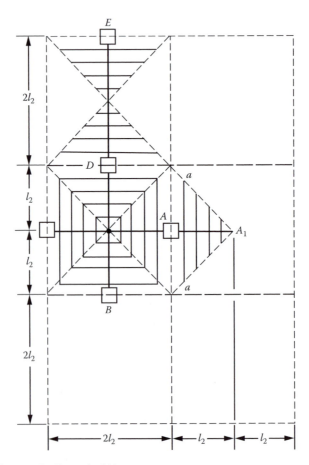

FIGURE 4.38 Service area for Example 4.11.

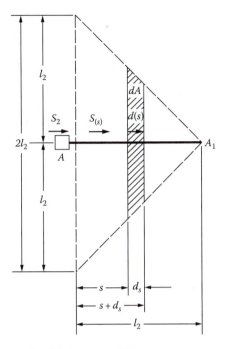

FIGURE 4.39 Linearly decreasing load for Example 4.11.

a. Find the differential area dA and the differential kilovolt-ampere load supplied $d(S)$ shown in Figure 4.41.
b. Find the kVA load flow in the main at any point s, that is, S_s. Express the S_s in terms of S_2, s, and l_2.
c. Find the differential voltage drop at point s and then show that the total load may be concentrated at $s = l_2/3$ for the purpose of computing the VD_2.
d. Suppose that this two-feeders-per-substation pattern is to be implemented with TL, that is, ampacity-loaded, feeders.

Assume that the load density is 500 kVA/mi², the line-to-line voltage is 12.47 kV, and the feeder mains are #4/0 AWG ACSR open-wire lines. Find the substation spacing, both ways, that is, $2l_2$, and the load on the substation transformers, that is, TS_2.

Solution

a. From Figure 4.41, the differential area is

$$dA = 2(l_2 - s)ds \text{ mi}^2 \tag{4.97}$$

Therefore, the differential kilovolt-ampere load supplied is

$$d(S) = 2D(l_2 - s)ds \text{ kVA} \tag{4.98}$$

b. The kilovolt-ampere load flow in the main at any point s is

$$S_s = 2(l_2 - s)^2 D$$

$$= 2(l_2 - s)^2 \times \frac{S_2}{2l_2^2} \text{ kVA} = \left(\frac{l_2 - s}{l_2}\right)^2 \times S_2 \text{ kVA} \tag{4.99}$$

c. The differential current at any point s is

$$I_s = \frac{S_s}{\sqrt{3} \times V_{L-L}} \tag{4.100}$$

Hence, the differential voltage drop at point s is

$$d(\text{VD})_s = I_s \times z_2 ds$$

$$= \frac{S_s}{\sqrt{3} \times V_{L-L}} \times z_2 ds$$

$$= \left(\frac{l_2 - s}{l_2}\right)^2 \left(\frac{S_2}{\sqrt{3} \times V_{L-L}}\right) z_2 ds$$

$$= \frac{S_2 \times z_2}{\sqrt{3} \times V_{L-L} \times l_2^2} \times (l_2 - s)^2 ds \tag{4.101}$$

The integration of either side of Equation 4.101 gives the voltage drop at point s:

$$\text{VD}_s = \int_0^s d(\text{VD})_s$$

$$= \int_0^s \frac{S_2 \times z_2}{\sqrt{3} \times V_{L-L} \times l_2^2}(l_2 - s)^2 ds$$

$$= \frac{S_2 \times z_2}{\sqrt{3} \times V_{L-L} \times l_2^2}\frac{l_2^3}{3} - \frac{S_2 \times z_2}{\sqrt{3} \times V_{L-L} \times l_2^2}\frac{(l_2 - s)^3}{3}$$

$$= \frac{S_2 \times z_2}{3\sqrt{3} \times V_{L-L} \times l_2^2}\left[l_2^3 - (l - s)^3\right] \tag{4.102}$$

When $s = l_2$, Equation 4.102 becomes

$$\text{VD}_2 = \frac{S_2 \times z_2 \times l_2^3}{3\sqrt{3} \times V_{L-L} \times l_2^2}$$

$$= \frac{S_2 \times z_2 \times l_2}{3\sqrt{3} \times V_{L-L}}$$

$$= \frac{S_2}{\sqrt{3} \times V_{L-L}} \times z_2 \times \frac{l_2}{3} \tag{4.103}$$

Therefore, the load has to be lumped at $l_2/3$.

d. From Table A.5, the conductor ampacity for #4/0 AWG ACSR conductor can be found as 340 A. Therefore

$$S_2 = \sqrt{3} \times 12.47\,\text{kV} \times 340\,\text{A}$$

$$\cong 7343.5\,\text{kVA}$$

Since

$$S_2 = D \times l_2^2$$

then

$$I_2 = \left(\frac{S_2}{D}\right)^{1/2}$$

$$= \left(\frac{7343.5\,\text{kVA}}{500\,\text{kVA/mi}^2}\right)$$

$$= 3.83\,\text{mi} \tag{4.104}$$

Therefore, the substation spacing, both ways, is

$$2I_2 = 2 \times 3.83$$

$$= 7.66\,\text{mi}$$

Total load supplied by one substation is

$$TS_2 = 2 \times S_2$$

$$= 2 \times 7343.5$$

$$= 14{,}687\,\text{kVA}$$

Example 4.12

Compare the method of service area coverage given in Example 4.11 with the four-feeders-per-substation pattern of Section 4.6 (see Figure 4.25). Use the same feeder main conductors so that $K_2 = K_4$ and the same line-to-line nominal operating voltage V_{L-L}. Here, let S_4 be the kilovolt-ampere input to one feeder main of the four-feeder substation, and let TS_4, A_4, VD_4, K_4, etc., all pertain similarly to the four-feeder substation. Investigate the VDL feeders and determine the following:

a. Ratio of substation spacings $= 2I_2/2I_4$
b. Ratio of areas covered per feeder main $= A_2/A_4$
c. Ratio of substation loads $= TS_2/TS_4$

Solution

a. Assuming the percent voltage drops and the K constants are the same in both cases,

$$\%VD_2 = \%VD_4$$

and

$$K_2 = K_4$$

where

$$\%VD_2 = \left(D \times I_2^2\right)(K_2)\left(\frac{1}{3}I_2\right)$$

$$= \frac{1}{3}K_2 \times D \times I_2^3$$

and

$$\%VD_4 = \left(D \times I_4^2\right)(K_4)\left(\frac{1}{3}I_4\right)$$

$$= \frac{1}{3}K_4 \times D \times I_4^3$$

Therefore,

$$I_2^3 = 2I_4^3$$

or the ratio of substation spacings is

$$\left(\frac{I_2}{I_4}\right)^3 = 2 \tag{4.105}$$

or, for both ways,

$$\frac{2I_2}{2I_4} \cong 2 \tag{4.106}$$

b. The ratio of areas covered per feeder main is

$$\frac{A_2}{A_4} = \frac{I_2^2}{I_4^2}$$

$$= \left(\frac{I_2}{I_4}\right)^2$$

$$= 2^{2/3}$$

$$\cong 1.59 \tag{4.107}$$

c. The ratio of substation loads is

$$\frac{TS_2}{TS_4} = \frac{2 \times D \times I_2^2}{4 \times D \times I_4^2}$$

$$= \frac{1}{2}\left(\frac{I_2}{I_4}\right)^2$$

$$\cong 0.8 \tag{4.108}$$

4.12 CAPABILITY OF FACILITIES

The capability of distribution substations to supply its service area load is usually determined by the capability of substation transformer banks. Occasionally, the capability of the transmission facilities supplying the substation or the capability of distribution feeders emanating from it will impose a lower limit on the amount of load the substation can supply.

Each substation transformer bank and each feeder have a normal capability, 100°F (40°C), and also an emergency capability that is usually higher. These capabilities are usually determined by the temperature rise limitations and the transformer and feeder components. Thus, they are higher in the warm interior area. In practice, normal and emergency capabilities in kVA of both existing and proposed banks should be computed by using a transformer capability assessment computer program.

Also, the component that limits the capability of a feeder may be the station breaker or the switches associated with it, the underground or overhead conductors, current transformers, metering, or the protective relay setting.

Each component should be checked to determine the amount of current it can carry under normal and emergency conditions. In some cases it will be possible to increase this capability at a relatively small cost by replacing the limiting component or modifying the feeder protective scheme.

According to practices of some utility companies, the capabilities of feeder circuit breakers and associated switches that are in good condition are 100% of their nameplate ratings for summer and winter normal conditions and summer emergency conditions and 110% of their nameplate ratings for winter emergency conditions. If the equipment is not in good condition, it may be necessary to establish lower limits or replace the equipment.

In general, it is a good practice to multiply the ampacities of overhead conductors, switches, and single-phase feeder regulators by 0.95 to permit for phase unbalance and, in cases where the substation is circuit limited, by the coincidence factor between feeders. All the cables in a duct share the heat buildup, so such a multiplier is unnecessary for cables in underground ducts and risers. Furthermore, the phase unbalance multiplier is not used for oil circuit breakers.

Having established the normal and emergency capabilities of feeders in amps, they can be converted to kVA by multiplying by specific factors. For example, nominal circuit voltages of 4,160, 4,800, 12,000, 17,000, and 20,780 V are multiplied by factors of 7.6, 8.73, 21.82, 30.92, and 37.8, respectively. These multiplying factors are based upon input voltage to the feeder of 126 V on a 120 V base. However, the multiplying factor of 0.95 to account for the effect of phase unbalance is not included.

4.13 SUBSTATION GROUNDING

4.13.1 ELECTRIC SHOCK AND ITS EFFECTS ON HUMANS

To properly design a grounding (called *equipment grounding*) for the high-voltage lines and/or substations, it is important to understand the electrical characteristics of the most important part of the circuit, the human body. In general, shock currents are classified based on the degree of severity of the shock they cause. For example, currents that produce direct physiological harm are called primary shock currents. Whereas currents that cannot produce direct physiological harm but may cause involuntary muscular reactions are called *secondary shock currents*. These shock currents can be either steady state or transient in nature. In alternating current (ac) power systems, steady-state currents are sustained currents of 60 Hz or its harmonics. The transient currents, on the other hand, are capacitive discharge currents whose magnitudes diminish rapidly with time.

Table 4.6 gives the possible effects of electrical shock currents on humans. Note that threshold value for a normally healthy person to be able to feel a current is about 1 mA. (Experiments have long ago established the well-known fact that *electrical shock effects are due to current, not voltage* [11].) This is the value of current at which a person is just able to detect a slight tingling sensation on the hands or fingers due to current flow. Currents of approximately 10–30 mA can cause lack of muscular control. In most humans, a current of 100 mA will cause ventricular fibrillation. Currents of higher magnitudes can stop the heart completely or cause severe electrical burns. The ventricular fibrillation is a condition where the heart beats in an abnormal and ineffective manner, with fatal results. Therefore, its threshold is the main concern in grounding design.

Currents of 1 mA or more but less than 6 mA are often defined as the secondary shock currents (*let-go currents*). The let-go current is the maximum current level at which a human holding an

TABLE 4.6

Effect of Electric Current (mA) on Men and Women

		DC		AC (60 Hz)	
Effects		**Men**	**Women**	**Men**	**Women**
1. No sensation on hand		1	0.6	0.4	0.3
2. Slight tingling; per caption threshold		5.2	3.5	1.1	0.7
3. Shock—not painful and muscular control not lost		9	6	1.8	1.2
4. Painful shock—painful but muscular control not lost		62	41	9	6
5. Painful shock—let-go threshold[a]		76	51	16	10.5
6. Painful and severe shock, muscular contractions, breathing difficult		90	60	23	15
7. Possible ventricular fibrillation from short shocks:					
a. Shock duration 0.03 s		1300	1300	1000	1000
b. Shock duration 3.0 s		500	500	100	100
c. Almost certain ventricular fibrillation (if shock duration over one heart beat interval)		1375	1375	275	275

[a] Threshold for 50% of the males and female tested.

energized conductor can control his muscles enough to release it. The 60 Hz minimum required body current leading to possible fatality through ventricular fibrillation can be expressed as

$$I = \frac{0.116}{\sqrt{t}} \, A \qquad (4.109)$$

where t is in seconds in the range from approximately 8.3 ms to 5 s.

The effects of an electric current passing through the vital parts of a human body depend on the duration, magnitude, and frequency of this current. The body resistance considered is usually between two extremities, either from one hand to both feet or from one foot to the other one.

Experiments have shown that the body can tolerate much more current flowing from one leg to the other than it can when current flows from one hand to the legs. Treating the foot as a circular plate electrode gives an approximate resistance of $3\rho_s$, where ρ_s is the soil resistivity. The resistance of the body itself is usually used as about 2300 Ω hand to hand or 1100 Ω hand to foot [12]. However, IEEE Std. 80-1976 [14] recommends the use of 1000 Ω as a reasonable approximation for body resistance. Therefore, the total branch resistance can be expressed as

$$R = 1000 + 1.5\rho_s \, \Omega \qquad (4.110)$$

for hand-to-foot currents and

$$R = 1000 + 6\rho_s \, \Omega \qquad (4.111)$$

for foot-to-foot currents, where ρ_s is the soil resistivity in ohm meters. If the surface of the soil is covered with a layer of crushed rock or some other high-resistivity material, its resistivity should be used in Equations 4.110 and 4.111.

Since it is much easier to calculate and measure potential than current, the fibrillation threshold, given by Equation 4.109, is usually given in terms of voltage. Therefore, the maximum allowable (or tolerable) touch and step potentials, respectively, can be expressed as

$$V_{touch} = \frac{0.116(1000 + 1.5\rho_s)}{\sqrt{t}} \, V \qquad (4.112)$$

TABLE 4.7

Resistivity of Different Soils

Ground Type	Resistivity, ρ_s
Seawater	0.01–1.0
Wet organic soil	10
Moist soil (average earth)	100
Dry soil	1000
Bedrock	10^4
Pure slate	10^7
Sandstone	10^9
Crushed rock	1.5×10^8

---

and

$$V_{\text{step}} = \frac{0.116(1000 + 6\rho_s)}{\sqrt{t}} \text{ V} \tag{4.113}$$

Table 4.7 gives typical values for various ground types. However, the resistivity of ground also changes as a function of temperature, moisture, and chemical content. Therefore, in practical applications, the only way to determine the resistivity of soil is by measuring it.

Example 4.13

Assume that a human body is part of a 60 Hz electric power circuit for about 0.25 s and that the soil type is average earth. Based on the IEEE Std. 80-1976, determine the following:

    a. Tolerable touch potential
    b. Tolerable step potential

Solution

    a. Using Equation 4.112,

$$V_{\text{touch}} = \frac{0.116(1000 + 1.5\rho_s)}{\sqrt{t}} = \frac{0.116(1000 + 1.5 \times 100)}{\sqrt{0.25}} \cong 267 \text{ V}$$

    b. Using Equation 4.113,

$$V_{\text{step}} = \frac{0.116(1000 + 6\rho_s)}{\sqrt{t}} = \frac{0.116(1000 + 6 \times 100)}{\sqrt{0.25}} \cong 371 \text{ V}$$

4.13.2 GROUND RESISTANCE

Ground is defined as a conducting connection, either intentional or accidental, by which an electric circuit or equipment becomes grounded. Therefore, *grounded* means that a given electric system, circuit, or device is connected to the earth serving in the place of the former with the purpose of establishing and maintaining the potential of conductors connected to it approximately at the potential of the earth and allowing for conducting electric currents from and to the earth of its equivalent.

Thus, a *safe grounding design* should provide the following:

1. A means to carry and dissipate electric currents into ground under normal and fault conditions without exceeding any operating and equipment limits or adversely affecting continuity of service
2. Assurance for such a degree of human safety so that a person working or walking in the vicinity of grounded facilities is not subjected to the danger of critic electrical shock

However, a low ground resistance is not, in itself, a guarantee of safety. For example, about three or four decades ago, a great many people assumed that any object grounded, however crudely, could be safely touched. This misconception probably contributed to many tragic accidents in the past. Since there is no simple relation between the resistance of the ground system as a whole and the maximum shock current to which a person might be exposed, a system or system component (e.g., substation or tower) of relatively low ground resistance may be dangerous under some conditions, whereas another system component with very high ground resistance may still be safe or can be made safe by careful design.

Ground potential rise (GPR) is a function of fault-current magnitude, system voltage, and ground (system) resistance. The current through the ground system multiplied by its resistance measured from a point remote from the substation determines the GPR with respect to remote ground.

The ground resistance can be reduced by using electrodes buried in the ground. For example, metal rods or *counterpoise* (i.e., buried conductors) are used for the lines, while the grid system made of copper-stranded copper cable and rods is used for the substations.

The grounding resistance of a buried electrode is a function of (1) the resistance of the electrode itself and connections to it, (2) contact resistance between the electrode and the surrounding soil, and (3) resistance of the surrounding soil, from the electrode surface outward. The first two resistances are very small with respect to soil resistance and therefore may be neglected in some applications. However, the third one is usually very large depending on the type of soil, chemical ingredients, moisture level, and temperature of the soil surrounding the electrode.

Table 4.8 gives typical resistivity values for various ground types. However, the resistivity of the ground also changes as a function of temperature, moisture, and chemical content. Therefore, in practical applications, the only way to determine the resistivity of soil is by measuring it.

Table 4.9 presents data indicating the effect of moisture contents on the soil resistivity. The resistance of the soil can be measured by using the three-electrode method or by using self-contained instruments such as the Biddle Megger Ground Resistance Tester.

The human body can tolerate slightly larger currents at 25 Hz and about five times larger at direct current (dc). Similarly, at frequencies of 1,000 or 10,000 Hz, even larger currents can be tolerated.

TABLE 4.8
Resistivity of Different Soils

Ground Type	Resistivity, ρ_s
Seawater	0.01–1.0
Wet organic soil	10
Moist soil (average earth)	100
Dry soil	1000
Bedrock	10^4
Pure slate	10^7
Sandstone	10^9
Crushed rock	1.5×10^8

TABLE 4.9

Effect of Moisture Content on Soil Resistivity

Moisture Content (Wt.%)	Resistivity (Ω-cm)	
	Topsoil	Sandy Loam
0	>10^9	>10^9
2.5	250,000	150,00
5	165,000	43,000
10	53,000	18,500
15	19,000	10,500
20	12,000	6,300
30	6,400	4,200

In the case of lighting surges, the human body appears to be able to tolerate very high currents, perhaps on the order of several hundreds of amperes [17].

When the human body becomes a part of the electric circuit, the current that passes through it can be found by applying Thévenin's theorem and Kirchhoff's current law (KCL), as illustrated in Figure 4.40. For dc and ac at 60 Hz, the human body can be substituted by a resistance in the equivalent circuits. The body resistance considered is usually between two extremities, either from one hand to both feet or from one foot to the other one.

Experiments have shown that the body can tolerate much more current flowing from one leg to the other than it can when current flows from one hand to the legs. Figure 4.40a shows a touch contact with current flowing from hand to feet. On the other hand, Figure 4.40b shows a step contact where current flows from one foot to the other. Note that in each case the body current I_b is driven by the potential difference between points A and B.

Currents of 1 mA or more but less than 6 mA are often defined as the secondary shock currents (*let-go currents*). The let-go current is the maximum current level at which a human holding an energized conductor can control his or her muscles enough to release it. For 99.5% of population, the 60 Hz minimum required body current, I_B, leading to possible fatality through ventricular fibrillation can be expressed as

$$I_b = \frac{0.116}{\sqrt{t_s}} \text{ A} \quad \text{for 50 kg body weight} \tag{4.114a}$$

or

$$I_b = \frac{0.157}{\sqrt{t_s}} \text{ A} \quad \text{for 70 kg body weight} \tag{4.114b}$$

where t is in seconds in the range from approximately 8.3 ms to 5 s.

The effects of an electric current passing through the vital parts of a human body depend on the duration, magnitude, and frequency of this current. The body resistance considered is usually between two extremities, either from one hand to both feet or from one foot to the other one. Figure 4.41 shows five basic situations involving a person and grounded facilities during fault.

Note that in the figure the *mesh voltage* is defined by the maximum touch voltage within a mesh of a ground grid. But the *metal-to-metal touch voltage* defines the difference in potential between metallic objects or structures within the substation site that may be bridged by direct hand-to-hand or hand-to-feet contact. However, the *step voltage* represents the difference in surface potential

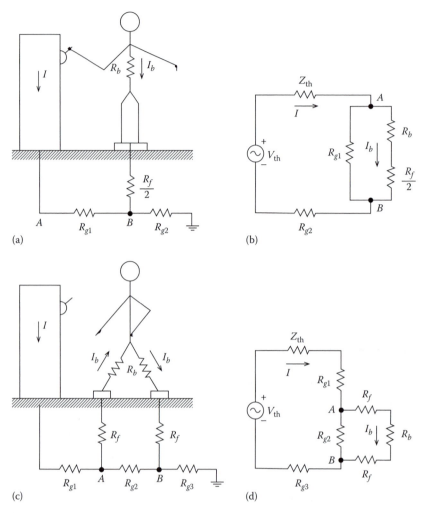

FIGURE 4.40 Typical electric shock hazard situations: (a) touch potential; (b) its equivalent circuit; (c) step potential; (d) its equivalent circuit.

experienced by a person bridging a distance of 1 m with the feet without contacting any other grounded object.

On the other hand, the *touch voltage* represents the potential difference between the GPR and the surface potential at the point where a person is standing while at the same time having a hand in contact with a grounded structure. The *transferred voltage* is a special case of the touch voltage where a voltage is transferred into or out of the substation from or to a remote point external to the substation site [12].

Finally, GPR is the maximum electrical potential that a substation grounding grid may have relative to a distant grounding point assumed to be at the potential of remote earth. This voltage, GPR, is equal to the maximum grid current times the grid resistance. Under normal conditions, the grounded electrical equipment operates at near-zero ground potential. That is, the potential of a grounded neutral conductor is nearly identical to the potential of remote earth. During a ground fault, the portion of fault current that is conducted by substation grounding grid into the earth causes the rise of the grid potential with respect to remote earth.

Exposure to touch potential normally poses a greater danger than exposure to step potential. The step potentials are usually smaller in magnitude (due to the greater corresponding body resistance),

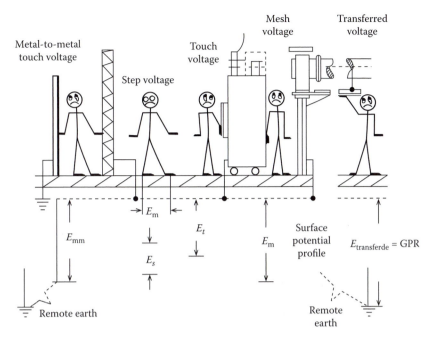

FIGURE 4.41 Possible basic shock situations. (From Keil, R.P., Substation grounding, in J.D. McDonald, ed., *Electric Power Substation Engineering*, 2nd edn., CRC Press, Boca Raton, FL, 2007, Chapter 11. Used with permission.)

and the allowable body current is higher than the touch contacts. In either case, the value of the body resistance is difficult to establish.

As said before, experiments have shown that the body can tolerate much more current flowing from one leg to the other than it can when current flows from one hand to the legs. Treating the foot as a circular plate electrode gives an approximate resistance of $3\rho_s$, where ρ_s is the soil resistivity. The resistance of the body itself is usually used as about 2300 Ω hand to hand or 1100 Ω hand to foot.

However, IEEE Std. 80-2000 [12] recommends the use of 1000 Ω as a reasonable approximation for body resistance. Therefore, the total branch resistance, for hand-to-foot currents, can be expressed as

$$R_b = 1000 + 1.5\rho_s \Omega \quad \text{for touch voltage} \tag{4.115a}$$

and, for foot-to-foot currents,

$$R_b = 1000 + 6\rho_s \Omega \quad \text{for step voltage} \tag{4.115b}$$

where ρ_s is the soil resistivity, Ω-m. If the surface of the soil has covered with a layer of crushed rock or some other high-resistivity material, its resistivity should be used in Equations 4.56 and 4.57. The *touch voltage limit* can be determined from

$$V_{\text{touch}} = \left(R_b + \frac{R_f}{2}\right)I_b \tag{4.116}$$

and

$$V_{\text{step}} = (R_b + 2R_f)I_b \tag{4.117}$$

where

$$R_f = 3C_s\rho_s \tag{4.118}$$

where

R_b is the resistance of human body, typically 1000 Ω for 50 and 60 Hz

R_f is the ground resistance of one foot

I_b is the rms magnitude of the current going through the body in A, per Equations 4.114a and 4.114b

C_s is the surface layer derating factor based on the thickness of the protective surface layer spread above the earth grade at the substation (per IEEE Std. 80-2000, if no protective layer is used, then $C_s = 1$)

Since it is much easier to calculate and measure potential than current, the fibrillation threshold, given by Equations 4.114a and 4.114b, is usually given in terms of voltage. Thus, if there is no protective surface layer, for a person with body weight of 50 or 70 kg, the *maximum allowable* (or *tolerable*) *touch voltages*, respectively, can be expressed as

$$V_{\text{touch}\,50} = \frac{0.116(1000 + 1.5\rho_s)}{\sqrt{t_s}} \text{ V} \quad \text{for 50 kg body weight} \tag{4.119a}$$

and

$$V_{\text{touch}\,70} = \frac{0.157(1000 + 1.5\rho_s)}{\sqrt{t_s}} \text{ V} \quad \text{for 70 kg body weight} \tag{4.119b}$$

If *the event of no protective surface layer is used, for the metal-to-metal touch* in V, Equations 4.119a and 4.119b become

$$V_{\text{mm-touch}\,50} = \frac{116}{\sqrt{t_s}} \text{ V} \quad \text{for 50 kg body weight} \tag{4.119c}$$

and

$$V_{\text{mm-touch}\,70} = \frac{157}{\sqrt{t_s}} \text{ V} \quad \text{for 70 kg body weight} \tag{4.119d}$$

If *a protective layer does exists*, then the *maximum allowable* (or *tolerable*) *step voltages*, for a person with body weight of 50 or 70 kg, are given, respectively, as

$$V_{\text{step}\,50} = \frac{0.116(1000 + 6C_s\rho_s)}{\sqrt{t_s}} \text{ V} \quad \text{for 50 kg body weight} \tag{4.120a}$$

and

$$V_{\text{step}\,70} = \frac{0.157(1000 + 6C_s\rho_s)}{\sqrt{t_s}} \text{ V} \quad \text{for 70 kg body weight} \tag{4.120b}$$

If a protective layer does exists, then the *maximum allowable* (or *tolerable*) *touch voltages*, for a person with body weight of 50 or 70 kg, are given, respectively, as

$$V_{touch\,50} = \frac{0.116(1000 + 1.5C_s\rho_s)}{\sqrt{t_s}} \text{ V } \text{ for 50 kg body weight} \qquad (4.120c)$$

$$V_{touch\,70} = \frac{0.157(1000 + 1.5C_s\rho_s)}{\sqrt{t_s}} \text{ V } \text{ for 70 kg body weight} \qquad (4.120d)$$

The earlier equations are applicable only in the event that a protection surface layer is used. For metal-to-metal contacts, use $\rho_s = 0$ and $C_s = 1$. For more detailed applications, see IEEE Std. 2000 [12]. Also, it is important to note that in using the earlier equations, it is assumed that they are applicable to 99.5% of the population. There are always exceptions.

4.13.3 REDUCTION OF FACTOR C_s

Note that according to IEEE Std. 80-2000, a *thin layer of highly resistive protective surface material* such as gravel spread across the earth at a substation greatly reduced the possible shock current at a substation. IEEE Std. 80-2000 gives the required equations to determine the ground resistance of one foot on a thin layer of surface material as

$$C_s = 1 + \frac{1,6b}{\rho_s} \sum_{n=1}^{\infty} K^n R_{m(2nh_s)} \qquad (4.121)$$

and

$$C_s = 1 - \frac{0.09(1 - (\rho/\rho_s))}{2h_s + 0.09} \qquad (4.122)$$

where

$$K = \frac{\rho - \rho_s}{\rho + \rho_s} \qquad (4.123)$$

where
   C_s is the surface layer derating factor (it can be considered as a corrective factor to compute the effective foot resistance in the presence of a finite thickness of surface material); see Figure 4.42
   ρ_s is the surface material resistivity, Ω-m
   K is the reflection factor between different material resistivity
   ρ is the resistivity of earth beneath the substation, Ω-m
   h_s is the thickness of the surface material, m
   b is the radius of circular metallic disk representing the foot, m
   $R_{m(2nhs)}$ is the mutual ground resistance between two similar, parallel, coaxial plates that are separated by a distance of $(2nh_s)$, Ω-m

Note that Figure 4.42 gives the exact value of C_s instead of using the empirical equation (4.64) for it. The empirical equation gives approximate values that are within 5% of the values that can be found in the equation.

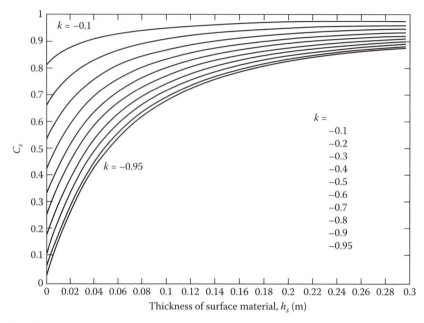

FIGURE 4.42 Surface layer derating factor C_s versus thickness of surface material in m. (From Keil, R.P., Substation grounding, in J.D. McDonald, ed., *Electric Power Substation Engineering*, 2nd edn., CRC Press, Boca Raton, FL, 2007, Chapter 11. Used with permission.)

Example 4.14

Assume that a human body is part of a 60 Hz electric power circuit for about 0.49 s and that the soil type is average earth. Based on the IEEE Std. 80-2000, determine the following:

a. Tolerable touch potential, for 50 kg body weight
b. Tolerable step potential, for 50 kg body weight
c. Tolerable touch voltage limit for metal-to-metal contact, if the person is 50 kg
d. Tolerable touch voltage limit for metal-to-metal contact, if the person is 70 kg

Solution

a. Using Equation 4.61a, for 50 kg body weight,

$$V_{touch50} = \frac{0.116(1000 + 1.5\rho_s)}{\sqrt{t_s}}$$

$$= \frac{0.116(1000 + 1.5 \times 100)}{\sqrt{0.49}}$$

$$\cong 191\,V$$

b. Using Equation 4.61b,

$$V_{step50} = \frac{0.116(1000 + 6\rho_s)}{\sqrt{t_s}}$$

$$= \frac{0.116(1000 + 6 \times 100)}{\sqrt{0.49}}$$

$$\cong 265\,V$$

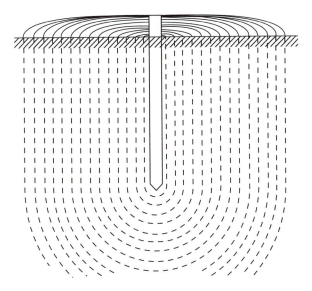

FIGURE 4.43   Resistance of earth surrounding an electrode.

c. Since $\rho_s = 0$,

$$V_{mm\text{-}touch50} = \frac{116}{\sqrt{t_s}} = \frac{116}{\sqrt{0.49}} = 165.7 \text{ V} \quad \text{for 50 kg body weight}$$

d. Since $\rho_s = 0$,

$$V_{mm\text{-}touch70} = \frac{157}{\sqrt{t_s}} = \frac{157}{\sqrt{0.49}} = 224.3 \text{ V} \quad \text{for 70 kg body weight}$$

Figure 4.43 shows a ground rod driven into the soil and conducting current in all directions. Resistance of the soil has been illustrated in terms of successive shells of the soil of equal thickness. With increased distance from the electrode, the soil shells have greater area and therefore lower resistance. Thus, the shell nearest the rod has the smallest cross section of the soil and therefore the highest resistance. Measurements have shown that 90% of the total resistance surrounding an electrode is usually with a radius of 6–10 ft.

The assumptions that have been made in deriving these formulas are that the soil is perfectly homogeneous and the resistivity is of the same known value throughout the soil surrounding the electrode. Of course, these assumptions are seldom true. The only way one can be sure of the resistivity of the soil is by actually measuring it at the actual location of the electrode and at the actual depth.

Figure 4.44 shows the variation of soil resistivity with depth for a soil having uniform moisture content at all depths [23]. In reality, however, deeper soils have greater moisture content, and the advantage of depth is more visible. Some nonhomogeneous soils can also be modeled by using the two-layer method [20].

The resistance of the soil can be measured by using the three-electrode method or by using self-contained instruments such as the Biddle Megger Ground Resistance Tester. Figure 4.45 shows the approximate ground resistivity distribution in the United States.

If the surface of the soil is covered with a layer of crushed rock or some other high-resistivity material, its resistivity should be used in the previous equations. Table 4.9 gives typical values for

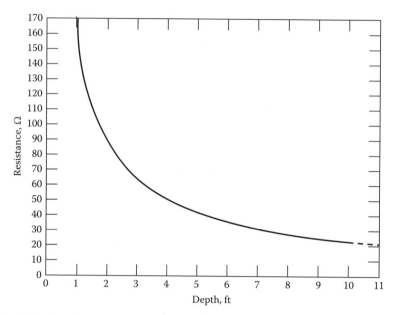

FIGURE 4.44 Variation of soil resistivity with depth for soil having uniform moisture content at all depths. (From National Bureau of Standards Technical Report 108, Department of Commerce, Washington DC, 1978.)

various ground types. However, the resistivity of ground also changes as a function of temperature, moisture, and chemical content. Thus, in practical applications, the only way to determine the resistivity of soil is by measuring it.

In general, soil resistivity investigations are required to determine the soil structure. Table 4.9 gives only very rough estimates. The soil resistivity can vary substantially with changes in temperature, moisture, and chemical content. To determine the soil resistivity of a specific site, it is required to take soil resistivity measurements. Since soil resistivity can change both horizontally and vertically, it is necessary to take more than one set of measurements. IEEE Std. 80-2000 [12] describes various measuring techniques in detail. There are commercially available computer programs that use the soil data and calculate the soil resistivity and provide a confidence level based on the test. There is also a graphical method that was developed by Sunde [20] to interpret the test results.

4.13.4 Soil Resistivity Measurements

Table 4.9 gives estimates on soil classification that are only an approximation of the actual resistivity of a given site. Actual resistivity tests therefore are crucial. They should be made at a number of places within the site. In general, substation sites where the soil has uniform resistivity throughout the entire area and to a considerable depth are seldom found.

4.13.4.1 Wenner Four-Pin Method

More often than not, there are several layers, each having a different resistivity. Furthermore, lateral changes also take place, however, with respect to the vertical changes; these changes usually are more gradual. Hence, soil resistivity tests should be made to find out if there are any substantial changes in resistivity with depth. If the resistivity varies considerably with depth, it is often desirable to use an increased range of probe spacing in order to get an estimate of the resistivity of deeper layers.

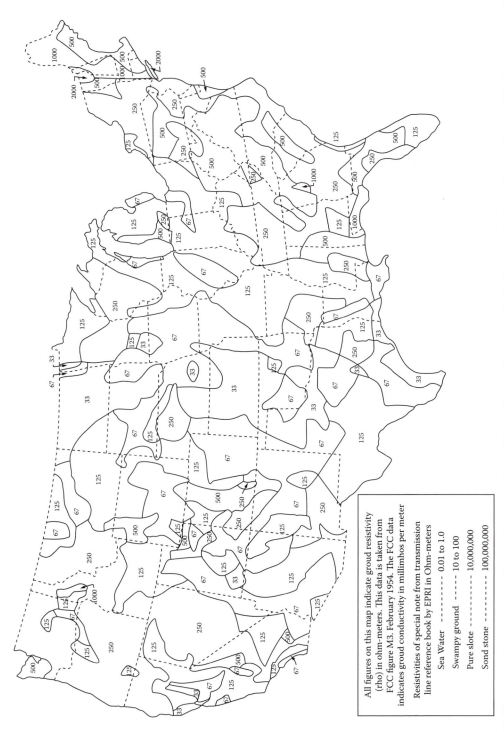

FIGURE 4.45 Approximate ground resistivity distribution in the United States. (From Farr, H. H., Transmission Line Design Manual, U.S. Department of Interior, Water and Power Resources Service, Denver, CO, 1980.)

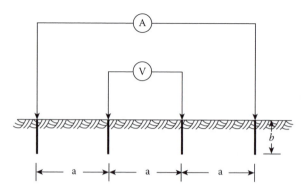

FIGURE 4.46 The Wenner four-pin method. (From Gonen, T., *Electric Power Transmission System Engineering*, 2nd edn., CRC Press, Boca Raton, FL, 2009.)

IEEE Std. 81-1983 describes a number of measuring techniques. The Wenner four-pin method is the most commonly used technique. Figure 4.46 illustrates this method. In this method, four probes (or pins) are driven into the earth along a straight line, at equal distances apart, driven to a depth b. The voltage between the two inner (i.e., potential) electrodes is then measured and divided by the current between the two outer (i.e., current) electrodes to give a value of resistance R. The apparent resistivity of soil is determined from

$$\rho_a = \frac{4\pi a R}{1 + \left(2a / \sqrt{a^2 + 4b^2}\right)} - \frac{a}{\sqrt{a^2 + b^2}} \tag{4.124}$$

where
  ρ_a is the apparent resistivity of the soil, Ω-m
  R is the measured resistivity, Ω
  a is the distance between adjacent electrodes, m
  b is the depth of the electrodes, m

In the event that b is small in comparison to a, then

$$\rho_a = 2\pi a R \tag{4.125}$$

The current tends to flow near the surface for the small probe spacing, whereas more of the current penetrates deeper soils for large spacing. Because of this fact, the previous two equations can be used to determine the apparent resistivity ρ_a at a depth a.

The Wenner four-pin method obtains the soil resistivity data for deeper layers without driving the test pins to those layers. No heavy equipment is needed to do the four-pin test. The results are not greatly affected by the resistance of the test pins or the holes created in driving the test pins into the soil. Because of these advantages, the Wenner method is the most popular method.

4.13.4.2 Three-Pin or Driven Ground Rod Method

IEEE Std. 81-1983 describes *a second method of measuring soil resistivity*. It is illustrated in Figure 4.47. In this method, the depth (L_r) of the driven rod located in the soil to be tested is varied. The other two rods are known as *reference rods*. They are driven to a shallow depth in a straight line. The location of the voltage rod is varied between the test rod and the current rod.

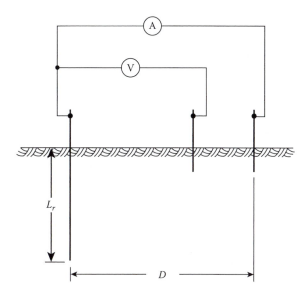

FIGURE 4.47 Circuit diagram for three-pin or driven ground rod method. (From Gonen, T., *Electric Power Transmission System Engineering*, 2nd edn., CRC Press, Boca Raton, FL, 2009.)

Alternatively, the voltage rod can be placed on the other side of the driven rod. The apparent resistivity is found from

$$\rho_a = \frac{2\pi L_r R}{\ln(8L_r/d) - 1} \qquad (4.126)$$

where
   L_r is the length of the driven rod, m
   d is the diameter of the rod, m
   R is the measured resistivity, Ω

A plot of the measured resistivity value ρ_a versus the rod length (L_r) provides a visual aid for finding out earth resistivity variations with depth. An advantage of the driven-rod method, even though not related necessarily to the measurements, is the ability to determine to what depth the ground rods can be driven. This knowledge can save the need to redesign the ground grid. Because of hard layers in the soil such as rock and hard clay, it becomes practically impossible to drive the test rod any further resulting in insufficient data.

A *disadvantage of the driven-rod method* is that when the test rod is driven deep in the ground, it usually losses contact with the soil due to the vibration and the larger diameter couplers resulting in higher measured resistance values. A ground grid designed with these higher soil resistivity values may be unnecessarily conservative. Thus, this method presents an uncertainty in the resistance value.

4.14 SUBSTATION GROUNDING

Grounding at substation has paramount importance: Again, the purpose of such a grounding system includes the following:

1. To provide the ground connection for the grounded neutral for transformers, reactors, and capacitors
2. To provide the discharge path for lightning rods, arresters, gaps, and similar devices

3. To ensure safety to operating personnel by limiting potential differences that can exist in a substation
4. To provide a means of discharging and de-energizing equipment in order to proceed with the maintenance of the equipment
5. To provide a sufficiently low-resistance path to ground to minimize rise in ground potential with respect to remote ground

A multigrounded, common neutral conductor used for a primary distribution line is always connected to the substation grounding system where the circuit originates and to all grounds along the length of the circuit. If separate primary and secondary neutral conductors are used, the conductors have to be connected together provided the primary neutral conductor is effectively grounded.

The substation grounding system is connected to every individual equipment, structure, and installation so that it can provide the means by which grounding currents are connected to remote areas. It is extremely important that the substation ground has a low ground resistance, adequate current-carrying capacity, and safety features for personnel. It is crucial to have the substation ground resistance very low so that the total rise of the ground system potential will not reach values that are unsafe for human contact.*

The substation grounding system normally is made of buried horizontal conductors and driven ground rods interconnected (by clamping, welding, or brazing) to form a continuous grid (also called mat) network. A continuous cable (usually it is 4/0 bare copper cable buried 12–18 in. below the surface) surrounds the grid perimeter to enclose as much ground as possible and to prevent current concentration and thus high gradients at the ground cable terminals. Inside the grid, cables are buried in parallel lines and with uniform spacing (e.g., about 10 × 20 ft).

All substation equipment and structures are connected to the ground grid with large conductors to minimize the grounding resistance and limit the potential between equipment and the ground surface to a safe value under all conditions. All substation fences are built inside the ground grid and attached to the grid in short intervals to protect the public and personnel. The surface of the substation is usually covered with crushed rock or concrete to reduce the potential gradient when large currents are discharged to ground and to increase the contact resistance to the feet of personnel in the substation.

IEEE Std. 80-1976 [13] provides a formula for a quick simple calculation of the grid resistance to ground after a minimum design has been completed. It is expressed as

$$R_{grid} = \frac{\rho_s}{4r} + \frac{\rho_s}{L_T} \qquad (4.127)$$

where
ρ_s is the soil resistivity, Ω-m
L is the total length of grid conductors, m
R is the radius of circle with area equal to that of grid, m

IEEE Std. 80-2000 [19] provides the following equation to determine the grid resistance after a minimum design has been completed:

$$R_{grid} = \frac{\rho_s}{4} \sqrt{\frac{\pi}{A}} \qquad (4.128)$$

---

* *Mesh voltage* is the worst possible value of a touch voltage to be found within a mesh of a ground grid if standing at or near the center of the mesh.

Also, IEEE Std. 80-2000 provides the following equation to determine *the upper limit for grid resistance to ground after a minimum design has been completed*:

$$R_{\text{grid}} = \frac{\rho_s}{4}\sqrt{\frac{\pi}{A}} + \frac{\rho_s}{L_T} \tag{4.129}$$

where
R_{grid} is the grid resistance, Ω
ρ is the soil resistance, Ω-m
A is the area of the ground, m^2
L_T is the total buried length of conductors, m

But, Equation 4.129 requires a uniform soil resistivity. Hence, a substantial engineering judgment is necessary for reviewing the soil resistivity measurements to decide the value of soil resistivity. However, it does provide a guideline for the uniform soil resistivity to be used in the ground grid design. Alternatively, Sverak [19] provides the following formula for the grid resistance:

$$R_{\text{grid}} = \rho_s \left[\frac{1}{L_T} + \frac{1}{\sqrt{20\,A}} \left(1 + \frac{1}{1 + h\sqrt{20/A}} \right) \right] \tag{4.130}$$

where
R_{grid} is the substation ground resistance, Ω
ρ_s is the soil resistivity, Ω-m
A is the area occupied by the ground grid, m^2
H is the depth of the grid, m
L_T is the total buried length of conductors, m

IEEE Std. 80-1976 also provides formulas to determine the effects of the grid geometry on the step and mesh voltage (which is the worst possible value of the touch voltage) in volts. *Mesh voltage is the worst possible value of a touch voltage to be found within a mesh of a ground grid if standing at or near the center of the mesh*. They can be expressed as

$$E_{\text{step}} = \frac{\rho_s \times K_s \times K_i \times I_G}{L_s} \tag{4.131}$$

and

$$E_{\text{mesh}} = \frac{\rho_s \times K_m \times K_i \times I_G}{L_m} \tag{4.132}$$

where
ρ_s is the average soil resistivity, Ω-m
K_s is the step coefficient
K_m is the mesh coefficient
K_i is the irregularity coefficient
I_G is the maximum rms current flowing between ground grid and earth, A
L_s is the total length of buried conductors, including cross connections, and (optionally) the total effective length of ground rods, m
L_m is the total length of buried conductors, including cross connections, and (optionally) the combined length of ground rods, m

Many utilities have computer programs for performing grounding grid studies. The number of tedious calculations that must be performed to develop an accurate and sophisticated model of a system is no longer a problem.

In general, in the event of a fault, overhead ground wires, neutral conductors, and directly buried metal pipes and cables conduct a portion of the ground fault current away from the substation ground grid and have to be taken into account when calculating the maximum value of the grid current. Based on the associated equivalent circuit and resultant current division, one can determine what portion of the total current flows into the earth and through other ground paths. It can be used to determine the approximate amount of current that did not use the ground as flow path. The fault-current division factor (also known as the *split factor*) can be expressed as

$$S_{\text{split}} = \frac{I_{\text{grid}}}{3I_{ao}} \tag{4.133}$$

where
S_{split} is the fault-current division factor
I_{grid} is the rms symmetrical grid current, A
I_{ao} is the zero-sequence fault current, A

The *split factor* is used to determine the approximate amount of current that did not use the ground flow path. Computer programs can determine the split factor easily, but it is also possible to determine the split factor through graphs. With the Y ordinate representing the split factor and the X axis representing the grid resistance, it is obvious that the grid resistance has to be known to determine the split factor.

As previously said, the split factor determines the approximate amount of current that does use the earth as return path. The amount of current that does enter the earth is found from the following equation. Hence, the design value of the *maximum grid current* can be found from

$$L_G = D_f \times I_{\text{grid}} \tag{4.134}$$

where
I_G is the maximum grid current, A
D_f is the decrement factor for the entire fault duration of t_f, s
I_{grid} is the rms symmetrical grid current, A

Here, Figure 4.48 illustrates the relationship between asymmetrical fault current, dc decaying component, and symmetrical fault current and the relationship between variables I_F, I_f, and D_f for the fault duration t_f.

The *decrement factor* is an adjustment factor that is used in conjunction with the symmetrical ground fault-current parameter in safety-oriented grounding calculations. It determines the rms equivalent of the asymmetrical current wave for a given fault duration, accounting for the effect of initial dc offset and its attenuation during the fault. The decrement factor can be calculated from

$$D_f = \sqrt{1 + \frac{T_a}{I_f}\left(1 - e^{-(2t_f/T_a)}\right)} \tag{4.135}$$

where t_f is the time duration of fault, s

$$T_a = \frac{X}{\omega R} \quad \text{dc offset time constant, s}$$

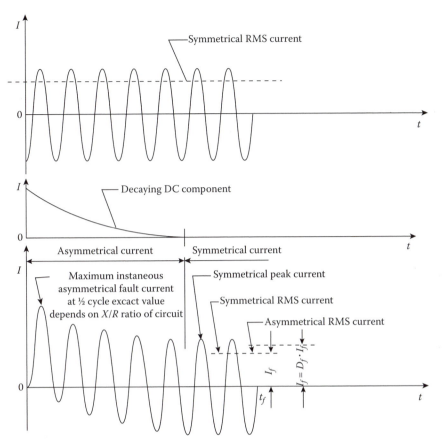

FIGURE 4.48 The relationship between asymmetrical fault current, dc decaying component, and symmetrical fault current.

Here, t_f should be chosen as the fastest clearing time. (The fastest clearing time includes breaker and relay time for transmission substations.) It is assumed here that the ac components do not decay with time.

The *symmetrical grid current* is defined as that portion of the symmetrical ground fault current that flows between the grounding grid and surrounding earth. It can be expressed as

$$I_{\text{grid}} = S_f \times I_f \qquad (4.136)$$

where
I_f is the rms value of symmetrical ground fault current, A
S_f is the fault-current division factor

IEEE Std. 80-2000 provides a series of current based on computer simulations for various values of ground grid resistance and system conditions to determine the grid current. Based on those split-current curves, one can determine the maximum grid current.

4.15 GROUND CONDUCTOR SIZING FACTORS

The flow of excessive currents will be very dangerous if the right equipments are not used to help dissipate the excessive currents. Ground conductors are means of providing a path for excessive currents from the substation to ground grid. Hence, the ground grid can then spread the current into the

TABLE 4.10

Material Constants of the Typical Grounding Material Used

Description	K_f	T_m (°C)	α_r Factor at 20°C (1/°C)	ρ_r at 20°C ($\mu\Omega \cdot$ cm)	K_0 at 0°C (0°C)	Fusing Temperature, T_m (0°C)	Material Conducting (%)	TCAP Thermal Capacity [J/cm³ · °C]
Copper-annealed Soft-drawn	7	1083	0.0393	1.72	234	1083	100	3.42
Copper-annealed Hard-drawn	1084	1084	0.00381	1.78	242	1084	97	3.42
Copper-clad Steel wire	1084	12.06	0.00378	5.86	245	1084	30	3.85
Stainless steel 304	1510	14.72	0.00130	15.86	749	1400	2.4	3.28
Zinc-coated Steel rod	28.96	28.96	0.0030	72	293	419	8.6	4.03

ground, creating a zero potential between the substation and the ground. Table 4.10 gives the list of possible conductors that can be used for such conductors.

In the United States, there are only two types of conductors that are used, namely, copper and/or copper-clad steel conductors that are used for this purpose. The copper one is mainly used due to its high conductivity and the high resistance to corrosion. The next step is to determine the size of ground conductor that needs to be buried underground.

Thus, based on the *symmetrical conductor current*, the required conductor size can be found from

$$I_f = A_{mm^2} \left[\left(\frac{TCAP \times 10^{-4}}{t_c \times \alpha_r \times \rho_r} \right) \ln \left(\frac{K_0 + T_{max}}{K_0 + T_{amb}} \right) \right]^{1/2} \tag{4.137}$$

if the conductor size should be in mm², it can be found from

$$A_{mm^2} = \frac{I_f}{\left[\left(\dfrac{TCAP \times 10^{-4}}{t_c \times \alpha_r \times \rho_r} \right) \ln \left(\dfrac{K_0 + T_{max}}{K_0 + T_{amb}} \right) \right]^{1/2}} \tag{4.138}$$

Alternatively, in the event that it should be in kcmil, since

$$A_{kcmil} = 1.974 \times A_{mm^2} \tag{4.139}$$

then Equation 4.130 can be expressed as

$$I_f = 5.07 \times 10^{-3} A_{kcmil} \left[\left(\frac{TCAP \times 10^{-4}}{t_c \times \alpha_r \times \rho_r} \right) \ln \left(\frac{K_0 + T_{max}}{K_0 + T_{amb}} \right) \right]^{1/2} \tag{4.140}$$

Note that both α_r and ρ_r can be found at the same reference temperature of T_r, °C. Also, note that Equations 4.137 and 4.140 can also be used to determine the short-time temperature rise in a ground conductor. Thus, taking other required conversions into account, the conductor size in kcmil can be found from

$$A_{kcmil} = \frac{197.4 \times I_f}{\left[\left(\frac{TCAP \times 10^{-4}}{t_c \times \alpha_r \times \rho_r} \right) \ln \left(\frac{K_0 + T_{max}}{K_0 + T_{amb}} \right) \right]^{1/2}} \tag{4.141}$$

where

I_f is the rms current (without dc offset), kA
A_{mm^2} is the conductor cross section, mm²
A_{kcmil} is the conductor cross section, kcmil
TCAP is the thermal capacity per unit volume, J/(cm³·°C) (it is found from Table 4.10, per IEEE Std. 80-2000)
t_c is the duration of current, s
α_r is the thermal coefficient of resistivity at reference temperature T_r, 1/°C (it is found from Table 4.10, per IEEE Std. 80-2000 for 20°C)
ρ_r is the resistivity of the ground conductor at reference temperature T_r, μΩ-cm (it is found from Table 4.10, per IEEE Std. 80-2000 for 20°C)
K_0 is $1/\alpha_0$ or $(1/\alpha_r) - T_r$, °C
T_{max} is the maximum allowable temperature, °C
T_{amb} is the ambient temperature, °C
I_f is the rms current (without dc offset), kA
A_{mm^2} is the conductor cross section, mm²
A_{kcmil} is the conductor cross section, kcmil

For a given conductor material, once the TCAP is found from Table 4.10 or calculated from

$$TCAP[J/(cm^3 \cdot °C)] = 4.184(J/cal) \times SH[cal/(g \cdot °C)] \times SW(g/cm^3) \tag{4.142}$$

where

SH is the specific heat, in cal/(g × °C), which is related to the thermal capacity per unit volume, J/(cm³ × °C)
SW is the specific weight, in g/cm³, which is related to the thermal capacity per unit volume, J/(cm³ × °C)

Thus, TCAP is defined by

$$TCAP[J/(cm^3 \cdot °C)] = 4.184(J/cal) \times SH[cal/(g \cdot °C)] \times SW(g/cm^3) \tag{4.143}$$

Asymmetrical fault currents consist of subtransient, transient, and steady-state ac components and the dc offset current component. To find the *asymmetrical fault current* (i.e., if the effect of the dc offset is needed to be included in the fault current), the equivalent value of the asymmetrical current I_F is found from

$$I_F = D_f \times I_f \tag{4.144}$$

where I_F is representing the rms value of an asymmetrical current integrated over the entire fault duration, t_c, which can be found as a function of X/R by using D_f, before using Equation 4.137 or 4.140 and where D_f is the decrement factor and is found from

$$D_f = \left[1 + \frac{T_a}{t_f} \left(1 - e^{(-2t_f/T_a)} \right) \right]^{1/2}$$ (4.145)

where

t_f is the time duration of fault, s
T_a is the dc offset time constant, s

Note that

$$T_a = \frac{X}{\omega R}$$ (4.146)

and for 60 Hz,

$$T_a = \frac{X}{120 \pi R}$$ (4.147)

The resulting I_F is always greater than I_f. However, if the X/R ratio is less than 5 and the fault duration is greater than 1 s, the effects of the dc offset are negligible.

4.16   MESH VOLTAGE DESIGN CALCULATIONS

If the GPR value exceeds the tolerable touch and step voltages, it is necessary to perform the mesh voltage design calculations to determine whether the design of a substation is safe. If the design is again unsafe, conductors in the form of ground rods are added to the design until the design is considered safe. The mesh voltage is the maximum touch voltage and it is found from

$$E_{mesh} = \frac{\rho \times K_m \times K_i \times I_G}{L_M}$$ (4.148)

where
ρ is soil resistivity, Ω-m
K_m is mesh coefficient
K_i is correction factor for grid geometry
I_G is maximum grid current that flows between ground grid and surrounding earth, A
L_m is length of $L_c + L_R$ for mesh voltage, m

The mesh coefficient K_m is determined from

$$K_m = \frac{1}{2\pi} \left[\ln \left(\frac{D^2}{16 \times h \times d} + \frac{(D+2 \times h)^2}{8 \times D \times d} - \frac{h}{4 \times d} \right) + \frac{K_{ii}}{K_h} \ln \left(\frac{8}{\pi(2 \times n - 1)} \right) \right]$$ (4.149)

where

    d is diameter of grid conductors, m

    D is spacing between parallel conductors, m

    K_{ii} is irregularity factor (*corrective weighting factor* that adjusts for the effects of inner conductors on the corner mesh)

    K_h is corrective weighting factor that highlight for the effects of grid depth

    n is geometric factor

    h is depth of ground grid conductors, m

As it can be observed from Equation 4.149, the geometric factor K_m has the following variables: (D_s) the spacing between the conductors, (n_s) the number of conductors, (d) the diameter of the conductors used, and (h) the depth of the grid. The effect of each variable on the K_m is different. Figure 4.49 shows the effect of the spacing (D) between conductors on K_m. Figure 4.50 shows the effect of the number of conductors (n) on the K_m. Figure 4.51 shows the relationship between the diameter of the conductor (d) and the K_m. Figure 4.52 shows the relationship between the depth of the conductor (h) and K_m [17].

Note that the value of K_{ii} depends on the following circumstances:

a. For the grids with ground rods existing in grid corners as well as perimeter:

$$K_{ii} = 1 \tag{4.150}$$

b. For the grids with no or few ground rods with none existing in corners or perimeter:

$$K_{ii} = \frac{1}{(2n)^{\frac{2}{n}}} \tag{4.151}$$

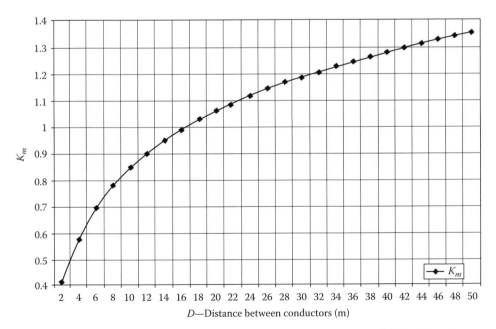

FIGURE 4.49 The effect of the spacing (D) between conductors on K_m. (From Keil, R.P., Substation grounding, in J.D. McDonald, ed., *Electric Power Substation Engineering*, 2nd edn., CRC Press, Boca Raton, FL, 2007, Chapter 11.)

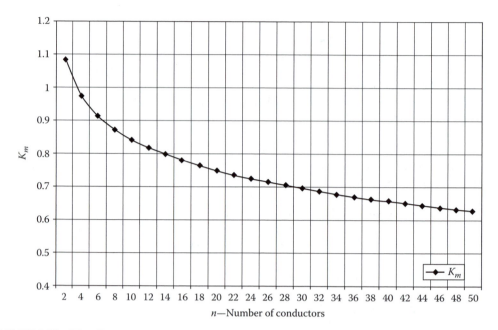

FIGURE 4.50 The effect of the number of conductors (n) on the K_m. (From Keil, R.P., Substation grounding, in J.D. McDonald, ed., *Electric Power Substation Engineering*, 2nd edn., CRC Press, Boca Raton, FL, 2007, Chapter 11.)

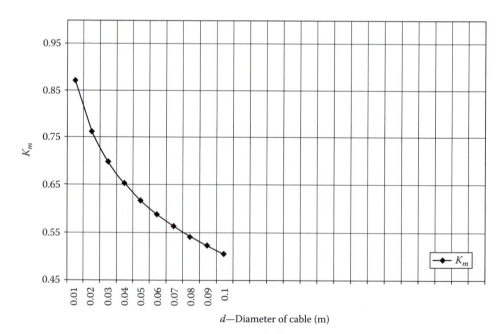

FIGURE 4.51 The relationship between the diameter of the conductor (d) and the K_m. (From Keil, R.P., Substation grounding, in J.D. McDonald, ed., *Electric Power Substation Engineering*, 2nd edn., CRC Press, Boca Raton, FL, 2007, Chapter 11.)

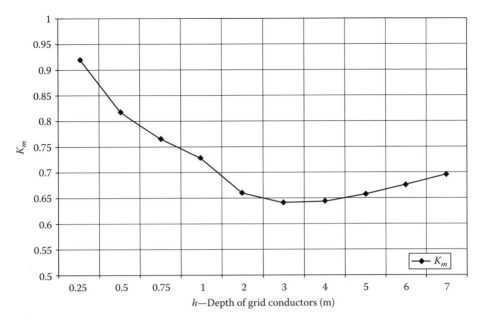

FIGURE 4.52 The relationship between the depth of the conductor (h) and K_m. (From Keil, R.P., Substation grounding, in J.D. McDonald, ed., *Electric Power Substation Engineering*, 2nd edn., CRC Press, Boca Raton, FL, 2007, Chapter 11.)

and

$$K_h = \sqrt{1 + \frac{h}{h_0}} \qquad (4.152)$$

where h_0 is grid reference depth = 1 m.

The effective number of parallel conductors (n) given in a given grid are found from

$$n = n_a \times n_b \times n_c \times n_d \qquad (4.153)$$

where
$$n_a = \frac{2L_c}{L_p}$$
n_b is 1, for square grids
n_c is 1, for square and rectangular grids
n_d is 1, for square, rectangular, and L-shaped grids

Otherwise, the following equations are used to determine the n_b, n_c, and n_d so that

$$n_b = \left[\frac{L_p}{4\sqrt{A}} \right] \qquad (4.154)$$

$$n_c = \left[\frac{L_X \times L_Y}{A} \right]^{\frac{0.7A}{L_X \times L_Y}} \qquad (4.155)$$

$$n_d = \frac{D_m}{\sqrt{L_X^2 + L_Y^2}} \tag{4.156}$$

where
  L_p is the peripheral length of the grid, m
  A is the area of the grid, m^2
  L_X is the maximum length of the grid in the x direction, m
  L_Y is the maximum length of the grid in the y direction, m
  D_m is the maximum distance between any two points on the grid, m

Note that the irregularity factor is determined from

$$K_{ii} = 0.644 + 0.148n \tag{4.157}$$

The effective buried length (L_M) for grids is as follows:

  a. With little or no ground rods but none located in the corners or along the perimeter of the grid:

$$L_M = L_C + L_R \tag{4.158}$$

  where
    L_R is the total length of all ground rods, m
    L_C is the total length of the conductor in the horizontal grid, m

  b. With ground rods in corners and along the perimeter and throughout the grid:

$$L_M = L_C + \left[1.55 + 1.22 \left(\frac{L_R}{\sqrt{L_X^2 + L_Y^2}} \right) \right] L_R \tag{4.159}$$

  where L_R is the length of each ground rod, m.

4.17   STEP VOLTAGE DESIGN CALCULATIONS

According to IEEE Std. 80-2000, in order for the ground system to be safe, step voltage has to be less than the tolerable step voltage. Furthermore, step voltages within the grid system designed for safe mesh voltages will be well within the tolerable limits, the reason for this is that both feet and legs are in series rather than in parallel and the current takes the path from one leg to the other rather than through vital organs. The step voltage is determined from

$$E_{step} = \frac{\rho \times K_s \times K_i \times I_G}{L_S} \tag{4.160}$$

where
  K_s is the step coefficient
  L_s is the buried conductor length, m

Again, for grids with or without ground rods,

$$L_S = 0.75L_C + 0.85L_R \tag{4.161}$$

so that the step coefficient can be found from

$$K_S = \frac{1}{\pi}\left[\frac{1}{2h} + \frac{2}{D+h} + \frac{1}{D}(1-0.5^{n-2})\right] \qquad (4.162)$$

where h is the depth of ground grid conductors in meters, usually in the range 0.25 m $< h <$ 2.5 m.

As shown in Equation 4.162, the geometric factor K_s is a function of D, n, d, π, and h. Figure 4.53 shows the relationship between the distance (D) between the conductors and the geometric factor K_s. Figure 4.54 shows the relationship between the number of conductors (n) and the geometric factor K_s. Figure 4.55 shows the relationship between the depth of grid conductors (D) in meter and the geometric factor K_s.

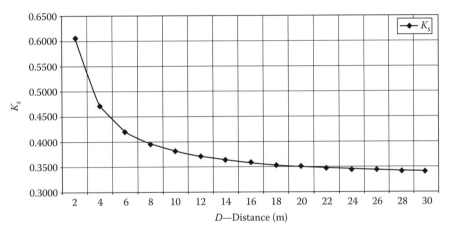

FIGURE 4.53 The relationship between the distance (D) between the conductors and the geometric factor K_s. (From Keil, R.P., Substation grounding, in J.D. McDonald, ed., *Electric Power Substation Engineering*, 2nd edn., CRC Press, Boca Raton, FL, 2007, Chapter 11. Used with permission.)

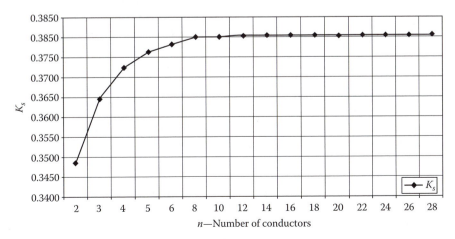

FIGURE 4.54 The relationship between the number of conductors (n) and the geometric factor K_s. (From Keil, R.P., Substation grounding, in J.D. McDonald, ed., *Electric Power Substation Engineering*, 2nd edn., CRC Press, Boca Raton, FL, 2007, Chapter 11. Used with permission.)

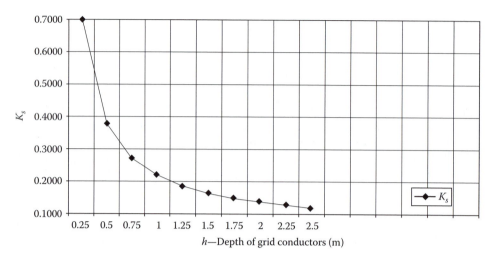

FIGURE 4.55 The relationship between the depth of grid conductors (h) in meter and the geometric factor K_s. (From Keil, R.P., Substation grounding, in J.D. McDonald, ed., *Electric Power Substation Engineering*, 2nd edn., CRC Press, Boca Raton, FL, 2007, Chapter 11. Used with permission.)

4.18 TYPES OF GROUND FAULTS

In general, it is difficult to determine which fault type and location will result in the greatest flow of current between the ground grid and surrounding earth because no simple rule applies. IEEE Std. 80-2000 recommends to not consider multiple simultaneous faults since their probability of occurrence is negligibly small. Instead, it recommends investigating single line-to-ground and line-to-line-to-ground faults.

4.18.1 LINE-TO-LINE-TO-GROUND FAULT

For a line-to-line-to-ground (i.e., double line-to-ground) fault, IEEE Std. 80-2000 gives the following equation to calculate the zero-sequence fault current:

$$I_{a0} = \frac{E(R_2 + jX_2)}{(R_1 + jX_1)[R_0 + R_2 + 3R_f + j(X_0 + X_2)] + (R_2 + jX_2)(R_0 + 3R_f + jX_0)} \qquad (4.163)$$

where
  I_{a0} is the symmetrical rms value of zero-sequence fault current, A
  E is the phase-to-neutral voltage, V
  R_f is the estimated resistance of the fault, Ω (normally it is assumed $R_f = 0$)
  R_1 is the positive-sequence system resistance, Ω
  R_2 is the negative-sequence system resistance, Ω
  R_0 is the zero-sequence system resistance, Ω
  X_1 is the positive-sequence system reactance (subtransient), Ω
  X_2 is the negative-sequence system reactance, Ω
  X_0 is the zero-sequence system reactance, Ω

The values of R_0, R_1, R_2, and X_0, X_1, X_2 are determined by looking into the system from the point of fault. In other words, they are determined from the Thévenin equivalent impedance at the fault point for each sequence.* Often, however, the resistance quantities given in the earlier equation is negligibly small. Hence,

$$I_{a0} = \frac{E \times X_2}{X_1(X_0 + X_2)(X_0 + X_2)} \tag{4.164}$$

4.18.2 SINGLE LINE-TO-GROUND FAULT

For a single line-to-ground fault, IEEE Std. 80-2000 gives the following equation to calculate the zero-sequence fault current:

$$I_{a0} = \frac{E}{3R_f + R_0 + R_1 + R_2 + j(X_0 + X_1 + X_2)} \tag{4.165}$$

Often, however, the resistance quantities in the earlier equation are negligibly small. Hence,

$$I_{a0} = \frac{E}{X_0 + X_1 + X_2} \tag{4.166}$$

4.19 GROUND POTENTIAL RISE

As said before in Section 4.8.2, the GPR is a function of fault-current magnitude, system voltage, and ground system resistance. The GPR with respect to remote ground is determined by multiplying the current flowing through the ground system by its resistance measured from a point remote from the substation. Here, the current flowing through the grid is usually taken as the maximum available line-to-ground fault current.

GPR is a function of fault-current magnitude, system voltage, and ground (system) resistance. The current through the ground system multiplied by its resistance measured from a point remote from the substation determines the GPR with respect to remote ground. Hence, GPR can be found from

$$V_{GPR} = I_G \times R_g \tag{4.167}$$

where
V_{GPR} is the GPR, V
R_g is the ground grid resistance, Ω

For example, if a ground fault current of 20,000 A is flowing into a substation ground grid due to a line-to-ground fault and the ground grid system has a 0.5 Ω resistance to the earth, the resultant *IR* voltage drop would be 10,000 V. It is clear that such 10,000 V *IR* voltage drop could cause serious problems to communication lines in and around the substation in the event that the communication equipment and facilities are not properly insulated and/or neutralized.

---

* It is often acceptable to use $X_1 = X_2$, especially if an appreciable percentage of the positive-sequence reactance to the point of fault is that of static equipment and transmission lines.

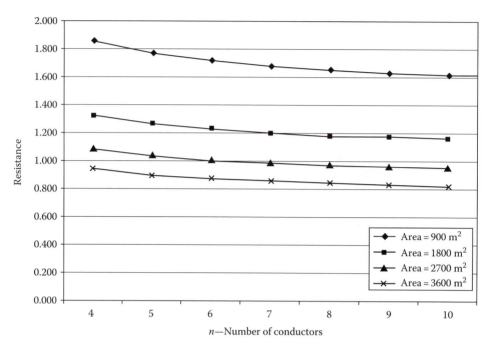

FIGURE 4.56 The effects of the number of grid conductors (n), without ground rods, on the ground grid resistance.

The ground grid resistance can be found from

$$R_g = \rho \left[\frac{1}{L_T} + \frac{1}{\sqrt{20A}} \left(1 + \frac{1}{1 + h\sqrt{20/A}} \right) \right] \tag{4.168}$$

where
  L_T is the total buried length of conductors, m
  h is the depth of the grid, m
  A is the area of substation ground surface, m^2

Figure 4.56 shows the effects of the number of grid conductors (n), without ground rods, on the ground grid resistance. It shows that the area (A) has a substantial influence on the grid resistance. Figure 4.57 shows the relationship between the burial depth of the grid (h), in meter, and the grid resistance. Here, the depth is varied from 0.5 to 2.5 m and the number of conductors from 4 to 10 [17].

In order to aid the substation grounding design engineer, the IEEE Standard 80-2000 includes a design procedure that has a 12-step process, as shown in Figure 4.58, in terms of substation grounding design procedure block diagram, based on a preliminary of a somewhat arbitrary area, that is, the standard suggests the grid be approximately the size of the distribution substation. But, some references state a common practice that is to extend the grid three meters beyond the perimeter of the substation fence.

Example 4.15

Let the starting grid be an 84.5 m by 49.6 m ground grid. Design a proper substation grounding to provide safety measures for anyone going near or working on a substation. Hence, use the IEEE 12-step process shown in Figure 4.58, then build a grid large enough to dissipate the ground fault current into the earth. (A large grounding grid extending far beyond the substation fence and made

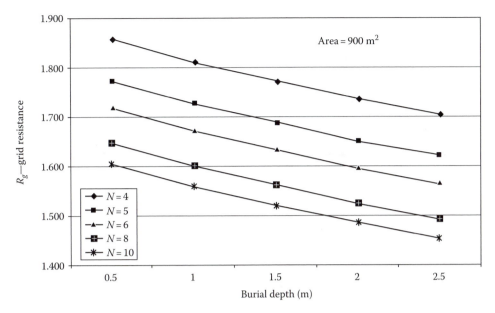

FIGURE 4.57 The effects of varying the depth of burial of the grid (*h*) from 0.5 to 2.5 m and the number of conductors from 4 to 10[17]. (From Fink, D.G. and H.W. Beaty, *Standard Handbook for Electrical Engineers*, 11th edn., McGraw-Hill, New York, 1978. Used with permission.)

of a single copper plate would have the most desirable effect for dispersing fault currents to remote earth and thereby ensure the safety of personnel at the surface. Unfortunately, a copper plate of such size is not an economically viable option.)

A grounding system is considered for this three-phase 230 kV system that feeds two step-down transformers that step down the voltage from 230 to 69 kV. The two transformers are connected in parallel with respect to each other. One of the transformers feeds a switchyard. The other one is connected to a transformer bank (which has three single-phase 4 MVA transformers) that steps down the 69 to 13.8 kV and feeds an industrial facility.

One alternative is to design a grid by using a series of horizontal conductors and vertical ground rods. Of course, the application of conductors and rods depends on the resistivity of the substation ground. Change the variables as necessary in order to meet specifications for grounding of the substation. The variables include the size of the grid, the size of the conductors used, the amount of conductors used, and the spacing of each grounding rod. Use 17,000 A as the maximum value fault current, a maximum clearing time of 1 s, and a conductor diameter of 210.5 kcmil, based on the given information. The soil resistivity is 50 Ω-m and the crushed rock resistivity on the surface of the substation is 2500 Ω-m. Assume that the incoming transmission line into substation has no shield wires and but there are four distribution neutrals. Design a grounding grid system by using a series of horizontal conductors and vertical ground rods, based on the resistivity of the soil.

Solution

Step 1: Field data

Assume that a uniform average soil resistivity of the substation ground is measured to be 50 Ω-m. The initial design parameters are given in Table 4.11.

Step 2: Conductor size

The analysis of the grounding grid should be based on the most conservative fault conditions. For example, the fault current $3I_{ao}$ is assumed maximum value, with all current dispersed through the grid (i.e., there is no alternative path for ground other than through the grid to remote earth). As said before, the maximum value of the fault current is given as 17,000 A; thus, the conductor size is selected based on this current and the duration of the fault. Thus, use 17,000 A as the maximum

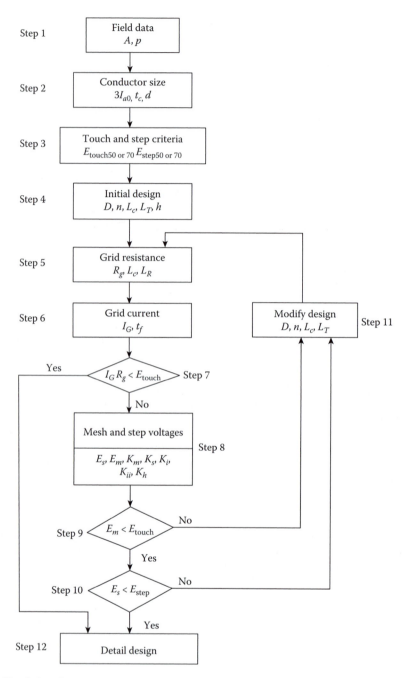

FIGURE 4.58 Substation grounding design procedure block diagram.

value fault current, a maximum clearing time of 0.5 s, and a conductor diameter of 210.5 kcmil, which is determined from the following calculation:

$$A_{kcmil} = I \times K \times \sqrt{t_c}$$

$$= 17 \times 10.45 \times \sqrt{1}$$

$$= 210.5 \, kcmil \tag{4.169}$$

TABLE 4.11

Initial Design Parameters

ρ	A	L_r	L_C	L_R	L_T	h	L_X	L_Y	D
50 Ω-m	4204.6 m²	3.048 m	1825	76	1901	1.524 m	84.6 m	49.6 m	4.97 m
t_c	h_s	D	$3I_{a0}$	ρ_s	D_f	L_p	n_c	n_d	t_f
1 s	0.11 m	0.018 m (for 500 kcmil)	17,000 A	2500 Ω-m	1.026	75 m	1	1	0.5 s

However, the conductor selected is 500 kcmil. This is based on the given guidelines so that the size is more than enough to handle the fault current. The diameter of the conductor can be found from Table A.1. Based on the selected conductor, the diameter (*d*) of the conductor is 0.018 m. The crushed rock resistivity is 2500 Ω-m. Surface derating factor is 0.714.

Step 3: Touch and step voltage criteria

In order to move to the third step in the design process, it is first needed to determine the surface layer derating factor C_s as

$$C_s = 1 - \frac{0.09(1 - (\rho/\rho_s))}{2h_s + 0.09}$$

$$= 1 - \frac{0.09(1 - (50/2500))}{2 \times 0.1524 + 0.09}$$

$$= 0.78$$

According to the federal law, all known hazards must be eliminated when GPR takes place for the safety of workers at a work site. In order to remove the hazards associated with GPR, a grounding grid is designed to reduce the hazardous potentials at the surface. First, it is necessary to determine what was not hazardous to the body. For two body types, the potential safe touch and step voltages a human could withstand before the fault is cleared need to be determined from Equations 4.120d and 4.120b, respectively, as

$$V_{touch70} = (1000 + 1.5C_s \times \rho_s)\frac{0.157}{\sqrt{t_s}}$$

$$= (1000 + 1.5 \times 0.78_s \times 2500)\frac{0.157}{\sqrt{0.5}}$$

$$= 871.5 \text{ V}$$

and

$$V_{step70} = (1000 + 6C_s \times \rho_s)\frac{0.157}{\sqrt{t_s}}$$

$$= (1000 + 1.6 \times 0.78 \times 2500)\frac{0.157}{\sqrt{0.5}}$$

$$= 2819.5 \text{ V}$$

Step 4: Initial design

Step 4 deals with the layout of the grounding conductors and the amount of conductors being used for the design. The initial design consists of factors obtained from the general knowledge of the substation. The preliminary size of the grounding grid system is largely based on the size of the substation to include all dimensions within the perimeter of the fence. To establish economic viability, the maximum area is considered and formed the shape of a square with an area of 4204.6 m². The spacing of conductors (D) is selected as 4.97 m.

The maximum lengths (L_x) of the conductor in the x direction and the y direction (L_y) are determined to be 84.6 m and 49.6 m, respectively. Based on the information given in this section, the total length of the grounding conductor is 1825 m. The length of the ground rods (L_r) is 3.048 m. A total of 25 ground rods are used, which gives the total length of the ground rods (L_R) to be approximately 76 m. Thus, the total conductor length that includes the conductor plus the ground rods is 1901 m.

The depth of the ground grid (h) is determined as 1.525 m below the surface. The next step is to take into account the geometry of the ground grid. Given the length of each side of the grounding grid, it is determined that the shape of the grid design will be a rectangle. The geometric factor can be calculated by determining n_a, n_b, n_c, and n_d as

$$n_a = \frac{2 \times L_c}{L_p} \quad .$$

$$= \frac{2 \times 1825}{268}$$

$$= 14$$

and

$$n_b = \sqrt{\frac{L_p}{4 \times \sqrt{A}}}$$

$$= \sqrt{\frac{268}{4\sqrt{4204.6}}}$$

$$= 1.03$$

and

$$n_c = \left[\frac{L_x \times L_y}{A} \right]^{\frac{0.7 \times A}{L_x \times L_y}}$$

$$= \left[\frac{84.6 \times 49.6}{4204.6} \right]^{\frac{0.7 \times 4204.6}{84.6 \times 49.6}}$$

$$= 1.00$$

and

$$n_d = \frac{D_m}{\sqrt{L_x^2 + L_y^2}}$$

$$= \frac{98.1}{\sqrt{84.6^2 + 49.6^2}}$$

$$= 1.00$$

The geometric factor is then calculated to be

$$n = n_a \times n_b \times n_c \times n_d$$

$$= 14 \times 1.03 \times 1 \times 1$$

$$= 14$$

Thus, they are approximately equal to 1 due to the shape of the grid.

Step 5: Grid resistance

A good grounding system provides a low resistance to remote earth in order to minimize the GPR. The next step is to evaluate the grid resistance by using Equation 4.103. All design parameters can be found in Table 4.11. Table 4.12 gives the approximate equivalent impedance of transmission line overhead shield wires and distribution feeder neutrals, according to their numbers. From Equation 4.168 for $L_T = 1901$ m, a grid area of $A = 4205$ m², $\rho = 50$ Ω-m, and $h = 1.524$ m, the grid resistance is

$$R_g = \rho \left[\frac{1}{L_T} + \frac{1}{\sqrt{20 \times A}} \left(1 + \frac{1}{1 + h \times \sqrt{20/A}} \right) \right]$$

$$= 50 \left[\frac{1}{1901} + \frac{1}{\sqrt{20 \times 4205}} \left(1 + \frac{1}{1 + 1.524 \times \sqrt{20/4205}} \right) \right]$$

$$= 0.35 \ \Omega$$

Step 6: Grid current

In step 6 of the logic flow diagram of the IEEE Std. 80-2000, the amount of current that flows within the designed grid (I_G) is determined. The GPR is determined as

$$V_{GPR} = I_G \times R_g$$

Determining the GPR and comparing it to the tolerable touch voltage is the first step to find out whether the grid design is a safe design for the people in and around the substation. The next step is to find the grid current I_G. But, it is first needed to determine the split factor from the following equation:

$$S_f = \left| \frac{Z_{eq}}{Z_{eq} + R_g} \right| \tag{4.170}$$

TABLE 4.12

Approximate Equivalent Impedance of Transmission Line Overhead Shield Wires and Distribution Feeder Neutrals

Number of Transmission Lines	Number of Distribution Neutrals	$R_{tg} = 15$ and $R_{d\phi} = 25$ $R + jX\ \Omega$	$R_{tg} = 15$ and $R_{d\phi} = 25$ $R + jX\ \Omega$
1	1	$0.91 + j0.485$ Ω	$3.27 + j0.652$ Ω
1	2	$0.54 + j0.33$ Ω	$2.18 + j0.412$ Ω
1	4	$0.295 + j0.20$ Ω	$1.32 + j0.244$ Ω
4	4	$0.23 + j0.12$ Ω	$0.817 + j0.16$ Ω
0	4	$0.322 + j0.242$ Ω	$1.65 + j0.291$ Ω

Since the substation has no impedance line shield wires and four distribution neutrals, from Table 4.12, the equivalent impedance can be found as $Z_{eq} = 0.322 + j0.242 \ \Omega$. Thus, $R_g = 1.0043 \ \Omega$ and a total fault current of $3I_{a0} = 17,000$ A, a decrement factor of $D_f = 1.026$. Thus, the current division factor (or the split factor) can be found as

$$S_f = \left| \frac{Z_{eq}}{Z_{eq} + R_g} \right|$$

$$= \left| \frac{(0.322 + j0.242)}{(0.322 + j0.242) + 1.0043} \right|$$

$$\cong 0.2548$$

since

$$I_g = S_f \times 3I_{a0}$$

$$= 0.2548 \times 17,000$$

$$= 4,331.6 \text{ A}$$

thus,

$$I_G = D_f \times I_g$$

$$= 1.026 \times 4,331.6$$

$$= 4,444.2 \text{ A}$$

Step 7: Determination of GPR

As said before, the product of I_G and R_g is the GPR. It is necessary to compare the GPR to the tolerable touch voltage, $V_{touch70}$. If the GPR is larger than the $V_{touch70}$, further design evaluations are necessary and the tolerable touch and step voltages should be compared to the maximum mesh and step voltages. Hence, first determine the GPR as

$$GPR = I_G \times R_g$$

$$= 4444.2 \times 0.35$$

$$= 1555.48 \text{ V}$$

Check to see whether

$$GPR > V_{touch70}$$

Indeed,

$$1555.2 \text{ V} > 871.5 \text{ V}$$

As it can be observed from the results, the GPR is much larger than the step voltage. Therefore, further design considerations are necessary and thus the step and mesh voltages must be calculated and compared to the tolerable touch and step voltage as follows.

Step 8: Mesh and step voltage calculations

a. Determination of the mesh voltage

In order to calculate the mesh voltage by using Equation 4.148, it is necessary first to calculate the variables K_h, K_m, and K_{ii}. Here, the correction factor that accounts for the depth of the grid (K_h) can be determined from Equation 4.152 as

$$K_h = \sqrt{1 + \frac{h}{h_0}}$$

$$= \sqrt{1 + \frac{1.524}{1}}$$

$$= 1.59$$

The corrective factor for grid geometry (K_{ii}) can be calculated from Equation 4.157 as

$$K_{ii} = 0.644 + 0.148 \times n$$

$$= 0.644 + 0.148 \times 14$$

$$= 2.716$$

b. Comparison of mesh voltage and allowable touch voltage

Using K_h, and K_{ii}, the spacing factor for mesh voltage (K_m) can be calculated. Here, the corrective weighting factor that can be used to adjust conductors on the corner mesh (K_{ii}) is considered to be 1.0 due to the shape of the grid being rectangular. Hence,

$$K_m = \frac{1}{2\pi}\left[\ln\left(\frac{D^2}{16 \times h \times d} + \frac{(D + 2 \times h)^2}{8 \times D \times d} - \frac{h}{4 \times d}\right) + \frac{K_{ii}}{K_h}\ln\left(\frac{8}{\pi(2 \times 14 - 1)}\right)\right]$$

$$= \frac{1}{2\pi}\left[\ln\left(\frac{4.97^2}{16 \times 1.524 \times 0.018} + \frac{(4.97 + 2 \times 1.524)^2}{8 \times 4.97 \times 0.018} - \frac{1.524}{4 \times 0.018}\right) + \frac{1}{1.589}\ln\left(\frac{8}{\pi(2 \times 14 - 1)}\right)\right]$$

$$= 0.53$$

Thus, the mesh voltage can now be calculated as

$$E_m = \frac{\rho \times I_G \times K_m \times K_{ii}}{L_C + \left[1.55 + 1.22\left(L_r/\sqrt{L_x^2 + L_y^2}\right)\right]L_R}$$

$$= \frac{50 \times 4,444.2 \times 0.53 \times 2.716}{1825 + \left[1.55 + 1.22\left(3.048/\sqrt{84.6^2 + 49.6^2}\right)\right]76.2}$$

$$= 164.39 \text{ V}$$

c. Determination of the step voltage

In order for the ground to be safe, step voltage has to be less than the tolerable step voltages. Also, step voltages within a grid system designed for mesh voltages will be well within the tolerable limits. To determine the step voltage (E_{step}), unknown variables of K_s and L_s are to be calculated. Thus, the spacing factor for step voltage (K_s) can be found from

$$K_s = \frac{1}{\pi}\left[\frac{1}{2\times h} + \frac{1}{D+h} + \frac{1}{D}(1-0.5^{n-2})\right]$$

$$= \frac{1}{\pi}\left[\frac{1}{2\times 1.524} + \frac{1}{3.97+1.524} + \frac{1}{4.97}(1-0.5^{14-2})\right]$$

$$= 0.22$$

The effective length (L_s) for the step voltage is

$$L_s = 0.75\times L_c + 0.85\times L_R$$

$$= 0.75\times 1825 + 0.85\times 76$$

$$= 1433.5\,\text{m}$$

Thus, the step voltage (E_{step}) determined as

$$E_{step} = \frac{\rho\times I_G\times K_s\times K_{ii}}{L_s}$$

$$= \frac{50\times 4,444.2\times 0.22_s\times 2.716}{1433.5}$$

$$= 92.62\,\text{V}$$

Step 9: Comparison of E_{mesh} versus V_{touch}

Here, the mesh voltage that is calculated in step 8 is compared with the tolerable touch voltages calculated in step 4. If the calculated mesh voltage E_{mesh} is greater than the tolerable $V_{touch70}$, further design evaluations are necessary. If the mesh voltage E_{mesh} is smaller than the $V_{touch70}$, then it can be moved to the next step and compare E_{step} with V_{step70}. Accordingly,

$$E_{mesh} < V_{touch70}(?)$$

$$164.39\,\text{V} < 871.5\,\text{V}$$

Here, the original grid design passes the second critical criteria in step 9. Hence, it can be moved to step 10 to find out whether the final criterion is met.

Step 10: Comparison of E_{step} versus V_{step70}

This is the final step that the design has to meet before the grounding system is considered safe. At this step, E_{step} is compared with the calculated tolerable step voltage V_{step70}. If

$$E_{step} > V_{step70}(?)$$

A refinement of the preliminary design is necessary and can be accomplished by decreasing the total grid resistance, closer grid spacing, adding more ground grid rods, if possible, and/or limiting the total fault current.

On the other hand, if

$$E_{step} < V_{step70}(?)$$

then the designed grounding grid is considerably safe. Since here,

$$92.62 \text{ V} < 2819.8 \text{ V}$$

then for the design,

$$E_{step} < V_{step70}$$

In summary, according to the calculations, the calculated mesh and step voltages are smaller than the tolerable touch and step voltages; therefore, in a typical shock situation, humans (weighting 70 kg) that become part of the circuit during a fault will have only what is considered a safe amount of current passing through their bodies.

There are many variables that can be changed in order to meet specifications for grounding a substation. Some variables include the size of the grid, the size of the conductors used, the amount of conductors used, and the spacing of each ground rod. After many processes an engineer has to go through, the project would then be put into construction if it is approved. Designing safe substation grounding is obviously not an easy task, but there are certain procedures that an engineer can follow to make the designing of substation grounding easier.

4.20 TRANSMISSION LINE GROUNDS

High-voltage transmission lines are designed and built to withstand the effects of lightning with a minimum damage and interruption of operation. If the lightning strikes an overhead ground wire (also called *static wire*) on a transmission line, the lightning current is conducted to ground through the ground wire installed along the pole or through the metal tower. The top of the line structure is raised in potential to a value determined by the magnitude of the lightning current and the surge impedance of the ground connection.

In the event that the impulse resistance of the ground connection is large, this potential can be in the magnitude of thousands of volts. If the potential is greater than the insulation level of the apparatus, a flashover will take place, causing an arc. The arc, in turn, will start the operation of protective relays, causing the line to be taken out of service. In the event that the transmission structure is well grounded and there is a sufficient coordination between the conductor insulation and the ground resistance, flashover can generally be avoided.

The transmission line grounds can be in various ways to achieve a low ground resistance. For example, a pole butt grounding plate or butt coil can be employed on wood poles. A butt coil is a spiral coil of bare copper wire installed at the bottom of a pole. The wire of the coil is extended up the pole as the ground wire lead. In practice, usually one or more ground rods are employed instead to achieve the required low ground resistance.

The sizes of the rods used are usually ⅝ or ¾ in. in diameter and 10 ft in length. The thickness of the rod does not play a major role in reducing the ground resistance as does the length of the rod. Multiple rods are usually used to provide the low ground resistance required by the high-capacity structures. But if the rods are moderately close to each other, the overall resistance will be more than if the same number of rods were spaced far apart. In other words, adding a second rod does not provide a total resistance of half that of a single rod unless the two are several rod lengths apart (actually infinite distance). Lewis [30] has shown that at 2 ft apart the resistance of two pipes (used as ground rods) in parallel is about 61% of the resistance of one of them, and at 6 ft apart it is about 55% of the resistance of one pipe.

Where there is bedrock near the surface or where sand is encountered, the soil is usually very dry and therefore has high resistivity. Such situations may require a grounding system known as the *counterpoise*, made of buried metal (usually galvanized steel wire) strips, wires, or cables. The counterpoise for an overhead transmission line consists of a special grounding terminal that reduces the surge impedance of the ground connection and increases the coupling between the ground wire and the conductors.

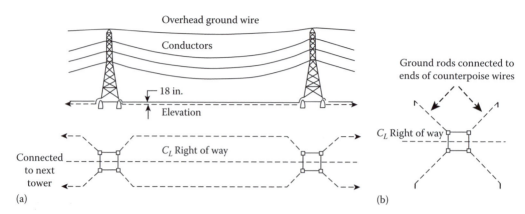

FIGURE 4.59 Two basic types of counterpoises: (a) continuous (parallel) and (b) radial. (From Gonen, T., *Electric Power Transmission System Engineering*, 2nd edn., CRC Press, Boca Raton, FL, 2009.)

The basic types of counterpoises used for transmission lines located in areas with sandy soil or rock close to the surface are the continuous type (also called the *parallel type*) and the radial (also called the *crowfoot type*), as shown in Figure 4.59. The continuous counterpoise is made of one or more conductors buried under the transmission line for its entire length.

The counterpoise wires are connected to the overhead ground (or *static*) wire at all towers or poles. But, the radial-type counterpoise is made of a number of wires and extends radially (in some fashion) from the tower legs. The number and length of the wires are determined by the tower location and the soil conditions. The counterpoise wires are usually installed with a cable plow at a length of 18 in. or more so that they will not be disturbed by cultivation of the land.

A multigrounded, common neutral conductor used for a primary distribution line is always connected to the substation grounding system where the circuit originates and to all grounds along the length of the circuit. If separate primary and secondary neutral conductors are used, the conductors have to be connected together provided that the primary neutral conductor is effectively grounded. The resistance of a single buried horizontal wire, when it is used as radial counterpoise, can be expressed as [16]

$$R = \frac{\rho}{\pi \ell}\left(\ln \frac{2\ell}{2(ad)^{1/2}} - 1 \right) \quad \text{when } d \ll \ell \tag{4.171}$$

where
ρ is the ground resistivity, Ω-m
ℓ is the length of wire, m
a is the radius of wire, m
d is the burial depth, m

It is assumed that the potential is uniform over the entire length of the wire. This is only true when the wire has ideal conductivity. If the wire is very long, such as with the radial counterpoise, the potential is not uniform over the entire length of the wire. Hence, Equation 4.141 cannot be used. Instead, the resistance of such a continuous counterpoise when $\ell(r/\rho)^{1/2}$ is large can be expressed as

$$R = (r\rho)^{1/2} \coth\left[\ell\left(\frac{r}{\rho} \right)^{1/2} \right] \tag{4.172}$$

where r is the resistance of wire, Ω-m.

If the lightning current flows through a counterpoise, the effective resistance is equal to the surge impedance of the wire. The wire resistance decreases as the surge propagates along the wire. For a given length counterpoise, the transient resistance will diminish to the steady-state resistance if the same wire is used in several shorter radial counterpoises rather than as a continuous counterpoise. Thus, the first 250 ft of counterpoise is most effective when it comes to grounding of lightning currents.

4.21 TYPES OF GROUNDING

In general, transmission and subtransmission systems are solidly grounded. Transmission systems are usually connected to a grounded wye, but subtransmission systems are often connected in delta. Delta systems may also be grounded through grounding transformers. In most high-voltage systems, the neutrals are solidly grounded, that is, connected directly to the ground. The advantages of such grounding are as follows:

1. Voltages to ground are limited to the phase voltage.
2. Intermittent ground faults and high voltages due to arcing faults are eliminated.
3. Sensitive protective relays operated by ground fault currents clear these faults at an early stage.

The grounding transformers used are normally either small distribution transformers (are connected normally in wye–delta, having their secondaries in delta) or small grounding autotransformers with interconnected wye or *zigzag* windings, as shown in Figure 4.60. The three-phase autotransformer has a single winding. If there is a ground fault on any line, the ground current flows equally in the three legs of the autotransformer. The interconnection offers the minimum impedance to the flow of the single-phase fault current.

The transformers are only used for grounding and carry little current except during a ground fault. Because of that, they can be fairly small. Their ratings are based on the stipulation that they carry current for no more than 5 min since the relays normally operate long before that. The grounding transformers are connected to the substation ground.

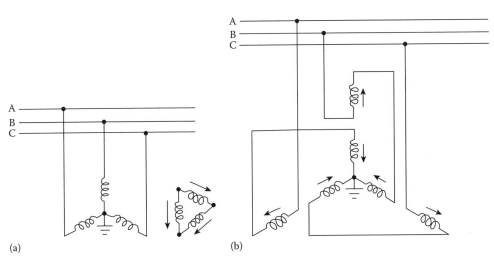

FIGURE 4.60 Grounding transformers used in delta-connected systems: (a) using grounded wye–delta-connected small distribution transformers or (b) using grounding autotransformers with interconnected wye or "zigzag" windings. (From Gonen, T., *Electric Power Transmission System Engineering*, 2nd edn., CRC Press, Boca Raton, FL, 2009.)

All substation equipment and structures are connected to the ground grid with large conductors to minimize the grounding resistance and limit the potential between equipment and the ground surface to a safe value under all conditions. All substation fences are built inside the ground grid and attached to the grid at short intervals to protect the public and personnel. Furthermore, the surface of the substation is usually covered with crushed rock or concrete to reduce the potential gradient when large currents are discharged to ground and to increase the contact resistance to the feet of personnel in the substation.

As said before, the substation grounding system is connected to every individual equipment, structure, and installation in order to provide the means by which grounding currents are conducted to remote areas. Thus, it is extremely important that the substation ground has a low ground resistance, adequate current-carrying capacity, and safety features for personnel.

It is crucial to have the substation ground resistance very low so that the total rise of the grounding system potential will not reach values that are unsafe for human contact. Therefore, the substation grounding system normally is made up of buried horizontal conductors and driven ground rods interconnected (by clamping, welding, or brazing) to form a *continuous grid* (also called *mat*) network.

Notice that a continuous cable (usually it is 4/0 bare stranded copper cable buried 12–18 in. below the surface) surrounds the grid perimeter to enclose as much ground as possible and to prevent current concentration and thus high gradients at the ground cable terminals. Inside the grid, cables are buried in parallel lines and with uniform spacing (e.g., about 10 × 20 ft).

Today, many utilities have computer programs for performing grounding grid studies. Thus, the number of tedious calculations that must be performed to develop an accurate and sophisticated model of a system is no longer a problem.

The GPR depends on grid burial depth, diameter, and length of conductors used, spacing between each conductor, fault-current magnitude, system voltage, ground system resistance, soil resistivity, distribution of current throughout the grid, proximity of the fault electrodes, and the system grounding electrodes to the conductors. IEEE Std. 80-1976 [14] provides a formula for a quick simple calculation of the grid resistance to ground after a minimum design has been completed. It is expressed as

$$R = \frac{\rho}{4r} + \frac{\rho}{L} \, \Omega \tag{4.173}$$

where
ρ is the soil resistivity, Ω-m
L is the total length of grid conductors, m
R is the radius of circle with area equal to that of grid, m

IEEE Std. 80-1976 also provides formulas to determine the effects of the grid geometry on the step and mesh voltage (which is the worst possible value of the touch voltage) in volts. They can be expressed as

$$V_{\text{step}} = \frac{K_s K_i \rho I_G}{L} \tag{4.174}$$

and

$$V_{\text{mesh}} = \frac{K_m K_i \rho I_G}{L} \tag{4.175}$$

where
K_s is the step coefficient
K_m is the mesh coefficient
K_i is the irregularity coefficient

Many utilities have computer programs for performing grounding grid studies. The number of tedious calculations that must be performed to develop an accurate and sophisticated model of a system is no longer a problem.

4.22 TRANSFORMER CLASSIFICATIONS

In power system applications, the single- or three-phase transformers with ratings up to 500 kVA and 34.5 kV are defined as distribution transformers, whereas those transformers with ratings over 500 kVA at voltage levels above 34.5 kV are defined as power transformers. Most distribution and power transformers are immersed in a tank of oil for better insulation and cooling purposes.

Today, various methods are in use in power transformers to get the heat pot of the tank more effectively. Historically, as the transformer sizes increased, the losses outgrew any means of self-cooling that was available at the time; thus, a water-cooling method was put into practice. This was done by placing metal coil tubing in the top oil, around the inside of the tank. Water was pumped through this cooling coil to get rid of the heat from oil.

Another method was circulating the hot oil through an external oil-to-water heat exchanger. This method is called *forced-oil-to-water cooling* (*FOW*). Today, the most common of these forced-oil-cooled power transformers uses an external bank of oil-to-air heat exchangers through which the oil is continuously pumped. It is known as type FOA.

In present practice, fans are automatically used for the first stage and pumps for the second, in triple-rated transformers that are designated as type *OA/FA/FOA*. These transformers carry up to about 60% of maximum nameplate rating (i.e., *FOA* rating) by natural circulation of the oil (*OA*) and 80% of maximum nameplate rating by forced cooling that consists of fans on the radiators (*FA*). Finally, at maximum nameplate rating (*FOA*), not only is oil forced to circulate through external radiators, but fans are also kept on to blow air onto the radiators as well as into the tank itself. In summary, the power transformer classes are as follows:

OA: Oil-immersed, self-cooled
OW: Oil-immersed, water-cooled
OA/FA: Oil-immersed, self-cooled/forced-air-cooled
OA/FA/FOA: Oil-immersed, self-cooled/forced-air-cooled/forced-oil-cooled
FOA: Oil-immersed, forced-oil-cooled with forced-air cooler
FOW: Oil-immersed, forced-oil-cooled with water cooler

In a distribution substation, power transformers are used to provide the conversion from subtransmission circuits to the distribution level. Most are connected in delta–wye grounded to provide ground source for the distribution neutral and to isolate the distribution grounding system from the subtransmission system.

Substation transformers can range from 5 MVA in smaller rural substations to over 80 MVA at urban stations (in terms of base ratings). As said earlier, power transformers have multiple ratings, depending on cooling methods. The base rating is the self-cooled rating, just due to the natural flow to the surrounding air through radiators. The transformer can supply more load with extra cooling turned on, as explained before.

However, the ANSI ratings were revised in the year 2000 to make them more consistent with IEC designations. This system has a four-letter code that indicates the cooling (IEEE C57.12.00-2000):

First letter—Internal cooling medium in contact with the windings:

O: Mineral oil or synthetic insulating liquid with fire point = 300°C
K: Insulating liquid with fire point >300°C
L: Insulating liquid with no measurable fire point

TABLE 4.13
Equivalent Cooling Classes

Year 2000 Designations	Designation prior to Year 2000
ONAN	OA
ONAF	FA
ONAN/ONAF/ONAF	OA/FA/FA
ONAN/ONAF/OFAF	OA/FA/FOA
OFAF	FOA
OFWF	FOW

Source: IEEE Std. C57.12.00-2000, IEEE Standard, General Requirements for
Liquid-Immersed Distribution, Power and Regulating Transformers, 2000.

Second letter—Circulation mechanism for internal cooling medium:

N: Natural convection flow through cooling equipment and in windings
F: Forced circulation through cooling equipment (i.e., *coolant pumps*); natural convection
flow in windings (also called *nondirected flow*)
D: Forced circulation through cooling equipment, directed from the cooling equipment into
at least the main windings

Third letter—External cooling medium:

A: Air
W: Water

Fourth letter—Circulation mechanism for external cooling medium:

N: Natural convection
F: Forced circulation: Fans (*air cooling*), pumps (*water cooing*)

Therefore, *OA/FA/FOA* is equivalent to *ONAA/ONAF/OFAF*. Each cooling level typically provides
an extra one-third capability: 21/28/35 MVA. Table 4.13 shows the equivalent cooling classes in old
and new naming schemes.

Utilities do not overload substation transformers as much as distribution transformers, but they
do not run them hot at times. As with distribution transformers, the tradeoff is loss of life versus the
immediate replacement cost of the transformer. Ambient conditions also affect loading. Summer
peaks are much worse than winter peaks. IEEE Std. C57.91-1995 provides detailed loading guidelines and also suggests an approximate adjustment of 1% of the maximum nameplate rating for
every degree C above or below 30°C.

The hottest-spot conductor temperature is the critical point where insulation degrades. Above the
hot-spot conductor temperature of 110°C, life expectancy of a transformer decreases exponentially.
The life of a transformer halves for every 8°C *increase in operating temperature*. Most of the time,
the hottest temperatures are nowhere near this. The impedance of substation transformers is normally
about 7%–10%. This is the impedance on the base rating, the self-cooled rating (OA or ONAN).

PROBLEMS

4.1   Verify Equation 4.17.
4.2   Derive Equation 4.44.
4.3   Prove that doubling feeder voltage level causes the percent voltage drop in the primary-feeder
circuit to be reduced to one-fourth of its previous value.

4.4 Repeat Example 4.2, parts (a) and (b), assuming a three-phase 34.5 kV wye-grounded feeder main that has 350 kcmil 19-strand copper conductors with an equivalent spacing of 37 in between phase conductors and a lagging-load power factor of 0.9.

4.5 Repeat part (a) of Problem 4.4, assuming 300 kcmil ACSR conductors.

4.6 Repeat Problem 4.5, assuming a lagging-load power factor of 0.7.

4.7 Repeat Problem 4.6, assuming AWG #4/0 conductors.

4.8 Repeat Example 4.3, assuming ACSR conductors.

4.9 Repeat Example 4.4, assuming ACSR conductors.

4.10 Repeat Example 4.5, assuming ACSR conductors.

4.11 Repeat Example 4.6, assuming ACSR conductors.

4.12 Repeat Example 4.8, assuming a 13.2/22.9 kV voltage level.

4.13 Repeat Example 4.9 for a load density of 1000 kVA/mil.

4.14 Repeat part d of Example 4.11 for a load density of 1000 kVA/mil.

4.15 A three-phase 34.5 kV wye-grounded feeder has 500 kcmil ACSR conductors with an equivalent spacing of 60 in. between phase conductors and a lagging-load power factor of 0.8. Use 25°C and 25 Hz and find the K constant in %VD per kVA per mile.

4.16 Assume a square-shaped distribution substation service area and that it is served by four three-phase 12.47 kV wye-grounded feeders. Feeder mains are of 2/0 copper conductors are made up of three-phase open-wire overhead lines having a geometric mean spacing of 37 in. between phase conductors. The percent voltage drop of the feeder is given as 0.0005 per kVA-mile. If the uniformly distributed load has a 4 MVA per square mile load density and a lagging-load factor of 0.9, and conductor ampacity is 360 A, find the following:
   (a)   Maximum load per feeder
   (b)   Substation size
   (c)   Substation spacing, both ways
   (d)   Total percent voltage drop from the feed point to the end of the main

4.17 Repeat Problem 4.15 for a load density of 1000 kVA/mi.

4.18 Assume that a 5 mile long feeder is supplying a 2000 kVA load of increasing load density starting at a substation. If the K constant of the feeder is given as 0.00001%VD per kVA·mi, determine the following:
   (a)   The percent voltage drop in the main.
   (b)   Repeat part (a) but assume that the load is a lumped-sum load and connected at the end of the feeder.
   (c)   Repeat part (a) but assume that the load is distributed uniformly along the main.

4.19 Consider the two-transformer bank shown in Figure P3.1 of Problem 3.3. Connect them in open-delta primary and open-delta secondary.
   (a)   Draw and label the voltage-phasor diagram required for the open-delta primary and open-delta secondary on the given 0° reference line.
   (b)   Show the connections required for the open-delta primary and open-delta secondary. Show the dot markings.

4.20 A three-phase 12.47 kV wye-grounded feeder main has 250 kcmil with 19-strand, copper conductors with an equivalent spacing of 54 in. between phase conductors, and a lagging-load power factor of 0.85. Use 50°C and 60 Hz, and compute the K constant.

4.21 Suppose that a human being is a part of a 60 Hz electric power circuit for about 0.25 s and that the soil type is dry soil. Based on the IEEE Std. 80-1976, determine the following:
   (a)   Tolerable touch potential
   (b)   Tolerable step potential

4.22 Consider the square-shaped distribution substation given in Example 4.10. The dimension of the area is 2 × 2 miles and served by a 12,470 V (line-to-line) feeder main and laterals. The load density is 1200 kVA/mi² and is uniformly distributed, having a lagging power factor of 0.9. A young distribution engineer is considering selection of 4/0 copper conductors with 19 strands and 1/0 copper conductors, operating 60 Hz and 50°C, for main and laterals, respectively. The geometric mean distances are 53 in. for main and 37 in. for the lateral. If the width of the service area of a lateral is 528 ft, determine the following:

    a. The percent voltage drop at the end of the last lateral, if the laterals are also three-phase four-wire wye grounded.

    b. The percent voltage drop at the end of the last lateral, if the laterals are single-phase two-wire wye grounded. Apply Morrison's approximation. (Explain what is right or wrong in the parameter selection in the problem mentioned earlier.) Any suggestions?

4.23 Resolve Example 3.8 by using MATLAB. Assume that all the quantities remain the same.

REFERENCES

1. Fink, D. G. and H. W. Beaty: *Standard Handbook for Electrical Engineers*, 11th edn., McGraw-Hill, New York, 1978.
2. Seely, H. P.: *Electrical Distribution Engineering*, 1st edn., McGraw-Hill, New York, 1930.
3. Van Wormer, F. C.: Some aspects of distribution load area geometry, *AIEE Trans.*, 73(2), December 1954, 1343–1349.
4. Denton, W. J. and D. N. Reps: Distribution substation and primary feeder planning, *AIEE Trans.*, 74(3), June 1955, 484–499.
5. Westinghouse Electric Corporation: *Electric Utility Engineering Reference Book—Distribution Systems*, Vol. 3, East Pittsburgh, PA, 1965.
6. Morrison, C.: A linear approach to the problem of planning new feed points into a distribution system, *AIEE Trans.*, pt. III (PAS), December 1963, 819–832.
7. Sciaca, S. C. and W. R. Block: Advanced SCADA concepts, *IEEE Comput Appl Power*, 8(1), January 1995, 23–28.
8. Gönen, T. et al.: Toward automated distribution system planning, *Proceedings of the IEEE Control of Power Systems Conference*, Texas A& M University, College Station, TX, March 19–21, 1979, pp. 23–30.
9. Gönen, T.: Power distribution, in *The Electrical Engineering Handbook*, 1st edn., Academic Press, New York, 2005, pp. 749–759, Chapter 6.
10. Bricker, S., L. Rubin, and T. Gönen: Substation automation techniques and advantages, *IEEE Comp. Appl. Power*, 14(3), July 2001, 31–37.
11. Ferris, L. P. et al.: Effects of electrical shock on the heart, *Trans. Am. Inst. Electric. Eng.*, 55, 1936, 498–515.
12. Gönen, T.: *Modern Power System Analysis*, Wiley, New York, 1988.
13. IEEE Standard 399-1980: *Recommended Practice for Industrial and Commercial Power System Analysis*, 1980, IEEE, New York.
14. IEEE Standard 80-1976: *IEEE Guide for Safety in AC Substation Grounding*, 1976, IEEE, New York.
15. ABB Power T & D Company, Inc.: *Introduction to Integrated Resource T & D Planning*, Cary, NC, 1994.
16. McDonald, D. J.: Substation integration and automation, in *Electric Power Substation Engineering*, 2nd edn., CRC Press, Boca Raton, FL, 2003, Chapter 7, 7-1–7-22.
17. Keil, R. P.: Substation grounding, in J. D. McDonald, ed., *Electric Power Substation Engineering*, 2nd edn., CRC Press, Boca Raton, FL, 2007, Chapter 11, 11-1–11-23.
18. Farr, H. H.: *Transmission Line Design Manual*, U.S. Department of Interior, Water and Power Resources Service, Denver, CO, 1980.
19. Institute of Electrical and Electronics Engineers, IEEE Std. 80-2000: *IEEE Guide for Safety in AC Substation Grounding*, IEEE, Piscataway, NJ, 2000.
20. Sunde, E. D.: *Earth Conduction Effects in Transmission Systems*, Macmillan, New York, 1968.
21. Gönen, T.: *Engineering Economy for Engineering Managers: With Computer Applications*, Wiley, New York, 1990.
22. Gonen, T.: *Electric Power Transmission System Engineering*, 2nd edn., CRC Press, Boca Raton, FL, 2009.
23. National Bureau of Standards Technical Report 108, Department of Commerce, Washington DC, 1978.

5 Design Considerations of Primary Systems

Imagination is more important than knowledge.

Albert Einstein

The great end of learning is nothing else but to seek for the lost mind.

Mencius, Works, 299 BC

Earn your ignorance! Learn something about everything before you know nothing about anything.

Turan Gönen

5.1 INTRODUCTION

The part of the electric utility system that is between the distribution substation and the distribution transformers is called the *primary* system. It is made of circuits known as *primary feeders* or *primary distribution feeders.*

Figure 5.1 shows a one-line diagram of a typical primary distribution feeder. A feeder includes a "main" or main feeder, which usually is a three-phase four-wire circuit, and branches or laterals, which usually are single-phase or three-phase circuits tapped off the main. Also sublaterals may be tapped off the laterals as necessary. In general, laterals and sublaterals located in residential and rural areas are single phase and consist of one-phase conductor and the neutral. The majority of the distribution transformers are single phase and are connected between the phase and the neutral through fuse cutouts.

A given feeder is sectionalized by reclosing devices at various locations in such a manner as to remove the faulted circuit as little as possible so as to hinder service to as few consumers as possible. This can be achieved through the coordination of the operation of all the fuses and reclosers.

It appears that, due to growing emphasis on the service reliability, the protection schemes in the future will be more sophisticated and complex, ranging from manually operated devices to remotely controlled automatic devices based on supervisory controlled or computer-controlled systems.

The congested and heavy-load locations in metropolitan areas are served by using underground primary feeders. They are usually radial three-conductor cables. The improved appearance and less frequent trouble expectancy are among the advantages of this method. However, it is more expensive, and the repair time is longer than the overhead systems. In some cases, the cable can be employed as suspended on poles. The cost involved is greater than that of open wire but much less than that of underground installation. There are various and yet interrelated factors affecting the selection of a primary-feeder rating. Examples are

1. The nature of the load connected
2. The load density of the area served
3. The growth rate of the load
4. The need for providing spare capacity for emergency operations
5. The type and cost of circuit construction employed

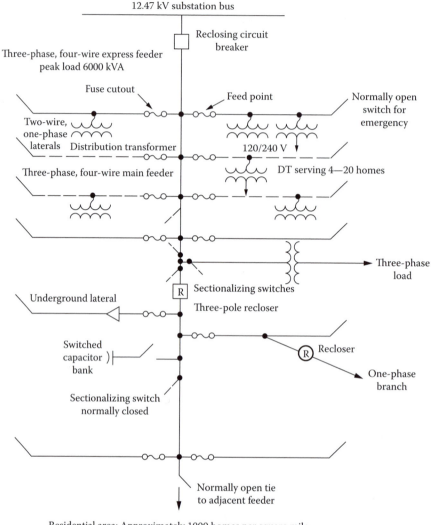

Residential area: Approximately 1000 homes per square mile
Feeder area: 1–4 min^2 depending on load density
15–30 single-phase laterals per feeder
150–500 MVA short-circuit available at substation bus

FIGURE 5.1   One-line diagram of typical primary distribution feeders. (From Fink, D.G. and Beaty, H.W., *Standard Handbook for Electrical Engineers*, 11th edn., McGraw-Hill, New York, 1978.)

6. The design and capacity of the substation involved
7. The type of regulating equipment used
8. The quality of service required
9. The continuity of service required

The voltage conditions on distribution systems can be improved by using shunt capacitors that are connected as near the loads as possible to derive the greatest benefit. The use of shunt capacitors also improves the power factor involved, which in turn lessens the voltage drops and currents, and therefore losses, in the portions of a distribution system between the capacitors and the bulk power buses. The capacitor ratings should be selected carefully to prevent the occurrence of excessive overvoltages at times of light loads due to the voltage rise produced by the capacitor currents.

The voltage conditions on distribution systems can also be improved by using series capacitors. But the application of series capacitors does not reduce the currents and therefore losses, in the system.

5.2 RADIAL-TYPE PRIMARY FEEDER

The simplest and the lowest cost and therefore the most common form of primary feeder is the radial-type primary feeder as shown in Figure 5.2. The main primary feeder branches into various primary laterals that in turn separates into several sublaterals to serve all the distribution transformers. In general, the main feeder and subfeeders are three-phase three- or four-wire circuits and the laterals are three phase or single phase. The current magnitude is the greatest in the circuit conductors that leave the substation. The current magnitude continually lessens out toward the end of the feeder as laterals and sublaterals are tapped off the feeder. Usually, as the current lessens, the size of the feeder conductors is also reduced. However, the permissible voltage regulation may restrict any feeder size reduction, which is based only on the thermal capability, that is, current-carrying capacity, of the feeder.

The reliability of service continuity of the radial primary feeders is low. A fault occurrence at any location on the radial primary feeder causes a power outage for every consumer on the feeder unless the fault can be isolated from the source by a disconnecting device such as a fuse, sectionalizer, disconnect switch, or recloser.

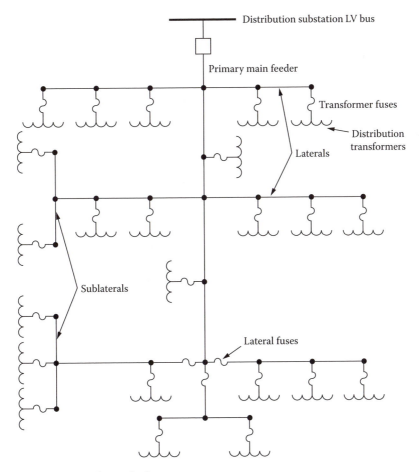

FIGURE 5.2 Radial-type primary feeder.

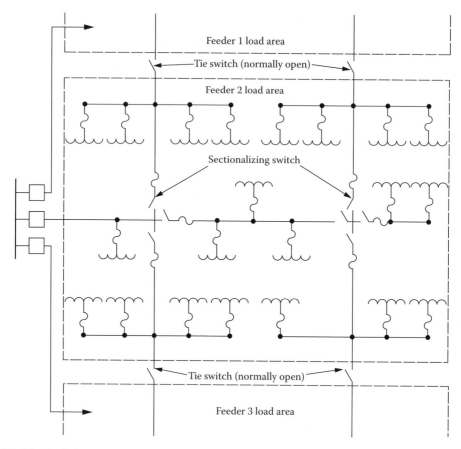

FIGURE 5.3 Radial-type primary feeder with tie and sectionalizing switches.

Figure 5.3 shows a modified radial-type primary feeder with tie and sectionalizing switches to provide fast restoration of service to customers by switching unfaulted sections of the feeder to an adjacent primary feeder or feeders. The fault can be isolated by opening the associated disconnecting devices on each side of the faulted section.

Figure 5.4 shows another type of radial primary feeder with express feeder and backfeed. The section of the feeder between the substation low-voltage bus and the load center of the service area is called an express feeder. No subfeeders or laterals are allowed to be tapped off the express feeder. However, a subfeeder is allowed to provide a backfeed toward the substation from the load center.

Figure 5.5 shows a radial-type phase-area feeder arrangement in which each phase of the three-phase feeder serves its own service area. In Figures 5.4 and 5.5, each dot represents a balanced three-phase load lumped at that location.

5.3 LOOP-TYPE PRIMARY FEEDER

Figure 5.6 shows a loop-type primary feeder that loops through the feeder load area and returns back to the bus. Sometimes the loop tie disconnect switch is replaced by a loop tie breaker due to the load conditions. In either case, the loop can function with the tie disconnect switches or breakers normally open (NO) or normally closed.

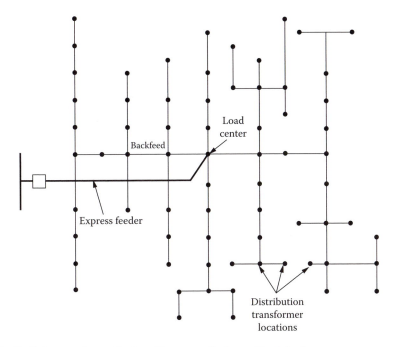

FIGURE 5.4 Radial-type primary feeder with express feeder and backfeed.

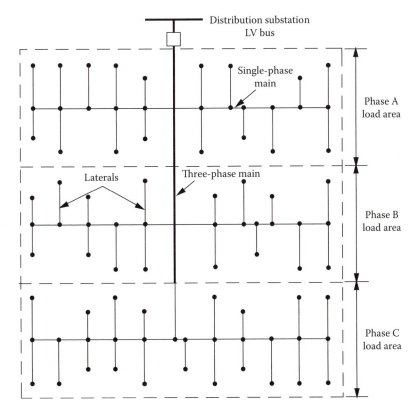

FIGURE 5.5 Radial-type phase-area feeder.

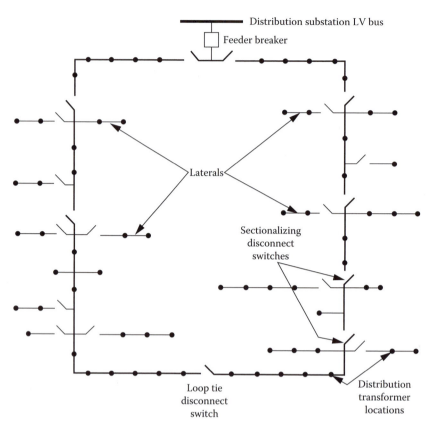

FIGURE 5.6 Loop-type primary feeder.

Usually, the size of the feeder conductor is kept the same throughout the loop. It is selected to carry its normal load plus the load of the other half of the loop. This arrangement provides two parallel paths from the substation to the load when the loop is operated with NO tie breakers or disconnect switches.

A primary fault causes the feeder breaker to be open. The breaker will remain open until the fault is isolated from both directions. The loop-type primary-feeder arrangement is especially beneficial to provide service for loads where high service reliability is important. In general, a separate feeder breaker on each end of the loop is preferred, despite the cost involved. The parallel feeder paths can also be connected to separate bus sections in the substation and supplied from separate transformers. In addition to main feeder loops, NO lateral loops are also used, particularly in underground systems.

5.4 PRIMARY NETWORK

As shown in Figure 5.7, a primary network is a system of interconnected feeders supplied by a number of substations. The radial primary feeders can be tapped off the interconnecting tie feeders. They can also be served directly from the substations. Each tie feeder has two associated circuit breakers at each end in order to have less load interrupted due to a tie-feeder fault.

The primary-network system supplies a load from several directions. Proper location of transformers to heavy-load centers and regulation of the feeders at the substation buses provide for adequate voltage at utilization points. In general, the losses in a primary network are lower than those in a comparable radial system due to load division.

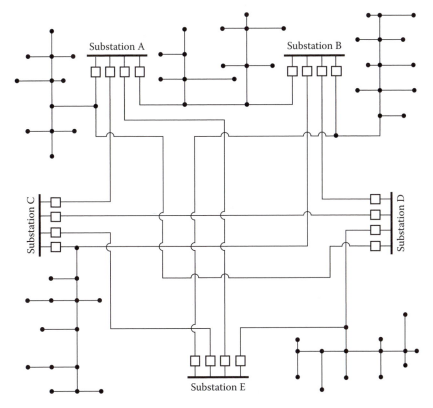

FIGURE 5.7 Primary network.

The reliability and the quality of service of the primary-network arrangement are much higher than the radial and loop arrangements. However, it is more difficult to design and operate than the radial or loop systems.

5.5 PRIMARY-FEEDER VOLTAGE LEVELS

The primary-feeder voltage level is the most important factor affecting the system design, cost, and operation. Some of the design and operation aspects affected by the primary-feeder voltage level are [2]

1. Primary-feeder length
2. Primary-feeder loading
3. Number of distribution substations
4. Rating of distribution substations
5. Number of subtransmission lines
6. Number of customers affected by a specific outage
7. System maintenance practices
8. The extent of tree trimming
9. Joint use of utility poles
10. Type of pole-line design and construction
11. Appearance of the pole line

There are additional factors affecting the decisions for primary-feeder voltage-level selection, as shown in Figure 5.8. Table 5.1 gives typical primary voltage levels used in the United States. Three-phase four-wire multigrounded common neutral primary systems, for example, 12.47Y/7.2 kV,

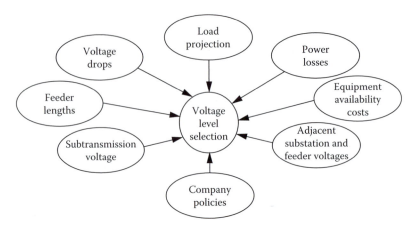

FIGURE 5.8 Factors affecting primary-feeder voltage-level selection decision.

TABLE 5.1
Typical Primary Voltage Levels

Class, kV	3φ Voltage	
2.5	2,300	3W-Δ
	2,400[a]	3W-Δ
5.0	4,000	3W-Δ or 3W-Y
	4,160[a]	4W-Y
	4,330	3W-Δ
	4,400	3W-Δ
	4,600	3W-Δ
	4,800	3W-Δ
8.66	6,600	3W-Δ
	6,900	3W-Δ or 4W-Y
	7,200[a]	3W-Δ or 4W-Y
	7,500	4W-Y
	8,320	4W-Y
15	11,000	3W-Δ
	11,500	3W-Δ
	12,000	3W-Δ or 4W-Y
	12,470[a]	4W-Y
	13,200[a]	3W-Δ or 4W-Y
	13,800[a]	3W-Δ
	14,400	3W-Δ
25	22,900[a]	4W-Y
	24,940[a]	4W-Y
34.5	34,500[a]	4W-Y

[a] Most common voltage in the individual classes.

24.9Y/14.4 kV, and 34.5Y/19.92 kV, are employed almost exclusively. The fourth wire is used as the multigrounded neutral for both the primary and secondary systems. The 15 kV-class primary voltage levels are most commonly used. The most common primary distribution voltage in use throughout North America is 12.47 kV. However, the current trend is toward higher voltages, for example, the 34.5 kV class is gaining rapid acceptance. The 5 kV class continues to decline in usage.

Some distribution systems use more than one primary voltage, for example, 12.47 and 34.5 kV. California is one of the few states that has three-phase three-wire primary systems. The four-wire system is economical, especially for underground residential distribution (URD) systems, since each primary lateral has only one insulated phase wire and the bare neutral instead of having two insulated wires.

Usually, primary feeders located in low-load density areas are restricted in length and loading by permissible voltage drop rather than by thermal restrictions, whereas primary feeders located in high-load density areas, for example, industrial and commercial areas, may be restricted by thermal limitations.

In general, for a given percent voltage drop, the feeder length and loading are direct functions of the feeder voltage level. This relationship is known as the *voltage-square rule*. For example, if the feeder voltage is doubled, for the same percent voltage drop, it can supply the same power four times the distance. However, as Lokay [2] explains it clearly, the feeder with the increased length feeds more load. Therefore, the advantage obtained by the new and higher-voltage level through the voltage-square factor, that is,

$$\text{Voltage-square factor} = \left(\frac{V_{L-N,\text{new}}}{V_{L-N,\text{old}}} \right)^2 \tag{5.1}$$

has to be allocated between the growth in load and in distance. Further, the same percent voltage drop will always result provided that the following relationship exists:

$$\text{Distance ratio} \times \text{Load ratio} = \text{Voltage-square factor} \tag{5.2}$$

where

$$\text{Distance ratio} = \frac{\text{New distance}}{\text{Old distance}} \tag{5.3}$$

and

$$\text{Load ratio} = \frac{\text{New feeder loading}}{\text{Old feeder loading}} \tag{5.4}$$

The relationship between the voltage-square factor rule and the feeder *distance-coverage principle* is further explained in Figure 5.9.

There is a relationship between the area served by a substation and the voltage rule. Lokay [2] defines it as the *area-coverage principle*. As illustrated in Figure 5.10, for a constant percent voltage drop and a uniformly distributed load, the feeder service area is proportional to

$$\left[\left(\frac{V_{L-N,\text{new}}}{V_{L-N,\text{old}}} \right)^2 \right]^{2/3} \tag{5.5}$$

provided that both dimensions of the feeder service area change by the same proportion. For example, if the new feeder voltage level is increased to twice the previous voltage level, the new load and area that can be served with the same percent voltage drop are

$$\left[\left(\frac{V_{L-N,\text{new}}}{V_{L-N,\text{old}}} \right)^2 \right]^{2/3} = (2^2)^{2/3} = 2.52 \tag{5.6}$$

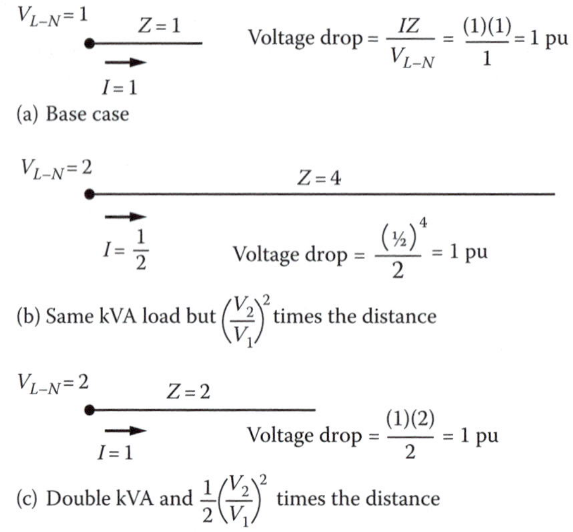

$V_{L-N} = 1$

$Z = 1$ Voltage drop $= \dfrac{IZ}{V_{L-N}} = \dfrac{(1)(1)}{1} = 1$ pu

$I = 1$

(a) Base case

$V_{L-N} = 2$

$Z = 4$

$I = \dfrac{1}{2}$ Voltage drop $= \dfrac{\left(\frac{1}{2}\right)^{4}}{2} = 1$ pu

(b) Same kVA load but $\left(\dfrac{V_2}{V_1}\right)^2$ times the distance

$V_{L-N} = 2$

$Z = 2$

$I = 1$ Voltage drop $= \dfrac{(1)(2)}{2} = 1$ pu

(c) Double kVA and $\dfrac{1}{2}\left(\dfrac{V_2}{V_1}\right)^2$ times the distance

FIGURE 5.9 Illustration of the voltage-square rule and the feeder distance-coverage principle as a function of feeder voltage level and a single load. (From Westinghouse Electric Corporation, *Electric Utility Engineering Reference Book-Distribution Systems*, Vol. 3, East Pittsburgh, Pittsburgh, PA, 1965.)

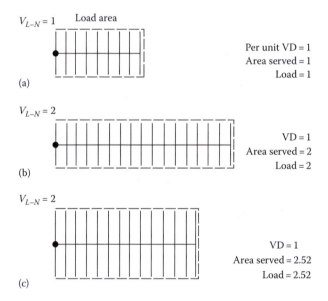

$V_{L-N} = 1$ Load area

Per unit VD $= 1$
Area served $= 1$
Load $= 1$

(a)

$V_{L-N} = 2$

VD $= 1$
Area served $= 2$
Load $= 2$

(b)

$V_{L-N} = 2$

VD $= 1$
Area served $= 2.52$
Load $= 2.52$

(c)

FIGURE 5.10 Feeder area-coverage principle as related to feeder voltage and a uniformly distributed load. (From Westinghouse Electric Corporation, *Electric Utility Engineering Reference Book-Distribution Systems*, Vol. 3, East Pittsburgh, Pittsburgh, PA, 1965.)

times the original load and area. If the new feeder voltage level is increased to three times the previous voltage level, the new load and area that can be served with the same percent voltage drop are

$$\left[\left(\frac{V_{L-N,\text{new}}}{V_{L-N,\text{old}}}\right)^2\right]^{2/3} = (3^2)^{2/3} = 4.32 \tag{5.7}$$

times the original load and area.

5.6 PRIMARY-FEEDER LOADING

Primary-feeder loading is defined as the loading of a feeder during peak-load conditions as measured at the substation [2]. Some of the factors affecting the design loading of a feeder are

1. The density of the feeder load
2. The nature of the feeder load
3. The growth rate of the feeder load
4. The reserve-capacity requirements for emergency
5. The service-continuity requirements
6. The service-reliability requirements
7. The quality of service
8. The primary-feeder voltage level
9. The type and cost of construction
10. The location and capacity of the distribution substation
11. The voltage regulation requirements

There are additional factors affecting the decisions for feeder routing, the number of feeders, and feeder conductor size selection, as shown in Figures 5.11 through 5.13.

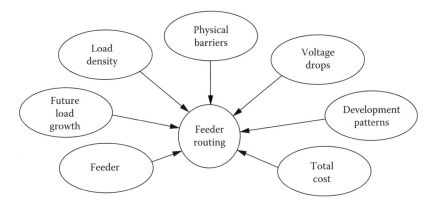

FIGURE 5.11   Factors affecting feeder routing decisions.

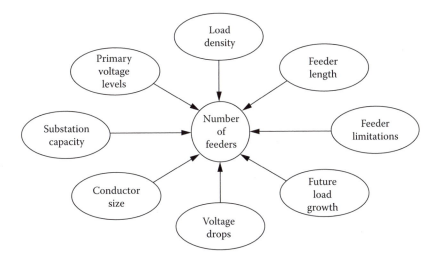

FIGURE 5.12   Factors affecting the number of feeders.

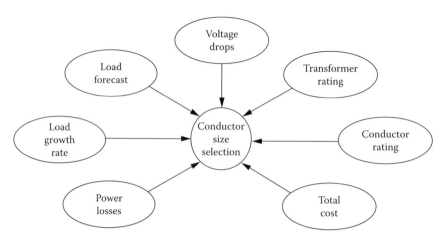

FIGURE 5.13   Factors affecting conductor size selection.

5.7   TIE LINES

A tie line is a line that connects two supply systems to provide emergency service to one system from another, as shown in Figure 5.14. Usually, a tie line provides service for area loads along its route as well as providing for emergency service to adjacent areas or substations. Therefore, tie lines are needed to perform either of the following two functions:

1. To provide emergency service for an adjacent feeder for the reduction of outage time to the customers during emergency conditions.
2. To provide emergency service for adjacent substation systems, thereby eliminating the necessity of having an emergency backup supply at every substation. Tie lines should be installed when more than one substation is required to serve the area load at one primary distribution voltage.

Usually the substation primary feeders are designed and installed in such an arrangement as to have the feeders supplied from the same transformer extend in opposite directions so that all required ties can be made with circuits supplied from different transformers. For example, a substation with two transformers and four feeders might have the two feeders from one transformer extending north and south. The two feeders from the other transformer may extend east and west. All tie lines should be made to circuits supplied by other transformers. This would make it much easier to restore service to an area that is affected by a transformer failure.

Disconnect switches are installed at certain intervals in main feeder tie lines to facilitate load transfer and service restoration. The location of disconnect switches needs to be selected carefully to obtain maximum operating flexibility. Not only the physical arrangement of the circuit but also the size and nature of loads between switches are important. Loads between the disconnect switches should be balanced as much as possible so that load transfers between circuits do not adversely affect circuit operation. The optimum voltage conditions are obtained only if the circuit is balanced as closely as possible throughout its length.

5.8   DISTRIBUTION FEEDER EXIT: RECTANGULAR-TYPE DEVELOPMENT

The objective of this section is to provide an example for a uniform area development plan to minimize the circuitry changes associated with the systematic expansion of the distribution system.

Assume that underground feeder exits are extended out of a distribution substation into an existing overhead system. Also assume that at the ultimate development of this substation, a 6 min^2 service

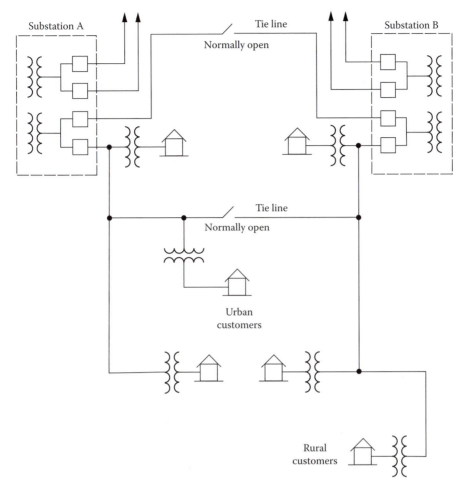

FIGURE 5.14 One-line diagram of typical two-substation area supply with tie lines.

area will be served with a total of 12 feeder circuits, four per transformer. Assuming uniform load distribution, each of the 12 circuits would serve approximately ½ mi² in a fully developed service area. This is called the *rectangular-type development* and illustrated in Figures 5.15 through 5.18.

In general, adjacent service areas are served from different transformer banks in order to provide for transfer to adjacent circuits in the event of transformer outages. The addition of new feeder circuits and transformer banks requires circuit number changes as the service area develops. The center transformer bank is always fully developed when the substation has eight feeder circuits. As the service area develops, the remaining transformer banks develop to full capacity. There are two basic methods of development, depending upon the load density of a service area, namely, the 1-2-4-8-12 *feeder circuit method* and the 1-2-4-6-8-12 *feeder circuit method*. The numbers shown for feeders and transformer banks in the following figures represent only the sequence of installation as the substation develops.

Method of development for high-load density areas: In service areas with high-load density, the adjacent substations are developed similarly to provide for adequate load-transfer capability and service continuity. Here, for example, a two-transformer-bank substation can carry a firm rating of the emergency rating of one bank plus circuit ties, plus reserve considerations. Since sufficient circuit ties must be available to support the loss of a large transformer unit, the 1-2-4-8-12 *feeder method* is especially desirable for a high-load density area. Figures 5.15 through 5.18 show the sequence of installing additional transformers and feeders.

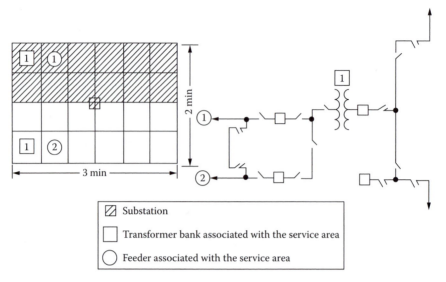

FIGURE 5.15 Rectangular-type development.

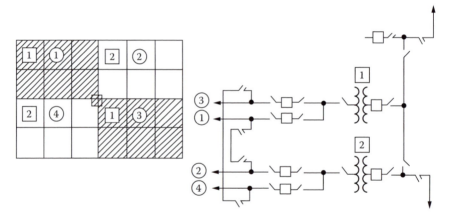

FIGURE 5.16 Rectangular-type development with two transformers, type 1.

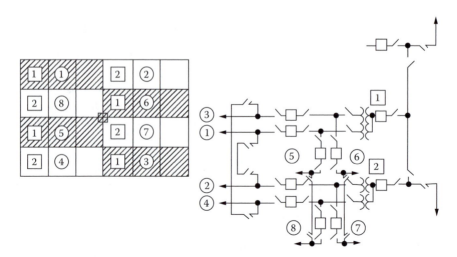

FIGURE 5.17 Rectangular-type development with two transformers, type 2.

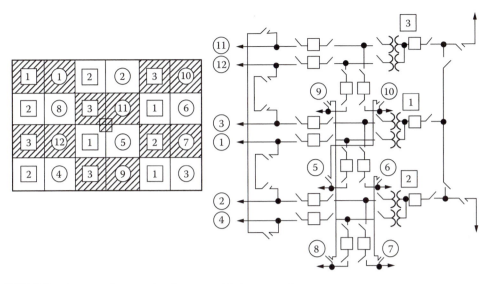

FIGURE 5.18 Rectangular-type development with three transformers.

Method of development for low-load density areas: In low-load density areas, where adjacent substations are not adequately developed and circuit ties are not available due to excessive distances between substations, the 1-2-4-6-8-12 *circuit-developing substation scheme* is more suitable. These large distances between substations generally limit the amount of load that can be transferred between substations without objectionable outage time due to circuit switching and guarantee that minimum voltage levels are maintained. This method requires the substation to have all three transformer banks before using the larger transformers in order to provide a greater firming capability within the individual substation.

As illustrated in Figures 5.19 through 5.23, once three, for example, 12/16/20-MVA, transformer units and six feeders are reached in the development of this type of substation, there are two alternatives for further expansion: (1) either remove one of the banks and increase the remaining two bank sizes to the larger, for example, 24/32/40 MVA, transformer units employing the low-side bays of

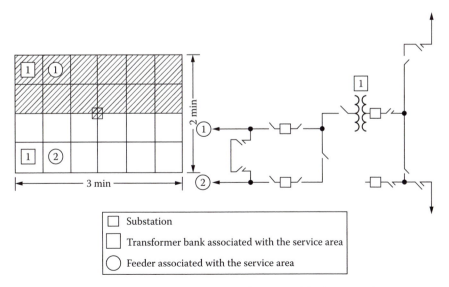

FIGURE 5.19 The sequence of installing additional transformers and feeders, type 1.

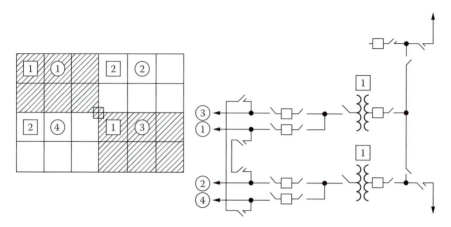

FIGURE 5.20 The sequence of installing additional transformers and feeders, type 2.

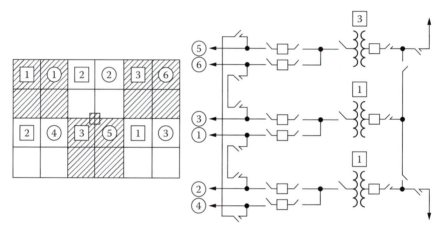

FIGURE 5.21 The sequence of installing additional transformers, type 3.

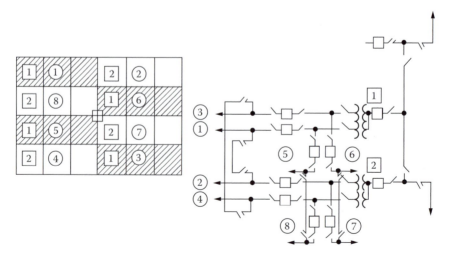

FIGURE 5.22 The sequence of installing additional transformers and feeders, type 4.

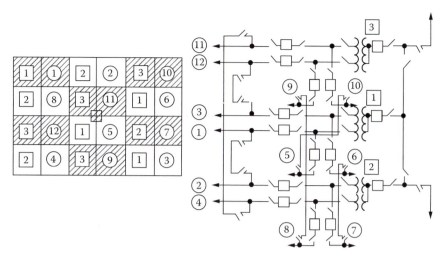

FIGURE 5.23 The sequence of installing additional transformers, type 5.

the third transformer as part of the circuitry in the development of the remaining two banks, or (2) completely ignore the third transformer-bank area and complete the development of two remaining sections similar to the previous method.

5.9 RADIAL-TYPE DEVELOPMENT

In addition to the rectangular-type development associated with overhead expansion, there is a second type of development that is due to the growth of URD subdivisions with underground feeders serving local load as they exit into the adjacent service areas. At these locations, the overhead feeders along the quarter section lines are replaced with underground cables, and as these underground lines extend outward from the substation, the area load is served. These underground lines extend through the platted service area developments and terminate usually on a remote overhead feeder along a section line. This type of development is called radial-type development, and it resembles a wagon wheel with the substation as the hub and the radial spokes as the feeders, as shown in Figure 5.24.

5.10 RADIAL FEEDERS WITH UNIFORMLY DISTRIBUTED LOAD

The single-line diagram, shown in Figure 5.25, illustrates a three-phase feeder main having the same construction, that is, in terms of cable size or open-wire size and spacing, along its entire length l. Here, the line impedance is $z = r + jx$ per unit length.

The load flow in the main is assumed to be perfectly balanced and uniformly distributed at all locations along the main. In practice, a reasonably good phase balance sometimes is realized when single-phase and open-wye laterals are wisely distributed among the three phases of the main.

Assume that there are many closely spaced loads and/or lateral lines connected to the main but not shown in Figure 5.25. Since the load is uniformly distributed along the main, as shown in Figure 5.26, the load current in the main is a function of the distance. Therefore, in view of the many closely spaced small loads, a differential tapped-off load current $d\overline{I}$, which corresponds to a dx differential distance, is to be used as an idealization. Here, l is the total length of the feeder and x is the distance of the point 1 on the feeder from the beginning end of the feeder. Therefore, the distance of point 2 on the feeder from the beginning end of the feeder is $x + dx$. $\overline{I}_s$ is the sending-end current at the feeder breaker, and $\overline{I}_r$ is the receiving-end current. $\overline{I}_{x1}$ and $\overline{I}_{x2}$ are the currents in the main at points 1 and 2, respectively. Assume that all loads connected to the feeder have the same power factor.

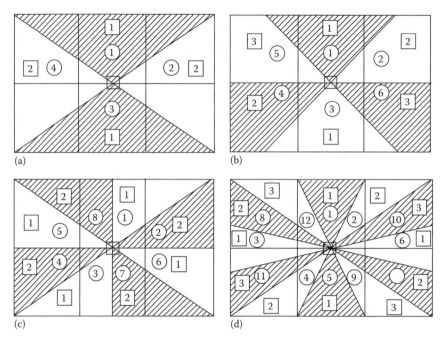

(a)   (b)

(c)   (d)

FIGURE 5.24 Radial-type development: (a) type 1, (b) type 2, (c) type 3, and (d) type 4.

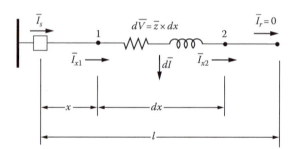

FIGURE 5.25 A radial feeder.

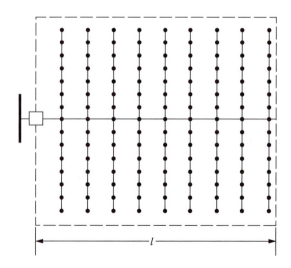

FIGURE 5.26 A uniformly distributed main feeder.

The following equations are valid both in per-unit or per-phase (line-to-neutral) dimensional variables. The circuit voltage is of either primary or secondary, and therefore shunt capacitance currents may be neglected. Since the total load is uniformly distributed from $x = 0$ to $x = \ell$,

$$\frac{d\bar{I}}{dx} = \bar{k} \tag{5.8}$$

which is a constant. Therefore $\bar{I}_x$, that is, the current in the main of some x distance away from the circuit breaker, can be found as a function of the sending-end current $\bar{I}_s$ and the distance x. This can be accomplished either by inspection or by writing a current equation containing the integration of $d\bar{I}$. Therefore, for dx distance,

$$\bar{I}_{x1} = \bar{I}_{x2} + d\bar{I} \tag{5.9}$$

or

$$\bar{I}_{x2} = \bar{I}_{x1} - d\bar{I} \tag{5.10}$$

From Equation 5.10,

$$\bar{I}_{x2} = \bar{I}_{x1} - d\bar{I}\,\frac{dx}{dx}$$

$$= \bar{I}_{x1} - \frac{d\bar{I}}{dx}\,dx \tag{5.11}$$

or

$$\bar{I}_{x2} = \bar{I}_{x1} - \bar{k}\,dx \tag{5.12}$$

where $\bar{k} = d\bar{I}/dx$ or, approximately,

$$\bar{I}_{x2} = \bar{I}_{x1} - k\,d\bar{I} \tag{5.13}$$

and

$$\bar{I}_{x1} = \bar{I}_{x2} + k\,d\bar{I} \tag{5.14}$$

Therefore, for the total feeder,

$$I_r = I_s - k \times l \tag{5.15}$$

and

$$I_s = I_r + k \times l \tag{5.16}$$

When $x = l$, from Equation 5.15,

$$I_r = I_s - k \times l = 0$$

hence

$$k = \frac{I_s}{l} \tag{5.17}$$

and since $x = l$,

$$I_r = I_s - k \times x \tag{5.18}$$

Therefore, substituting Equation 5.17 into Equation 5.18,

$$I_r = I_s\left(1 - \frac{x}{l}\right) \tag{5.19}$$

For a given x distance,

$$I_x = I_r$$

thus Equation 5.19 can be written as

$$I_x = I_s\left(1 - \frac{x}{l}\right) \tag{5.20}$$

which gives the current in the main at some x distance away from the circuit breaker. Note that from Equation 5.20,

$$I_s = \begin{cases} I_r = 0 & \text{at } x = l \\ I_r = I_s & \text{at } x = 0 \end{cases}$$

The differential series voltage drop $d\overline{V}$ and the differential power loss dP_{LS} due to I^2R losses can also be found as a function of the sending-end current I_s and the distance x in a similar manner.

Therefore, the differential series voltage drop can be found as

$$d\overline{V} = I_x \times z dx \tag{5.21}$$

or substituting Equation 5.20 into Equation 5.21,

$$d\overline{V} = I_s \times z\left(1 - \frac{x}{\ell}\right)dx \tag{5.22}$$

Also, the differential power loss can be found as

$$dP_{LS} = I_x^2 \times r dx \tag{5.23}$$

or substituting Equation 5.20 into Equation 5.23,

$$dP_{LS} = \left[I_s\left(1 - \frac{x}{\ell}\right)\right]^2 r dx \tag{5.24}$$

The series voltage drop VD_x due to I_x current at any point x on the feeder is

$$VD_x = \int_0^x dV \qquad (5.25)$$

Substituting Equation 5.22 into Equation 5.25,

$$VD_x = \int_0^x I_s \times z \left(1 - \frac{x}{\ell}\right) dx \qquad (5.26)$$

or

$$VD_x = I_s \times z \times x \left(1 - \frac{x}{2l}\right) \qquad (5.27)$$

Therefore, the total series voltage drop $\sum VD_x$ on the main feeder when $x = l$ is

$$\sum VD_x = I_s \times z \times \ell \left(1 - \frac{\ell}{2\ell}\right)$$

or

$$\sum VD_x = \frac{1}{2} z \times \ell \times I_s \qquad (5.28)$$

The total copper loss per phase in the main due to I^2R losses is

$$\sum P_{LS} = \int_0^l dP_{LS} \qquad (5.29)$$

or

$$\sum P_{LS} = \frac{1}{3} I_s^2 \times r \times \ell \qquad (5.30)$$

Therefore, from Equation 5.28, the distance x from the beginning of the main feeder at which location the total load current I_s may be concentrated, that is, lumped for the purpose of calculating the total voltage drop, is

$$x = \frac{\ell}{2}$$

whereas, from Equation 5.30, the distance x from the beginning of the main feeder at which location the total load current I_s may be lumped for the purpose of calculating the total power loss is

$$x = \frac{\ell}{3}$$

5.11 RADIAL FEEDERS WITH NONUNIFORMLY DISTRIBUTED LOAD

The single-line diagram, shown in Figure 5.27, illustrates a three-phase feeder main, which has the tapped-off load increasing linearly with the distance x. Note that the load is zero when $x = 0$. The plot of the sending-end current vs. the x distance along the feeder main gives the curve shown in Figure 5.28.

From Figure 5.28, the negative slope can be written as

$$\frac{dI_x}{dx} = -k \times I_s \times x \qquad (5.31)$$

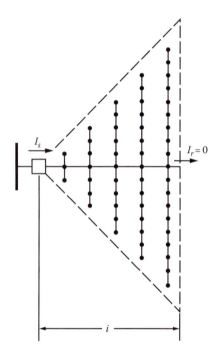

FIGURE 5.27 A uniformly increasing load.

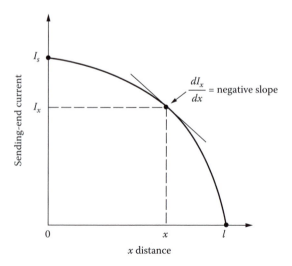

FIGURE 5.28 The sending-end current as a function of the distance along a feeder.

Here, the constant k can be found from

$$I_s = \int_{x=0}^{l} -dIx$$

$$= \int_{x=0}^{l} k \times I_s \times x dx \tag{5.32}$$

or

$$I_s = k \times I_s \times \frac{\ell^2}{2} \tag{5.33}$$

From Equation 5.33, the constant k is

$$k = \frac{2}{\ell^2} \tag{5.34}$$

Substituting Equation 5.34 into Equation 5.31,

$$\frac{dI_x}{dx} = -2I_s \times \frac{x}{\ell^2} \tag{5.35}$$

Therefore, the current in the main at some x distance away from the circuit breaker can be found as

$$I_x = I_s \left(1 - \frac{x^2}{\ell^2} \right) \tag{5.36}$$

Hence, the differential series voltage drop is

$$d\overline{V} = I_x \times z dx \tag{5.37}$$

or

$$d\overline{V} = I_s \times z \left(1 - \frac{x^2}{\ell^2} \right) dx \tag{5.38}$$

Also, the differential power loss can be found as

$$dP_{LS} = I_x^2 \times r dx \tag{5.39}$$

or

$$dP_{LS} = I_s^2 \times r \left(1 - \frac{x^2}{\ell^2} \right)^2 dx \tag{5.40}$$

The series voltage drop due to I_x current at any point x on the feeder is

$$VD_x = \int_0^x dV \tag{5.41}$$

Substituting Equation 5.38 into Equation 5.41 and integrating the result,

$$VD_x = I_s \times z \times x \left(1 - \frac{x^2}{3\ell^2}\right) \tag{5.42}$$

Therefore, the total series voltage drop on the main feeder when $x = 1$ is

$$\sum VD_x = \frac{2}{3} z \times \ell \times I_s \tag{5.43}$$

The total copper loss per phase in the main due to I^2R losses is

$$\sum P_{LS} = \int_0^l dP_{LS} \tag{5.44}$$

or

$$\sum P_{LS} = \frac{8}{15} I_s^2 \times r \times \ell \tag{5.45}$$

5.12 APPLICATION OF THE A, B, C, D GENERAL CIRCUIT CONSTANTS TO RADIAL FEEDERS

Assume a single-phase or balanced three-phase transmission or distribution circuit characterized by the $\bar{A}$, $\bar{B}$, $\bar{C}$, $\bar{D}$ general circuit constants, as shown in Figure 5.29. The mixed data assumed to be known, as commonly encountered in system design, are $|\bar{V}_S|$, P_r, and $\cos\theta$. Assume that all data represent either per-phase dimensional values or per unit values.

As shown in Figure 5.30, taking phasor $\bar{V}_r$ as the reference,

$$\bar{V}_r = V_r \angle 0° \tag{5.46}$$

$$\bar{V}_s = V_s \angle \delta \tag{5.47}$$

FIGURE 5.29 A symbolic representation of a line.

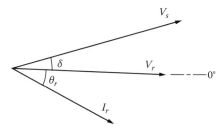

FIGURE 5.30 Phasor diagram.

$$\bar{I}_r = I_r \angle -\theta_r \tag{5.48}$$

where
$\bar{V}_r$ = receiving-end voltage phasor
$\bar{V}_s$ = sending-end voltage phasor
$\bar{I}_r$ = receiving-end current phasor

The sending-end voltage in terms of the general circuit constants can be expressed as

$$\bar{V}_s = \bar{A} \times \bar{V}_r + \bar{B} \times \bar{I}_r \tag{5.49}$$

where

$$\bar{A} = \bar{A}_1 + j\bar{A}_2 \tag{5.50}$$

$$\bar{B} = \bar{B}_1 + j\bar{B}_2 \tag{5.51}$$

$$\bar{I}_r = I_r(\cos\theta_r - j\sin\theta_r) \tag{5.52}$$

$$\bar{V}_r = V_r\angle 0° = V_r \tag{5.53}$$

$$\bar{V}_s = V_s(\cos\delta + j\sin\delta) \tag{5.54}$$

Therefore, Equation 5.49 can be written as

$$V_s\cos\delta + jV_s\sin\delta = (A_1 + jA_2)V_r + (B_1 + jB_2)(I_r\cos\theta_r - jI_r\sin\theta_r)$$

from which

$$V_s\cos\delta = A_1V_r + B_1I_r\cos\theta_r + B_2I_r\sin\theta_r \tag{5.55}$$

and

$$V_s\sin\delta = A_2V_r + B_2I_r\cos\theta_r - B_1I_r\sin\theta_r \tag{5.56}$$

By taking squares of Equations 5.55 and 5.56, and adding them side by side,

$$V_s^2 = (A_1 V_r + B_1 I_r \cos\theta_r + B_2 I_r \sin\theta_r)^2 + (A_2 V_r + B_2 I_r \cos\theta_r - B_1 I_r \sin\theta_r)^2 \qquad (5.57)$$

or

$$V_s^2 = V_r^2 \left(A_1^2 + A_2^2 \right) + 2 V_r I_r \cos\theta_r (A_1 B_1 + A_2 B_2) + B_1^2 \left(V_r^2 \cos^2\theta_r + I_r^2 \sin^2\theta_r \right)$$

$$+ B_2^2 \left(I_r^2 \sin^2\theta_r + I_r^2 \cos^2\theta_r \right) + 2 V_r I_r \sin\theta_r (A_1 B_2 - B_1 A_2) \qquad (5.58)$$

Since

$$P_r = V_r I_r \cos\theta_r \qquad (5.59)$$

$$Q_r = V_r I_r \sin\theta_r \qquad (5.60)$$

and

$$Q_r = P_r \tan\theta \qquad (5.61)$$

Equation 5.58 can be rewritten as

$$V_r^2 \left(A_1^2 + A_2^2 \right) + \left(B_1^2 + B_2^2 \right)\left(1 + \tan^2\theta_r \right)\frac{P_r^2}{V_r^2} = V_s^2 - 2 P_r [(A_1 B_1 + A_2 B_2) + (A_1 B_2 - B_1 A_2)\tan\theta_r] \quad (5.62)$$

Let

$$\widehat{K} = V_s^2 - 2 P_r [(A_1 B_1 + A_2 B_2) + (A_1 B_2 - B_1 A_2)\tan\theta_r] \qquad (5.63)$$

Then Equation 8.62 becomes

$$V_r^2 \left(A_1^2 + A_2^2 \right) + \left(B_1^2 + B_2^2 \right)\left(1 + \tan^2\theta_r \right)\frac{P_r^2}{V_r^2} - \widehat{K} = 0 \qquad (5.64)$$

or

$$V_r^2 \left(A_1^2 + A_2^2 \right) + \left(B_1^2 + B_2^2 \right)\left(\sec^2\theta_r \right)\frac{P_r^2}{V_r^2} - \widehat{K} = 0 \qquad (5.65)$$

Therefore, from Equation 5.65, the receiving-end voltage can be found as

$$V_r = \left\{ \frac{\widehat{K} \pm \left[\widehat{K}^2 - 4\left(A_1^2 + A_2^2 \right)\left(B_1^2 + B_2^2 \right) P_r^2 \sec^2\theta_r \right]^{1/2}}{2\left(A_1^2 + A_2^2 \right)} \right\}^{1/2} \qquad (5.66)$$

Also, from Equations 5.55 and 5.56,

$$V_s \sin \delta = A_2 V_r + B_2 I_r \cos \theta_r - B_1 I_r \sin \theta_r$$

and

$$V_s \cos \delta = A_1 V_r + B_1 I_r \cos \theta_r - B_2 I_r \sin \theta_r$$

where

$$I_r = \frac{P_r}{V_r \cos \theta_r} \tag{5.67}$$

Therefore,

$$V_s \sin \delta = A_2 V_r + \frac{B_2 P_r}{V_r} - \frac{B_1 P_r}{V_r} \tan \theta_r \tag{5.68}$$

and

$$V_s \cos \delta = A_1 V_r + \frac{B_1 P_r}{V_r} + \frac{B_2 P_r}{V_r} \tan \theta_r \tag{5.69}$$

By dividing Equation 5.68 by Equation 5.69,

$$\tan \delta = \frac{A_2 V_r^2 + B_2 P_r - B_1 P_r \tan \theta_r}{A_1 V_r^2 + B_1 P_r + B_2 P_r \tan \theta_r} \tag{5.70}$$

or

$$\tan \delta = \frac{A_2 V_r^2 + P_r(B_2 - B_1 \tan \theta_r)}{A_1 V_r^2 + P_r(B_1 + B_2 \tan \theta_r)} \tag{5.71}$$

Equations 5.66 and 5.71 are found for a general transmission system. They could be adapted to the simpler transmission consisting of a short primary-voltage feeder where the feeder capacitance is usually negligible, as shown in Figure 5.31.

To achieve the adaptation, Equations 5.63, 5.66, and 5.71 can be written in terms of R and X. Therefore, for the feeder shown in Figure 5.31,

$$[\bar{I}] = [\bar{Y}][\bar{V}] \tag{5.72}$$

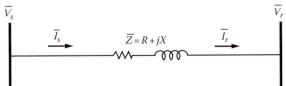

FIGURE 5.31   A radial feeder.

or

$$\begin{bmatrix} \bar{I}_s \\ \bar{I}_r \end{bmatrix} = \begin{bmatrix} \bar{Y}_{11} & \bar{Y}_{12} \\ \bar{Y}_{21} & \bar{Y}_{22} \end{bmatrix} \begin{bmatrix} \bar{V}_s \\ \bar{V}_r \end{bmatrix} \tag{5.73}$$

where

$$\bar{Y}_{11} = \frac{1}{\bar{Z}} \tag{5.74}$$

$$\bar{Y}_{21} = \bar{Y}_{12} - \frac{1}{\bar{Z}} \tag{5.75}$$

$$\bar{Y}_{22} = \frac{1}{\bar{Z}} \tag{5.76}$$

Therefore,

$$A_1 = -\frac{\bar{Y}_{22}}{\bar{Y}_{21}} = 1 \tag{5.77}$$

or

$$A_1 + jA_2 = 1 \tag{5.78}$$

where

$$A_1 = 1 \tag{5.79}$$

and

$$A_2 = 0 \tag{5.80}$$

Similarly,

$$\bar{B}_1 = -\frac{1}{\bar{Y}_{21}} = \bar{Z} \tag{5.81}$$

or

$$B_1 + jB_2 = R + jX \tag{5.82}$$

where

$$B_1 = R \tag{5.83}$$

and

$$B_2 = X \tag{5.84}$$

Substituting Equations 5.79, 5.80, 5.83, and 5.84 into Equation 5.66,

$$V_r = \left\{ \frac{\widehat{K} \pm \left[\widehat{K}^2 - 4(R^2 + X^2)P_r^2 \sec^2 \theta_r \right]^{1/2}}{2} \right\}^{1/2} \tag{5.85}$$

or

$$V_r = \left(\frac{\widehat{K}}{2} \left\{ 1 \pm \left[1 - \frac{4(R^2 + X^2)P_r^2}{\widehat{K}^2 \cos^2 \theta_r} \right]^{1/2} \right\} \right)^{1/2} \tag{5.86}$$

or

$$V_r = \left(\frac{\widehat{K}}{2} \left\{ 1 \pm \left[1 - \left(\frac{2ZP_r}{\widehat{K} \cos \theta_r} \right) \right]^{1/2} \right\} \right)^{1/2} \tag{5.87}$$

where

$$\widehat{K} = V_s^2 - 2 \times P_r(R + X \times \tan \theta_r) \tag{5.88}$$

Also, from Equation 5.71,

$$\tan \delta = \frac{P_r(X - R \times \tan \theta_r)}{V_r^2 + P_r(R + X \times \tan \theta_r)} \tag{5.89}$$

Example 5.1

Assume that the radial express feeder, shown in Figure 5.31, is used on rural distribution and is connected to a lumped-sum (or concentrated) load at the receiving end. Assume that the feeder impedance is $0.10 + j0.10$ pu, the sending-end voltage is 1.0 pu, P_r is 1.0 pu constant power load, and the power factor at the receiving end is 0.80 lagging. Use the given data and the exact equations for K, P_r, and $\tan \delta$ given previously and determine the following:

   a. Compute V_r and δ by using the exact equations and find also the corresponding values of the I_r and I_s currents.
   b. Verify the numerical results found in part (a) by using those results in

$$\overline{V_s} = \overline{V_r} + (R + jX)\overline{I_r} \tag{5.90}$$

Solution
   a. From Equation 5.88,

$$\widehat{K} = V_s^2 - 2 \times P_r(R + X \times \tan \theta_r)$$

$$= 1.0^2 - 2 \times 1[0.10 + 0.1 \times \tan(\cos^{-1}0.80)]$$

$$= 0.65 \text{ pu}$$

From Equation 5.87,

$$V_r = \left(\frac{\hat{K}}{2} \left\{ 1 \pm \left[1 - \left(\frac{2ZP_r}{\hat{K}\cos\theta_r} \right) \right]^{1/2} \right\} \right)^{1/2}$$

$$= \left(\frac{0.65}{2} \left\{ 1 \pm \left[1 - \left(\frac{2 \times 0.141 \times 1.0}{0.65 \times 0.8} \right) \right]^{1/2} \right\} \right)^{1/2}$$

$$= 0.7731\, pu$$

From Equation 5.89,

$$\tan\delta = \frac{P_r(X - R \times \tan\theta_r)}{V_r^2 + P_r(R + X \times \tan\theta_r)}$$

$$= \frac{1.0\,[0.10 - 0.10 \times \tan(\cos^{-1}0.80)]}{0.7731^2 + 1.0\,[0.10 + 0.10 \times \tan(\cos^{-1}0.80)]}$$

$$= 0.0323$$

therefore

$$\delta \cong 1.85°$$

$$\bar{I}_r = \bar{I}_s = \frac{P_r}{V_r \cos\theta_r} \angle - \theta_r$$

$$= \frac{1.0}{0.7731 \times 0.80} \angle - 36.8°$$

$$= 1.617 \angle - 36.8°\, pu$$

b. From the given equation,

$$\bar{V}_r = \bar{V}_s - (R + jX)\bar{I}_r$$

$$= 1.0 \angle 1.85° - (0.10 + j0.10)(1.617 \angle - 36.8°)$$

$$\cong 0.7731 \angle 0°\, pu$$

5.13 DESIGN OF RADIAL PRIMARY DISTRIBUTION SYSTEMS

The radial primary distribution systems are designed in several different ways: (1) overhead primaries with overhead laterals or (2) URD, for example, with mixed distribution of overhead primaries and underground laterals.

5.13.1 OVERHEAD PRIMARIES

For the sake of illustration, Figure 5.32 shows an arrangement for overhead distribution, which includes a main feeder and 10 laterals connected to the main with sectionalizing fuses. Assume that the distribution substation, shown in the figure, is arbitrarily located; it may also serve a second

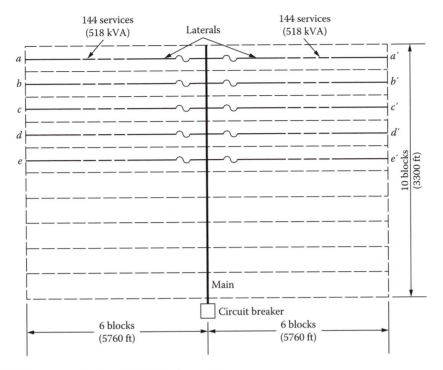

FIGURE 5.32 An overhead radial distribution system.

area, which is not shown in the figure, that is equal to the area being considered and, for example, located "below" the shown substation site.

Here, the feeder mains are three phase and of 10 short block length or less. The laterals, on the other hand, are all of six long block length and are protected with sectionalizing fuses. In general, the laterals may be either single phase, open wye grounded, or three phase.

Here, in the event of a permanent fault on a lateral line, only a relatively small fraction of the total area is outaged. Ordinarily, permanent faults on the overhead line can be found and repaired quickly.

5.13.2 UNDERGROUND RESIDENTIAL DISTRIBUTION

Even though a URD costs somewhere between 1.25 and 10 times more than a comparable overhead system, due to its certain advantages, it is used commonly [4,5]. Among the advantages of the underground system are the following:

1. The lack of outages caused by the abnormal weather conditions such as ice, sleet, snow, severe rain and storms, and lightning
2. The lack of outages caused by accidents, fires, and foreign objects
3. The lack of tree trimming and other preventative maintenance tasks
4. The aesthetic improvement

For the sake of illustration, Figure 5.33 shows a URD for a typical overhead and underground primary distribution system of the two-way feed type. The two arbitrarily located substations are assumed to be supplied from the same subtransmission line, which is not shown in the figure, so that the low-voltage buses of the two substations are nominally in phase. In the figure, the two overhead primary-feeder mains carry the total load of the area being considered, that is, the area of

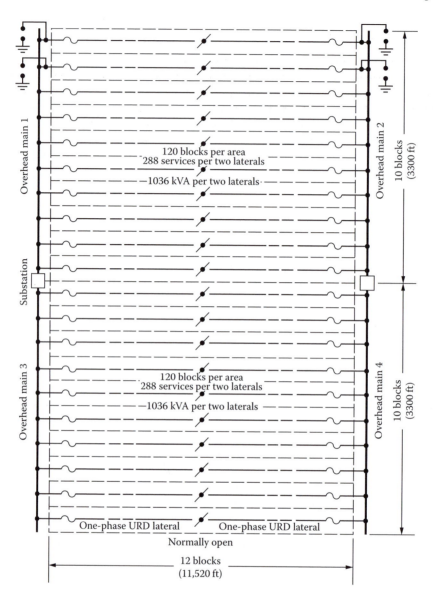

FIGURE 5.33 A two-way feed-type underground residential distribution system.

the 12 block by 10 block. The other two overhead feeder mains carry the other equally large area. Therefore, in this example, each area has 120 blocks.

The laterals, in residential areas, typically are single phase and consist of directly buried (rather than located in ducts) concentric neutral-type cross-linked polyethylene (XLPE)-insulated cable. Such cables usually insulated for 15 kV line-to-line solidly grounded neutral service and the commonly used single-phase line-to-neutral operating voltages are nominally 7200 or 7620 V.

The installation of long lengths of cable capable of being plowed directly into the ground or placed in narrow and shallow trenches, without the need for ducts and manholes, naturally reduces installation and maintenance costs. The heavy three-phase feeders are overhead along the periphery of a residential development, and the laterals to the pad-mount transformers are buried about 40 in. deep. The secondary service lines then run to the individual dwellings at a depth of about 24 in. and come up into the dwelling meter through a conduit. The service conductors run along easements and do not cross adjacent property lines.

The distribution transformers now often used are of the *pad-mounted* or *submersible* type. The pad-mounted distribution transformers are completely enclosed in strong, locked sheet metal enclosures and mounted on grade on a concrete slab. The submersible-type distribution transformers are placed in a cylindrical excavation that is lined with a concrete bituminized fiber or corrugated sheet metal tube. The tubular liner is secured after near-grade level with a locked cover.

Ordinarily, each lateral line is operated NO at or near the center as Figure 5.33 suggests. An excessive amount of time may be required to locate and repair a fault in a directly buried URD cable. Therefore, it is desirable to provide switching so that any one run of primary cable can be de-energized for cable repair or replacement while still maintaining service to all (or nearly all) distribution transformers.

Figure 5.34 shows apparatus, suggested by Lokay [2], which is or has been used to accomplish the desired switching or sectionalizing. The figure shows a single-line diagram of loop-type primary-feeder circuit for a low-cost underground distribution system in residential areas. Figure 5.34a shows it with a disconnect switch at each transformer, whereas Figure 5.34b shows the similar setup without a disconnect switch at each transformer. In Figure 5.34a, if the cable "above" C is faulted, the switch at C and the switch or cutout "above" C are opened, and, at the same time, the sectionalizing switch at B is closed. Therefore, the faulted cable above C and the distribution transformer at C are then out of service.

Figure 5.35 shows a distribution transformer with internal high-voltage fuse and with stick-operated plug-in type of high-voltage load-break connectors. Some of the commonly used plug-in types of load-break connector ratings include 8.66 kV line-to-neutral, 200 A continuous 200 A load break, and 10,000 A symmetrical fault close-in rating.

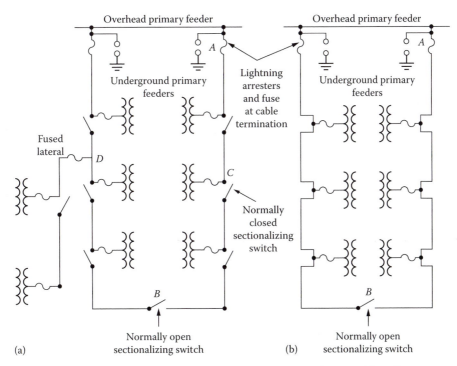

FIGURE 5.34 Single-line diagram of loop-type primary-feeder circuits: (a) with a disconnect switch at each transformer and (b) without a disconnect switch at each transformer. (From Westinghouse Electric Corporation, *Electric Utility Engineering Reference Book-Distribution Systems*, Vol. 3, East Pittsburgh, Pittsburgh, PA, 1965.)

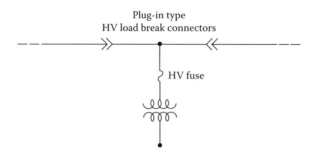

FIGURE 5.35   A distribution transformer with internal high-voltage fuse and load-break connectors.

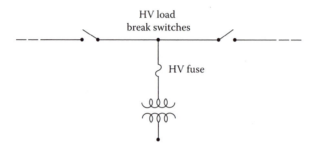

FIGURE 5.36   A distribution transformer with internal high-voltage fuses and load-break switches.

Figure 5.36 shows a distribution transformer with internal high-voltage fuse and with stick-operated high-voltage load-break switches that can be used in Figure 5.34a to allow four modes of operation, namely, the following:

1. The transformer is energized and the loop is closed
2. The transformer is energized and the loop is open to the right
3. The transformer is energized and the loop is open to the left
4. The transformer is de-energized and the loop is open

In Figure 5.33, note that, in case of trouble, the open may be located near one of the underground feed points. Therefore, at least in this illustrative design, the single-phase underground cables should be at least ampacity sized for the load of 12 blocks, not merely six blocks.

In Figure 5.33, note further the difficulty in providing abundant overvoltage protection to cable and distribution transformers by placing lightning arresters at the open cable ends. The location of the open moves because of switching, whether for repair purposes or for load balancing.

Example 5.2

Consider the layout of the area and the annual peak demands shown in Figure 5.32. Note that the peak demand per lateral is found as

$$144 \text{ customers} \times 3.6 \text{ kVA/customer} \cong 518 \text{ kVA}$$

Assume a lagging-load power factor of 0.90 at all locations in all primary circuits at the time of the annual peak load. For purposes of computing voltage drop in mains and in three-phase laterals, assume that the single-phase load is perfectly balanced among the three phases. Idealize the voltage-drop calculations further by assuming uniformly distributed load along all laterals. Assume nominal operating voltage when computing current from the kilovoltampere load.

For the open-wire overhead copper lines, compute the percent voltage drops, using the precalculated percent voltage drop per kilovoltampere–mile curves given in Chapter 4. Note that $D_m = 37$ in. is assumed.

The joint EEI-NEMA report [6] defines *favorable* voltages at the point of utilization, inside the buildings, to be from 110 to 125 V. Here, for illustrative purposes, the lower limit is arbitrarily raised to 116 V at the meter, that is, at the end of the service-drop cable. This allowance may compensate for additional voltage drops, not calculated, due to the following:

1. Unbalanced loading in three-wire single-phase secondaries
2. Unbalanced loading in four-wire three-phase primaries
3. Load growth
4. Voltage drops in building wiring

Therefore, the voltage criteria that are to be used in this problem are

$$V_{max} = 125 \text{ V} = 1.0417 \text{ pu}$$

and

$$V_{min} = 116 \text{ V} = 0.9667 \text{ pu}$$

at the meter. *The maximum voltage drop, from the low-voltage bus of the distribution substation to the most remote meter, is 7.50%.* It is assumed that a 3.5% maximum steady-state voltage drop in the secondary distribution system is reasonably achievable. Therefore, *the maximum allowable primary voltage drop for this problem is limited to 4.0%.*

Assume open-wire overhead primaries with three-phase four-wire laterals, and that the nominal voltage is used as the base voltage and is equal to 2400/4160 V for the three-phase four-wire grounded-wye primary system with copper conductors and $D_m = 37$ in. *Consider only the "longest" primary circuit, consisting of a 3300 ft main and the two most remote laterals,* like the laterals a and a' of Figure 5.32. *Use ampacity-sized conductors but in no case smaller than AWG #6 for reasons of mechanical strength.* Determine the following:

a. The percent voltage drops at the ends of the laterals and the main.
b. If the 4% maximum voltage-drop criterion is exceeded, find a reasonable combination of larger conductors for main and for lateral that will meet the voltage-drop criterion.

Solution

a. Figure 5.37 shows the "longest" primary circuit, consisting of the 3300 ft main and the most remote laterals a and a'. In Figure 5.37, the signs //// indicate that there are three phase and one neutral conductors in that portion of the one-line diagram. The current in the lateral is

$$I_{lateral} = \frac{S_l}{\sqrt{3} \times V_{L-L}}$$

$$= \frac{518}{\sqrt{3} \times 4.16} \cong 72 \text{ A} \tag{5.91}$$

Thus, from Table A.1, AWG #6 copper conductor with 130-A ampacity is selected for the laterals. The current in the main is

$$I_{main} = \frac{S_m}{\sqrt{3} \times V_{L-L}}$$

$$= \frac{1036}{\sqrt{3} \times 4.16} \cong 144 \text{ A} \tag{5.92}$$

---

* Note that the whole area is *not* considered here, but only the last two laterals, for practice.

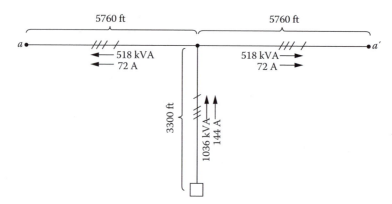

FIGURE 5.37   The "longest" primary circuit.

Hence, from Table A.1, AWG #4 copper conductor with 180-A ampacity is selected for the mains. Here, note that the AWG #5 copper conductors with 150-A ampacity are not selected due to the resultant too-high total voltage drop.

From Figure 4.17, the K constants for the AWG #6 laterals and the AWG #4 mains can be found as 0.015 and 0.01, respectively. Therefore, since the load is assumed to be uniformly distributed along the lateral,

$$\% \, VD_{lateral} = \frac{\ell}{2} \times K \times S$$

$$= \frac{1}{2} \times \frac{5760 \text{ ft}}{5280 \text{ ft/min}} \times 0.015 \times 518 \text{ kVA}$$

$$= 4.24 \tag{5.93}$$

and since the main is considered to have a lumped-sum load of 1036 kVA at the end of its length,

$$\% VD_{main} = \ell \times K \times S$$

$$= \frac{3300 \text{ ft}}{5280 \text{ ft/min}} \times 0.01 \times 1036 \text{ kVA}$$

$$= 6.48 \tag{5.94}$$

Therefore, the total percent primary voltage drop is

$$\sum \% \, VD = \% \, VD_{main} + \% \, VD_{lateral}$$

$$= 6.48 + 4.24$$

$$= 10.72 \tag{5.95}$$

which *exceeds* the maximum primary voltage-drop criterion of 4.00%.

Here, note that if single-phase laterals were used instead of the three-phase laterals, *according to Morrison* [7], the percent voltage drop of a single-phase circuit is approximately four times that for a three-phase circuit, assuming the use of the same-size conductors. Hence, for the laterals,

$$\sum \% \, VD_{1\phi} = 4(\% \, VD_{3\phi})$$

$$= 4 \times 4.24$$

$$= 16.96 \tag{5.96}$$

Therefore, from Equation 5.95, the new total percent voltage drop would be

$$\sum \% \, VD = \% \, VD_{main} + \% \, VD_{lateral}$$

$$= 6.48 + 16.96$$

$$= 23.44$$

which would be far *exceeding* the maximum primary voltage-drop criterion of 4.00%.

b. Therefore, to meet the maximum primary voltage-drop criterion of 4.00%, from Table A.1, select 4/0 and AWG #1 copper conductors with ampacities of 480 A and 270 A for the main and laterals, respectively. Hence, from Equation 5.93,

$$\% \, VD_{lateral} = \frac{\ell}{2} \times K \times S$$

$$= \frac{1}{2} \times \frac{5760 \text{ ft}}{5280 \text{ ft/min}} \times 0.006 \times 518 \text{ kVA}$$

$$= 1.695$$

and from Equation 5.94,

$$\% VD_{main} = \ell \times K \times S$$

$$= \frac{3300 \text{ ft}}{5280 \text{ ft/min}} \times 0.003 \times 1036 \text{ kVA}$$

$$= 1.943$$

Therefore, from Equation 5.95,

$$\sum \% VD = \% VD_{main} + \% VD_{lateral}$$

$$= 1.943 + 1.695$$

$$= 3.638$$

which *meets* the maximum primary voltage-drop criterion of 4.00%.

Example 5.3

Repeat Example 5.2 but assume that, instead of the open-wire overhead primary system, a self-supporting aerial messenger cable with aluminum conductors is being used. This is to be considered one step toward the improvement of the aesthetics of the overhead primary system, since, in general, very few crossarms are required.

Consider again only the "longest" primary circuit, consisting of a 3300 ft main and the two most remote laterals, * like the laterals a and a' of Figure 5.32. For the voltage-drop calculations in the self-supporting aerial messenger cable, use Table A.23 for its resistance and reactance values. For *ampacities*, use Table 5.2, which gives data for XLPE-insulated aluminum conductor, grounded neutral +3/0 aerial cables. These ampacities are based on 40°C ambient and 90°C conductor temperatures and are taken from the General Electric Company's Publication No. PD-16.

---

* Note that the whole area is *not* considered here again, but only the last two laterals, for practice.

TABLE 5.2
Current-Carrying Capacity of XLPE
Aerial Cables

	Ampacity, A	
Conductor Size	5 kV Cable	15 kV Cable
6 AWG	75	
4 AWG	99	
2 AWG	130	135
1 AWG	151	155
1/0 AWG	174	178
2/0 AWG	201	205
3/0 AWG	231	237
4/0 AWG	268	273
250 kcmil	297	302
350 kcmil	368	372
500 kcmil	459	462

Solution

a. The voltage drop, due to the uniformly distributed load, at the lateral is

$$VD_{lateral} = I(r \times \cos\theta + x_L \times \sin\theta)\frac{\ell}{2} \text{ V} \qquad (5.97)$$

where
  $I = 72$ A, from Example 5.2
  $r = 4.13$ Ω/min, for AWG #6 aluminum conductors from Table A.23
  $x_L = 0.258$ Ω/min, for AWG #6 aluminum conductors from Table A.23
  $\cos\theta = 0.90$
  $\sin\theta = 0.436$

Therefore,

$$VD_{lateral} = 72(4.13 \times 0.9 + 0.258 \times 0.436)\frac{5760 \text{ ft}}{5280 \text{ ft/min}} \times \frac{1}{2}$$

$$= 150.4 \text{ V}$$

or, in percent,

$$\% VD_{lateral} = \frac{150.4 \text{ V}}{2400 \text{ V}}$$

$$= 6.27$$

The voltage drop due to the lumped-sum load at the end of main is

$$VD_{main} = I(r \times \cos\theta + x_L \times \sin\theta)\ell \text{ V} \qquad (5.98)$$

where
  $I = 144$ A, from Example 5.2
  $r = 1.29$ Ω/min, for AWG #1 aluminum conductors from Table A.23
  $x_L = 0.211$ Ω/min, for AWG #1 aluminum conductors from Table A.23

Therefore,

$$VD_{main} = 144(1.29 \times 0.9 + 0.211 \times 0.436)\frac{3300 \text{ ft}}{5280 \text{ ft/min}}$$

$$\cong 112.8 \text{ V}$$

or, in percent,

$$\% VD_{main} = \frac{112.8 \text{ V}}{2400 \text{ V}}$$

$$= 4.7$$

Thus, from Equation 5.95, the total percent primary voltage drop is

$$\sum \% VD = \% VD_{main} + \% VD_{lateral}$$

$$= 4.7 + 6.27$$

$$= 10.97$$

which *far exceeds* the maximum primary voltage-drop criterion of 4.00%.
b. Therefore, to meet the maximum primary voltage-drop criterion of 4.00%, from Tables 5.2 and A.23, select 4/0 and 1/0 aluminum conductors with ampacities of 268 A and 174 A for the main and laterals, respectively.
   Hence, from Equation 5.97,

$$VD_{lateral} = 72(1.03 \times 0.9 + 0.207 \times 0.436)\frac{5760 \text{ ft}}{5280 \text{ ft/min}} \times \frac{1}{2}$$

$$= 39.95 \text{ V}$$

or, in percent,

$$\% VD_{lateral} = \frac{39.95 \text{ V}}{2400 \text{ V}}$$

$$= 1.66$$

From Equation 5.98,

$$VD_{main} = 144(0.518 \times 0.9 + 0.191 \times 0.436)\frac{3300 \text{ ft}}{5280 \text{ ft/min}} = 49.45 \text{ V}$$

or, in percent,

$$\% VD_{main} = \frac{49.45 \text{ V}}{2400 \text{ V}}$$

$$= 2.06$$

Thus, from Equation 5.95, the total percent primary voltage drop is

$$\sum \% VD = 2.06 + 1.66$$

$$= 3.72$$

which *meets* the maximum primary voltage-drop criterion of 4.00%.

Example 5.4

Repeat Example 5.2, but assume that the nominal operating voltage is used as the base voltage and is equal to 7,200/12,470 V for the three-phase four-wire grounded-wye primary system with copper conductors. Use $D_m = 37$ in. although $D_m = 53$ in. is more realistic for this voltage class. This simplification allows the use of the precalculated percent voltage drop per kilovoltampere–mile curves given in Chapter 4.

Consider serving the *total area* of $12 \times 10 = 120$-block area, shown in Figure 5.32, with *two feeder mains* so that the longest of the two feeders would consist of a 3300 ft main and 10 laterals, that is, the laterals a through e and the laterals a′ through e′. Use ampacity-sized conductors, but not smaller than AWG #6, and determine the following:

a. Repeat part (a) of Example 5.2.
b. Repeat part (b) of Example 5.2.
c. The deliberate use of too-small D leads to small errors in what and why?

Solution

a. The assumed load on the longer feeder is

$$\frac{518\,\text{kVA}}{\text{Lateral}} \times \frac{10\,\text{Laterals}}{\text{Feeder}} = 5180\,\text{kVA}$$

Therefore, the current in the main is

$$I_{main} = \frac{5180\,\text{kVA}}{\sqrt{3} \times 12.47\,\text{kV}} = 240.1\,\text{A}$$

Thus, from Table A.1, AWG #2, three-strand copper conductor, is selected for the mains. The current in the lateral is

$$I_{lateral} = \frac{518\,\text{kVA}}{\sqrt{3} \times 12.47\,\text{kV}} = 24.1\,\text{A}$$

Hence, from Table A.1, AWG #6 copper conductor is selected for the laterals.

From Figure 4.17, the K constants for the AWG #6 laterals and the AWG #2 mains can be found as 0.00175 and 0.0008, respectively. Therefore, since the load is assumed to be uniformly distributed along the lateral, from Equation 5.93,

$$\% \text{VD}_{lateral} = \frac{1}{2} \times K \times S$$

$$= \frac{1}{2} \times \frac{5760\,\text{ft}}{5280\,\text{ft/min}} \times 0.00175 \times 518\,\text{kVA}$$

$$= 0.50$$

and since, due to the peculiarity of this new problem, one-half of the main has to be considered as an express feeder and the other half is connected to a uniformly distributed load of 5180 kVA,

$$\% \text{VD}_{main} = \frac{3}{4} \times l \times K \times S$$

$$= \frac{3}{4} \times \frac{3300\,\text{ft}}{5280\,\text{ft/min}} \times 0.0008 \times 5180\,\text{kVA}$$

$$= 1.94 \tag{5.99}$$

Therefore, from Equation 5.95, the total percent primary voltage drop is

$$\sum \% \, VD = 1.94 + 0.50$$

$$= 2.44$$

b. It *meets* the maximum primary voltage-drop criterion of 4.00%.
c. Since the inductive reactance of the line is

$$x_L = 0.1213 \times \ln \frac{1}{D_s} + 0.1213 \times \ln D_m \, \Omega/\text{min}$$

or

$$x_L = x_a + x_d \, \Omega/\text{min}$$

when $D_m = 37$ in.,

$$x_d = 0.1213 \times \ln \frac{37 \text{ in.}}{12 \text{ in./ft}}$$

$$= 0.1366 \, \Omega/\text{min}$$

and when $D_m = 53$ in.,

$$x_d = 0.1213 \times \ln \frac{53 \text{ in.}}{12 \text{ in./ft}}$$

$$= 0.1802 \, \Omega/\text{min}$$

Hence, there is a difference of

$$\Delta x_d = 0.0436 \, \Omega/\text{min}$$

which calculates a voltage-drop value smaller than it really is.

Example 5.5

Consider the layout of the area and the annual peak demands shown in Figure 5.33. The primary distribution system in the figure is a mixed system with overhead mains and a URD system. Assume that open-wire overhead mains are used with 7,200/12,470 V three-phase four-wire grounded-wye ACSR conductors and that $D_m = 53$ in. Also assume that concentric neutral XLPE-insulated underground cable with aluminum conductors is used for single-phase and 7200 V underground cable laterals.

For voltage-drop calculations and ampacity of concentric neutral XLPE-insulated URD cable with aluminum conductors, use Table 5.3.

The foregoing data are for a currently used 15 kV solidly grounded neutral class of cable construction consisting of (1) Al phase conductor, (2) extruded semiconducting conductor shield, (3) 175 mils thickness of cross-linked PE insulation, (4) extruded semiconducting sheath and insulation shield, and (5) bare copper wires spirally applied around the outside to serve as the current-carrying grounded neutral. The data given are for a cable intended for single-phase service, hence the number and the size of concentric neutral are selected to have "100% neutral" ampacity. When three such cables are to be installed to make a three-phase circuit, the number and/or size of copper concentric neutral strands on each cable are reduced to 33% (or less) neutral ampacity per cable.

Another type of insulation in current use is high-molecular-weight PE (HMWPE). It is rated for only 75°C conductor temperature and, therefore, provides a little less ampacity than XLPE

TABLE 5.3
15 kV Concentric Neutral XLPE-Insulated Al URD Cable

| Al Conductor Size | Cu Neutral | Ω/1000 ft[a] | | Ampacity, A | |
		r[b]	XL	Direct Burial	In Duct
4 AWG	6-#14	0.526	0.0345	128	91
2 AWG	104 14	0.331	0.0300	168	119
1 AWG	13-#14	0.262	0.0290	193	137
1/0 AWG	16-114	0.208	0.0275	218	155
2/0 AWG	134 12	0.166	0.0260	248	177
3/0 AWG	16-#12	0.132	0.0240	284	201
4/0 AWG	20-#12	0.105	0.0230	324	230
250 kcmil	25-112	0.089	0.0220	360	257
300 kcmil	18-1110	0.074	0.0215	403	291
350 kcmil	204 10	0.063	0.0210	440	315

Source: Data abstracted from Rome Cable Company, *URD Technical Manual*, 4th edn.
[a] For single-phase circuitry.
[b] At 90°C conductor temperature.

insulation on the same conductor size. The HMWPE requires 220 mils insulation thickness in lieu of 175 mil. Cable reactances are, therefore, slightly higher when HMWPE is used. However, the Δx_L is negligible for ordinary purposes.

The determination of correct $r + jx_L$ values of these relatively new concentric neutral cables is a subject of current concern and research. A portion of the neutral current remains in the bare concentric neutral conductors; the remainder returns to the earth (Carson's equivalent conductor). More detailed information about this matter is available in Refs. [8,9]. Use the given data and determine the following:

a. Size each of the overhead mains 1 and 2, of Figure 5.33, *with enough ampacity to serve the entire 12 × 10 block area.* Size each single-phase lateral URD cable with ampacity for the load of 12 blocks.
b. Find the percent voltage drop at the ends of the most remote laterals under *normal operation,* that is, all laterals open at the center, and both mains are energized.
c. Find the percent voltage drop at the most remote lateral under *the worst possible emergency operation,* that is, one main is outaged, and all laterals are fed full length from the one energized main.
d. Is the voltage-drop criterion met for normal operation and for the worst emergency operation?

Solution

a. Since *under the emergency operation* the remaining energized main supplies the doubled number of laterals, the assumed load is

$$\frac{2 \times 518 \text{ kVA}}{\text{Lateral}} \times \frac{10 \text{ Laterals}}{\text{Feeder}} = 10,360 \text{ kVA}$$

Therefore, the current in the main is

$$I_{main} = \frac{10,360 \text{ kVA}}{\sqrt{3} \times 12.47 \text{ kV}}$$

$$= 480.2 \text{ A}$$

Thus, from Table A.5, 300 kcmil ACSR conductors, with 500 A ampacity, are selected for the mains. Since under the emergency operation, due to doubled load, the current in the lateral is doubled,

$$I_{lateral} = \frac{2 \times 518 \, kVA}{7.2 \, kV}$$

$$= 144 \, A$$

Therefore, from Table 5.3, AWG #2 XLPE Al URD cable, with 168 A ampacity, is selected for the laterals.

b. *Under normal operation*, all laterals are open at the center, and both mains are energized. Thus the voltage drop, due to uniformly distributed load, at the main is

$$VD_{main} = I[r \times \cos\theta + x_L \times \sin\theta]\frac{l}{2} \, V \tag{5.100}$$

or

$$VD_{main} = I[r \times \cos\theta + (x_a + x_d) \times \sin\theta]\frac{l}{2} \, V \tag{5.101}$$

where
  $I = 480.2/2 = 240.1$ A
  $r = 0.342$ Ω/min for 300-kcmil ACSR conductors from Table A.5
  $x_a = 0.458$ Ω/min for 300-kcmil ACSR conductors from Table A.5
  $x_d = 0.1802$ Ω/min for $D_m = 53$ in. from Table A.10
  $\cos\theta = 0.90$
  $\sin\theta = 0.436$

Therefore,

$$VD_{main} = 240.1[0.342 \times 0.9 + (0.458 + 0.1802)0.436]\frac{3300 \, ft}{5280 \, ft/min} \times \frac{1}{2}$$

$$\cong 44 \, V$$

or, in percent,

$$\% VD_{main} = \frac{44 \, V}{7200 \, V}$$

$$= 0.61$$

The voltage drop at the lateral, due to the uniformly distributed load, from Equation 5.97 is

$$VD_{lateral} = I(r \times \cos\theta + x \times \sin\theta)\frac{l}{2} \, V$$

where
  $I = 144/2 = 72$ A
  $r = 0.331$ Ω/1000 ft for AWG #2 XLPE Al URD cable from Table 5.3
  $x_L = 0.0300$ ft/1000 ft for AWG #2 XLPE Al URD cable from Table 5.3

Therefore,

$$VD_{lateral} = 72(0.331 \times 0.9 + 0.0300 \times 0.436)\frac{5760 \text{ ft}}{1000 \text{ ft}} \times \frac{l}{2}$$

$$= 64.5 \text{ V}$$

or, in percent,

$$\%VD_{lateral} = \frac{64.5 \text{ V}}{7200 \text{ V}}$$

$$= 0.9$$

Thus, from Equation 5.95, the total percent primary voltage drop is

$$\sum\% \text{ VD} = 0.61 + 0.9$$

$$= 1.51$$

c. *Under the worst possible emergency operation*, one main is outaged and all laterals are supplied full length from the remaining energized main. Thus the voltage drop in the main, due to uniformly distributed load, from Equation 5.101 is

$$VD_{main} = 480.2(0.3078 + 0.2783)\frac{3300 \text{ ft}}{5280 \text{ ft/min}} \times \frac{1}{2}$$

$$= 88 \text{ V}$$

or, in percent,

$$\%VD_{main} = 1.22$$

The voltage drop at the lateral, due to uniformly distributed load, from Equation 5.97 is

$$VD_{lateral} = 144(0.331 \times 0.9 + 0.03 \times 0.435)\frac{5760 \text{ ft}}{1000 \text{ ft}}$$

$$= 258 \text{ V}$$

or, in percent,

$$\%VD_{lateral} = \frac{258 \text{ V}}{7200 \text{ V}}$$

$$= 3.5$$

Therefore, from Equation 5.95, the total percent primary voltage drop is

$$\sum\%VD = 1.22 + 3.5$$

$$= 4.72$$

d. The primary voltage-drop criterion is *met for normal operation* but is *not met for the worst emergency operation.*

5.14 PRIMARY SYSTEM COSTS

Based on the 1994 prices, construction of three-phase, overhead, wooden pole crossarm-type feeders of normal large conductor (e.g., 600 kcmil per phase) at about 12.47 kV voltage level costs about $150,000 per mile. However, cost can vary greatly due to variations in labor, filing, and permit costs among utilities, as well as differences in design standards, and very real differences in terrain and geology. The aforementioned feeder would be rated with a thermal capacity of about 15 MVA and a recommended economic peak loading of about 10 MVA peal, depending on losses and other costs. At $150,000 per mile, this provides a cost of $10–$15 per kW-mile. Underground construction of three-phase primary is more expensive, requiring buried ductwork and cable, and usually works out to a range of $30–$50 per kW-mile.

The costs of lateral lines vary between about $5 and $15 per kW-mole overhead. The underground lateral lines cost between $5 and $15 per kW-mile for direct buried cables and $30 and $100 per kW-mile for ducted cables. Costs of other distribution equipment, including regulators, capacitor banks and their switches, sectionalizers, and line switches, vary greatly depending on specifics to each application. In general, the cost of the distribution system will vary between $10 and $30 per kW-mile.

PROBLEMS

5.1 Repeat Example 5.2, assuming a 30 min annual maximum demand of 4.4 kVA per customer.

5.2 Repeat Example 5.3, assuming the nominal operating voltage to be 7,200/12,470 V.

5.3 Repeat Example 5.3, assuming a 30 min annual maximum demand of 4.4 kVA per customer for a 12.47 kV system.

5.4 Repeat Example 5.4 and find the exact solution by using $D_m = 53$ in.

5.5 Repeat Example 5.5, assuming a lagging-load power factor of 0.80 at all locations.

5.6 Assume that a radial express feeder used in rural distribution is connected to a concentrated and static load at the receiving end. Assume that the feeder impedance is $0.15 + j0.30$ pu, the sending-end voltage is 1.0 pu, and the constant power load at the receiving end is 1.0 pu with a lagging power factor of 0.85. Use the given data and the exact equations for $\bar{K}$, $\bar{V}_r$, and tan δ given in Section 5.12 and determine the following:
   a. The values $\bar{V}_r$ and δ by using the exact equations
   b. The corresponding values of the $\bar{I}_r$ and $\bar{I}_s$ currents

5.7 Use the results found in Problem 5.6 and Equation 5.90 and determine the receiving-end voltage $\bar{V}_r$.

5.8 Assume that a three-phase 34.5 kV radial express feeder is used in rural distribution and that the receiving-end voltages at full load and no load are 34.5 and 36.9 kV, respectively. Determine the percent voltage regulation of the feeder.

5.9 A three-phase radial express feeder has a line-to-line voltage of 22.9 kV at the receiving end, a total impedance of $5.25 + j10.95$ Ω/phase, and a load of 5 MW with a lagging power factor of 0.90. Determine the following:
   a. The line-to-neutral and line-to-line voltages at the sending end
   b. The load angle

5.10 Use the results of Problem 5.9 and determine the percent voltage regulation of the feeder.

5.11 Assume that a wye-connected three-phase load is made up of three impedances of $50\angle 25°$ Ω each and that the load is supplied by a three-phase four-wire primary express feeder. The balanced line-to-neutral voltages at the receiving end are

$$\bar{V}_{an} = 7630\angle 0° \, \text{V}$$

$$\bar{V}_{bn} = 7630\angle 240° \text{ V}$$

$$\bar{V}_{cn} = 7630\angle 120' \text{ V}$$

Determine the following:
a. The phasor currents in each line
b. The line-to-line phasor voltages
c. The total active and reactive power supplied to the load

5.12 Repeat Problem 5.11, if the same three load impedances are connected in a delta connection.

5.13 Assume that the service area of a given feeder is increasing as a result of new residential developments. Determine the new load and area that can be served with the same percent voltage drop if the new feeder voltage level is increased to 34.5 kV from the previous voltage level of 12.47 kV.

5.14 Assume that the feeder in Problem 5.13 has a length of 2 min and that the new feeder uniform loading has increased to three times the old feeder loading. Determine the new maximum length of the feeder with the same percent voltage drop.

5.15 Consider a 12.47 kV three-phase four-wire grounded-wye overhead radial distribution system, similar to the one shown in Figure 5.32. The uniformly distributed area of 12 × 10 = 120 – block area is served by one main located in the middle of the service area. There are 10 laterals (6 blocks each) on each side of the main. The lengths of the main and the laterals are 3300 and 5760 ft, respectively. From Table A.1, arbitrarily select 4/0 copper conductor with 12 strands for the main and AWG # 6 copper conductor for the lateral. The K constants for the main and lateral are 0.0032% and 0.00175% VD per kVA-min, respectively. If the maximum diversified demand per lateral is 518.4 kVA, consider the total service area and determine the following:
a. The total load of the main feeder in kVA.
b. The amount of current in the main feeder.
c. The amount of current in the lateral.
d. The percent voltage drop at the end of the lateral.
e. The percent voltage drop at the end of the main.
f. The total voltage drop for the last lateral. Is it acceptable if the 4% maximum voltage-drop criterion is used?

5.16 After solving Problem 5.15, use the results obtained, but assume that the main is made up of 500 kcmil, 19-strand copper conductors with $D_m = 37$ in. and determine the following:
a. The percent voltage drop at the end of the main.
b. The total voltage drop to the end of last lateral. Is it acceptable and why?

5.17 After solving Problem 5.15, use the results obtained, but assume that the main is made up of 350 kcmil, 12-strand copper conductors with $D_m = 37$ in. and determine the following:
a. The percent voltage drop at the end of the main.
b. The total voltage drop to the end of last lateral. Is it acceptable and why?

5.18 After solving Problem 5.15, use the results obtained, but assume that the main is made up of 250 kcmil, 12-strand copper conductors with $D_m = 37$ in. and determine the following:
a. The percent voltage drop at the end of the main.
b. The total voltage drop to the end of last lateral. Is it acceptable and why?

5.19 Resolve Example 5.2 by using MATLAB®. Use the same selected conductors and their parameters.

5.20 Resolve Example 5.3 by using MATLAB, assuming the nominal operating voltage to be 7,200/12,470 V. Use the same selected conductors and their parameters.

REFERENCES

1. Fink, D. G. and H. W. Beaty: *Standard Handbook for Electrical Engineers*, 11th edn., McGraw-Hill, New York, 1978.
2. Westinghouse Electric Corporation: *Electric Utility Engineering Reference Book-Distribution Systems*, Vol. 3, East Pittsburgh, Pittsburgh, PA, 1965.
3. Gönen, T. et al.: *Development of Advanced Methods for Planning Electric Energy Distribution Systems*, US Department of Energy, October 1979. Available from the National Technical Information Service, US Department of Commerce, Springfield, VA.
4. Edison Electric Institute: *Underground Systems Reference Book*, 2nd edn., New York, 1957.
5. Andrews, F. E.: Residential underground distribution adaptable, *Electr. World*, December 12, 1955, 107–113.
6. EEI-NEMA: *Preferred Voltage Ratings for AC Systems and Equipment*, EEI Publication No. R-6, NEMA Publication No. 117, May 1949.
7. Morrison, C.: A linear approach to the problem of planning new feed points into a distribution system, *AIEE Trans.*, pt. III (PAS), December 1963, 819–832.
8. Smith, D. R. and J. V. Barger: Impedance and circulating current calculations_ for URD multi-wire concentric neutral circuits, *IEEE Trans. Power Appar. Syst.*, PAS-91(3), May/June 1972, 992–1006.
9. Stone, D. L.: Mathematical analysis of direct buried rural distribution cable impedance, *IEEE Trans. Power Appar. Syst.*, PAS-91(3), May/June 1972, 1015–1022.
10. Gönen, T.: High-temperature superconductors, in *McGraw-Hill Encyclopedia of Science and Technology*, 7th edn., Vol. 7, 1992, pp. 127–129.
11. Gönen, T. and D. C. You: A comparative analysis of distribution feeder costs, *Proceeding of the Southwest Electrical Exposition and IEEE Conference*, Houston, TX, January 22–24, 1080.
12. Gönen, T.: Power distribution, Chapter 6, in *The Electrical Engineering Handbook*, 1st edn., Academic Press, New York, 2005, pp. 749–759.
13. Rome Cable Company, *URD Technical Manual*, 4th edn., Rome, NY, 1962.

6 Design Considerations of Secondary Systems

Egyptian Proverb: The worst things:
To be in bed and sleep not,
To want for one who comes not,
To try to please and please not.

Francis Scott Fitzgerald, *Notebooks***, 1925**

6.1 INTRODUCTION

A realistic view of the power distribution systems should be based on "*gathering*" functions rather than on "*distributing*" since the size and locations of the customer demands are not determined by the distribution engineer but by the customers. Customers install all types of energy-consuming devices that can be connected in every conceivable combination and at times of customers' choice. This concept of distribution starts with the individual customers and loads and proceeds through several gathering stages where each stage includes various groups of increasing numbers of customers and their loads. Ultimately the generating stations themselves are reached through services, secondaries, distribution transformers, primary feeders, distribution substations, subtransmission and bulk power stations, and transmission lines.

In designing a system, distribution engineers should consider not only the immediate, that is, short-range, factors but also the long-range problems. The designed system should not only solve the problems of economically building and operating the systems to serve the loads of today but also require a long-range projection into the future to determine the most economical distribution system components and practices to serve the higher levels of the customers' demands, which will then exist. Therefore, the present design practice should be influenced by the requirements of the future system.

Distribution engineers, who have to consider the many factors, variables, and alternative solutions of the complex distribution design problems, need a technique that will enable them to select the most economical size combination of distribution transformers, secondary conductors, and service drops (SDs).

The recent developments in high-speed digital computers, through the use of computer programs, have provided (1) the fast and economic consideration of many feasible alternatives and (2) the economic and engineering evaluation of these alternatives as they evolve with different strategies throughout the study period. The strategies may include, for example, cutting the secondary, changing the transformers, and possibly adding capacitors.

Naturally, each designed system should meet a specified performance criterion throughout the study period. The most optimum, that is, most economical, system design that corresponds to a load-growth projection schedule can be selected. Also, through the periodic use of the programs, distribution engineers can determine whether strategies adopted continue to be desirable or whether they require some modification as a result of some changes in economic considerations and load-growth projections.

To minimize the secondary-circuit lengths, distribution engineers locate the distribution transformers close to the load centers and try to have the secondary SDs to the individual customers as short as possible.

Since only a small percentage of the total service interruptions are due to failures in the secondary system, distribution engineers, in their system design decisions of the secondary distribution, are primarily motivated by the considerations of economy, copper losses (I^2R) in the transformer and secondary circuit, permissible voltage drops, and voltage flicker of the system. Of course, there are some other engineering and economic factors affecting the selection of the distribution transformer and the secondary configurations, such as permissible transformer loading, balanced phase loads for the primary system, investment costs (ICs) of the various secondary system components, cost of labor, cost of capital, and inflation rates.

Distribution transformers represent a significant part of the secondary system cost. Therefore, one of the major concerns of distribution engineers is to minimize the investment in distribution transformers. In general, the present practice in the power industry is to plan the distribution transformer loading on the basis that there should not be excessive spare capacity installed, and transformers should be exchanged, or banked, as the secondary load grows.

Usually, a transformer load management (TLM) system is desirable for consistent loading practices and economical expansion plans. Distribution engineers, recognizing the impracticality of obtaining complete demand information on all customers, have attempted to combine a limited amount of demand data with the more complete, and readily available, energy consumption data available in the customer account files. A typical demand curve is scaled according to the energy consumed, and the resultant information is used to estimate the peak loading on specific pieces of equipment, such as distribution transformers, in which case it is known as TLM, feeders, and substations [3–6].

However, in general, residential, commercial, and industrial customers are categorized in customer files by rate classification only; that is, potentially useful and important subclassifications are not distinguished. Therefore, the demand data are generally collected for the purpose of generating typical curves only for each rate classification.

6.2 SECONDARY VOLTAGE LEVELS

Today, the standard (or preferred) voltage levels for the electric power systems are given by the American National Standards Institute's (ANSI) Standard C84.1-1977 entitled *Voltage Ratings for Electric Power Systems and Equipment* (60 Hz).

Accordingly, the standard voltage level for single-phase residential loads is 120/240 V. It is supplied through three-wire single-phase services, from which both 120 V lighting and 240 V single-phase power connections are made to large household appliances such as ranges, clothes dryers, and water heaters. For grid- or mesh-type secondary-network systems, used usually in the areas of commercial and residential customers with high-load densities, the voltage level is 208Y/120 V. It is also supplied through three-wire single-phase services, from which both 120 V lighting and 208 V single-phase power connections are made. For "spot" networks used in downtown areas for high-rise buildings with superhigh-load densities and also for areas of industrial and/or commercial customers, the voltage level is 480Y/277 V. It is supplied through four-wire three-phase services, from which both 277 V for fluorescent lighting and other single-phase loads and 480 V three-phase power connections are made.

Today, one can also find other voltage levels in use contrary to the ANSI standards, for example, 120/240 V four wire three phase, 240 V three wire three phase, 480 V three wire three phase, 240/416 V four wire three phase, or 240/480 V four wire three phase.

To increase the service reliability for critical loads, such as hospitals, computer centers, and crucial industrial loads, some backup systems, for example, emergency generators and/or batteries, with automatic switching devices are provided.

6.3 PRESENT DESIGN PRACTICE

The part of the electric utility system that is between the primary system and the consumer's property is called the *secondary system*. Secondary distribution systems include step-down distribution transformers, secondary circuits (secondary mains), consumer services (or SDs), and meters to measure consumer energy consumption.

Generally, the secondary distribution systems are designed in single phase for areas of residential customers and in three phase for areas of industrial or commercial customers with high-load densities. The types of the secondary distribution systems include the following:

1. The separate-service system for each consumer with separate distribution transformer and secondary connection
2. The radial system with a common secondary main, which is supplied by one distribution transformer and feeding a group of consumers
3. The secondary-bank system with a common secondary main that is supplied by several distribution transformers, which are all fed by the same primary feeder
4. The secondary-network system with a common grid-type main that is supplied by a large number of the distribution transformers, which may be connected to various feeders for their supplies

The separate-service system is seldom used and serves the industrial- or rural-type service areas. Generally speaking, most of the secondary systems for serving residential, rural, and light-commercial areas are radial designed. Figure 6.1 shows the one-line diagram of a radial secondary system. It has a low cost and is simple to operate.

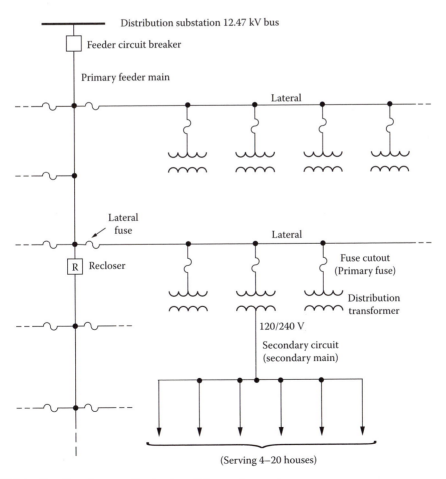

FIGURE 6.1   One-line diagram of a simple radial secondary system.

6.4 SECONDARY BANKING

The "banking" of the distribution transformers, that is, parallel connection, or, in other words, *interconnection*, of the secondary sides of two or more distribution transformers, which are supplied from the same primary feeder, is sometimes practiced in residential and light-commercial areas where the services are relatively close to each other, and therefore, the required spacing between transformers is little.

However, many utilities prefer to keep the secondary of each distribution transformer separate from all others. In a sense, secondary banking is a special form of network configuration on a radial distribution system. The advantages of the banking of the distribution transformers include the following:

1. Improved voltage regulation
2. Reduced voltage dip or light flicker due to motor starting, by providing parallel supply paths for motor-starting currents
3. Improved service continuity or reliability
4. Improved flexibility in accommodating load growth, at low cost, that is, possible increase in the average loading of transformers without corresponding increase in the peak load

Banking the secondaries of the distribution transformers allows us to take advantage of the load diversity existing among the greater number of consumers, which, in turn, induces a savings in the required transformer kilovolt-amperes. These savings can be as large as 35% according to Lokay [7], depending upon the load types and the number of consumers.

Figure 6.2 shows two different methods of banking secondaries. The method illustrated in Figure 6.2a is commonly used and is generally preferred because it permits the use of a lower-rated fuse on the high-voltage side of the transformer, and it prevents the occurrence of cascading the fuses. This method also simplifies the coordination with primary-feeder sectionalizing fuses by having a lower-rated fuse on the high side of the transformer. Furthermore, it provides the most economical system.

Figure 6.3 gives two other methods of banking secondaries. The method shown in Figure 6.3a is the oldest one and offers the least protection, whereas the method shown in Figure 6.3b offers

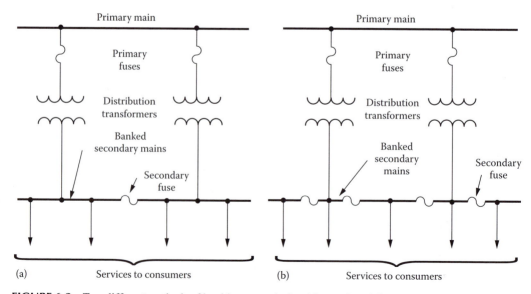

FIGURE 6.2   Two different methods of banking secondaries: (a) type 1 and (b) type 2.

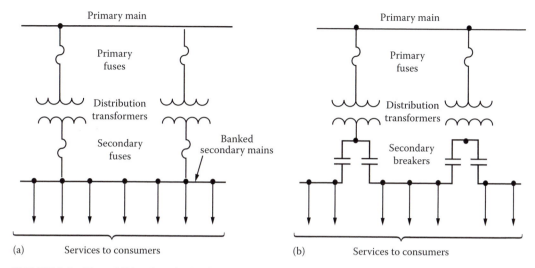

FIGURE 6.3   Two additional methods of banking secondaries: (a) type 3 and (b) type 4.

the greatest protection. Therefore, the methods illustrated in Figures 6.2a and b and 6.3a have some definite disadvantages, which include the following:

1. The requirement for careful policing of the secondary system of the banked transformers to detect blown fuses.
2. The difficulty in coordination of secondary fuses.
3. Furthermore, the method illustrated in Figure 6.2b has the additional disadvantage of being difficult to restore service after a number of fuses on adjacent transformers have been blown.

Today, due to the aforementioned difficulties, many utilities prefer the method given in Figure 6.3b. The special distribution transformer known as the *completely self-protecting-bank* (CSPB) *transformer* has, in its unit, a built-in high-voltage protective link, secondary breakers, signal lights for overload warnings, and lightning protection.

CSPB transformers are built in both single phase and three phase. They have two identical secondary breakers that trip independently of each other upon excessive current flows. In case of a transformer failure, the primary protective links and the secondary breakers will both open. Therefore, the service interruption will be minimum and restricted only to those consumers who are supplied from the secondary section that is in fault.

However, all the methods of secondary banking have an inherent disadvantage: the difficulty in performing TLM to keep up with changing load conditions. The main concern when designing a banked secondary system is the equitable load division among the transformers. It is desirable that transformers whose secondaries are banked in a straight line be within one size of each other.

For other types of banking, transformers may be within two sizes of each other to prevent excessive overload in case the primary fuse of an adjacent larger transformer should blow. Today, in general, the banking is applied to the secondaries of single-phase transformers, and all transformers in a bank must be supplied from the same phase of the primary feeder.

6.5   SECONDARY NETWORKS

Generally speaking, most of the secondary systems are radial designed except for some specific service areas (e.g., downtown areas or business districts, some military installations, hospitals) where the reliability and service-continuity considerations are far more important than the cost and economic considerations. Therefore the secondary systems may be designed in grid- or mesh-type network configurations in those areas.

The low-voltage secondary networks are particularly well justified in the areas of high-load density. They can also be built in underground to avoid overhead (OH) congestion. The OH low-voltage secondary networks are economically preferable over underground low-voltage secondary networks in the areas of medium-load density. However, the underground secondary networks give a very high degree of service reliability. In general, where the load density justifies an underground system, it also justifies a secondary-network system.

Figure 6.4 shows a one-line diagram of a small segment of a secondary network supplied by three primary feeders. In general, the usually low-voltage (208Y/120 V) grid- or mesh-type secondary-network system is supplied through network-type transformers by two or more primary feeders to increase the service reliability. In general, these are radial-type primary feeders. However, the loop-type primary feeders are also in use to a very limited extent. The primary feeders are interlaced in a way to prevent the supply to any two adjacent transformer banks from the same feeder. As a result of this arrangement, if one primary feeder is out of service for any reason (*single contingency*), the remaining feeders can feed the load without overloading and without any objectionable voltage drop. The primary-feeder voltage levels are in the range of 4.16–34.5 kV. However, there is a tendency toward the use of higher primary voltages.

Currently, the 15 kV class is predominating. The secondary network must be designed in such a manner as to provide at least one of the primary feeders as a spare capacity together with its transformers. To achieve even load distribution between transformers and minimum voltage drop in the network, the network transformers must be located accordingly throughout the secondary network.

As explained previously, the smaller secondary networks are designed based upon single contingency, that is, the outage of one primary feeder. However, larger secondary-network systems must be designed based upon *double contingency* or *second contingency*, that is, having two feeder

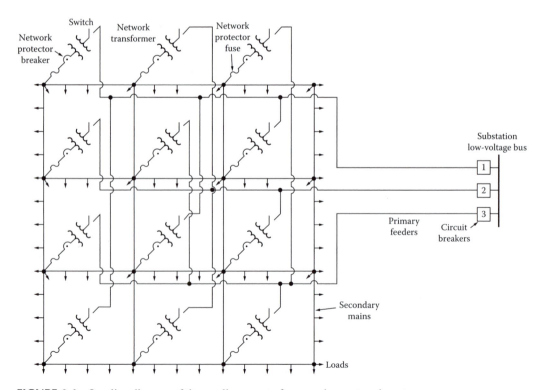

FIGURE 6.4 One-line diagram of the small segment of a secondary-network system.

outages simultaneously. According to Reps [7], the factors affecting the probability of occurrence of double outages are as follows:

1. The total number of primary feeders
2. The total mileage of the primary-feeder outages per year
3. The number of accidental feeder outages per year
4. The scheduled feeder-outage time per year
5. The time duration of a feeder outage

Even though theoretically the primary feeders may be supplied from different sources such as distribution substations, bulk power substations, or generating plants, it is generally preferred to have the feeders supplied from the same substation to prevent voltage magnitude and phase-angle differences among the feeders, which can cause a decrease in the capacities of the associated transformers due to improper load division among them. Also, during light-load periods, the power flow in a reverse direction in some feeders connected to separate sources is an additional concern.

6.5.1 SECONDARY MAINS

Seelye [8] suggested that the proper size and arrangement of the secondary mains should provide for the following:

1. The proper division of the normal load among the network transformers
2. The proper division of the fault current among the network transformers
3. Good voltage regulation to all consumers
4. Burning off short circuits or grounds at any point without interrupting service

All secondary mains (underground or OH) are routed along the streets and are three phase four wire wye connected with solidly grounded neutral conductor. In the underground networks, the secondary mains usually consist of single-conductor cables, which may be either metallic or nonmetallic sheathed. Secondary cables commonly have been rubber insulated, but PE cables are now used to a considerable extent. They are installed in duct lines or duct banks. Manholes at the street intersections are constructed with enough space to provide for various cable connections and limiters and to permit any necessary repair activities by workers.

The secondary mains in the OH secondary networks usually are open-wire circuits with weatherproof conductors. The conductor sizes depend upon the network-transformer ratings. For a grid-type secondary main, the minimum conductor size must be able to carry about 60% of the full-load current to the largest network transformer. This percentage is much less for the underground secondary mains. The most frequently used cable sizes for secondary mains are 4/0 or 250 kcmil and, to a certain extent, 350 and 500 kcmil.

The selection of the sizes of the mains is also affected by the consideration of burning faults clear. In case of a phase-to-phase or phase-to-ground short circuit, the secondary network is designed to burn itself clear without using sectionalizing fuses or other overload protective devices. Here, *"burning clear"* of a faulted secondary-network cable refers to a burning away of the metal forming the contact between phases or from phase to ground until the low voltage of the secondary network can no longer support the arc.

To achieve fast clearing, the secondary network must be able to provide for high current values to the fault. The larger the cable, the higher the short-circuit current value needed to achieve the burning clear of the faulted cable. Therefore, conductors of 500 kcmil are about the largest conductors used for secondary-network mains.

The conductor size is also selected keeping in mind the voltage-drop criterion, so that the voltage drop along the mains under normal load conditions does not exceed a maximum of 3%.

6.5.2 Limiters

Most of the time, the method permitting secondary-network conductors to burn clear, especially in 120/208 V, gives good results without loss of service. However, under some circumstances, particularly at higher voltages, for example, 480 V, this method may not clear the fault due to insufficient fault current, and, as a result, extensive cable damage, manhole fires, and service interruptions may occur.

To have fast clearing of such faults, the so-called limiters are used. The limiter is a high-capacity fuse with a restricted copper section, and it is installed in each phase conductor of the secondary main at each junction point. The limiter's fusing or time–current characteristics are designed to allow the normal network load current to pass without melting but to operate and clear a faulted section of main before the cable insulation is damaged by the heat generated in the cable by the fault current.

The fault should be cleared away by the limiters rapidly, before the network protector (NP) fuses blow. Therefore, the time–current characteristics of the selected limiters should be coordinated with the time–current characteristics of the NPs and the insulation-damage characteristics of the cable.

The distribution engineer's decision of using limiters should be based upon two considerations: (1) minimum service interruption and (2) whether the saving in damage to cables pays more than the cost of the limiters. Figure 6.5 shows the time–current characteristics of limiters used in 120/208 V systems and the insulation-damage characteristics of the underground-network cables (paper or rubber insulated).

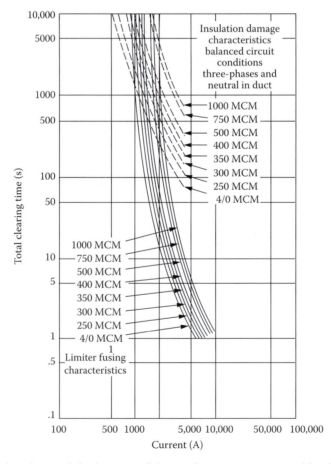

FIGURE 6.5 Limiter characteristics in terms of time to fuse versus current and insulation-damage characteristics of the underground-network cables. (From Westinghouse Electric Corporation, *Electric Utility Engineering Reference Book-Distribution Systems*, Vol. 3, East Pittsburgh, PA, 1965.)

6.5.3 NETWORK PROTECTORS

As shown in Figure 6.4, the network transformer is connected to the secondary network through an NP. The NP consists of an air circuit breaker with a closing and tripping mechanism controlled by a network master and phasing relay and backup fuses.

All these are enclosed in a metal case, which may be mounted on the transformer or separately mounted. The fuses provide backup protection to disconnect the network transformer from the network if the NP fails to do so during a fault. The functions of an NP include the following:

1. To provide automatic isolation of faults occurring in the network transformer or in the primary feeder. For example, when a fault occurs in one of the high-voltage feeders, it causes the feeder circuit breaker, at the substation, to be open. At the same time, a current flows to the feeder fault point from the secondary network through the network transformers normally supplied by the faulted feeder. This reverse power flow triggers the circuit breakers of the NPs connected to the faulty feeder to open. Therefore, the fault becomes isolated without any service interruption to any of the consumers connected to the network.
2. To provide automatic closure under the predetermined conditions, that is, when the primary-feeder voltage magnitude and the phase relation with respect to the network voltage are correct. For example, the transformer voltage should be slightly higher (about 2 V) than the secondary-network voltage in order to achieve power flow from the network transformer to the secondary-network system and not the reverse. Also, the low-side transfer voltage should be in phase with, or leading, the network voltage.
3. To provide its reverse power relay to be adequately sensitive to trip the circuit breaker with currents as small as the exciting current of the transformer. For example, this is important for the protection against line-to-line faults occurring in ungrounded three-wire primary feeders feeding network transformers with delta connections.
4. To provide protection against the reverse power flow in some feeders connected to separate sources. For example, when a network is fed from two different substations, under certain conditions, the power may flow from one substation to the other through the secondary network and network transformers. Therefore, the NPs should be able to detect this reverse power flow and to open. Here, the best protection is not to employ more than one substation as the source.

As previously explained, each network contains backup fuses, one per phase. These fuses provide backup protection for the network transformer if the NP breakers fail to operate.

Figure 6.6 illustrates an ideal coordination of secondary-network protective apparatus. The *coordination* is achieved by proper selection of time delays for the successive protective devices placed in series. Table 6.1 indicates the required action or operation of each protective equipment under different fault conditions associated with the secondary-network system. For example, in case of a fault in a given secondary main, only the associated limiters should isolate the fault, whereas in case of a transformer internal fault, both the NP breaker and the substation breaker should trip. Figures 6.4 and 6.7 show three-position switches electrically located at the high-voltage side of the network transformers. They are physically mounted on one end of the network transformer.

6.5.4 HIGH-VOLTAGE SWITCH

As shown in Figure 6.7, position 2 is for normal operation, position 3 is for disconnecting the network transformer, and position 1 is for grounding the primary circuit. In any case, the switch is manually operated and is not designed to interrupt current. The first step is to open the primary-feeder circuit breaker at the substation before opening the switch and taking the network unit out of service. After taking the unit out, the feeder circuit breaker may be closed to reestablish service to the rest of the network.

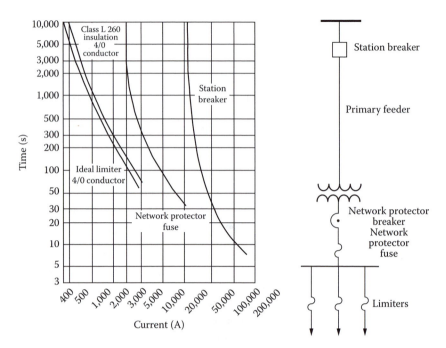

FIGURE 6.6 An ideal coordination of secondary-network overcurrent protection devices. (From Westinghouse Electric Corporation, *Electric Utility Engineering Reference Book-Distribution Systems*, Vol. 3, East Pittsburgh, PA, 1965.)

TABLE 6.1
Required Operation of the Protective Apparatus

Fault Type	Limiter	NP Fuse	NP Breaker	Substation Circuit Breaker
Mains	Yes	No	No	No
Low-voltage bus	Yes	Yes	No	No
Transformer internal fault	No	No	Trips	Trips
Primary feeder	No	No	Trips	Trips

However, the switch cannot be operated, due to an electric interlock system, unless the network transformer is first de-energized. The grounding position provides safety for the workers during any work on the de-energized primary feeders.

To facilitate the disconnection of the transformer from an energized feeder, sometimes a special disconnecting switch that has an interlock with the associated NP is used, as shown in Figure 6.7. Therefore, the switch cannot be opened unless the load is first removed by the NP from the network transformer.

6.5.5 NETWORK TRANSFORMERS

In the OH secondary networks, the transformers can be mounted on poles or platforms, depending on their sizes. For example, the small ones (75 or 150 kVA) can be mounted on poles, whereas larger transformers (300 kVA) are mounted on platforms. The transformers are either single-phase or three-phase distribution transformers.

In the underground secondary networks, the transformers are installed in vaults. The NP is mounted on one side of the transformer and the three-position high-voltage switch on the other side. This type of arrangement is called a *network unit*.

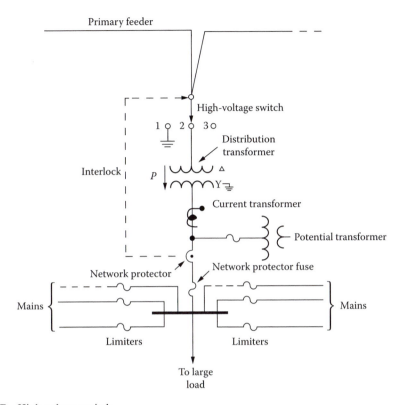

FIGURE 6.7 High-voltage switch.

A typical network transformer is three phase, with a low-voltage rating of 216Y/125 V, and can be as large as 1000 kVA. Table 6.2 gives standard ratings for three-phase transformers, which are used as secondary-network transformers. Because of the savings in vault space and in installation costs, network transformers are now built as three-phase units.

In general, the network transformers are submersible and oil or askarel cooled. However, because of environmental concerns, askarel is not used as an insulating medium in new installations anymore. Depending upon the locale of the installation, the network transformers can also be ventilated dry type or sealed dry type and submersible.

6.5.6 TRANSFORMER APPLICATION FACTOR

Reps [7] defines the application factor as "the ratio of installed network transformer to load." Therefore, by the same token,

$$\text{Application factor} = \frac{\sum S_T}{\sum S_L} \tag{6.1}$$

where
$\sum S_T$ is the total capacity of network transformers
$\sum S_L$ is the total load of secondary network

TABLE 6.2
Standard Ratings for Three-Phase Secondary-Network Transformers Transformer High Voltage

Preferred Nominal System Voltage	Rating	BIL (kV)	Taps		Standard kVA Ratings for Low-Voltage Rating of 216Y/125 V
			Above	Below	
2400/4160Y	4160[a]	60	None	None	300, 500, 750
	4160Y/2400[a,b]		None	None	
	4330		None	None	
	4330Y/2500[b]		None	None	
4800	5000	60	None	4875/4750/4625/4500	300, 500, 750
7200	7200[a]	75	None	7020/6840/6660/6480	300, 500, 750
	7500		None	7313/7126/6939/6752	
7200	11,500	95	None	11,213/10,926/10,639/10,w352	300, 500, 750, 1000
12,000	12,000[a]	95	None	11,700/11,400/11,100/10,800	300, 500, 750, 1000
	12,500		None	12,190/11,875/11,565/11,250	
7200/12,470Y	13,000Y/7500[b]	95	None	12,675/12,350/12,025/11,700	300, 500, 750, 1000
13,200	13,200[a]	95	None	12,870/12,540/12,210/11,880	300, 500, 750, 1000
7620/13,200Y	13,200Y/7620[a,b]		None	12,870/12,540/12,210/11,880	
	13,750		None	13,406/13,063/12,719/12,375	
	13,750Y/7940[b]		None	13,406/13,063/12,719/12,375	
14,440	14,400[a]	95	None	14,040/13,680/13,320/12,960	300, 500, 750, 1000
23,000	22,900[a]	150	24,100/23,500	22,300/21,700	500, 750, 1000
	24,000		25,200/24,600	23,400/22,800	

Source: Westinghouse Electric Corporation, *Electric Utility Engineering Reference Book Distribution Systems*, Vol. 3, East Pittsburgh, PA, 1965. With permission.

Note: All windings are delta connected unless otherwise indicated.

[a] Preferred ratings that should be used when establishing new networks.

[b] High-voltage and low-voltage neutrals are internally connected by a removable link.

The application factor is based upon single contingency, that is, the loss of one of the primary feeders. According to Reps [7], the application factor is a function of the following:

1. The number of primary feeders used
2. The ratio of Z_M/Z_T, where Z_M is the impedance of each section of secondary main and Z_T is the impedance of the secondary-network transformer
3. The extent of nonuniformity in load distribution among the network transformers under the single contingency

Figure 6.8 gives the plots of the transformer application factor versus the ratio of Z_M/Z_T for different numbers of feeders. For a given number of feeders and a given Z_M/Z_T ratio, the required capacity of network transformers to supply a given amount of load can be found by using Figure 6.8.

6.6 SPOT NETWORKS

A spot network is a special type of network that may have two or more network units feeding a common bus from which services are tapped. The transformer capacity utilization is better in the spot networks than in the distributed networks due to equal load division among the transformers regardless of a single-contingency condition.

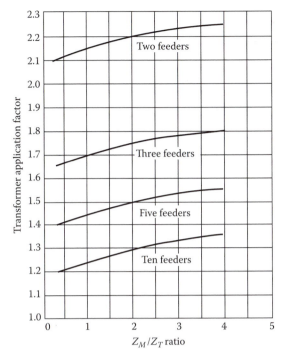

FIGURE 6.8 Network-transformer application factors as a function of Z_M/Z_T ratio and number of feeders used. (From Westinghouse Electric Corporation, *Electric Utility Engineering Reference Book-Distribution Systems*, Vol. 3, East Pittsburgh, PA, 1965.)

The impedance of the secondary main, between transformers, is zero in the spot networks. The spot networks are likely to be found in new high-rise commercial buildings. Even though spot networks with light loads can utilize 208Y/120 V as the nominal low voltage, the commonly used nominal low voltage of the spot networks is 480Y/277 V. Figure 6.9 shows a one-line diagram of the primary system for the John Hancock Center.

6.7 ECONOMIC DESIGN OF SECONDARIES

In this section, a method for (at least approximately) minimizing the *total annual cost* (TAC) of owning and operating the secondary portion of a three-wire single-phase distribution system in a residential area is presented. The method can be applied either to OH or *underground residential distribution* (URD) construction. Naturally, it is hoped that a design for satisfactory voltage-drop and voltage-dip performance will agree at least reasonably well with the design that yields minimum TAC.

6.7.1 PATTERNS AND SOME OF THE VARIABLES

Figure 6.10 illustrates the layout and one particular pattern having one span of secondary line (SL) each way from the distribution transformer. The system is assumed to be built in a straight line along an alley or along rear lot lines. The lots are assumed to be of uniform width *d* so that each span of SL is of length 2*d*. If SLs are not used, then there is a distribution transformer on every pole and OH construction, and every transformer supplies four SDs.

The primary line, which obviously must be installed along the alley, is not shown in Figure 6.10. The number of spans of SLs each way from a transformer is an important variable. Sometimes no

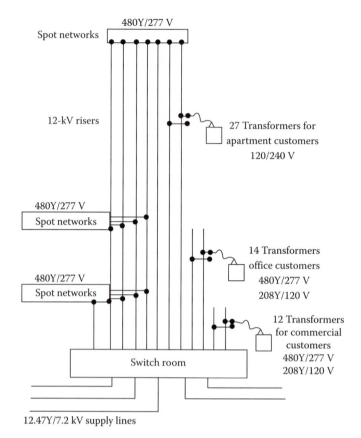

FIGURE 6.9 One-line diagram of the multiple primary system for the John Hancock Center. (From Westinghouse Electric Corporation, *Electric Utility Engineering Reference Book-Distribution Systems*, Vol. 3, East Pittsburgh, PA, 1965.)

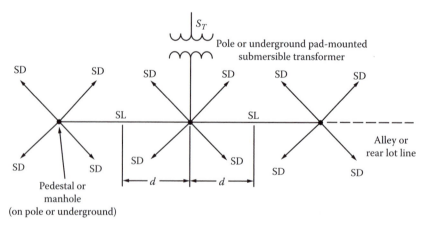

FIGURE 6.10 Illustration of a typical pattern.

SL is used in high-load density areas. In light-load density areas, three or more spans of SL each way from the transformer may be encountered in practice.

If Figure 6.10 represents an OH system, the transformer, with its arrester(s) and fuse cutout(s), is pole mounted. The SL and the SD may be of either open-wire or triplex cable construction. If Figure 6.10 represents a typical URD design, the transformer is grade mounted on a concrete slab and completely enclosed in a grounded metal housing, or else it is submersibly installed in a hole lined with concrete, Transite, or equivalent material. Both SL and SDs are triplexed or twin concentric neutral direct-burial cable laid in narrow trenches, which are backfilled after the installation of the cable. The distribution transformers have the parameters defined in the following:

S_T is the transformer capacity, continuously rated kVA.
I_{exc} is the per unit exciting current (based on S_T).
$P_{T,Fe}$ is the transformer core loss at rated voltage and rated frequency, kW.
$P_{T,Cu}$ is the transformer copper loss at rated kVA load, kW.

The SL has the parameters defined in the following:

A_{SL} is the conductor area, kcmil.
ρ is the conductor resistivity, $(\Omega \cdot \text{cmil})/\text{ft}$.
$= 20.5$ at 65°C for aluminum cable.

The SDs have the parameters A_{SD} and ρ with meanings that correspond to those given for SLs.

6.7.2 FURTHER ASSUMPTIONS

1. All secondaries and services are single phase three wire and nominally 120/240 V.
2. Perfectly balanced loading obtains in all three-wire circuits.
3. The system is energized 100% of the time, that is, 8760 h/year.
4. The annual loss factor is estimated by using Equation 2.40a, that is,

$$F_{LS} = 0.3F_{LD} + 0.7F_{LD}^2 \qquad (2.40a)$$

5. The annual peak-load kilovolt-ampere loading in any element of the pattern, that is, SD, section of SL, or transformer, is estimated by using the maximum diversified demand of the particular number of customers located downstream from the circuit element in question. This point is illustrated later.
6. Current flows are estimated in kilovolt-amperes and nominal operating voltage, usually 240 V.
7. All loads have the same (and constant) power factor.

6.7.3 GENERAL TAC EQUATION

The TAC of owning and operating one pattern of the secondary system is a summation of *investment (fixed) costs* (ICs) and *operating (variable) costs* (OCs). The costs to be considered are contained in the following equation:

$$TAC = \sum IC_T + \sum IC_{SL} + \sum IC_{SD} + \sum IC_{PH} + \sum OC_{exc}$$
$$+ \sum OC_{T,Fe} + \sum OC_{T,Cu} + \sum OC_{SL,Cu} + \sum OC_{SD,Cu} \qquad (6.2)$$

The summations are to be taken for the one standard pattern being considered, like Figure 6.10, but modified appropriately for the number of spans of SL being considered. It is apparent that the TAC so found may be divided by the number of customers per pattern so that the TAC can be allocated on a *per customer* basis.

6.7.4 Illustrating the Assembly of Cost Data

The following cost data are sufficient for illustrative purposes but not necessarily of the accuracy required for engineering design in commercial practice. Some of the cost data given may be quite inaccurate because of recent, severe inflation. The data are intended to represent an OH system using three-conductor triplex aluminum cable for both SLs and SDs. The important aspect of the following procedures is the finding of equations for all costs so that analytical methods can be employed to minimize the TAC:

1. IC_T is the annual installed cost of the distribution transformer + associated protective equipment

$$= (250 + 7.26 \times S_T) \times i \ \$/\text{transformer} \tag{6.3}$$

where

$$15\,\text{kVA} \le S_T \le 100\,\text{kVA}$$

S_T is the transformer-rated kVA

2. IC_{SL} is the annual installed cost of triplex aluminum SL cable

$$= (60 + 4.50 \times A_{SL}) \times i \ \$/1000 \ \text{ft} \tag{6.4}$$

where
A_{SL} is the conductor area, kcmil
i is the pu fixed charge rate on investment

Note that this cost is 1000 ft of cable, that is, 3000 ft of conductor.
3. IC_{SD} is the annual installed cost of triplex aluminum SD cable

$$= (60 + 4.50 \times A_{SD}) \times i \ \$/1000 \ \text{ft} \tag{6.5}$$

In this example, Equations 6.4 and 6.5 are alike because the same material, that is, triplex aluminum cable, is assumed to be used for both SL and SD construction.
4. IC_{PH} is the annual installed cost of pole and hardware on it but excluding transformer and transformer protective equipment

$$= \$160 \times i \ \$/\text{pole} \tag{6.6}$$

In case of URD design, the cost item IC_{PH} would designate the annual installed cost of a secondary pedestal or manhole.
5. OC_{exc} is the annual operating cost of transformer exciting current

$$= I_{exc} \times S_T \times IC_{cap} \times i \ \$/\text{transformer} \tag{6.7}$$

where
IC_{cap} is the total installed cost of primary-voltage shunt capacitors = \$5.00/kvar
I_{exc} is the an average value of transformer exciting current based on S_T kVA rating = 0.015 pu

6. $OC_{T,Fe}$ is the annual operating cost of transformer due to core (iron) losses

$$= (IC_{sys} \times i + 8760 \times EC_{off})P_{T,Fe} \text{ \$/transformer} \qquad (6.8)$$

where
IC_{sys} is the average investment cost of power system upstream, that is, toward generator, from distribution transformers
$= \$350/kVA$
EC_{off} is the incremental cost of electric energy (off-peak)
$= \$0.008/kWh$
$P_{T,Fe}$ is the annual transformer core loss, kW
$= 0.004 \times S_T \ 15 \text{ kVA} \leq S_T \leq 100 \text{ kVA}$

7. $OC_{T,Cu}$ is the annual operating cost of transformer due to copper losses

$$= (IC_{sys} \times i + 8760 \times EC_{on} \times F_{LS})\left(\frac{S_{max}}{S_T}\right)^2 \times P_{T,Cu} \text{ \$/transformer} \qquad (6.9)$$

where
EC_{on} is the incremental cost of electric energy (on-peak)
$= \$0.010/kWh$
S_{max} is the annual maximum kVA demand on transformer
$P_{T,Cu}$ is the transformer copper loss, kW at rated kVA load

$$= 0.073 + 0.00905 \times S_T \quad \text{where } 15 \text{ kVA} \leq S_T \leq 100 \text{ kVA} \qquad (6.10)$$

F_{LS} is the annual loss factor

8. $OC_{SL,Cu}$ is the annual operating cost of copper loss in a unit length of SL

$$= (IC_{sys} \times i + 8760 \times EC_{on} \times F_{LS})P_{SL,Cu} \qquad (6.11)$$

where
$P_{SL,Cu}$ is the power loss in a unit of SL at time of annual peak load due to copper losses, kW
$P_{SL,Cu}$ is an I^2R loss, and it must be related to conductor area A_{SL} with $R = \rho L/A_{SL}$

One has to decide carefully whether L should represent length of conductor or length of cable. When establishing $\sum OC_{SL,Cu}$ for the particular pattern being used, one has to remember that different sections of SLs may have different values of current and, therefore, different $P_{SL,Cu}$.

9. $OC_{SD,Cu}$ is the annual operating cost of copper loss in a unit length of SD. $OC_{SD,Cu}$ is handled like $OC_{SL,Cu}$ as described in Equation 6.11. When developing $\sum OC_{SD,Cu}$, it is important to relate $P_{SD,Cu}$ properly to the total length of SDs in the entire pattern.

6.7.5 Illustrating the Estimation of Circuit Loading

The simplifying assumptions (5) and (6) earlier describe one method for estimating the loading of each element of the pattern. It is important to find reasonable estimates for the current loads in each SD, in each section of SL, and in the transformer so that reasonable approximations will be used for the copper-loss costs $OC_{T,Cu}$, $\sum OC_{SL,Cu}$, and $\sum OC_{SD,Cu}$.

TABLE 6.3
Illustrative Load Data

No. of Customers Being Diversified	Ann. Max. Demand (kVA/Customer)
1	5.0
2	3.8
4	3.0
8	2.47
10	2.2
20	2.1
30	2.0
100	1.8

Source: Lawrence, R.F. et al., *AIEE Trans.*, PAS-79(pt. III), 199, June 1960, Fig. 3.

To proceed, it is necessary to have data for the annual maximum diversified kilovolt-ampere demand per customer versus the number of customers being diversified. The illustrative data tabulated in Table 6.3 have been taken from Lawrence, Reps, and Patton's paper entitled *Distribution System Planning through Optimized Design, I-Distribution Transformers and Secondaries* [13, Fig. 3]. As explained in that paper, the maximum diversified demand data were developed with the appliance diversity curves and the hourly variation factors.

It is apparent that the data could be plotted and the demand per customer for intermediate numbers of customers could then be read from the curve. Alternately, if a digital computer is programmed to perform the work described here, a linear interpolation might reasonably be used to estimate the per customer demand for intermediate numbers of customers.

Figure 6.11 shows a pattern having two SLs each way from the transformer. The reader can apply the foregoing data and with linear interpolation find the flows shown in Figure 6.11. The nominal voltage used is 240 V.

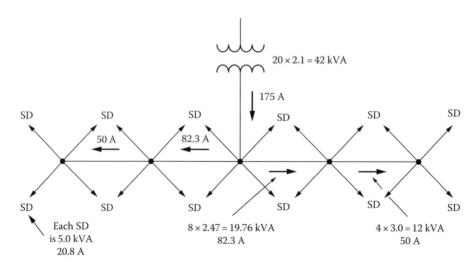

FIGURE 6.11 Estimated circuit loading for copper-loss determinations.

6.7.6 Developed Total Annual Cost Equation

Upon expanding all the cost items (1)–(9) in Section 6.7.4, taking the correct summations for the pattern being used, and introducing the results into Equation 6.2, one finds that

$$TAC = A + \frac{B}{S_T^2} + \frac{C}{S_T} + D \times S_T + E \times A_{SD} + \frac{F}{A_{SD}} + G \times A_{SL} + \frac{H}{A_{SL}} \tag{6.12}$$

where the coefficients A to H are numerical constants. It is important to note that TAC has been reduced to a function of three design variables, that is,

$$TAC = f(S_T, A_{SD}, \text{ and } A_{SL}) \tag{6.13}$$

However, one has to remember that many parameters, such as the fixed charge rate i, transformer core and copper losses, and installed costs of poles and lines, are contained in coefficients A to H. It should be further noted that the variables S_T, A_{SD}, and A_{SL} are in fact discrete variables. They are not continuous variables. For example, if theory indicates that $S_T = 31$ kVA is the optimum transformer size, the designer must choose rather arbitrarily between the standard commercial sizes of 25 and 37.5 kVA. The same ideas apply to conductor sizes for A_{SL} and A_{SD}.

6.7.7 Minimization of the Total Annual Costs

One may commence by using Equation 6.12, taking three partial derivatives, and setting each derivative to zero:

$$\frac{\partial(TAC)}{\partial S_T} = 0 \tag{6.14}$$

$$\frac{\partial(TAC)}{\partial A_{SL}} = 0 \tag{6.15}$$

$$\frac{\partial(TAC)}{\partial A_{SD}} = 0 \tag{6.16}$$

The work required by Equation 6.14 is formidable. The roots of a cubic must be found. At this point, one has the minimum TAC if only S_T is varied and similarly for only A_{SL} and A_{SD} variables. There is no assurance that the true, grant minimum of TAC will be achieved if the results of Equations 6.14 through 6.16 are applied simultaneously.

Having in fact discrete variables in this problem, one now discards continuous variable methods. The results of Equations 6.14 through 6.16 are used henceforth merely as indicators of the region that contains the minimum TAC achievable with standard commercial equipment sizes. The problem is continued by computing TAC for the standard commercial sizes of equipment nearest to the results of Equations 6.14 through 6.16 and then for one (or more?) standard sizes both larger and smaller than those indicated by Equations 6.14 through 6.16.

The results at this point are a reasonable number of computed TAC values, all close to the idealized, continuous variable TAC. Designers can easily scan these final few TAC results and select the $(S_T, A_{SL}, \text{ and } A_{SD})$ combinations that they think best.

6.7.8 Other Constraints

There are additional criteria that must be met in the total design of the distribution system, whether or not minimum TAC is realized. The further criteria involve quality of utility service. Minimum TAC designs may be encountered, which will violate one or more of the commonly used criteria:

1. A minimum allowable steady-state voltage at the most remote service entrance may have been set by law, public utility commission order, or company policy.
2. A maximum allowable motor-starting voltage dip at the most remote service entrance similarly may have been established.
3. Ordinarily, the ampacity of no section of SLs or SDs should be exceeded by the designer.
4. The maximum allowable distribution transformer loading, in per unit of the transformer continuous rating, should not be exceeded by the designer.

Example 6.1

This example deals with the costs of a single-phase OH secondary distribution system in a residential area. Figures 6.12 and 6.13 show the layouts and the service arrangement to be considered. Note that equal lot widths, hence uniform load spacings, are assumed. All SDs are assumed to be 70 ft long. The calculations should be done *for one block* of the residential area.

In case of OH secondary distribution system, assume that *there are* 12 *services per transformers*, that is, there are two transformers per block that are at poles 2 and 5, as shown in Figure 6.12.

Subsequent problems of succeeding chapters will deal with the voltage-drop constraints, which are used to set a minimum standard of quality of service. Naturally, it is hoped that a design for satisfactory voltage-drop performance will agree at least reasonably well with the design for minimum TAC.

Table 6.4 gives load data to be used in this example problem. Use 30 min annual maximum demands for customer class 2 for this problem.

Use the following data and assumptions:

1. All secondaries and services are single phase three wire, nominally 120/240 V.
2. Assume perfectly balanced loading in all single-phase three-wire circuits.
3. Assume that the system is energized 100% of the time, that is, 8760 h/year.
4. Assume the annual load factor to be $F_{LD} = 0.35$.
5. Assume the annual loss factor to be

$$F_{LS} = 0.3F_{LD} + 0.7F_{LD}^2$$

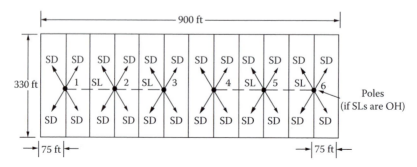

FIGURE 6.12   Residential area lot layout and service arrangement.

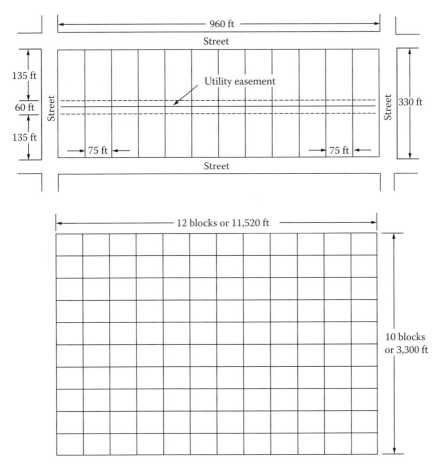

FIGURE 6.13 Residential area lot layout and utility easement arrangement.

TABLE 6.4
Load Data for Example 6.1

No. of Customers Being Diversified	30 Min Ann. Max. Demands (kVA/Customer)		
	Class 1	Class 2	Class 3
1	18.0	10.0	2.5
2	14.4	7.6	1.8
4	12.0	6.0	1.5
12	10.0	4.4	1.2
100	8.4	3.6	1.1

Source: Fink, D.G. and Beaty, H.W., *Standard Handbook for Electrical Engineers*, 11th edn., McGraw-Hill, New York, 1978, Figure 3.

Note: The kilovolt-ampere demands cited have been doubled arbitrarily in an effort to modernize the data. It is explained in the reference cited that the original maximum demand data were developed from appliance diversity curves and hourly variation factors.

6. Assume that the annual peak-load copper losses are properly evaluated $\left(\sum I^2 R\right)$ by applying the given class 2 loads as
   a. One consumer per SD
   b. Four consumers per section of SL
   c. Twelve consumers per transformer

   Here, $P_{SL,Cu}$ is an I^2R loss, and it must be related to conductor area A_{SL} with

$$R = \frac{\rho \times L}{1000 \times A_{SL}}$$

where
   A_{SL} is the conductor area, kcmil
   ρ is 20.5 ($\Omega \cdot$ cmil)/ft at 65°C for aluminum cable
   L is the length of conductor wire involved (not cable length)

(The designer must be careful to establish a correct relation between $\sum OC_{SL,Cu}$, that is, the annual OC per block, and the amount of SL for which $P_{SL,Cu}$ is evaluated.)
7. Assume nominal operating voltage of 240 V when computing currents.
8. Assume a 90% power factor for all loads.
9. Assume a fixed charge (capitalization) rate of 0.15.

Using the given data and assumptions, develop a numerical TAC *equation applicable to one block of these residential areas for the case of 12 services per transformer, that is, two transformers per block.* The equation should contain the variables of S_T, A_{SD}, and A_{SL}. Also determine the following:

a. The most economical SD size (A_{SD}) and the nearest larger standard AWG wire size
b. The most economical SL size (A_{SL}) and the nearest larger standard AWG wire size
c. The most economical distribution transformer size (S_T) and the nearest larger standard transformer size
d. The TAC per block for the theoretically most economical sizes of equipment
e. The TAC per block for the nearest larger standard commercial sizes of equipment
f. The TAC per block for the nearest larger transformer size and for the second larger sizes of A_{SD} and A_{SL}
g. Fixed charges per customer per month for the design using the nearest larger standard commercial sizes of equipment
h. The variable (operating) costs per customer per month for the design using the nearest larger standard commercial sizes of equipment

Solution

From Equation 6.2, the TAC is

$$TAC = \sum IC_T + \sum IC_{SL} + \sum IC_{SD} + \sum IC_{PH} + \sum OC_{exc}$$
$$+ \sum OC_{T,Fe} + \sum OC_{T,Cu} + \sum OC_{SL,Cu} + \sum OC_{SD,Cu} \qquad (6.2)$$

Since there are two transformers per block and 12 services per transformer, from Equation 6.3, the annual installed cost of the two distribution transformers and associated protective equipment is

$$IC_T = 2(250 + 7.26 \times S_T) \times i$$

$$= 2(250 + 7.26 \times S_T) \times 0.15$$

$$= 75 + 2.178\, S_T \ \$/block \qquad (6.17)$$

From Equation 6.4, the annual installed cost of the triplex aluminum cable used for 300 ft per transformer (since there is 150 ft SL at each side of each transformer) in the SLs is

$$IC_{SL} = 2(60 + 4.50 \times A_{SL}) \times i$$

$$= 2(60 + 4.50 \times A_{SL}) \times 0.15 \times \frac{300\, \text{ft/transformer}}{1000\, \text{ft}}$$

$$= 5.4 + 0.405\, A_{SL}\ \$/\text{block} \tag{6.18}$$

From Equation 6.5, the annual installed cost of triplex aluminum 24 SDs per block (each SD is 70 ft long) is

$$IC_{SD} = 2(60 + 4.50 \times A_{SD}) \times i$$

$$= 2(60 + 4.50 \times A_{SD}) \times 0.15 \times \frac{12 \times 70\, \text{ft/SD}}{1000\, \text{ft}}$$

$$= 15.12 + 1.134 \times A_{SD}\ \$/\text{block} \tag{6.19}$$

From Equation 6.6, the annual cost of pole and hardware for the six poles per block is

$$IC_{PH} = \$160 \times i \times 6\ \text{poles/block}$$

$$= \$160 \times 0.15 \times 6$$

$$= \$144/\text{block} \tag{6.20}$$

From Equation 6.7, the annual OC of transformer exciting current per block is

$$OC_{exc} = 2I_{exc} \times S_T \times IC_{cap} \times i$$

$$= 2(0.015) \times S_T \times \$5/\text{kvar} \times 0.15$$

$$= 0.0225\, S_T\ \$/\text{block} \tag{6.21}$$

From Equation 6.8, the annual OC of core (iron) losses of the two transformers per block is

$$OC_{T,Fe} = 2(IC_{sys} \times i + 8760 \times EC_{off})\, 0.004 \times S_T$$

$$= 2(\$350/\text{kVA} \times 0.15 + 8760 \times \$0.008/\text{kWh})\, 0.004 \times S_T$$

$$= 0.98 S_T\ \$/\text{block} \tag{6.22}$$

From Equation 6.9, the annual OC of transformer copper losses of the two transformers per block is

$$OC_{T,Cu} = (IC_{sys} \times i + 8760 \times EC_{on} \times F_{LS}) \left(\frac{S_{max}}{S_T}\right)^2 \times P_{T,Cu}$$

where

$$F_{LS} = 0.3F_{LD} + 0.7F_{LD}^2$$

$$= 0.3(0.35) + 0.7(0.35)^2$$

$$= 0.1904$$

$$S_{max} = 12 \text{ customers/transformer} \times 4.4 \text{ kVA/customer}$$

$$= 52.8 \text{ kVA/transformer}$$

Here, the figure of 4.4 kVA/customer is found from Table 6.4 for 12 class 2 customers.

From Equation 6.10, the transformer copper loss in kilowatts at rated kilovolt-ampere load is found as

$$P_{T,Cu} = 0.073 + 0.00905S_T$$

Therefore,

$$OC_{T,Cu} = 2[(\$350/\text{kVA}) \times 0.15 + 8760 \times (\$0.01/\text{kWh}) \times 0.1904]$$

$$\times \left(\frac{52.8\,\text{kVA/transformer}}{S_T}\right)^2 (0.073 + 0.00905 \times S_T)$$

$$= \frac{28,170}{S_T^2} + \frac{3492}{S_T} \$/\text{block} \tag{6.23}$$

From Equation 6.11, the annual OC of copper losses in the four SLs is

$$OC_{SL,Cu} = 2(IC_{sys} \times i + 8760 \times EC_{on} \times F_{LS})P_{LS,Cu}$$

where $P_{LS,Cu}$ is the copper losses in two SLs at time of annual peak load, kW/transformer (see Figure 6.14) = $I^2 \times R$

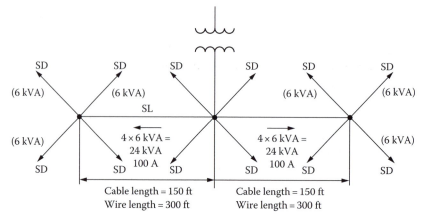

FIGURE 6.14   Illustration of the SLs.

where

$$R = \frac{\rho \times L}{1000 \times A_{SL}}$$

$$= \frac{20.5(\Omega \cdot \text{cmil})/\text{ft} \times 300 \text{ ft wire} \times 2}{1000 \times A_{SL}}$$

$$= \frac{12.3}{A_{SL}}(\Omega \cdot \text{kcmil})/\text{transformer}$$

$$P_{LS,Cu} = \left(\frac{24\text{kVA}}{240\text{V}}\right)^2 \times \frac{12.3}{A_{SL}} \times \frac{1}{1000}$$

$$= \frac{123}{A_{SL}} \text{kW/transformer}$$

Thus,

$$OC_{SL,Cu} = 2[(\$350/\text{kVA}) \times 0.15 + 8760 \times (\$0.01/\text{kWh}) \times 0.1904]\frac{123}{A_{SL}}$$

$$= \frac{17,018}{A_{SL}} \$/\text{block} \tag{6.24}$$

Also from Equation 6.11, the annual OC of copper losses in the 24 SDs is

$$OC_{SD,Cu} = (IC_{sys} \times i + 8760 \times EC_{on} \times F_{LS})P_{SD,Cu}$$

$$= (69.179)P_{SD,Cu}$$

where $P_{SD,Cu}$ is the copper losses in the 24 secondary drops at the time of annual peak load, kW

$$= I^2 \times R$$

where

$$R = \frac{\rho \times L}{1000 \times A_{SD}}$$

$$= \frac{20.5(\Omega \cdot \text{cmil/ft})(70 \text{ ft}) \times (24 \text{ SD/block}) \times (2 \text{ wires/SD})}{1000 \times A_{SD}}$$

$$= \frac{68.88}{A_{SD}}(\Omega \cdot \text{kcmil})/\text{block}$$

From Table 6.4, the 30 min annual maximum demand for one SD per one class 2 customer can be found as 10 kVA. Therefore,

$$P_{SD,Cu} = \left(\frac{10\text{kVA}}{0.240\text{kV}}\right)^2 \times \frac{68.88}{A_{SD}} \times \frac{1}{1000}$$

$$= \frac{119.58}{A_{SD}} \text{kW/block}$$

Thus,

$$OC_{SD,Cu} = 69.179 \times \frac{119.58}{A_{SD}}$$

$$\cong \frac{8273}{A_{SD}} \, \$/\text{block} \tag{6.25}$$

Substituting Equations 6.17 through 6.25 into Equation 6.2, the TAC equation can be found as

$$TAC = (75 + 2.178 \times S_T) + (5.4 + 0.405 \times A_{SL}) + (15.12 + 1.134 \times A_{SD}) + (144 + 0.0225 \times S_T)$$

$$+ (0.98 \times S_T) + \left(\frac{28,170}{S_T^2} + \frac{3,492}{S_T}\right) + \frac{17,108}{A_{SL}} + \frac{8,273}{A_{SD}}$$

After simplifying,

$$TAC = 239.52 + 3.1805 \times S_T + \frac{3,492}{S_T} + \frac{28,170}{S_T^2} + 0.405 \times A_{SL}$$

$$+ \frac{17,018}{A_{SL}} + 1.134 \times A_{SD} + \frac{8,273}{A_{SD}} \tag{6.26}$$

a. By partially differentiating Equation 6.26 with respect to A_{SD} and equating the resultant to zero,

$$\frac{\partial(TAC)}{\partial A_{SD}} = 1.134 - \frac{8273}{A_{SD}^2} = 0$$

from which the most economical service-drop size can be found as

$$A_{SD} = \left(\frac{8273}{1.134}\right)^{1/2}$$

$$= 85.41\,\text{kcmil}$$

Therefore, the nearest larger standard AWG wire size can be found from the copper-conductor table (see Table A.1) as 1/0, that is, 106,500 cmil.
b. Similarly, the most economical SL size can be found from

$$\frac{\partial(TAC)}{\partial A_{SL}} = 0.405 - \frac{17,018}{A_{SL}^2} = 0$$

as

$$A_{SL} = \left(\frac{17,018}{0.405}\right)^{1/2}$$

$$= 204.99\,\text{kcmil}$$

Therefore, the nearest larger AWG wire size is 4/0, that is, 211.6 kcmil.
c. The most economical distribution transformer size can be found from

$$\frac{\partial(TAC)}{\partial S_T} = 3.1805 - \frac{3,492}{S_T^2} - \frac{56,340}{S_T^3} = 0$$

or

$$S_T \cong 39 \text{kVA}$$

Therefore, the nearest larger standard transformer size is 50 kVA.

d. By substituting the found values of A_{SD}, A_{SL}, and S_T into Equation 6.26, the TAC per block for the theoretically most economical sizes of equipment can be found as

$$TAC = 239.52 + 3.1805 \times (39) + \frac{3,492}{(39)} + \frac{28,170}{(39^2)} + 0.405 \times (204.99)$$

$$+ \frac{17,018}{(204.99)} + 1.134 \times (85.41) + \frac{8,273}{(85.41)} \cong \$838/\text{block}$$

e. By substituting the found standard values of A_{SD}, A_{SL}, and S_T into Equation 6.26, the TAC per block for the nearest larger standard commercial sizes of equipment can be found as

$$TAC = 239.52 + 3.1805 \times (50) + \frac{3,492}{(50)} + \frac{28,170}{(50^2)} + 0.405 \times (211.6)$$

$$+ \frac{17,018}{(211.6)} + 1.134 \times (106.5) + \frac{8,273}{(106.5)} \cong \$844/\text{block}$$

f. The second larger sizes of A_{SD} and A_{SL} are 133.1 kcmil and 250 kcmil, respectively. Therefore,

$$TAC = 239.52 + 3.1805 \times (50) + \frac{3492}{(50)} + \frac{28,170}{(50^2)} + 0.405 \times (250)$$

$$+ \frac{17,018}{(250)} + 1.134 \times (133.1) + \frac{8,273}{(133.1)} \cong \$862/\text{block}$$

g. The fixed charges per customer per month for the design using the nearest larger standard commercial sizes of equipment is

$$TAC = \left(\sum IC_T + \sum IC_{SL} + \sum IC_{SD} + \sum IC_{PH} \right) \times \frac{1}{24 \text{ customers/block} \times 12 \text{ month/year}}$$

$$\cong \$1.9225/\text{customer/month}$$

h. The variable (operating) cost per customer per month for the design using the nearest larger standard commercial sizes of equipment is

$$TAC = \left(\sum OC_{exc} + \sum OC_{T,Fe} + \sum OC_{T,Cu} + \sum OC_{SL,Cu} + \sum OC_{SD,Cu} \right) \frac{1}{24 \times 12}$$

$$= \left[0.0225(50) + 0.98(50) + \frac{28,170}{(50^2)} + \frac{3,492}{50} + \frac{17,018}{211.6} + \frac{8,273}{106.5} \right] \frac{1}{24 \times 12}$$

$$= \$1.0084/\text{customer/month}$$

Note that the fixed charges are larger than the OCs.

6.8 UNBALANCED LOAD AND VOLTAGES

A single-phase three-wire circuit is regarded as unbalanced if the neutral current is not zero. This happens when the loads connected, for example, between line and neutral, are not equal. The result is unsymmetrical current and voltages and a nonzero current in the neutral line. In that case, the necessary calculations can be done by using the method of symmetrical components.

Example 6.2

This example and Examples 6.3 and 6.4 deal with the computation of voltages in unbalanced single-phase three-wire secondary circuits, as shown in Figure 6.15. Here, both the mutual-impedance methods and the flux-linkage methods are applicable as alternative methods for computing the voltage drops in the SL. This example deals with the computation of the complex linkages due to the line currents in the conductors a, b, and n. Assume that the distribution transformer used for this single-phase three-wire distribution is rated as 7200/120–240 V, 25 kVA, and 60 Hz, and the n_1 and n_2 turn ratios are 60 and 30. As Figure 6.15 suggests, the two halves of the low-voltage winding of the distribution transformer are independently loaded with unequal secondary loads. Therefore, the single-phase three-wire secondaries are unbalanced. The vertical spacing between the secondary wires is as illustrated in Figure 6.16. Assume that the secondary wires are

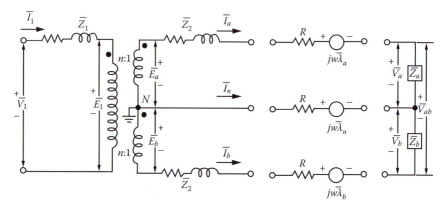

FIGURE 6.15 An unbalanced single-phase three-wire secondary circuit.

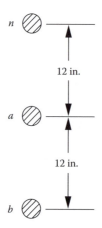

FIGURE 6.16 Vertical spacing between the secondary wires.

made of #4/0 seven-strand hard-drawn aluminum conductors and 400 ft of line length. Use 50°C resistance in finding the line impedances.

Furthermore, assume that (1) the load impedances $\bar{Z}_a$ and $\bar{Z}_b$ are independent of voltage, (2) the primary-side voltage is $\bar{V}_1 = 7272$ V and is maintained constant, and (3) the line capacitances and transformer exciting current are negligible. Use the given information and develop numerical equations for the phasor expressions of the flux linkages $\bar{\lambda}_a$, $\bar{\lambda}_b$, and $\bar{\lambda}_n$ in terms of $\bar{I}_a$ and $\bar{I}_b$. In other words, find the coefficient matrix, numerically, in the equation

$$\begin{bmatrix} \bar{\lambda}_a \\ \bar{\lambda}_b \\ \bar{\lambda}_c \end{bmatrix} = \begin{bmatrix} \text{coefficient} \\ \text{matrix} \end{bmatrix} \begin{bmatrix} \bar{I}_a \\ \bar{I}_b \end{bmatrix} \tag{6.27}$$

Solution

The phasor expressions of the complex flux linkages $\bar{\lambda}_a$, $\bar{\lambda}_b$, and $\bar{\lambda}_n$ due to the line currents in the conductors a, b, and n can be written as*

$$\bar{\lambda}_a = 2 \times 10^{-7} \left(\bar{I}_a \times \ln \frac{1}{D_{aa}} + \bar{I}_b \times \ln \frac{1}{D_{ab}} + \bar{I}_n \times \ln \frac{1}{D_{an}} \right) \frac{\text{Wb} \cdot \text{T}}{\text{m}} \tag{6.28}$$

$$\bar{\lambda}_b = 2 \times 10^{-7} \left(\bar{I}_a \times \ln \frac{1}{D_{ab}} + \bar{I}_b \times \ln \frac{1}{D_{aa}} + \bar{I}_n \times \ln \frac{1}{D_{bn}} \right) \frac{\text{Wb} \cdot \text{T}}{\text{m}} \tag{6.29}$$

$$\bar{\lambda}_n = 2 \times 10^{-7} \left(\bar{I}_a \times \ln \frac{1}{D_{na}} + \bar{I}_b \times \ln \frac{1}{D_{nb}} + \bar{I}_n \times \ln \frac{1}{D_{nn}} \right) \frac{\text{Wb} \cdot \text{T}}{\text{m}} \tag{6.30}$$

Since

$$\bar{I}_a + \bar{I}_b + \bar{I}_n = 0 \tag{6.31}$$

the current in the neutral conductor can be written as

$$\bar{I}_n = -\bar{I}_a - \bar{I}_b \tag{6.32}$$

Thus, substituting Equation 6.32 into Equations 6.28 through 6.30,

$$\bar{\lambda}_a = 2 \times 10^{-7} \left(\bar{I}_a \times \ln \frac{D_{an}}{D_{aa}} + \bar{I}_b \times \ln \frac{D_{an}}{D_{ab}} \right) \frac{\text{Wb} \cdot \text{T}}{\text{m}} \tag{6.33}$$

$$\bar{\lambda}_b = 2 \times 10^{-7} \left(\bar{I}_a \times \ln \frac{D_{bn}}{D_{ab}} + \bar{I}_b \times \ln \frac{D_{bn}}{D_{bb}} \right) \frac{\text{Wb} \cdot \text{T}}{\text{m}} \tag{6.34}$$

$$\bar{\lambda}_c = 2 \times 10^{-7} \left(\bar{I}_a \times \ln \frac{D_{nn}}{D_{na}} + \bar{I}_b \times \ln \frac{D_{nn}}{D_{ab}} \right) \frac{\text{Wb} \cdot \text{T}}{\text{m}} \tag{6.35}$$

---

* The notation "ln" is used for "log to the base e."

Therefore, from Equations 6.33 through 6.35,

$$
\begin{bmatrix} \bar{\lambda}_a \\ \bar{\lambda}_b \\ \bar{\lambda}_c \end{bmatrix} = \begin{bmatrix} 2\times10^{-7}\times\ln\dfrac{D_{an}}{D_{aa}} + 2\times10^{-7}\times\ln\dfrac{D_{an}}{D_{ab}} \\ 2\times10^{-7}\times\ln\dfrac{D_{bn}}{D_{ab}} + 2\times10^{-7}\times\ln\dfrac{D_{bn}}{D_{bb}} \\ 2\times10^{-7}\times\ln\dfrac{D_{nn}}{D_{na}} + 2\times10^{-7}\times\ln\dfrac{D_{nn}}{D_{ab}} \end{bmatrix} \begin{bmatrix} \bar{I}_a \\ \bar{I}_b \end{bmatrix} \dfrac{Wb\cdot T}{m}
$$
(6.36)

Thus, from Equation 6.36, the coefficient matrix can be found numerically as

$$
\begin{bmatrix} coefficient \\ matrix \end{bmatrix} = \begin{bmatrix} 2\times10^{-7}\times\ln\dfrac{1}{0.01577} & 2\times10^{-7}\times\ln\dfrac{1}{1} \\ 2\times10^{-7}\times\ln\dfrac{2}{1} & 2\times10^{-7}\times\ln\dfrac{2}{0.01577} \\ 2\times10^{-7}\times\ln\dfrac{0.01577}{1} & 2\times10^{-7}\times\ln\dfrac{0.01577}{2} \end{bmatrix}
$$

$$
= \begin{bmatrix} 8.2992\times10^{-7} & 0 \\ 1.3862\times10^{-7} & 9.6855\times10^{-7} \\ -8.2992\times10^{-7} & -9.6855\times10^{-7} \end{bmatrix} \dfrac{Wb\cdot T}{m}
$$

Note that the elements in the coefficient matrix can be converted to weber-teslas per foot if they are multiplied by 0.3048 m/ft.

Example 6.3

Assume that, in Example 6.2, $\bar{I}_a$, $\bar{I}_b$, and $\bar{V}_1$ are specified but not the load impedances $\bar{Z}_a$ and $\bar{Z}_b$. Develop symbolic equations that will give solutions for the load voltages $\bar{V}_a$, $\bar{V}_a$, and $\bar{V}_{ab}$ in terms of the voltage $\bar{V}_1$, the impedances, and the flux linkages.

Solution

Since the transformation ratio of the distribution transformer is

$$
n = \frac{E_1}{E_a} = \frac{E_1}{E_b}
$$

$$
= \frac{7200\,V}{120\,V}
$$

$$
= 60
$$

the primary-side current can be written as

$$
\bar{I}_1 = \frac{\bar{I}_a - \bar{I}_b}{n}
$$
(6.37)

Here,

$$
\bar{E}_1 = \bar{V}_1 - \bar{I}_1\bar{Z}_1
$$
(6.38)

Substituting Equation 6.37 into Equation 6.38,

$$\bar{E}_1 = \bar{V}_1 - \bar{Z}_1 \frac{\bar{I}_a - \bar{I}_b}{n} \tag{6.39}$$

Also,

$$\bar{E}_a = \bar{E}_b = \frac{\bar{E}_1}{n} \tag{6.40}$$

Substituting Equation 6.39 into Equation 6.40,

$$\bar{E}_a = \bar{E}_b = \frac{\bar{V}_1}{n} - \frac{\bar{Z}_1}{n^2}(\bar{I}_a - \bar{I}_b) \tag{6.41}$$

By writing a loop equation for the secondary side of the equivalent network of Figure 6.15,

$$-\bar{E}_a + \bar{Z}_2\bar{I}_a + R\bar{I}_a + j\omega\bar{\lambda}_a + \bar{V}_a - j\omega\bar{\lambda}_n + R(\bar{I}_a + \bar{I}_b) = 0 \tag{6.42}$$

Substituting Equation 6.41 into Equation 6.42,

$$-\frac{\bar{V}_1}{n} + \frac{\bar{Z}_1}{n^2}(\bar{I}_a - \bar{I}_b) + \bar{Z}_2\bar{I}_a + R\bar{I}_a + j\omega\bar{\lambda}_a + \bar{V}_a - j\omega\bar{\lambda}_n + R(\bar{I}_a + \bar{I}_b) = 0$$

or

$$\bar{V}_a = \frac{\bar{V}_1}{n} + \left(\frac{\bar{Z}_1}{n^2} - R\right)\bar{I}_b - \left(\frac{\bar{Z}_1}{n^2} + \bar{Z}_2 + 2R\right)\bar{I}_a - j\omega(\bar{\lambda}_a - \bar{\lambda}_n) \tag{6.43a}$$

Also, by writing a second loop equation,

$$\bar{E}_b + \bar{Z}_2\bar{I}_b + R\bar{I}_b + j\omega\bar{\lambda}_b - \bar{V}_b + R(\bar{I}_a + \bar{I}_b) - j\omega\bar{\lambda}_n = 0 \tag{6.44}$$

Substituting Equation 6.41 into Equation 6.45,

$$\bar{V}_b = \frac{\bar{V}_1}{n} - \left(\frac{\bar{Z}_1}{n^2} - R\right)\bar{I}_a + \left(\frac{\bar{Z}_1}{n^2} + \bar{Z}_2 + 2R\right)\bar{I}_b + j\omega(\bar{\lambda}_b - \bar{\lambda}_n) \tag{6.45}$$

However, from Figure 6.15,

$$\bar{V}_{ab} = \bar{V}_a + \bar{V}_b \tag{6.46}$$

Therefore, substituting Equations 6.43 and 6.45 into Equation 6.46,

$$\bar{V}_{ab} = 2\frac{\bar{V}_1}{n} - \left(\frac{2\bar{Z}_1}{n^2} + R + \bar{Z}_2\right)\bar{I}_a + \left(\frac{2\bar{Z}_1}{n^2} + R + \bar{Z}_2\right)\bar{I}_b + j\omega(\bar{\lambda}_b - \bar{\lambda}_a) \tag{6.47}$$

Example 6.4

Assuming that in Example 6.3 the given voltages are

$$\bar{V}_1 = 7272\angle 0° \, V$$

$$\bar{E}_a = 120\angle 0° \, V$$

$$\bar{E}_b = 120\angle 0° \, V$$

and the load impedances are

$$\bar{Z}_a = 0.80 + j0.60 \, \Omega$$

$$\bar{Z}_b = 0.80 + j0.60 \, \Omega$$

and

$$\bar{Z}_1 = 14.5152 + j19.90656 \, \Omega$$

$$\bar{Z}_2 = 0.008064 + j0.0027648 \, \Omega$$

determine the following:

a. The secondary currents $\bar{I}_a$ and $\bar{I}_b$
b. The secondary neutral current $\bar{I}_n$
c. The secondary voltages $\bar{V}_a$ and $\bar{V}_b$
d. The secondary voltage $\bar{V}_{ab}$

Solution

From Equation 6.43,

$$\bar{V}_a = \bar{I}_a \bar{Z}_a = \frac{\bar{V}_1}{n} + \left(\frac{\bar{Z}_1}{n^2} - R \right) \bar{I}_b - \left(\frac{\bar{Z}_1}{n^2} + \bar{Z}_2 + 2R \right) \bar{I}_a - j\omega(\bar{\lambda}_a - \bar{\lambda}_n) \tag{6.43b}$$

or

$$\frac{\bar{V}_1}{n} = \left(\frac{\bar{Z}_1}{n^2} + \bar{Z}_2 + 2R + \bar{Z}_a \right) \bar{I}_a - \left(\frac{\bar{Z}_1}{n^2} - R \right) \bar{I}_b + j\omega(\bar{\lambda}_a - \bar{\lambda}_n) \tag{6.48}$$

Similarly, from Equation 6.45,

$$\bar{V}_b = -\bar{I}_b \bar{Z}_b = \frac{\bar{V}_1}{n} - \left(\frac{\bar{Z}_1}{n^2} - R \right) \bar{I}_a + \left(\frac{\bar{Z}_1}{n^2} + \bar{Z}_2 + 2R \right) \bar{I}_b + j\omega(\bar{\lambda}_b - \bar{\lambda}_n)$$

or

$$\frac{\bar{V}_1}{n} = \left(\frac{\bar{Z}_1}{n^2} - R \right) \bar{I}_a + \left(\frac{\bar{Z}_1}{n^2} + \bar{Z}_2 + \bar{Z}_b + 2R \right) \bar{I}_b + j\omega(\bar{\lambda}_b - \bar{\lambda}_n) \tag{6.49}$$

Substituting the given values into Equation 6.48,

$$\frac{7272}{60} = \overline{I_a}\left[\frac{14.5152}{60^2} + j\frac{19.90656}{60^2} + 0.008064 + j0.027648 + 0.8 + j0.6\right.$$

$$+ \frac{2(400)(0.486)}{5280} + \overline{I_b}\left[\frac{(400)(0.486)}{5280} - \frac{14.5152}{60^2} - j\frac{19.90656}{60^2}\right]$$

$$+ j377(0.3048)(400) \times 10^{-7}(8.299\overline{I_a} + 8.299\overline{I_a} + 9.686\overline{I_b})$$

or

$$121.2 = \overline{I_a}(0.8857 + j0.6846) + \overline{I_b}(0.03279 + j0.03899) \tag{6.50}$$

Also, substituting the given values into Equation 6.49,

$$\frac{7272}{60} = \overline{I_a}\left[\frac{14.5152}{60^2} + j\frac{19.90656}{60^2} - \frac{(400)(0.486)}{60^2}\right]$$

$$+ \overline{I_b}\left[-0.8 + j0.6 - \frac{14.5152}{60^2} - j\frac{19.90656}{60^2} - 0.008064 - j0.027648 - \frac{2(400)(0.486)}{5280}\right]$$

$$- j377(0.3048)(400) \times 10^{-7} \times (1.386\overline{I_a} + 9.686\overline{I_b} + 8.299\overline{I_a} + 9.686\overline{I_b})$$

or

$$121.2 = \overline{I_a}(-0.03279 - j0.03899) + \overline{I_b}(-0.88574 + j0.50267) \tag{6.51}$$

Therefore, from Equations 6.50 and 6.51,

$$\begin{bmatrix} 121.2 \\ 121.2 \end{bmatrix} = \begin{bmatrix} 0.8857 + j0.6846 & 0.03279 + j0.03899 \\ -0.03279 - j0.03899 & -0.88574 + j0.50267 \end{bmatrix}\begin{bmatrix} \overline{I_a} \\ \overline{I_b} \end{bmatrix} \tag{6.52}$$

By solving Equation 6.52,

$$\begin{bmatrix} \overline{I_a} \\ \overline{I_b} \end{bmatrix} = \begin{bmatrix} 89.8347 & -j62.393 \\ -107.387 & -j62.5885 \end{bmatrix} A \tag{6.53}$$

a. From Equation 6.53, the secondary currents are

$$\overline{I_a} = 89.8347 - j62.393$$

$$= 109.376 \angle -34.78°A$$

and

$$\overline{I_b} = -107.387 - j62.5885$$

$$= 124.295 \angle 210.24°A$$

  b. Therefore, the secondary neutral current is

$$\overline{I}_n = -\overline{I}_a - \overline{I}_b$$

$$= 17.5523 + j124.9815 \text{ A}$$

c. The secondary voltages are

$$\overline{V}_a = \overline{I}_a \times \overline{Z}_a$$

$$= (109.376\angle{-34.78°})(1\angle{36.87°})$$

$$= 109.376\angle{2.09°} \text{ V}$$

and

$$\overline{V}_b = -\overline{I}_b \times \overline{Z}_b$$

$$= -(124.295\angle{210.24°})(1\angle{-36.87°})$$

$$= 124.295\angle{-6.63°} \text{ V}$$

d. Therefore, the secondary voltage $\overline{V}_{ab}$ is

$$\overline{V}_{ab} = \overline{V}_a + \overline{V}_b$$

$$= 109.376\angle{2.09°} + 124.295\angle{-6.63°}$$

$$= 232.997\angle{-2.55°} \text{ V}$$

Example 6.5

Figure 6.17 shows an ac secondary network, which has been adapted from Ref. [7]. The loads shown in Figure 6.17 are in three-phase kilowatts and kilovars, with a lagging power factor of 0.85. The nominal voltage is 208 V. All distribution transformers are rated 500 kVA three phase, with 4160 V delta high voltage and 125/216 V wye grounded low voltage. They have leakage imped-ance Z_T of 0.0086 + j0.0492 pu based on transformer ratings.

All secondary underground mains have copper 3-#4/0 per phase and 3-#3/0 neutral cables in nonmagnetic conduits. The positive-sequence impedance Z_M of 500 ft of main is 0.181 + j0.115 pu on a 1000 kVA base.

All primary-feeder circuits are 1.25 min long. Three single-conductor 500 kcmil 5 kV shielded-copper PE-insulated underground cables are used at 90° conductor temperature. Their imped-ances within the small area of the network are neglected. The positive-sequence impedance Z_F of the feeder cable is 0.01 + j0.017 pu on a 1000 kVA base for 1.25 min long feeders. The approximate ampacities are 473 A for one circuit per duct bank and 402 A for four equally loaded circuits per duct bank.

The bases used are (1) three-phase power base of 1000 kVA; (2) for secondaries, 125/216 V, 2666.7 A, and 0.04687 Ω; and (3) for primaries, 2400/4160 V, 138.9 A, and 17.28 Ω.

The standard 125/216 V network-capacitor sizes used are 40, 80, and 120 kvar. In this study, these capacitors are not switched. Ordinarily, it is desired that distribution circuits not get into leading power factor operation during off-peak load periods. Therefore, the total magnetizing vars generated by unswitched shunt capacitors should not exceed the total magnetizing vars taken by the off-peak load. In this example, the total reactive load is 3150 kvar at peak load, and it is assumed that off-peak load is one-third of peak load, or 1050 kvar. Therefore, a total capacitor size of 960 kvar has been used. It has been distributed arbitrarily throughout the network in standard sizes but with the larger capacitor banks generally being located at the larger-load buses and at the ends of radial stubs from the network.

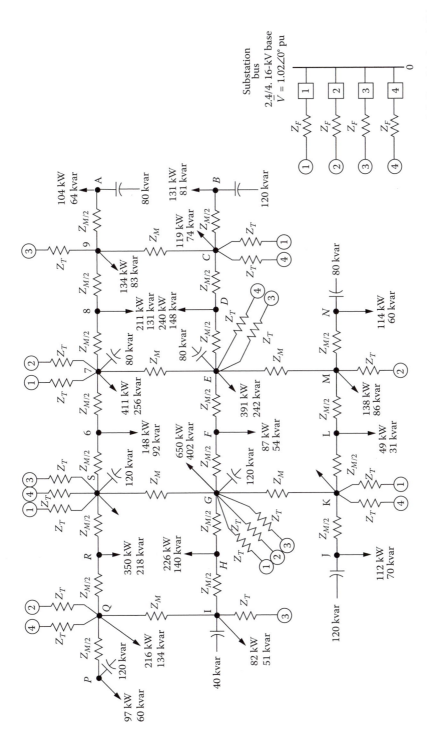

FIGURE 6.17 (Adapted from Westinghouse Electric Corporation, *Electric Utility Engineering Reference Book-Distribution Systems*, Vol. 3, East Pittsburgh, PA, 1965.)

TABLE 6.5
Bus Voltage Value (pu)

Buses	Case 1	Case 2	Case 3	Case 4
A	0.951	0.967	0.954	0.915
B	0.958	0.975	0.955	0.860
C	0.976	0.986	0.966	0.873
J	0.959	0.976	0.954	0.864
K	0.974	0.984	0.962	0.875
N	0.958	0.973	0.963	0.924
P	0.960	0.977	0.966	0.926
R	0.945	0.954	0.938	0.890
S	0.964	0.972	0.951	0.898

Using the given data, four separate load flow solutions have been obtained for the following operating conditions in the example secondary network:

Case 1: *Normal switching*: Normal loads and all shunt capacitors are off.
Case 2: *Normal switching*: Normal loads and all shunt capacitors are on.
Case 3: *First-contingency outage*: Primary feeder 1 is out. Normal loads and all shunt capacitors are on.
Case 4: *Second-contingency outage*: Primary feeders 1 and 4 are out. Normal loads and all shunt capacitors are on. Note that this second-contingency outage is very severe, causing the largest load (at bus 5) to lose two-thirds of its transformer capacity.

To make a voltage study, Table 6.5 has been developed based on the load flow studies for the four cases. The values given in the table are per unit bus voltage values. Here, the buses selected for the study are the ones located at the ends of radials or else the ones that are badly disturbed by the second-contingency outage of case 4.

Use the given data and determine the following:

a. If the lowest *"favorable"* and the lowest *"tolerable"* voltages are defined as 114 and 111 V, respectively, what are the pu voltages, based on 125 V, that correspond to the lowest favorable voltage and the lowest tolerable voltage for nominally 120/208Y systems?
b. List the buses given in Table 6.5, for the first-contingency outage, that have (1) less than favorable voltage and (2) less than tolerable voltage.
c. List the buses given in Table 6.5, for the second-contingency outage, that have (1) less than favorable voltage and (2) less than tolerable voltage.
d. Find Z_M/Z_T, $1/2(Z_M/Z_T)$, and using Figure 6.8, find the value of the *"application factor"* for this example network and make an approximate judgment about the sufficiency of the design of this network.

Solution

a. The lowest favorable voltage in per unit is 114 V:

$$\frac{114\,V}{125\,V} = 0.912\,pu$$

and the lowest tolerable voltage in per unit is

$$\frac{111\,V}{125\,V} = 0.888\,pu$$

b. There are no buses in Table 6.5, for the first-contingency outage, that have (1) less than favorable voltage or (2) less than tolerable voltage.

c. For the second-contingency outage, the buses in Table 6.5 that have (1) less than favorable voltage are B, C, J, K, R, and S and (2) less than tolerable voltage are B, C, J, and K.

d. The given transformer impedance of 0.0086 + j0.0492 pu is based on 500 kVA. Therefore, it corresponds to

$$Z_T = 0.0172 + j0.0984 \text{ pu } \Omega$$

that is based on 1000 kVA. Therefore, the ratios are

$$\frac{Z_M}{Z_T} = \frac{0.181 + j0.115}{0.0172 + j0.0984}$$

$$= 2.147$$

or

$$\frac{1}{2}\left(\frac{Z_M}{Z_T}\right) = 1.0735$$

Thus, from Figure 6.8, the corresponding average transformer application factor for four feeders can be found as 1.6. To verify this value for the given design, the actual application factor can be recalculated as

$$\text{Actual application factor} = \frac{\text{Total installed network-transformer capacity}}{\text{Total load}}$$

$$= \frac{19 \text{ transformers} \times 500 \text{ kVA/transformer}}{5096 + j3158}$$

$$= 1.5846$$

Therefore, the design of this network is sufficient.

6.9 SECONDARY SYSTEM COSTS

As discussed previously, the secondary system consists of the service transformers that convert primary voltage to utilization voltage, the secondary circuits that operate at utilization voltage, and the SDs that feed power directly to each customer. Many utilities develop cost estimates for this equipment on a per customer basis. The annual costs of operating, maintenance, and taxes for a secondary system are typically between 1/8 and 1/30 of the capital cost.

In general, it costs more to upgrade given equipment to a higher capacity than to build to that capacity in the first place. Upgrading an existing SL entails removing the old conductor and installing new. Usually, new hardware is required, and sometimes poles and crossarms must be replaced. Therefore, usually, the cost of this conversion greatly exceeds the cost of building to the higher-capacity design in the first place. Because of this, T&D engineers have an incentive to look at long-term needs carefully and to install extra capacity for future growth.

Example 6.6

It has been estimated that a 12.47 kV OH, three-phase feeder with 336 kcmil costs $120,000/mile. It has been also estimated that to build the feeder with 600 kcmil conductor instead and a 15 MVA capacity would cost about $150,000/mile. Upgrading the existing 9 MVA capacity line later to

15 MVA capacity entails removing the old conductor and installing new. The cost of upgrade is $200,000/mile. Determine the following:

  a. The cost of building the 9 MVA capacity line in dollars per kVA-mile
  b. The cost of building the 15 MVA capacity line in dollars per kVA-mile
  c. The cost of the upgrade in dollars per kVA-mile

Solution

  a. The cost of building the 9 MVA capacity line is

$$\text{Cost}_{9\,\text{MVA line}} = \frac{\$120,000}{9,000\,\text{kVA}} = \$13.33/\text{kVA-mile}$$

  b. The cost of building the 15 MVA capacity is

$$\text{Cost}_{15\,\text{MVA line}} = \frac{\$150,000}{15,000\,\text{kVA}} = \$10/\text{kVA-mile}$$

  c. The cost of the upgrade is

$$\text{Cost}_{\text{upgrade}} = \frac{\$200,000}{(15,000 - 9,000)\,\text{kVA}} = \$33.33/\text{kVA-mile}$$

As it can be seen earlier, when judged against the additional capacity (15 − 9 MVA), the upgrade option is very costly, that is, over $33/kVA-mile.

PROBLEMS

6.1 Repeat Example 6.1. Assume that there are four services per transformer, that is, one transformer on each pole so that there are six transformers per block.

6.2 Repeat Example 6.1. Assume that the annual load factor is 0.35.

6.3 Repeat Problem 6.1. Assume that the annual load factor is 0.65.

6.4 Consider Problem 6.1 and find the following:
  a. The most economical service-drop size (A_{SD}) and the nearest larger commercial wire size
  b. The most economical SL size (A_{SL}) and the nearest larger standard transformer size
  c. The TAC per block for the nearest larger standard sizes of equipment

6.5 Repeat Example 6.4, assuming the load impedances are

$$\bar{Z}_a = 1.0 + j0.0 \ \Omega$$

and

$$\bar{Z}_b = 1.5 + j0.0 \ \Omega$$

6.6 Repeat Example 6.4, assuming the load impedances are

$$\bar{Z}_a = 1.0 + j0.0 \ \Omega$$

and

$$\bar{Z}_b = 3.0 + j0.0 \ \Omega$$

6.7 Repeat Example 6.4, assuming the load impedances are

$$\bar{Z}_a = 0.80 + j0.60 \ \Omega$$

and

$$\bar{Z}_b = 1.5 + j0.0 \ \Omega$$

6.8 The following table gives the total real and reactive power losses for the secondary network given in Example 6.5. Explain the circumstances that cause minimum and maximum losses. Bear in mind that the total $P + jQ$ power delivered to the loads is identical in all cases.

Case No.	$\sum P_L$, MW	$\sum Q_L$, Mvar
1	0.16379	0.38807
2	0.14160	0.33142
3	0.19263	0.46648
4	0.36271	0.82477

6.9 The following table gives the primary-feeder circuit loading for the primary feeders given in Example 6.5.

	$P + jQ$, pu MVA			
Case No.	Feeder 1	Feeder 2	Feeder 3	Feeder 4
1	$1.3575 - j0.9012$	$1.186 - j0.8131$	$1.3822 - j0.9381$	$1.3341 - j0.8857$
2	$1.3496 - j0.6540$	$1.1854 - j0.5894$	$1.375 - j0.6936$	$1.3278 - j0.6308$
3	Out	$1.5965 - j0.8468$	$1.8427 - j0.952$	$1.8495 - j0.9354$
4	Out	$2.5347 - j1.4587$	$2.924 - j1.7285$	Out

Determine the ampere loads of each feeder and complete the following table.

	Percent of Ampacity Rating			
Case No.	Feeder 1	Feeder 2	Feeder 3	Feeder 4
1	—	—	—	—
2	—	—	—	—
3	Out	—	—	—
4	Out	—	—	—

6.10 Assume that the following table gives the transformer loading for transformers 1, 3, and 4, using bus S data, for Example 6.5.

	Transformer Loading, kVA		
Case No.	Transformer 1	Transformer 3	Transformer 4
1	380.365	374.00	385.450
2	358.475	352.31	363.375
3		509.42	508.921
4		812.61	

Complete the following table. Note that bus S not only has the largest load but also loses two-thirds of its transformer capacity in the event of the second-contingency outage being considered here.

| | Load-In Percent of Transformer Rating | | |
Case No.	Transformer 1	Transformer 3	Transformer 4
1			
2			
3			
4			

6.11 Assume that the following table gives the loading of the secondary ins close to bus S in Example 6.5.

| | Loading of Secondary Mains, pu MVA | | | | |
Case No.	S–R	R–Q	S–6	6–7	S–G
1	0.1715	0.2516	0.0699	0.1065	0.0361
2	0.1662	0.2560	0.0692	0.1072	0.0364
3	0.1252	0.3110	0.0816	0.0945	0.0545
4	0.0872	0.3778	0.0187	0.1901	0.1430

Determine the ampere loading of the mains close to bus S and also complete the following table.

| | Loading of Secondary Mains, % of Rated Ampacity | | | | |
Case No.	S–R	R–Q	S–6	6–7	S–G
1					
2					
3					
4					

6.12 Resolve Example 6.2 by using MATLAB®. Assume that all the quantities remain the same.

REFERENCES

1. Gönen, T. et al.: *Development of Advanced Methods For Planning Electric Energy Distribution Systems*, US Department of Energy, October 1979. Available from the National Technical Information Service, US Department of Commerce, Springfield, VA.
2. Westinghouse Electric Corporation: *Electrical Transmission and Distribution Reference Book*, East Pittsburgh, PA, 1964.
3. Davey, J. et al.: Practical application of weather sensitive load forecasting to system planning, *Proceedings, the IEEE PES Summer Meeting*, San Francisco, CA, July 9–14, 1972.
4. Chang, N. E.: Loading distribution transformers, *Transmission and Distribution*, 26(8), August 1974, 58–59.
5. Chang, N. E.: Determination and evaluation of distribution transformer losses of the electric system through transformer load monitoring, *IEEE Trans. Power Appar. Syst.*, PAS-89, July/August 1970, 1282–1284.

6. Electric Power Research Institute: *Analysis of Distribution R&D Planning*, EPRI Report 329, Palo Alto, CA, October 1975.
7. Westinghouse Electric Corporation: *Electric Utility Engineering Reference Book-Distribution Systems*, Vol. 3, East Pittsburgh, PA, 1965.
8. Seelye, H. P.: *Electrical Distribution Engineering*, 1st edn., McGraw-Hill, New York, 1930.
9. Fink, D. G. and H. W. Beaty: *Standard Handbook for Electrical Engineers*, 11th edn., McGraw-Hill, New York, 1978.
10. Chang, S. H.: Economic design of secondary distribution system by computer, MS thesis, Iowa State University, Ames, IA, 1974.
11. Robb, D. D.: *ECDES Program User Manual. Power System Computer Service*, Iowa State University, Ames, IA, 1975.
12. Edison Electric Institute-National Electric Manufacturers Association: *EEI-NEMA Standards for Secondary Network Transformers*, EEI Publication no. 57-7, NEMA Publication No. TR4-1957, 1968.
13. Lawrence, R. F., D. N. Reps, and A. D. Patton: Distribution system planning through optimal design, I-distribution transformers and secondaries, *AIEE Trans.*, PAS-79(pt. III), June 1960, 199–204.
14. Gönen, T.: *Engineering Economy for Engineering Managers: With Computer Applications*, Wiley, New York, 1990.

7 Voltage-Drop and Power-Loss Calculations

Any man may make a mistake; none but a fool will stick to it.

M.T. Cicero, 51 B.C.

Time is the wisest counselor.

Pericles, 450 B.C.

When others agree wth me, I wonder what is wrong!

Author Unknown

7.1 THREE-PHASE BALANCED PRIMARY LINES

As discussed in Chapter 5, a utility company strives to achieve a well-balanced distribution system in order to improve system voltage regulation by means of equally loading each phase. Figure 7.1 shows a primary system with either a three-phase three-wire or a three-phase four-wire main. The laterals can be either (1) three-phase three-wire, (2) three-phase four-wire, (3) single phase with line-to-line voltage, ungrounded, (4) single phase with line-to-neutral voltage, grounded, or (5) two-phase plus neutral, open wye.

7.2 NON-THREE-PHASE PRIMARY LINES

Usually there are many laterals on a primary feeder that are not necessarily in three phase, for example, single phase, which causes the voltage drop and power loss due to load current not only in the phase conductor but also in the return path.

7.2.1 SINGLE-PHASE TWO-WIRE LATERALS WITH UNGROUNDED NEUTRAL

Assume that an overloaded single-phase lateral is to be changed to an equivalent three-phase three-wire and balanced lateral, holding the load constant. Since the power input to the lateral is the same as before,

$$S_{1\phi} = S_{3\phi} \tag{7.1}$$

(From Gonen's book Electric Power Distribution System Engineering, Figure 7.1 on page 324.) where the subscripts 1ϕ and 3ϕ refer to the single-phase and three-phase circuits, respectively. Equation 7.1 can be rewritten as

$$\left(\sqrt{3} \times V_s\right)I_{1\phi} = 3V_s I_{3\phi} \tag{7.2}$$

where V_s is the line-to-neutral voltage. Therefore, from Equation 7.2,

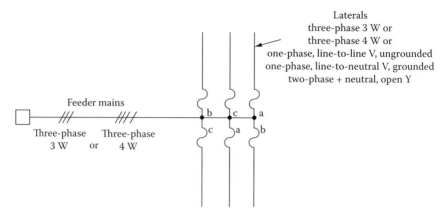

FIGURE 7.1 Various lateral types that exist in the United States.

$$I_{1\phi} = \sqrt{3} \times I_{3\phi} \tag{7.3}$$

which means that the current in the single-phase lateral is 1.73 times larger than the one in the equivalent three-phase lateral. The voltage drop in the three-phase lateral can be expressed as

$$VD_{3\phi} = I_{3\phi}(R\cos\theta + X\sin\theta) \tag{7.4}$$

and in the single-phase lateral as

$$VD_{1\phi} = I_{1\phi}(K_R R\cos\theta + K_X X\sin\theta) \tag{7.5}$$

where
K_R and K_X are conversion constants of R and X and are used to convert them from their three-phase values to the equivalent single-phase values
$K_R = 2.0$
$K_X = 2.0$ when underground (UG) cable is used
$K_X \cong 2.0$ when overhead line is used, with approximately a ±10% accuracy

Therefore, Equation 7.5 can be rewritten as

$$VD_{1\phi} = I_{1\phi}(2R\cos\theta + 2X\sin\theta) \tag{7.6}$$

or substituting Equation 7.3 into Equation 7.6,

$$VD_{1\phi} = 2\sqrt{3} \times I_{3\phi}(R\cos\theta + X\sin\theta) \tag{7.7}$$

By dividing Equation 7.7 by Equation 7.4 side by side,

$$\frac{VD_{1\phi}}{VD_{3\phi}} = 2\sqrt{3} \tag{7.8}$$

which means that *the voltage drop in the single-phase ungrounded lateral is approximately 3.46 times larger than the one in the equivalent three-phase lateral.* Since base voltages for the single-phase and three-phase laterals are

$$V_{B(1\phi)} = \sqrt{3} \times V_{s,L-N} \tag{7.9}$$

and

$$V_{B(3\phi)} = V_{s,L-N} \tag{7.10}$$

Equation 7.8 can be expressed in per units as

$$\frac{\mathrm{VD}_{pu,1\phi}}{\mathrm{VD}_{pu,3\phi}} = 2.0 \tag{7.11}$$

which means that *the per unit voltage drop in the single-phase ungrounded lateral is two times larger than the one in the equivalent three-phase lateral.* For example, if the per unit voltage drop in the single-phase lateral is 0.10, it would be 0.05 in the equivalent three-phase lateral.

The power losses due to the load currents in the conductors of the single-phase lateral and the equivalent three-phase lateral are

$$P_{\mathrm{LS},1\phi} = 2 \times I_{1\phi}^2 R \tag{7.12}$$

and

$$P_{\mathrm{LS},3\phi} = 3 \times I_{3\phi}^2 R \tag{7.13}$$

respectively. Substituting Equation 7.3 into Equation 7.12,

$$P_{\mathrm{LS},1\phi} = 2\left(\sqrt{3} \times I_{3\phi}\right)^2 R \tag{7.14}$$

and dividing the resultant Equation 7.14 by Equation 7.13 side by side,

$$\frac{P_{\mathrm{LS},1\phi}}{P_{\mathrm{LS},3\phi}} = 2.0 \tag{7.15}$$

which means that the *power loss due to the load currents in the conductors of the single-phase lateral is two times larger than the one in the equivalent three-phase lateral.*

Therefore, one can conclude that *by changing a single-phase lateral to an equivalent three-phase lateral, both the per unit voltage drop and the power loss due to copper losses in the primary line are approximately halved.*

7.2.2 Single-Phase Two-Wire Ungrounded Laterals

In general, this system is presently not used due to the following disadvantages. There is no earth current in this system. It can be compared to a three-phase four-wire balanced lateral in the following manner. Since the power input to the lateral is the same as before,

$$S_{1\phi} = S_{3\phi} \tag{7.16}$$

or

$$V_s \times I_{1\phi} = 3 \times V_s \times I_{3\phi} \tag{7.17}$$

from which

$$I_{1\phi} = 3 \times I_{3\phi} \tag{7.18}$$

The voltage drop in the three-phase lateral can be expressed as

$$VD_{3\phi} = I_{3\phi}(R\cos\theta + X\sin\theta) \tag{7.19}$$

and in the single-phase lateral as

$$VD_{1\phi} = I_{1\phi}(K_R R\cos\theta + K_X X\sin\theta) \tag{7.20}$$

where
  $K_R = 2.0$ when a full-capacity neutral is used, that is, if the wire size used for neutral conductor
    is the same as the size of the phase wire
  $K_R > 2.0$ when a reduced-capacity neutral is used
  $K_X \cong 2.0$ when overhead line is used

Therefore, if $K_R = 2.0$ and $K_X = 2.0$, Equation 7.20 can be rewritten as

$$VD_{1\phi} = I_{1\phi}(2R\cos\theta + 2X\sin\theta) \tag{7.21}$$

or substituting Equation 7.18 into Equation 7.21,

$$VD_{1\phi} = 6 \times I_{3\phi}(R\cos\theta + X\sin\theta) \tag{7.22}$$

Dividing Equation 7.22 by Equation 7.19 side by side,

$$\frac{VD_{1\phi}}{VD_{3\phi}} = 6.0 \tag{7.23a}$$

or

$$\frac{VD_{pu,1\phi}}{VD_{pu,1\phi}} = 2\sqrt{3} = 3.46 \tag{7.23b}$$

which means that *the voltage drop in the single-phase two-wire ungrounded lateral with full-capacity neutral is six times larger than the one in the equivalent three-phase four-wire balanced lateral.*
    The power losses due to the load currents in the conductors of the single-phase two-wire unigrounded lateral with full-capacity neutral and the equivalent three-phase four-wire balanced lateral are

$$P_{LS,1\phi} = I_{1\phi}^2(2R) \tag{7.24}$$

and

$$P_{LS,3\phi} = 3 \times I_{3\phi}^2 R \tag{7.25}$$

respectively. Substituting Equation 7.18 into Equation 7.24,

$$P_{LS,1\phi} = (3 \times I_{3\phi})^2 (2R) \tag{7.26}$$

and dividing Equation 7.26 by Equation 7.25 side by side,

$$\frac{P_{LS,1\phi}}{P_{LS,3\phi}} = 6.0 \tag{7.27}$$

Therefore, *the power loss due to load currents in the conductors of the single-phase two-wire ungrounded lateral with full-capacity neutral is six times larger than the one in the equivalent three-phase four-wire lateral.*

7.2.3 Single-Phase Two-Wire Laterals with Multigrounded Common Neutrals

Figure 7.2 shows a single-phase two-wire lateral with multigrounded common neutral. As shown in the figure, the neutral wire is connected in parallel (i.e., multigrounded) with the ground wire at various places through ground electrodes in order to reduce the current in the neutral. I_a is the current in the phase conductor, I_w is the return current in the neutral wire, and I_d is the return current in Carson's equivalent ground conductor. According to Morrison [1], the return current in the neutral wire is

$$I_n = \zeta_1 I_a \quad \text{where } \zeta_1 = 0.25\text{--}0.33 \tag{7.28}$$

and it is almost independent of the size of the neutral conductor.

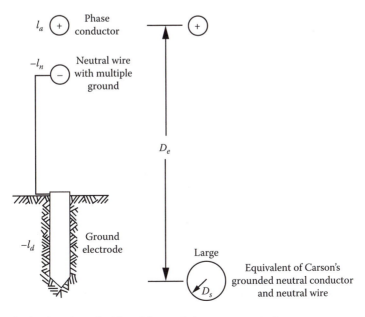

FIGURE 7.2 A single-phase lateral with multigrounded common neutral.

In Figure 7.2, the constant K_R is less than 2.0 and the constant K_X is more or less equal to 2.0 because of conflictingly large D_m (i.e., mutual geometric mean distance or geometric mean radius) of Carson's equivalent ground (neutral) conductor. Therefore, Morrison's data [1] (probably empirical) indicate that

$$\text{VD}_{\text{pu},1\phi} = \zeta_2 \times \text{VD}_{\text{pu},3\phi} \quad \text{where } \zeta_2 = 3.8\text{--}4.2 \tag{7.29}$$

and

$$P_{\text{LS},1\phi} = \zeta_3 \times P_{\text{LS},3\phi} \quad \text{where } \zeta_3 = 3.5\text{--}3.75 \tag{7.30}$$

Therefore, assuming that the data from Morrison [1] are accurate,

$$K_R < 2.0 \quad \text{and} \quad K_X < 2.0$$

the per unit voltage drops and the power losses due to load currents can be approximated as

$$\text{VD}_{\text{pu},1\phi} \cong 4.0 \times \text{VD}_{\text{pu},3\phi} \tag{7.31}$$

and

$$P_{\text{LS},1\phi} \cong 3.6 \times P_{\text{LS},3\phi} \tag{7.32}$$

for the illustrative problems.

7.2.4 Two-Phase Plus Neutral (Open-Wye) Laterals

Figure 7.3 shows an open-wye-connected lateral with two phase and neutral. The neutral conductor can be unigrounded or multigrounded, but because of disadvantages, the unigrounded neutral is generally not used. If the neutral is unigrounded, all neutral current is in the neutral conductor itself. Theoretically, it can be expressed that

$$\mathbf{V} = \mathbf{ZI} \tag{7.33}$$

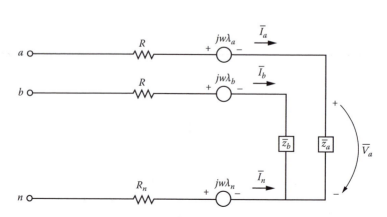

FIGURE 7.3 An open-wye connected lateral.

where

$$\bar{V}_a = \bar{Z}_a \bar{I}_a \tag{7.34}$$

$$\bar{V}_b = \bar{Z}_b \bar{I}_b \tag{7.35}$$

It is correct for equal load division between the two phases.

Assuming equal load division among phases, the two-phase plus neutral lateral can be compared to an equivalent three-phase lateral, holding the total kilovoltampere load constant. Therefore,

$$S_{2\phi} = S_{3\phi} \tag{7.36}$$

or

$$2V_s I_{2\phi} = 3V_s I_{3\phi} \tag{7.37}$$

from which

$$I_{2\phi} = \frac{3}{2} I_{3\phi} \tag{7.38}$$

The voltage-drop analysis can be performed depending upon whether the neutral is unigrounded or multigrounded. *If the neutral is unigrounded and the neutral conductor impedance (Z_n) is zero, the voltage drop in each phase is*

$$VD_{2\phi} = I_{2\phi}(K_R R \cos\theta + K_X X \sin\theta) \tag{7.39}$$

where
  $K_R = 1.0$
  $K_X = 1.0$

Therefore,

$$VD_{2\phi} = I_{2\phi}(R \cos\theta + X \sin\theta) \tag{7.40}$$

or substituting Equation 7.38 into Equation 7.40,

$$VD_{2\phi} = \frac{3}{2} I_{3\phi}(R \cos\theta + X \sin\theta) \tag{7.41}$$

Dividing Equation 7.41 by Equation 7.19, side by side,

$$\frac{VD_{2\phi}}{VD_{3\phi}} = \frac{3}{2} \tag{7.42}$$

However, *if the neutral is unigrounded and the neutral conductor impedance (Z_n) is larger than zero,*

$$\frac{VD_{2\phi}}{VD_{3\phi}} > \frac{3}{2} \tag{7.43}$$

Therefore, in this case, some unbalanced voltages are inherent.

However, *if the neutral is multigrounded and $Z_n > 0$*, the data from Morrison [1] indicate that the per unit voltage drop in each phase is

$$VD_{pu,2\phi} = 2.0 \times VD_{pu,3\phi} \tag{7.44}$$

when a full-capacity neutral is used and

$$VD_{pu,2\phi} = 2.1 \times VD_{pu,3\phi} \tag{7.45}$$

when a reduced-capacity neutral (i.e., when the neutral conductor employed is one or two sizes smaller than the phase conductors) is used.

The power loss analysis also depends upon whether the neutral is unigrounded or multigrounded. *If the neutral is unigrounded*, the power loss is

$$P_{LS,2\phi} = I_{2\phi}^2 (K_R R) \tag{7.46}$$

where
 $K_R = 3.0$ when a full-capacity neutral is used
 $K_R > 3.0$ when a reduced-capacity neutral is used

Therefore, if $K_R = 3.0$,

$$\frac{P_{LS,2\phi}}{P_{LS,3\phi}} = \frac{3I_{2\phi}^2 R}{3I_{3\phi}^2 R} \tag{7.47}$$

or

$$\frac{P_{LS,2\phi}}{P_{LS,3\phi}} = 2.25 \tag{7.48}$$

On the other hand, *if the neutral is multigrounded*,

$$\frac{P_{LS,2\phi}}{P_{LS,3\phi}} < 2.25 \tag{7.49}$$

Based on the data from Morrison [1], the approximate value of this ratio is

$$\frac{P_{LS,2\phi}}{P_{LS,3\phi}} \cong 1.64 \tag{7.50}$$

which means that the *power loss* due to load currents in the conductors of the *two-phase three-wire lateral with multigrounded neutral is approximately 1.64 times larger than the one in the equivalent three-phase lateral.*

Example 7.1

Assume that a uniformly distributed area is served by a three-phase four-wire multigrounded 6-mile-long main located in the middle of the service area. There are six laterals on each side of the main. Each lateral is 1 mile apart with respect to each other, and the first lateral is located on the main 1 mile away from the substation so that the total three-phase load on the main is 6000 kVA. Each lateral is 10 mi long and is made up of #6 AWG copper conductors and serving a uniformly distributed peak load of 500 kVA, at 7.2/12.47 kV. The K constant of a #6 AWG copper conductor is 0.0016/kVA-mi. Determine the following:

a. The maximum voltage drop to the end of each lateral, if the lateral is a three-phase lateral with multigrounded common neutrals
b. The maximum voltage drop to the end of each lateral, if the lateral is a two-phase plus full-capacity multigrounded neutral (open-wye) lateral
c. The maximum voltage drop to the end of each lateral, if the lateral is a single-phase two-wire lateral with multigrounded common neutrals

Solution

a. For the three-phase four-wire lateral with multigrounded common neutrals,

$$\%VD_{3\phi} = \frac{1}{2} \times K \times S$$

$$= \left(\frac{10\,\text{mi}}{2}\right)\left(0.0016\frac{\%\,VD}{kVA - mi}\right)(500\,kVA) = 4$$

b. For the two-phase plus full-capacity multigrounded neutral (open-wye) lateral, according to the results of Morrison,

$$\%\,VD_{2\phi} = 2(\%\,VD_{3\phi})$$

$$= 2(4\%) = 8$$

c. For the single-phase two-wire lateral with multigrounded common neutrals, according to the results of Morrison,

$$\%\,VD_{1\phi} = 4(\%\,VD_{3\phi})$$

$$= 4(4\%) = 16$$

Example 7.2

A three-phase express feeder has an impedance of $6 + j20$ ohms per phase. At the load end, the line-to-line voltage is 13.8 kV, and the total three-phase power is 1200 kW at a lagging power factor of 0.8. By using the *line-to-neutral* method, determine the following:

a. The line-to-line voltage at the sending end of the feeder (i.e., at the substation low-voltage bus)
b. The power factor at the sending end
c. The copper loss (i.e., the transmission loss) of the feeder
d. The power at the sending end in kW

Solution

Since in an express feeder, the line current is the same at the beginning or at the end of the line,

$$I_L = I_S = I_R = \frac{P_{R(3\phi)}}{\sqrt{3}V_{R(L-L)}\cos\theta}$$

$$= \frac{1,200\,kW}{\sqrt{3}(13.8\,kV)0.8} = 62.75\,A$$

and

$$V_{R(L-N)} = \frac{V_{R(L-L)}}{\sqrt{3}}$$

$$= \frac{13,800 \text{ V}}{\sqrt{3}} = 7,976.4 \text{ V}.$$

using this *as the reference voltage*, the sending-end voltage is found from

$$\bar{V}_{S(L-N)} = \bar{V}_{R(L-N)} + \bar{I}_L \bar{Z}_L$$

where

$$\bar{V}_{R(L-N)} = 7,976.9 \angle 0° \text{ V}$$

$$\bar{I}_L = \bar{I}_S = \bar{I}_R = I_L (\cos\theta_R - \sin\theta_R)$$

$$= 62.83(0.8 - j0.6) = 62.83 \angle -36.87° \text{ A}$$

$$\bar{Z}_L = 6 + j20 = 20.88 \angle 73.3° \text{ }\Omega$$

a. $$\bar{V}_{S(L-N)} = 7,976.9 \angle 0° + (62.83 \angle -36.87°)(20.88 \angle 73.7°)$$

$$= 9,065.95 \angle 4.93° \text{ V}$$

and

$$\bar{V}_{S(L-L)} = \sqrt{3}\bar{V}_{S(L-N)}$$

$$= \sqrt{3}(9,065.95) \angle 4.93° + 30°$$

$$= 15,684.09 \angle 34.93° \text{ V}$$

b. $$\theta_S = \left|\theta_{\bar{V}_{S(L-N)}}\right| - \left|\theta_{\bar{I}_S}\right| = 4.9° - \left|-36.87°\right| = 41.8° \quad \text{and} \quad \cos\theta_S = 0.745 \text{ lagging}$$

c. $$P_{\text{loss}(3\phi)} = 3I_L^2 R = 3(62.83)^2 \times 6$$

$$= 71,056.96 \text{ W} \cong 71.057 \text{ kW}$$

d. $$P_{S(3\phi)} = P_{R(3\phi)} + P_{\text{loss}(3\phi)}$$

$$= 1,200 + 71.057 = 1,271.057 \text{ kW}$$

or

$$P_{S(3\phi)} = \sqrt{3}V_{S(L-L)}I_S \cos\theta_S$$

$$= \sqrt{3}(15,684.09)(62.83)0.745 \cong 1,270.073 \text{kW}$$

Example 7.3

Repeat Example 7.2 by using the *single-phase equivalent* method.

Solution

Here, the single-phase equivalent current is found from

$$I_{eq(1\phi)} = \frac{P_{3\phi}}{V_{R(L-L)} \cos\theta}$$

$$= \frac{1200 \text{ kW}}{(13.8 \text{ kV})(0.8)} = 108.7 \text{ A}$$

where

$$I_{eq(1\phi)} = \sqrt{3} I_{3\phi}$$

or

$$I_{3\phi} = I_L = \frac{I_{eq(1\phi)}}{\sqrt{3}}$$

$$= \frac{108.7 \text{ A}}{\sqrt{3}} = 62.8 \text{ A}$$

a.
$$\bar{V}_{S(L-L)} = \bar{V}_{R(L-L)} + \bar{I}_{eq(1\phi)} \bar{Z}_L$$

$$= 13,800\angle 0° + (108.7\angle -36.9°)(20.88\angle 73.3°)$$

$$= 15,684.76\angle 4.93° + 30° \text{ V}$$

b.
$$\theta_S = \theta_{\bar{V}_{S(L-N)}} - \theta_{\bar{I}_S} = 41.8° \quad \text{so that} \quad \cos\theta_S = 0.745 \text{ lagging}$$

c.
$$P_{loss(3\phi)} = I_{eq(1\phi)}^2 R$$

$$= 108.7^2 \times 6 = 70.89 \text{kW}$$

d.
$$P_{S(3\phi)} = P_{R(3\phi)} + P_{loss(3\phi)}$$

$$= 1,200 + 70.89 = 1,270.89 \text{kW}$$

7.3 FOUR-WIRE MULTIGROUNDED COMMON NEUTRAL DISTRIBUTION SYSTEM

Figure 7.4 shows a typical four-wire multigrounded common neutral distribution system. Because of the economic and operating advantages, this system is used extensively. The assorted secondaries can be, for example, either (1) 120/240 V single-phase three wire, (2) 120/240 V three-phase four wire connected in delta, (3) 120/240 V three-phase four-wire connected in open delta, or (4) 120/208 V three-phase four wire connected in grounded wye. Where primary and secondary systems are both existent, the same conductor is used as the *common* neutral for both systems. The neutral is grounded at each distribution transformer, at various places where no transformers are connected and to water pipes or driven ground electrodes at each user's service entrance. The secondary neutral is also grounded at the distribution transformer and the service drops (SDs).

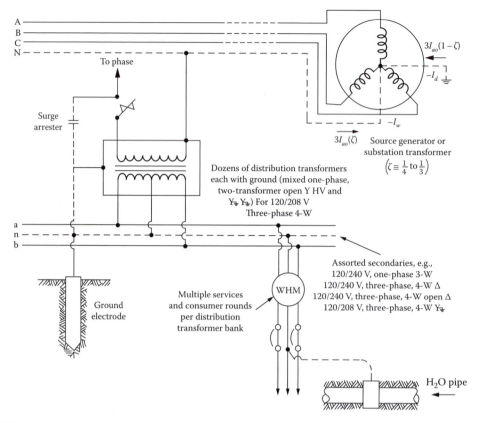

FIGURE 7.4 A four-wire multigrounded common neutral distribution system.

Typical values of the resistances of the ground electrodes are 5, 10, or 15 Ω. Under no circumstances should they be larger than 25 Ω. Usually, a typical metal water pipe system has a resistance value of less than 3 Ω. A part of the unbalanced, or zero sequence, load current flows in the neutral wire, and the remaining part flows in the ground and/or the water system. Usually the same conductor size is used for both phase and neutral conductors.

Example 7.4

Assume that the circuit shown in Figure 7.5 represents a single-phase circuit if dimensional variables are used; it represents a balanced three-phase circuit if per unit variables are used. $R + jX$ represents the total impedance of lines and/or transformers. The power factor of the load is $\cos\theta = \cos(\theta_{V_R} - \theta_T)$. Find the load power factor for which the voltage drop is maximum.

Solution

The line voltage drop is

$$VD = I(R\cos\theta + X\sin\theta)$$

By taking its partial derivative with respect to the θ angle and equating the result to zero,

$$\frac{\partial(VD)}{\partial\theta} = I(R\cos\theta + X\sin\theta) = 0$$

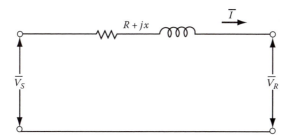

FIGURE 7.5   A single-phase circuit.

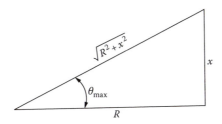

FIGURE 7.6   Impedance triangle.

or

$$\frac{X}{R} = \frac{\sin\theta}{\cos\theta} = \tan\theta$$

therefore

$$\theta_{max} = \tan^{-1}\frac{X}{R}$$

and from the impedance triangle shown in Figure 7.6, the load power factor for which the voltage drop is maximum is

$$PF = \cos\theta_{max} = \frac{R}{(R^2 + X^2)^{1/2}} \tag{7.51}$$

also

$$\cos\theta_{max} = \cos\left(\tan^{-1}\frac{X}{R}\right) \tag{7.52}$$

Example 7.5

Consider the three-phase four-wire 416-V secondary system with balanced per-phase loads at A, B, and C as shown in Figure 7.7. Determine the following:

   a. Calculate the total voltage drop, or as it is sometimes called, *voltage regulation*, in one phase of the lateral by using the approximate method.
   b. Calculate the real power per phase for each load.
   c. Calculate the reactive power per phase for each load.
   d. Calculate the total (*three-phase*) kilovoltampere output and load power factor of the distribution transformer.

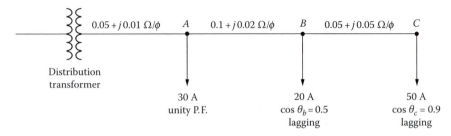

FIGURE 7.7   One-line diagram of a three-phase four-wire secondary system.

Solution

a. Using the approximate voltage-drop equation, that is,

$$VD = I(R\cos\theta + X\sin\theta)$$

the voltage drop for each load can be calculated as

$$VD_A = 30(0.05 \times 1.0 + 0.01 \times 0) = 1.5\,V$$

$$VD_B = 20(0.15 \times 0.5 + 0.03 \times 0.866) = 2.02\,V$$

$$VD_C = 50(0.20 \times 0.9 + 0.08 \times 0.436) = 10.744\,V$$

Therefore, the total voltage drop is

$$\sum VD = VD_A + VD_B + VD_C$$

$$= 1.5 + 2.02 + 10.744$$

$$= 14.264\ V$$

or

$$\frac{14.264\,V}{240\,V} = 0.0594\,pu\,V$$

b. The per-phase real power for each load can be calculated from

$$P = VI\cos\theta$$

or

$$P_A = 240 \times 30 \times 1.0 = 7.2\,kW$$

$$P_B = 240 \times 20 \times 0.5 = 2.4\,kW$$

$$P_C = 240 \times 50 \times 0.9 = 10.8\,kW$$

Therefore, the total per-phase real power is

$$\sum P = P_A + P_B + P_C$$

$$= 7.2 + 2.4 + 10.8$$

$$= 20.4\,kW$$

c. The reactive power per phase for each load can be calculated from

$$Q = VI \sin\theta$$

or

$$Q_A = 240 \times 30 \times 0 = 0 \, \text{kvar}$$

$$Q_B = 240 \times 20 \times 0.866 = 4.156 \, \text{kvar}$$

$$Q_C = 240 \times 50 \times 0.436 = 5.232 \, \text{kvar}$$

Therefore, the total per-phase reactive power is

$$\sum Q = Q_A + Q_B + Q_C$$

$$= 0 + 4.156 + 5.232$$

$$= 9.389 \, \text{kvar}$$

d. Therefore, the per-phase kilovoltampere output of the distribution transformer is

$$S = (P^2 + Q^2)^{1/2}$$

$$= (20.4^2 + 9.389^2)^{1/2}$$

$$\cong 22.457 \, \text{kVA/phase}$$

Thus the total (or three-phase) kilovoltampere output of the distribution transformer is

$$3 \times 22.457 = 67.37 \, \text{kVA}$$

Hence, the load power factor of the distribution transformer is

$$\cos\theta = \frac{\sum P}{S}$$

$$= \frac{20.4 \, \text{kW}}{22.457 \, \text{kVA}}$$

$$= 0.908 \, \text{lagging}$$

Example 7.6

This example is a continuation of Example 6.1. It deals with voltage drops in the secondary distribution system. In this and the following examples, a single-phase three-wire 120/240 V directly buried underground residential distribution (URD) secondary system will be analyzed, and calculations will be made for motor-starting voltage dip (VDIP) and for steady-state voltage drops at the time of annual peak load. Assume that the cable impedances given in Table 7.2 are correct for a typical URD secondary cable.

Transformer data. The data given in Table 7.1 are for modern single-phase 65°C OISC distribution transformers of the 7200-120/240 V class. The data were taken from a recent catalog of a manufacturer. All given per unit values are based on the transformer-rated kilovoltamperes and voltages.

TABLE 7.1
Single-Phase 7200-120/240-V Distribution Transformer Data at 65°C

Rated (kVA, kW)	Core Loss[a] (kW)	Copper Loss[b] (kW)	R (pu)	X (pu)	Excitation Current (A)
15	0.083	0.194	0.0130	0.0094	0.014
25	0.115	0.309	0.0123	0.0138	0.015
37.5	0.170	0.400	0.0107	0.0126	0.014
50	0.178	0.537	0.0107	0.0139	0.014
75	0.280	0.755	0.0101	0.0143	0.014
100	0.335	0.975	0.0098	0.0145	0.014

[a] At rated voltage and frequency.
[b] At rated voltage and kilovoltampere load.

The 2400 V-class transformers of the sizes being considered have about 15% less R and about 7% less X than the 7200 V transformers. Ignore the small variation of impedance with rated voltage and assume that voltage drop calculated with the given data will suffice for whichever primary voltage is used.

URD secondary cable data. Cable insulations and manufacture are constantly being improved, especially for high-voltage cables. Therefore, any cable data soon become obsolete. The following information and data have been abstracted from recent cable catalogs.

Much of the 600 V-class cables now commonly used for secondary lines (SLs) and services have Al conductor and cross-linked PE insulation, which can stand 90°C conductor temperature. The triplexed cable assembly shown in Figure 7.8 (quadruplexed for three-phase four-wire service) has three or four insulated conductors when aluminum is used. When copper is used, the one grounded neutral conductor is bare. The neutral conductor typically is two AWG sizes smaller than the phase conductors.

The twin concentric cable assembly shown in Figure 7.9 has two insulated copper or aluminum phase conductors plus several spirally served small bare copper binding conductors that act as the current-carrying grounded neutral. The number and size of the spiral neutral wires vary so

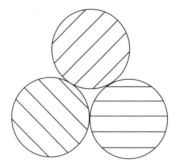

FIGURE 7.8   Triplexed cable assembly.

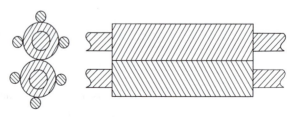

FIGURE 7.9   Twin concentric cable assembly.

TABLE 7.2
Twin Concentric Al/Cu XLPE 600 V Cable Data

| | R (Ω/1000 ft) per Conductor | | | | Direct Burial $\tilde{K}^a$ | |
| | Phase | Neutral | X (Ω/1000 ft) per | | | |
Size	Conductor 90°C	Conductor 80°C	Phase Conductor	Ampacity (A)	90% PF	50% PF
2 AWG	0.334	0.561	0.0299	180	0.02613	0.01608
1 AWG	0.265	0.419	0.0305	205	0.02098	0.01324
1/0 AWG	0.210	0.337	0.0297	230	0.01683	0.01089
2/0 AWG	0.167	0.259	0.0290	265	0.01360	0.00905
3/0 AWG	0.132	0.211	0.0280	300	0.01092	0.00752
4/0 AWG	0.105	0.168	0.0275	340	0.00888	0.00636
250 kcmil	0.089	0.133	0.0280	370	0.00769	0.00573
350 kcmil	0.063	0.085	0.0270	445	0.00571	0.00458
500 kcmil	0.044	0.066	0.0260	540	0.00424	0.00371

[a] Per unit voltage drop per 10^4 A · ft (amperes per conductor times feet of cable) based on 120 V line-to-neutral or 240 V line to line. Valid for the two power factors shown and for perfectly balanced three-wire loading.

that the ampacity of the neutral circuit is equivalent to two AWG wire sizes smaller than the phase conductors. Table 7.2 gives data for twin concentric aluminum/copper XLPE 600 V-class cable.

The triplex and twin concentric assemblies obviously have the same resistance for a given size of phase conductors. The triplex assembly has very slightly higher reactance than the concentric assembly. The difference in reactances is too small to be noted unless precise computations are undertaken for some special purpose. The reactances of those cables should be increased by about 25% if they are installed in iron conduit. The reactances given in the following text are valid only for balanced loading (where the neutral current is zero).

The triplex assembly has about 15% smaller ampacity than the concentric assembly, but the exact amount of reduction varies with wire size. The ampacities given are for 90°C conductor temperature, 20°C ambient earth temperature, direct burial in earth, and 10% daily load factor. When installed in buried duct, the ampacities are about 70% of those listed later. For load factors less than 100%, consult current literature or cable standards. The increased ampacities are significantly large.

Arbitrary criteria

1. Use the approximate voltage-drop equation, that is,

$$VD = I(R \cos\theta + X \sin\theta)$$

   and adapt it to per unit data when computing transformer voltage drops and adapt it to ampere and ohm data when computing SD and SL voltage drops. Obtain all voltage-drop answers in per unit based on 240 V.
2. Maximum allowable motor-starting VDIP = 3% = 0.03 pu = 3.6 V based on 120 V. This figure is arbitrary; utility practices vary.
3. Maximum allowable steady-state voltage drop in the secondary system (transformer + SL + SD) = 3.50% = 0.035 pu = 4.2 V based on 120 V. This figure also is quite arbitrary; regulatory commission rules and utility practices vary. More information about favorable and tolerable amounts of voltage drop will be discussed in connection with subsequent examples, which will involve voltage drops in the primary lines.
4. The loading data for the computation of steady-state voltage drop is given in Table 7.3.
5. As loading data for transient motor-starting VDIP, assume an air-conditioning compressor motor located most unfavorably. It has a 3 hp single-phase 240 V 80 A locked rotor current, with a 50% PF locked rotor.

TABLE 7.3
Load Data

Circuit Element	Load (kV)
SD	1 class 2 load (10 kVA)
SL	1 class 2 load (10 kVA) + 3 diversified class 2 loads (6.0 kVA each)
Transformer	1 class 2 load (10 kVA) + either 3 diversified class 2 loads (6.0 kVA each) or 11 diversified class 2 loads (4.4 kVA each)

Source: Lawrence, R.F. et al., *AIEE Trans.*, pt. III, PAS-79, 199, 1960.

Assumptions

1. Assume perfectly balanced loading in all three-wire single-phase circuits.
2. Assume nominal operating voltage of 240 V when computing currents from kilovoltampere loads.
3. Assume 90% lagging power factor for all loads.

Using the given data and assumptions, calculate the $\tilde{K}$ constant for any one of the secondary cable sizes, hoping to verify one of the given values in Table 7.2.

Solution

Let the secondary cable size be #2 AWG, arbitrarily. Also let the I current be 100 A and the length of the SL be 100 ft. Using the values from Table 7.2, the resistance and reactance values for 100 ft of cable can be found as

$$R = 0.334\,\Omega/1000\,\text{ft} \times \frac{100\,\text{ft}}{1000\,\text{ft}}$$

$$= 0.0334\,\Omega$$

and

$$X = 0.0299\,\Omega/1000\,\text{ft} \times \frac{100\,\text{ft}}{1000\,\text{ft}}$$

$$= 0.00299\,\Omega$$

Therefore, using the approximate voltage-drop equation,

$$\text{VD} = I(R\cos\theta + X\sin\theta)$$

$$= 100(0.0334 \times 0.9 + 0.00299 \times 0.435)$$

$$= 3.136\,\text{V}$$

or, in per unit volts,

$$\frac{3.136\,V}{120\,V} = 0.0261\,pu\,V$$

which is very close to the value given in Table 7.2 for the $\tilde{K}$ constant, that is, 0.02613 pu V/(10^4 A · ft) of cable.

Example 7.7

Use the information and data given in Examples 6.1 and 7.3. Assume a URD system. Therefore, the SLs shown in Figure 6.12 are made of UG secondary cables. Assume 12 services per distribution transformer and two transformers per block that are at the locations of poles 2 and 5, as shown in Figure 6.12. Service pedestals are at the locations of poles 1, 3, 4, and 6. Assume that the selected equipment sizes (for S_T, A_{SL}, A_{SD}) are of the nearest standard size that are larger than the theoretically most economical sizes and determine the following:

    a. Find the steady-state voltage drop in per units at the most remote consumer's meter for the annual maximum system loads given in Table 7.3.
    b. Find the VDIP in per units for motor starting at the most unfavorable location.
    c. If the voltage-drop and/or VDIP criteria are not met, select larger equipment and find a design that will meet these arbitrary criteria. Do not, however, immediately select the largest sizes of ST, ASL, and ASD equipment and call that a worthwhile design. In addition, contemplate the data and results and attempt to be wise in selecting ASL or ASD (or both) for enlarging to meet the voltage criteria.

Solution

    a. Due to the diversity factors involved, the load values given in Table 7.3 are different for SDs, SLs, and transformers. For example, the load on the transformer is selected as

$$\text{Transformer load} = 10 + 11 \times 4.4$$

$$= 58.4\,kVA$$

Therefore, selecting a 50 kVA transformer,

$$I = \frac{58.4\,kVA/240\,V}{S_r/240\,V}$$

$$= \frac{58.4\,kVA}{50\,kVA}$$

$$= 1.168\,pu\,A$$

Thus, the per unit voltage drop in the transformer is

$$VD_T = I(R\cos\theta + X\sin\theta)$$

$$= (1.168\,pu\,A)(0.0107 \times 0.9 + 0.0139 \times 0.435)$$

$$= 0.0183\,pu\,V$$

    As shown in Figure 7.10, the load on each SL (that portion of the wiring between the transformer and the service pedestal) is similarly calculated as

$$\text{SL load} = 10 + 3 \times 6$$

$$= 28\,kVA$$

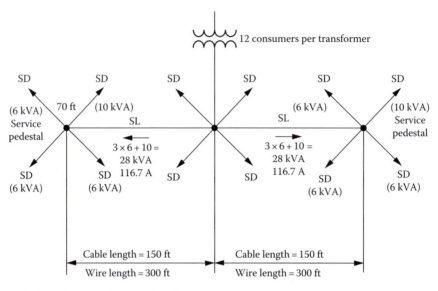

FIGURE 7.10 Circulation of the secondary-line currents.

or 116.7 A. If the SL is selected to be #4/0 AWG with the $\tilde{K}$ constant of 0.0088 from Table 7.2, the per unit voltage drop in each SL is

$$VD_{SL} = \tilde{K}\left(\frac{I \times \ell}{10^4}\right)$$

$$= 0.0088\left(\frac{116.7 \times 150\,\text{ft}}{10^4}\right)$$

$$= 0.01554\,\text{pu V}$$

The load on each SD is given to be 10 kVA or 41.6 A from Table 7.3. If each SD of 70 ft length is selected to be #1/0 AWG with the $\tilde{K}$ constant of 0.01683 from Table 7.2, the per unit voltage drop in each SD is

$$VD_{SD} = \tilde{K}\left(\frac{I \times \ell}{10^4}\right)$$

$$= 0.01683\left(\frac{41.6\,\text{A} \times 70\,\text{ft}}{10^4}\right)$$

$$= 0.0049\,\text{pu V}$$

Therefore, the total steady-state voltage drop in per units at the most remote consumer's meter is

$$\sum VD = VD_T + VD_{SL} + VD_{SD}$$

$$= 0.0183 + 0.01554 + 0.0049$$

$$= 0.0388\,\text{pu V}$$

which exceeds the given criterion of 0.035 pu V.

b. To find the VDIP in per units for motor starting at the most unfavorable location, the given starting current of 80 A can be converted to a kilovoltampere load of 19.2 kVA (80 A × 240 V). Therefore, the per unit VDIP in the 50 kVA transformer is

$$VDIP_T = (R\cos\theta + X\sin\theta)\left(\frac{19.2\,kVA}{50\,kVA}\right)$$

$$= (0.0107 \times 0.5 + 0.0139 \times 0.866)\left(\frac{19.2\,kVA}{50\,kVA}\right)$$

$$= 0.00668\,pu\,V$$

The per unit VDIP in the SL of #4/0 AWG cable is

$$VDIP_{SL} = \tilde{K}\left(\frac{80\,A \times 150\,ft}{10^4}\right)$$

$$= 0.00636\left(\frac{80 \times 150}{10^4}\right)$$

$$= 0.00763\,pu\,V$$

The per unit VDIP in the SD of #1/0 AWG cable is

$$VDIP_{SD} = \tilde{K}\left(\frac{80\,A \times 70\,ft}{10^4}\right)$$

$$= 0.01089\left(\frac{80 \times 70}{10^4}\right)$$

$$= 0.0061\,pu\,V$$

Therefore, the total VDIP in per units due to motor starting at the most unfavorable location is

$$\sum VDIP = VDIP_T + VDIP_{SL} + VDIP_{SD}$$

$$= 0.00668 + 0.00763 + 0.0061$$

$$= 0.024\,pu\,V$$

which meets the given criterion of 0.03 pu V.
c. Since in part (a) the voltage-drop criterion has not been met, select the SL cable size to be one size larger than the previous #4/0 AWG size, that is, 250 kcmil, keeping the size of the transformer the same. Therefore, the new per unit voltage drop in the SL becomes

$$VD_{SL} = 0.00769\left(\frac{116.7\,A \times 150\,ft}{10^4}\right)$$

$$= 0.01347\,pu\,V$$

Also, selecting one-size-larger cable, that is, #2/0 AWG, for the SD, the new per unit voltage drop in the SD becomes

$$VD_{SD} = 0.0136\left(\frac{41.6\,A \times 70\,ft}{10^4}\right)$$

$$= 0.00396\,pu\,V$$

Therefore, the new total steady-state voltage drop in per units at the most remote consumer's meter is

$$VD = VD_T + VD_{SL} + VD_{SD}$$

$$= 0.0183 + 0.01347 + 0.00396$$

$$= 0.03573\,pu\,V$$

which is still larger than the criterion. Thus, select 350 kcmil cable size for the SLs and #2/0 AWG cable size for the SDs to meet the criteria.

Example 7.8

Figure 7.11 shows a residential secondary distribution system. Assume that the distribution transformer capacity is 75 kVA (use Table 7.1), all secondaries and services are single-phase three wire, nominally 120/240 V, and all SLs are of #2/0 A1/Cu XLPE cable, and SDs are of #1/0 Al/Cu XLPE cable (use Table 7.2). All SDs are 100 ft long, and all SLs are 200 ft long. Assume an average lagging-load power factor of 0.9 and 100% load diversity factors and determine the following:

a. Find the total load on the transformer in kilovoltamperes and in per units.
b. Find the total steady-state voltage drop in per units at the most remote and severe customer's meter for the given annual maximum system loads.

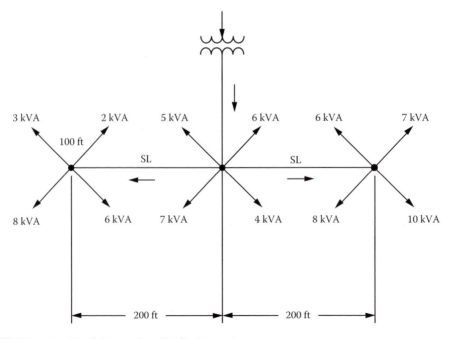

FIGURE 7.11　A residential secondary distribution system.

Solution

a. Assuming a diversity factor of 100%, the total load on the transformer is

$$S_T = (3+2+8+6)+(5+6+7+4)+(6+7+8+10)$$

$$= 19+22+31$$

$$= 72 \text{ kVA}$$

or, in per units,

$$I = \frac{S_r}{S_B}$$

$$= \frac{72 \text{kVA}}{75 \text{kVA}}$$

$$= 0.96 \text{ pu A}$$

b. To find the total voltage drop in per units at the most remote and severe customer's meter, calculate the per unit voltage drops in the transformer, the service line, and the SD of the most remote and severe customer. Therefore,

$$VD_T = I(R\cos\theta + X\sin\theta)$$

$$= 0.96(0.0101\times0.90+0.0143\times0.4359)$$

$$= 0.0147 \text{ pu V}$$

$$VD_{SL} = \tilde{K}\left(\frac{I\times l}{10^4}\right)$$

$$= 0.0136\left(\frac{129.17\,\text{A}\times200\,\text{ft}}{10^4}\right)$$

$$= 0.03513 \text{ pu V}$$

$$VD_{SD} = \tilde{K}\left(\frac{I\times l}{10^4}\right)$$

$$= 0.01683\left(\frac{41.67\,\text{A}\times100\,\text{ft}}{10^4}\right)$$

$$= 0.0070 \text{ pu V}$$

Therefore, the total voltage drop is

$$\sum VD = VD_T + VD_{SL} + VD_{SD}$$

$$= 0.0147+0.03513+0.0070$$

$$= 0.0568 \text{ pu V}$$

Example 7.9

Figure 7.12 shows a three-phase four-wire grounded-wye distribution system with multigrounded neutral, supplied by an express feeder and mains. In the figure, d and s are the width and length of a primary lateral, where s is much larger than d. Main lengths are equal to $cb = ce = s/2$. The number of the primary laterals can be found as s/d. The square-shaped service area (s^2) has a uniformly distributed load density, and all loads are presumed to have the same lagging power factor. Each primary lateral, such as ba, serves an area of length s and width d. Assume that

D = uniformly distributed load density, kVA/(unit length)^2
V_{L-L} = nominal operating voltage, which is also the base voltage, line-to-line kV
$r_m + jx_m$ = impedance of three-phase express and mains, Ω/(phase · unit length)
$r_l + jx_l$ = impedance of a three-phase lateral line, Ω/(phase · unit length)

Use the given information and data and determine the following:

1. Assume that the laterals are in three phase and find the per unit voltage-drop expressions for
   a. The express feeder fc, that is, $\text{VD}_{pu,fc}$
   b. The main cb, that is, $\text{VD}_{pu,cb}$
   c. The primary lateral ba, that is, $\text{VD}_{pu,ba}$
      Note that the equations to be developed should contain the constants D, s, d, impedances, θ, load power-factor angle, V_{L-L}, etc., but not current variable I.

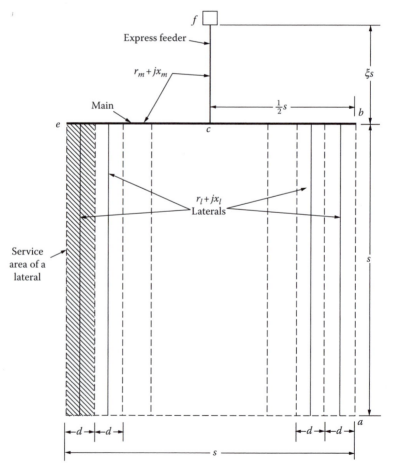

FIGURE 7.12   The distribution system of Example 7.9.

2. Change all the laterals from the three-phase four-wire system to an open-wye system so that investment costs will be reduced, but three-phase secondary service can still be rendered where needed. Assume that the phasing connections of the many laterals are well balanced on the mains. Use Morrison's approximations and modify the equations derived in part (1).

Solution

1.

$$\text{Total kVA load served} = D \times s^2 \text{ kVA} \tag{7.53}$$

$$\text{Current at point } f = \frac{D \times s^2}{\sqrt{3} \times V_{L-L}} \tag{7.54}$$

$$\text{Voltage drop} = I \times z \times l_{eff} \tag{7.55}$$

Therefore,

a.
$$VD_{pu,fc} = \frac{D \times s^2}{\sqrt{3} \times V_{L-L}} (r_m \cos\theta + x_m \sin\theta) \frac{\sqrt{3}}{1000 \times V_{L-L}} (\zeta \times s)$$

$$= \frac{\zeta \times D \times s^3}{1000 \times V_{L-L}^2} (r_m \cos\theta + x_m \sin\theta) \tag{7.56}$$

b.
$$VD_{pu,cb} = \frac{\frac{1}{2} D \times s^2}{\sqrt{3} \times V_{L-L}} (r_m \cos\theta + x_m \sin\theta) \frac{\sqrt{3}}{1000 \times V_{L-L}} \left(\frac{1}{4} \times s\right)$$

$$= \frac{D \times s^3}{8000 \times V_{L-L}^2} (r_m \cos\theta + x_m \sin\theta) \tag{7.57}$$

c
$$VD_{pu,ba} = \frac{D(d \times s)}{\sqrt{3} \times V_{L-L}} (r_l \cos\theta + x_l \sin\theta) \frac{\sqrt{3}}{1000 \times V_{L-L}} \left(\frac{1}{2} \times s\right)$$

$$= \frac{D \times d \times s^2}{2000 \times V_{L-L}^2} (r_l \cos\theta + x_l \sin\theta) \tag{7.58}$$

2. There would not be any change in the equations given in part (1).

Example 7.10

Figure 7.13 shows a square-shaped service area ($A = 4$ mi²) with a uniformly distributed load density of D kVA/mi² and 2 mi of #4/0 AWG copper overhead main from a to b. There are many closely spaced primary laterals that are not shown in the square-shaped service area of the figure. In this voltage-drop study, use the precalculated voltage-drop curves of Figure 4.17 when applicable. Use the nominal primary voltage of 7,620/13,200 V for a three-phase four-wire wye-grounded system. Assume that at peak loading, the load density is 1000 kVA/mi² and the lumped load is 2000 kVA, and that at off-peak loading, the load density is 333 kVA/mi² and the lumped load is still 2000 kVA. The lumped load is of a small industrial plant working three shifts a day. The substation bus voltages are 1.025 pu V of 7620 base volts at peak load and 1.000 pu V during off-peak load.

The transformer located between buses c and d has a three-phase rating of 2000 kVA and a delta-rated high voltage of 13,200 V and grounded-wye-rated low voltage of 277/480 V. It has

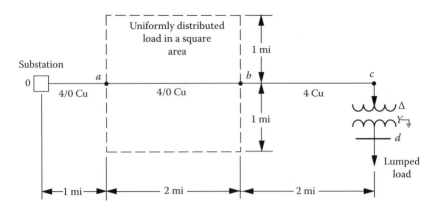

FIGURE 7.13 A square-shaped service area and a lumped-sum load.

$0 + j0.05$ per unit impedance based on the transformer ratings. It is tapped up to raise the low voltage 5.0% relative to the high voltage, that is, the equivalent turns ratio in use is $(7620/277) \times 0.95$. Use the given information and data for peak loading and determine the following:

a. The percent voltage drop from the substation to point a, from a to b, from b to c, and from c to d on the main
b. The per unit voltages at the points a, b, c, and d on the main
c. The line-to-neutral voltages at the points a, b, c, and d

Solution

a. The load connected in the square-shaped service area is

$$S_n = D \times A_n$$

$$= 1000 \times 4$$

$$= 4000 \, kVA$$

Thus the total kilovoltampere load on the main is

$$S_m = 4000 + 2000$$

$$= 6000 \, kVA$$

From Figure 4.17, for #4/0 copper, the K constant is found to be 0.0003. Therefore, the percent voltage drop from the substation to point a is

$$\% \, VD_{0a} = K \times S_m \times 1$$

$$= 0.0003 \times 6000 \times 1$$

$$= 1.48\% \, V \quad or \quad 0.018 \, pu \, V$$

The percent voltage drop from point a to point b is

$$\% \, VD_{ab} = K \times S_n \times 1 + K \times S_{lump} \times 1$$

$$= 0.0003 \times 4000 \times 1 + 0.0003 \times 2000 \times 2$$

$$= 2.4\% \, V \quad or \quad 0.024 \, pu \, V$$

The percent voltage drop from point b to point c is

$$\% \, VD_{bc} = K \times S_{lump} \times l$$

$$= 0.0009 \times 2000 \times 2$$

$$= 3.6\% \, V \quad \text{or} \quad 0.036 \, pu \, V$$

To find the percent voltage drop from point c to bus d,

$$l = \frac{2000 \, kVA}{\sqrt{3} \times V_{L-L} \text{ at point } c}$$

$$= \frac{2000 \, kVA}{\sqrt{3} \times (0.947 \times 13.2 \, kV)}$$

$$= 92.373 \, A$$

$$l_B = \frac{2000 \, kVA}{\sqrt{3} \times 13.2 \, kV}$$

$$= 87.477 \, A$$

$$l_{pu} = \frac{l}{l_B}$$

$$= 1.056 \, pu \, A$$

Note that usually in a simple problem like this, the reduced voltage at point c is ignored. In that case, for example, the per unit current would be 1.0 pu A rather than 1.056 pu A. Since

$$Z_{T, \, pu} = 0 + j0.05 \, pu \, \Omega$$

and

$$\cos \theta = 0.9 \quad \text{or} \quad \theta = 25.84° \text{ lagging}$$

therefore

$$l_{pu} = 1.056 \angle 25.84° \, pu \, A$$

Thus, to find the percent voltage drop at bus d, first it can be found in per unit as

$$VD_{cd} = \frac{l(R \cos \theta + X \sin \theta)}{V_B} \, pu \, V$$

but since the low voltage has been tapped up 5%,

$$VD_{cd} = \frac{l(R \cos \theta + X \sin \theta)}{V_B} - 0.05 \, pu \, V$$

Therefore,

$$VC_{cd} = \frac{1.056(0 \times 0.9 + 0.05 \times 0.4359)}{1.0} - 0.05$$

$$= -0.0267 \, pu \, V$$

or

$$\% VD_{cd} = -2.67\% \, V$$

Here, the negative sign of the voltage drop indicates that it is in fact a voltage rise rather than a voltage drop.

b. The per unit voltages at the points *a*, *b*, *c*, and *d* on the main are

$$V_a = V_0 - V_{0a}$$

$$= 1.025 - 0.018$$

$$= 1.007 \, pu \, V \quad \text{or} \quad 100.7\% \, V$$

$$V_b = V_a - V_{ab}$$

$$= 1.007 - 0.024$$

$$= 0.983 \, pu \, V \quad \text{or} \quad 98.3\% \, V$$

$$V_c = V_b - V_{bc}$$

$$= 0.983 - 0.036$$

$$= 0.947 \, pu \, V \quad \text{or} \quad 94.7\% \, V$$

$$V_d = V_c - V_{cd}$$

$$= 0.947 - (-0.0267)$$

$$= 0.9737 \, pu \, V \quad \text{or} \quad 97.37\% \, V$$

c. The line-to-neutral voltages are

$$V_a = 7620 \times 1.007$$

$$= 7673.3 \, V$$

$$V_b = 7620 \times 0.983$$

$$= 7490.5 \, V$$

$$V_c = 7620 \times 0.947$$

$$= 7216.1 \, V$$

$$V_d = 277 \times 0.9737$$

$$= 269.7 \, V$$

Example 7.11

Use the relevant information and data given in Example 7.10 for off-peak loading and repeat Example 7.10, and find the V_d voltage at bus d in line-to-neutral volts. Also write the necessary codes to solve the problem in MATLAB.

Solution

a. At off-peak loading, the load connected in the square-shaped service area is

$$S_n = D \times A_n$$

$$= 333 \times 4$$

$$= 1332 \text{kVA}$$

Thus the total kilovoltampere load on the main is

$$S_m = 1332 + 2000$$

$$= 3332 \text{kVA}$$

Therefore, the percent voltage drop from the substation to point a is

$$\% \text{VD}_{0a} = K \times S_m \times 1$$

$$= 0.0003 \times 3332 \times 1$$

$$= 1.0\% \text{ V} \quad \text{or} \quad 0.01 \text{pu V}$$

The percent voltage drop from point a to point b is

$$\% \text{VD}_{ab} = K \times S_n \times l/2 + K \times S_{\text{lump}} \times 1$$

$$= 0.003 \times 1332 \times 1 + 0.0003 \times 2000 \times 2$$

$$= 1.6\% \text{ V} \quad \text{or} \quad 0.016 \text{pu V}$$

The percent voltage drop from point b to point c is

$$\text{VD}_{bc} = K \times S_{\text{lump}} \times l$$

$$= 0.0009 \times 2000 \times 2$$

$$= 3.6\% \text{ V} \quad \text{or} \quad 0.036 \text{pu V}$$

To find the percent voltage drop from point c to bus d, the percent voltage drop at bus d can be found as before

$$\% \text{VD}_{cd} = 0.0267 \text{pu V} \quad \text{or} \quad -2.67\% \text{ V}$$

b. The per unit voltages at points a, b, c, and d on the main are

$$V_a = V_0 - V_{0a}$$

$$= 1.0 - 0.01$$

$$= 0.99 \text{pu V} \quad \text{or} \quad 99\% \text{ V}$$

$$V_b = V_a - V_{ab}$$

$$= 0.99 - 0.016$$

$$= 0.974 \, pu \, V \quad or \quad 97.4\% \, V$$

$$V_c = V_b - V_{bc}$$

$$= 0.974 - 0.036$$

$$= 0.938 \, pu \, V \quad or \quad 93.8\% \, V$$

$$V_d = V_c - V_{cd}$$

$$= 0.938 - (-0.0267)$$

$$= 0.9647 \, pu \, V \quad or \quad 96.47\% \, V$$

c. The line-to-neutral voltages are

$$V_a = 7620 \times 0.99$$

$$= 7543.8 \, V$$

$$V_b = 7620 \times 0.974$$

$$= 7421.9 \, V$$

$$V_c = 7620 \times 0.938$$

$$= 7147.6 \, V$$

$$V_d = 277 \times 0.9647$$

$$= 267.2 \, V$$

Note that the voltages at bus *d* during peak and off-peak loading are nearly the same.

Here is the MATLAB script:

```
clc
clear

% System parameters
St = 2000;% in kVA
D = 1000;% in kVA/mi^2
An = 4;% in mi^2
K40 = 0.0003;% from Figure 4.17 for 4/0 AWG
K4 = 0.0009;% from Figure 4.17 for 4 AWG
L1 = 1;% distanced from substation to point a in miles
L2 = 2;% distanced from point a to b in miles
kV = 13.2;
Xt = 0.05;
```

```
PF = 0.9;
Vopu = 1.025;
VBp = 7620;% Voltage base primary
VBs = 277;% Voltage base secondary

% Solution for part a
Sn = D*An

% Total kVA on main
Sm = Sn + St

% Per unit voltage drop from substation to point a
VDoapu = (K40*Sm*L1)/100

% Per unit voltage drop from point a to point b
VDabpu = (K40*Sn*(L2/2)+K40*St*L2)/100

% Per unit voltage drop from point b to point c
VDbcpu = (K4*St*L2)/100
I = St/(sqrt(3)*0.947*kV);
IB = St/(sqrt(3)*kV);
Ipu = I/IB

% Per unit voltage drop from point c to point d
VDcdpu = Ipu*(Xt*sin(acos(PF)))-0.05

% Solution for part b in per units
Vapu = Vopu-VDoapu
Vbpu = Vapu-VDabpu
Vcpu = Vbpu-VDbcpu
Vdpu = Vcpu-VDcdpu

% Solution for part c in per units
Va = Vapu*VBp
Vb = Vbpu*VBp
Vc = Vcpu*VBp
Vd = Vdpu*VBs
```

Example 7.12

Figure 7.14 shows that a large number of small loads are closely spaced along the length l. If the loads are single phase, they are assumed to be well balanced among the three phases.

A three-phase four-wire wye-grounded 7.62/13.2 kV primary line is to be built along the length l and fed through a distribution substation transformer from a high-voltage transmission line. Assume that the uniform (or linear) distribution of the connected load along the length l is

$$\frac{S_{\text{connected load}}}{l} = 0.45\,\text{kVA/ft}$$

The 30-min annual demand factor of all loads is 0.60, the diversity factor (F_D) among all loads is 1.08, and the annual loss factor (F_{LS}) is 0.20. Assuming a lagging power factor of 0.9 for all loads and a 37 in. geometric mean spacing of phase conductors, use Figure 4.17 for voltage-drop calculations for copper conductors. Use the relevant tables in Appendix A for additional data about copper and ACSR conductors and determine the following:

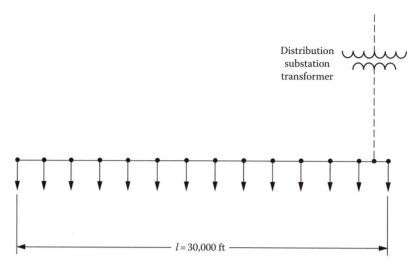

FIGURE 7.14 The distribution system of Example 7.12.

Locate the distribution substation where you think it would be the most economical, considering only the 13.2 kV system, and then find

a. The minimum ampacity-sized copper and ACSR phase conductors
b. The percent voltage drop at the location having the lowest voltage at the time of the annual peak load, using the ampacity-sized copper conductor found in part (a)
c. Also write the necessary codes to solve the problem in MATLAB

Solution

To achieve the minimum voltage drop, the substation should be located at the middle of the line 1, and therefore

a. From Equation 2.13, the diversified maximum demand of the group of the load is

$$D_g = \frac{\sum_{i-1}^{n} TCD_i \times DF_i}{F_D}$$

$$= \frac{0.45\,kVA/ft \times 0.60}{1.08}$$

$$= 0.250\,kVA/ft$$

Thus the peak load of each main on the substation transformer is

$$S_{PK} = 0.250\,kVA/ft \times 15,000\,ft$$

$$= 3750\,kVA$$

or

$$3750\ kVA\ = \sqrt{3} \times 13.2\ kV \times I$$

hence

$$I = \frac{3750\,\text{kVA}}{\sqrt{3} \times 13.2\,\text{kV}}$$

$$= 164.2\,\text{A}$$

in each main out of the substation. Therefore, from the tables of Appendix A, it can be recommended that either #4 AWG copper conductor or #2 AWG ACSR conductor be used.

b. Assuming that #4 AWG copper conductor is used, the percent voltage drop at the time of the annual peak load is

$$\%\text{VD} = [K\,\%\,\text{VD}/(\text{kVA} \cdot \text{mi})] \times [S_{PK}\text{kVA}] \times \frac{l\,\text{ft}}{2} \frac{1}{5280\,\text{ft/mi}}$$

$$= 0.0009 \times 3750 \times \frac{15,000}{2 \times 5280}$$

$$= 5.3\%\,\text{V}$$

c. Here is the MATLAB script:

```
clc
clear

% System parameters
D = 333;% off-peak load density in kVA/mi^2
An = 4;% in mi^2
K40 = 0.0003;% from Figure 4.17 for 4/0 AWG
K4 = 0.0009;% from Figure 4.17 for 4 AWG
L1 = 1;% distanced from substation to point a in miles
L2 = 2;% distanced from point a to b in miles
St = 2000;% in kVA
Vopu = 1.0;
VBp = 7620;
VBs = 277;

% Solution for part a
Sn = D*An

% Total kVA on main
Sm = Sn + St

% Per unit voltage drop from substation to point a
VDoapu = (K40*Sm*L1)/100

% Per unit voltage drop from point a to point b
VDabpu = (K40*Sn*(L2/2)+K40*St*L2)/100

% Per unit voltage drop from point b to point c
VDbcpu = (K4*St*L2)/100
VDcdpu = -0.027% as before
```

```
% Solution for part b in per units
Vapu = Vopu-VDoapu
Vbpu = Vapu-VDabpu
Vcpu = Vbpu-VDbcpu
Vdpu = Vcpu-VDcdpu

% Solution for part c in per units
Va = Vapu*VBp
Vb = Vbpu*VBp
Vc = Vcpu*VBp
Vd = Vdpu*VBs
```

Example 7.13

Now suppose that the line in Example 7.12 is arbitrarily constructed with #4/0 AWG ACSR phase conductor and that the substation remains where you put it in part (a). Assume 50°C conductor temperature and find the total annual I^2R energy loss (TAEL$_{Cu}$), in kilowatthours, in the entire line length.

Solution

The total I^2R loss in the entire line length is

$$\sum I^2R = 3I^2\left(r\times\frac{l}{2}\right)$$

$$= 3(164.2)^2(0.592\ \Omega/\text{mi})\frac{30,000\,\text{ft}}{2\times5280\,\text{ft/mi}}$$

$$= 136,033.729\,\text{W}$$

Therefore, the total I^2R energy loss is

$$\text{TAEL}_{Cu} = \left[\left(\sum I^2R\right)F_{LS}\right](8760\ \text{h/year})$$

$$= \frac{136,033.729}{10^3}\times0.20\times8760$$

$$= 238,331.09\ \text{kWh}$$

Example 7.14

Figure 7.15 shows a single-line diagram of a simple three-phase four-wire wye-grounded primary feeder. The nominal operating voltage and the base voltage are given as 7,200/12,470 V. Assume that all loads are balanced three phase and all have 90% power factor, lagging. The given values of the constant K in Table 7.4 are based on 7,200/12,470 V. There is a total of a 3000 kVA uniformly distributed load over a 4 mi line between b and c. Use the given data and determine the following:

a. Find the total percent voltage drop at points a, b, c, and d.
b. If the substation bus voltages are regulated to 7,300/12,650 V, what are the line-to-neutral and line-to-line voltages at point a?

Solution

a. The total load flowing through the line between points 0 and a is

$$\sum S = 2000\,\text{kVA} + 3000\,\text{kVA} = 5000\,\text{kVA}$$

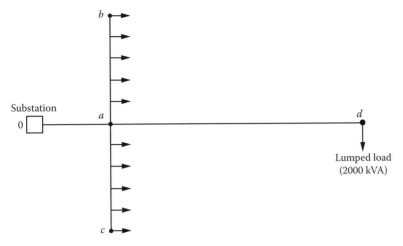

FIGURE 7.15   The distribution system of Example 7.14.

TABLE 7.4
***K* Constants**

Run	Conductor Type	Distance (mi)	*K*, % VD/(kVA·mi)
Sub. to a	#4/0 ACSR	1.0	0.0005
a to b	# 1 ACSR	2.0	0.0010
a to c	#1 ACSR	2.0	0.0010
a to d	#1 ACSR	2.0	0.0010

---

therefore the percent voltage drop at point *a* is

$$\% \text{VD}_a = K\left(\sum S\right)l$$

$$= 0.0005 \times 5000 \times 1.0$$

$$= 2.5\% \, \text{V}$$

Similarly, the load flowing through the line between points *a* and *b* is

$$S = 1500 \, \text{kVA}$$

Therefore,

$$\% \, \text{VD}_b = K \times S \times l/2 + \% \, \text{VD}_a$$

$$= 0.0010 \times 1500 \times 1 + 2.5\%$$

$$= 4\% \, \text{V}$$

$$\% \, \text{VD}_c = \% \, \text{VD}_b = 4\% \, \text{V}$$

$$\% \, VD_d = K \times S_{\text{lump}} \times l + \% \, VD_a$$

$$= 0.0010 \times 2000 \times 2 + 2.5\%$$

$$= 6.5\% \, V$$

b. If the substation bus voltages are regulated to 7,300/12,650 V at point a, the line-to-neutral voltage is

$$V_{a,L-N} = 7300 - VD_{a,L-N}$$

$$= 7300 - 7300 \times 0.025$$

$$= 7117.5 \, V$$

and the line-to-line voltage is

$$V_{a,L-L} = 12{,}650 - VD_{a,L-L}$$

$$= 12{,}650 - 12{,}650 \times 0.025$$

$$= 12{,}333.8 \, V$$

7.4 PERCENT POWER (OR COPPER) LOSS

The percent power (or conductor) loss of a circuit can be expressed as

$$\% \, I^2 R = \frac{P_{\text{LS}}}{P_r} \times 100$$

$$= \frac{I^2 R}{P_r} \times 100 \tag{7.59}$$

where
P_{LS} is the power loss of a circuit, kW
$= I^2 R$
P_r is the power delivered by the circuit, kW

The conductor $I^2 R$ losses at a load factor of 0.6 can readily be found from Table 7.5 for various voltage levels.

At times, in ac circuits, the ratio of percent power, or conductor, loss to percent voltage regulation can be used, and it is given by the following approximate expression:

$$\frac{\% \, I^2 R}{\% \, VD} = \frac{\cos \phi}{\cos \theta \times \cos(\phi - \theta)} \tag{7.60}$$

where
$\% \, I^2 R$ is the percent power loss of a circuit
$\% \, VD$ is the percent voltage drop of the circuit
ϕ is the impedance angle $= \tan^{-1}(X/R)$
θ is the power-factor angle

TABLE 7.5
Conductor I^2R Losses, kWh/(mi year), at 7.2/12.5 kV and a Load Factor of 0.6

Annual Peak	Single Phase				"V" Phase				Three Phase					
	8 Copper	6 Copper	4 Copper	2 Copper	8 Copper	6 Copper	4 Copper	2 Copper	6 Copper	4 Copper	2 Copper	1 Copper	1/0 Copper	2/0 Copper
Load (kW)		4 ACSR	2 ACSR	1/0 ACSR		4 ACSR	2 ACSR	1/0 ACSR	4 ACSR	2 ACSR	1/0 ACSR	2/0 ACSR	3/0 ACSR	4/0 ACSR
20	124	82	55	37	62	41	27	19	25	16	10			
40	495	329	218	149	248	164	109	75	99	63	39	31		
60	1,110	740	491	335	557	370	246	168	224	141	88	70	56	
80	1,980	1,320	873	596	990	658	437	298	398	250	157	125	99	78
100	3,100	2,060	1,360	932	1,550	1,030	682	466	621	391	245	195	154	122
120	4,460	2,960	1,960	1,340	2,230	1,480	982	671	895	563	353	280	222	176
140	6,070	4,030	2,670	1,830	3,030	2,010	1,340	913	1,220	766	481	382	302	240
160	7,920	5,260	3,490	2,390	3,960	2,630	1,750	1,190	1,590	1,000	628	498	395	313
180	10,000	6,660	4,420	3,020	5,010	3,330	2,210	1,510	2,010	1,270	795	631	500	396
200	12,400	8,220	5,460	3,730	6,190	4,110	2,730	1,860	2,490	1,560	982	779	617	489
225	15,700	10,400	6,910	4,720	7,830	5,200	3,450	2,360	3,150	1,980	1240	986	780	619
250	19,300	12,800	8,530	5,820	9,670	6,420	4,260	2,910	3,880	2,440	1530	1220	964	764
275	23,400	15,500	10,300	7,050	11,700	7,770	5,160	3,520	4,700	2,960	1860	1470	1170	925
300		18,500	12,300	8,390	13,900	9,250	6,140	4,190	5,590	3,520	2210	1750	1390	1100
325		21,700	14,400	9,840	16,300	10,900	7,210	4,920	6,560	4,130	2590	2060	1630	1280
350			16,700	11,400	18,900	12,600	8,360	5,710	7,610	4,790	3010	2380	1890	1500
375			19,200	13,100	21,800	14,400	9,590	6,550	8,740	5,500	3450	2740	2170	1720
400			21,800	14,900	24,800	16,400	10,900	7,450	9,940	6,260	3930	3120	2470	1960
450				18,900		20,800	13,800	9,430	12,600	7,920	4970	3940	3120	2480
500				23,300		25,700	17,100	11,600	15,500	9,780	6140	4870	3850	3060
550							20,600	14,100	18,800	11,800	7420	5890	4660	3700
600							24,600	16,800	22,400	14,100	8840	7010	5550	4400

Source: Rural Electrification Administration, *U.S. Department of Agriculture: Economic Design of Primary Lines for Rural Distribution Systems,* REA Bulletin, 60, 1960.

Note: This table is calculated for a PF of 90%. To adjust for a different PF, multiply these values by the factor of $k = (90)^2/(\text{PF})^2$.

For 7.62/13.2 kV, multiply these values by 0.893; for 14.4/24.9 kV, multiply by 0.25.

7.5 METHOD TO ANALYZE DISTRIBUTION COSTS

To make any meaningful feeder-size selection, the distribution engineer should make a cost study associated with feeders in addition to the voltage-drop and power-loss considerations. The cost analysis for each feeder size should include (1) investment cost of the installed feeder, (2) cost of energy lost due to I^2R losses in the feeder conductors, and (3) cost of demand lost, that is, the cost of useful system capacity lost (including generation, transmission, and distribution systems), in order to maintain adequate system capacity to supply the I^2R losses in the distribution feeder conductors. Therefore, the total annual feeder cost of a given size feeder can be expressed as

$$\text{TAC} = \text{AIC} + \text{AEC} + \text{ADC} \ \$/\text{mi} \tag{7.61}$$

where
    TAC is the total annual equivalent cost of feeder, \$/mi
    AIC is the annual equivalent of investment cost of installed feeder, \$/mi
    AEC is the annual equivalent of energy cost due to I^2R losses in feeder conductors, \$/mi
    ADC is the annual equivalent of demand cost incurred to maintain adequate system capacity to
        supply I^2R losses in feeder conductors, \$/mi

7.5.1 ANNUAL EQUIVALENT OF INVESTMENT COST

The annual equivalent of investment cost of a given size feeder can be expressed as

$$\text{AIC} = \text{IC}_F \times i_F \ \$/\text{mi} \tag{7.62}$$

where
    AIC is the annual equivalent of investment cost of a given size feeder, \$/mi
    IC_F is the cost of installed feeder, \$/mi
    i_F is the annual fixed charge rate applicable to feeder

The general utility practice is to include cost of capital, depreciation, taxes, insurance, and operation and maintenance (O&M) expenses in the *annual fixed charge rate* or so-called *carrying charge rate*. It is given as a decimal.

7.5.2 ANNUAL EQUIVALENT OF ENERGY COST

The annual equivalent of energy cost due to I^2R losses in feeder conductors can be expressed as

$$\text{AEC} = 3I^2R \times \text{EC} \times \text{F}_{\text{LL}} \times \text{F}_{\text{LSA}} \times 8760 \ \$/\text{mi} \tag{7.63}$$

where
    AEC is the annual equivalent of energy cost due to I^2R losses in feeder conductors, \$/mi
    EC is the cost of energy, \$/kWh
    F_{LL} is the load-location factor
    F_{LS} is the loss factor
    F_{LSA} is the loss-allowance factor

The load-location factor of a feeder with uniformly distributed load can be defined as

$$F_{\text{LL}} = \frac{s}{\ell} \tag{7.64}$$

where
F_{LL} is the load-location factor in decimal
s is the distance of point on feeder where total feeder load can be assumed to be concentrated for the purpose of calculating I^2R losses
ℓ is the total feeder length, mi

The loss factor can be defined as the ratio of the average annual power loss to the peak annual power loss and can be found approximately for urban areas from

$$F_{LS} = 0.3F_{LD} + 0.7\,F_{LD}^2 \tag{7.65}$$

and for rural areas [6],

$$F_{LS} = 0.16\,F_{LD} + 0.84\,F_{LD}^2$$

The loss-allowance factor is an allocation factor that allows for the additional losses incurred in the total power system due to the transmission of power from the generating plant to the distribution substation.

7.5.3 ANNUAL EQUIVALENT OF DEMAND COST

The annual equivalent of demand cost incurred to maintain adequate system capacity to supply the I^2R losses in the feeder conductors can be expressed as

$$\text{ADC} = 3I^2R \times F_{LL} \times F_{PR} \times F_R$$
$$\times F_{LSA}[(C_G \times i_G) + (C_T \times i_T) + (C_S \times i_S)]\ \$/\text{mi} \tag{7.66}$$

where
ADC is the annual equivalent of demand cost incurred to maintain adequate system capacity to supply 12R losses in feeder conductors, $/mi
F_{LL} is the load-location factor
F_{PR} is the peak-responsibility factor
F_R is the reserve factor
F_{LSA} is the loss-allowance factor
C_G is the cost of (peaking) generation system $/kVA
C_T is the cost of transmission system, $/kVA
C_S is the cost of distribution substation, $/kVA
i_G is the annual fixed charge rate applicable to generation system
i_T is the annual fixed charge rate applicable to transmission system
i_S is the annual fixed charge rate applicable to distribution substation

The reserve factor is the ratio of total generation capability to the total load and losses to be supplied. The peak-responsibility factor is a per unit value of the peak feeder losses that are coincident with the system peak demand.

7.5.4 LEVELIZED ANNUAL COST

In general, the costs of energy and demand and even O&M expenses vary from year to year during a given time, as shown in Figure 7.16a; therefore, it becomes necessary to *levelize* these costs over the expected economic life of the feeder, as shown in Figure 7.16b.

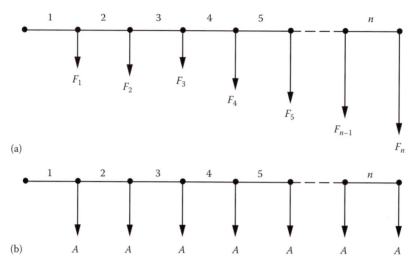

FIGURE 7.16 Illustration of the levelized annual cost concept: (a) unlevelized annual cost flow diagram and (b) levelized cost flow diagram.

Assume that the costs occur discretely at the end of each year, as shown in Figure 7.16a. The *levelized annual* cost* of equal amounts can be calculated as

$$A = \left[F_1 \left(\frac{P}{F} \right)_1^i + F_2 \left(\frac{P}{F} \right)_2^i + F_3 \left(\frac{P}{F} \right)_3^i + \cdots + F_n \left(\frac{P}{F} \right)_n^i \right] \left(\frac{A}{P} \right)_n^i \qquad (7.67)$$

or

$$A = \left[\sum_{j=1}^{n} F_i \left(\frac{P}{F} \right)_j^i \right] \left(\frac{A}{P} \right)_n^i \qquad (7.68)$$

where
  A is the levelized annual cost, $/year
  F_i is the unequal (or actual or unlevelized) annual cost, $/year
  n is the economic life, year
  i is the interest rate
  $(P/F)_n^i$ is the present worth (or present equivalent) of a future sum factor (with i interest rate and n years of economic life), also known as *single-payment discount factor*
  $(A/P)_n^i$ is the uniform series worth of a present sum factor, also known as *capital-recovery factor*

The single-payment discount factor and the capital-recovery factor can be found from the compounded-interest tables or from the following equations, respectively,

$$\left(\frac{P}{F} \right)_n^i = \frac{1}{(1+i)^n} \qquad (7.69)$$

---

* Also called the *annual equivalent* or *annual worth*.

and

$$\left(\frac{A}{P}\right)^i_n = \frac{i(1+i)^n}{(1+i)^n - 1}$$ (7.70)

Example 7.15

Assume that the following data have been gathered for the system of the NL&NP Company.

Feeder length = 1 mi
Cost of energy = 20 mills/kWh (or $0.02/kWh)
Cost of generation system = $200/kW
Cost of transmission system = $65/kW
Cost of distribution substation = $20/kW
Annual fixed charge rate for generation = 0.21
Annual fixed charge rate for transmission = 0.18
Annual fixed charge rate for substation = 0.18
Annual fixed charge rate for feeders = 0.25
Interest rate = 12%
Load factor = 0.4
Loss-allowance factor = 1.03
Reserve factor = 1.15
Peak-responsibility factor = 0.82

Table 7.6 gives cost data for typical ACSR conductors used in rural areas at 12.5 and 24.9 kV. Table 7.7 gives cost data for typical ACSR conductors used in urban areas at 12.5 and 34.5 kV. Using the given data, develop nomographs that can be readily used to calculate the total annual equivalent cost of the feeder in dollars per mile.

TABLE 7.6
Typical ACSR Conductors Used in Rural Areas

Conductor Ground Wire	Conductor Ground Wire			Installation Cost and Hardware Installed		Total Cost and Hardware Installed
Size	Size	Weight (lb)	Weight (lb)	$/lb	Cost ($)	Feeder Cost ($)
At 12.5 kV						
#4	#4	356	356	0.6	6,945.6	7,800
1/0	#2	769	566	0.6	7,176.2	8,900
3/0	1/0	1223	769	0.6	7,737.2	10,400
4/0	1/0	1542	769	0.6	8,563	11,800
266.8 kcmil	1/0	1802	769	0.6	9,985	13,690
477 kcmil	1/0	3642	769	0.6	10,967	17,660
At 24.9 kV						
#4	#4	356	356	0.6	7,605.6	8,460
1/0	#2	769	566	0.6	7,856.2	9,580
3/0	1/0	1223	769	0.6	8,217.2	10,880
4/0	1/0	1542	769	0.6	8,293	11,530
266.8 kcmil	1/0	1802	769	0.6	9,615	13,320
477 kcmil	1/0	3462	769	0.6	11,547	18,240

TABLE 7.7

Typical ACSR Conductors Used in Urban Areas

| Conductor Ground Wire | | Conductor Ground Wire | | Cost | Installation and Hardware Installed | Total |
Size	Size	Weight (lb)	Weight (lb)	$/lb	Cost ($)	Feeder Cost ($)
At 12.5 kV						
#4	#4	356	356	0.6	21,145.6	22,000
1/0	#4	769	356	0.6	22,402.2	24,000
3/0	#4	1223	356	0.6	24,585	27,000
477 kcmil	1/0	3462	769	0.6	28,307	35,000
At 34.5 kV						
#4	#4	356	356	0.6	21,375.6	22,230
1/0	#4	769	356	0.6	22,632.2	24,230
3/0	#4	1223	356	0.6	24,815	27,230
477 kcmil	1/0	3462	769	0.6	28,537	35,230

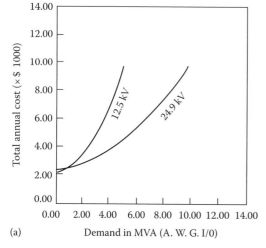

(a) Demand in MVA (A. W. G. I/0)

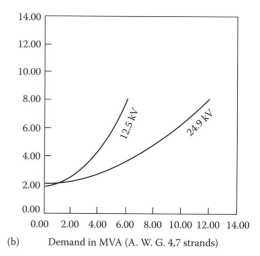

(b) Demand in MVA (A. W. G. 4,7 strands)

FIGURE 7.17 Total annual equivalent cost of ACSR feeders for *rural* areas in thousands of dollars per mile: (a) 477 cmil, 26 strands, (b) 266.8 cmil, 6 strands, AWG 4/0, and AWG 3/0.

Solution

Using the given and additional data and appropriate equations from Section 7.5, the following nomographs have been developed. Figures 7.17 and 7.18 give nomographs to calculate the total annual equivalent cost of ACSR feeders of various sizes for rural and urban areas, respectively, in thousands of dollars per mile.

Example 7.16

The NL&NP power and light company is required to serve a newly developed residential area. There are two possible routes for the construction of the necessary power line. Route *A* is 18 miles long and goes around a lake. It has been estimated that the required overhead power line will cost $8000 per mile to build and $350 per mile per year to maintain. Its salvage value will be $1500 per mile at the end of its useful life of 20 years.

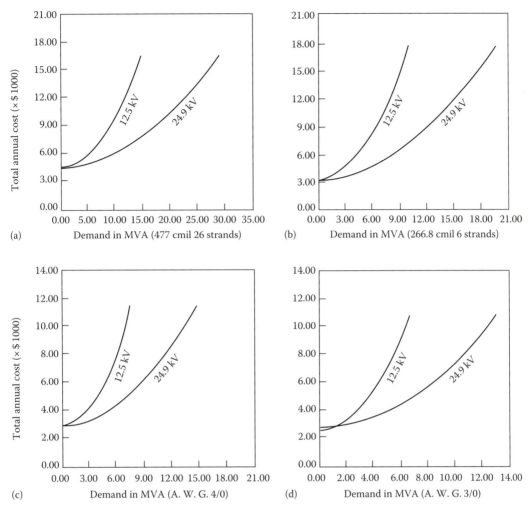

FIGURE 7.18 Total annual equivalent cost of ACSR feeders for *urban* areas in thousands of dollars per mile: (a) 477 cmil, 26 strands, (b) AWG 3/0, (c) AWG 1/0, and (d) AWG 4, 7 strands.

On the other hand, route B is 6 miles long and is an underwater line that goes across the lake. It has been estimated that the required underwater line using submarine power cables will cost $21,000 per mile to build and $1,200 per mile per year to maintain. Its salvage value will be $6000 per mile at the end of 20 years. Assume that the fixed charge rate is 10% and that the annual ad valorem (property) taxes are 3% of the first cots of each power line. Use any engineering economy interest tables and determine the economically preferable alternative.

Solution

Route A: The first cost of the overhead power line is

$$P = (\$8,000/\text{mile})(18\,\text{miles}) = \$144,000$$

and its estimated salvage value is

$$F = (\$1,500/\text{mile})(18\,\text{miles}) = \$27,000$$

The annual equivalent cost of capital invested in the line is

$$A_1 = \$144,000\left(\frac{A}{P}\right)_{20}^{10\%} - \$27,000\left(\frac{A}{F}\right)_{20}^{10\%}$$

$$= \$144,000(0.11746) - \$27,000(0.01746) = \$16,443$$

The annual equivalent cost of the tax and maintenance is

$$A_2 = (\$144,000)(3\%) + (\$350/\text{mile})(18\,\text{miles}) = \$10,620$$

Route B: The first cost of the submarine power line is

$$P = (\$21,000/\text{mile})(6\,\text{miles}) = \$126,000$$

and its estimated salvage value is

$$F = (\$6,000/\text{mile})(6\,\text{miles}) = \$36,000$$

Its annual equivalent cost of capital invested is

$$A_1 = \$126,000\left(\frac{A}{P}\right)_{20}^{10\%} - \$36,000\left(\frac{A}{F}\right)_{20}^{10\%}$$

$$= \$14,171$$

The annual equivalent cost of the tax and maintenance is

$$A_2 = (\$126,000)(3\%) + (\$1,200/\text{mile})(6\,\text{miles}) = \$10,980$$

The total annual equivalent cost of the submarine power line is

$$A = A_1 + A_2$$

$$= \$14,171 + \$10,980 = \$25,151$$

Hence, the *economically preferable alternative* is route B. Of course, if the present worth of the costs are calculated, the conclusion would still be the same. For example, the present worth of costs for A and B are

$$PW_A = \$27,063\left(\frac{P}{A}\right)_{20}^{10\%} = \$230,414$$

and

$$PW_B = \$25,151\left(\frac{P}{A}\right)_{20}^{10\%} = \$214,136$$

Thus, route B is still the preferred route.

Example 7.17

Use the data given in Example 6.6 and assume that the fixed charge rate is 0.15, and zero salvage values are expected at the end of useful lives of 30 years for each alternative. But the salvage value for 9 MVA capacity line is $2000 at the end of the 10th year. Use a study period of 30 years and determine the following:

a. The annual equivalent cost of 9 MVA capacity line.
b. The annual equivalent cost of 15 MVA capacity line.
c. The annual equivalent cost of the upgrade option if the upgrade will take place at the end of 10 years. Use an average value of $5000 at the end of 20 years for the new 15 MVA upgrade line.

Solution

a. The annual equivalent cost of 9 MVA capacity line is

$$A_1 = \$120,000\left(\frac{A}{P}\right)_{30}^{15\%} = \$120,000(0.15230) = \$18,276 \text{ per mile per year}$$

b.
$$A_2 = \$150,000\left(\frac{A}{P}\right)_{30}^{15\%} = \$150,000(0.15230) = \$22,845$$

c. The annual equivalent cost of 15 MVA capacity line is

$$A_2 = \left[\$120,000 - \$2,000\left(\frac{P}{F}\right)_{10}^{15\%} + \$200,000\left(\frac{P}{F}\right)_{10}^{15\%} - \$5,000\left(\frac{P}{F}\right)_{30}^{15\%}\right]\left(\frac{A}{P}\right)_{30}^{15\%}$$

$$= [\$120,000 - \$2,000(0.2472) + \$200,000(0.2472) - \$5,000(0.0151)](0.15230)$$

$$= \$25,718.92$$

As it can be seen, the upgrade option is still the bad option. Furthermore, if one considers the 9 MVA vs. 15 MVA capacities, building the 15 MVA capacity line from the start is still the best option.

7.6 ECONOMIC ANALYSIS OF EQUIPMENT LOSSES

Today, the substantially escalating plant, equipment, energy, and capital costs make it increasingly more important to evaluate losses of electric equipment (e.g., power or distribution transformers) before making any final decision for purchasing new equipment and/or replacing (or retiring) existing ones. For example, nowadays it is not uncommon to find out that a transformer with lower losses but higher initial price tag is less expensive than the one with higher losses but lower initial price when total cost over the life of the transformer is considered.

However, in the replacement or retirement decisions, the associated cost savings in O&M costs in a given *life cycle analysis** or *life cycle cost study* must be greater than the total purchase price of the more efficient replacement transformer. Based on the "*sunk cost*" concept of engineering economy, the carrying charges of the existing equipment do not affect the retirement decision, regardless of the age of the existing unit. In other words, the fixed, or carrying, charges of an existing equipment must be amortized (written off) whether the unit is retired or not.

---

* These phrases are used by some governmental agencies and other organizations to specifically require that bid evaluations or purchase decisions be based not just on first cost but on all factors (such as future operating costs) that influence the alternative that is more economical.

The transformer cost study should include the following factors:

1. Annual cost of copper losses
2. Annual cost of core losses
3. Annual cost of exciting current
4. Annual cost of regulation
5. Annual cost of fixed charges on the first cost of the installed equipment

These annual costs may be different from year to year during the economical lifetime of the equipment. Therefore, it may be required to levelize them, as explained in Section 7.5.4. Read Section 6.7 for further information on the cost study of the distribution transformers. For the economic replacement study of the power transformers, the following simplified technique may be sufficient. Dodds [10] summarizes the economic evaluation of the cost of losses in an old and a new transformer step by step as given in the following text:

1. Determine the power ratings for the transformers as well as the peak and average system loads.
2. Obtain the load and no-load losses for the transformers under rated conditions.
3. Determine the original cost of the old transformer and the purchase price of the new one.
4. Obtain the carrying charge rate, system capital cost rate, and energy cost rate for your particular utility.
5. Calculate the transformer carrying charge and the cost of losses for each transformer. The cost of losses is equal to the system carrying charge plus the energy charge.
6. Compare the total cost per year for each transformer. The total cost is equal to the sum of the transformer carrying charge and the cost of losses.
7. Compare the total cost per year of the old and new transformers. If the total cost per year of the new transformer is less, replacement of the old transformer can be economically justified.

PROBLEMS

7.1 Consider Figure P7.1 and repeat Example 7.5.

7.2 Repeat Example 7.7, using a transformer with 75 kVA capacity.

7.3 Repeat Example 7.7, assuming four services per transformer. Here, omit the UG SL. Assume that there are six transformers per block, that is, one transformer at each pole location.

7.4 Repeat Problem 7.3, using a 75 kVA transformer.

7.5 Repeat Example 7.8, using a 100 kVA transformer and #3/0 AWG and #2 AWG cables for the SLs and SDs, respectively.

7.6 Repeat Example 7.10. Use the nominal primary voltage of 19,920/34,500 V and assume that the remaining data are the same.

7.7 Assume that a three-conductor dc overhead line with equal conductor sizes (see Figure P7.7) is considered to be employed to transmit three-phase three-conductor ac energy at 0.92 power factor.

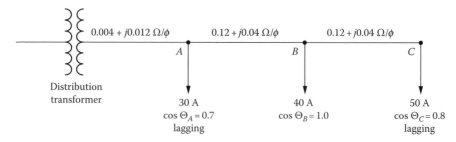

FIGURE P7.1 One-line diagram for Problem 7.1.

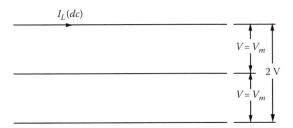

FIGURE P7.7 Illustration for Example 7.7.

If voltages to ground and transmission line efficiencies are the same for both direct and alternating currents, and the load is balanced, determine the change in the power transmitted in percent.

7.8 Assume that a single-phase feeder circuit has a total impedance of $1 + j3\ \Omega$ for lines and/or transformers. The receiving-end voltage and load current are $2400\angle0°$ V and $50\ \angle{-}30°$ A, respectively. Determine the following:
a. The power factor of the load.
b. The load power factor for which the voltage drop is maximum, using Equation 7.51.
c. Repeat part (b), using Equation 7.52.

7.9 An unbalanced three-phase wye-connected and grounded load is connected to a balanced three-phase four-wire source. The load impedance Z_a, Z_b, and Z_c are given as $70\ \angle30°$, $85\ \angle{-}40°$, and $50\ \angle35°\ \Omega$/phase, respectively, and the phase a line voltage has an effective value of 13.8 kV: Use the line-to-neutral voltage of phase a as the reference and determine the following:
a. The line and neutral currents
b. The total power delivered to the loads

7.10 Consider Figure P7.1 and assume that the impedances of the three line segments from left to right are $0.1 + j0.3$, $0.1 + j0.1$, and $0.08 + j0.12\ \Omega$/phase, respectively. Also assume that this three-phase three-wire 480-V secondary system supplies balanced loads at A, B, and C. The loads at A, B, and C are represented by 50 A with a lagging power factor of 0.85, 30 A with a lagging power factor of 0.90, and 50 A with a lagging power factor of 0.95, respectively. Determine the following:
a. The total voltage drop in one phase of the lateral using the approximate method
b. The real power per phase for each load
c. The reactive power per phase for each load
d. The kilovoltampere output and load power factor of the distribution transformer

7.11 *Assume* that bulk power substation 1 supplies substations 2 and 3, as shown in Figure P7.11, through three-phase lines. Substations 2 and 3 are connected to each other over a tie line, as shown. Assume that the line-to-line voltage is 69 kV and determine the following:
a. The voltage difference between substations 2 and 3 when tie line 23 is open-circuited
b. The line currents when all three lines are connected as shown in the figure

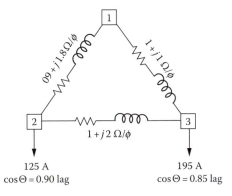

FIGURE P7.11 Distribution system for Problem 7.11.

c. The total power loss in part (a)

d. The total power loss in part (b).

7.12 Repeat Example 7.6, assuming 50% lagging power factor for all loads.

7.13 Resolve Example 7.4 by using MATLAB and assuming four services per transformer. Here, omit the UG SL. Assume that there are six transformers per block, that is, one transformer at each pole location.

REFERENCES

1. Morrison, C.: A Linear approach to the problem of planning new feed points into a distribution system, *AIEE Trans.*, 82, pt. III (PAS) December 1963, 819–832.

2. Westinghouse Electric Corporation: *Electric Utility Engineering Reference Book-Distribution Systems*, Vol. 3, Westinghouse Electric Corporation, East Pittsburgh, PA, 1965.

3. Fink, D. G. and H. W. Beaty: *Standard Handbook for Electrical Engineers*. 11th edn., McGraw-Hill, New York, 1978.

4. Gönen, T. et al.: *Development of Advanced Methods for Planning Electric Energy Distribution Systems*, US Department of Energy, October 1979. Available from the National Technical Information Service, US Department of Commerce, Springfield, VA.

5. Gönen, T. and D. C. Yu: A comparative analysis of distribution feeder costs, *Southwest Electrical Exposition and IEEE Conference Proceedings*, Houston, TX, January 22–24, 1980.

6. Rural Electrification Administration: *U.S. Department of Agriculture: Economic Design of Primary Lines for Rural Distribution Systems*, REA Bulletin, May, 1960, pp. 60–69.

7. Gönen, T. and D. C. Yu: A distribution system planning model, *Control of Power Systems Conference Proceedings*, Oklahoma City, OK, March 17–18, 1980, pp. 28–34.

8. Schlegel, M. C.: New selection method reduces conductor losses, *Electr. World*, February 1, 1977, 43–44.

9. Light, J.: An economic approach to distribution conductor size selection, paper presented at the *Missouri Valley Electric Association 49th Annual Engineering Conference*, Kansas City, MO, April 12–14, 1978.

10. Dodds, T. H.: Costs of losses can economically justify replacement of an old transformer with a new one, *The Line*, 80, 2, July 1980, 25–28.

11. Smith, R. W. and D. J. Ward: Does early distribution transformer retirement make sense? *Electric. Forum*, 6, 3, 1980, 6–9.

12. Delaney, M. B.: Economic analysis of electrical equipment losses, *The Line*, 74, 4, 1974, 7–8.

13. Klein, K. W.: Evaluation of distribution transformer losses and loss ratios, *Elec. Light Power*, July 15, 1960, 56–61.

14. Jeynes, P. H.: Evaluation of capacity differences in the economic comparison of alternative facilities, *AIEE Trans.*, pt. III (PAS), January 1952, 62–80.

8 Application of Capacitors to Distribution Systems

Who neglects learning in his youth, loses the past and is dead for the future.

Euripides, 438 BC

Where is there dignity unless there is honesty?

Cicero

8.1 BASIC DEFINITIONS

Capacitor element: an indivisible part of a capacitor consisting of electrodes separated by a dielectric material

Capacitor unit: an assembly of one or more capacitor elements in a single container with terminals brought out

Capacitor segment: a single-phase group of capacitor units with protection and control system

Capacitor module: a three-phase group of capacitor segments

Capacitor bank: a total assembly of capacitor modules electrically connected to each other

8.2 POWER CAPACITORS

At a casual look, a capacitor seems to be a very simple and unsophisticated apparatus, that is, two metal plates separated by a dielectric insulating material. It has no moving parts but instead functions by being acted upon by electric stress.

In reality, however, a power capacitor is a highly technical and complex device in that very thin dielectric materials and high electric stresses are involved, coupled with highly sophisticated processing techniques. Figure 8.1 shows a cutaway view of a power factor correction capacitor. Figure 8.2 shows a typical capacitor utilization in a switched pole-top rack.

In the past, most power capacitors were constructed with two sheets of pure aluminum foil separated by three or more layers of chemically impregnated kraft paper. Power capacitors have been improved tremendously over the last 30 years or so, partly due to improvements in the dielectric materials and their more efficient utilization and partly due to improvements in the processing techniques involved. Capacitor sizes have increased from the 15–25 kvar range to the 200–300 kvar range (capacitor banks are usually supplied in sizes ranging from 300 to 1800 kvar).

Nowadays, power capacitors are much more efficient than those of 30 years ago and are available to the electric utilities at a much lower cost per kilovar. In general, capacitors are getting more attention today than ever before, partly due to a new dimension added in the analysis: changeout economics. Under certain circumstances, even replacement of older capacitors can be justified on the basis of lower-loss evaluations of the modern capacitor design.

Capacitor technology has evolved to extremely low-loss designs employing the all-film concept; as a result, the utilities can make economic loss evaluations in choosing between the presently existing capacitor technologies.

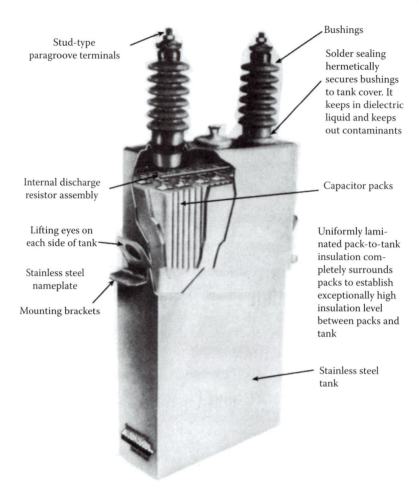

Stud-type
paragroove terminals

Bushings

Solder sealing
hermetically
secures bushings
to tank cover. It
keeps in dielectric
liquid and keeps
out contaminants

Internal discharge
resistor assembly

Capacitor packs

Lifting eyes on
each side of tank

Uniformly lami-
nated pack-to-tank
insulation com-
pletely surrounds
packs to establish
exceptionally high
insulation level
between packs and
tank

Stainless steel
nameplate

Mounting brackets

Stainless steel
tank

FIGURE 8.1 A cutaway view of a power factor correction capacitor. (From McGraw-Edison Company, *The ABC of Capacitors*, Bulletin R230-90-1, 1968.)

FIGURE 8.2   A typical utilization in a switched pole-top rack.

8.3 EFFECTS OF SERIES AND SHUNT CAPACITORS

As mentioned earlier, the fundamental function of capacitors, whether they are series or shunt, installed as a single unit or as a bank, is to regulate the voltage and reactive power flows at the point where they are installed. The shunt capacitor does it by changing the power factor of the load, whereas the series capacitor does it by directly offsetting the inductive reactance of the circuit to which it is applied.

8.3.1 SERIES CAPACITORS

Series capacitors, that is, *capacitors connected in series with lines*, have been used to a very limited extent on distribution circuits due to being a more specialized type of apparatus with a limited range of application. Also, because of the special problems associated with each application, there is a requirement for a large amount of complex engineering investigation. Therefore, in general, utilities are reluctant to install series capacitors, especially of small sizes.

As shown in Figure 8.3, a series capacitor compensates for inductive reactance. In other words, a series capacitor is a negative (capacitive) reactance in series with the circuit's positive (inductive) reactance with the effect of compensating for part or all of it. Therefore, the primary effect of the series capacitor is to minimize, or even suppress, the voltage drop caused by the inductive reactance in the circuit.

At times, a series capacitor can even be considered as a voltage regulator that provides for a voltage boost that is proportional to the magnitude and power factor of the through current. Therefore, a series capacitor provides for a voltage rise that increases automatically and instantaneously as the load grows.

Also, a series capacitor produces more net voltage rise than a shunt capacitor at lower power factors, which creates more voltage drop. However, a series capacitor betters the system power factor much less than a shunt capacitor and has little effect on the source current.

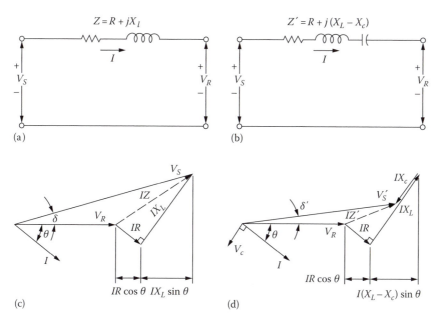

FIGURE 8.3 Voltage phasor diagrams for a feeder circuit of lagging power factor: (a) and (c) without and (b) and (d) with series capacitors.

Consider the feeder circuit and its voltage phasor diagram as shown in Figure 8.3a and c. The voltage drop through the feeder can be expressed approximately as

$$\text{VD} = IR\cos\theta + IX_L\sin\theta \tag{8.1}$$

where
  R is the resistance of the feeder circuit
  X_L is the inductive reactance of the feeder circuit
  $\cos\theta$ is the receiving-end power factor
  $\sin\theta$ is the sine of the receiving-end power factor angle

As can be observed from the phasor diagram, the magnitude of the second term in Equation 8.1 is much larger than the first. The difference gets to be much larger when the power factor is smaller and the ratio of R/X_L is small.

However, when a series capacitor is applied, as shown in Figure 8.3b and d, the resultant lower voltage drop can be calculated as

$$\text{VD} = IR\cos\theta + I(X_L - X_c)\sin\theta \tag{8.2}$$

where X_c is the capacitive reactance of the series capacitor.

8.3.1.1 Overcompensation

Usually, the series-capacitor size is selected for a distribution feeder application in such a way that the resultant capacitive reactance is smaller than the inductive reactance of the feeder circuit. However, in certain applications (where the resistance of the feeder circuit is larger than its inductive reactance), the reverse might be preferred so that the resultant voltage drop is

$$\text{VD} = IR\cos\theta - I(X_c - X_L)\sin\theta \tag{8.3}$$

The resultant condition is known as *overcompensation*. Figure 8.4a shows a voltage phasor diagram for overcompensation at normal load. At times, when the selected level of overcompensation is strictly based on normal load, the resultant overcompensation of the receiving-end voltage may not be pleasing at all because the lagging current of a large motor at start can produce an extraordinarily large voltage rise, as shown in Figure 8.4b, which is especially harmful to lights (shortening their lives) and causes light flicker, resulting in consumers' complaints.

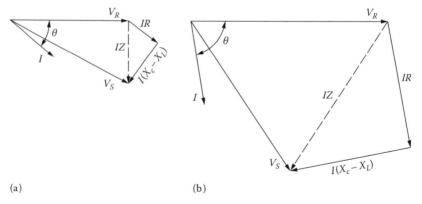

(a)   (b)

FIGURE 8.4 Overcompensation of the receiving-end voltage: (a) at normal load and (b) at the start of a large motor.

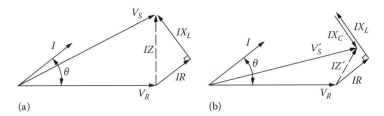

FIGURE 8.5 Voltage phasor diagram with leading power factor: (a) without series capacitors and (b) with series capacitors.

8.3.1.2 Leading Power Factor

To decrease the voltage drop considerably between the sending and receiving ends by the application of a series capacitor, the load current must have a lagging power factor. As an example, Figure 8.5a shows a voltage phasor diagram with a leading-load power factor without having series capacitors in the line. Figure 8.5b shows the resultant voltage phasor diagram with the same leading-load power factor but this time with series capacitors in the line. As can be seen from the figure, the receiving-end voltage is reduced as a result of having series capacitors.

When $\cos \theta = 1.0$, $\sin \theta \cong 0$, and therefore,

$$I(X_L - X_c)\sin\theta \cong 0$$

hence, Equation 8.2 becomes

$$VD \cong IR \tag{8.4}$$

Thus, in such applications, series capacitors practically have no value.

Because of the aforementioned reasons and others (e.g., ferroresonance in transformers, subsynchronous resonance during motor starting, shunting of motors during normal operation, and difficulty in protection of capacitors from system fault current), series capacitors do not have large applications in distribution systems.

However, they are employed in subtransmission systems to modify the load division between parallel lines. For example, often a new subtransmission line with larger thermal capability is parallel with an already existing line. It may be very difficult, if not impossible, to load the subtransmission line without overloading the old line. Here, series capacitors can be employed to offset some of the line reactance with greater thermal capability. They are also employed in subtransmission systems to decrease the voltage regulation.

8.3.2 Shunt Capacitors

Shunt capacitors, that is, *capacitors connected in parallel with lines*, are used extensively in distribution systems. Shunt capacitors supply the type of reactive power or current to counteract the out-of-phase component of current required by an inductive load. In a sense, shunt capacitors modify the characteristic of an inductive load by drawing a leading current that counteracts some or all of the lagging component of the inductive load current at the point of installation. Therefore, a shunt capacitor has the same effect as an overexcited synchronous condenser, generator, or motor.

As shown in Figure 8.6, by the application of shunt capacitor to a feeder, the magnitude of the source current can be reduced, the power factor can be improved, and consequently the voltage drop between the sending end and the load is also reduced. However, shunt capacitors do not affect current or power factor beyond their point of application. Figure 8.6a and c shows the single-line diagram of a line and its voltage phasor diagram before the addition of the shunt capacitor, and Figure 8.6b and d shows them after the addition.

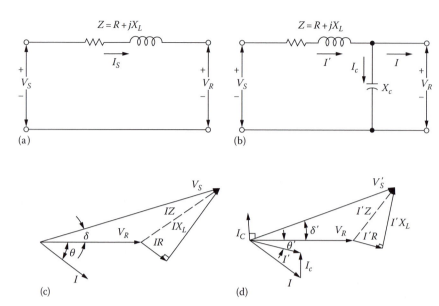

FIGURE 8.6 Voltage phasor diagrams for a feeder circuit of lagging power factor: (a) and (c) without and (b) and (d) with shunt capacitors.

Voltage drop in feeders, or in short transmission lines, with lagging power factor can be approximated as

$$\text{VD} = I_R R + I_X X_L \tag{8.5}$$

where
R is the total resistance of the feeder circuit, Ω
X_L is the total inductive reactance of the feeder circuit, Ω
I_R is the real power (or in-phase) component of the current, A
I_X is the reactive (or out-of-phase) component of the current lagging the voltage by 90°, A

Example 8.1

Consider the right-angle triangle shown in Figure 8.7b. Determine the power factor of the load on a 460 V three-phase system, if the ammeter reads 100 A and the wattmeter reads 70 kW.

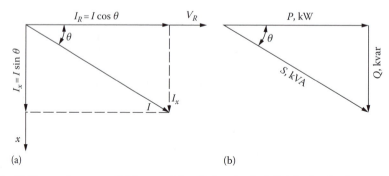

FIGURE 8.7 (a) Phasor diagram and (b) power triangle for a typical distribution load.

Solution

$$S = \frac{\sqrt{3}(V)(I)}{1000}$$

$$= \frac{\sqrt{3}(460 \text{ V})(100 \text{ A})}{1000}$$

$$\cong 79.67 \text{ kVA}$$

Thus,

$$PF = \cos\theta = \frac{P}{S}$$

$$= \frac{70 \text{ kW}}{79.67 \text{ kVA}}$$

$$\cong 0.88 \quad \text{or} \quad 88\%$$

When a capacitor is installed at the receiving end of the line, as shown in Figure 8.6b, the resultant voltage drop can be calculated approximately as

$$VD = I_R R_R + I_X X_L - I_c X_L \tag{8.6}$$

where I_c is the reactive (or out-of-phase) component of current leading the voltage by 90°, A.

The difference between the voltage drops calculated by using Equations 8.5 and 8.6 is the voltage rise due to the installation of the capacitor and can be expressed as

$$VR = I_c X_L \tag{8.7}$$

8.4 POWER FACTOR CORRECTION

8.4.1 GENERAL

A typical utility system would have a reactive load at 80% power factor during the summer months. Therefore, in typical distribution loads, the current lags the voltage, as shown in Figure 8.7a. The cosine of the angle between current and sending voltage is known as the *power factor* of the circuit. If the in-phase and out-of-phase components of the current I are multiplied by the receiving-end voltage V_R, the resultant relationship can be shown on a triangle known as the *power triangle*, as shown in Figure 8.7b. Figure 8.7b shows the triangular relationship that exists between kilowatts, kilovoltamperes, and kilovars.

Note that, by adding the capacitors, the reactive power component Q of the apparent power S of the load can be reduced or totally suppressed. Figures 8.8a and 8.9 illustrate how the reactive power component Q increases with each 10% change of power factor. Figure 8.8a also illustrates how a portion of lagging reactive power Q_{old} is cancelled by the leading reactive power of capacitor Q_c.

Note that, as illustrated in Figure 8.8, even an 80% power factor of the reactive power (kilovar) size is quite large, causing a 25% increase in the total apparent power (kilovoltamperes) of the line. At this power factor, 75 kvar of capacitors is needed to cancel out the 75 kvar of the lagging component.

As previously mentioned, the generation of reactive power at a power plant and its supply to a load located at a far distance is not economically feasible, but it can easily be provided by capacitors (or overexcited synchronous motors) located at the load centers. Figure 8.10 illustrates the power factor correction for a given system. As illustrated in the figure, capacitors draw leading reactive

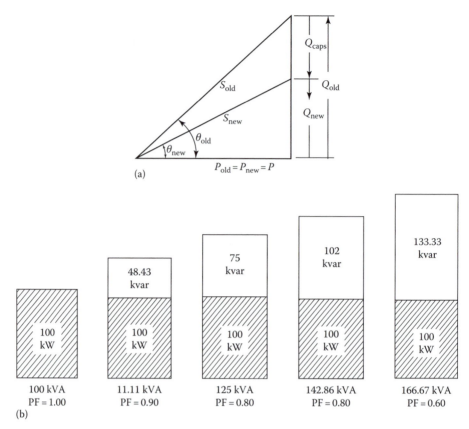

FIGURE 8.8 Illustration of (a) the use of a power triangle for power factor correction by employing capacitive reactive power and (b) the required increase in the apparent and reactive powers as a function of the load power factor, holding the real power of the load constant.

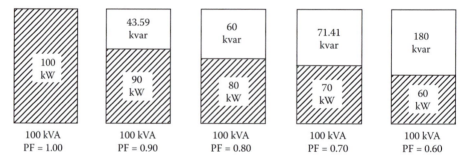

FIGURE 8.9 Illustration of the change in the real and reactive powers as a function of the load power factor, holding the apparent power of the load constant.

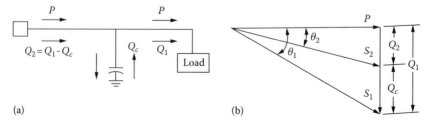

FIGURE 8.10 Illustration of power factor correction.

power from the source; that is, they supply lagging reactive power to the load. Assume that a load is supplied with a real power P, lagging reactive power Q_1, and apparent power S_1 at a lagging power factor of

$$\cos\theta_1 = \frac{P}{S_1}$$

or

$$\cos\theta_1 = \frac{P}{\left(P^2 + Q_1^2\right)^{1/2}} \tag{8.8}$$

When a shunt capacitor of Q_c kVA is installed at the load, the power factor can be improved from $\cos\theta_1$ to $\cos\theta_2$, where

$$\cos\theta_2 = \frac{P}{S_2}$$

$$= \frac{P}{\left(P^2 + Q_2^2\right)^{1/2}}$$

or

$$\cos\theta_2 = \frac{P}{\left[P^2 + (Q_1 - Q_c)^2\right]^{1/2}} \tag{8.9}$$

8.4.2 CONCEPT OF LEADING AND LAGGING POWER FACTORS

Many consider that the terms "lagging" and "leading" power factor are somewhat confusing, and they are meaningless, if the directions of the flows of real and reactive powers are not known. In general, for a given load, the power factor is *lagging* if the load withdraws reactive power; on the other hand, it is *leading* if the load supplies reactive power.

Hence, an induction motor has a lagging power factor since it withdraws reactive power from the source to meet its magnetizing requirements. But a capacitor (or an overexcited synchronous motor) supplies reactive power and thus has a leading power factor, as shown in Figure 8.11 and indicated in Table 8.1.

On the other hand, an underexcited synchronous motor withdraws both the real and reactive power from the source, as indicated. The use of varmeters instead of power factor meters avoids the confusion about the terms "lagging" and "leading." Such a varmeter has a zero center point with scales on either side, one of them labeled "in" and the other one "out."

8.4.3 ECONOMIC POWER FACTOR

As can be observed from Figure 8.10b, the apparent power and the reactive power are decreased from S_1 to S_2 kVA and from Q_1 to Q_2 kvar (by providing a reactive power of Q), respectively. The reduction of reactive current results in a reduced total current, which in turn causes less power losses.

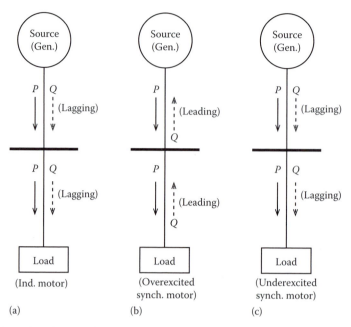

FIGURE 8.11 Examples of some of the sources of leading and lagging reactive power at the load.

TABLE 8.1
Power Factor of Load and Source

Load Type	At Load			At Generator		
	P	**Q**	**Power Factor[a]**	**P**	**Q**	**Power Factor[b]**
Induction motor	In	Out	Lagging			
Induction generator				Out	Out	Lagging
Synchronous motor (Underexcited)	In	In	Lagging	Out	Out	Lagging
Synchronous motor (Overexcited)	In	Out	Leading	Out	In	Leading

[a] Power factor measured at the load.
[b] Power factor measured at the generator.

Thus, the power factor correction produces economic savings in capital expenditures and fuel expenses through a release of kilovoltamperage capacity and reduction of power losses in all the apparatus between the point of installation of the capacitors and the power plant source, including distribution lines, substation transformers, and transmission lines.

The economic power factor is the point at which the economic benefits of adding shunt capacitors just equal the cost of the capacitors. In the past, this economic power factor was around 95%. Today's high plant and fuel costs have pushed the economic power factor toward unity.

However, as the corrected power factor moves nearer to unity, the effectiveness of capacitors in improving the power factor, decreasing the line kilovoltamperes transmitted, increasing the load capacity, or reducing line copper losses by decreasing the line current sharply decreases. Therefore, the correction of power factor to unity becomes more expensive with regard to the marginal cost of capacitors installed.

8.4.4 Use of a Power Factor Correction Table

Table 8.2 is a power factor correction table to simplify the calculations involved in determining the capacitor size necessary to improve the power factor of a given load from original to desired value. It gives a multiplier to determine the kvar requirement. It is based on the following formula:

$$Q = P(\tan \theta_{\text{orig}} - \tan \theta_{\text{new}})$$ (8.10)

or

$$Q = P\left(\sqrt{\frac{1}{PF_{\text{orig}}^2} - 1} - \sqrt{\frac{1}{PF_{\text{new}}^2} - 1} \right)$$ (8.11)

where
 Q is the required compensation in kvar
 P is the real power kW
 PF_{orig} is the original power factor
 PF_{new} is the desired power factor

8.4.5 Alternating Cycles of a Magnetic Field

Furthermore, in order to understand how the power factor of a device can be improved, one has to understand what is taking place electrically. Consider an induction motor that is being supplied by the real power P and the reactive power Q. The real power P is lost, whereas the reactive power Q is not lost. But instead it is used to store energy in the magnetic field of the motor.

Since the current is alternating, the magnetic field undergoes cycles of building up and breaking down. As the field is building up, the reactive current flows from the supply or source to the motor. As the field is breaking down, the reactive current flows out of the motor back to the supply or source. In such application, what is needed is some type of device that can be used as a temporary storage area for the reactive power when the magnetic field of the motor breaks down.

The ideal device for this is a *capacitor* that also stores energy. However, this energy is stored in an electric field. By connecting a capacitor *in parallel with the supply line of the load*, the cyclic flow of reactive power takes place between the motor and the capacitor. Here, the supply lines carry only the current supplying real power to the motor. *This is only applicable for a unity power factor condition.* For other power factors, the supply lines would carry some reactive power.

8.4.6 Power Factor of a Group of Loads

In general, the power factor of a single load is known. However, it is often that the power factor of a group of various loads needs to be determined. This is accomplished based on the known power relationship.

Example 8.2

Assume that a substation supplies three different kinds of loads, mainly, incandescent lights, induction motors, and synchronous motors, as shown in Figure 8.12. The substation power factor is found from the total reactive and real powers of the various loads that are connected. Based on the given data in Figure 8.12, determine the following:

 a. The apparent, real, and kvars of each load
 b. The total apparent, real, and reactive powers of the power that should be supplied by the substation
 c. The total power factor of the substation

TABLE 8.2
Determination of kW Multiplies to Calculate kvar Requirement for Power Factor Correction

Reactive Factor	Original Power Factor (%)	Correcting Factor / Desired Power Factor (%)																				
		80	81	82	83	84	85	86	87	88	89	90	91	92	93	94	95	96	97	98	99	100
0.800	60	0.584	0.610	0.636	0.662	0.688	0.714	0.741	0.767	0.794	0.822	0.850	0.878	0.905	0.939	0.971	1.005	1.043	1.083	1.130	1.192	1.334
0.791	61	0.549	0.575	0.601	0.627	0.653	0.679	0.706	0.732	0.759	0.787	0.815	0.843	0.870	0.904	0.936	0.970	1.008	1.048	1.096	1.157	1.299
0.785	62	0.515	0.541	0.567	0.593	0.619	0.645	0.672	0.698	0.725	0.753	0.781	0.809	0.836	0.870	0.902	0.936	0.974	1.014	1.062	1.123	1.265
0.776	63	0.483	0.509	0.535	0.561	0.587	0.613	0.640	0.666	0.693	0.721	0.749	0.777	0.804	0.838	0.870	0.904	0.942	0.982	1.030	1.091	1.233
0.768	64	0.450	0.476	0.502	0.528	0.554	0.580	0.607	0.633	0.660	0.688	0.716	0.744	0.771	0.805	0.837	0.871	0.909	0.949	0.997	1.058	1.200
0.759	65	0.419	0.445	0.471	0.497	0.523	0.549	0.576	0.602	0.629	0.657	0.685	0.713	0.740	0.774	0.806	0.840	0.878	0.918	0.966	1.027	1.169
0.751	66	0.388	0.414	0.440	0.466	0.492	0.518	0.545	0.571	0.598	0.626	0.654	0.682	0.709	0.743	0.775	0.809	0.847	0.887	0.935	0.996	1.138
0.744	67	0.358	0.384	0.410	0.436	0.462	0.488	0.515	0.541	0.568	0.596	0.624	0.652	0.679	0.713	0.745	0.779	0.817	0.857	0.905	0.966	1.108
0.733	68	0.329	0.355	0.381	0.407	0.433	0.459	0.486	0.512	0.539	0.567	0.595	0.623	0.650	0.684	0.716	0.750	0.788	0.828	0.876	0.937	1.079
0.725	69	0.299	0.325	0.351	0.377	0.403	0.429	0.456	0.482	0.509	0.537	0.565	0.593	0.620	0.654	0.686	0.720	0.758	0.798	0.840	0.907	1.049
0.714	70	0.270	0.296	0.322	0.348	0.374	0.400	0.427	0.453	0.480	0.508	0.536	0.564	0.591	0.625	0.657	0.691	0.729	0.769	0.811	0.878	1.020
0.704	71	0.242	0.268	0.294	0.320	0.346	0.372	0.399	0.425	0.452	0.480	0.508	0.536	0.563	0.597	0.629	0.663	0.700	0.741	0.783	0.850	0.992
0.694	72	0.213	0.239	0.265	0.291	0.317	0.343	0.370	0.396	0.423	0.451	0.479	0.507	0.534	0.568	0.600	0.634	0.672	0.712	0.754	0.821	0.963
0.682	73	0.186	0.212	0.238	0.264	0.290	0.316	0.343	0.369	0.396	0.424	0.452	0.480	0.507	0.541	0.573	0.607	0.645	0.685	0.727	0.794	0.936
0.673	74	0.159	0.185	0.211	0.237	0.263	0.289	0.316	0.342	0.369	0.397	0.425	0.453	0.480	0.514	0.546	0.580	0.618	0.658	0.700	0.767	0.909
0.661	75	0.132	0.158	0.184	0.210	0.236	0.262	0.289	0.315	0.342	0.370	0.398	0.426	0.453	0.487	0.519	0.553	0.591	0.631	0.673	0.740	0.882
0.650	76	0.105	0.131	0.157	0.183	0.209	0.235	0.262	0.288	0.315	0.343	0.371	0.399	0.426	0.460	0.492	0.526	0.564	0.604	0.652	0.713	0.855
0.637	77	0.079	0.105	0.131	0.157	0.183	0.209	0.236	0.262	0.289	0.317	0.345	0.373	0.400	0.434	0.466	0.500	0.538	0.578	0.620	0.687	0.829
0.626	78	0.053	0.079	0.105	0.131	0.157	0.183	0.210	0.236	0.263	0.291	0.319	0.347	0.374	0.408	0.440	0.474	0.512	0.552	0.594	0.661	0.803

		80	81	82	83	84	85	86	87	88	89	90	91	92	93	94	95	96	97	98	99	
79	0.613	0.026	0.052	0.078	0.104	0.130	0.156	0.183	0.209	0.236	0.264	0.292	0.320	0.347	0.381	0.413	0.447	0.485	0.525	0.567	0.634	0.776
80	0.600	0.000	0.026	0.052	0.078	0.104	0.130	0.157	0.183	0.210	0.238	0.266	0.294	0.321	0.355	0.387	0.421	0.459	0.499	0.541	0.608	0.750
81	0.588		0.000	0.026	0.052	0.078	0.104	0.131	0.157	0.184	0.212	0.240	0.268	0.295	0.329	0.361	0.395	0.433	0.473	0.515	0.528	0.724
82	0.572			0.000	0.026	0.052	0.078	0.105	0.131	0.158	0.186	0.214	0.242	0.269	0.303	0.335	0.369	0.407	0.447	0.489	0.556	0.698
83	0.559				0.000	0.026	0.052	0.079	0.105	0.132	0.160	0.188	0.216	0.243	0.277	0.309	0.343	0.381	0.421	0.463	0.530	0.672
84	0.543					0.000	0.026	0.053	0.079	0.106	0.134	0.162	0.190	0.217	0.251	0.283	0.317	0.355	0.395	0.437	0.504	0.646
85	0.529						0.000	0.027	0.053	0.080	0.108	0.136	0.164	0.191	0.225	0.257	0.291	0.329	0.369	0.417	0.478	0.620
86	0.510							0.000	0.026	0.053	0.081	0.109	0.137	0.167	0.198	0.230	0.265	0.301	0.342	0.390	0.451	0.593
87	0.497								0.000	0.027	0.055	0.083	0.111	0.141	0.172	0.204	0.239	0.275	0.316	0.364	0.425	0.567
88	0.475									0.000	0.028	0.056	0.083	0.113	0.144	0.176	0.211	0.247	0.288	0.336	0.397	0.540
89	0.455										0.000	0.028	0.055	0.086	0.117	0.149	0.183	0.221	0.262	0.309	0.370	0.512
90	0.443											0.000	0.028	0.058	0.089	0.121	0.155	0.193	0.234	0.281	0.342	0.484
91	0.427												0.000	0.030	0.061	0.093	0.127	0.165	0.206	0.253	0.314	0.456
92	0.392													0.000	0.031	0.063	0.097	0.135	0.176	0.223	0.284	0.426
93	0.386														0.000	0.032	0.066	0.104	0.145	0.192	0.253	0.395
94	0.341															0.000	0.035	0.072	0.113	0.160	0.221	0.363
95	0.327																0.000	0.036	0.078	0.125	0.186	0.328
96	0.280																	0.000	0.041	0.089	0.150	0.292
97	0.242																		0.000	0.048	0.109	0.251
98	0.199																			0.000	0.061	0.203
99	0.137																				0.000	0.142

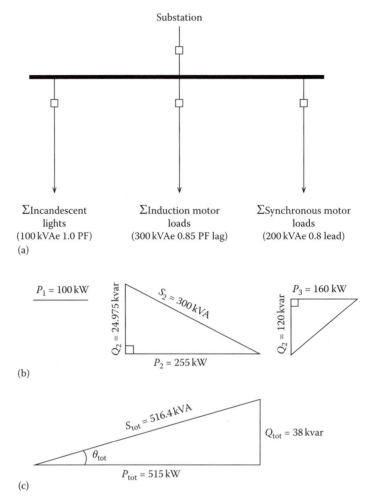

FIGURE 8.12 For Example 8.2: (a) connection diagram, (b) phasor diagrams of individual loads, and (c) phasor diagram of combined loads.

Solution

a.
1. *For a 100 kVA lighting load*

   Since incandescent lights are basically a unity power factor load, it is assumed that all the current is kilowatt current. Hence,

   $$S_1 = P_1$$

   $$100 \text{ kVA} \cong 100 \text{ kW}$$

2. *For 400 hp of connected induction motor loads*

   Assume that for the motor loads,

   $$\text{kVA load} = 0.75 \times (\text{Connected motor horse power})$$

   with an opening power factor of 85% lagging:

   $$S_2 = \left(0.75 \, \frac{\text{kW}}{\text{hp}}\right)(400 \text{ hp}) = 300 \text{ kVA}$$

$$P_3 = (0.75 \times 400) \times 0.85 = 255 \text{ kW}$$

$$Q_2 = \sqrt{(300)^2 - (255)^2}$$

$$= \sqrt{90,000 - 65,025}$$

$$= 24,975$$

$$\cong 158 \text{ kvar}$$

3. *200 hp motor with a 0.8 leading power factor*
   At full load, assume kVA = motor-hp rating = 200 kVA:

$$P_3 = (200 \text{ kVA}) \cos\theta$$

$$= 200 \times 0.8$$

$$= 160 \text{ kW}$$

$$Q_2 = \sqrt{(200 \text{ kVA})^2 - (160 \text{ kW})^2}$$

$$= \sqrt{40,000 - 25,600}$$

$$= \sqrt{14,400}$$

$$= 120 \text{ kvar}$$

b. At the substation, the total real power is

$$P_{\text{total}} = P_{\text{lights}} + P_{\text{ind.mot.}} + P_{\text{sync.mot.}}$$

$$= 100 + 255 + 160$$

$$= 515 \text{ kW}$$

The total reactive power is

$$Q_{\text{total}} = Q_{\text{lights}} + Q_{\text{ind. mot.}}$$

$$= 0 + 158$$

$$= 158 \text{ kvar}$$

Thus, an overexcited synchronous motor operating without the mechanical load connected to its shaft can supply the leading reactive power. Hence, the net lagging reactive power that must be supplied by the substation is the difference between the reactive power supplied by the synchronous motor and the reactive power required by the induction motor loads:

Induction motor load required = 158 kvar
Synchronous motor supplied = 120 kvar
Substation must supply = 38 kvar

c. The kVA of the substation is

$$S_{\text{total}} = \sqrt{P_{\text{tot}}^2 + Q_{\text{tot}}^2} \qquad (8.12)$$

or

$$S_{total} = \sqrt{515^2 + 38^2}$$

$$= \sqrt{266,669}$$

$$= 516.4\,\text{kVA}$$

The power factor of the substation is

$$PF = \text{power factor} = \frac{P}{S}$$

$$= \frac{515\,\text{kW}}{516.4\,\text{kVA}}$$

$$= 0.997\,\text{lagging}$$

8.4.7 Practical Methods Used by the Power Industry for Power Factor Improvement Calculations

It is often that the formulas that are used by the power industry contains kW, kVA, or kvar instead of the symbols of P, S, Q, which are the correct form and used in the academia. However, there are certain advantages of using them since one does not have to think which one is P, S, or Q.

From the right-triangle relationship, several simple and useful mathematical expressions may be written as

$$PF = \cos\theta = \frac{kW}{kVA} \tag{8.13}$$

$$\tan\theta = \frac{kvar}{kW} \tag{8.14}$$

$$\sin\theta = \frac{kvar}{kVA} \tag{8.15}$$

Since the kW component normally stays the same (the kVA and kvar components change with power factor), it is convenient to use Equation 8.11 involving the kW component. The relationship can be reexpressed as

$$kvar = kW \times \tan\theta \tag{8.16}$$

For instance, if it is necessary to determine the capacitor rating to improve the load's power factor, one would use the following relationships:

$$kvar\ at\ original\ PF = kW \times \tan\theta_1 \tag{8.17}$$

$$kvar\ at\ improved\ PF = kW \times \tan\theta_2 \tag{8.18}$$

Thus, the capacitor rating required to improve the power factor can be expressed as

$$ckvar^* = kW \times (\tan\theta_1 - \tan\theta_2) \tag{8.19}$$

or

$$\Delta \tan\theta = \tan\theta_1 - \tan\theta_2 \tag{8.20}$$

then

$$\text{ckvar*} = \text{kW} \times \Delta \tan\theta \tag{8.21}$$

Table 8.2 has a "kW multiplier" for determining the capacitor based on the previously mentioned expression. Also, note that the prefix "c" in ckvar is employed to designate the capacitor kvar in order to differentiate it from load kvar.

To find irrespective currents of kVA, kW, and kvar, use the following relationships:

$$\text{kVA} = \sqrt{(\text{kW})^2 + (\text{kvar})^2} \tag{8.22}$$

$$\text{kW} = \sqrt{(\text{kVA})^2 - (\text{kvar})^2} \tag{8.23}$$

$$\text{kvar} = \sqrt{(\text{kVA})^2 - (\text{kW})^2} \tag{8.24}$$

Example 8.3

Assume that a load withdraws 80 kW and 60 kvar at a 0.8 power factor. It is required that its power factor is to be improved from 80% to 90% by using capacitors. Determine the amount of the reactive power to be provided by using capacitors.

Solution

Without capacitors at PF = 0.8

$$\text{kW} = 80$$

$$\text{kvar} = 60$$

Thus, the kVA requirement of the load is

$$\text{kVA} = (80^2 + 60^2)^{1/2} = 100 \text{ kVA}$$

With capacitors at PF = 0.9

$$\text{kW} = 80$$

$$\text{kVA} \cong 88.9$$

$$\text{Line kvar} = (88.9^2 - 80^2)^{1/2} (7903 - 6400)^{1/2} = 38.7$$

Hence, the line supplies 38.7 kvar and the load needs 60 kvar, and the capacitor supplies the difference, or

$$\text{ckvar} = 60 - 38.7 = 21.3 \text{ kvar}$$

as it is illustrated in Figure 8.13.

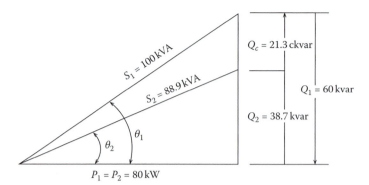

FIGURE 8.13 Illustration of power factor correction using a shunt capacitor in Example 8.3.

Example 8.4

Determine the capacitor rating in Example 8.3 by using Table 8.2.

Solution

From Table 8.2, the "kW multiplier" or $\Delta \tan \theta$ is read as 0.266. Therefore,

$$\text{ckvar} = \text{kW} \times \Delta \tan \theta$$

$$= (80 \text{ kW}) \times (0.266)$$

$$= 21.3 \text{ kvar}$$

which is the same value determined by the calculation in Example 8.3.

Example 8.5

Assume that a certain load withdraws a kilowatt current of 2 A and kilovar current of 2 A. Determine the amount of total current that it withdraws.

Solution

The answer is not the following!

$$(2 \text{ A}) + (2 \text{ A}) = 4 \text{ A}$$

The correct answer can be found from the following right-triangle relationships:

$$(\text{kvar current})^2 + (\text{kW current})^2 = (\text{total current})^2$$

$$(2 \text{ A})^2 + (2 \text{ A})^2 = (\text{total current})^2$$

or

$$4 + 4 = (\text{total current})^2$$

Hence,

$$\text{Total current} = 8^{1/2} = 2.83 \text{ A}$$

Thus,

$$2 + 2 \neq 4!$$

Figure 8.14 shows the component current diagram.

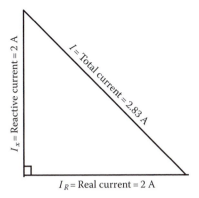

FIGURE 8.14 Component current diagram.

Example 8.6

Assume that a 460 V cable circuit is rated at 240 A but is carrying a load of 320 A at 0.65 power factor. Determine the kvar of capacitor that is needed to reduce the current to 240 A.

Solution

$$kVA = \frac{\sqrt{3} \times (460 \text{ V}) \times (320 \text{ A})}{1000}$$

$$= 254.96 \text{ kVA}$$

$$kW = (254.96 \text{ kVA})0.65$$

$$= 165.72 \text{ kW}$$

The kVA corresponding to 240 A is

$$kVA = \frac{\sqrt{3} \times (460 \text{ V})(240 \text{ A})}{1000}$$

$$= 191.2 \text{ kVA}$$

Thus, the operating power factor corresponding to the new load is

$$PF_2 = \cos\theta_2 = \frac{P}{S_2}$$

$$= \frac{165.72 \text{ kW}}{191.2 \text{ kVA}}$$

$$= 0.8667$$

The capacitor kvar required is

$$ckvar = (165.72 \text{ kW})\tan(\cos^{-1}0.8667)$$

$$= (165.72 \text{ kW})\tan(29.92°)$$

$$\cong (165.72)(0.5755)$$

$$\cong 95.38$$

8.4.8 Real Power-Limited Equipment

Certain equipments such as turbogenerator (i.e., turbine generators) and engine generator sets have a real power (P) limit of the prime mover as well as a kVA limit of the generator. Usually the real power limit corresponds to the generator S rating, and the set is rated at that P value at *unity power* factor operation.

Other real power (P) values that correspond to the lesser power factor operations are determined by the *power factor and real power (S) rating at the generator in order that the P and S ratings of the load do not exceed the S rating* of the generator. Any improvement of the power factor can release both P and S capacities.

Example 8.7

Assume that a 1000 kW turbine unit (turbogenerator set) has a turbine capability of 1250 kW. It is operating at a rated load of 1250 kVA at 0.85 power factor. An additional load of 150 kW at 0.85 power factor is to be added. Determine the value of capacitors needed in order not to overload the turbine nor the generator.

Solution

Original load

$$P = 1000 \text{ kW}$$

$$Q = \sqrt{(kVA)^2 - (kW)^2}$$

$$= \sqrt{(1250)^2 - (1000)^2}$$

$$= 750 \text{ kvar}$$

Additional load

$$P = kW = 150 \text{ kW}$$

$$S = kVA$$

$$= \frac{150 \text{ kW}}{0.85}$$

$$= 200 \text{ kVA}$$

$$Q = \sqrt{(200)^2 - (1000)^2}$$

$$= 132.29 \text{ kvar}$$

Total load

$$P_{tot} = kW = 1000 + 150$$

$$= 1150 \text{ kW}$$

$$Q_{tot} = 750 + 132.29$$

$$= 882.29 \text{ kvar}$$

The minimum operating power factor for a load of 1150 kW and not exceeding the kVA rating of the generator is

$$PF = \cos\theta = \frac{1150 \text{ kW}}{1250 \text{ kVA}}$$

$$= 0.92$$

The maximum load kvar for this situation is

$$Q = (1150\ \text{kW})\tan^{-1}\theta$$

$$= 1150 \times \tan^{-1}(23.073918°)$$

$$\cong 489.9\ \text{kvar}$$

where 0.426 is the tangent corresponding to the maximum power factor of 0.935.

Thus, the capacitors must provide the difference between the total load kvar and the permissible generator kvar, or

$$\text{ckvar} = 882.29 - 489.9$$

$$= 392.39\ \text{kvar}$$

Example 8.8

Assume that a 700 k VA load has a 65% power factor. It is desired to improve the power factor to 92%. Using Table 8.2, determine the following:

a. The correction factor required.
b. The capacitor size required.
c. What would be the resulting power factor if the next higher standard capacitor size is used?

Solution

a. From Table 8.2, the correction factor required can be found as 0.74.
b. The 700 kVA load at 65% power factor is

$$P_L = S_L \times \cos\theta$$

$$= 700 \times 0.65$$

$$= 455\ \text{kW} \tag{8.25}$$

The capacitor size necessary to improve the power factor from 65% to 92% can be found as

$$\text{Capacitor size} = P_L(\text{correlation factor})$$

$$= 455(0.74)$$

$$= 336.7\ \text{kvar} \tag{8.26}$$

c. Assume that the next higher standard capacitor size (or rating) is selected to be 360 kvar. Therefore, the resulting new correction factor can be found from

$$\text{New correction factor} = \frac{\text{Standard capacitor rating}}{P_L}$$

$$= \frac{360\ \text{kvar}}{455\ \text{kW}}$$

$$= 0.7912 \tag{8.27}$$

From the table by linear interpolation, the resulting corrected percent power factor, with an original power factor of 65% and a correction factor of 0.7912, can be found as

$$\text{New corrected \% power factor} = 93 + \frac{172}{320}$$

$$\cong 93.5$$

8.4.9 Computerized Method to Determine the Economic Power Factor

As suggested by Hopkinson [1], a load flow digital computer program can be employed to determine the kilovoltamperes, kilovolts, and kilovars at annual peak level for the whole system (from generation through the distribution substation buses) as the power factor is varied.

As a start, shunt capacitors are applied to each substation bus for correcting to an initial power factor, for example, 90%. Then, a load flow run is performed to determine the total system kilovoltamperes, and kilowatt losses (from generator to load) at this level and capacitor kilovars are noted. Later, additional capacitors are applied to each substation bus to increase the power factor by 1%, and another load flow run is made. This process of iteration is repeated until the power factor becomes unity.

As a final step, the benefits and costs are calculated at each power factor. The economic power factor is determined as the value at which benefits and costs are equal. After determining the economic power factor, the additional capacitor size required can be calculated as

$$\Delta Q_c = P_{PK}(\tan\phi - \tan\theta) \tag{8.28}$$

where
ΔQ_c is the required capacitor size, kvar
P_{PK} is the system demand at annual peak, kW
$\tan\phi$ is the tangent of original power factor angle
$\tan\theta$ is the tangent of economic power factor angle

An illustration of this method is given in Example 8.12.

8.5 APPLICATION OF CAPACITORS

In general, capacitors can be applied at almost any voltage level. As illustrated in Figure 8.15, individual capacitor units can be added in parallel to achieve the desired kilovar capacity and can be added in series to achieve the required kilovolt voltage. They are employed at or near rated voltage for economic reasons.

The cumulative data gathered for the whole utility industry indicate that approximately 60% of the capacitors is applied to the feeders, 30% to the substation buses, and the remaining 10% to the transmission system [1].

The application of capacitors to the secondary systems is very rare due to small economic advantages. Zimmerman [3] has developed a nomograph, shown in Figure 8.16, to determine the economic justification, if any, of the secondary capacitors considering only the savings in distribution transformer cost.

Example 8.9

Assume that a three-phase 500 hp 60 Hz 4160 V wye-connected induction motor has a full-load efficiency (η) of 88% and a lagging power factor of 0.75 and is connected to a feeder. If it is desired to correct the power factor of the load to a lagging power factor of 0.9 by connecting three capacitors at the load, determine the following:

a. The rating of the capacitor bank, in kilovars
b. The capacitance of each unit if the capacitors are connected in delta, in microfarads
c. The capacitance of each unit if the capacitors are connected in wye, in microfarads

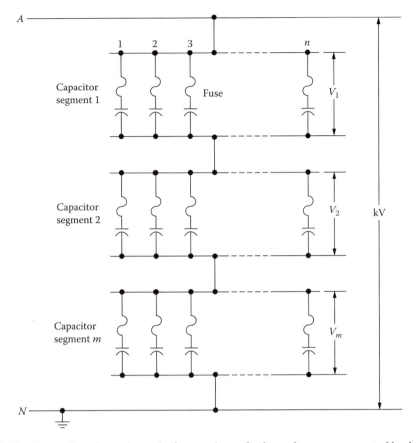

FIGURE 8.15 Connection of capacitor units for one phase of a three-phase wye-connected bank.

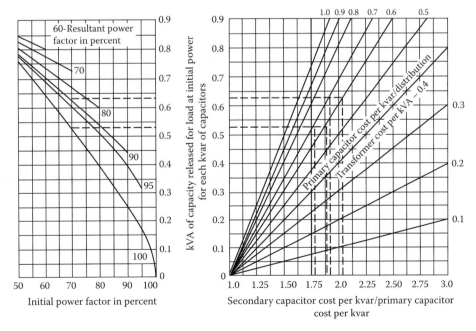

FIGURE 8.16 Secondary capacitor economics considering only savings in distribution transformer cost. (From Zimmerman, R.A., *AIEE Trans.*, 72, 694, Copyright 1953 IEEE. Used with permission.)

Solution

a. The input power of the induction motor can be found as

$$P = \frac{(HP)(0.7457 \text{ kW/hp})}{\eta}$$

$$= \frac{(500 \text{ hp})(0.7457 \text{ kW/hp})}{0.88}$$

$$= 423.69 \text{ kW}$$

The reactive power of the motor at the uncorrected power factor is

$$Q_1 = P \tan \theta_1$$

$$= 423.69 \tan(\cos^{-1} 0.75)$$

$$= 423.69 \times 0.8819$$

$$= 373.7 \text{ kvar}$$

The reactive power of the motor at the corrected power factor is

$$Q_2 = P \tan \theta_2$$

$$= 423.69 \tan (\cos^{-1} 0.90)$$

$$= 423.69 \times 0.4843$$

$$= 205.2 \text{ kvar}$$

Therefore, the reactive power provided by the capacitor bank is

$$Q_c = Q_1 - Q_2$$

$$= 373.7 - 205.2$$

$$= 168.5 \text{ kvar}$$

Hence, assuming the losses in the capacitors are negligible, the rating of the capacitor bank is 168.5 kvar.

b. If the capacitors are connected in delta as shown in Figure 8.17a, the line current is

$$I_L = \frac{Q_c}{\sqrt{3} \times V_{L-L}}$$

$$= \frac{168.5}{\sqrt{3} \times 4.16}$$

$$= 23.39 \text{ A}$$

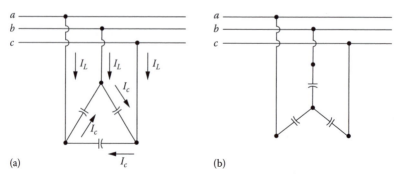

(a)   (b)

FIGURE 8.17 Capacitor connected (a) in delta and (b) in wye.

and therefore,

$$I_c = \frac{I_L}{\sqrt{3}}$$

$$= \frac{23.39}{\sqrt{3}}$$

$$= 13.5 \, A$$

Thus, the reactance of each capacitor is

$$X_c = \frac{V_{L-L}}{I_c}$$

$$= \frac{4160}{13.5}$$

$$= 308.11 \, \Omega$$

and hence, the capacitance of each unit,* if the capacitors are connected in delta, is

$$C = \frac{10^6}{\omega X_c} \, \mu F$$

or

$$C = \frac{10^6}{\omega X_c}$$

$$= \frac{10^6}{2\pi \times 60 \times 308.11}$$

$$= 8.61 \, \mu F$$

c. If the capacitors are connected in wye as shown in Figure 8.17b,

$$I_c = I_L = 23.39 \, A$$

and therefore,

$$X_c = \frac{V_{L-N}}{I_c}$$

$$= \frac{4160}{\sqrt{3} \times 23.39}$$

$$= 102.70 \, \Omega$$

Thus, the capacitance of each unit, if the capacitors are connected in wye, is

$$C = \frac{10^6}{\omega X_c}$$

$$= \frac{10^6}{2\pi \times 60 \times 102.70}$$

$$= 25.82 \, \mu F$$

---

* Note that $C = 1/(\omega X_c)$ F. If the equation is divided by 10^6 side by side, then $C = 10^6/(\omega X_c)(10^{-6} \, \text{F})$ or $C = 10^6/(\omega X_c) \, \mu\text{F}$:

$$C/10^6 = (1/\omega X_c)/10^6 \, \text{F}$$

Example 8.10

Assume that a 2.4 kV single-phase circuit feeds a load of 360 kW (measured by a wattmeter) at a lagging load factor and the load current is 200 A. If it is desired to improve the power factor, determine the following:

   a. The uncorrected power factor and reactive load.
   b. The new corrected power factor after installing a shunt capacitor unit with a rating of 300 kvar.
   c. Also write the necessary codes to solve the problem in MATLAB®.

Solution

   a. Before the power factor correction,

$$S_1 = V \times I$$

$$= 2.4 \times 200$$

$$= 480 \text{ kVA}$$

therefore, the uncorrected power factor can be found as

$$\cos \theta_1 = \frac{P}{S_1}$$

$$= \frac{360 \text{ kW}}{480 \text{ kVA}}$$

$$= 0.75$$

and the reactive load is

$$Q_1 = S_1 \times \sin (\cos^{-1} \theta_1)$$

$$= 480 \times 0.661$$

$$= 317.5 \text{ kvar}$$

   b. After the installation of the 300 kvar capacitors,

$$Q_2 = Q_1 - Q_c$$

$$= 317.5 - 300$$

$$= 17.5 \text{ kvar}$$

and therefore, the new power factor can be found from Equation 8.9 as

$$\cos \theta_2 = \frac{P}{[P^2 + (Q_1 - Q_c)^2]^{1/2}}$$

$$= \frac{360}{(360^2 + 17.5^2)^{1/2}}$$

$$= 0.9989 \quad \text{or} \quad 99.89\%$$

   c. Here is the MATLAB script:

~~~~~~

```
clc
clear

% System parameters
HP = 500;
PFold = 0.75;
PFnew = 0.90;
effi = 0.88;
kVL = 4.16;
VL = 4160;

% Solution for part a

% Solve for input power of motor in kW
P = (HP*0.7457)/effi

% Solve for reactive power of motor in kvar
Q1 = P*tan(acos(PFold))

% Reactive power of motor at corrected power factor in kvar
Q2 = P*tan(acos(PFnew))

% Reactive power provided by capacitor bank in kvar
Qc = Q1 - Q2

% Solution for part b when capacitors are in delta

% Line current when capacitors are in delta, in Amps
IL = Qc/(sqrt(3)*kVL)

% Capacitor current in Amps
Ic = IL/sqrt(3)

% Reactance of each capacitor in ohms
Xc = VL/Ic

% Capacitance of each unit in micro farads
C = (10^6)/(120*pi*Xc)

% Solution for part c when capacitors are in wye

% Capacitance current is equal to line current when capacitors are in wye

% Reactance of each capacitor in ohms
Xcwye = VL/(sqrt(3)*IL)

% Capacitance of each unit in micro farads
Cwye = (10^6)/(120*pi*Xcwye)
```

Example 8.11

Assume that the Riverside Substation of the NL&NP Company has a bank of three 2000 kVA transformers that supplies a peak load of 7800 kVA at a lagging power factor of 0.89. All three transformers have a thermal capability of 120% of the nameplate rating. It has already been planned to install 1000 kvar of shunt capacitors on the feeder to improve the voltage regulation.

Determine the following:

a. Whether or not to install additional capacitors on the feeder to decrease the load to the thermal capability of the transformer
b. The rating of the additional capacitors

Solution

a. Before the installation of the 1000 kvar capacitors,

$$P = S_1 \times \cos \theta$$

$$= 7800 \times 0.89$$

$$= 6942 \text{ kW}$$

and

$$Q_1 = S_1 \times \sin \theta$$

$$= 7800 \times 0.456$$

$$= 3556.8 \text{ kvar}$$

Therefore, after the installation of the 1000 kvar capacitors,

$$Q_2 = Q_1 - Q_c$$

$$= 3556.8 - 1000$$

$$= 2556.8 \text{ kvar}$$

and using Equation 8.9,

$$\cos \theta_2 = \frac{P}{[P^2 + (Q_1 - Q_c)^2]^{1/2}}$$

$$= \frac{6942}{(6942^2 + 2556.8^2)^{1/2}}$$

$$= 0.938 \quad \text{or} \quad 93.8\%$$

and the corrected apparent power is

$$S_2 = \frac{P}{\cos \theta_2}$$

$$= \frac{6942}{0.938}$$

$$= 7397.9 \text{ kVA}$$

On the other hand, the transformer capability is

$$S_T = 6000 \times 1.20$$

$$= 7200 \text{ kVA}$$

Therefore, the capacitors installed to improve the voltage regulation are not adequate; additional capacitor installation is required.

b. The new or corrected power factor required can be found as

$$PF_{2,new} = \cos\theta_{2,new} = \frac{P}{S_T}$$

$$= \frac{6942}{7200}$$

$$= 0.9642 \quad \text{or} \quad 96.42\%$$

and thus, the new required reactive power can be found as

$$Q_{2,new} = P \times \tan\theta_{2,new}$$

$$= P \times \tan(\cos^{-1}PF_{2,new})$$

$$= 6942 \times 0.2752$$

$$= 1910 \text{ kvar}$$

Therefore, the rating of the additional capacitors required is

$$Q_{c,add} = Q_2 - Q_{2,new}$$

$$= 2556.8 - 1910$$

$$= 646.7 \text{ kvar}$$

Example 8.12

If a power system has 10,000 kVA capacity and is operating at a power factor of 0.7 and the cost of a synchronous capacitor (i.e., synchronous condenser) to correct the power factor is $10 per kVA, find the investment required to correct the power factor to

    a. 0.85 lagging power factor
    b. Unity power factor

Solution

At original cost

$$\theta_{old} = \cos^{-1}PF = \cos^{-1}0.7 = 45.57°$$

$$P_{old} = S\cos\theta_{old} = (10,000 \text{ kVA})0.7 = 7,000 \text{ kW}$$

$$Q_{old} = S\sin\theta_{old} = (10,000 \text{ kVA})\sin45.57° = 7,141.43 \text{ kvar}$$

a. *For PF = 0.85 lagging*

$$P_{new} = P_{old} = 7000 \text{ kW} \quad (\text{as before})$$

$$S_{new} = \frac{P_{new}}{\cos\theta_{new}} = \frac{7000 \text{ kW}}{0.85} = 8235.29 \text{ kVA}$$

$$Q_{new} = S_{new}\sin(\cos^{-1}PF) = (8235.29 \text{ kVA})\sin(\cos^{-1}0.85) = 4338.21 \text{ kvar}$$

$$Q_c = Q_{required} = Q_{old} - Q_{new} = 7141.43 - 4338.21 = 2803.22 \text{ kvar correction needed}$$

Hence, the *theoretical* cost of the synchronous capacitor is

$$\text{Cost}_{\text{capacitor}} = (2,803.22 \text{ kVA})\left(\frac{\$10}{\text{kVA}}\right) = \$28,032.20$$

Note that it is customary to give the cost of capacitors in dollars per kVA rather than in dollars per kvar.

b. *For PF = 1.0*

$$Q_c = Q_{\text{required}} = Q_{\text{old}} - Q_{\text{new}} = 7141.43 - 0.0 = 7141.43 \text{ kvar}$$

Thus, the *theoretical* cost of the synchronous capacitor is

$$\text{Cost}_{\text{capacitor}} = (7,141.43 \text{ kVA})\left(\frac{\$10}{\text{kVA}}\right) = \$71,414.30$$

Note that $P_{\text{new}} = 7000$ kW is the same as before.

Example 8.13

If a power system has 15,000 kVA capacity, operating at a 0.65 lagging power factor, and the cost of synchronous capacitors to correct the power factor is \$12.5/kVA, determine the costs involved and also develop a table showing the required (*leading*) reactive power to increase the power factor to

a. 0.85 lagging power factor
b. 0.95 lagging power factor
c. Unity power factor

Solution
At original power factor or 0.65

$$P = S\cos\theta = (15,000 \text{ kVA})0.65 = 9750 \text{ kW at a power factor angle of } 49.46°$$

$$Q = S\sin\theta = (15,000 \text{ kVA})\sin(\cos^{-1} 0.65) = 11,399 \text{ kvar}$$

The following table shows the amount of reactive power that is required to improve the power factor from one level to the next at 0.05 increments.

PF	P (kW)	Q (kvar)	Q to Correct from Next Lower PF (kvar)	Cumulative Q Required for Correction (kvar)
0.65	9,750	11,399	—	—
0.70	10,500	10,712	687	687
0.75	11,250	9,922	790	1,477
0.80	12,000	9,000	922	2,399
0.85	12,750	7,902	1098	3,497
0.90	13,500	6,538	1364	4,861
0.95	14,250	4,684	1854	6,715
1.00	15,000	0	4684	11,399

a. *For PF = 0.85 lagging*
$P = S \cos \theta = (15,000 \text{ kVA}) \times 0.65 = 9,750 \text{ kW}$. It will be the same at a power factor of 0.85.

$$S = \frac{P}{\cos\theta} = \frac{9,750 \text{ kW}}{0.85} = 11,470 \text{ kVA}$$

and

$$Q = S \sin\theta = (11,470 \text{ kVA}) \sin(\cos^{-1}0.85) = 6,042 \text{ kvar}$$

The amount of additional reactive power correction required is

$$\text{Additional var correction} = 11,399 - 6,042 = 5,357 \text{ kvar}$$

The cost of this correction is

$$\text{Cost of correction} = (5,357 \text{ kVA})\left(\frac{\$12.5}{\text{kVA}}\right) = \$66,962.50$$

b. *For PF = 0.95 lagging*:

$$S = \frac{9,750 \text{ kW}}{0.95} = 10,263 \text{ kVA}$$

and

$$Q = (10,263 \text{ kVA}) \sin(\cos^{-1}0.95) = 3,204 \text{ kvar}$$

The amount of additional reactive power correction required is

$$\text{Additional var correction} = 11,399 - 3,204 = 8,195 \text{ kvar}$$

The cost of this correction is

$$\text{Cost of correction} = (8,195 \text{ kVA})\left(\frac{\$12.5}{\text{kVA}}\right) = \$102,438$$

c. *For unity PF*
The amount of additional reactive power correction required is

$$\text{Additional var correction} = 11,399 \text{ kvar}$$

The cost of this correction is

$$\text{Cost of correction} = (11,399 \text{ kVA})\left(\frac{\$12.5}{\text{kVA}}\right) = \$142,487.50$$

8.5.1 CAPACITOR INSTALLATION TYPES

In general, capacitors installed on feeders are pole-top banks with necessary group fusing. The fusing applications restrict the size of the bank that can be used. Therefore, the maximum sizes used are about 1800 kvar at 15 kV and 3600 kvar at higher voltage levels. Usually, utilities do not install more than four capacitor banks (of equal sizes) on each feeder.

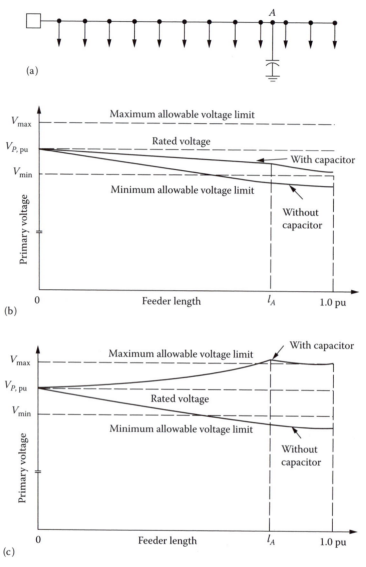

FIGURE 8.18 The effects of a fixed capacitor on the voltage profile of (a) feeder with uniformly distributed load (b) at heavy load and (c) at light load.

Figure 8.18 illustrates the effects of a fixed capacitor on the voltage profiles of a feeder with uniformly distributed load at heavy load and light load. If only fixed-type capacitors are installed, as can be observed in Figure 8.18c, the utility will experience an excessive leading power factor and voltage rise at that feeder. Therefore, as shown in Figure 8.19, some of the capacitors are installed as *switched capacitor banks* so they can be switched off *during light-load conditions*.

Thus, the *fixed capacitors* are sized for light load and connected permanently. As shown in the figure, the switched capacitors can be switched as a block or in several consecutive steps as the reactive load becomes greater from light-load level to peak load and sized accordingly.

However, in practice, the number of steps or blocks is selected to be much less than the ones shown in the figure due to the additional expenses involved in the installation of the required switchgear and control equipment.

A system survey is required in choosing the type of capacitor installation. As a result of load flow program runs or manual load studies on feeders or distribution substations, the system's

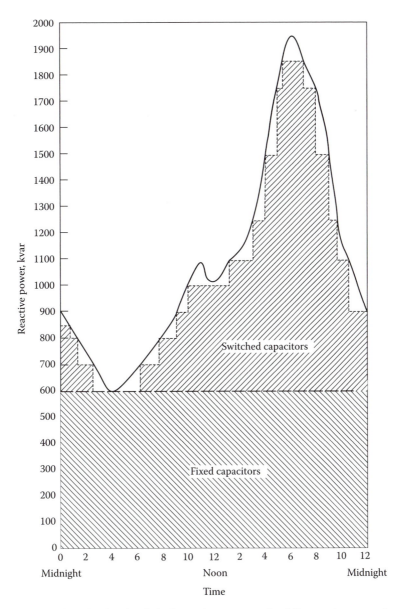

FIGURE 8.19 Sizing of the fixed and switched capacitors to meet the daily reactive power demands.

lagging reactive loads (i.e., power demands) can be determined and the results can be plotted on a curve as shown in Figure 8.19. This curve is called the *reactive load–duration curve* and is the cumulative sum of the reactive loads (e.g., fluorescent lights, household appliances, and motors) of consumers and the reactive power requirements of the system (e.g., transformers and regulators). Once the daily reactive load–duration curve is obtained, then by visual inspection of the curve, the size of the fixed capacitors can be determined to meet the minimum reactive load. For example, from Figure 8.19 one can determine that the size of the fixed capacitors required is 600 kvar.

The remaining kilovar demands of the loads are met by the generator or preferably by the switched capacitors. However, since meeting the kilovar demands of the system from the generator is too expensive and may create problems in the system stability, capacitors are used. Capacitor sizes are selected to match the remaining load characteristics from hour to hour.

Many utilities apply the following rule of thumb to determine the size of the switched capacitors: Add switched capacitors until

$$\frac{\text{kvar from switched} + \text{fixed capacitors}}{\text{kvar of peak reactive feeder load}} \geq 0.70 \tag{8.29}$$

From the voltage regulation point of view, the kilovars needed to raise the voltage at the end of the feeder to the maximum allowable voltage level at minimum load (25% of peak load) are the size of the fixed capacitors that should be used. On the other hand, if more than one capacitor bank is installed, the size of each capacitor bank at each location should have the same proportion, that is,

$$\frac{\text{kvar of load center}}{\text{kvar of total feeder}} = \frac{\text{kVA of load center}}{\text{kVA of total feeder}} \tag{8.30}$$

However, the resultant voltage rise must not exceed the light-load voltage drop. The approximate value of the percent voltage rise can be calculated from

$$\%\text{VR} = \frac{Q_{c,3\phi} \times x \times l}{10 \times V_{L-L}^2} \tag{8.31}$$

where
%VR is the percent voltage rise
$Q_{c,3\phi}$ is the three-phase reactive power due to fixed capacitors applied, kvar
x is the line reactance, Ω/min
l is the length of feeder from sending end of feeder to fixed capacitor location, min
V_{L-L} is the line-to-line voltage, kV

The percent voltage rise can also be found from

$$\%\text{VR} = \frac{I_c \times x \times l}{10 \times V_{L-L}} \tag{8.32}$$

where

$$I_c = \frac{Q_{c,3\phi}}{\sqrt{3} \times V_{L-L}}$$

$$= \text{current drawn by fixed-capacitor bank} \tag{8.33}$$

If the fixed capacitors are applied to the end of the feeder and if the percent voltage rise is already determined, the maximum value of the fixed capacitors can be determined from

$$\text{Max } Q_{c,3\phi} = \frac{10(\%\text{VR})V_{L-L}^2}{x \times l} \text{ kvar} \tag{8.34}$$

Equations 8.31 and 8.32 can also be used to calculate the percent voltage rise due to the switched capacitors. Therefore, once the percent voltage rises due to both fixed and switched capacitors, the total percent voltage rise can be calculated as

$$\sum \%\text{VR} = \%\text{VR}_{\text{NSW}} + \%\text{VR}_{\text{SW}} \tag{8.35}$$

where

$\sum$ %VR is the total percent voltage rise

%VR$_{NSW}$ is the percent voltage rise due to fixed (or nonswitched) capacitors

%VR$_{SW}$ is the percent voltage rise due to switched capacitors

Some utilities use the following rule of thumb: *The total amount of fixed and switched capacitors for a feeder is the amount necessary to raise the receiving-end feeder voltage to maximum at 50% of the peak feeder load.*

Once the kilovars of capacitors necessary for the system are determined, there remains only the question of proper location. *The rule of thumb for locating the fixed capacitors on feeders with uniformly distributed loads is to locate them approximately at two-thirds of the distance from the substation to the end of the feeder.*

For the uniformly decreasing loads, fixed capacitors are located approximately halfway out on the feeder. On the other hand, the location of the switched capacitors is basically determined by the voltage regulation requirements, and it usually turns out to be the last one-third of the feeder away from the source.

8.5.2 Types of Controls for Switched Shunt Capacitors

The switching process of capacitors can be done by manual control or by automatic control using some type of control intelligence. Manual control (at the location or as remote control) can be employed at distribution substations. The intelligence types that can be used in automatic control include time–switch, voltage, current, voltage–time, voltage–current, and temperature.

The most popular types are the time–switch control, voltage control, and voltage––current control. The time–switch control is the least-expensive one. Some combinations of these controls are also used to follow the reactive load–duration curve more closely, as illustrated in Figure 8.20.

8.5.3 Types of Three-Phase Capacitor-Bank Connections

A three-phase capacitor bank on a distribution feeder can be connected in (1) delta, (2) grounded wye, or (3) ungrounded wye. The type of connection used depends upon the following:

1. System type, that is, whether it is a grounded or an ungrounded system
2. Fusing requirements
3. Capacitor-bank location
4. Telephone interference considerations

A *resonance condition* may occur in delta and ungrounded-wye (floating neutral) banks when there is a one- or two-line open-type fault that occurs on the source side of the capacitor bank due to the maintained voltage on the open phase that backfeeds any transformers located on the load side of the open conductor through the series capacitor. As a result of this condition, the single-phase distribution transformers on four-wire systems may be damaged. Therefore, *ungrounded-wye capacitor banks are not recommended under the following conditions*:

1. On feeders with light load where the minimum load per phase beyond the capacitor bank does not exceed 150% of the per phase rating of the capacitor bank
2. On feeders with single-phase breaker operation at the sending end
3. On fixed capacitor banks
4. On feeder sections beyond a sectionalizing-fuse or single-phase recloser
5. On feeders with emergency load transfers

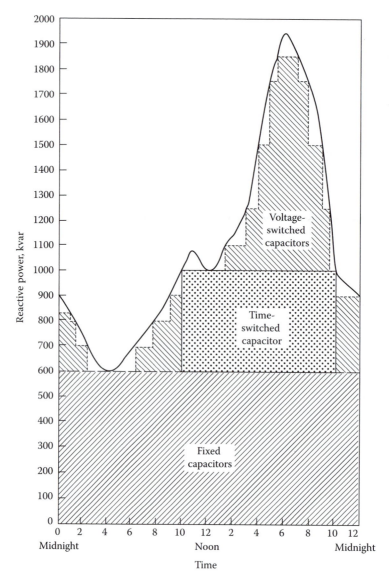

FIGURE 8.20 Meeting the reactive power requirements with fixed, voltage-controlled, and time-controlled capacitors.

However, the *ungrounded-wye capacitor banks are recommended if one or more of the following conditions exist*:

1. Excessive harmonic currents in the substation neutral can be precluded.
2. Telephone interferences can be minimized.
3. Capacitor-bank installation can be made with two single-phase switches rather than with three single-pole switches.

Usually, *grounded-wye capacitor banks* are used only on four-wire three-phase primary systems. Otherwise, if a grounded-wye capacitor bank is used on a three-phase three-wire ungrounded-wye or delta system, it furnishes a ground current source that may disturb sensitive ground relays.

8.6 ECONOMIC JUSTIFICATION FOR CAPACITORS

Loads on electric utility systems include two components: active power (measured in kilowatts) and reactive power (measured in kilovars). Active power has to be generated at power plants, whereas reactive power can be provided by either power plants or capacitors. It is a well-known fact that shunt power capacitors are the most economical source to meet the reactive power requirements of inductive loads and transmission lines operating at a lagging power factor.

When reactive power is provided only by power plants, each system component (i.e., generators, transformers, transmission and distribution lines, switchgear, and protective equipment) has to be increased in size accordingly. Capacitors can mitigate these conditions by decreasing the reactive power demand all the way back to the generators. Line currents are reduced from capacitor locations all the way back to generation equipment. As a result, losses and loadings are reduced in distribution lines, substation transformers, and transmission lines.

Depending upon the uncorrected power factor of the system, the installation of capacitors can increase generator and substation capability for additional load at least 30% and can increase individual circuit capability, from the voltage regulation point of view, approximately 30%–100%.

Furthermore, the current reduction in transformer and distribution equipment and lines reduces the load on these kilovoltampere-limited apparatus and consequently delays the new facility installations. In general, the economic benefits force capacitor banks to be installed on the primary distribution system rather than on the secondary.

It is a *well-known rule of thumb that the optimum amount of capacitor kilovars to employ is always the amount at which the economic benefits obtained from the addition of the last kilovar exactly equal the installed cost of the kilovars of capacitors.*

The methods used by the utilities to determine the economic benefits derived from the installation of capacitors vary from company to company, but the determination of the total installed cost of a kilovar of capacitors is easy and straightforward.

In general, the *economic benefits that can be derived from capacitor installation* can be summarized as follows:

1. Released generation capacity
2. Released transmission capacity
3. Released distribution substation capacity
4. Additional advantages in distribution system
   a. Reduced energy (copper) losses
   b. Reduced voltage drop and consequently improved voltage regulation
   c. Released capacity of feeder and associated apparatus
   d. Postponement or elimination of capital expenditure due to system improvements and/or expansions
   e. Revenue increase due to voltage improvements

8.6.1 BENEFITS DUE TO RELEASED GENERATION CAPACITY

The released generation capacity due to the installation of capacitors can be calculated approximately from

$$
\Delta S_G = \begin{cases} \left[\left(1 - \dfrac{Q_c^2 \times \cos^2 \theta}{S_G^2} \right)^{1/2} + \dfrac{Q_c \times \sin \theta}{S_G} - 1 \right] S_G & \text{when } Q_c > 0.10 S_G \\[4mm] Q_c \times \sin \theta & \text{when } Q_c \le 0.10 S_G \end{cases} \tag{8.36}
$$

where

ΔS_G is the released generation capacity beyond maximum generation capacity at original power factor, kVA

S_G is the generation capacity, kVA

Q_c is the reactive power due to corrective capacitors applied, kvar

$\cos \theta$ is the original (or uncorrected or old) power factor before application of capacitors

Therefore, the annual benefits due to the released generation capacity can be expressed as

$$\Delta\$_G = \Delta S_G \times C_G \times i_G \tag{8.37}$$

where

$\Delta\$_G$ is the annual benefits due to released generation capacity, \$/year

ΔS_G is the released generation capacity beyond maximum generation capacity at original power factor, kVA

C_G is the cost of (peaking) generation, \$/kW

i_G is the annual fixed charge rate* applicable to generation

8.6.2 Benefits due to Released Transmission Capacity

The released transmission capacity due to the installation of capacitors can be calculated approximately as

$$\Delta S_T = \begin{cases} \left[\left(1 - \dfrac{Q_c^2 \times \cos^2 \theta}{S_T^2} \right)^{1/2} + \dfrac{Q_c \times \sin \theta}{S_T} - 1 \right] S_T & \text{when } Q_c > 0.10 S_T \\[4mm] Q_c \times \sin \theta & \text{when } Q_c \le 0.10 S_T \end{cases} \tag{8.38}$$

where

ΔS_T is the released transmission capacity[†] beyond maximum transmission capacity at original power factor, kVA

S_T is the transmission capacity, kVA

Thus, the annual benefits due to the released transmission capacity can be found as

$$\Delta\$_T = \Delta S_T \times C_T \times i_T \tag{8.39}$$

where

$\Delta\$_T$ is the annual benefits due to released transmission capacity, \$/year

ΔS_T is the released transmission capacity beyond maximum transmission capacity at original power factor, kVA

C_T is the cost of transmission line and associated apparatus, \$/kVA

i_T is the annual fixed charge rate applicable to transmission

---

* Also called *carrying charge rate*. It is defined as that portion of the annual revenue requirements that results from a plant investment. Total carrying charges include (1) return (on equity and debt), (2) book depreciation, (3) taxes (including amount paid currently and amounts deferred to future years), (4) insurance, and (5) operations and maintenance. It is expressed as a decimal.

[†] Note that the symbol S_T now stands for transmission capacity rather than transformer capacity.

8.6.3 BENEFITS DUE TO RELEASED DISTRIBUTION SUBSTATION CAPACITY

The released distribution substation capacity due to the installation of capacitors can be found approximately from

$$
\Delta S_S = \begin{cases} \left[\left(1 - \dfrac{Q_c^2 \times \cos^2 \theta}{S_S^2} \right)^{1/2} + \dfrac{Q_c \times \sin \theta}{S_S} - 1 \right] S_S & \text{when } Q_c > 0.10 S_S \\[2mm] Q_c \times \sin \theta & \text{when } Q_c \leq 0.10 S_S \end{cases} \tag{8.40}
$$

where
  ΔS_S is the released distribution substation capacity beyond maximum substation capacity at original power factor, kVA
  S_S is the distribution substation capacity, kVA

Hence, the annual benefits due to the released substation capacity can be calculated as

$$
\Delta \$_S = \Delta S_S \times C_S \times i_s \tag{8.41}
$$

where
  $\Delta \$_S$ is the annual benefits due to the released substation capacity, \$/year
  ΔS_S is the released substation capacity, kVA
  C_S is the cost of substation and associated apparatus, \$/kVA
  i_S is the annual fixed charge rate applicable to substation

8.6.4 BENEFITS DUE TO REDUCED ENERGY LOSSES

The annual energy losses are reduced as a result of decreasing copper losses due to the installation of capacitors. The conserved energy can be expressed as

$$
\Delta \text{ACE} = \frac{Q_{c,3\phi} R (2 S_{L,3\phi} \sin \theta - Q_{c,3\phi}) 8760}{1000 \times V_{L-L}^2} \tag{8.42}
$$

where
  ΔACE is the annual conserved energy, kWh/year
  $Q_{c,3\phi}$ is the three-phase reactive power due to corrective capacitors applied, kvar
  R is the total line resistance to load center, Ω
  $Q_{L,3\phi}$ is the original, that is, uncorrected, three-phase load, kVA
  $\sin \theta$ is the sine of original (uncorrected) power factor angle
  V_{L-L} is the line-to-line voltage, kV

Therefore, the annual benefits due to the conserved energy can be calculated as

$$
\Delta \$_{\text{ACE}} = \Delta \text{ACE} \times \text{EC} \tag{8.43}
$$

where
  ΔACE is the annual benefits due to conserved energy, \$/year
  EC is the cost of energy, \$/kWh

8.6.5 BENEFITS DUE TO REDUCED VOLTAGE DROPS

The following advantages can be obtained by the installation of capacitors into a circuit:

1. The effective line current is reduced, and consequently, both IR and IX_L voltage drops are decreased, which results in improved voltage regulation.
2. The power factor improvement further decreases the effect of reactive line voltage drop.

The percent voltage drop that occurs in a given circuit can be expressed as

$$\%VD = \frac{S_{L,3\phi}(r\cos\theta + x\sin\theta)l}{10 \times V_{L-L}^2} \qquad (8.44)$$

where
   $\%VD$ is the percent voltage drop
   $S_{L,3\phi}$ is the three-phase load, kVA
   r is the line resistance, 0/min
   x is the line reactance, 0/min
   l is the length of conductors, min
   V_{L-L} is the line-to-line voltage, kV

The voltage drop that can be calculated from Equation 8.44 is the basis for the application of the capacitors. After the application of the capacitors, the system yields a voltage rise due to the improved power factor and the reduced effective line current. Therefore, the voltage drops due to IR and IX_L are minimized. The approximate value of the percent voltage rise along the line can be calculated as

$$\%VR = \frac{Q_{c,3\phi} \times x \times l}{10 \times V_{L-L}^2} \qquad (8.45)$$

Furthermore, an additional voltage-rise phenomenon through every transformer from the generating source to the capacitors occurs due to the application of capacitors. It is independent of load and power factor of the line and can be expressed as

$$\%VR_T = \left(\frac{Q_{c,3\phi}}{S_{T,3\phi}}\right)x_T \qquad (8.46)$$

where
   $\%VR_T$ is the percent voltage rise through the transformer
   $S_{T,3\phi}$ is the total three-phase transformer rating, kVA
   x_T is the percent transformer reactance (approximately equal to the transformer's nameplate impedance).

8.6.6 BENEFITS DUE TO RELEASED FEEDER CAPACITY

In general, feeder capacity is restricted by allowable voltage drop rather than by thermal limitations (as seen in Chapter 4). Therefore, the installation of capacitors decreases the voltage drop and consequently increases the feeder capacity.

Without including the released regulator or substation capacity, this additional feeder capacity can be calculated as

$$\Delta S_F = \frac{(Q_{c,3\phi})x}{x\sin\theta + r\cos\theta}\, \text{kVA} \tag{8.47}$$

Therefore, the annual benefits due to the released feeder capacity can be calculated as

$$\Delta \$_F = \Delta S_F \times C_F \times i_F \tag{8.48}$$

where
  $\Delta \$_F$ is the annual benefits due to released feeder capacity, \$/year
  ΔS_F is the released feeder capacity, kVA
  C_F is the cost of installed feeder, \$/kVA
  i_F is the annual fixed charge rate applicable to the feeder

8.6.7 FINANCIAL BENEFITS DUE TO VOLTAGE IMPROVEMENT

The revenues to the utility are increased as a result of increased kilowatthour energy consumption due to the voltage rise produced on a system by the addition of the corrective capacitor banks. This is especially true for residential feeders.

The increased energy consumption depends on the nature of the apparatus used. For example, energy consumption for lighting increases as the square of the voltage used. As an example, Table 8.3 gives the additional kilowatthour energy increase (in percent) as a function of the ratio of the average voltage after the addition of capacitors to the average voltage before the addition of capacitors (based on a typical load diversity).

Thus, the increase in revenues due to the increased kilowatthour energy consumption can be calculated as

$$\Delta \$_{\text{BEC}} = \Delta \text{BEC} \times \text{BEC} \times \text{EC} \tag{8.49}$$

where
  $\Delta \$_{\text{BEC}}$ is the additional annual revenue due to increased kWh energy consumption, \$/year
  ΔBEC is the additional kWh energy consumption increase
  BEC is the original (or base) annual kWh energy consumption, kWh/year

TABLE 8.3
Additional kWh Energy Increase
After Capacitor Addition

$\dfrac{V_{av,after}}{V_{av,before}}$	ΔkWh Increase, %
1.00	0
1.05	8
1.10	16
1.15	25
1.20	34
1.25	43
1.30	52

8.6.8 Total Financial Benefits due to Capacitor Installations

Therefore, the total benefits due to the installation of capacitor banks can be summarized as

$$\sum \Delta\$ = (\text{Demand reduction}) + (\text{Energy reduction}) + (\text{Revenue increase})$$

$$= (\Delta\$_G + \Delta\$_T + \Delta\$_S + \Delta\$_F) + \Delta\$_{ACE} + \Delta\$_{BEC} \qquad (8.50)$$

The total benefits obtained from Equation 8.50 should be compared against the annual equivalent of the total cost of the installed capacitor banks. The total cost of the installed capacitor banks can be found from

$$\Delta EIC_c = \Delta Q_c \times IC_c \times i_c \qquad (8.51)$$

where
ΔEIC_c is the annual equivalent of the total cost of installed capacitor banks, \$/year
ΔQ_c is the required amount of capacitor-bank additions, kvar
IC_c is the cost of installed capacitor banks, \$/kvar
i_c is the annual fixed charge rate applicable to capacitors

In summary, capacitors can provide the utility industry with a very effective cost reduction instrument. With plant costs and fuel costs continually increasing, electric utilities benefit whenever new plant investment can be deferred or eliminated and energy requirements reduced.

Thus, capacitors aid in minimizing operating expenses and allow the utilities to serve new loads and customers with a minimum system investment. Today, utilities in the United States have approximately 1 kvar of power capacitors installed for every 2 kW of installed generation capacity in order to take advantage of the economic benefits involved [4].

Example 8.19*

Assume that a large power pool is presently operating at 90% power factor. It is desired to improve the power factor to 98%. To improve the power factor to 98%, a number of load flow runs are made, and the results are summarized in Table 8.4.

Assume that the average fixed charge rate is 0.20, average demand cost is \$250/kW, energy cost is \$0.045/kWh, the system loss factor is 0.17, and an average capacitor cost is \$4.75/kvar. Use

TABLE 8.4
For Example 8.19

Comment	At 90% PF	At 98% PF
Total loss reduction due to capacitors applied to substation buses, kW	495,165	491,738
Additional loss reduction due to capacitors applied to feeders, kW	85,771	75,342
Total demand reduction due to capacitors applied to substation buses and feeders, kVA	22,506,007	21,172,616
Total required capacitor additions at buses and feeders, kvar	9,810,141	4,213,297

* Based on Ref. [3].

responsibility factors of 1.0 and 0.9 for capacitors installed on the substation buses and on feeders, respectively. Determine the following:

a. The resulting additional savings in kilowatt losses at the 98% power factor when all capacitors are applied to substation buses.
b. The resulting additional savings in kilowatt losses at the 98% power factor when some capacitors are applied to feeders.
c. The total additional savings in kilowatt losses.
d. The additional savings in the system kilovoltampere capacity.
e. The additional capacitors required, kvars.
f. The total annual savings in demand reduction due to additional capacitors applied to substation buses and feeders, $/year.
g. The annual savings due to the additional released transmission capacity, $/year.
h. The total annual savings due to the energy loss reduction, $/year.
i. The total annual cost of the additional capacitors, $/year.
j. The total net annual savings, $/year.
k. Is the 98% power factor the economic power factor?

Solution

a. From Table 8.4, the resulting additional savings in kilowatt losses due to the power factor improvement at the substation buses is

$$\Delta P_{LS} = 495,165 - 491,738 = 3427 \text{ kW}$$

b. From Table 8.4 for feeders,

$$\Delta P_{LS} = 85,771 - 75,342$$

$$= 10,429 \text{ kW}$$

c. Therefore, the total additional kilowatt savings is

$$\Delta P_{LS} = 3,427 + 10,429$$

$$= 13,856 \text{ kW}$$

As can be observed, the additional kilowatt savings due to capacitors applied to the feeders is more than three times that of capacitors applied to the substation buses. This is due to the fact that power losses are larger at the lower voltages.

d. From Table 8.4, the additional savings in the system kilovoltampere capacity is

$$\Delta S_{sys} = 22,506,007 - 21,172,616$$

$$= 1,333,391 \text{ kVA}$$

e. From Table 8.4, the additional capacitors required are

$$\Delta Q_c = 9,810,141 - 4,213,297$$

$$= 5,596,844 \text{ kvar}$$

f. The annual savings in demand reduction due to capacitors applied to distribution substation buses is approximately

$$(3427 \text{ kW})(1.0)(\$250/\text{kW})(0.20/\text{year}) = \$171,350/\text{year}$$

and due to capacitors applied to feeders is

$$(10,429 \text{ kW})(0.9)(\$250/\text{kW})(0.20/\text{year}) = \$469,305/\text{year}$$

Therefore, the total annual savings in demand reduction is

$$\$171,350 + \$469,305 = \$640,655/\text{year}$$

g. The annual savings due to the additional released transmission capacity is

$$(1,333,391 \text{ kVA})(\$27/\text{kVA})(0.20/\text{year}) = \$7,200,311/\text{year}$$

h. The total annual savings due to the energy loss reduction is

$$(\$13,856 \text{ kW})(8760 \text{ h/year})(0.17)(\$0.045/\text{kWh}) = \$928,546/\text{year}$$

i. The total annual cost of the additional capacitors is

$$(5,596,844 \text{ kvar})(\$4.75/\text{kvar})(0.20/\text{year}) = \$5,317,002/\text{year}$$

j. The total annual savings is summation of the savings in demand, capacity, and energy:

$$\$640,655 + \$7,200,311 + \$928,5466 = \$8,769,512/\text{year}$$

Therefore, the total net annual savings is

$$\$8,769,512 - \$5,317,002 = \$3,452,510/\text{year}$$

k. No, since the total net annual savings is not zero.

8.7 PRACTICAL PROCEDURE TO DETERMINE THE BEST CAPACITOR LOCATION

In general, the best location for capacitors can be found by optimizing power loss and voltage regulation. A feeder voltage profile study is performed to warrant the most effective location for capacitors and the determination of a voltage that is within recommended limits.

Usually, a 2 V rise on circuits used in urban areas and a 3 V rise on circuits used in rural areas are approximately the maximum voltage changes that are allowed when a switched capacitor bank is placed into operation. The general iteration process involved is summarized in the following steps:

1. Collect the following circuit and load information:
   a. Any two of the following for each load: kilovoltamperes, kilovars, kilowatts, and load power factor
   b. Desired corrected power of circuit
   c. Feeder-circuit voltage
   d. A feeder-circuit map that shows locations of loads and presently existing capacitor banks
2. Determine the kilowatt load of the feeder and the power factor.
3. From Table 8.2, determine the kilovars per kilowatts of load (i.e., the correction factor) necessary to correct the feeder-circuit power factor from the original to the desired power factor. To determine the kilovars of capacitors required, multiply this correction factor by the total kilowatts of the feeder circuit.
4. Determine the individual kilovoltamperes and power factor for each load or group of loads.
5. To determine the kilovars on the line, multiply individual load or groups of loads by their respective reactive factors that can be found from Table 8.2.
6. Develop a nomograph to determine the line loss in W/1000 ft due to the inductive loads tabulated in steps 4 and 5. Multiply these line losses by their respective line lengths in thousands of feet. Repeat this process for all loads and line sections and add them to find the total inductive line loss.

7. In the case of having presently existing capacitors on the feeder, perform the same calculations as in step 6, but this time subtract the capacitive line loss from the total inductive line loss. Use the capacitor kilovars determined in step 3 and the nomograph developed for step 6 and find the line loss in each line section due to capacitors.
8. To find the distance to capacitor location, divide the total inductive line loss by capacitive line loss per thousand feet. If this quotient is greater than the line section length
   a. Divide the remaining inductive line loss by the capacitive line loss in the next line section to find the location
   b. If this quotient is still greater than the line section length, repeat step 8a
9. Prepare a voltage profile by hand calculations or by using a computer program for voltage profile and load analysis to determine the circuit voltages. If the profile shows that the voltages are inside the recommended limits, then the capacitors are installed at the location of minimum loss. If not, then use engineering judgment to locate them for the most effective voltage control application.

8.8 MATHEMATICAL PROCEDURE TO DETERMINE THE OPTIMUM CAPACITOR ALLOCATION

The optimum application of shunt capacitors on distribution feeders to reduce losses has been studied in numerous papers such as those by Neagle and Samson [5], Schmidt [7], Maxwell [8,9], Cook [10], Schmill [11], Chang [12–14], Bae [15], Gönen and Djavashi [17], and Grainger et al. [19,21–23]. Figure 8.21 shows a realistic representation of a feeder that contains a number of line segments with a combination of concentrated (or lumped-sum) and uniformly distributed loads, as suggested by Chang [13]. Each line segment represents a part of the feeder between sectionalizing devices, voltage regulators, and other points of significance. For the sake of convenience, the load or line current and the resulting I^2R loss can be assumed to have two components, namely, (1) those due to the in-phase or active component of current and (2) those due to the out-of-phase or reactive

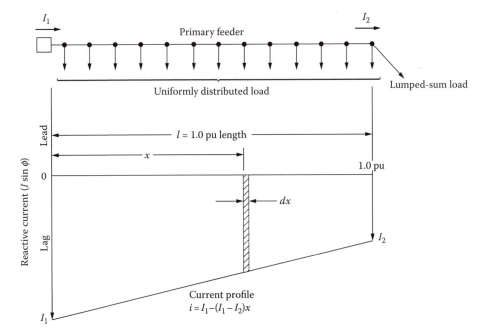

FIGURE 8.21 Primary feeder with lumped-sum (or concentrated) and uniformly distributed loads and reactive current profile before adding the capacitor.

component of current. Since losses due to the in-phase or active component of line current are not significantly affected by the application of shunt capacitors, they are not considered. This can be verified as follows.

Assume that the I^2R losses are caused by a lagging line current I flowing through the circuit resistance R. Therefore, it can be shown that

$$I^2R = (I \cos\phi)^2 R + (I \sin\phi)^2 R \tag{8.52}$$

After adding a shunt capacitor with current I_c, the resultants are a new line current I_1 and a new power loss I_1^2R. Hence,

$$I_1^2R = (I \cos\phi)^2 R + (I \sin\phi - I_c)^2 R \tag{8.53}$$

Therefore, the loss reduction as a result of the capacitor addition can be found as

$$\Delta P_{LS} = I^2R - I_1^2R \tag{8.54}$$

or by substituting Equations 8.56 and 8.57 into Equation 8.58,

$$\Delta P_{LS} = 2(I \sin\phi)I_c R - I_c^2 R \tag{8.55}$$

Thus, only the out-of-phase or reactive component of line current, that is, $I \sin\theta$, should be taken into account for I^2R loss reduction as a result of a capacitor addition.

Assume that the length of a feeder segment is 1.0 pu length, as shown in Figure 8.21. The current profile of the line current at any given point on the feeder is a function of the distance of that point from the beginning of the feeder. Therefore, the differential I^2R loss of a dx differential segment located at a distance x can be expressed as

$$dP_{LS} = 3[I_1 - (I_1 - I_2)x]^2 R\, dx \tag{8.56}$$

Therefore, the total I^2R loss of the feeder can be found as

$$P_{LS} = \int_{x=0}^{1.0} dP_{LS}$$

$$= 3 \int_{x=0}^{1.0} [I_1 - (I_1 - I_2)x]^2 R\, dx$$

$$= \left(I_1^2 + I_1 I_2 + I_2^2\right) R \tag{8.57}$$

where

P_{LS} is the total I^2R loss of the feeder before adding the capacitor
I_1 is the reactive current at the beginning of the feeder segment
I_2 is the reactive current at the end of the feeder segment
R is the total resistance of the feeder segment
x is the per unit distance from the beginning of the feeder segment

8.8.1   Loss Reduction due to Capacitor Allocation

8.8.1.1   Case 1: One Capacitor Bank

The insertion of one capacitor bank on the primary feeder causes a break in the continuity of the reactive load profile, modifies the reactive current profile, and consequently reduces the loss, as shown in Figure 8.22.

Therefore, the loss equation after adding one capacitor bank can be found as before:

$$P'_{LS} = 3 \int_{x=0}^{x_1} [I_1 - (I_1 - I_2)x - I_c]^2 R\,dx + 3 \int_{x=x_1}^{1.0} [I_1 - (I_1 - I_2)x]^2 R\,dx \qquad (8.58)$$

or

$$P'_{LS} = \left(I_1^2 + I_1 I_2 + I_2^2\right) R + 3x_1 \left[(x_1 - 2)I_1 I_c - x_1 I_2 I_c + I_c^2\right] R \qquad (8.59)$$

Thus, the per unit power loss reduction as a result of adding one capacitor bank can be found from

$$\Delta P_{LS} = \frac{P_{LS} - P'_{LS}}{P_{LS}} \qquad (8.60)$$

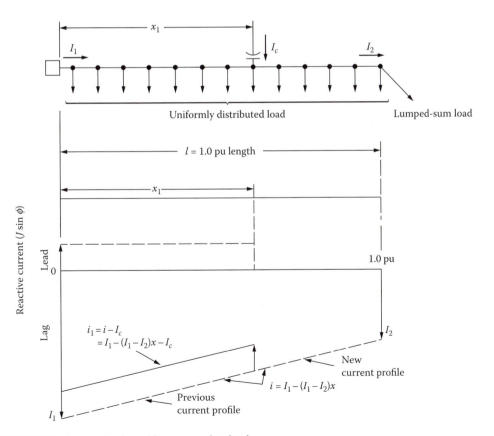

FIGURE 8.22   Loss reduction with one capacitor bank.

or substituting Equations 8.57 and 8.58 into Equation 8.60,

$$\Delta P_{LS} = \frac{-3x_1 \left[(x_1 - 2)I_1 I_c - x_1 I_2 I_c + I_c^2 \right] R}{\left(I_1^2 + I_1 I_2 + I_2^2 \right) R} \tag{8.61}$$

or rearranging Equation 8.61 by dividing its numerator and denominator by I_1^2 so that

$$\Delta P_{LS} = \frac{3x_1}{1 + (I_2/I_1) + (I_2/I_1)^2} \left[(2 - x_1)\left(\frac{I_c}{I_1} \right) + x_1 \left(\frac{I_2}{I_1} \right)\left(\frac{I_c}{I_1} \right) - \left(\frac{I_c}{I_1} \right)^2 \right] \tag{8.62}$$

If c is defined as the ratio of the capacitive kilovoltamperes (ckVAs) of the capacitor bank to the total reactive load, that is,

$$c = \frac{\text{ckVA of capacitor installed}}{\text{Total reactive load}} \tag{8.63}$$

then

$$c = \frac{I_c}{I_1} \tag{8.64}$$

and if λ is defined as the ratio of the reactive current at the end of the line segment to the reactive current at the beginning of the line segment, that is,

$$\lambda = \frac{\text{Reactive current at the end of line segment}}{\text{Reactive current at the beginning of line segment}} \tag{8.65}$$

then

$$\lambda = \frac{I_2}{I_1} \tag{8.66}$$

Therefore, substituting Equations 8.64 and 8.66 into Equation 8.62, the per unit power loss reduction can be found as

$$\Delta P_{LS} = \frac{3x_1}{1 + \lambda + \lambda^2} [(2 - x_1)c + x_1 \lambda c - c^2] \tag{8.67}$$

or

$$\Delta P_{LS} = \frac{3cx_1}{1 + \lambda + \lambda^2} [(2 - x_1) + x_1 \lambda - c] \tag{8.68}$$

where x_1 is the per unit distance of the capacitor-bank location from the beginning of the feeder segment (between 0 and 1.0 pu).

If α is defined as the reciprocal of $1 + \lambda + \lambda^2$, that is,

$$\alpha = \frac{1}{1 + \lambda + \lambda^2} \tag{8.69}$$

then Equation 8.68 can also be expressed as

$$\Delta P_{LS} = 3\alpha c x_1 [(2 - x_1) + \lambda x_1 - c] \tag{8.70}$$

Figures 8.23 through 8.27 give the loss reduction that can be accomplished by changing the location of a single capacitor bank with any given size for different capacitor compensation ratios along the feeder for different representative load patterns, for example, uniformly distributed loads ($\lambda = 0$), concentrated or lumped-sum loads ($\lambda = 1$), or a combination of concentrated and uniformly distributed loads ($0 < \lambda < 1$). To use these nomographs for a given case, the following factors must be known:

1. Original losses due to the reactive current
2. Capacitor compensation ratio
3. The location of the capacitor bank

As an example, assume that the load on the line segment is uniformly distributed and the desired compensation ratio is 0.5. From Figure 8.23, it can be found that the maximum loss reduction can be obtained if the capacitor bank is located at 0.75 pu length from the source. The associated loss reduction is 0.85 pu or 85%. If the bank is located anywhere else on the feeder, however, the loss reduction would be less than the 85%.

In other words, there is only one location for any given-size capacitor bank to achieve the maximum loss reduction. Table 8.5 gives the optimum location and percent loss reduction for a given-size capacitor bank located on a feeder with uniformly distributed load ($\lambda = 0$). From the table it can be observed that the maximum loss reduction can be achieved by locating the single capacitor bank at the two-thirds length of the feeder away from the source. Figure 8.28 gives the loss reduction for a given capacitor bank of any size and located at the optimum location on a feeder with various combinations of load types based on Equation 8.70. Figure 8.29 gives the loss reduction due to an optimum-sized capacitor bank located on a feeder with various combinations of load types.

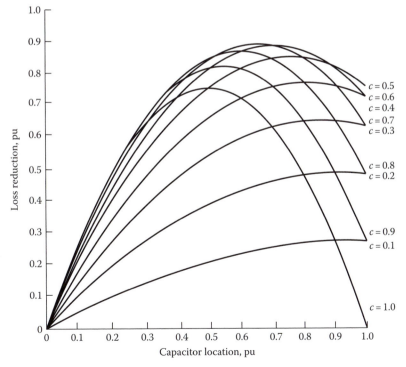

FIGURE 8.23   Loss reduction as a function of the capacitor-bank location and capacitor compensation ratio for a line segment with uniformly distributed loads ($\lambda = 0$).

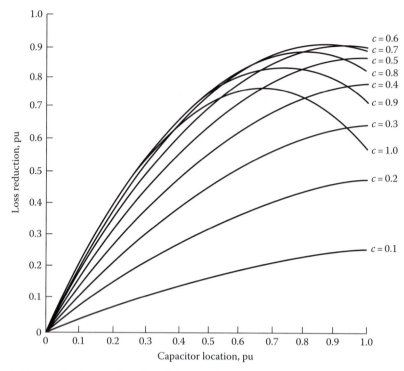

FIGURE 8.24 Loss reduction as a function of the capacitor-bank location and capacitor compensation ratio for a line segment with a combination of concentrated and uniformly distributed loads ($\lambda = 1/4$).

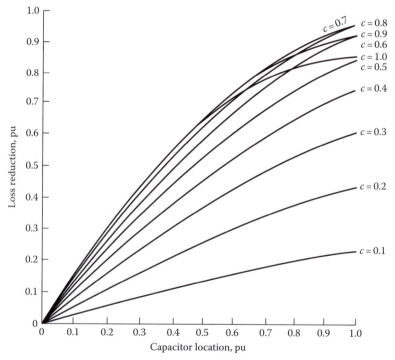

FIGURE 8.25 Loss reduction as a function of the capacitor-bank location and capacitor compensation ratio for a line segment with a combination of concentrated and uniformly distributed loads ($\lambda = 1/2$).

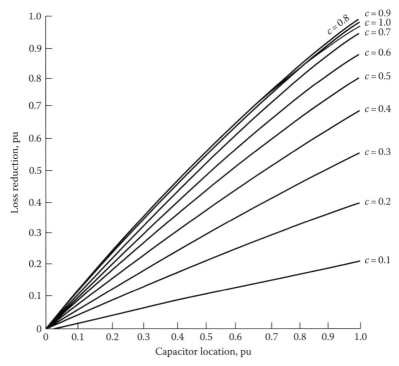

FIGURE 8.26 Loss reduction as a function of the capacitor-bank location and capacitor compensation ratio for a line segment with a combination of concentrated and uniformly distributed loads ($\lambda = 3/4$).

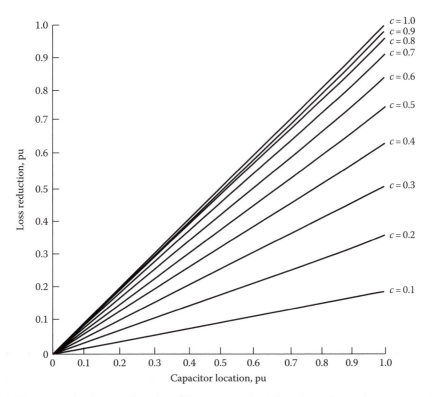

FIGURE 8.27 Loss reduction as a function of the capacitor-bank location and capacitor compensation ratio for a line segment with concentrated loads ($\lambda = 1$).

TABLE 8.5
Optimum Location and Optimum Loss Reduction

Capacitor-Bank Rating, pu	Optimum Location, pu	Optimum Loss Reduction, %
0.0	1.0	0
0.1	0.95	27
0.2	0.90	49
0.3	0.85	65
0.4	0.80	77
0.5	0.75	84
0.6	0.70	88
0.7	0.65	89
0.8	0.60	86
0.9	0.55	82
1.0	0.50	75

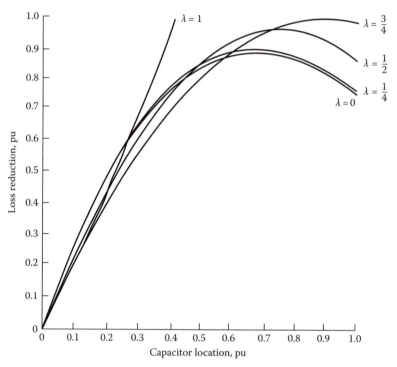

FIGURE 8.28   Loss reduction due to a capacitor bank located at the optimum location on a line section with various combinations of concentrated and uniformly distributed loads.

8.8.1.2   Case 2: Two Capacitor Banks

Assume that two capacitor banks of equal size are inserted on the feeder, as shown in Figure 8.30. The same procedure can be followed as before, and the new loss equation becomes

$$P'_{LS} = 3 \int_{x=0}^{x_1} [I_1 - (I_1 - I_2)x - 2I_c]^2 R\, dx + 3 \int_{x=x_1}^{x_2} [I_1 - (I_1 I_2)x - I_c]^2 R\, dx + 3 \int_{x=x_2}^{1.0} [I_1 - (I_1 - I_2)x]^2 R\, dx$$

$$(8.71)$$

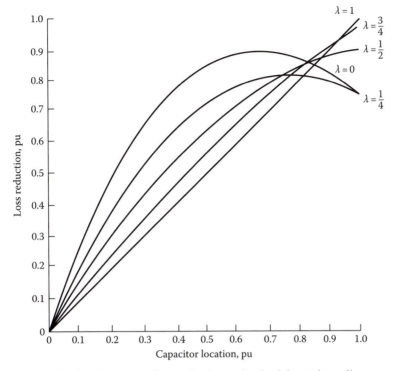

FIGURE 8.29 Loss reduction due to an optimum-sized capacitor bank located on a line segment with various combinations of concentrated and uniformly distributed loads.

Therefore, substituting Equations 8.57 and 8.71 into Equation 8.60, the new per unit loss reduction equation can be found as

$$\Delta P_{LS} = 3\alpha c x_1[(2-x_1)+\lambda x_1 - 3c] + 3\alpha c x_2[(2-x_2)+\lambda x_2 - c] \tag{8.72}$$

or

$$\Delta P_{LS} = 3\alpha c\{x_1[(2-x_1)+\lambda x_1 - 3c] + x_2[(2-x_2)+\lambda x_2 - c]\} \tag{8.73}$$

8.8.1.3 Case 3: Three Capacitor Banks

Assume that three capacitor banks of equal size are inserted on the feeder, as shown in Figure 8.31. The relevant per unit loss reduction equation can be found as

$$\Delta P_{LS} = 3\alpha c\{x_1[(2-x_1)+\lambda x_1 - 5c] + x_2[(2-x_2)+\lambda x_2 - 3c] + x_3[(2-x_3)+\lambda x_3 - c]\} \tag{8.74}$$

8.8.1.4 Case 4: Four Capacitor Banks

Assume that four capacitor banks of equal size are inserted on the feeder, as shown in Figure 8.32. The relevant per unit loss reduction equation can be found as

$$\Delta P_{LS} = 3\alpha c\{x_1[(2-x_1)+\lambda x_1 - 7c] + x_2[(2-x_2)+\lambda x_2 - 5c]$$
$$+ x_3[(2-x_3)+\lambda x_3 - 3c] + x_4[(2-x_4)+\lambda x_4 - c]\} \tag{8.75}$$

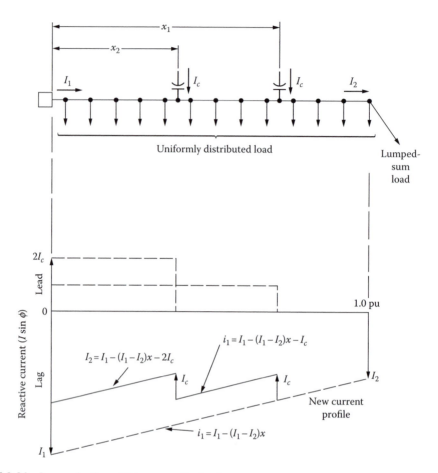

FIGURE 8.30 Loss reduction with two capacitor banks.

8.8.1.5 Case 5: *n* Capacitor Banks

As the aforementioned results indicate, the per unit loss reduction equations follow a definite pattern as the number of capacitor banks increases. Therefore, the general equation for per unit loss reduction, for an *n* capacitor-bank feeder, can be expressed as

$$P_{LS} = 3\alpha c \sum_{i=1}^{n} x_i[(2 - x_i) + \lambda x_i - (2i - 1)c] \tag{8.76}$$

where
    c is the capacitor compensation ratio at each location (determined from Equation 8.63)
    x_i is the per unit distance of the *i*th capacitor-bank location from the source
    n is the total number of capacitor banks

8.8.2 OPTIMUM LOCATION OF A CAPACITOR BANK

The optimum location for the *i*th capacitor bank can be found by taking the first-order partial derivative of Equation 8.76 with respect to x_i and setting the resulting expression equal to zero. Therefore,

$$x_{i,\text{opt}} = \frac{1}{1 - \lambda} - \frac{(2i - 1)c}{2(1 - \lambda)} \tag{8.77}$$

where $x_{i,\text{opt}}$ is the optimum location for the *i*th capacitor bank in per unit length.

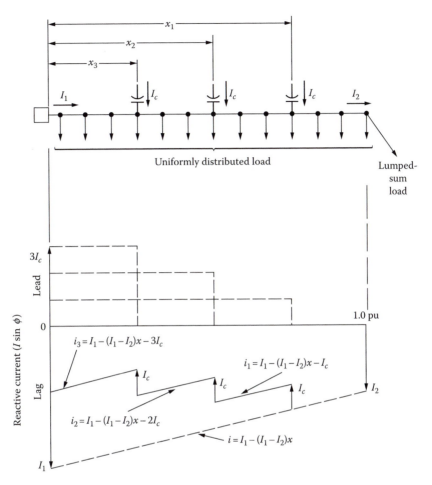

FIGURE 8.31　Loss reduction with three capacitor banks.

By substituting Equation 8.81 into Equation 8.80, the optimum loss reduction can be found as

$$P_{LS, opt} = 3\alpha c \sum_{i=1}^{n} \left[\frac{1}{1-\lambda} - \frac{(2i-1)c}{(1-\lambda)} + \frac{i^2 c^2}{1-\lambda} - \frac{c^2}{4(1-\lambda)} - \frac{ic^2}{1-\lambda} \right] \tag{8.78}$$

Equation 8.78 is an infinite series of algebraic form that can be simplified by using the following relations:

$$\sum_{i=1}^{n} (2i-1) = n^2 \tag{8.79}$$

$$\sum_{i=1}^{n} i = \frac{n(n+1)}{2} \tag{8.80}$$

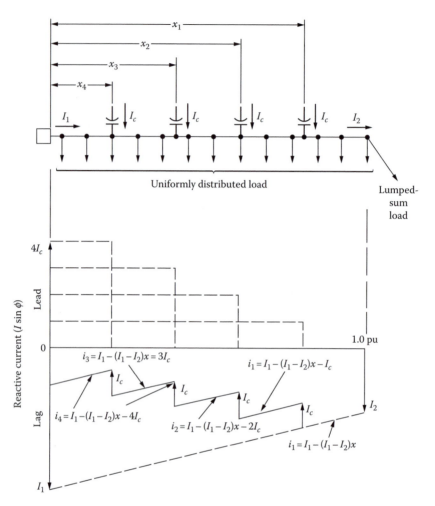

FIGURE 8.32   Loss reduction with four capacitor banks.

$$\sum_{i=1}^{n} i^2 = \frac{n(n+1)(2n+1)}{6} \tag{8.81}$$

$$\sum_{i=1}^{n} \frac{1}{1-\lambda} = \frac{n}{1-\lambda} \tag{8.82}$$

Therefore,

$$P_{LS,\,opt} = 3\alpha c \sum_{i=1}^{n} \left[\frac{n}{1-\lambda} - \frac{n^2 c}{(1-\lambda)} + \frac{nc^2(n+1)(2n+1)}{6} - \frac{nc^2}{4(1-\lambda)} - \frac{nc^2(n+1)}{2(1-\lambda)} \right] \tag{8.83}$$

$$P_{LS,\,opt} = \frac{3\alpha c}{1-\lambda} \left[n - cn^2 + \frac{c^2 n(4n^2 - 1)}{12} \right] \tag{8.84}$$

The capacitor compensation ratio at each location can be found by differentiating Equation 8.88 with respect to c and setting it equal to zero as

$$c = \frac{2}{2n+1} \tag{8.85}$$

Equation 8.86 can be called the $2/(2n + 1)$ *rule*. For example, for $n = 1$, the capacitor rating is two-thirds of the total reactive load that is located at

$$x_1 = \frac{2}{3(1-\lambda)} \tag{8.86}$$

of the distance from the source to the end of the feeder, and the peak loss reduction is

$$\Delta P_{LS,opt} = \frac{2}{3(1-\lambda)} \tag{8.87}$$

For a feeder with a uniformly distributed load, the reactive current at the end of the line is zero (i.e., $I_2 = 0$); therefore,

$$\lambda = 0 \quad \text{and} \quad \alpha = 1$$

Thus, for the optimum loss reduction of

$$\Delta P_{LS,opt} = \frac{8}{9} \text{ pu} \tag{8.88}$$

the optimum value of x_1 is

$$x_1 = \frac{2}{3} \text{ pu} \tag{8.89}$$

and the optimum value of c is

$$c = \frac{2}{3} \text{ pu} \tag{8.90}$$

Figure 8.33 gives a maximum loss reduction comparison for capacitor banks, with various total reactive compensation levels and located optimally on a line segment that has uniformly distributed load ($\lambda = 0$), based on Equation 8.84. The given curves are for one, two, three, and infinite number of capacitor banks.

For example, from the curve given for one capacitor bank, it can be observed that a capacitor bank rated two-thirds of the total reactive load and located at two-thirds of the distance out on the feeder from the source provides for a loss reduction of 89%. In the case of two capacitor banks, with four-fifths of the total reactive compensation, located at four-fifths of the distance out on the feeder, the maximum loss reduction is 96%. Figure 8.34 gives similar curves for a combination of concentrated and uniformly distributed loads ($\lambda = 1/4$).

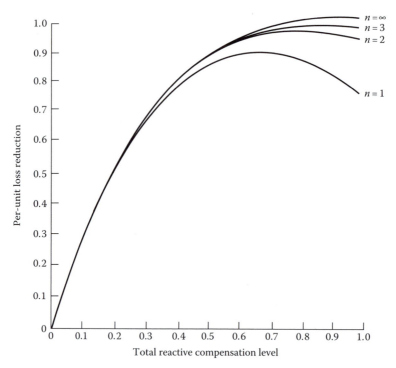

FIGURE 8.33 Comparison of loss reduction obtainable from $n = 1, 2, 3,$ and ∞ number of capacitor banks, with $\lambda = 0$.

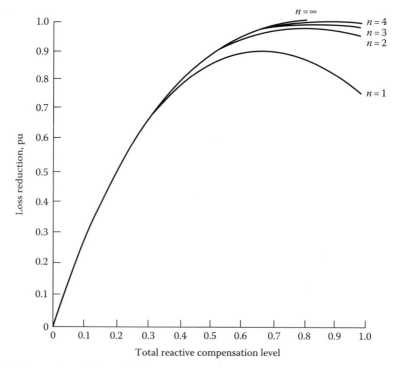

FIGURE 8.34 Comparison of loss reduction obtainable from $n = 1, 2, 3, 4,$ and ∞ number of capacitor banks, with $\lambda = 1/4$.

8.8.3 ENERGY LOSS REDUCTION DUE TO CAPACITORS

The per unit energy loss reduction in a three-phase line segment with a combination of concentrated and uniformly distributed loads due to the allocation of fixed shunt capacitors is

$$\Delta EL = 3\alpha c \sum_{i=1}^{n} x_i[(2 - x_i)F'_{LD} + x_i\lambda F'_{LD} - (2i - 1)c]T \tag{8.91}$$

where
F'_{LD} is the reactive load factor = Q/S
T is the total time period during which fixed-shunt-capacitor banks are connected
ΔEL is the energy loss reduction, pu

The optimum locations for the fixed shunt capacitors for the maximum energy loss reduction can be found by differentiating Equation 8.91 with respect to x_i and setting the result equal to zero. Therefore,

$$\frac{\partial(\Delta EL)}{\partial x_i} = 3\alpha c[2F'_{LD}(\lambda - 1)x_i + 2F'_{LD} - (2i - 1)c] \tag{8.92}$$

$$\frac{\partial^2(\Delta EL)}{\partial x_i^2} = -2F'_{LD}(1 - \lambda) < 0 \tag{8.93}$$

The optimum capacitor location for the maximum energy loss reduction can be found by setting Equation 8.92 to zero, so that

$$x_{i,\text{opt}} = \frac{1}{1 - \lambda} - \frac{(2i - 1)c}{2(1 - \lambda)F'_{LD}} \tag{8.94}$$

Similarly, the optimum total capacitor rating can be found as

$$C_T = \frac{2n}{2n + 1}F'_{LD} \tag{8.95}$$

From Equation 8.95, it can be observed that if the total number of capacitor banks approaches infinity, then the optimum total capacitor rating becomes equal to the reactive load factor.

If only one capacitor bank is used, the optimum capacitor rating to provide for the maximum energy loss reduction is

$$C_T = \frac{2}{3}F'_{LD} \tag{8.96}$$

This equation gives the well-known *two-thirds rule for fixed shunt capacitors*. Figure 8.35 shows the relationship between the total capacitor compensation ratio and the reactive load factor, in order to achieve maximum energy loss reduction, for a line segment with uniformly distributed load where $\lambda = 0$ and $\alpha = 1$.

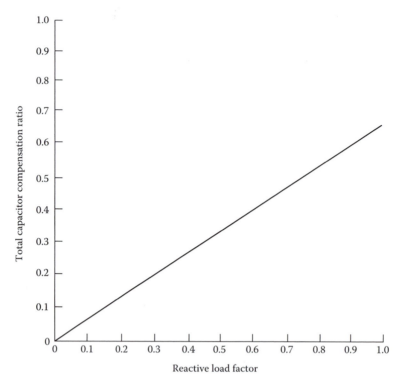

FIGURE 8.35 Relationship between the total capacitor compensation ratio and the reactive load factor for uniformly distributed load ($\lambda = 0$ and $\alpha = 1$).

By substituting Equation 8.94 into Equation 8.95, the optimum energy loss reduction can be found as

$$
\begin{aligned}
\Delta\mathrm{EL}_{\mathrm{opt}} &= \frac{3\alpha c}{1-\lambda}\left[nF'_{\mathrm{LD}} - cn^2 + \frac{c^2 n(4n^2-1)}{12F'_{\mathrm{LD}}}\right]T \\
&= \frac{3\alpha cn}{1-\lambda}\left[F'_{\mathrm{LD}} - cn + \frac{c^2 n^2(4n^2-1)}{12n^2 F'_{\mathrm{LD}}}\right]T \\
&= \frac{3\alpha C_T}{1-\lambda}\left[F'_{\mathrm{LD}} - C_T + \frac{C_T^2(4n^2-1)}{12n^2 F'_{\mathrm{LD}}}\right]T
\end{aligned}
\tag{8.97}
$$

where C_T is the total reactive compensation level $= cn$.

Based on Equation 8.97, the optimum energy loss reductions with any size capacitor bank located at the optimum location for various reactive load factors have been calculated, and the results have been plotted in Figures 8.36 through 8.40. It is important to note the fact that, for all values of λ, when reactive load factors are 0.2 or 0.4, the use of a fixed capacitor bank with corrective ratios of 0.4 and 0.8, respectively, gives a zero energy loss reduction.

Figures 8.41 through 8.45 show the effects of various reactive load factors on the maximum energy loss reductions for a feeder with different load patterns.

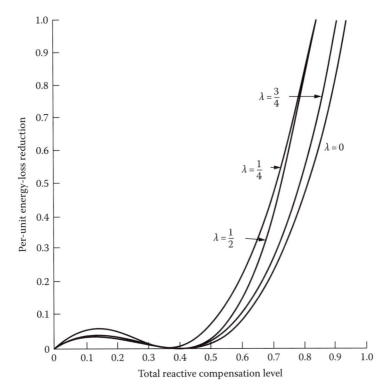

FIGURE 8.36 Energy loss reduction with any capacitor-bank size, located at optimum location ($F'_{LD} = 0.2$).

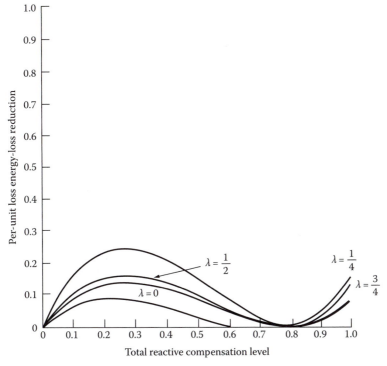

FIGURE 8.37 Energy loss reduction with any capacitor-bank size, located at the optimum location ($F'_{LD} = 0.4$).

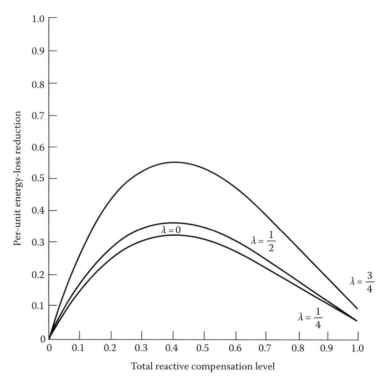

FIGURE 8.38   Energy loss reduction with any capacitor-bank size, located at the optimum location ($F'_{LD} = 0.6$).

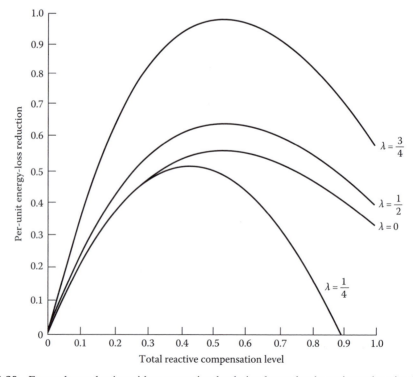

FIGURE 8.39   Energy loss reduction with any capacitor-bank size, located at the optimum location ($F'_{LD} = 0.8$).

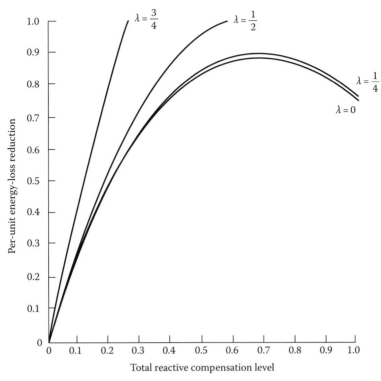

FIGURE 8.40   Energy loss reduction with any capacitor-bank size, located at the optimum location ($F'_{LD} = 1.0$).

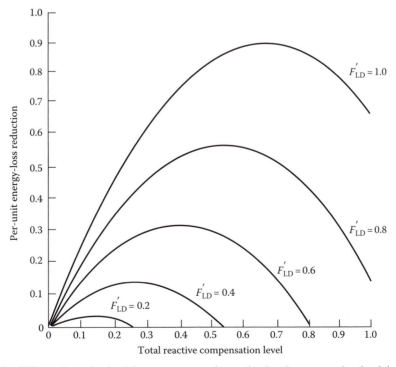

FIGURE 8.41   Effects of reactive load factors on energy loss reduction due to capacitor-bank installation on a line segment with uniformly distributed load ($\lambda = 0$).

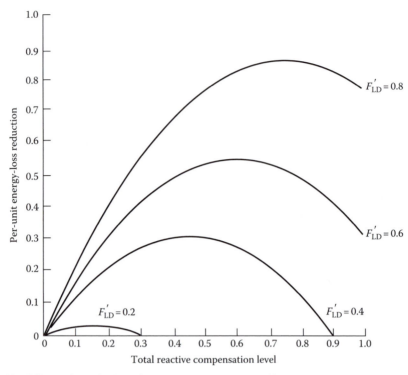

FIGURE 8.42 Effects of reactive load factors on energy loss reduction due to capacitor-bank installation on a line segment with a combination of concentrated and uniformly distributed loads ($\lambda = 1/4$).

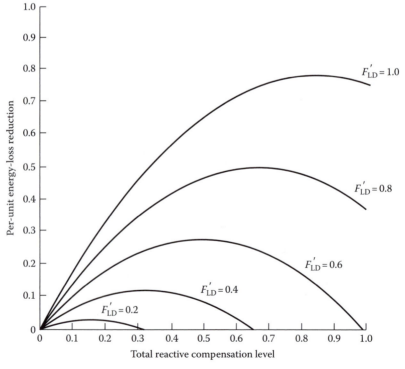

FIGURE 8.43 Effects of reactive load factors on energy loss reduction due to capacitor-bank installation on a line segment with a combination of concentrated and uniformly distributed loads ($\lambda = 1/2$).

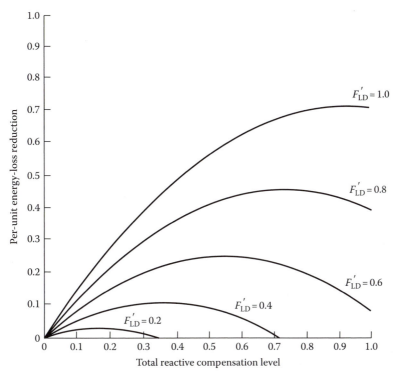

FIGURE 8.44 Effects of reactive load factors on loss reduction due to capacitor-bank installation on a line segment with a combination of concentrated and uniformly distributed loads ($\lambda = 3/4$).

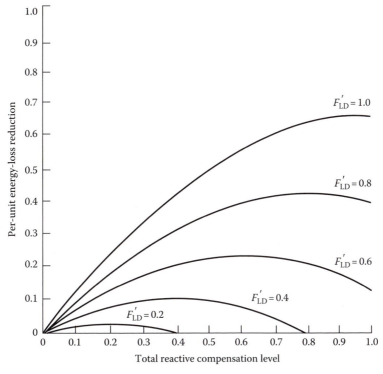

FIGURE 8.45 Effects of reactive load factors on energy loss reduction due to capacitor-bank installation on a line segment with a concentrated load ($\lambda = 1$).

8.8.4 Relative Ratings of Multiple Fixed Capacitors

The total savings due to having two fixed-shunt-capacitor banks located on a feeder with uniformly distributed load can be found as

$$\sum \$ = 3c_1\left(1 - c_1 + \frac{c_1^2}{4}\right)K_2 + 3c_2\left(1 - c_2 + \frac{c_2^2}{4}\right)K_2 + 3c_1\left(F'_{LD} - c_1 + \frac{c_1^2}{4F'_{LD}}\right)K_1T$$

$$+ 3c_2\left(F'_{LD} - c_2 + \frac{c_2^2}{4F'_{LD}}\right)K_1T \tag{8.98}$$

or

$$\sum \$ = 3[(c_1 + c_2)(K_1 + K_2T\,F'_{LD})] - \left(c_1^2 + c_2^2\right)(K_2 + K_1T)$$

$$+ \frac{1}{4}\left(c_1^3 + c_2^3\right)\left(K_2 + \frac{K_1T}{F'_{LD}}\right) \tag{8.99}$$

where
K_1 is the a constant to convert energy loss savings to dollars, \$/kWh
K_2 is the a constant to convert power loss savings to dollars, \$/kWh

Since the total capacitor-bank rating is equal to the sum of the ratings of the capacitor banks,

$$C_T = c_1 + c_2 \tag{8.100}$$

or

$$c_1 = C_T - c_2 \tag{8.101}$$

By substituting Equation 8.101 into Equation 8.99,

$$\sum \$ = 3\left[C_T(K_1 + K_2T\,F'_{LD}) - \left(C_T^2 + 2c_2^2 - 2c_2C_T\right)(K_1T + K_2) \right.$$

$$\left. + \frac{1}{4}\left(C_T^3 - 3c_2C_T^2 + 3c_2^2C_T\right)\left(K_2 + \frac{K_1T}{F'_{LD}}\right) \right] \tag{8.102}$$

The optimum rating of the second fixed capacitor bank as a function of the total capacitor-bank rating can be found by differentiating Equation 8.106 with respect to c_2, so that

$$\frac{\partial\left(\sum \$\right)}{\partial c_2} = -3(4c_2 - 2C_T)(K_2 + K_1T) + \frac{3}{4}\left(-3C_T^2 + 6c_2C_T\right)\left(K_2 + \frac{K_1T}{F'_{LD}}\right) \tag{8.103}$$

and setting the resultant equation equal to zero,

$$2c_2 = C_T \tag{8.104}$$

and since

$$C_T = c_1 + c_2 \tag{8.105}$$

then

$$c_1 = c_2 \tag{8.106}$$

The result shows that if multiple fixed-shunt-capacitor banks are to be employed on a feeder with uniformly distributed loads, in order to receive the maximum savings, all capacitor banks should have the same rating.

8.8.5 General Savings Equation for Any Number of Fixed Capacitors

From Equations 8.76 and 8.92, the total savings equation in a three-phase primary feeder with a combination of concentrated and uniformly distributed loads can be found as

$$\sum \$ = 3K_1\alpha c \sum_{i=1}^{n} x_i[(2 - x_i)F'_{LD} + x_i\lambda F'_{LD} - (2i - 1)c]T$$

$$+ 3K_2\alpha c \sum_{i=1}^{n} x_i[(2 - x_i) + x_i\lambda - (2i - 1)c] - K_3 C_T \tag{8.107}$$

where
K_1 is the constant to convert energy loss savings to dollars, \$/kWh
K_2 is the constant to convert power loss savings to dollars, \$/kWh
K_3 is the constant to convert total fixed capacitor ratings to dollars, \$/kvar
x_i is the ith capacitor location, pu length
n is the total number of capacitor banks
F'_{LD} is the reactive load factor
C_T is the total reactive compensation level
c is the capacitor compensation ratio at each location
λ is the ratio of reactive current at the end of the line segment to the reactive load current at the beginning of the line segment

$$\alpha = 1/(1 + \lambda + \lambda^2)$$

T is the total time period during which fixed-shunt-capacitor banks are connected

By taking the first- and second-order partial derivatives of Equation 8.107 with respect to x_i,

$$\frac{\partial\left(\sum \$\right)}{\partial x_i} = 3\alpha c[2x_i(K_2 + K_1 T F'_{LD})(\lambda - 1) + 2(K_2 + K_1 T F'_{LD}) - (2i - 1)c(K_2 + K_1 T)] \tag{8.108}$$

and

$$\frac{\partial^2\left(\sum \$\right)}{\partial x_i^2} = -6\alpha c(1 - \lambda)(K_2 + K_1 T F'_{LD}) < 0 \tag{8.109}$$

Setting Equation 8.108 equal to zero, the optimum location for any fixed capacitor bank with any rating can be found as

$$x_i = \frac{1}{1-\lambda} - \frac{(2i-1)c}{1-\lambda} \frac{K_2 + K_1 T}{K_2 + K_1 T F'_{LD}}$$ (8.110)

where $0 \leq x_i \leq 1.0$ pu length. Setting the capacitor bank anywhere else on the feeder would decrease rather than increase the savings from loss reduction.

Some of the *cardinal rules* that can be derived for the application of capacitor banks include the following:

1. The location of fixed shunt capacitors should be based on the average reactive load.
2. There is only one location for each size of capacitor bank that produces maximum loss reduction.
3. One large capacitor bank can provide almost as much savings as two or more capacitor banks of equal size.
4. When multiple locations are used for fixed-shunt-capacitor banks, the banks should have the same rating to be economical.
5. For a feeder with a uniformly distributed load, a fixed capacitor bank rated at two-thirds of the total reactive load and located at two-thirds of the distance out on the feeder from the source gives an 89% loss reduction.
6. The result of the two-thirds rule is particularly useful when the reactive load factor is high. It can be applied only when fixed shunt capacitors are used.
7. In general, particularly at low reactive load factors, some combination of fixed and switched capacitors gives the greatest energy loss reduction.
8. In actual situations, it may be difficult, if not impossible, to locate a capacitor bank at the optimum location; in such cases the permanent location of the capacitor bank ends up being suboptimum.

8.9 FURTHER THOUGHTS ON CAPACITORS AND IMPROVING POWER FACTORS

In a given circuit, when the relative current is reduced, the total current is reduced. If the real current does not change, which is usually the case, the power factor will improve as reactive current is reduced. When the reactive current becomes zero, all the current becomes the real current, and thus, the power factor will be equal to 1.0, that is, unity, or 100%. For instance, in Figure 8.46a, if a capacitor is installed to supply the total reactive power, or 60 kvar, the line power factor may be improved by supplying the load reactive power requirements by a capacitor.

In an induction load or equipment (e.g., induction motor) the real power is supplied and lost, whereas the reactive power is supplied and not lost. Here, the reactive power is used to store energy in the motor's magnetic energy field. Since the current is alternating, the magnetic field undergoes cycles of building up and breaking down. When the current is breaking down, the reactive current flows out of the motor back to the supply, as shown in Figure 8.46a. Note that the flow of the real power is unidirectional, that is, from the source to the motor. However, it is possible to provide only real power from the power source and provide the necessary reactive power from a capacitor.

Note that contrary to the real power flow (which is unidirectional), the reactive power flow is cyclic in nature, that is, flows out of capacitor to the motor and then flows back to the capacitor, as illustrated in Figure 8.46b. Here, the capacitor acts as a temporary storage area for the reactive power by storing energy units in electric field.

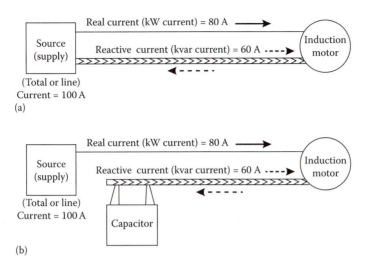

FIGURE 8.46 Schematic diagram for the use of capacitors to reduce total line current by supplying reactive power locally: (a) without the use of capacitors and (b) with the use of capacitors.

Figure 8.46 illustrates how to provide reactive power alternatively by the use of capacitors locally. Notice that the working load requires 80 A, but due to the magnetic requirements of the motor of 60 A, the supply circuit must carry 100 A. After the installation of the capacitor, the necessary magnetic requirements of the motor are supplied by the capacitor. Thus, the supply circuit needs to deliver only 80 A to do exactly the same work. The supply circuit is now carrying only kilowatts, so no system capacity is wasted in carrying the so-called *non-working current*.

Also, it can be observed from the power triangle that the simple subtraction of real power from total apparent power never equals the reactive power except at unity power factor.

However, in actual practice, it is generally not necessary nor economical to improve the power factor to 100%. Because of this fact, the utility companies provide the power factor of about 0.95 at their power plants. The remainder is supplied by using overexcited synchronous motors orating with no loads, or capacitor banks.

8.10 CAPACITOR TANK–RUPTURE CONSIDERATIONS

When the total energy input to the capacitor is larger than the strength of the tank's envelope to withstand such input, the tank of the capacitor ruptures. This energy input could happen under a wide range of current–time conditions. Through numerous testing procedures, capacitor manufacturers have generated tank–rupture curves as a function of fault current available.

The resulting tank–rupture time–current characteristic curves with which fuse selection is coordinated have furnished comparatively good protection against tank rupture. Figure 8.47 shows the results of tank–rupture tests conducted on all-film capacitors. Figure 8.48 shows the capacitor reliability cycle during its lifetime. The longer it takes for the dielectric material to wear out due to the forces generated by the combination of electric stress and temperature, the greater is its reliability. In other words, the wear-out process or time to failure is a measure of life and reliability. Currently, there are numerous methods that can be used to detect the capacitor tank ruptures. Burrage [24] categorizes them as follows:

1. Detection of sound produced by the rupture
2. Observation of smoke and/or vapor from the capacitor tank upon rupture
3. Observation of ultraviolet light generated by the arc getting outside the capacitor tank
4. Measurement of the change in arc voltage when the capacitor tank is breached
5. Detection of a sudden reduction in internal pressure
6. Measurement of the distortion generated by gas pressure within the capacitor tank

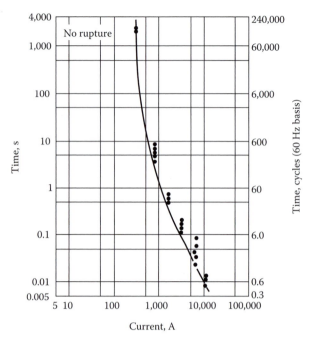

FIGURE 8.47 Time-to-rupture characteristics for 200 kvar 7.2 kV all-film capacitors. (From McGraw-Edison Company, *The ABC of Capacitors*, Bulletin R230-90-1, 1968.)

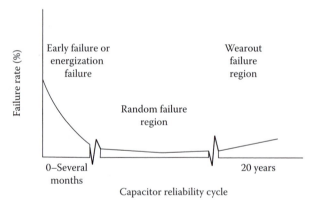

FIGURE 8.48 Capacitor reliability cycle. (From McGraw-Edison Company, *The ABC of Capacitors*, Bulletin R230-90-1, 1968.)

8.11 DYNAMIC BEHAVIOR OF DISTRIBUTION SYSTEMS

The characteristics of the distribution system dynamic behavior include (1) fault effects and transient recovery voltage (TRV), (2) switching and lightning surges, (3) inrush and cold-load current transients, (4) ferroresonance, and (5) harmonics.

In the event of a fault on a distribution system, there will be a substantial change in current magnitude on the faulted phase. It is also possible for the current to have a dc offset that is a function of the voltage wave at the time of the fault and the X/R ratio of the circuit. This may cause a voltage rise on the unfaulted phases due to neutral shift that results in saturation of transformers and increased load current magnitudes.

On the other hand, it may cause a reduction in voltage and load current on the faulted phase. When a circuit recloser clears the fault at a current zero, a higher-frequency transient voltage is superimposed on the power frequency recovery voltage. The resultant voltage is called the TRV. It is possible to have its crest magnitudes be two or three times the nominal system voltage. This may cause failure to clear or restrike that may produce substantial switching surges.

Switching surges are generated when loads, station capacitor banks, or feeders are energized or reenergized; or when faults are initiated, cleared, and reinitiated. The factors affecting the magnitude and duration of the resultant voltage transients include (1) the system impedance characteristics, (2) the amount of capacitive kilovars connected at the time of switching, (3) the location of the capacitor bank on the system, (4) the type of breaker, and (5) the breaker pole-closing angles. In general, the switching surges on distribution systems have not been taken seriously so far.

However, if the current trend toward higher-distribution voltage levels and reduced insulation levels continues, this may change. But voltage surges resulting from lightning strokes to the distribution line will usually require the most severe design requirements. The factors affecting the lightning surge include (1) the system configuration and the system grounding and shielding, (2) the stroke characteristics and stroke location, (3) the sparkover of arresters remote from the converters, (4) the amount of the connected capacitive kilovars in the surge path, and (5) the loss mechanism (corona, skin effect) in the surge path.

The energization of motors, transformers, capacitors, feeders, and loads generates current transients. For example, when motors and other loads draw high starting currents, capacitors draw a high-frequency inrush based on the instantaneous voltage and the circuit inductance as well as the capacitance, whereas in a transformer the magnitude of this inrush depends upon the voltage wave at the time of energization and the residual flux in the core.

It is important to recognize the fact that low voltage during inrush can harm equipment involved and stop the circuit from recovering without sectionalizing. Furthermore, protective devices may operate incorrectly or not operate due to the high-magnitude and high-frequency currents.

8.11.1 Ferroresonance

Ferroresonance is an oscillatory phenomenon caused by the interaction of system capacitance with the nonlinear inductance of a transformer. These capacitive and inductive elements make a series-resonant circuit that can generate high transient or sustained overvoltages that can damage system equipment.

These overvoltages are more likely to take place where a considerable length of cable is connected to an overloaded three-phase transformer (or bank) and single-phase switching is done at a point remote from the transformer (e.g., riser pole). Serious damage to equipment may be prevented by recognizing the conditions that increase the probability of these overvoltages and taking appropriate preventive measures.

The more serious overvoltages may be evidenced by (1) flashover or damage to lightning arresters, (2) transformer humming with only one phase closed, (3) damage to transformers and other equipment, (4) three-phase motor reversals, and (5) high secondary voltages.

Even though the ferroresonant phenomenon has been recognized for some time, until recently it has not been considered as a serious operating problem on electric distribution systems. Changes in the characteristics of distribution systems and in transformer design have resulted in the increased probability of ferroresonant overvoltages when switching three-phase transformer installations.

For example, the capacitance of cable is much greater (i.e., capacitive reactance lower) than that of open wire, and present trends are toward a greater use of underground cables due to the esthetic considerations. Also, system operation at higher than nominal voltages and trends in transformer design have led to the operation of distribution transformer cores at higher saturation.

Furthermore, the use of higher-distribution voltages results in distribution transformers with greater magnetizing reactance. At the same time, underground system capacitance will be greater (capacitive reactance lower).

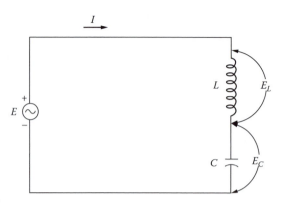

FIGURE 8.49   The *LC* circuit for ferroresonance.

Consider the *LC* circuit shown in Figure 8.49. Note that the resistance is neglected for the sake of simplicity. If the inductive reactance X_L of the inductor is equal in magnitude to the capacitive reactance X_C of the capacitor, the circuit is in resonance. The voltage E across the inductor is 180° out of phase with the voltage E_C across the capacitor. The voltages E_L and E_C can be expressed as

$$E_L = \frac{E}{jX_L - jX_C}(jX_L)$$

$$= \frac{E}{1 - X_C/X_L} \tag{8.111}$$

and

$$E_C = \frac{E}{jX_L - jX_C}(-jX_C)$$

$$= \frac{E}{1 - X_L/X_C} \tag{8.112}$$

Example 8.20

Consider a downtown 34.5 kV underground cable, and for the purpose of illustration, assume that $X_L/X_C = 0.9$ in the *LC* circuit given by Figure 8.49, and therefore, $X_C/X_L = 1.1111$. If the line-to-line voltage is 34.5 kV, determine

    a. The voltages E_L
    b. The voltages E_C

Solution

    a.
$$E_L = \frac{E}{1 - X_C/X_L} = \frac{E}{1 - 1.1111} = -9E$$

    or

$$E_L = -9 \times \left(\frac{34.5 \text{ kV}}{\sqrt{3}}\right)$$

$$\cong 179.3 \text{ kV}$$

In practice, such situation often takes place in downtown underground networks when a 34.5 kV underground cable feeds a transformer. As a result of the ferroresonance (because of the X_L of the transformer and X_C of the cable), the 34.5 kV underground cable breaks down due to such high voltages, causing a short-circuit condition. The fault usually takes place at the entrance terminal to the transformer. It has often puzzled some of the protection engineers. Thus, in such situations, it is a good idea to check for possible ferroresonance conditions.

b.
$$E_C = \frac{E}{1 - X_L/X_C} = \frac{E}{1 - 0.9} = 10E$$

or

$$E_C = 10 \times \left(\frac{34.5\,\text{kV}}{\sqrt{3}} \right)$$

$$= \frac{345\,\text{kV}}{\sqrt{3}}$$

$$\cong 199.2\,\text{kV}$$

Therefore, in this case the voltage across the capacitor is 10 times the source voltage. The nearer the circuit is to the actual resonance, the greater will be the overvoltage.

Although the previous text is a relatively simple example of a resonant circuit, the basic concept is very similar to ferroresonance with one notable exception. In a ferroresonant circuit the capacitor is in series with a nonlinear (iron-core) inductor. A plot of the voltampere or impedance characteristic of an iron-core reactor would have the same general shape as the BH curve of the iron core. If the iron-core reactor is operating at a point near saturation, a small increase in voltage can cause a large decrease in the effective inductive reactance of the reactor. Therefore, the value of inductive reactance can vary widely and resonance can occur over a range of capacitance values. The effects of ferroresonance can be minimized by such measures as

1. Using grounded-wye–grounded-wye transformer connection
2. Using open-wye–open-delta transformer connection
3. Using switches rather than fuses at the riser pole
4. Using single-pole devices only at the transformer location and three-pole devices for remote switching
5. Avoiding switching an unloaded transformer bank at a point remote from the transformers
6. Keeping X_C/X_M ratios high (10 or more)
7. Installing neutral resistance
8. Using dummy loads to suppress ferroresonant overvoltages
9. Assuring load is present during switching
10. Using larger transformers
11. Limiting the length of cable serving the three-phase installation
12. Using only three-phase switching and sectionalizing devices at the terminal pole
13. Temporarily grounding the neutral of a floating-wye primary during switching operations

8.11.2 HARMONICS ON DISTRIBUTION SYSTEMS

The power industry has recognized the problem of power system harmonics since the 1920s when distorted voltage and current waveforms were observed on power lines. However, the levels of harmonics on distribution systems have generally been insignificant in the past.

Today, it is obvious that the levels of harmonic voltages and currents on distribution systems are becoming a serious problem. Some of the most important power system operational problems caused by harmonics have been reported to include the following [25]:

1. Capacitor-bank failure from dielectric breakdown or reactive power overload
2. Interference with ripple control and power-line carrier systems, causing misoperation of systems that accomplish remote switching, load control, and metering
3. Excessive losses in—and heating of—induction and synchronous machines
4. Overvoltages and excessive currents on the system from resonance to harmonic voltages or currents on the network
5. Dielectric instability of insulated cables resulting from harmonic overvoltages on the system
6. Inductive interference with telecommunication systems
7. Errors in induction watthour meters
8. Signal interference and relay malfunction, particularly in solid-state and microprocessor-controlled systems
9. Interference with large motor controllers and power plant excitation systems (reported to cause motor problems as well as nonuniform output)

These effects depend, of course, on the harmonic source, its location on the power system, and the network characteristics that promote propagation of harmonics. There are numerous sources of harmonics. In general, the harmonic sources can be classified as (1) previously known harmonic sources and (2) new harmonic sources. The previously known harmonic sources include the following:

1. Tooth ripples or ripples in the voltage waveform of rotating machines
2. Variations in air-gap reluctance over synchronous machine pole pitch
3. Flux distortion in the synchronous machine from sudden load changes
4. Nonsinusoidal distribution of the flux in the air gap of synchronous machines
5. Transformer magnetizing currents
6. Network nonlinearities from loads such as rectifiers, inverters, welders, arc furnaces, voltage controllers, frequency converters

While the established sources of harmonics are still present on the system, the power network is also subjected to new harmonic sources:

1. Energy conservation measures, such as those for improved motor efficiency and load matching, which employ power semiconductor devices and switching for their operation. These devices often produce irregular voltage and current waveforms that are rich in harmonics.
2. Motor control devices such as speed controls for traction.
3. High-voltage dc power conversion and transmission.
4. Interconnection of wind and solar power converters with distribution systems.
5. Static-var compensators that have largely replaced synchronous condensors as continuously variable-var sources.
6. The development and potentially wide use of electric vehicles that require a significant amount of power rectification for battery charging.
7. The potential use of direct energy conversion devices, such as magnetohydrodynamics, storage batteries, and fuel cells, that require dc/ac power converters.

The presence of harmonics causes the distortion of the voltage or current waves. The distortions are measured in terms of the voltage or current harmonic factors. The IEEE Standard 519-1981 [26] defines the harmonic factors as the ratio of the root-mean-square value of all the harmonics to the root-mean-square value of the fundamental. Therefore, the voltage harmonic factor HF_v can be expressed as

$$HF_V = \frac{\left(E_3^2 + E_5^2 + E_7^2 + \cdots\right)^{1/2}}{E_1} \tag{8.113}$$

and the current harmonic factor HF_I can be expressed as

$$HF_I = \frac{\left(I_3^2 + I_5^2 + I_7^2 + \cdots\right)^{1/2}}{I_1} \tag{8.114}$$

The presence of the voltage distortion results in harmonic currents. Figure 8.50 shows harmonic analysis of a peaked no-load current wave.

The characteristics of harmonics on a distribution system are functions of both the harmonic source and the system response. For example, utilities are presently installing more and larger transformers to meet ever-increasing power demands. Each transformer is a source of harmonics to the distribution system.

Furthermore, these transformers are being operated closer to the saturation point. Transformer saturation results in a nonsinusoidal exciting current in the iron core when a sinusoidal voltage is applied. The level of transformer saturation is affected by the magnitude of the applied voltage.

When the applied voltage is above the rated voltage, the harmonic components of the exciting current increase dramatically. Owen [27] has demonstrated this for a typical substation power transformer, as shown in Figure 8.51. Also, some utility companies are overexciting distribution transformers as a matter of policy and practice, which compounds the harmonic problem.

The current harmonics of consequence that are produced by transformers are generally in the order of the third, fifth, and seventh. Table 8.6 gives a summary of the conditions obtaining third harmonics with different connections of double-wound three-phase transformers. The table is prepared for third harmonics in double-wound single-phase core- and shell-type transformers and in three-phase shell-type transformers for three-phase service. Figure 8.52 shows the shape of the resultant waves obtained when combining the fundamental and third harmonic along with different positions of the harmonic.

Note that at harmonic frequencies the phase angles (due to the various harmonic impedances of each load) can be anything between 0° and 360°. Also, as the harmonic order increases, the

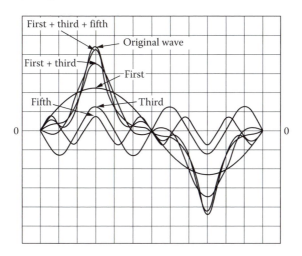

FIGURE 8.50　Harmonic analysis of peaked no-load current.

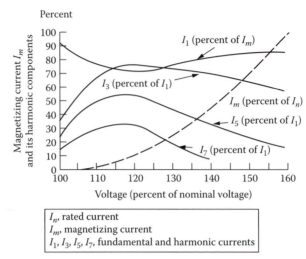

FIGURE 8.51 Harmonic components of transformer exciting current. (From Owen, R.E., Distribution system harmonics: Effects on equipment and operation, *Pacific Coast Electrical Association Engineering and Operating Conference*, Los Angeles, CA, March 15–16, 1979.)

power-line impedance itself plays the role of a controlling factor, and therefore, the harmonic current will have different phase angles at different locations.

The impact of harmonics on transformers is numerous. For example, voltage harmonics result in increased iron losses; current harmonics result in increased copper losses and stray flux losses. The losses may in turn cause the transformer to be overheated.

The harmonics may also cause insulation stresses and resonances between transformer windings and line capacitances at the harmonic frequencies. The total eddy-current losses are proportional to the harmonic frequencies and can be expressed as

$$\text{TECL} = \text{ECL}_1 \sum_{h=1}^{\infty} \left(\frac{h \times I_h}{I_1} \right)^2 \tag{8.115}$$

where

TECL is the total eddy-current loss
ECL_1 is the eddy-current loss at rated fundamental current
h is the harmonic order
I_1 is the rated fundamental current
I_h is the harmonic current

Capacitor-bank sizes and locations are critical factors in a distribution system's response to harmonic sources. The combination of capacitors and the system reactance causes both series- and parallel-resonant frequencies for the circuit.

The possibility of resonance between a shunt capacitor bank and the rest of the system, at a harmonic frequency, may be determined by calculating equal order of harmonic h at which resonance may take place [29]. This equal order of harmonic is found from

$$h = \left(\frac{S_{sc}}{Q_{cap}} \right)^{1/2} \tag{8.116}$$

TABLE 8.6
The Influence of Three-Phase Transformer Connections on Third Harmonics

Connections[a]	Primary					Secondary			
	Currents		Voltages		Flux	Currents		Voltages	
	No-Load	Line	Line	Phase		No-Load	Line	Line	Phase
1. Wye I.N./wye I.N.	Sine	Sine	Sine	Contains 3d h(P)	Contains 3d h(FT)		Sine	Sine	Contains 3d h(P)
2. Wye N. to G./wye I.N.	Contains 3d h(P)[b]	Contains 3d h(P)[b]	Sine	Contains 3d h(P)[b]	Contains 3d h(FT)[b]		Sine	Sine	Contains 3d h(P)[b]
3. Wye I.N./wye, four-wire	Sine	Sine	Sine	Contains 3d h(P)[b]	Contains 3d h(FT)[b]	Contains 3d h(P)[b]	Contains 3d h(P)[b]	Sine	Contains 3d h(P)[b]
4. Wye I.N. tertiary delta/wye I.N.	Sine in star, 3d h in delta (P)	Sine	Sine	Sine	Sine		Sine	Sine	Sine
5. Wye I.N./delta	Sine	Sine	Sine	Sine	Sine	Contains 3d h(P)	Sine	Sine	Sine
6. Wye N. to G./delta	Contains 3d h(P)[b]	Contains 3d h(P)[b]	Sine	Sine	Sine	Contains 3d h(P)[b]	Sine	Sine	Sine
7. Wye I.N./interconnected wye I.N.	Sine	Sine	Sine	Contains 3d h(P)	Contains 3d h(FT)		Sine	Sine	Sine
8. Wye I.N./interconnected wye, four-wire	Sine	Sine	Sine	Contains 3d h(P)	Contains 3d h(FT)		Sine	Sine	Sine
9. Delta/wye I.N.	Contains 3d h(P)	Sine	Sine	Sine	Sine		Sine	Sine	Sine
10. Delta/wye, four-wire	Contains 3d h(P)	Sine	Sine	Sine	Sine	Contains 3d h(P)	Contains 3d h(P)	Sine	Sine
11. Delta/delta	Contains 3d h(P)	Sine	Sine	Sine	Sine	Contains 3d h(P)[b]	Sine	Sine	Sine

Source: Stigant, S.A. and Franklin, A.C., *The J&P Transformer Book*, Butterworth, London, U.K., 1973.

[a] I.N., isolated neutral; N. to G., transformer primary neutral connected to generator neutral; (P), peaked wave; (FT), flat-top wave.

[b] In all these cases the third-harmonic component is less than it otherwise would be if (1) the circulating third-harmonic current flowed through a closed delta winding only or (2) the neutral point was isolated.

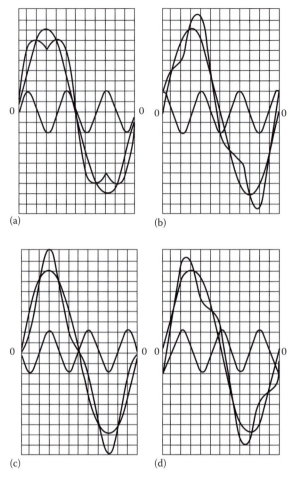

FIGURE 8.52 Combinations of fundamental and third-harmonic waves: (a) harmonic in phase, (b) harmonic 90° leading, (c) harmonic in opposition, and (d) harmonic 90° lagging. (From Stigant, S.A. and Franklin, A.C., *The J&P Transformer Book*, Butterworth, London, U.K., 1973.)

where

S_{sc} is the short-circuit power of a system at the point of application, MVA
Q_{cap} is the capacitor-bank size, Mvar

The parallel-resonant frequency f_p can be expressed as

$$f_p = f_1 \times h \tag{8.117}$$

Substituting Equation 8.120 into Equation 8.121,

$$f_p = f_1 \left(\frac{S_{sc}}{Q_{cap}} \right)^{1/2} \tag{8.118}$$

or

$$f_p = f_1 \left(\frac{X_{cap}}{X_{sc}} \right)^{1/2}$$

(8.119)

or

$$f_p = \frac{1}{2\pi} \left(\frac{1}{L_{sc} \times C} \right)^{1/2}$$

(8.120)

where
f_1 is the fundamental frequency, Hz
X_{cap} is the reactance of capacitor bank, pu or Ω
X_{sc} is the reactance of power system, pu or Ω
L_{sc} is the inductance of power system, H
C is the capacitance of capacitor bank, F

The effects of the harmonics on the capacitor bank include (1) overheating of the capacitors, (2) overvoltage at the capacitor bank, (3) changed dielectric stress, and (4) losses in capacitors. According to Kimbark [29], the increase of losses in capacitors due to harmonics can be expressed as

$$LCDH = \sum C(\tan \delta)_h w_h V_h^2$$

(8.121)

where
LCDH is the losses in capacitors due to harmonics
C is the capacitance
$(\tan \delta)_h$ is the loss factor at frequency of the hth harmonic
$w_h = 2\pi$ times frequency of the hth harmonic
V_h is the rms voltage of the hth harmonic

The harmonic control techniques include (1) locating the capacitor banks strategically, (2) selecting capacitor-bank sizes properly, (3) ungrounding or deleting the capacitor bank, (4) using shielded cables, (5) controlling grounds properly, and (6) using harmonic filters.

PROBLEMS

8.1 Assume that a feeder supplies an industrial consumer with a cumulative load of (1) induction motors totaling 300 hp that run at an average efficiency of 89% and a lagging average power factor of 0.85, (2) synchronous motors totaling 100 hp with an average efficiency of 86%, and (3) a heating load of 100 kW. The industrial consumer plans to use the synchronous motors to correct its overall power factor. Determine the required power factor of the synchronous motors to correct the overall power factor at peak load to
a. Unity
b. 0.96 lagging

8.2   A 2.4/4.16 kV wye-connected feeder serves a peak load of 300 A at a lagging power factor of 0.7 connected at the end of the feeder. The minimum daily load is approximately 135 A at a power factor of 0.62. If the total impedance of the feeder is $0.50 + j1.35$ Ω, determine the following:

   a.   The necessary kilovar rating of the shunt capacitors located at the load to improve the peak-load power factor to 0.96
   b.   The reduction in kilovoltamperes and line current due to the capacitors
   c.   The effects of the capacitors on the voltage regulation and voltage drop in the feeder
   d.   The power factor at minimum daily load level

8.3   Assume that a locked-rotor starting current of 90 A at a lagging load factor of 0.30 is supplied to a motor that is operated discontinuously. A normal operating current of 15 A, at a lagging power factor of 0.80, is drawn by the motor from the 2.4/4.16 kV feeder of Problem 8.2. Assume that a series capacitor is desired to be installed in the feeder to improve the voltage regulation and limit lamp flicker from the intermittent motor starting, and determine the following:

   a.   The voltage dip due to the motor starting, before the installation of the series capacitor
   b.   The necessary size of the capacitor to restrict the voltage dip at motor start to not more than 3%

8.4   Assume that a three-phase distribution substation transformer has a nameplate rating of 7250 kVA and a thermal capability of 120%. If the connected load is 8816 kVA with a 0.85 lagging power factor, determine the following:

   a.   The kilovar rating of the shunt capacitor bank required to decrease the kilovoltampere load on the transformer to its capability level
   b.   The power factor of the corrected load
   c.   The kilovar rating of the shunt capacitor bank required to correct the load power factor to unity
   d.   The corrected kilovoltampere load at this unity power factor

8.5   Assume that the NP&NL Utility Company is presently operating at 90% power factor. It is desired to improve the power factor to 98%. To study the power factor improvement, a number of load flow runs have been made and the results are summarized in the following table. Using the relevant additional information given in Table P8.5, repeat Example 8.19.

8.6   Assume that a manufacturing plant has a three-phase in-plant generator to supply only three-phase induction motors totaling 1200 hp at 2.4 kV with a lagging power factor and efficiency of 0.82 and 0.93, respectively. Using the given information, determine the following:

   a.   Find the required line current to serve the 1200 hp load and the required capacity of the generator.
   b.   Assume that 500 hp of the 1200 hp load is produced by an overexcited synchronous motor operating with a leading power factor and efficiency of 0.90 and 0.93, respectively. Find the required new total line current and the overall power factor.
   c.   Find the required size of shunt capacitors to be installed to achieve the same overall power factor as found in part b by replacing the overexcited synchronous motor.

TABLE P8.5
Summary of Load Flows

Comment	At 90% PF	At 99% PF
Total loss reduction due to capacitors applied to substation buses, kW	496	488
Additional loss reduction due to capacitors applied to feeders, kW	84	72
Total demand reduction due to capacitors applied to substation buses and feeders, kVA	21,824	19,743
Total required capacitor additions at buses and feeders, kvar	9,512	2,785

8.7 Verify that the loss reduction with two capacitor banks is

$$\Delta L = 3\alpha c\{x_1[(2-x_1)+\lambda x_1 -3c]+x_2[(2-x_2)+\lambda x_2 -c]\}$$

8.8 Derive Equation 8.96 from Equation 8.95.
8.9 Verify that the optimum loss reduction is

$$\Delta L_{opt} = 3\alpha c \sum_{i=1}^{n}\left[\frac{1}{1-\lambda}-\frac{(2i-1)c}{1-\lambda}+\frac{i^2 c^2}{1-\lambda}-\frac{c^2}{4(1-\lambda)}-\frac{ic^2}{1-\lambda}\right]$$

8.10 Derive Equation 8.105 from Equation 8.96.
8.11 Verify Equation 8.106.
8.12 If a power system has a 15,000 kVA capacity and is operating at a power factor of 0.65 lagging and the cost of synchronous capacitors is $15/kVA, find the investment required to correct the power factor to
   a. 0.85 lagging power factor
   b. Unity power factor
8.13 If a power system has a 20,000 kVA capacity and is operating at a power factor of 0.6 lagging and the cost of synchronous capacitors is $12.50/kVA, find the investment required to correct the power factor to
   a. 0.85 lagging power factor
   b. Unity power factor
8.14 If a power system has a 20,000 kVA capacity and is operating at a power factor of 0.6 lagging and the cost of synchronous capacitors is $17.50/kVA, develop a table showing the required (leading) reactive power to correct the power factor to
   a. 0.85 lagging power factor
   b. 0.95 lagging power factor
   c. Unity power factor
8.15 If a power system has a 25,000 kVA capacity and is operating at a power factor of 0.7 lagging and the cost of synchronous capacitors is $12.50/kVA, develop a table showing the required (leading) reactive power to correct the power factor to
   a. 0.85 lagging power factor
   b. 0.95 lagging power factor
   c. Unity power factor
8.16 If a power system has an 8000 kVA capacity and is operating at a power factor of 0.7 lagging and the cost of synchronous capacitors is $15/kVA, find the investment required to correct the power factor to
   a. Unity power factor
   b. 0.85 lagging power factor
8.17 Assume that a feeder supplies an industrial consumer with a cumulative load of (1) induction motors totaling 200 hp that run at an average efficiency of 90% and a lagging average power factor of 0.80, (2) synchronous motors totaling 200 hp with an average efficiency of 80%, and a lagging average power factor of 0.80 and (3) a heating load of 50 kW. The industrial consumer plans to use the synchronous motors to correct its overall power factor. Determine the required power factor of synchronous motors to correct the overall power factor at peak load to unity power factor.
8.18 Resolve Example 8.9 by using MATLAB for a 300 kW lagging load. Assume that all the other quantities remain the same.
8.19 Resolve Example 8.11 by using MATLAB. Assume that all the quantities remain the same.

REFERENCES

1. Hopkinson, R. H.: Economic power factor-key to kvar supply, *Electr. Forum*, 6(3), 1980, 20–22.
2. McGraw-Edison Company: *The ABC of Capacitors*, Bulletin R230-90-1, 1968.
3. Zimmerman, R. A.: Economic merits of secondary capacitors, *AIEE Trans.*, 72, 1953, 694–697.
4. Wallace, R. L.: Capacitors reduce system investment and losses, *The Line*, 76(1), 1976, 15–17.
5. Neagle, N. M. and D. R. Samson: Loss reduction from capacitors installed on primary feeders, *AIEE Trans.*, 75(pt. III), October 1956, 950–959.
6. Baum, W. U. and W. A. Frederick: A method of applying switched and fixed capacitors for voltage control. *IEEE Trans. Power Appar. Syst.*, PAS-84(1), January 1965, 42–48.
7. Schmidt, R. A.: DC circuit gives easy method of determining value of capacitors in reducing 12R losses, *AIEE Trans.*, 75(pt. III), October 1956, 840–848.
8. Westinghouse Electric Corporation: *Electric Utility Engineering Reference Book—Distribution Systems*, vol. 3, East Pittsburgh, PA, 1965.
9. Maxwell, N.: The economic application of capacitors to distribution feeders, *AIEE Trans.*, 79(pt. III), August 1960, 353–359.
10. Cook, R. F.: Optimizing the application of shunt capacitors for reactive voltampere control and loss reduction, *AIEE Trans.*, 80(pt. III), August 1961, 430–444.
11. Schmill, J. V.: Optimum size and location of shunt capacitors on distribution systems, *IEEE Trans. Power Apper. Syst.*, PAS-84(9), September 1965, 825–832.
12. Chang, N. E.: Determination of primary-feeder losses, *IEEE Trans. Power Appar. Syst.*, Pas-87(12), December 1968, 1991–1994.
13. Chang, N. E.: Locating shunt capacitors on primary feeder for voltage control and loss reduction, *IEEE Trans. Power Appar. Syst.*, PAS-88(10), October 1969, 1574–1577.
14. Chang, N. E.: Generalized equations on loss reduction with shunt capacitors, *IEEE Trans. Power Appar. Syst.*, PAS-91(5), September/October 1972, 2189–2195.
15. Bae, Y. G.: Analytical method of capacitor allocation on distribution primary feeders, *IEEE Trans. Power Appar. Syst.*, PAS-97(4), July/August 1978, 1232–1238.
16. Gönen, T. and F. Djavashi: Optimum loss reduction from capacitors installed on primary feeders, *IEEE Midwest Power Symposium*, Purdue University, West Lafayette, IN, October 27–28, 1980.
17. Gönen, T. and F. Djavashi: Optimum shunt capacitor allocation on primary feeders, *IEEE MEXICON-80 International Conference*, Mexico City, Mexico, October 22–25, 1980.
18. Lapp, J.: The impact of technical developments on power capacitors, *The Line*, 80(2), July 1980, 19–24.
19. Grainger, J. J. and S. H. Lee.: Optimum size and location of shunt capacitors for reduction of losses on distribution feeders, *IEEE Trans. Power Appar. Syst.*, PAS-100, March 1981, 1105–1118.
20. Oklahoma Gas and Electric Company: *Engineering Guides*, Oklahoma City, OK, January 1981.
21. Lee, S. H. and J. J. Grainger: Optimum placement of fixed and switched capacitors on primary distribution feeders, *IEEE Trans. Power Appar. Syst.*, PAS-100, January 1981, 345–351.
22. Grainger, J. J., A. A. El-Kib, and S. H. Lee: Optimal capacitor placement on three-phase primary feeders: Load and feeder unbalance effects, Paper 83 WM 160-9, *IEEE PES Winter Meeting*, New York, January 30–February 4, 1983.
23. Grainger, J. J., S. Civanlar, and S. H. Lee: Optimal design and control scheme for continuous capacitive compensation of distribution feeders, Paper 83 WM 159-1, *IEEE PES Winter Meeting*, New York, January 30–February 4, 1983.
24. Burrage, L. M.: Capacitor tank ruptures studied, *The Line*, 76(2), 1976, 2–5.
25. Mahmoud, A. A., R. E. Owen, and A. E. Emanuel: Power system harmonics: An overview, *IEEE Trans. Power Appar. Svst.*, PAS-102(8), August 1983, 2455–2460.
26. IEEE: *IEEE Guide for Harmonic Control and Reactive Compensation of Static Power Converters*, IEEE Std. 519-1981, 1981.
27. Owen, R. E.: Distribution system harmonics: Effects on equipment and operation, *Pacific Coast Electrical Association Engineering and Operating Conference*, Los Angeles, CA, March 15–16, 1979.
28. Stigant, S. A. and A. C. Franklin: *The J&P Transformer Book*, Butterworth, London, U.K., 1973.
29. Kimbark, E. W.: *Direct Current Transmission*, Wiley, New York, 1971.
30. Electric Power Research Institute: *Study of Distribution System Surge and Harmonic Characteristics*, Final Report, EPRI EL-1627, Palo Alto, CA, November 1980.
31. Owen, R. E., M. F. McGranaghan, and J. R. Vivirito: Distribution system harmonics: Controls for large power converters, Paper 81 SM 482-9, *IEEE PES Summer Meeting*, Portland, OR, July 26–31, 1981.

32. McGranaghan, M. F., R. C. Dugan, and W. L. Sponsler: Digital simulation of distribution system frequency-response characteristics, Paper 80 SM 665-0, *IEEE PES Summer Meeting*, Minneapolis, MN, July 13–18,1980.
33. McGranaghan, M. F., J. H. Shaw, and R. E. Owen: Measuring voltage and current harmonics on distribution systems, Paper 81 WM 126-2, *IEEE PES Winter Meeting*, Atlanta, GA, February 1–6, 1981.
34. Szabados, B., E. J. Burgess, and W. A. Noble: Harmonic interference corrected by shunt capacitors on distribution feeders, *IEEE Trans. Power Appar. Syst.*, PAS-96(1), January/February 1977, 234–239.
35. Gönen, T. and A. A. Mahmoud: Bibliography of power system harmonics, Part I, *IEEE Trans. Power Appar. Syst.*, PAS-103(9), September 1984, 2460–2469.
36. Gönen, T. and A. A. Mahmoud: Bibliography of power system harmonics, Part II, *IEEE Trans. Power Appar. Syst.*, PAS-103(9), September 1984, 2470–2479.

9 Distribution System Voltage Regulation

Nothing is so firmly believed as what we least know.

M.E. De Montaigne, Essays, 1580

Talk sense to a fool and he calls you foolish.

Euripides, The Bacchae, 407 B. C.

But talk nonsense to a fool and he calls you a genius.

Turan Gönen

9.1 BASIC DEFINITIONS

Voltage regulation: The percent voltage drop of a line (e.g., a feeder) with respect to the receiving-end voltage. Therefore,

$$\% \text{ regulation} = \frac{|V_s| - |V_r|}{|V_r|} \times 100 \tag{9.1}$$

Voltage drop: The difference between the sending-end and the receiving-end voltages of a line

Nominal voltage: The nominal value assigned to a line or apparatus or a system of a given voltage class

Rated voltage: The voltage at which performance and operating characteristics of apparatus are referred

Service voltage: The voltage measured at the ends of the service-entrance apparatus

Utilization voltage: The voltage measured at the ends of an apparatus

Base voltage: The reference voltage, usually 120 V

Maximum voltage: The largest 5-min average voltage

Minimum voltage: The smallest 5-min voltage

Voltage spread: The difference between the maximum and minimum voltages, without voltage dips due to motor starting

9.2 QUALITY OF SERVICE AND VOLTAGE STANDARDS

In general, performance of distribution systems and quality of the service provided are measured in terms of freedom from interruptions and maintenance of satisfactory voltage levels at the customer's premises that are within limits appropriate for this type of service. Due to economic considerations, an electric utility company cannot provide each customer with a constant voltage matching exactly the nameplate voltage on the customer's utilization apparatus.

Therefore, a common practice among the utilities is to stay with preferred voltage levels and ranges of variation for satisfactory operation of apparatus as set forth by the American National Standards Institute (ANSI) Standard [2]. In many states, the ANSI standard is the basis for the state regulatory commission rulings on setting forth voltage requirements and limits for various classes of electric service.

In general, based on experience, too-high steady-state voltage causes reduced light bulb life, reduced life of electronic devices, and premature failure of some types of apparatus. On the other hand, too-low steady-state voltage causes lowered illumination levels, shrinking of TV pictures, slow heating of heating devices, difficulties in motor starting, and overheating and/or burning out of motors. However, most equipment and appliances operate satisfactorily over some range of voltage so that a reasonable tolerance is allowable.

The nominal voltage standards for a majority of the electric utilities in the United States to serve residential and commercial customers are

1. 120/240 V three-wire single phase
2. 240/120 V four-wire three-phase delta
3. 208Y/120 V four-wire three-phase wye
4. 480Y/277 V four-wire three-phase wye

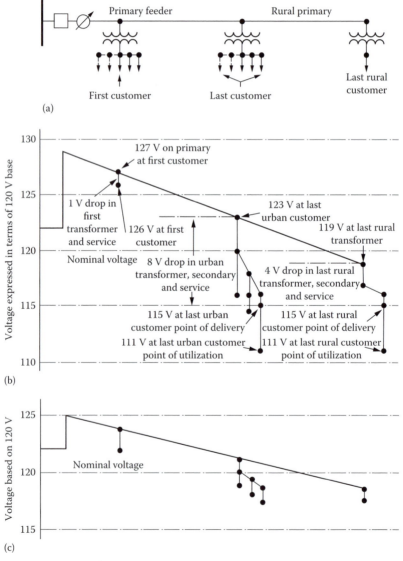

FIGURE 9.1 Illustration of voltage spread on a radial primary feeder: (a) one-line diagram of a feeder circuit, (b) voltage profile at peak-load conditions, and (c) voltage profile at light-load conditions.

As shown in Figure 9.1, *the voltage on a distribution circuit varies from a maximum value at the customer nearest to the source (first customer) to a minimum value at the end of the circuit (last customer)*. For the purpose of illustration, Table 9.1 gives typical secondary voltage standards applicable to residential and commercial customers. These voltage limits may be set by the state regulatory commission as a guide to be followed by the utility.

As can be observed in Table 9.1, for any given nominal voltage level, the actual operating values can vary over a large range. This range has been segmented into three zones, namely, (1) the *favorable zone* or *preferred zone*, (2) the *tolerable zone*, and (3) the *extreme zone*. The favorable zone includes the majority of the existing operating voltages and the voltages within this zone (i.e., range A) to produce satisfactory operation of the customer's equipment.

The distribution engineer tries to keep the voltage of every customer on a given distribution circuit within the favorable zone. Figure 9.1 illustrates the results of such efforts on urban and rural circuits. The tolerable zone contains a band of operating voltages slightly above and below the favorable zone. The operating voltages in the tolerable zone (i.e., range B) are usually acceptable for most purposes. For example, in this zone, the customer's apparatus may be expected to operate satisfactorily, although its performance may perhaps be less than warranted by the manufacturer.

However, if the voltage in the tolerable zone results in unsatisfactory service of the customer's apparatus, the voltage should be improved. The extreme or emergency zone includes voltages on the fringes of the tolerable zone, usually within 2% or 3% above or below the tolerable zone. They may or may not be acceptable depending on the type of application. At times, the voltage that usually stays within the tolerable zone may infrequently exceed the limits because of some extraordinary conditions. For example, failure of the principal supply line, which necessitates the use of alternative routes or voltage regulators being out of service, can cause the voltages to reach the emergency limits.

However, if the operating voltage is held within the extreme zone under these conditions, the customer's apparatus may still be expected to provide dependable operation, even though not the standard performance. However, voltages outside the extreme zone should not be tolerated under any conditions and should be improved right away. Usually, *the maximum voltage drop in the customer's wiring between the point of delivery and the point of utilization is accepted as* 4 V *based on* 120 V.

TABLE 9.1

Typical Secondary Voltage Standards Applicable to Residential and Commercial Customers

	Voltage Limits		
	At Point of Delivery		At Point of Utilization
Nominal Voltage Class	Maximum	Minimum	Minimum
120/240 V 1ϕ and 240/120 V 3ϕ			
Favorable zone, range A	126/252	114/228	110/220
Tolerable zone, range B	127/254	110/220	106/212
Extreme zone, emergency	130/260	108/216	104/208
208Y/120 V 3ϕ			
Favorable zone, range A	218Y/126	197Y/114	191Y/110
Tolerable zone, range B	220Y/127	191Y/110	184Y/106
Extreme zone, emergency	225Y/130	187Y/108	180Y/104
408Y/277 V 3ϕ			
Favorable zone, range A	504Y/291	456Y/263	440Y/254
Tolerable zone, range B	508Y/293	440Y/254	424Y/245
Extreme zone, emergency	520Y/300	432Y/249	416Y/240

9.3 VOLTAGE CONTROL

To keep distribution-circuit voltages within permissible limits, means must be provided to control the voltage, that is, to increase the circuit voltage when it is too low and to reduce it when it is too high. There are numerous ways to improve the distribution system's overall voltage regulation. The complete list is given by Lokay [1] as

1. Use of generator voltage regulators
2. Application of voltage-regulating equipment in the distribution substations
3. Application of capacitors in the distribution substation
4. Balancing of the loads on the primary feeders
5. Increasing of feeder conductor size
6. Changing of feeder sections from single phase to multiphase
7. Transferring of loads to new feeders
8. Installing of new substations and primary feeders
9. Increasing of primary voltage level
10. Application of voltage regulators out on the primary feeders
11. Application of shunt capacitors on the primary feeders
12. Application of series capacitors on the primary feeders

The selection of a technique or techniques depends upon the particular system requirement. However, *automatic voltage regulation is always provided by* (1) *bus regulation at the substation*, (2) *individual feeder regulation in the substation*, and (3) *supplementary regulation along the main by regulators mounted on poles.* Distribution substations are equipped with *load-tap-changing* (LTC) *transformers* that operate automatically under load or with separate voltage regulators that provide bus regulation.

Voltage-regulating apparatus are designed to maintain automatically a predetermined level of voltage that would otherwise vary with the load. As the load increases, the regulating apparatus boosts the voltage at the substation to compensate for the increased voltage drop in the distribution feeder. In cases where customers are located at long distances from the substation or where voltage drop along the primary circuit is excessive, additional regulators or capacitors, located at selected points on the feeder, provide supplementary regulation. Many utilities have experienced that the most economical way of regulating the voltage within the required limits is to apply both step voltage regulators and shunt capacitors.

Capacitors are installed out on the feeders and on the substation bus in adequate quantities to accomplish the economic power factor. Many of these installations have sophisticated controls designed to perform automatic switching. A fixed capacitor is not a voltage regulator and cannot be directly compared to regulators, but, in some cases, automatically switched capacitors can replace conventional step-type voltage regulators for voltage control on distribution feeders.

9.4 FEEDER VOLTAGE REGULATORS

Feeder voltage regulators are used extensively to regulate the voltage of each feeder separately *to maintain a reasonable constant voltage at the point of utilization.* They are either the induction type or the step type. However, since today's modern step-type voltage regulators have practically replaced induction-type regulators, only step-type voltage regulators will be discussed in this chapter.

Step-type voltage regulators can be either (1) *station type*, which can be single or three phase and which can be used in substations for bus voltage regulation (BVR) or individual feeder voltage regulation; or (2) *distribution type*, which can be only single phase and used pole-mounted out on overhead primary feeders. Single-phase step-type voltage regulators are available in sizes from 25 to 833 kVA, whereas three-phase step-type voltage regulators are available in sizes from 500 to 2000 kVA.

For some units, the standard capacity ratings can be increased by 25%–33% by forced-air cooling. Standard voltage ratings are available from 2,400 to 19,920 V, allowing regulators to be used on distribution circuits from 2,400 to 34,500 V grounded-wye/19,920 V multigrounded-wye. Station-type step voltage regulators for BVR can be up to 69 kV.

A step-type voltage regulator is fundamentally an autotransformer with many taps (or steps) in the series winding. Most regulators are designed to correct the line voltage from 10% boost to 10% buck (i.e., ±10%) in 32 steps, with a ⅝% voltage change per step.

(Note that the full voltage regulation range is 20%, and therefore if the 20% regulation range is divided by the 32 steps, a percent regulation per step is found.) If two internal coils of a regulator are connected in series, the regulator can be used for ±10% regulation; when they are connected in parallel, the current rating of the regulator would increase to 160%, but the regulation range would decrease to ±5%. Figure 9.2 shows a typical single-phase 32-step pole-type voltage regulator; Figure 9.3 shows its application on a feeder with essential components. Figure 9.4 shows typical

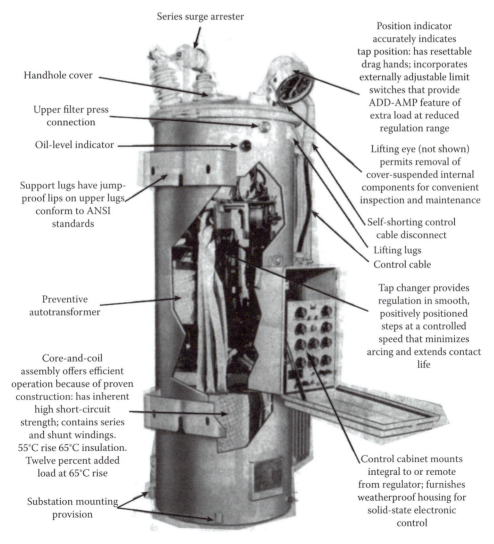

FIGURE 9.2 Typical single-phase 32-step pole-type voltage regulator used for 167 kVA or below. (McGraw-Edison Company, Belleville, NJ.)

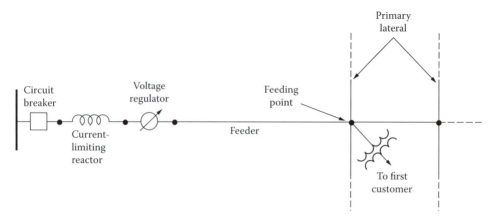

FIGURE 9.3 One-line diagram of a feeder, indicating the sequence of essential components.

FIGURE 9.4 Typical platform-mounted voltage-regulators. (Siemens-Allis Company.)

platform-mounted voltage regulators. Individual feeder regulation for a large utility can be provided at the substation by a bank of distribution voltage regulators, as shown in Figure 9.5.

   In addition to its autotransformer component, a step-type regulator also has two other major components, namely, the tap-changing mechanism and the control mechanism, as shown in Figure 9.2. Each voltage regulator ordinarily is equipped with the necessary controls and accessories so that the taps are changed automatically under load by a tap changer that responds to a voltage-sensing control to maintain a predetermined output voltage. By receiving its inputs from potential transformer (PT) and current transformer (CT), the control mechanism provides control of voltage level and bandwidth (BW).

FIGURE 9.5 Individual feeder voltage regulation provided by a bank of distribution voltage regulators. (Siemens-Allis Company.)

One of such control mechanisms is a *voltage regulating relay* (VRR), which controls tap changes. As illustrated in Figure 9.6, this relay has the following three basic settings that control tap changes:

1. *Set voltage*: It is the desired output of the regulator. It is also called the *set point* or *band center*.
2. *Bandwidth*: Voltage regulator controls monitor the difference between the measured voltage and the set voltage. Only when the difference exceeds one half of the BW will a tap change start.
3. *Time delay (TD)*: It is the waiting time between the time when the voltage goes out of the band and when the controller initiates the tap change. Longer TDs reduce the number of tap changes. Typical TDs are 10–120 s.

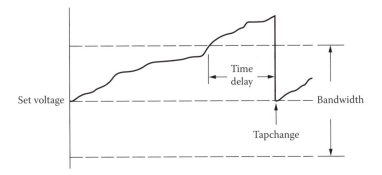

FIGURE 9.6 Regulator tap controls based on the set voltage, bandwidth, and time delay.

    Furthermore, the control mechanism also provides the ability to adjust line-drop compensation by selecting the resistance and reactance settings, as shown in Figure 9.7. Figure 9.8 shows a standard direct-drive tap changer. Figure 9.9 shows four-step *auto-booster regulators*. Auto-boosters basically are single-phase regulating autotransformers, which provide four-step feeder voltage regulation without the high degree of sophistication found in 32-step regulators. They can be used on circuits rated 2.4 to 12 kV delta and 2.4/4.16 to 19.92/34.5 kV multigrounded-wye. The auto-booster unit can have a continuous current rating of either 50 or 100 A. Each step represents either 1½% or 2½% voltage change depending on whether the unit has a 6% or 10% regulation range, respectively. They cost much less than the standard voltage regulators.

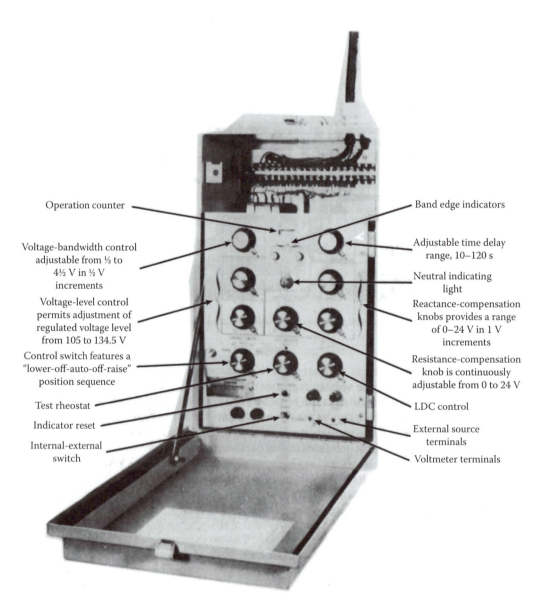

FIGURE 9.7 Features of the control mechanism of a single-phase 32-step voltage regulator. (McGraw-Edison Company, Belleville, NJ.)

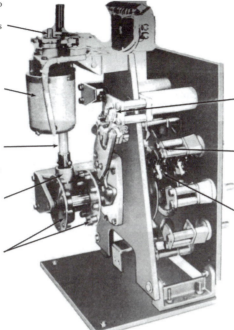

Geneva pinion transmits motion to drive; shaft; makes one complete revolution for every six revolutions of the pinion

Drive motor is reversible single type initiates capacitor type; initiates drive mechanism action; once a tap change is initiated, the holding switch provides an uninterruptible power supply through a complete tap change

Drive shaft transmits motion through gearing to scroll cam; operates through 180° per tap change

Scroll cam imparts; motion to roller plates designed to produce smooth tap changes

Roller plates are individual and independent for imparting motion through shaft to a particular movable contact

Reversing switch has sturdy, arc-resistant, copper-tungsten contacts that make solid connection to series winding

Stationary contacts have arc-resistant copper-tungsten tips on low-resistance copper base to ensure long life

Movable contacts slide on stationary contacts in a strong wiping movement that exerts firm pressure on two surfaces of stationary contact: are constructed of arc-resistant copper-tungsten: have individual collector rings

FIGURE 9.8 Standard direct-drive tap changer used through 150 kV BIL, above 219 A. (McGraw-Edison Company, Belleville, NJ.)

(a)

(b)

FIGURE 9.9 Four-step auto-booster regulators: (a) 50 A unit and (b) 100 A unit. (McGraw-Edison Company, Belleville, NJ.)

9.5 LINE-DROP COMPENSATION

Voltage regulators located in the substation or on a feeder are used to keep the voltage constant at a fictitious regulation or regulating point (RP) without regard to the magnitude or power factor of the load. The regulation point is usually selected to be somewhere between the regulator and the end of the feeder. This automatic voltage maintenance is achieved by dial settings of the adjustable resistance and reactance elements of a unit called the *line-drop compensator* (LDC) located on the control panel of the voltage regulator. Figure 9.10 shows a simple schematic diagram and phasor diagram of the control circuit and LDC circuit of a step or induction voltage regulator. Determination of the appropriate dial settings depends upon whether or not any load is tapped off the feeder between the regulator and the regulation point.

If no load is tapped off the feeder between the regulator and the regulation point, the R dial setting of the LDC can be determined from

$$R_{set} = \frac{CT_P}{PT_N} \times R_{eff} \ \Omega \tag{9.2}$$

where
CT_P is the rating of the current transformer's primary
PT_N is the potential transformer's turns ratio = V_{pri}/V_{sec}
R_{eff} is the effective resistance of a feeder conductor from regulator station to regulation point, Ω

$$R_{eff} = r_a \times \frac{\ell - s_1}{2} \ \Omega \tag{9.3}$$

where
r_a is the resistance of a feeder conductor from regulator station to regulation point, Ω/mi per conductor
s_1 is the length of three-phase feeder between regulator station and substation, mi (multiply length by 2 if feeder is in single phase)
ℓ is the primary feeder length, mi

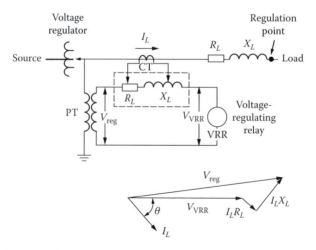

FIGURE 9.10 Simple schematic diagram and phasor diagram of the control circuit and line-drop compensator circuit of a step or induction voltage regulator. (From Westinghouse Electric Corporation, *Electric Utility Engineering Reference Book-Distribution Systems*, Vol. 3, Westinghouse Electric Corporation, East Pittsburgh, PA, 1965.)

Also, the *X dial setting of the LDC* can be determined from

$$X_{\text{set}} = \frac{\text{CT}_P}{\text{PT}_N} \times X_{\text{eff}} \; \Omega \tag{9.4}$$

where X_{eff} is the effective reactance of a feeder conductor from regulator to regulation point, Ω

$$X_{\text{eff}} = x_L \times \frac{\ell - s_1}{2} \; \Omega \tag{9.5}$$

and

$$x_L = x_a + x_d \; \Omega/\text{mi} \tag{9.6}$$

where

x_a is the inductive reactance of individual phase conductor of feeder at 12-in spacing, Ω/mi
x_d is the inductive-reactance spacing factor, Ω/mi
x_L is the inductive reactance of feeder conductor, Ω/mi

Note that since the *R* and *X* settings are determined for the total connected load, rather than for a small group of customers, the resistance and reactance values of the transformers are not included in the effective resistance and reactance calculations.

If load is tapped off the feeder between the regulator station and the regulation point, the *R* dial setting of the LDC can still be determined from Equation 9.2, but the determination of the R_{eff} is somewhat more involved. Lokay [1] gives the following equations to calculate the effective resistance:

$$R_{\text{eff}} = \frac{\sum_{i=1}^{n} |\text{VD}_R|_i}{|I_L|} \; \Omega \tag{9.7}$$

and

$$\sum_{i=1}^{n} |\text{VD}_R|_i = |I_{L,1}| \times r_{a,1} \times l_1 + |I_{L,2}| \times r_{a,2} \times l_2 + \cdots + |I_{L,n}| \times r_{a,n} \times l_n \tag{9.8}$$

where

$\|\text{VD}_R\|_i$ is the voltage drop due to line resistance of *i*th section of feeder between regulator station and regulation point, V/section
$\sum_{i=1}^{n} |\text{VD}_R|_i$ is the total voltage drop due to line resistance of feeder between regulator station and regulation point, V
$\|I_L\|$ is the magnitude of load current at regulator location, A
$\|I_{L,i}\|$ is the magnitude of load current in *i*th feeder section, A
$r_{a,i}$ is the resistance of a feeder conductor in *i*th section of feeder, Ω/mi
ℓ_i is the length of *i*th feeder section, mi

Also, the *X dial setting* of the LDC can still be determined from Equation 9.4, but the determination of the X_{eff} is again somewhat more involved. Lokay [1] gives the following equations to calculate the effective reactance:

$$X_{\text{eff}} = \frac{\sum_{i=1}^{n} |\text{VD}_X|_i}{|I_L|} \; \Omega \tag{9.9}$$

and

$$\sum_{i=1}^{n}\left|VD_X\right|_i = \left|I_{L,1}\right| \times X_{L,1} \times I_1 + \left|I_{L,2}\right| \times X_{L,2} \times I_2 + \cdots + \left|I_{L,n}\right| \times X_{L,n} \times I_n \tag{9.10}$$

where
   $\left\|VD_X\right\|_i$ is the voltage drop due to line reactance of ith section of feeder between regulator station
      and regulation point, V/section
   $\sum_{i=1}^{n}\left|VD_X\right|_i$ is the total voltage drop due to line reactance of feeder between regulator station
      and regulation point, V
   $X_{L,1}$ is the inductive reactance [as defined in Equation 9.6] of ith section of feeder, Ω/mi

   Since the methods just described to determine the effective R and X are rather involved, Lokay
[1] suggests as *an alternative and practical method to measure the current (I_L) and voltage at the
regulator location and the voltage at the RP*. The difference between the two voltage values is the
total voltage drop between the regulator and the regulation point, which can also be defined as

$$VD = \left|I_L\right| \times R_{eff} \times \cos\theta + \left|I_L\right| \times X_{eff} \times \sin\theta \tag{9.11}$$

from which the R_{eff} and X_{eff} values can be determined easily if the load power factor of the feeder
and the average R/X ratio of the feeder conductors between the regulator and the RP are known.
   Figure 9.11 gives an example for determining the voltage profiles for the peak and light loads.
Note that the primary-feeder voltage values are based on a 120 V base.
   It represents one-line diagram and voltage profiles of a feeder with distributed load beyond a
voltage regulator location: (a) one-line diagram and (b) peak- and light-load profiles showing fictitious RP for LDC settings. It is assumed that the conductor size between regulator and first distribution transformer is #2/0 copper conductor with 44-in flat spacing with resistance and reactance of

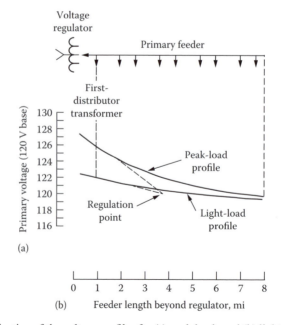

FIGURE 9.11   Determination of the voltage profiles for (a) peak loads and (b) light loads.

0.481 and 0.718 2/mi, respectively. The PT and CT ratios of the voltage regulator are 7960: 120 and 200: 5, respectively. Distance to fictitious RP is 3.9 mi. LDC settings are

$$R_{set} = 200 \times \frac{120}{7960} \times 0.481 \times 3.9 = 5.656$$

$$X_{set} = 200 \times \frac{120}{7960} \times 0.718 \times 3.9 = 8.4428$$

Voltage-regulating relay setting is 120.1 V. (From [1].)

Example 9.1

This example investigates the use of step-type voltage regulation (control) to improve the voltage profile of distribution systems. Figure 9.12 illustrates the elements of a distribution substation that is supplied from a subtransmission loop and feeds several radial primary feeders.

The substation LTC transformer can be used to regulate the primary distribution voltage (V_P) bus, holding V_P constant as both the subtransmission voltage (V_{ST}) and the IZ_T voltage drop in the substation transformer vary with load. If the typical primary-feeder main is voltage-drop-limited, it can be extended further and/or loaded more heavily if a feeder voltage regulator bank is used wisely. In Figure 9.12, the feeder voltage regulator, indicated with the symbol shaped as a 0 with an arrow going through it, is located at the point $s = s_1$, and it varies its boost and buck automatically to hold a set voltage at the RP, that is, at $s = s_{RP}$.

Typical LTC and feeder regulator data: The abbreviation VRR *stands for voltage-regulating relay* (or solid-state equivalent thereof), and it is adjustable within the approximate range from 110 to 125 V. The VRR measures the voltage at the RP, that is, V_{RP}, by means of the LDC.

The LDC has R and X settings, which are both adjustable within the approximate range from 0 to 24 Ω (often called *volts* because the CTs used with regulators have 1.A secondaries).

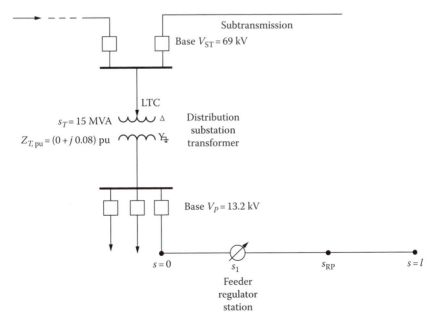

FIGURE 9.12 The elements of a distribution substation for Example 9.1.

The BW of the VRR is adjustable within the approximate range from ±¾ to ±1½ V based on 120 V. The TD is adjustable between about 10 and 120 s.

The location of the RP *is controlled by the R and X settings of the* LDC. If the R and X settings are set to be zero, the regulator regulates the voltage at its local terminal to the setting of the VRR ± BW. In this example, $s_{RP} = s_1$.

Overloading of step-type feeder regulators ANSI standards provide for regulator overload capacity as listed in Table 9.2 in case the full 10% range of regulation is not required. All modern regulators are provided with adjustments to reduce the range to which the motor can drive the tap-changer switching mechanism.

Good advantage sometimes can be taken of this designed *overload* type of limited-range operation. However, if load growth occurs, both a larger range of regulation and a larger regulator size (kilovoltamperes or current) can be expected to be needed. Table 9.3 gives some typical single-phase regulator sizes.

Substation data Make the following assumptions:

Base MVA$_{3\phi}$ = 15 MVA
Subtransmission base V_{L-L} = 69 kV
Primary base V_{L-L} = 13.2 kV

TABLE 9.2
Overloading of Step-Type Feeder Regulators

Reduced Range of Regulation (%)	Percent of Normal Load Current
±10.00	100
±8.75	110
±7.50	120
±6.25	135
±5.00	160

TABLE 9.3
Some Typical Single-Phase Regulator Sizes

Single Phase kVA	Volts	Amps	CTP[a]	PTN[b]
25	2500	100	100	20
⋮	⋮	⋮	⋮	⋮
125	2500	500	500	20
38.1	7620	50	50	63.5
57.2	7620	75	75	63.5
76.2	7620	100	100	63.5
114.3	7620	150	150	63.5
167	7620	219	250	63.5
250	7620	328	400	63.5

[a] Ratio of the current transformer contained within the regulator. (Here, the ratio is the high-voltage-side ampere rating because the low-voltage rating is 1.0 A.)

[b] Ratio of the potential transformer contained within the regulator. (All potential transformer secondaries are 120 V.)

The substation transformer is rated 15 MVA, 69–7.62/13.2 kV grounded-wye and has a per unit impedance ($Z_{T,pu}$) of $0 + j0.08$ based on its ratings. Its three-phase LTC can regulate ±10% voltage in 32 steps of 5/8% each.

Load flow data: Assume that the maximum subtransmission voltage (max V_{ST}) is 72.45 kV or 1.05 pu, which occurs during the off-peak period at which the off-peak kilovoltamperage is 0.25 pu with a leading power factor of 0.95. The minimum subtransmission voltage (min V_{ST}) is 69 kV or 1.00 pu, which occurs during the peak period at which the peak kilovoltamperage is 1.00 pu with a lagging power factor of 0.85.

Voltage data and voltage criteria: Assume that the maximum secondary voltage is 125 V or 1.0417 pu V (based on 120 V) and the minimum secondary voltage is 116 V or 0.9667 pu V, and that the maximum voltage drop in secondaries is 0.035 pu V.

Assume that the maximum primary voltage (max V_P) is 1.0417 pu V at zero load, and that at annual peak load, the maximum primary voltage is 1.0767 pu V (1.0417 + 0.035) considering the nearest secondary to the regulator and the minimum primary voltage is 1.0017 pu V (0.9667 + 0.035) considering the most remote secondary.

Feeder data: Assume that the annual peak load is 4000 kVA, at a lagging power factor of 0.85, and is distributed uniformly along the 10-mi-long feeder main. The main has 266.8 kcmil all-aluminum conductors (AACs) with 37 strands and 53-in geometric mean spacing. Use 3.88×10^{-6} pu VD/(kVA · mi) at 0.85 lagging power factor as the constant K.

Assume that the substation transformer LTC is used for BVR. Use a BW of ±1.0 V or (1 V/120 V) = 0.0083 pu V. Also use rounded figures of 1.075 and 1.000 pu V for the maximum and minimum primary voltages at peak load, respectively.

a. Specify the setting of the VRR for the highest allowable primary voltage (V_P), BW being considered, then round the setting to a convenient number.
b. Find the maximum number of steps of buck and boost that will be required.
c. Sketch voltage profiles of the feeder being considered for zero load and for the annual peak load. Label the significant voltage values on the curves.

Solution

a. Since the LDC of the regulator is not used,

$$R_{set} = 0 \quad \text{and} \quad X_{set} = 0$$

Therefore, the setting of the VRR for the highest allowable primary voltage, BW being considered, occurs at the zero load and is

$$VRR = (V_P)_{max} - BW$$

$$= 1.0417 - 0.0083$$

$$= 1.0334 \text{ pu V}$$

$$\cong 1.035 \text{ pu V}$$

$$= 124.2 \text{ V}$$

b. To find the maximum number of buck and boost that will be required, the highest allowable primary voltages at off-peak and on-peak have to be found. Therefore, *at off-peak*,

$$\bar{V}_{P,pu} = \bar{V}_{ST,pu} - \bar{I}_{P,pu} \times \bar{Z}_{T,pu} \tag{9.12}$$

where
$V_{ST,pu}$ is the per unit subtransmission voltage at primary side of the substation transformer
$\quad = 1.05\angle 0° \text{ pu V}$
$I_{P,pu}$ is the per unit no-load primary current at substation (transformer) = 0.2381 pu
$\quad$ A = per unit impedance of substation transformer

$$Z_{T,pu} = 0 + j0.08 \text{ pu } \Omega$$

Therefore,

$$V_{P,pu} = 1.05 - (0.2381)(\cos\theta + j\sin\theta)(0 + j0.08)$$
$$= 1.05 - (0.2381)(0.95 + j0.3118)(0 + j0.08)$$
$$= 1.0589 \text{ pu V}$$

whereas *at on-peak*,

$$V_{P,pu} = 1.0 - (1.00)(0.85 - j0.53)(0 + j0.08)$$
$$= 0.9602 \text{ pu V}$$

Since the LTC of the substation can regulate ±10% voltage in 32 steps of 5/8% volts (or 0.00625 pu V) each, the maximum number of steps of buck required, *at off-peak*, is

$$\text{No. of steps} = \frac{V_{P,pu} - VRR_{pu}}{0.00625}$$
$$= \frac{1.0589 - 1.035}{0.00625}$$
$$\cong 3 \quad \text{or} \quad 4 \text{ steps} \tag{9.13}$$

and the maximum number of steps of boost required, *at peak*, is

$$\text{No. of steps} = \frac{V_{P,pu} - VRR_{pu}}{0.00625}$$
$$= \frac{1.035 - 0.9602}{0.00625}$$
$$\cong 12 \text{ steps} \tag{9.14}$$

c. To sketch voltage profiles of the primary feeder for the annual peak load, the total voltage drop of the feeder has to be known. Therefore,

$$\sum VD_{pu} = K \times S \times \frac{\ell}{2}$$
$$= (3.88 \times 10^{-6})(4000 \text{ kVA})\left(\frac{10 \text{ mi}}{2}\right)$$
$$= 0.0776 \text{ pu V} \tag{9.15}$$

and thus the minimum primary-feeder voltage at the end of the 10 mi feeder, as shown in Figure 9.13, is

$$\text{Min } V_{P,pu} = VRR_{pu} - \sum VD_{pu}$$
$$= 1.035 - 0.0776$$
$$= 0.9574 \text{ pu V} \tag{9.16}$$

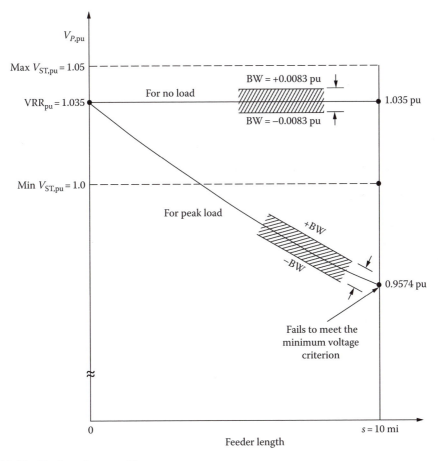

FIGURE 9.13 Feeder voltage profile.

At the annual peak load, the rounded voltage criteria are

$$\text{Max } V_{P,\text{pu}} = 1.075 - \text{BW}$$

$$= 1.075 - 0.0083$$

$$= 1.0667 \text{ pu V}$$

and

$$\text{Min } V_{P,\text{pu}} = 1.00 + \text{BW}$$

$$= 1.00 + 0.0083$$

$$= 1.0083 \text{ pu V}$$

At no-load, the rounded voltage criteria are

$$\text{Max } V_{P,\text{pu}} = 1.0417 - \text{BW}$$

$$= 1.0417 - 0.0083$$

$$= 1.035 \text{ pu V}$$

and

$$\text{Min } V_{P,pu} = 1.0083 \text{ pu V}$$

As can be seen from Figure 9.13, the minimum primary-feeder voltage at the end of the 10 mi feeder *fails to meet the minimum voltage criterion* at the annual peak load. Therefore, a voltage regulator has to be used.

Example 9.2

Use the information and data given in Example 9.1 and locate the voltage regulator, that is, determine the s_1 distance at which the regulator must be located as shown in Figure 9.12, for the following two cases, where the peak-load primary-feeder voltage ($V_{P,pu}$) at the input to the regulator is

a. $V_{P,pu} = 1.010$ pu V.
b. $V_{P,pu} = 1.000$ pu V.
c. What is the advantage of part *a* over part *b*, or vice versa?

Solution

a. When $V_{P,pu} = 1.010$ pu V, the associated voltage drop at the distance s_1, as shown in Figure 9.14, is

$$VD_{s1} = VRR_{pu} - V_{P,pu}$$

$$= 1.035 - 1.01$$

$$= 0.025 \text{ pu V} \tag{9.17}$$

From Example 9.1, the total voltage drop of the feeder is

$$\sum VD_{pu} = 0.0776 \text{ pu V}$$

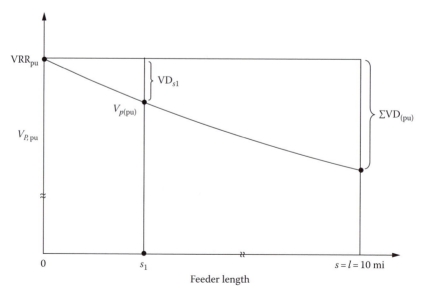

FIGURE 9.14   Voltage profile for Example 9.2.

Therefore, the distance s_1 can be found from the following parabolic formula for the uniformly distributed load:

$$\frac{VDs_1}{\sum VD_{pu}} = \frac{s_1}{\ell}\left(2 - \frac{s_1}{\ell}\right) \tag{9.18}$$

or

$$\frac{0.025}{0.0776} = \frac{s_1}{10}\left(2 - \frac{s_1}{10}\right)$$

from which the following quadratic equation can be obtained:

$$s_1^2 - 20s_1 + 32.2165 = 0$$

which has two solutions, namely, 1.75 and 18.23 mi. Therefore, the distance s_1, taking the acceptable answer, is 1.75 mi.

b. When $V_{P,pu}$ = 1.00 pu V, the associated voltage drop at the distance s_1 is

$$VD_{s1} = VRR_{pu} - V_{P,pu}$$

$$= 1.035 - 1.00$$

$$= 0.035 \text{ pu V}$$

Therefore, from Equation 9.18,

$$\frac{0.035}{0.0776} = \frac{s_1}{10}\left(2 - \frac{s_1}{10}\right)$$

or

$$s_1^2 - 20s_1 + 45.1031 = 0$$

which has two solutions, namely, 2.6 and 17.4 mi. Thus, taking the acceptable answer, the distance s_1 is 2.6 mi.

c. The advantage of part (a) over part (b) is that it can compensate for future growth. Otherwise, the VP, pu might be less than 1.00 pu V in the future.

Example 9.3

Assume that the peak-load primary-feeder voltage at the input to the regulator is 1.010 pu V as given in Example 9.2. Determine the necessary minimum kilovoltampere size of each of three single-phase feeder regulators.

Solution

From Example 9.2, the distance s_1 is found to be 1.75 mi. Previously, the annual peak load and the standard regulation range have been given as 4000 kVA and ±10%, respectively.

The *uniformly distributed three-phase load* at s_1 is

$$S_{3\phi}\left(1 - \frac{s_1}{\ell}\right) = 4000\left(1 - \frac{1.75}{10.00}\right)$$

$$= 3300 \text{ kVA}$$

Therefore, the single-phase load at s_1 is

$$\frac{3300\text{ kVA}}{3} = 1100\text{ kVA}$$

Since the single-phase regulator kilovoltampere rating is given by

$$S_{reg} = \frac{(\%R_{max})S_{ckt}}{100} \tag{9.19}$$

where S_{ckt} is the circuit kilovoltamperage, then

$$S_{reg} = \frac{10 \times 1100\text{ kVA}}{100} = 110\text{ kVA}$$

Thus, from Table 9.3, the corresponding minimum kilovoltampere size of the regulator size can be found as 114.3 kVA.

Example 9.4

Use the distance of $s_1 = 1.75$ mi found in Example 9.2 and assume that the distance of the RP is equal to s_1, that is, $s_{RP} = s_1$, or, in other words, *the RP is located at the regulator station*, and determine the following:

    a. Specify the best settings for the LDC's R and X, and for the VRR.
    b. Sketch voltage profiles for zero load and for the annual peak load. Label significant voltage values on the curves.
    c. Are the primary-feeder voltage ($V_{P,pu}$) criteria met?

Solution

    a. The $x_{RP} = s_1$ means that the *RP is located at the feeder regulator station*. Therefore, the *best settings for the* LDC *of the regulator are when settings for both R and X are zero* and

$$\text{VRR}_{pu} = V_{RP,pu} = 1.035\text{ pu V}$$

    b. The voltage drop occurring in the feeder portion *between the RP and the end of the feeder* is

$$\text{VD}_{pu} = K \times S \times \frac{\ell}{2}$$

$$= (3.88 \times 10^{-6})(3300)\left(\frac{8.25}{2}\right)$$

$$= 0.0528\text{ pu V}$$

Thus the *primary-feeder voltage at the end of the feeder* for the annual peak load is

$$V_{P,10mi} = 1.035 - 0.0528$$

$$= 0.9809\text{ pu V}$$

Note that the $V_{P,pu}$ used at the regulator point is the *no-load value* rather than the annual peak-load value. If, instead, the 1.0667 pu value is used, then, for example, television sets of those customers located at the vicinity of the RP might be damaged during the off-peak periods because of the too-high V_{RP} value.

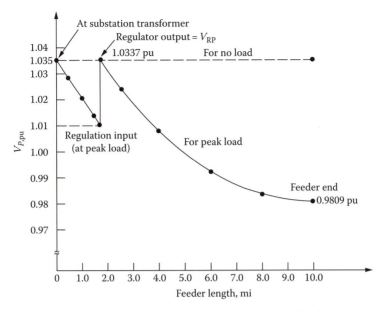

FIGURE 9.15 Feeder voltage profiles for zero load and for the annual peak load.

As can be seen from Figure 9.15, the peak-load voltage profile is not in linear but in parabolic shape. *The voltage-drop value for any given point s between the substation and the regulator station* can be calculated from

$$VD_s = K\left(S_{3\phi} - \frac{S_{3\phi} \times s}{\ell}\right)s + K\left(\frac{S_{3\phi} \times s}{\ell}\right)\frac{s}{2} \text{ pu V} \tag{9.20}$$

where
  K is the percent voltage drop per kilovoltampere-mile characteristic of feeder
  $S_{3\phi}$ is the uniformly distributed three-phase annual peak load, kVA
  ℓ is the primary feeder length, mi
  s is the distance from substation, mi

Therefore, from Equation 9.20,

$$VD_s = 3.88 \times 10^{-6}\left(4000 - \frac{4000s}{10}\right)s + 3.88 \times 10^{-6}\left(\frac{4000s}{10}\right)\frac{s}{2} \text{ pu V} \tag{9.21}$$

For various values of s, the associated values of the voltage drops and $V_{P,pu}$ can be found, as given in Table 9.4.

TABLE 9.4
For Annual Peak Load

s (mi)	VD$_s$ (pu V)	$V_{P,\,pu}$ (pu V)
0.0	0.0	1.035
0.5	0.0076	1.0274
1.0	0.0071	1.0203
1.5	0.0068	1.0135
1.75	0.025	1.010

The *voltage-drop value for any given point s between the substation and the regulator station* can also be calculated from

$$VD_s = I(r \times \cos\theta + x \times \sin\theta)s\left(1 - \frac{s}{2\ell}\right) \qquad (9.22)$$

where
I_L is the load current in feeder at substation end

$$= \frac{S_{3\phi}}{\sqrt{3} \times V_{L-L}} \qquad (9.23)$$

r is the resistance of feeder main, Ω/mi per phase
x is the reactance of feeder main, Ω/mi per phase

Therefore, the voltage drop in per units can be found as

$$VD_s = \frac{VD_s}{V_{L-N}} \text{ pu V} \qquad (9.24)$$

The *voltage-drop value for any given point s between the regulator station and the end of the feeder* can be calculated from the following equation:

$$VD_s = K\left(S'_{3\phi} - \frac{S'_{3\phi} \times s}{\ell - s}\right)s + K\left(\frac{S'_{3\phi} \times s}{\ell - s}\right)\frac{s}{2} \text{ pu V} \qquad (9.25)$$

where $S'_{3\phi}$ = uniformly distributed three-phase annual peak load at distance s_1, kVA

$$= S_{3\phi}\left(1 - \frac{s_1}{\ell}\right) \text{ kVA}$$

s_1 is the distance of feeder regulator station from substation, mi
Therefore, from Equation 9.25,

$$VD_s = 3.88 \times 10^{-6}\left(3300 - \frac{3300s}{8.25}\right)s + 3.88 \times 10^{-6}\left(\frac{3300s}{8.25}\right)\frac{s}{2} \text{ pu V} \qquad (9.26)$$

For various values of s, the corresponding values of the voltage drops and $V_{P,pu}$ can be found, as given in Table 9.5.

TABLE 9.5
For Annual Peak Load

s (mi)	VD$_s$ (pu V)	V$_{P,pu}$ (pu V)
0.00	0.00	1.0337
0.75	0.0092	1.0245
2.25	0.0157	1.0088
4.25	0.0155	0.9933
6.25	0.0093	0.9840
8.25	0.0031	0.9809

The voltage profiles for the annual peak load can be obtained by plotting the $V_{P,pu}$ values from Tables 9.4 and 9.5. Since there is no voltage drop at zero load, the $V_{P,pu}$ remains constant at 1.035 pu. Therefore, the voltage profile for the zero load is a horizontal line (with zero slope).

c. The minimum $V_{P,pu}$ criterion of 1.0083 pu V is not met even though the regulator voltage has been set as high as possible without exceeding the maximum voltage criterion of 1.035 pu V.

Example 9.5

Assume that the regulator station is located at a distance s_1 as found in part (a) of Example 9.2, but the RP has been moved to the end of the feeder so that $s_{RP} = l = 10$ mi.

a. Determine good settings for the values of VRR, R, and X so that all $V_{P,pu}$ voltage criteria will be met, if possible.
b. Sketch voltage profiles and label the values of significant voltages, in per unit volts.

Solution

a. From Table A.4 of Appendix A, the resistance at 50°C and the reactance of the 266.8 kcmil AAC with 37 strands are 0.386 and 0.4809 Ω/mi, respectively. From Table A.10, the inductive-reactance spacing factor for the 53-in geometric mean spacing is 0.1802 Ω/mi. Therefore, from Equation 9.6, the inductive reactance of the feeder conductor is

$$x_L = x_a + x_d$$

$$= 0.4809 + 0.1802$$

$$= 0.6611 \, \Omega/\text{mile}$$

From Equations 9.3 and 9.5,

$$R_{\text{eff}} = r_a \times \frac{\ell - s_1}{2}$$

$$= 0.386 \times \frac{8.25}{2}$$

$$= 1.5923 \, \Omega$$

and

$$X_{\text{eff}} = x_L \times \frac{\ell - s_1}{2}$$

$$= 0.6611 \times \frac{8.25}{2}$$

$$= 2.7270 \, \Omega$$

From Table 9.3, for the regulator size of 114.3 kVA found in Example 9.3, the primary rating of the CT and the PT ratio are 150 and 63.5, respectively. Therefore, from Equations 9.2 and 9.4, the R and X dial settings can be found as

$$R_{\text{set}} = \frac{CT_P}{PT_N} \times R_{\text{eff}} \, \Omega$$

$$= \frac{150}{63.5} \times 1.5923$$

$$= 3.761 \, \text{V} \quad \text{or} \quad 0.0313 \, \text{pu V}$$

based on 120 V and

$$X_{set} = \frac{CT_P}{PT_N} \times X_{eff} \ \Omega$$

$$= \frac{150}{63.5} \times 2.727$$

$$= 6.442 \ V \quad \text{or} \quad 0.0537 \ pu \ V$$

Assume that the voltage at the RP (V_{RP}) is arbitrarily set to be 1.0138 pu V using the R and X settings of the LDC of the regulator so that the V_{RP} is always the same for zero load or for the annual peak load. Therefore, the output voltage of the regulator for the annual peak load can be found from

$$V_{reg} = V_{RP} + \frac{\dfrac{S_{1\phi}}{V_{L-N}}(R_{set} \times \cos\theta + X_{set} \times \sin\theta)}{CT_P \times V_B} \ pu \ V$$

$$= 1.0138 + \frac{(1100/7.62)(3.761 \times 0.85 + 6.442 \times 0.527)}{150 \times 120}$$

$$= 1.0666 \ pu \ V \tag{9.27}$$

Here, note that the regulator regulates the regulator output voltage automatically according to the load at any given time in order to maintain the RP voltage at the predetermined voltage value.

Table 9.6 gives the $V_{P,pu}$ values for the purpose of comparing the actual voltage values against the established voltage criteria for the annual peak and for zero load.

As can be observed from Table 9.6, the primary voltage criteria are met by using the R and X settings.

b. The voltage profiles for the annual peak load and zero load can be obtained by plotting the $V_{P,pu}$ values from Tables 9.6 and 9.7 (based on Equation 9.26), as shown in Figure 9.16.

TABLE 9.6
Actual Voltages vs. Voltage Criteria at Peak and Zero Loads

Voltage	Actual Voltage (pu V)		Voltage Criteria (pu V)	
	At Peak Load	At Zero Load	At Peak Load	At Zero Load
Max $V_{P,pu}$	1.0666	1.0138	1.0667	1.0337
Min $V_{P,pu}$	1.0138	1.0138	1.0083	1.0083

TABLE 9.7
Values Obtained

s (mi)	VD_s (pu V)	$V_{P,pu}$ (pu V)
0.00	0.00	1.0666
0.75	0.0092	1.0574
2.25	0.0157	1.0417
4.25	0.0155	1.0262
6.25	0.0093	1.0169
8.25	0.0031	1.0138

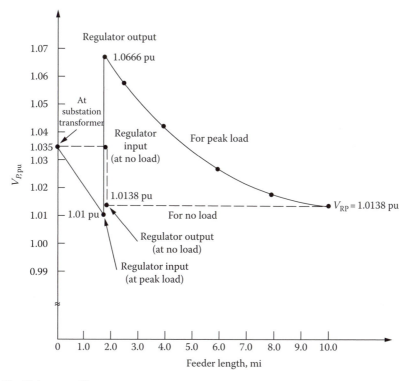

FIGURE 9.16 Voltage profiles.

Example 9.6

Consider the results of Examples 9.4 and 9.5 and determine the following:

  a. The number of steps of buck and boost the regulators will achieve in Example 9.4.
  b. The number of steps of buck and boost the regulators will achieve in Example 9.5.

Solution

  a. For Example 9.4, the number of steps of buck is

$$No. of steps = \frac{1.035 - 1.0337}{0.00625}$$

$$= 0.208$$

thus it is either zero or one step. The number of boost is

$$No. of steps = \frac{1.0337 - 1.010}{0.00625}$$

$$= 3.79$$

therefore it is either three or four steps.
  b. For Example 9.5, the number of steps of buck is

$$No. of steps = \frac{1.035 - 1.0138}{0.00625}$$

$$= 3.39$$

hence, it is either three or four steps. The number of steps of boost is

$$\text{No. of steps} = \frac{1.0666 - 1.010}{0.00625}$$

$$= 9.06$$

therefore, it is either 9 or 10 steps.

Example 9.7

Consider the results of Examples 9.4 and 9.5 and answer the following:

a. Can reduced range of regulation be used gainfully in Example 9.4? Explain.
b. Can reduced range of regulation be used gainfully in Example 9.5? Explain.

Solution

a. Yes, the reduced range of regulation can be used gainfully in Example 9.4 since the next-smaller-size regulator, that is, 76.2 kVA, at ±5% regulation range can be selected. This ±5% regulation range would allow the capacity of the regulator to be increased to 160% (see Table 9.2) so that

$$1.6 \times 76.2 \text{ kVA} = 121.92 \text{ kVA}$$

which is much larger than the required capacity of 110 kVA. It would allow the user ±8 steps of buck and boost, which is more than the required one step of buck and four steps of boost.

b. No, the reduced range of regulation cannot be used gainfully in Example 9.5 since the required steps of buck and boost are 4 and 10, respectively. The reduced range of regulation at ±6.25% would provide the ±10 steps of buck and boost, but it would allow the capacity of the regulator to be increased only up to 135% (see Table 9.2) so that

$$1.35 \times 76.2 \text{ kVA} = 102.87 \text{ kVA}$$

which is smaller than the required capacity of 110 kVA.

Example 9.8

Figure 9.17 shows a one-line diagram of a primary feeder supplying an industrial customer. The nominal voltage at the utility substation low-voltage bus is 7.2/13.2 kV three-phase wye-grounded. The voltage regulator bank is made up of three single-phase step-type voltage regulators with a PT ratio of 63.5 (7620:120).

The industrial customer's bus is located at the end of a 3 mi primary line with a resistance of 0.30 Ω/mi and an inductive reactance of 0.80 Ω/mi.

The customer's transformer is rated 5000 kVA in three phase with a 12,800 V primary connected in delta (taps in use) and a 2400/4160 V secondary connected in grounded-wye. The transformer impedance is $0 + j0.05$ pu Ω based on the rated kilovoltamperes and tap voltages in use. Assume that the bases to be used are 5,000 kVA, 2,400/4,160 V, and 7,390/12,800 V.

Assume that the customer asks that the low-voltage bus be regulated to 2450/4244 V and determine the following:

a. Find the necessary setting of the voltage-setting dial of the VRR of each single-phase regulator in use.
b. Assume that the ratio of the CT in each regulator is 250:1 A and find the necessary R and X dial settings of LDCs.

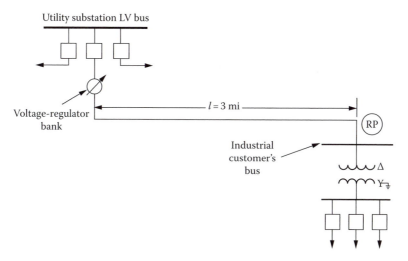

Utility substation LV bus

Voltage-regulator
bank

$l = 3$ mi

RP

Industrial
customer's
bus

FIGURE 9.17   One-line diagram of a primary feeder supplying an industrial customer.

Solution

a. The voltage at the RP, *which is located at the customer's bus*, is

$$V_{RP} = \frac{2450\,V}{2400\,V}$$

$$= 1.02083\,pu\,V$$

Therefore,

$$VRR = \frac{7390}{7620} \times 1.02083$$

$$= 0.99\,pu\,V \quad or \quad 7620\,V$$

Thus,

$$VRR = \frac{7620}{PT_N} \times 0.99$$

$$= \frac{7620}{63.5} \times 0.99$$

$$= 120 \times 0.99$$

$$= 118.8\,V$$

or, alternatively,

$$VRR_{set} = V_{RP} \times V_{B,sec}$$

$$= 1.02083 \times \frac{12,800}{\sqrt{3} \times 63.5}$$

$$\cong 118.8\,V \tag{9.28}$$

b. The applicable impedance base is

$$Z_B = \frac{(kV_{L-L})^2}{MVA}$$

$$= \frac{(12.8\,kV)^2}{5\,MVA}$$

$$= 32.768\,\Omega$$

therefore, the transformer impedance is

$$Z_T = Z_{T,pu} \times Z_B$$

$$= (0 + j0.05) \times 32.768$$

$$= 0 + j1.6384\,\Omega \qquad\qquad (9.29)$$

Since here the R and X settings are determined for only one customer, the resistance and reactance values of the customer's transformer have to be included in the effective resistance and reactance calculations. Therefore,

$$R_{eff} = r \times l + R_T$$

$$= (0.3\,\Omega/mi)(3\,mi) + 0$$

$$= 0.9\,\Omega \qquad\qquad (9.30)$$

and

$$X_{eff} = x \times l + X_T$$

$$= (0.8\,\Omega/mi)(3\,mi) + 1.6384\,\Omega$$

$$= 4.0384\,\Omega \qquad\qquad (9.31)$$

Thus, the R dial setting of the LDC is

$$R_{set} = \frac{CT_P}{PT_N} \times R_{eff}$$

$$= \frac{250}{63.5} \times 0.9$$

$$= 3.5433\,\Omega$$

and the X dial setting is

$$X_{set} = \frac{CT_P}{PT_N} \times X_{eff}$$

$$= \frac{250}{63.5} \times 4.0384$$

$$= 15.8992\,\Omega$$

Example 9.9

Consider the 10 mi feeder of Example 9.1. Assume that the substation has a bus voltage regulator (BVR) with transformer LTC and that the primary feeder voltage (V_p) has been set on the VRR to be 1.035 pu V.

Assume that the main feeder is made up of 266.8 kcmil with 37 strands and 53-in geometric mean spacing. It has been found in Example 9.5 that the main feeder has an inductive reactance of 0.661 Ω/mi per conductor. Assume that the annual peak load is 4000 kVA at a lagging power factor of 0.85 and distributed uniformly along the main or, in other words, the uniformly distributed load is

$$S_{3\phi} = P_L + jQ_L = 3400 + j2100 \text{ kVA}$$

Assume that, as found in Example 9.1, the total voltage drop of the feeder is

$$\sum VD_{pu} = VD_{1,pu}$$

$$= 0.0776 \text{ pu V}$$

and the reactive load factor is 0.40. Use the given data and determine the following:

a. Design a fixed, that is, nonswitched (NSW)-capacitor bank for the maximum loss reduction.
b. Sketch the voltage profiles when there is no-capacitor (N/C) bank installed and when there is a fixed-capacitor bank, that is, Q_{NSW} installed.
c. Add a switched-capacitor bank for voltage control on the feeder. Locate the switched-capacitor bank, that is, Q_{sw}, at the end of the feeder for the feeder at the annual peak load is 1.000 pu V. Sketch the associated voltage profiles.

Solution

a. From Figure 8.31, the corrective ratio (CR) for the given reactive load factor of 0.40 is found to be 0.27. Therefore, the required size of the NSW-capacitor bank is

$$Q_{NSW} = CR \times Q_L$$

$$= (0.27)(2100 \text{ kvar})$$

$$= 567 \text{ kvar per three phase} \tag{9.32}$$

Thus, two single-phase standard 100-kvar-size capacitor units are required to be used on each phase and located on the feeder at a distance of

$$s = \frac{2}{3} \times l$$

$$= \frac{2}{3} \times 10$$

$$= 6.67 \text{ mi}$$

for the optimum result, as shown in Figure 9.18.

Therefore, the per unit voltage rise (VRP_{pu}) due to the *NSW-capacitor bank* is

$$VR_{pu} = \frac{Q_{NSW} \times \frac{2}{3} X_L}{1000(kV_{L-L})^2}$$

$$= \frac{600 \times 4.41}{1000 \times 13.2^2}$$

$$= 0.0152 \text{ pu V} \tag{9.33}$$

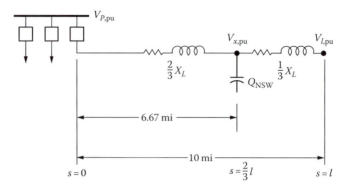

FIGURE 9.18 Optimum location of a capacitor bank.

b. When there is *N/C bank installed,* the voltage drop for the uniformly distributed load at a distance of $s = 2/3\ \ell$ can be found from Equation 9.18 as

$$\frac{VDs_{pu}}{\sum VD_{pu}} = \frac{s}{\ell}\left(2 - \frac{s}{\ell}\right) \tag{9.34}$$

or

$$\frac{VDs_{pu}}{0.0776} = \frac{6.67}{10}\left(2 - \frac{6.67}{10}\right)$$

from which

$$VD_{s,pu} = 0.069 \text{ pu V}$$

Therefore, the feeder voltage at the $2/3\ \ell$ distance is

$$V_{s,pu} = V_{P,pu} - VD_{s,pu}$$

$$= 1.035 - 0.069$$

$$= 0.966 \text{ pu V}$$

When there is a *fixed-capacitor bank installed,* the new voltage at the $2/3\ \ell$ distance due to the voltage rise is

$$\text{New } V_{s,pu} = V_{s,pu} + VR_{pu}$$

$$= 0.966 + 0.0152$$

$$= 0.9812 \text{ pu V} \tag{9.35}$$

When there is *N/C bank installed,* the voltage at the end of the feeder is

$$V_{l,pu} = V_{P,pu} - \sum VD_{pu}$$

$$= 1.035 - 0.0776$$

$$= 0.9574 \text{ pu V} \tag{9.36}$$

When there is a *fixed-capacitor bank installed*, the new voltage at the end of the feeder due to the voltage rise is

$$\text{New } V_{l,\text{pu}} = V_{l,\text{pu}} + \text{VR}_{\text{pu}}$$

$$= 0.9574 + 0.0152$$

$$= 0.9726 \text{ pu V} \qquad (9.37)$$

The associated voltage profiles are shown in Figure 9.19.

c. Since the new voltage at the end of the feeder due to the Q_{sw} installation is 1.000 pu V at the annual peak load, the required voltage rise is

$$\text{VR}_{\text{pu}} = V_{l,\text{pu}} - \text{new} V_{l,\text{pu}}$$

$$= 1.000 - 0.9726$$

$$= 0.0274 \text{ pu V} \qquad (9.38)$$

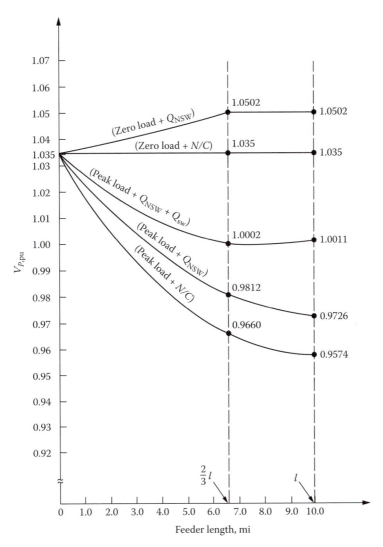

FIGURE 9.19   Voltage profiles.

Therefore, the *required size of the switched-capacitor bank* can be found from

$$VR_{pu} = \frac{Q_{3\phi,SW} \times X_L}{1000(kV_{L-L})^2} \text{ pu V}$$ (9.39)

or

$$Q_{3\phi,SW} = \frac{1000(kV_{L-L})^2 VR_{pu}}{X_L}$$ (9.40)

$$= \frac{1000 \times 13.2^2 \times 0.0274}{6.611}$$

$$= 722.2 \text{ kvar}$$

Hence, the possible combinations of the single-phase standard-size capacitor units to make up the capacitor bank are
 i. Fifteen single-phase standard 50 kvar capacitor units, for a total of 750 kvar
 ii. Six single-phase standard 100 kvar capacitor units, for a total of 600 kvar
 iii. Nine single-phase standard 100 kvar capacitor units, for a total of 900 kvar
For example, assume that the first combination, that is,

$$Q_{3\phi,SW} = 750 \text{ kvar}$$

is selected. The *resultant new voltage rises* at the distance of *l* and $s = s = 2/3\,\ell$ are

$$VR_{l,pu} = VR_{pu} \times \frac{\text{selected } Q_{SW}}{\text{required } Q_{SW}} \text{ pu V}$$ (9.41)

$$= 0.0274 \times \frac{750 \text{ kvar}}{722.2 \text{ kvar}}$$

$$\cong 0.0285 \text{ pu V}$$

and

$$VR_{s,pu} = \frac{2}{3} VR_{l,pu} \text{ pu V}$$ (9.42)

$$= \frac{2}{3} \times 0.0285$$

$$= 0.0190 \text{ pu V}$$

Therefore, *at the peak load when both the NSW-* (i.e., fixed) and the *switched-capacitor banks* are on, the voltage at two-thirds of the line and at the end of the line are

$$V_{s,pu} = \text{new } V_{s,pu} + VR_{s,pu} \text{ pu V}$$ (9.43)

$$= 0.9812 + 0.0190$$

$$= 1.0002 \text{ pu V}$$

and

$$V_{s,pu} = newV_{l,pu} + VR_{l,pu} \text{ pu V} \tag{9.44}$$

$$= 0.9726 + 0.0285$$

$$= 1.0011 \text{ pu V}$$

respectively. At zero load when there is N/C bank installed, the voltages at two-thirds of the line and at the end of the line are the same and equal to 1.035 pu V. The associated voltage profiles are shown in Figure 9.19.

Example 9.10

Consider Example 9.8 and assume that the industrial load at the annual peak is 5000 kVA at 80% lagging power factor. Assume that the customer wishes to add some additional load, is currently paying a monthly power-factor penalty, and the single-phase voltage regulators are approaching full boost. Select a proper three-phase capacitor-bank size (in terms of the multiples of three-phase 150 kvar capacitor units) to be connected to the 4 kV bus that will (1) produce a voltage rise of at least 0.020 pu V on the 4 kV bus and (2) raise the on-peak power factor of the present load to at least 88% lagging power factor.

Solution

The presently existing load is

$$S_{3\phi} = 5000\angle 36.87° \text{ kVA}$$

or

$$S_{3\phi} = 4000 + j3000 \text{ kVA}$$

at 80% lagging power factor. When a properly sized capacitor bank is connected to the bus to improve the on-peak power factor to 88%, the real power portion will be the same but the reactive power portion will be different. In other words,

$$|S|\angle 28.36° = 4000 + jQ_{L,new}$$

from which

$$\tan 28.36° = \frac{Q_{L,new}}{4000}$$

therefore

$$Q_{L,new} = 4000 \times \tan 28.36°$$

$$= 4000 \times 0.5397$$

$$= 2158.97 \text{ kvar}$$

and hence the magnitude of the new apparent power is

$$|S| = 4545.45 \text{ kVA}$$

Therefore, the minimum size of the capacitor bank required to raise the load power factor to 0.88 is

$$Q_{3\phi} = 3000 - 2158.97$$

$$= 841.03 \text{ kvar}$$

Thus, if a 900-kvar-capacity bank is used, the resultant voltage rise from Equation 9.39 is

$$VR_{pu} = \frac{Q_{3\phi} \times X_L}{1000(kV_{B,L-L})^2}$$

where

$$X_L = X_{line} + X_T$$

$$= (0.83 \ \Omega/\text{mi})(3 \ \text{mi}) + 1.6384 \ \Omega$$

$$= 4.0384 \Omega$$

hence

$$VR_{pu} = \frac{900 \times 4.0384}{1000 \times 12.8^2}$$

$$= 0.0222 \text{ pu V}$$

which is larger than the given voltage-rise criterion of 0.020 pu V. Therefore, it is proper to install six 150 kvar three-phase units as the capacitor bank to meet the criteria.

9.6 DISTRIBUTION CAPACITOR AUTOMATION

Today, intelligent customer meters can now monitor voltage at key customer sites and communicate this information to the utility company. Thus, system information can be fine-tuned based on the actual measured values at the end point, rather than on projected values, combined with var information integrated into the control scheme.

The distributed capacitor automation takes advantage of distributed-processing capabilities of electronic meters, capacitor controllers, radios, and substation processors. It uses an algorithm to switch field and substations' capacitors on and off remotely, using voltage information from meters located at key customer sites, and var information from the substation, as illustrated in Figure 9.20.

In the past, capacitors on distribution system were switched on and off mainly by stand-alone controllers that monitored circuit voltage at the capacitor. Various control strategies, including temperature and/or time bias settings on capacitor controllers, were used to ensure operation during predicted peak loading conditions. While this system provided adequate peak voltage/var support, it necessarily involved overcompensation to ensure that all customers were receiving adequate voltage service. Also, capacitors operated independently and were not integrated into a system-wide control scheme.

Distribution capacitor automation integrates field and substation capacitors into a closed-loop control scheme, within a structure that operates as follows:

1. Intelligent customer meters provide exception reporting on voltages out of set BWs. They also report 5-min average voltages when polled.
2. Meters communicate via power-line carrier to the nearest packed radio, and via radio-frequency packet communication, to the designed capacitor controller.

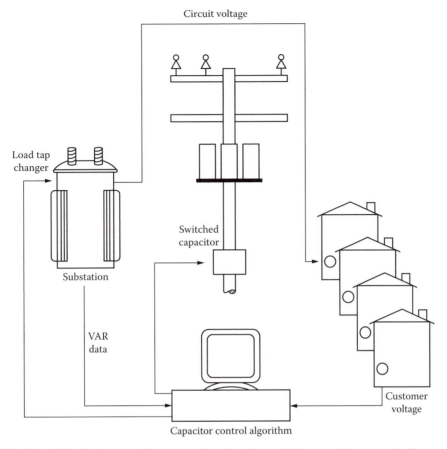

FIGURE 9.20 A distribution capacitor automation algorithm switches capacitors on and off remotely and automatically, using voltage information from customer meters and var information from the substation.

3. Each capacitor controller-automation programmed to receive meter voltages (received from several meters) to a substation processor.
4. The distribution capacitor automation program algorithm, running on an industrial-grade processor at the substation, determines the optimal capacitor-switching pattern and communicates control instructions to capacitor controllers.

Customer meters are strategically placed to provide a consistent sample of lower voltage customers. The system aims to maintain every customer's voltage within a tighter BW targeted at a minimum of 114 V.

Substation reactive power flow is optimized by using control BW set points in the processor. For example, the operator may set a desired power factor as measured at the substation transformer, and the algorithm would act accordingly, choosing the pattern of capacitor switching to both maintain minimum customer voltage and at the same time meet the substation var requirements. In cases where distribution substations have load-tap changing (LTC) transformers, the control algorithm calculates optimal bus voltage in order to produce unity power factor, and the processor issues commands to the LTC controller to hold to this optimal voltage level [19].

To control subtransmission reactive power flow, the transmission substation processor interfaces with distribution substation processors to derive a subtransmission voltage level for minimum var flow and customer voltages.

9.7 VOLTAGE FLUCTUATIONS

In general, voltage fluctuations and lamp flicker on distribution systems are caused by a customer's utilization apparatus. Lamp flicker can be defined as a sudden change in the intensity of illumination due to an associated abrupt change in the voltage across the lamp. Most flickers are caused by the starting of motors. The large momentary inrush of starting current creates a sudden dip in the illumination level provided by incandescent and/or fluorescent lamps since the illumination is a function of voltage.

Therefore, a utility company tries not to endanger other customers, from the quality-of-service point of view, in the process of serving a new customer who could generate excessive flicker by the company's standards. Thus, the distribution circuits are checked to determine whether or not the flicker caused by the new customer's load in addition to the existing flicker-generating loads will meet the company's voltage-fluctuating standards.

The decision to serve such a customer is based on the load location, load type, service voltage, frequency of the motor starts, the motor's horsepower rating, and the motor's NEMA code considerations. Momentary, or pulsating, loads are considered for both their starting requirements and the change in power requirements per unit time. Often, more severe flickers result due to the running operation of pulsating loads than the starting loads. Usually, the pulsating loads, such as grinders, hammer mills, rock crushers, reciprocating pumps, and arc welders, require additional study.

The annoyance created by lamp flicker is a very subjective matter and differs from person to person. However, in certain cases, the flicker can be very objectionable and can create great discomfort. In general, the degree of objection to lamp flicker is a function of the frequency of its occurrence and the rate of change of the voltage dip. The voltage changes resulting in lamp flicker can be either cyclic or noncyclic in nature. Usually, cyclic flicker is more objectionable.

Figure 9.21 shows a typical curve used by the utilities to determine the amount of voltage flicker to be allowed on their system. As indicated in the figure, flicker values located above the curve are likely to be objectionable to lighting customers. For example, from the figure, it can be observed that 5 V dips, based on 120 V, are satisfactory to lighting customers as long as the number of dips does not exceed 3/h.

Thus, more frequent dips of this magnitude are in the objectionable-flicker zone, that is, they are objectionable to the lighting customers. The curve for sinusoidal flicker should be used for the sinusoidal voltage change caused by pump compressors and equipment of similar characteristics. Each utility company develops its own voltage-flicker-limit curve based on its own experiences with customer complaints in the past.

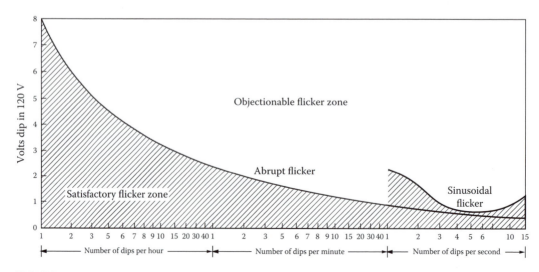

FIGURE 9.21 Permissible voltage-flicker-limit curve.

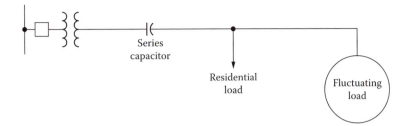

FIGURE 9.22 Installation of series capacitor to reduce the flicker voltage caused by a fluctuating load.

Distribution engineers strive to keep voltage flickers in the satisfactory zone by securing compliance with the company's flicker standards and requirements, and by designing new extensions and rebuilds that will provide service within the satisfactory-flicker zone. Flicker due to motor starting can be reduced by the following remedies:

1. Using a motor that requires less kilovoltamperes per horsepower to start
2. Choosing a low-starting-torque motor if the motor starts under light load
3. Replacing the large-size motor with a smaller-size motor or motors
4. Employing motor starters to reduce the motor inrush current at the start
5. Using shunt or series capacitors to correct the power factor

As mentioned at the start of this section, distribution engineers try every reasonable means to satisfy the motor-start flicker requirement. After exhausting other alternatives, they may choose to satisfy the flicker condition by installing shunt or series capacitors.

Shunt capacitors compensate for the low power factor of the motor during start. They are removed from the circuit when the motor reaches nominal running speed. At start and for a very short time, not to exceed 10 s, capacitors rated at line-to-neutral voltage are often connected line-to-line, that is, at a voltage greater than their rating by a factor of $\sqrt{3}$. Thus, the momentary effective kilovar rating of the capacitors becomes equal to three times the rated kilovars since $(V_{L-L}/V_{L-N})^2 = 3$.

If series capacitors are used, they should be installed between the substation transformer and the residential or lighting tap, as shown in Figure 9.22. Installing the capacitor between the residential load and the fluctuating load would not reduce the flicker voltage since it would not reduce the impedance between the source and the lighting bus.

Since series capacitors are permanently installed in the primary feeder, they require special devices to protect them against overvoltage and resonance conditions. Therefore, a typical series-capacitor installation costs three or four times as much as shunt-motor-start capacitors. Series capacitors correct the power factor of the system, not the motor. The position of the series capacitors in the circuit is especially important if there are other customers along the line.

9.7.1 Shortcut Method to Calculate the Voltage Dips due to a Single-Phase Motor Start

If the starting kilovoltamperage of a single-phase motor is known, the motor's starting current can be found as

$$I_{start} = \frac{S_{start}}{V_{L-N}} \text{ A} \tag{9.45}$$

and the voltage dip based on 120 V can be calculated from

$$VDIP = \frac{120 \times I_{start} \times Z_G}{V_{L-N}} \text{ V} \tag{9.46}$$

where

$$Z_G = \frac{V_{L-N}}{I_{f,L-G}} \; \Omega \qquad (9.47)$$

Substituting Equations 9.45 and 9.47 into Equation 9.46, the voltage dip can be expressed as

$$\text{VDIP} = \frac{120 \times S_{\text{start}}}{I_{f,L-G} \times V_{L-N}} \; V \qquad (9.48)$$

where

VDIP is the voltage dip due to single-phase motor start expressed in terms of 120 V base, V
S_{start} is the starting kVA of single-phase motor, kVA
$I_{f,L-G}$ is the line-to-ground fault current available at point of installation and obtained from fuse coordination, A
V_{L-N} is the line-to-neutral voltage, kV

Example 9.11

Assume that a 10-hp single-phase 7.2 kV motor with NEMA code letter "G" starting 15 times per hour is to be served at a certain location. If the starting kilovoltamperes per horsepower for this motor is given by the manufacturer as 6.3, and the line-to-ground fault at the installation location calculated to be 1438 A, determine the following:

  a. The voltage dip due to the motor start, in volts
  b. Whether or not the resultant voltage dip is objectionable

Solution

  a. Since the starting kilovoltamperes per horsepower is given as 6.3, the starting kilovoltamperes can be found as

$$S_{\text{start}} = (\text{kVA/hp})_{\text{start}} \times \text{hp}_{\text{motor}} \; \text{kVA}$$

$$= 6.3 \; \text{kVA/hp} \times 10 \; \text{hp}$$

$$= 63 \; \text{kVA} \qquad (9.49)$$

Therefore, the voltage dip due to the motor start, from Equation 9.48, can be calculated as

$$\text{VDIP} = \frac{120 \times S_{\text{start}}}{I_{f,L-G} \times V_{L-N}}$$

$$= \frac{120 \times 63 \; \text{kVA}}{1438 \; \text{A} \times 7.2 \; \text{kV}}$$

$$= 0.73 \; \text{V}$$

  b. From Figure 9.21, it can be found that the voltage dip of 0.73 V with a frequency of 15 times/h is in the satisfactory-flicker zone and therefore is not objectionable to the immediate customers.

9.7.2 SHORTCUT METHOD TO CALCULATE THE VOLTAGE DIPS DUE TO A THREE-PHASE MOTOR START

If the starting kilovoltamperes of a three-phase motor is known, its starting current can be found as

$$I_{start} = \frac{S_{start}}{\sqrt{3} \times V_{L-L}} \quad A \tag{9.50}$$

and the voltage dip based on 120 V can be calculated from

$$VDIP = \frac{120 \times I_{start} \times Z_l}{V_{L-N}} \quad V \tag{9.51}$$

where

$$Z_l = \frac{V_{L-N}}{I_{3\phi}} \quad \Omega \tag{9.52}$$

Substituting Equations 9.50 and 9.52 into Equation 9.51, the voltage dip can be expressed as

$$VDIP = \frac{69.36 \times S_{start}}{I_{3\phi} \times V_{L-L}} \quad V \tag{9.53}$$

where
  VDIP is the voltage dip due to three-phase motor start expressed in terms of 120 V base, V
  S_{start} is the starting kVA of three-phase motor, kVA
  $I_{3\phi}$ is the three-phase fault current available at the point of installation and obtained from fuse coordination, A
  V_{L-L} is the line-to-line voltage, kV

Example 9.12

Assume that a 100-hp three-phase 12.47 kV motor with NEMA code letter "F" starting three times per hour is to be served at a certain location. If the starting kilovoltampere per horsepower for this motor is given by the manufacturer as 5.6, and the three-phase fault current at the installation location calculated to be 1765 A, determine the following:

  a. The voltage dip due to the motor start
  b. Whether or not the resultant voltage dip is objectionable

Solution

  a. Since the starting kilovoltampere per horsepower is given as 5.6, the starting kilovoltampere can be found from Equation 9.49 as

$$S_{start} = (kVA/hp)_{start} \times hp_{motor}$$

$$= 5.6 \, kVA/hp \times 100 \, hp$$

$$= 560 \, kVA$$

Therefore, the voltage dip due to the motor start, from Equation 9.53, can be calculated as

$$\text{VDIP} = \frac{69.36 \times S_{\text{start}}}{I_{3\phi} \times V_{L-L}}$$

$$= \frac{69.36 \times 560 \text{ kVA}}{1765 \text{ A} \times 12.47 \text{ kV}}$$

$$= 1.76 \text{ V}$$

b. From Figure 9.21, it can be found that the voltage dip of 1.72 V with a frequency of three times per hour is in the satisfactory-flicker zone and therefore is not objectionable to the immediate customers.

PROBLEMS

9.1 Derive, or prove, Equation 9.18.

9.2 Derive, or prove, Equation 9.20.

9.3 Repeat Example 9.1, assuming 336.4 kcmil ACSR conductors and annual peak load of 5000 kVA at a lagging-load power factor of 0.90.

9.4 Repeat Example 9.2, assuming 336.4 kcmil ACSR conductors and annual peak load of 5000 kVA at a lagging-load power factor of 0.90.

9.5 Repeat Example 9.3, assuming 336.4 kcmil ACSR conductors and annual peak load of 5000 kVA at a lagging-load power factor of 0.90.

9.6 Repeat Example 9.4, assuming 336.4 kcmil ACSR conductors and annual peak load of 5000 kVA at a lagging-load power factor of 0.90.

9.7 Repeat Example 9.5, assuming 336.4 kcmil ACSR conductors and annual peak load of 5000 kVA at a lagging-load power factor of 0.90.

9.8 Repeat Example 9.6, assuming 336.4 kcmil ACSR conductors and annual peak load of 5000 kVA at a lagging-load power factor of 0.90.

9.9 Assume that a subtransmission line is required to be designed to carry a contingency peak load of $2 \times \text{SIL}$. A 60% series compensation is to be used, that is, the capacitive reactance (X_c) of the capacitor bank required to be installed is equal to 60% of the total series inductive reactance per phase of the transmission line. Assume that each phase of the series-capacitor bank is to be made up of series and parallel groups of two-bushing 12 kV 150 kvar shunt power-factor-correction capacitors. Assume that the three-phase SIL of the line is 416.5 MVA and its inductive line reactance is 117.6 Ω/phase. Specify the necessary series–parallel arrangement of capacitors for each phase.

9.10 In this problem, design improvements of the designer choice to correct the undervoltage conditions are investigated on the radial system shown in Figure P9.10.

The voltage at the distribution substation low-voltage bus is kept at 1.04 pu V with bus-voltage regulation. The per unit voltages at annual peak-load values at the points *a*, *b*, *c*, *d*, *e*, and *f* are 1.0049, 0.9815, 0.9605, 0.8793, 0.8793, and 0.8793, respectively. Use the nominal operating voltage of 7,200/12,470 V of the three-phase four-wire wye-grounded system as the base voltage. Assume that all given kilovoltampere loads are annual peak values at 85% lagging-load power factor. The load between the substation bus and point *a* is a uniformly distributed load of 2000 kVA. The loads on the laterals *c*, *d*, *c–e*, and *c–f* are also uniformly distributed, each with 400 kVA. There is a lumped load of 800 kVA at point *b*.

The line data for the #4/0 and 4 ACSR conductors are given in Table P9.10.

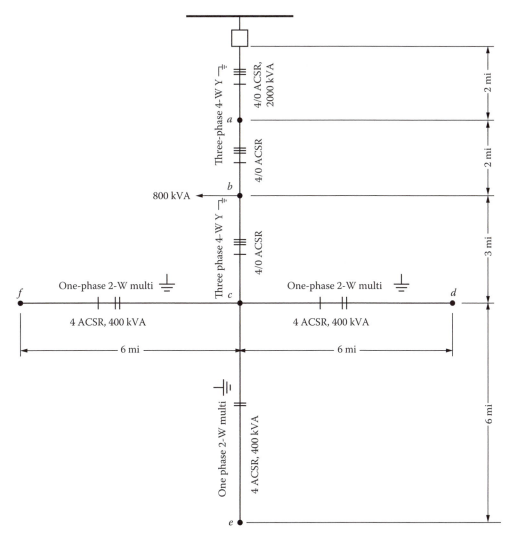

FIGURE P9.10 Figure for Problem 9.10.

TABLE P9.10
Table for Problem P9.10

Conductor Size	R, Ω/(phase · mi)	X, Ω/(phase · mi)	K, pu VD/(kVA · mi)
4/0	0.592	0.761	5.85×10^{-6}
4	2.55	0.835	1.69×10^{-5}

To improve voltage conditions, consider any or all combinations of the following design remedies:

1. Installation of shunt capacitor bank(s)
2. Installation of 32-step voltage regulators with a maximum regulation range of ±10%
3. Addition of new phase conductors

Using these remedies attempt to meet the following primary-voltage criteria:
1. Maximum primary voltage must be 1.040 pu V at zero load.
2. Maximum primary voltage must be 1.07 pu V at peak load.
3. Minimum primary voltage must be 1.00 pu V at peak load.

If the installation of the capacitor alternative is chosen, determine the following:
a. The rating of the capacitor bank(s) in three-phase kilovars.
b. The location of the capacitor bank(s) on the given system.
c. Whether or not voltage-controlled automatic switching is required.

If the installation of the voltage-regulator alternative is chosen, determine the following:
a. The location of the regulator bank(s)
b. The standard kilovoltampere rating of each single-phase regulator
c. The location of the RP on the system
d. The setting of the voltage-regulating relay (VRR)
e. The R and X settings of the LDC

9.11 Repeat Example 9.9. Assume that the annual peak load is 5000 kVA at a lagging power factor of 0.80 and that the reactive load factor is 0.60.

9.12 Figure P9.12 shows an open-wire primary line with many laterals and uniformly distributed load. The voltage at the distribution substation low-voltage bus is held at 1.03 pu V with BVR. When there is N/C bank installed on the feeder, that is, $Q_{NSW} = 0$, the per unit voltage at the end of the line at annual peak load is 0.97. Use the nominal operating voltage of 7.97/13.8 kV of the three-phase four-wire wye-grounded system as the base voltage. Assume that the off-peak load of the system is about 25% of the on-peak load. Also assume that the line reactance is 0.80 Ω/(phase · mi), but the line resistance is not given and determine the following:
a. When the shunt-capacitor bank is not used, find the V_x voltages at the times of peak load and off-peak load.
b. Apply an NSW-capacitor bank and locate it at the point of $x = 4$ mi on the line, and size the capacitor bank to yield the per unit voltage of 1.05 at point x at the time of zero load. Find the size of the capacitor (Q_{NSW}) in three-phase kilovars. Also find the per unit voltage of V_x and V_l at the time of peak load.

9.13 Figure P9.13 shows a system that has a load connected at the end of a 3.5 mi #4 ACSR open-wire primary line. The load belongs to an important scientific equipment installation, and it varies from nearly zero to 1000 kVA. The load requires a closely regulated voltage

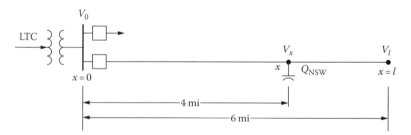

FIGURE P9.12 Figure for Problem 9.12.

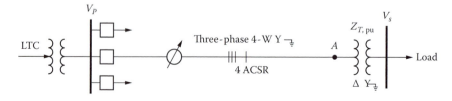

FIGURE P9.13 Figure for Problem 9.13.

at the V_s bus. The consumer requests the voltage V_s to be equal to 1.000 ± 0.010 pu and offers to compensate properly the supplying utility company for such high-quality voltage-regulated service.

A junior engineer proposes to build the 3.5 mi #4 ACSR line to a nearby distribution sub-station and to place the feeder voltage regulators there in order to render the service requested. His wire size is generous for ampacity. He proposes a BW setting of ±1.0 V based on 120 V.

There is BVR at the substation, but at times the BVR equipment is disconnected and bypassed for maintenance and repair. Therefore, the substation bus voltage $V_{P,\mathrm{pu}}$ is as follows. When BVR is in use,

$$V_{P,\mathrm{pu}} = 1.030 \text{ pu V}$$

When BVR is out,

$$\text{Max } V_{P,\mathrm{pu}} = 1.060 \text{ pu V}$$

$$\text{Min } V_{P,\mathrm{pu}} = 0.970 \text{ pu V}$$

The nominal and base voltage at the distribution substation low-voltage bus is 7,200/12,470 V for the three-phase four-wire wye-grounded system. The nominal and base voltage at the consumer's bus is 277/480 V for the three-phase four-wire wye-grounded service. Assume that the regulator bank is made up of three single-phase 32-step feeder voltage regulators with ±10% regulation range. The feeder impedance is given as $2.55 + j0.835 \ \Omega/(\mathrm{mi} \cdot \mathrm{phase})$. Assume that the precalculated K constant of the line is 1.69×10^{-5} pu VD/(kVA · mi) at 85% power factor. The consumer's transformer is rated as 1000 kVA, three phase, with 12,470 V high-voltage rating. It has an impedance of $0 + j0.055$ pu based on the transformer ratings.

Using the given information and data, determine and state whether or not the young engineer's proposed design will meet the consumer's requirements. (Check for both cases, that is, when BVR is in use and BVR is not in use.)

9.14 Repeat Example 9.11, assuming 20 starts per hour and a line-to-ground current of 350 A.

9.15 Repeat Example 9.12, assuming 10 starts per hour and a three-phase fault current of 750 A.

9.16 Resolve parts (a) and (b) of Example 9.1 by using MATLAB. Assume that all the quantities remain the same.

REFERENCES

1. Westinghouse Electric Corporation: *Electric Utility Engineering Reference Book-Distribution Systems*, Vol. 3, Westinghouse Electric Corporation, East Pittsburgh, PA, 1965.
2. ANSI C84.1-1977: *Voltage Ratings for Electric Power Systems and Equipment*, American National Standards Institute.
3. McCrary, M. R.: Regulating voltage on a major power system, *The Line*, 81(1), 1981, 2–8.
4. Hopkinson, R. H.: Recap $ computer program aids voltage regulation studies. *Electr. Forum*. 4(4), 1978, 20–23.
5. Fink, D. G. and H. W. Beaty: *Standard Handbook for Electrical Engineers*, 11th edn., McGraw-Hill, New York, 1978.
6. Gönen, T. et al.: *Development of Advanced Methods for Planning Electric Energy Distribution Systems*, U.S. Department of Energy, October 1979. Available from the National Technical Information Service, U.S. Department of Commerce, Springfield, VA.
7. Oklahoma Gas and Electric Company: *Engineering Guides*, Oklahoma Gas and Electric Company, Oklahoma City, OK, January 1981.
8. Bovenizer, W. N.: New 'Simplified' regulator for lower-cost distribution voltage regulation, *The Line*, 75(1), 1975, 25–28.
9. Bovenizer, W. N.: Paralleling voltage regulators, *The Line*, 75(2), 1975, 6–9.

10. Sealey, W. C.: Increased current ratings for step regulators, *AIEE Trans.*, 74(pt. III), August 1955, 737–742.

11. Lokay, H. E. and D. N. Reps: Distribution system primary-feeder voltage control: Part I-A new approach using the digital computer, *AIEE Trans.*, 77(pt. III), October 1958, 845–855.

12. Reps, D. N. and G. J. Kirk, Jr.: Distribution system primary-feeder voltage control: Part II-digital computer program, *AIEE Trans.*, 77(pt. III), October 1958, 856–865.

13. Amchin, H. K., R. J. Bentzel, and D. N. Reps: Distribution system primary-feeder voltage control: Part III-computer program application, *AIEE Trans.*, 77(pt. III), October 1958, 865–879.

14. Reps, D. N. and R. F. Cook: Distribution system primary-feeder voltage control: Part IV-A supplementary computer program for main-circuit analysis, *AIEE Trans.*, 78(pt. III), October 1958, 904–913.

15. Chang, N. E.: Locating shunt capacitors on primary feeder for voltage control and loss reduction, *IEEE Trans. Power Appar. Syst.*, PAS-88(10), October 1969, 1574–1577.

16. Gönen, T. and F. Djavashi: Optimum shunt capacitor allocation on primary feeders, *IEEE MEXICON-80 International Conference*, Mexico City, Mexico, October 22–25, 1980.

17. Ku, W. S.: Economic comparison of switched capacitors and voltage regulators for system voltage control, *AIEE Trans.*, 76(pt. III), December 1957, 891–906.

18. Gönen, T. and F. Djavashi: Optimum loss reduction from capacitors installed on primary feeders, *The Midwest Power Symposium*, Purdue University, West Lafayette, IN, October 27–28, 1980.

19. Williams, B.R. and D.G. Walden: Distribution automation strategy for the future, *IEEE Comp. Appl. Power*, 7(3), July 1994, 16–21.

10 Distribution System Protection

It is curious that physical courage should be so common in the world and moral courage so rare.

Mark Twain

A lie can travel halfway around the world while the truth is putting on its shoes.

Mark Twain

10.1 BASIC DEFINITIONS

Switch: A device for making, breaking, or changing the connection in an electric current

Disconnect switch: A switch designed to disconnect power devices at no-load conditions

Load-break switch: A switch designed to interrupt load currents but not (greater) fault currents

Circuit breaker: A switch designed to interrupt fault currents

Automatic circuit reclosers: An overcurrent protective device that trips and recloses a preset number of times to clear transient faults or to isolate permanent faults

Automatic line sectionalizer: An overcurrent protective device used only with backup circuit breakers or reclosers but not alone

Fuse: An overcurrent protective device with a circuit-opening fusible member directly heated and destroyed by the passage of overcurrent through it in the event of an overload or short-circuit condition

Relay: A device that responds to variations in the conditions in one electric circuit to affect the operation of other devices in the same or in another electric circuit

Lightning arrester: A device put on electric power equipment to reduce the voltage of a surge applied to its terminals

10.2 OVERCURRENT PROTECTION DEVICES

The overcurrent protective devices applied to distribution systems include relay-controlled circuit breakers, automatic circuit reclosers, fuses, and automatic line sectionalizers.

10.2.1 FUSES

A *fuse* is an overcurrent device with a circuit-opening fusible member (i.e., fuse link) directly heated and destroyed by the passage of overcurrent through it in the event of an overload or short-circuit condition. Therefore, the purpose of a fuse is to clear a permanent fault by removing the defective segment of a line or equipment from the system.

A fuse is designed to blow within a specified time for a given value of fault current. The time–current characteristics (TCCs) of a fuse are represented by two curves: (1) the *minimum-melting curve* and (2) the *total-clearing curve*. The *minimum-melting curve of a fuse* is a plot of the minimum time versus current required to melt the fuse link. The *total-clearing curve* is a plot of the maximum time versus current required to melt the fuse link and extinguish the arc.

Fuses designed to be used above 600 V are categorized as *distribution cutouts* (also known as *fuse cutouts*) or *power fuses*. Figure 10.1 gives a detailed classification of high-voltage fuses.

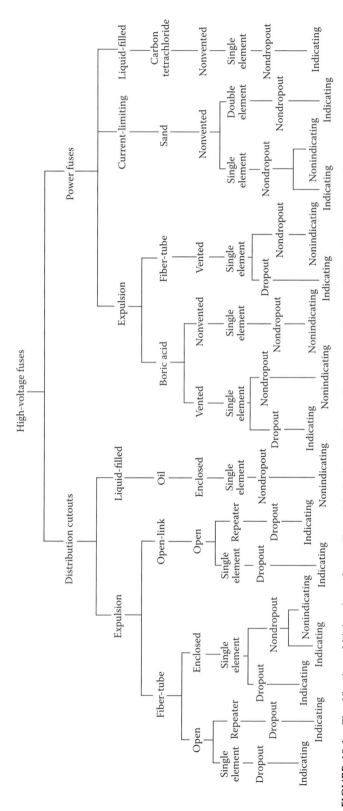

FIGURE 10.1 Classification of high-voltage fuses. (From Westinghouse Electric Corporation, *Electric Utility Engineering Reference Book-Distribution Systems,* vol. 3, East Pittsburgh, PA, 1965.)

The liquid-filled (oil-filled) cutouts are mainly used in underground installations and contain the fusible elements in an oil-filled and sealed tank. The expulsion-type distribution cutouts are by far the most common type of protective device applied to overhead primary distribution systems. In these cutouts, the melting of the fuse link causes heating of the fiber fuse tube, which, in turn, produces deionizing gases to extinguish the arc.

Expulsion-type cutouts are classified according to their external appearance and operation methods as (1) enclosed-fuse cutouts, (2) open-fuse cutouts, and (3) open-link-fuse cutouts.

The ratings of the distribution fuse cutouts are based on continuous current-carrying capacity, nominal and maximum design voltages, and interrupting capacity. In general, the *fuse cutouts are selected* based upon the following data:

1. The type of system for which they are selected, for example, overhead or underground and delta or grounded-wye system
2. The system voltage for which they are selected
3. The maximum available fault current at the point of application
4. The *X/R* ratio at the point of application
5. Other factors, for example, safety, load growth, and changing duty requirements

The use of symmetrical ratings simplified the selection of cutouts as a simple comparison of the calculated system requirements with the available fuse cutout ratings. In spite of that, fuse cutouts still have to be able to interrupt asymmetrical currents, which are, in turn, subject to the *X/R* ratios of the circuit. Therefore, symmetrical cutout rating tables are prepared on the basis of assumed maximum *X/R* ratios. Table 10.1 gives the interrupting ratings of open-fuse cutouts. Figure 10.2 shows a typical open-fuse cutout in pole-top style for 7.2/14.4 kV overhead distribution. Figure 10.3 shows a typical application of open-fuse cutouts in 7.2/14.4 kV overhead distribution.

TABLE 10.1

Interrupting Ratings of Open-Fuse Cutouts

Rating of Cutout			Interrupting Rating in Root-Mean-Rating of Cutout Square Amperes at				Interrupting Rating
Continuous Current, A	Nominal Voltage, kV	Maximum Design Voltage, kV	5.2 kV	7.8 kV	15 kV	27 kV	Nomenclature
100	5.0	5.2	3,000	—	—	—	Normal duty
100	5.0	5.2	5,000	—	—	—	Heavy duty
100	5.0	5.2	10,000	—	—	—	Extra heavy duty
200	5.0	5.2	4,000	—	—	—	Normal duty
200	5.0	5.2	12,000	—	—	—	Heavy duty
100	7.5	7.8	—	3,000	—	—	Normal duty
100	7.5	7.8	—	5,000	—	—	Heavy duty
100	7.5	7.8	—	10,000	—	—	Extra heavy duty
200	7.5	7.8	—	4,000	—	—	Normal duty
200	7.5	7.8	—	12,000	—	—	Heavy duty
100	15	15	—	—	2,000	—	Normal duty
100	15	15	—	—	4,000	—	Heavy duty
100	15	15	—	—	8,000	—	Extra heavy duty
200	15	15	—	—	4,000	—	Normal duty
200	15	15	—	—	10,000	—	Heavy duty
100	25	27	—	—	—	1200	Normal duty

Source: Westinghouse Electric Corporation, *Electric Utility Engineering Reference Book—Distribution Systems*, vol. 3, East Pittsburgh, PA, 1965. With permission.

FIGURE 10.2 Typical open-fuse cutout in pole-top style for 7.2/14.4 kV overhead distribution. (Courtesy of *S&C Electric Company*, Chicago, IL.)

FIGURE 10.3 Typical application of open-fuse cutouts in 7.2/14.4 kV overhead distribution. (Courtesy of S&C Electric Company, Chicago, IL.)

In 1951, a joint study by the Edison Electric Institute (EEI) and National Electrical Manufacturers Association (NEMA) established standards specifying *preferred* and *nonpreferred* current ratings for fuse links of distribution fuse cutouts and their associated TCCs in order to provide interchangeability for fuse links.

The reason for stating certain ratings to be preferred or nonpreferred is based on the fact that the ordering sequence of the current ratings is set up such that a preferred-size fuse link will protect the next higher preferred size. This is also true for the nonpreferred sizes. The current ratings of fuse links for preferred sizes are given as 6, 10, 15, 25, 40, 65, 100, 140, and 200 A and for nonpreferred sizes as 8, 12, 20, 30, 50, and 80 A.

Furthermore, the standards also classify the fuse links as (1) type K (*fast*) and (2) type T (*slow*). The difference between these two fuse links is in the relative melting time, which is defined by the speed ratio as

$$\text{Speed ratio} = \frac{\text{Melting current at } 0.1\,\text{s}}{\text{Melting current at } 300 \text{ or } 600\,\text{s}}$$

where
The 0.1 and 300 s are for fuse links rated 6–100 A
The 0.1 and 600 s are for fuse links rated 140–200 A

Therefore, the speed ratios for type *K* and type *T* fuse links are between 6 and 8, and 10 and 13, respectively. Figure 10.4 shows typical fuse links. Figure 10.5 shows minimum-melting TCC curves for typical (fast) fuse links.

Power fuses are employed where the system voltage is 34.5 kV or higher and/or the interrupting requirements are greater than the available fuse cutout ratings. They are different from fuse cutouts in terms of (1) higher interrupting ratings, (2) larger range of continuous current ratings, (3) applicable not only for distribution but also for subtransmission and transmission systems, and (4) designed and built usually for substation mounting rather than pole and crossarm mounting. A power fuse is made of a fuse mounting and a fuse holder. Its fuse link is called the *refill unit*. In general, they are designed and built as (1) expulsion (boric acid or other solid material [SM]) type, (2) current-limiting (silver-sand) type, or (3) liquid-filled type.

Power fuses are identified by the letter "E" (e.g., 200E or 300E) to specify that their TCCs comply with the interchangeability requirements of the standard. Figure 10.6 shows a typical transformer protection application of 34.5 kV SM-type power fuses. Figure 10.7 shows a feeder protection application of 34.5 kV SM-type power fuses. Figure 10.8 shows a cutaway view of a typical 34.5 kV SM-type refill unit.

10.2.2 AUTOMATIC CIRCUIT RECLOSERS

The automatic circuit recloser is an overcurrent protective device that automatically trips and recloses a preset number of times to clear temporary faults or isolate permanent faults. It also has provisions for manually opening and reclosing the circuit that is connected.

Reclosers can be set for a number of different operation sequences such as (1) two instantaneous (trip and reclose) operations followed by two time-delay trip operations prior to lockout, (2) one instantaneous plus three time-delay operations, (3) three instantaneous plus one time-delay operations, (4) four instantaneous operations, or (5) four time-delay operations. The instantaneous and time-delay characteristics of a recloser are a function of its rating. Recloser ratings range from 5 to 1120 A for the ones with series coils and from 100 to 2240 A for the ones with nonseries coils.

The minimum pickup for all ratings is usually set to trip instantaneously at two times the current rating. The reclosers must be able to interrupt asymmetrical fault currents related to their

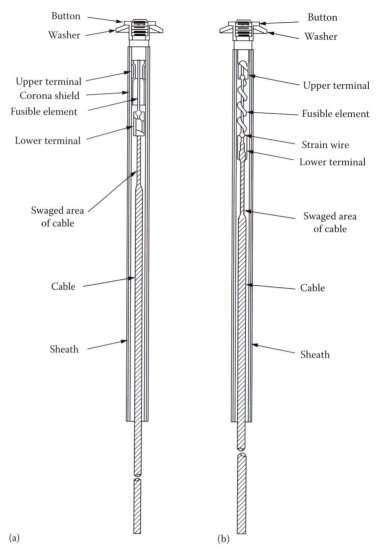

(a)   (b)

FIGURE 10.4 Typical fuse links used on outdoor distribution: (a) fuse link rated less than 10 A and (b) fuse link rated 10–100 A. (Courtesy of S&C Electric Company, Chicago, IL.)

symmetrical rating. The root-mean-square (rms) asymmetrical current ratings can be determined by multiplying the symmetrical ratings by the asymmetrical factor, from Table 10.2, corresponding to the specified X/R circuit ratio. Note that the asymmetrical factors given in Table 10.2 are the ratios of the asymmetrical to the symmetrical rms fault currents at 0.5 cycle after fault initiation for different circuit X/R ratios.

A generally accepted rule of thumb is to assume that the X/R ratios on distribution feeders are not to surpass 5 and therefore the corresponding asymmetry factor is to be about 1.25. However, the asymmetry factor for other parts of the system is assumed to be approximately 1.6.

Line reclosers are often installed at points on the circuit to reduce the amount of exposure on the substation equipment. For example, a feeder circuit serving both urban and rural load would probably have reclosers on the main line serving the rural load. Therefore, the installation of line reclosers will depend on the amount of exposure and operating experience. The maximum fault current available is always an important consideration in the application of line reclosers.

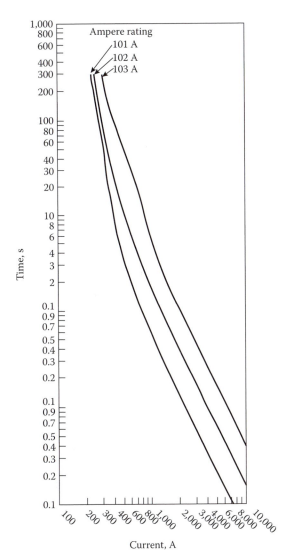

FIGURE 10.5 Minimum-melting-TCC curves for typical (fast) fuse links. Curves are plotted to minimum test points, so all variations should be +20% in current. (Courtesy of S&C Electric Company, Chicago, IL.)

In a sense, a recloser fulfills the same task as the combination of a circuit breaker, overcurrent relay, and reclosing relay. Fundamentally, a recloser is made of an interrupting chamber and the related main contacts that operate in oil, a control mechanism to trigger tripping and reclosing, an operator integrator, and a lockout mechanism.

Reclosers are designed and built in either single-phase or three-phase units. Single-phase reclosers inherently result in better service reliability as compared to three-phase reclosers. If the three-phase primary circuit is wye connected, either a three-phase recloser or three single-phase reclosers are used. However, if the three-phase primary circuit is delta connected, the use of two single-phase reclosers is adequate for protecting the circuit against either single- or three-phase faults. Figure 10.9 shows a typical single-phase hydraulically controlled automatic circuit recloser. Figures 10.10 and 10.11 show typical three-phase hydraulically controlled and electronically controlled automatic circuit reclosers, respectively. Single-phase reclosers inherently result in better service reliability as compared to three-phase reclosers.

FIGURE 10.6 Typical transformer protection application of 34.5 kV SM-type power fuses. (Courtesy of S&C Electric Company, Chicago, IL.)

10.2.3 AUTOMATIC LINE SECTIONALIZERS

The automatic line sectionalizer is an overcurrent protective device installed only with backup circuit breakers or reclosers. It counts the number of interruptions caused by a backup automatic interrupting device and opens during dead circuit time after a preset number (usually two or three) of tripping operations of the backup device.

Zimmerman [1] summarizes the operation modes of a sectionalizer as follows:

1. If the fault is cleared while the reclosing device is open, the sectionalizer counter will reset to its normal position after the circuit is reclosed.
2. If the fault persists when the circuit is reclosed, the fault-current counter in the sectionalizer will again prepare to count the next opening of the reclosing device.
3. If the reclosing device is set to go to lockout on the fourth trip operation, the sectionalizer will be set to trip during the open-circuit time following the third tripping operation of the reclosing device.

FIGURE 10.7 Feeder protection application of 34.5 kV SM-type power fuses. (Courtesy of S&C Electric Company, Chicago, IL.)

Contrary to expulsion-type fuses, a sectionalizer provides coordination (without inserting an additional time–current coordination) with the backup devices associated with very high fault currents and consequently provides an additional sectionalizing point on the circuit. On overhead distribution systems, they are usually installed on poles or crossarms. The application of sectionalizers entails certain requirements:

1. They have to be used in series with other protective devices but not between two reclosers.
2. The backup protective device has to be able to sense the minimum fault current at the end of the sectionalizer's protective zone.
3. The minimum fault current has to be greater than the minimum actuating current of the sectionalizer.
4. Under no circumstances should the sectionalizer's momentary and short-time ratings be exceeded.

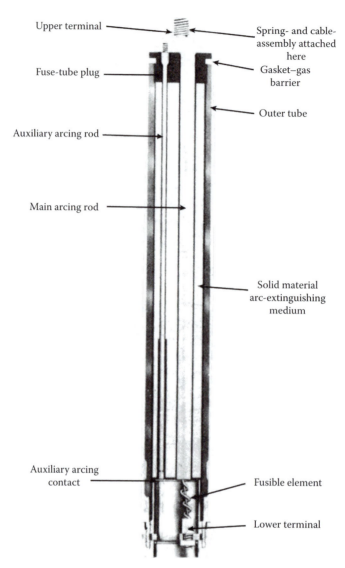

FIGURE 10.8 Cutaway view of typical 34.5 kV SM-type refill unit. (Courtesy of S&C Electric Company, Chicago, IL.)

TABLE 10.2

Asymmetrical Factors as Function of *X/R* Ratios

X/R	Asymmetrical Factor
2	1.06
4	1.20
8	1.39
10	1.44
12	1.48
14	1.51
25	1.60

FIGURE 10.9 Typical single-phase hydraulically controlled automatic circuit recloser: (a) type H, 4H, V4H, or L and (b) type D, E, 4E, or DV. (Courtesy of McGrawEdison Company, Belleville, NJ.)

FIGURE 10.10 Typical three-phase hydraulically controlled automatic circuit reclosers: (a) type 6H or V6H and (b) type RV, RVE, RX, RXE, etc. (Courtesy of McGraw-Edison Company, Belleville, NJ.)

FIGURE 10.11 Typical three-pole automatic circuit recloser. (From Westinghouse Electric Corporation, *Westinghouse Transmission and Distribution Reference Book*, East Pittsburgh, PA, 1964; Westinghouse Electric Corporation, *Electric Utility Engineering Reference Book-Distribution Systems*, vol. 3, East Pittsburgh, PA, 1965.)

5. If there are two backup protective devices connected in series with each other and located ahead of a sectionalizer toward the source, the first and second backup devices should be set for four and three tripping operations, respectively, and the sectionalizer should be set to open during the second dead circuit time for a fault beyond the sectionalizer.
6. If there are two sectionalizers connected in series with each other and located after a backup protective device that is close to the source, the backup device should be set to lock-out after the fourth operation, and the first and second sectionalizers should be set to open following the third and second counting operations, respectively.

The standard continuous current ratings for the line sectionalizers range from 10 to 600 A. Figure 10.12 shows typical single- and three-phase automatic line sectionalizers.

The advantages of using automatic line sectionalizers are as follows:

1. When employed as a substitute for reclosers, they have a lower initial cost and demand less maintenance.
2. When employed as a substitute for fused cutouts, they do not show the possible coordination difficulties experienced with fused cutouts due to improperly sized replacement fuses.
3. They may be employed for interrupting or switching loads within their ratings.

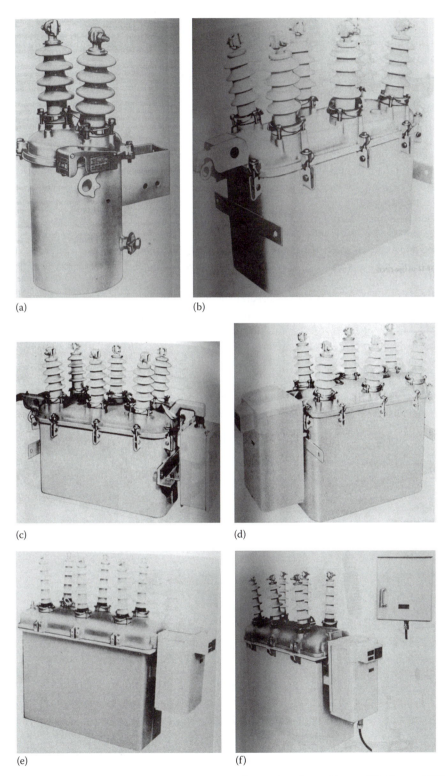

FIGURE 10.12   Typical single- and three-phase automatic line sectionalizers: (a) type GH, (b) type GN3, (c) type GN3E, (d) type GV, (e) type GW, and (f) type GWC. (Courtesy of McGraw-Edison Company, Belleville, NJ.)

The disadvantages of using automatic line sectionalizers are as follows:

1. When employed as a substitute for fused cutouts, they are more costly initially and demand more maintenance.
2. In general, in the past, their failure rate has been greater than that of fused cutouts.

10.2.4 AUTOMATIC CIRCUIT BREAKERS

Circuit breakers are automatic interrupting devices that are capable of breaking and reclosing a circuit under all conditions, that is, faulted or normal operating conditions. The primary task of a circuit breaker is to extinguish the arc that develops due to the separation of its contacts in an arc-extinguishing medium, for example, in air, as is the case for air circuit breakers; in oil, as is the case for oil circuit breakers (OCBs); in SF_6 (sulfur hexafluoride); or in vacuum. In some types, the arc is extinguished by a blast of compressed air, as is the case for magnetic blowout circuit breakers. The circuit breakers used at distribution system voltages are of the air circuit breaker or OCB type. For low-voltage applications, molded-case circuit breakers are available.

OCBs controlled by protective relays are usually installed at the source substations to provide protection against faults on distribution feeders.

Figures 10.13 and 10.14 show typical oil and vacuum circuit breakers, respectively.

Currently, circuit breakers are rated on the basis of rms symmetrical current. Usually, circuit breakers used in the distribution systems have minimum operating times of five cycles. In general, relay-controlled circuit breakers are preferred to reclosers due to their greater flexibility, accuracy, design margins, and esthetics. However, they are much more expensive than reclosers.

The relay, or fault-sensing device, that opens the circuit breaker is generally an overcurrent induction type with inverse, very inverse, or extremely inverse TCCs, for example, the CO relays by Westinghouse or the IAC relays by General Electric. Figure 10.15 shows a typical IAC single-phase overcurrent-relay unit. Figure 10.16 shows typical TCCs of overcurrent relays. Figure 10.17 shows time–current curves of typical overcurrent relays with inverse characteristics.

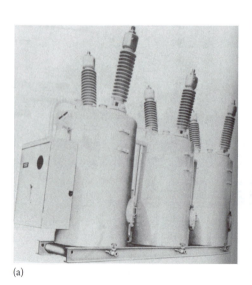
(a)

(b)

FIGURE 10.13   (a) and (b) Typical oil circuit breakers. (Courtesy of McGraw-Edison Company.)

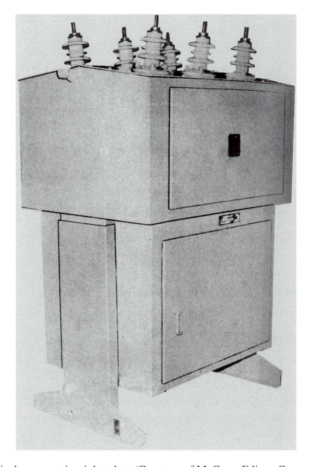

FIGURE 10.14 Typical vacuum circuit breaker. (Courtesy of McGraw-Edison Company.)

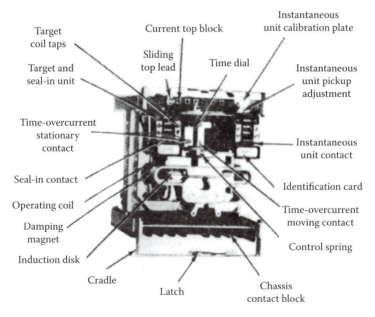

FIGURE 10.15 A typical IAC single-phase overcurrent-relay unit.

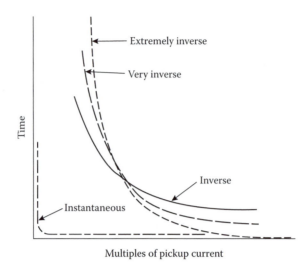

FIGURE 10.16 TCC of overcurrent relays. (From General Electric Company, *Distribution System Feeder Overcurrent Protection*, Application Manual GET-6450, 1979.)

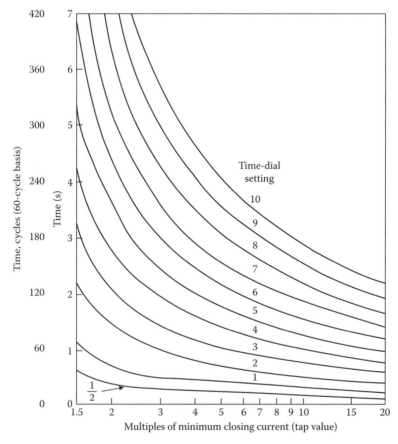

FIGURE 10.17 Time–current curves of IAC overcurrent relays with inverse characteristics. (From General Electric Company, *Overcurrent Protection for Distribution Systems*, Application Manual GET-1751A, 1962; General Electric Company, *Distribution System Feeder Overcurrent Protection*, Application Manual GET-6450, 1979.)

10.3 OBJECTIVE OF DISTRIBUTION SYSTEM PROTECTION

The main objectives of distribution system protection are (1) to minimize the duration of a fault and (2) to minimize the number of consumers affected by the fault.

The secondary objectives of distribution system protection are (1) to eliminate safety hazards as fast as possible, (2) to limit service outages to the smallest possible segment of the system, (3) to protect the consumers' apparatus, (4) to protect the system from unnecessary service interruptions and disturbances, and (5) to disconnect faulted lines, transformers, or other apparatus.

The overhead distribution systems are subject to two types of electric faults, namely, transient (or temporary) faults and permanent faults.

Depending on the nature of the system involved, approximately 75%–90% of the total number of faults are temporary in nature [2]. Usually, *transient faults* occur when phase conductors electrically contact other phase conductors or ground momentarily due to trees, birds or other animals, high winds, lightning, flashovers, etc. Transient faults are cleared by a service interruption of sufficient length of time to extinguish the power arc. Here, the fault duration is minimized, and unnecessary fuse blowing is prevented by using instantaneous or high-speed tripping and automatic reclosing of a relay-controlled power circuit breaker or the automatic tripping and reclosing of a circuit recloser. The breaker speed, relay settings, and recloser characteristics are selected in a manner to interrupt the fault current before a series fuse (i.e., the nearest source-side fuse) is blown, which would cause the transient fault to become permanent.

Permanent faults are those that require repairs by a repair crew in terms of (1) replacing burned-down conductors, blown fuses, or any other damaged apparatus; (2) removing tree limbs from the line; and (3) manually reclosing a circuit breaker or recloser to restore service. Here, the number of customers affected by a fault is minimized by properly selecting and locating the protective apparatus on the feeder main, at the tap point of each branch, and at critical locations on branch circuits.

Permanent faults on overhead distribution systems are usually sectionalized by means of fuses. For example, permanent faults are cleared by fuse cutouts installed at submain and lateral tap points. This practice limits the number of customers affected by a permanent fault and helps locate the fault point by reducing the area involved. In general, the only part of the distribution circuit not protected by fuses is the main feeder and feeder tie line. The substation is protected from faults on feeder and tie lines by circuit breakers and/or reclosers located inside the substation.

Most of the faults are permanent on an underground distribution system, thereby requiring a different protection approach. Even though the number of faults occurring on an underground system is relatively much less than that on the overhead systems, they are usually permanent and can affect a larger number of customers. Faults occurring in the underground residential distribution (URD) systems are cleared by the blowing of the nearest sectionalizing fuse or fuses. Faults occurring on the feeder are cleared by tripping and lockout of the feeder breaker.

Figure 10.18 shows a protection scheme of a distribution feeder circuit. As shown in the figure, each distribution transformer has a fuse that is located either externally, that is, in a fuse cutout next to the transformer, or internally, that is, inside the transformer tank as is the case for a completely self-protected (CSP) transformer.

As shown in Figure 10.18, it is a common practice to install a fuse at the head of each lateral (or branch). The fuse must carry the expected load, and it must coordinate with load-side transformer fuses or other devices. It is customary to select the rating of each lateral fuse adequately large so that it is protected from damage by the transformer fuses on the lateral. Furthermore, the lateral fuse is usually expected to clear faults occurring at the ends of the lateral. If the fuse does not clear the faults, then one or more additional fuses may be installed on the lateral.

As shown in the figure, a recloser, or circuit breaker A with reclosing relays, is located at the substation to provide a backup protection. It clears the temporary faults in its protective zone. At the limit of the protective zone, the minimum available fault current, determined by calculation,

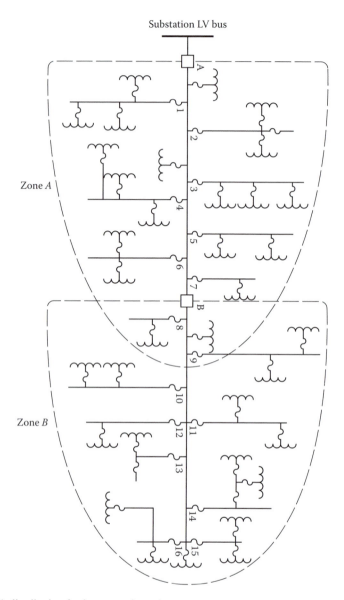

FIGURE 10.18   A distribution feeder protection scheme.

is equal to the smallest value of current (called *minimum pickup current*), which will trigger the recloser, or circuit breaker, to operate. However, a fault beyond the limit of this protection zone may not trigger the recloser, or circuit breaker, to operate. Therefore, this situation may require that a second recloser, with a lower pickup current rating, be installed at location B, as shown in the figure.

The major factors that play a role in making a decision to choose a recloser over a circuit breaker are (1) the costs of equipment and installation and (2) the reliability. Usually, a comparable recloser can be installed for approximately one-third less than a relay-controlled OCB. Even though a circuit breaker provides a greater interrupting capability, this excess capacity is not always required. Also, some distribution engineers prefer reclosers because of their flexibility, due to the many extras that are available with reclosers but not with circuit breakers.

10.4   COORDINATION OF PROTECTIVE DEVICES

The process of selecting overcurrent protection devices with certain time–current settings and their appropriate arrangement in series along a distribution circuit in order to clear faults from the lines and apparatus according to a preset sequence of operation is known as *coordination*. When two protective apparatus installed in series have characteristics that provide a specified operating sequence, they are said to be *coordinated* or *selective*. Here, the device that is set to operate first to isolate the fault (or interrupt the fault current) is defined as the *protecting device*. It is usually the apparatus closer to the fault.

The apparatus that furnishes backup protection but operates only when the protecting device fails to operate to clear the fault is defined as the *protected device*. Properly coordinated protective devices help (1) to eliminate service interruptions due to temporary faults, (2) to minimize the extent of faults in order to reduce the number of customers affected, and (3) to locate the fault, thereby minimizing the duration of service outages.

Since coordination is primarily the selection of protective devices and their settings to develop zones that provide temporary fault protection and limit an outage area to the minimum size possible if a fault is permanent, to coordinate protective devices, in general, the distribution engineer must assemble the following data:

1. Scaled feeder-circuit configuration diagram (map)
2. Locations of the existing protective devices
3. TCC curves of protective devices
4. Load currents (under normal and emergency conditions)
5. Fault currents or megavoltamperes (under minimum and maximum generation conditions) at every point where a protective apparatus might be located

Usually, these data are not readily available and therefore must be brought together from numerous sources. For example, the TCCs of protective devices are gathered from the manufacturers; the values of the load currents and fault currents are usually taken from computer runs called the *load flow* (or more correctly, *power flow*) *studies* and *fault studies*, respectively.

In general, manual techniques for coordination are still employed by most utilities, especially where distribution systems are relatively small or simple and therefore only a small number of protective devices are used in series.

However, some utilities have established standard procedures, tables, or other means to aid the distribution engineer and field personnel in coordination studies. Some utilities employ semiautomated, computerized coordination programs developed either by the protective device manufacturers or by the company's own staff.

A general coordination procedure, regardless of whether it is manual or computerized, can be summarized as follows [3,4]:

1. Gather the required and aforementioned data.
2. Select initial locations on the given distribution circuit for protective (i.e., sectionalizing) devices.
3. Determine the maximum and minimum values of fault currents (specifically for three-phase, line-to-line [L–L], and line-to-ground faults) at each of the selected locations and at the end of the feeder mains, branches, and laterals.
4. Pick out the necessary protective devices located at the distribution substation in order to protect the substation transformer properly from any fault that might occur in the distribution circuit.
5. Coordinate the protective devices from the substation outward or from the end of the distribution circuit back to the substation.

6. Reconsider and change, if necessary, the initial locations of the protective devices.
7. Reexamine the chosen protective devices for current-carrying capacity, interrupting capacity, and minimum pickup rating.
8. Draw a composite TCC curve showing the coordination of all protective devices employed, with curves drawn for a common base voltage (this step is optional).
9. Draw a circuit diagram that shows the circuit configuration, the maximum and minimum values of the fault currents, and the ratings of the protective devices employed.

There are also some additional factors that need to be considered in the coordination of protective devices (i.e., fuses, reclosers, and relays) such as (1) the differences in the TCCs and related manufacturing tolerances, (2) preloading conditions of the apparatus, (3) ambient temperature, and (4) effect of reclosing cycles. These factors affect the adequate margin for selectivity under adverse conditions.

10.5 FUSE-TO-FUSE COORDINATION

The selection of a fuse rating to provide adequate protection to the circuit beyond its location is based upon several factors. First of all, the selected fuse must be able to carry the expanded load current, and, at the same time, it must be sufficiently selective with other protective apparatus in series. Furthermore, it must have an adequate reach; that is, it must have the capability to clear a minimum fault current within its zone in a predetermined time duration.

A fuse is designed to blow within a specified time for a given value of fault current. The TCCs of a fuse are represented by two curves: the minimum-melting curve and the total-clearing curve, as shown in Figure 10.19. The minimum-melting curve of a fuse that represents the minimum

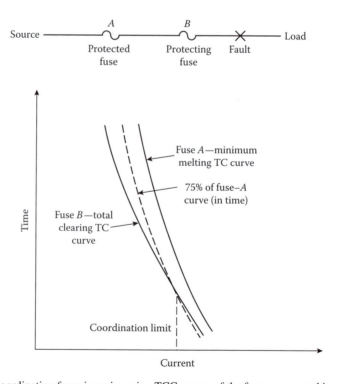

FIGURE 10.19 Coordinating fuses in series using TCC curves of the fuses connected in series.

time, and therefore it is the plot* of the minimum time versus current required to melt the fuse. The total-clearing (time) curve represents the total time, and therefore it is the plot of the maximum time versus current required to melt the fuse and extinguish the arc, plus manufacturing tolerance. It is also a standard procedure to develop "damaging" time curves from the minimum-melting-time curves by using a safety factor of 25%. Therefore, the damaging curve (due to the partial melting) is developed by taking 75% of the minimum-melting time of a specific-size fuse at various current values. The time unit used in these curves is seconds.

Fuse-to-fuse coordination, that is, the coordination between fuses connected in series, can be achieved by two methods:

1. Using the TCC curves of the fuses
2. Using the coordination tables prepared by the fuse manufacturers

Furthermore, some utilities employ certain rules of thumb as a third type of fuse-to-fuse coordination method.

In the first method, the coordination of the two fuses connected in series, as shown in Figure 10.19, is achieved by comparing the total-clearing-time–current curve of the *"protecting fuse,"* that is, fuse *B*, with the damaging-time curve of the *"protected fuse,"* that is, fuse *A*. Here, it is necessary that the total-clearing time of the protecting fuse not exceed 75% of the minimum-melting time of the protected fuse.

The 25% margin has been selected to take into account some of the operating variables, such as preloading, ambient temperature, and the partial melting of a fuse link due to a fault current of short duration. If there is no intersection between the aforementioned curves, a complete coordination in terms of selectivity is achieved. However, if there is an intersection of the two curves, the associated current value at the point of the intersection gives the coordination limit for the partial coordination achieved.

In the second method of fuse-to-fuse coordination, coordination is established by using the fuse sizes from coordination tables developed by the fuse link manufacturers. Tables 10.3 and 10.4 are such tables developed by the General Electric Company for fast and slow fuse links, respectively.

These tables give the maximum fault currents to achieve coordination between various fuse sizes and are based upon the 25% margin described in the first method. Here, the determination of the total-clearing curve is not necessary since the maximum value of fault current to which each combination of series fuses can be subjected with guaranteed coordination is given in the tables, depending upon the type of fuse link selected.

10.6   RECLOSER-TO-RECLOSER COORDINATION

The need for recloser-to-recloser coordination may arise due to any of the following situations that may exist in a given distribution system:

1. Having two three-phase reclosers
2. Having two single-phase reclosers
3. Having a three-phase recloser at the substation and a single-phase recloser on one of the branches of a given feeder

---

* TCC curves of overcurrent protective devices are plotted on *log–log coordinate* paper. The use of this standard-size transparent paper allows the comparison of curves by superimposing one sheet over another.

TABLE 10.3

Coordination Table for GE Type "K" (Fast) Fuse Links Used in GE 50, 100, or 200 A Expulsion Fuse Cutouts and Connected in Series

Type "K" Ratings of Protecting Fuse Links (B in Diagram), A	Type "K" Ratings of Protected Fuse Links (A in Diagram), A																		
	6 K	8 K	10 K	12 K	15 K	20 K	25 K	30 K	40 K	50 K	65 K	80 K	52ᵃ	100 K	101ᵃ	140 K	200 K	102ᵃ	103ᵃ
	Maximum short-circuit RMS Amperes to which fuse links will be protected																		
1 K	135	215	300	395	530	660	820	1100	1370	1720	2200	2750	3250	3600	5800	6000	9700	9500	16,000
2 K	110	195	300	395	530	660	820	1100	1370	1720	2200	2750	3250	3600	5800	6000	9700	9500	16,000
3 K	80	165	290	395	530	660	820	1100	1370	1720	2200	2750	3250	3600	5800	6000	9700	9500	16,000
5 A series hi-surge	14	133	270	395	530	660	820	1100	1370	1720	2200	2750	3250	3600	5800	6000	9700	9500	16,000
6 K		37	145	270	460	620	820	1100	1370	1720	2200	2750	3250	3600	5800	6000	9700	9500	16,000
8 K			133	170	390	560	820	1100	1370	1720	2200	2750	3250	3600	5800	6000	9700	9500	16,000
10 A series hi-surge			24	260	530	660	820	1100	1370	1720	2200	2750	3250	3600	5800	6000	9700	9500	16,000
10 K				38	285	470	720	1100	1370	1720	2200	2750	3250	3600	5800	6000	9700	9500	16,000
12 K					140	360	660	1100	1370	1720	2200	2750	3250	3600	5800	6000	9700	9500	16,000
15 K						95	410	960	1370	1720	2200	2750	3250	3600	5800	6000	9700	9500	16,000
20 K							70	700	1200	1720	2200	2750	3250	3600	5800	6000	9700	9500	16,000
25 K								140	580	1300	2200	2750	3250	3600	5800	6000	9700	9500	16,000
30 K									215	700	1800	2750	3250	3600	5800	6000	9700	9500	16,000
40 K										170	1200	2750	3250	3600	5800	6000	9700	9500	16,000
50 K											195	1600	3250	3600	5800	6000	9700	9500	16,000
65 K												330		2300	5800	6000	9700	9500	16,000
52ᵃ														290	5500	6000	9700	9500	16,000
80 K														580	5800	6000	9700	9500	16,000
100 K															300	4300	9700	9500	16,000
101ᵃ																385	7500	9500	16,000
140 K																	2800		16,000
102ᵃ																	1250		16,000

Source: General Electric Company, *Overcurrent Protection for Distribution Systems,* Application Manual GET-1751A, 1962. With permission.

RMS, root-mean-square.

ᵃ GE coordinating fuse links.

TABLE 10.4

Coordination Table for GE Type "T" (Slow) Fuse Links Used in GE 50, 100, or 200 A Expulsion Fuse Cutouts and Connected in Series

Type "T" Ratings of Protecting Fuse Links (B in Diagram), A	Type "T" Ratings of Protected Fuse Links (A in Diagram), A															
	6T	8T	10T	12T	15T	20T	25T	30T	40T	50T	65T	80T	100T	140T	200T	103
	Maximum short-circuit RMS Amperes to which fuse links will be protected															
1N[a]	250	395	540	710	950	1220	1500	1930	2500	3100	3950	4950	6300	9600	15,000	16,000
2N[a]	250	395	540	710	950	1220	1500	1930	2500	3100	3950	4950	6300	9600	15,000	16,000
3N[a]	250	395	540	710	950	1220	1500	1930	2500	3100	3950	4950	6300	9600	15,000	16,000
6T		33	365	650	950	1220	1500	1930	2500	3100	3950	4950	6300	9600	15,000	16,000
8T			125	480	850	1220	1500	1930	2500	3100	3950	4950	6300	9600	15,000	16,000
10 A series hi-surge		19	540	710	950	1220	1500	1930	2500	3100	3950	4950	6300	9600	15,000	16,000
10T				74	620	1130	1500	1930	2500	3100	3950	4950	6300	9600	15,000	16,000
12T					135	770	1400	1930	2500	3100	3950	4950	6300	9600	15,000	16,000
15T						100	880	1750	2500	3100	3950	4950	6300	9600	15,000	16,000
20T							105	1150	2300	3100	3950	4950	6300	9600	15,000	16,000
25T								190	1500	3100	3950	4950	6300	9600	15,000	16,000
30T									115	1900	3950	4950	6300	9600	15,000	16,000
40T										310	2350	4950	6300	9600	15000	16,000
50T											150	3400	6300	9600	15000	16,000
65T												270	4300	9600	15000	16,000
80T													660	9200	15,000	16,000
100T														6000	15,000	16,000
140T															6600	

RMS, root-mean-square.

[a] The 1N, 2N, and 3N ampere ratings of the GE 5 A series hi-surge fuse links have TCCs closely approaching those established by the American Standards for 1T, 2T, and 3T ampere ratings, respectively. Hence, they are recommended for applications requiring 1T, 2T, or 3T fuse links.

The required coordination between the reclosers can be achieved by using one of the following remedies:

1. Employing different recloser types and some mixture of coil sizes and operating sequences
2. Employing the same recloser type and operating sequence but using different coil sizes
3. Employing the same recloser type and coil sizes but using different operating sequences

In general, the utility industry prefers to use the first remedy over the other two. However, there may be some circumstances, for example, having two single-phase reclosers of the same type, where the second remedy can be applied.

When the TCC curves of the two reclosers are less than 12 cycles separate from each other, the reclosers may do their instantaneous or fast operations at the same time. To achieve coordination between the delayed-tripping curves of two reclosers, at least a minimum time margin of 25% must be applied.

10.7   RECLOSER-TO-FUSE COORDINATION

In Figure 10.20, curves represent the instantaneous, time-delay, and extended time-delay (as an alternative) tripping characteristics of a conventional automatic circuit recloser. Here, curves A and B symbolize the first and second openings and the third and fourth openings of the recloser, respectively.

To provide protection against permanent faults, fuse cutouts (or power fuses) are installed on overhead feeder taps and laterals. The use of an automatic reclosing device as a backup protection against temporary faults eliminates many unnecessary outages that occur when using fuses only. Here, the backup recloser can be either the substation feeder recloser, usually with an operating sequence of one fast- and two delayed-tripping operations, or a branch feeder recloser, with two fast- and two delayed-tripping operations. The recloser is set to trip for a temporary fault before any of the fuses can blow and then reclose the circuit. However, if the fault is a permanent one, it is cleared by the correct fuse before the recloser can go on time-delay operation following one or two instantaneous operations.

Figure 10.21 shows a portion of a distribution system where a recloser is installed ahead of a fuse. The figure also shows the superposition of the TCC curve of the fuse C on the fast and delayed

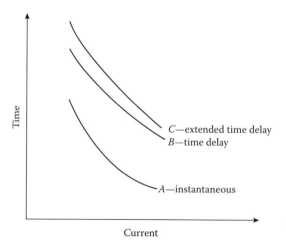

FIGURE 10.20   Typical recloser tripping characteristics.

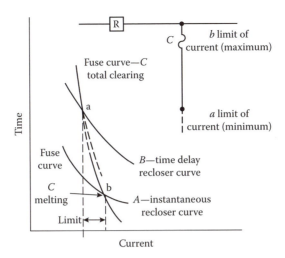

FIGURE 10.21 Recloser TCC curves superimposed on fuse TCC curves. (From General Electric Company, *Distribution System Feeder Overcurrent Protection*, Application Manual GET-6450, 1979.)

TCC curves of the recloser R. If the fault beyond fuse *C* is temporary, the instantaneous tripping operations of the recloser protect the fuse from any damage. This can be observed from the figure by the fact that the instantaneous recloser curve *A* lies below the fuse TCC for currents less than that associated with the intersection point *b*.

However, if the fault beyond fuse *C* is a permanent one, the fuse clears the fault as the recloser goes through a delayed operation *B*. This can be observed from the figure by the fact that the time-delay curve *B* of the recloser lies above the total-clearing-curve portion of the fuse TCC for currents greater than that associated with the intersection point *a*. The distance between the intersection points *a* and *b* gives the coordination range for the fuse and recloser.

Therefore, a proper coordination of the trip operations of the recloser and the total-clearing time of the fuse prevents the fuse link from being damaged during instantaneous trip operations of the recloser. The required coordination between the recloser and the fuse can be achieved by comparing the respective time–current curves and taking into account other factors, for example, preloading, ambient temperature, curve tolerances, and accumulated heating and cooling of the fuse link during the fast-trip operations of the recloser.

Figure 10.22 illustrates the temperature cycle of a fuse link during recloser operations. As can be observed from the figure, each of the first two (instantaneous) operations takes only 2 cycles, but each of the last two (delayed) operations lasts 20 cycles. After the fourth operation, the recloser locks itself open.

Therefore, the recloser-to-fuse coordination method illustrated in Figure 10.21 is an approximate one since it does not take into account the effect of the accumulated heating and cooling of the fuse link during recloser operation. Thus, it becomes necessary to compute the heat input to the fuse during, for example, two instantaneous recloser operations if the fuse is to be protected from melting during these two openings.

Figure 10.23 illustrates a practical yet sufficiently accurate method of coordination. Here, the maximum coordinating current is found by the intersection (at point *b'*) of two curves, the fuse-damage curve (which is defined as 75% of the minimum-melting-time curve of the fuse) and the maximum-clearing-time curve of the recloser's fast-trip operation (which is equal to $2 \times A$ "*in time*," since there are two fast trips). Similarly, point *a'* is found from the intersection of the fuse total-clearing curve with the shifted curve *B'* (which is equal to $2 \times A + 2 \times B$ "*in time*," since in addition to the two fast trips, there are two delayed trips).

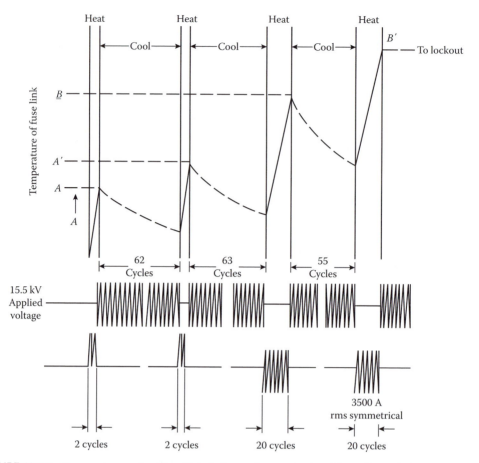

FIGURE 10.22 Temperature cycle of fuse link during recloser operation. (From General Electric Company, *Distribution System Feeder Overcurrent Protection*, Application Manual GET-6450, 1979.)

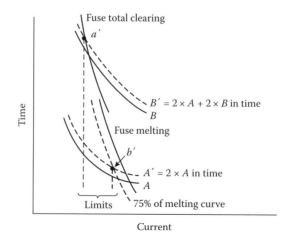

FIGURE 10.23 Recloser-to-fuse coordination (corrected for heating and cooling cycle). (From General Electric Company, *Distribution System Feeder Overcurrent Protection*, Application Manual GET-6450, 1979.)

TABLE 10.5
Automatic Recloser and Fuse Ratings

Recloser Rating, rms A (Continuous)	Fuse Link Ratings, rma A	2N[a]	3N[a]	Ratings of GE Type T Fuse Links, A						
				6T	8T	10T	12T	15T	20T	25T
				Range of Coordination, rms A						
5	Min	14	17.5	68						
	Max	55	55	123						
10	Min			31	45	75	200			
	Max[b]			110	152	220	300, 250			
15	Min			30	34	59	84	200	380	
	Max[b]			105	145	210	280	375	450[c]	
25	Min			50	50	50	68	105	145	300
	Max			89	130	190	265	360	480	610

Source: Fink, D.G. and H.W. Beaty, *Standard Handbook for Electrical Engineers*, 11th edn., McGraw-Hill, New York, 1978. With permission.

[a] The 1N, 2N, and 3N ampere ratings of the GE 5 A series hi-surge fuse links have TCCs closely approaching those established by the EEI–NEMA Standards for IT, 2T, and 3T ampere ratings, respectively. Hence, they are recommended for applications requiring 1T, 2T, or 3T fuse links.

[b] Where maximum lines have two values, the smaller value is for the 50 A frame, single-phase recloser. The larger value is for all others: 50 A frame, three phase; 140 A frame, single phase and three phase.

[c] Coordination with 50 A frame size single-phase recloser not possible since maximum interrupting capacity is less than minimum value.

Some distribution engineers *use rule-of-thumb methods*, based upon experience, to allow extra margin in the coordination scheme.

As shown in Table 10.5, there are also coordination tables developed by the manufacturers to coordinate reclosers with fuse links in a simpler way.

10.8   RECLOSER-TO-SUBSTATION TRANSFORMER HIGH-SIDE FUSE COORDINATION

Usually, a power fuse, located at the primary side of a delta–wye-connected substation transformer, provides protection for the transformer against the faults in the transformer or at the transformer terminals and also provides backup protection for feeder faults. These fuses have to be coordinated with the reclosers or reclosing circuit breakers located on the secondary side of the transformer to prevent the fuse from any damage during the sequential tripping operations. The effects of the accumulated heating and cooling of the fuse element can be taken into account by adjusting the delayed-tripping time of the recloser.

To achieve a coordination, the adjusted tripping time is compared to the minimum-melting time of the fuse element, which is plotted for a phase-to-phase fault that might occur on the secondary side of the transformer. If the minimum-melting time of the backup fuse is greater than the adjusted tripping time of the recloser, a coordination between the fuse and recloser is achieved.

The coordination of a substation circuit breaker with substation transformer primary fuses dictates that the total-clearing time of the circuit breaker (i.e., relay time plus breaker interrupting time) be less than 75%–90% of the minimum-melting time of the fuses at all values of current up to the maximum fault current.

The selected fuse must be able to carry 200% of the transformer full-load current continuously in any emergency in order to be able to carry the transformer *"magnetizing"* inrush current (which is usually 12–15 times the transformer full-load current) for 0.1 s [5].

10.9 FUSE-TO-CIRCUIT-BREAKER COORDINATION

The fuse-to-circuit-breaker (*overcurrent-relay*) coordination is somewhat similar to the fuse-to-recloser coordination. In general, the reclosing time intervals of a circuit breaker are greater than those of a recloser. For example, 5 s is usually the minimum reclosing time interval for a circuit breaker, whereas the minimum reclosing time interval for a recloser can be as small as 1/2 s.

Therefore, when a fuse is used as the backup or protected device, there is no need for heating and cooling adjustments. Thus, in order to achieve a coordination between a fuse and circuit breaker, the minimum-melting-time curve of the fuse is plotted for a phase-to-phase fault on the secondary side.

If the minimum-melting time of the fuse is approximately 135% of the combined time of the circuit breaker and related relays, the coordination is achieved. However, when the fuse is used as the protecting device, the coordination is achieved if the relay operating time is 150% of the total-clearing time of the fuse.

In summary, when the circuit breaker is tripped instantaneously, it has to clear the fault before the fuse is blown. However, the fuse has to clear the fault before the circuit breaker trips on time-delay operations.

Therefore, it is necessary that the relay characteristic curve, at all values of current up to the maximum current available at the fuse location, lie above the total-clearing characteristic curve of the fuse. Thus, it is usually customary to leave a margin between the relay and fuse characteristic curves to include a safety factor of 0.1 to 0.3 + 0.1 s for relay overtravel time.

A sectionalizing fuse installed at the riser pole to protect underground cables does not have to coordinate with the instantaneous trips since underground lines are usually not subject to transient faults. On looped circuits, the fuse size selected is usually the minimum size required to serve the entire load of the loop, whereas on lateral circuits, the fuse size selected is usually the minimum size required to serve the load and coordinate with the transformer fuses, keeping in mind the cold pickup load.

10.10 RECLOSER-TO-CIRCUIT-BREAKER COORDINATION

The reclosing relay recloses its associated feeder-circuit breaker at predetermined intervals (e.g., 15, 30, or 45 s cycles) after the breaker has been tripped by overcurrent relays. If desired, the reclosing relay can provide an instantaneous initial reclosure plus three time-delay reclosures.

However, if the fault is permanent, the reclosing relay recloses the breaker the predetermined number of times and then goes to the lockout position. Usually, the initial reclosing is so fast that customers may not even realize that service has been interrupted.

The crucial factor in coordinating the operation of a recloser and a circuit breaker (better yet, the relay that trips the breaker) is the reset time of the overcurrent relays during the tripping and reclosing sequence.

If the relay used is of an electromechanical type, rather than a solid-state type, it starts to travel in the trip direction during the operation of the recloser.

If the reset time of the relay is not adjusted properly, the relay can accumulate enough movement (or travel) in the trip direction, during successive recloser operations, to trigger a false tripping.

Example 10.1

Figure 10.24 gives an example* for proper recloser-to-relay coordination. In the figure, curves *A* and *B* represent, respectively, the instantaneous and time-delay TCCs of the 35 A reclosers. Curve *C* represents the TCC of the extremely inverse-type IAC overcurrent relay set on the number 1.0

---

* For further information, see Ref. [5].

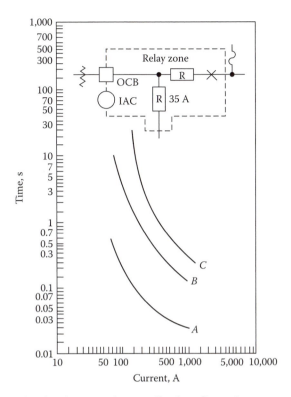

FIGURE 10.24 An example of recloser-to-relay coordination. Curve *A* represents TCCs of one instantaneous recloser opening. Curve *B* represents TCCs of one extended time-delay recloser opening. Curve *C* represents TCCs of the IAC relay. (Courtesy of Sec Company, Chicago, IL.)

time-dial adjustment and 4 A tap (160 A primary with 200:5 current transformer [CT]). Assume a permanent fault current of 700 A located at point *X* in the figure. Determine the necessary relay and recloser coordination.

Solution

From Figure 10.24, the operating time of the relay and recloser can be found as the following:

For recloser: Instantaneous (from curve *A*) = 0.03 s
Time delay (from curve *B*) = 0.17 s
For relay: Pickup (from curve *C*) = 0.42 s

$$\text{Reset} = \frac{1.0}{10} \times 60 = 6.0\,\text{s}$$

assuming a 60 s reset time for the relay with a number 10 time-dial setting [5].

Using the signs (+) for trip direction and (−) for reset direction, the percent of total relay travel, during the operation of the recloser, can be calculated in the following manner. During the instantaneous operation (from curve *A*) of the recloser,

$$\text{Relay-closing travel} = \frac{\text{Recloser-instantaneous time}}{\text{Relay-pickup time}}$$

$$= \frac{0.03}{0.42}$$

$$= 0.0714 \quad \text{or} \quad 7.14\% \tag{10.1}$$

Assuming that the recloser is open for 1 s,

$$\text{Relay-reset travel} = \frac{(-)\,\text{Recloser-open time}}{\text{Relay-reset time}}$$

$$= \frac{-1}{6.0}$$

$$= -0.1667 \quad \text{or} \quad -16.67\% \tag{10.2}$$

From the results, it can be seen that

$$|\text{Relay-closing travel}| < |\text{relay-reset travel}|$$

or

$$|7.14\%| < |16.67\%|$$

and therefore the relay will completely reset during the time that the recloser is open following each instantaneous opening.

Similarly, the travel percentages during the delayed-tripping operations can be calculated in the following manner. During the first time-delay trip operation (from curve B) of the recloser,

$$\text{Relay-closing travel} = \frac{\text{Recloser time-delay}}{\text{Recloser-pickup time}}$$

$$= \frac{0.17}{0.42}$$

$$\cong 0.405 \quad \text{or} \quad 40.5\% \tag{10.3}$$

Assuming that the recloser opens for 1 s,

$$\text{Relay-reset travel} = (-)\frac{\text{Recloser-open time}}{\text{Relay-reset time}}$$

$$= -\frac{1.0}{6.0}$$

$$= -16.67\%$$

During the second time-delay trip of the recloser,

$$\text{Relay-closing travel} = 40.5\%$$

Therefore, the net total relay travel is 64.3%:

$$(= +40.5\% - 16.67\% + 40.5\%)$$

Since this net total relay travel is less than 100%, the desired recloser-to-relay coordination is accomplished. In general, a 0.15–0.20 s safety margin is considered to be adequate for any possible errors that might be involved in terms of curve readings, etc.

Some distribution engineers use *a rule-of-thumb method* to determine whether the recloser-to-relay coordination is achieved or not. For example, if the operating time of the relay at any given fault-current value is less than twice the delayed-tripping time of the recloser, assuming a recloser operation sequence that includes two time-delay trips, there will be a possible lack of coordination.

Whenever there is a lack of coordination, either the time-dial or pickup settings of the relay must be increased or the recloser has to be relocated until the coordination is achieved.

In general, the reclosers are located at the end of the relay reach. The rating of each recloser must be such that it will carry the load current, have sufficient interrupting capacity for that location, and coordinate with both the relay and load-side apparatus. If there is a lack of coordination with the load-side apparatus, then the recloser rating has to be increased. After the proper recloser ratings are determined, each recloser has to be checked for reach. If the reach is insufficient, additional series reclosers may be installed on the primary main.

10.11 FAULT-CURRENT CALCULATIONS*

There are four possible fault types that might occur in a given distribution system:

1. Three-phase grounded or ungrounded fault (3ϕ)
2. Phase-to-phase (or $L–L$) ungrounded fault
3. Phase-to-phase (or double line-to-ground, 2LG) grounded fault
4. Phase-to-ground (or single line-to-ground, SLG) fault

The first type of fault can take place only on three-phase circuits, and the second and third on three-phase or two-phase (i.e., vee or open-delta) circuits. However, even on these circuits usually only SLG faults will take place due to the multigrounded construction. The relative numbers of the occurrences of different fault types depend upon various factors, for example, circuit configuration, the height of ground wires, voltage class, method of grounding, relative insulation levels to ground and between phases, speed of fault clearing, number of stormy days per year, and atmospheric conditions. Based on Ref. [6], the probabilities of prevalence of the various types of faults are[†]

SLG faults	= 0.70
$L–L$ faults	= 0.15
2LG faults	= 0.10
3ϕ faults	= 0.05
Total	= 1.00

The actual fault current is usually less than the bolted three-phase value. (Here, the term bolted means that there is no fault impedance [or fault resistance] resulting from fault arc, i.e., $Z_f = 0$.) However, the SLG fault often produces a greater fault current that the 3ϕ fault especially (1) where the associated generators have solidly grounded neutrals or low-impedance neutral impedances and (2) on the wye-grounded side of delta–wye-grounded transformer banks [7].

Therefore, for a given system, each fault at each fault location must be calculated based on actual circuit conditions. When this is done, according to Anderson [8], it is usually the case that the SLG fault is the most severe, with the 3ϕ, 2LG, and $L–L$ following in that order. In general, since the 2LG fault value is always somewhere in between the maximum and minimum, it is usually neglected in the distribution system fault calculations [3].

In general, the maximum and minimum fault currents are both calculated for a given distribution system. The *maximum fault current* is calculated based on the following assumptions:

1. All generators are connected, that is, in service.
2. The fault is a bolted one; that is, the fault impedance is zero.
3. The load is maximum, that is, on-peak load.

---

* More rigorous and detailed treatment of the subject is given by Anderson [3,8].
[†] One should keep in mind that these probabilities may differ substantially from one system to another in practice.

The *minimum current* is calculated based on the following assumptions:

1. The number of generators connected is minimum.
2. The fault is not a bolted one, that is, the fault impedance is not zero but has a value somewhere between 0 and 40 Ω.
3. The load is minimum, that is, off-peak load.

On 4 kV systems, the value of the minimum fault current available may be taken as 60%–70% of the calculated maximum line-to-ground fault current.

In general, these fault currents are calculated for each sectionalizing point, including the substation, and for the ends of the longest sections. The calculated maximum fault-current values are used in determining the required interrupting capacities (i.e., ratings) of the fuses, circuit breakers, or other fault-clearing apparatus; the calculated minimum fault-current values are used in coordinating the operations of fuses, reclosers, and relays.

To calculate the fault currents, one has to determine the zero-, positive-, and negative-sequence Thévenin impedances of the system* at the high-voltage side of the distribution substation transformer looking into the system. These impedances are usually readily available from transmission system fault studies. Therefore, for any given fault on a radial distribution circuit, one can simply add the appropriate impedances to the Thévenin impedances as the fault is moved away from the substation along the circuit. The most common types of distribution substation transformer connections are (1) delta–wye solidly grounded and (2) delta–delta.

10.11.1 THREE-PHASE FAULTS

Since this fault type is completely balanced, there are no zero- or negative-sequence currents. Therefore, when there is no fault impedance,

$$I_{f,3\phi} = I_{f,a} = I_{f,b} = I_{f,c}$$

$$= \left| \frac{\overline{V}_{L-N}}{\overline{Z}_1} \right| \tag{10.4}$$

and when there is a fault impedance,

$$I_{f,3\phi} = \left| \frac{\overline{V}_{L-N}}{\overline{Z}_1 + \overline{Z}_f} \right| \tag{10.5}$$

where
$\overline{I}_{f,3\phi}$ is the three-phase fault current, A
$\overline{V}_{L-N}$ is the line-to-neutral distribution voltage, V
$\overline{Z}_1$ is the total positive-sequence impedance, Ω
$\overline{Z}_f$ is the fault impedance, Ω
$I_{f,3\phi} = I_{f,a} = I_{f,b} = I_{f,c}$ are the fault currents in a, b, and c phases

Since the total positive-sequence impedance can be expressed as

$$\overline{Z}_1 = \overline{Z}_{1,\text{sys}} + \overline{Z}_{1,T} + \overline{Z}_{1,\text{ckt}} \tag{10.6}$$

---

* Anderson [3] recommends letting the positive-sequence. Thevenin impedance of the system, that is, $Z_{1,\text{sys}}$ be equal to zero, if there is no exact system information available and, at the same time, the substation transformer is small. Of course, by using this assumption, the system is treated as an infinitely large system.

where

$\bar{Z}_{1,sys}$ is the positive-sequence Thévenin-equivalent impedance of the system (or source) referred to distribution voltage,* Ω

$\bar{Z}_{1,T}$ is the positive-sequence transformer impedance referred to distribution voltage,† Ω

$\bar{Z}_{1,ckt}$ is the positive-sequence impedance of faulted segment of distribution circuit, Ω

substituting Equation 10.6 into Equations 10.4 and 10.5, the three-phase fault current can be expressed as

$$I_{f,3\phi} \left| \frac{\bar{V}_{L-N}}{\bar{Z}_{1,sys} + \bar{Z}_{1,T} + \bar{Z}_{1,ckt}} \right| A \tag{10.7}$$

$$I_{f,3\phi} \left| \frac{\bar{V}_{L-N}}{\bar{Z}_{1,sys} + \bar{Z}_{1,T} + \bar{Z}_{1,ckt} + \bar{Z}_f} \right| A \tag{10.8}$$

Equations 10.7 and 10.8 are applicable whether the source connection is wye grounded or delta. At times, it might be necessary to reflect a three-phase fault on the distribution system as a three-phase fault on the subtransmission system. This can be accomplished by using

$$I_{F,3\phi} = \frac{V_{L-L}}{V_{ST,L-L}} \times I_{f,3\phi} \ A \tag{10.9}$$

where

$I_{F,3\phi}$ is the three-phase fault current referred to subtransmission voltage, A

$I_{f,3\phi}$ is the three-phase fault current based on distribution voltage, A

V_{L-L} is the line-to-line distribution voltage, V

$V_{ST,L-L}$ is the line-to-line subtransmission voltage, V

10.11.2 Line-to-Line Faults

Assume that an *L–L* fault exists between phases *b* and *c*. Therefore, if there is no fault impedance,

$$I_{f,a} = 0$$

$$I_{f,L-L} = I_{f,c} = -I_{f,b}$$

$$= \left| \frac{j\sqrt{3} \times \bar{V}_{L-N}}{\bar{Z}_1 + \bar{Z}_2} \right| \tag{10.10}$$

where

$I_{f,L-L}$ is the line-to-line fault current, A

$\bar{Z}_2$ is the total negative-sequence impedance, Ω

However,

$$\bar{Z}_1 = \bar{Z}_2$$

---

* Remember that an impedance can be converted from one base voltage to another by using $Z_2 = Z_1(V_2/V_1)^2$, where Z_1 = impedance on V_1 base and Z_2 = impedance on V_2 base.

† Note that there has been a shift in notation and the symbol $Z_{1,T}$ stands for Z_T.

thus,

$$I_{f,L-L} = \left| \frac{j\sqrt{3} \times \bar{V}_{L-N}}{2\bar{Z}_1} \right| \tag{10.11}$$

or substituting Equation 10.6 into Equation 10.11,

$$I_{f,L-L} = \left| \frac{j\sqrt{3} \times \bar{V}_{L-N}}{2(\bar{Z}_{1,sys} + \bar{Z}_{1,T} + \bar{Z}_{1,ckt})} \right| \tag{10.12}$$

However, if there is a fault impedance,

$$I_{f,L-L} = \left| \frac{j\sqrt{3} \times \bar{V}_{L-N}}{2(\bar{Z}_{1,sys} + \bar{Z}_{1,T} + \bar{Z}_{1,ckt}) + \bar{Z}_f} \right| \tag{10.13}$$

By comparing Equation 10.11 with Equation 10.5, one can determine a relationship between the three-phase fault and *L–L* fault currents as

$$I_{f,L-L} = \frac{\sqrt{3}}{2} \times I_{f,3\phi} = 0.866 \times I_{f,3\phi} \tag{10.14}$$

that is applicable to any point on the distribution system. The equations derived in this section are applicable whether the source connection is wye grounded or delta.

10.11.3 SINGLE LINE-TO-GROUND FAULTS

Assume that an SLG fault exists on phase *a*. If there is no fault impedance,

$$I_{f,L-G} = \left| \frac{\bar{V}_{L-N}}{\bar{Z}_G} \right| \tag{10.15}$$

where
   $I_{f,L-G}$ is the line-to-ground fault current, A
   $\bar{Z}_G$ is the impedance to ground, Ω
   $\bar{V}_{L-N}$ is the line-to-neutral distribution voltage, V

However,

$$\bar{Z}_G = \frac{\bar{Z}_1 + \bar{Z}_2 + \bar{Z}_0}{3} \tag{10.16}$$

or

$$\bar{Z}_G = \frac{2\bar{Z}_1 + \bar{Z}_0}{3} \tag{10.17}$$

since

$$\bar{Z}_1 = \bar{Z}_2$$

Therefore, by substituting Equation 10.17 into Equation 10.15,

$$I_{f,L-G} = \left| \frac{\overline{V}_{L-N}}{\frac{1}{3}(2\overline{Z}_1 + \overline{Z}_0)} \right| \tag{10.18}$$

However, if there is a fault impedance,

$$I_{f,L-G} = \left| \frac{\overline{V}_{L-N}}{\frac{1}{3}(2\overline{Z}_1 + \overline{Z}_0) + \overline{Z}_f} \right| \tag{10.19}$$

where $\overline{Z}_0$ is the total zero-sequence impedance, Ω.

Equations 10.18 and 10.19 are only applicable if the source connection is wye grounded. If the source connection is delta, they are not applicable since the fault current would be equal to zero due to the zero-sequence impedance being infinite.

If the primary distribution feeders are supplied by a delta–wye solidly grounded substation transformer, an SLG fault on the distribution system is reflected as an L–L fault on the subtransmission system. Therefore, the low-voltage-side fault current may be referred to the high-voltage side by using the equation

$$I_{f,L-L} = \frac{V_{L-L}}{\sqrt{3} \times V_{ST,L-L}} \times I_{f,L-G} \tag{10.20}$$

where
  $I_{f,L-G}$ is the single line-to-ground fault current based on distribution voltage, A
  $I_{F,L-L}$ is the single line-to-ground fault current reflected as a line-to-line fault current on the sub-
    transmission system, A
  V_{L-L} is the line-to-line distribution voltage, V
  $V_{ST,L-L}$ is the line-to-line subtransmission voltage, V

In general, the zero-sequence impedance Z_0 of a distribution circuit with multigrounded neutral is very hard to determine precisely, but it is usually larger than its positive-sequence impedance Z_1. However, some empirical approaches are possible. For example, Anderson [3] gives the following relationship between the zero- and positive-sequence impedances of a distribution circuit with multigrounded neutral:

$$\overline{Z}_0 = K_0 \times \overline{Z}_1 \tag{10.21}$$

where
  Z_0 is the zero-sequence impedance of distribution circuit, Ω
  Z_1 is the positive-sequence impedance of distribution circuit, Ω
  K_0 is a constant

Table 10.6 gives various possible values for the constant K_0. If the earth has a very bad conducting characteristic, the constant K_0 is totally established by the neutral-wire impedance. Anderson [3] suggests using an average value of 4 where exact conditions are not known.

TABLE 10.6

Estimated Values of the K_0 Constant for Various Conditions

Condition	K_0
Perfectly conducting earth (e.g., a system with multiple water-pipe grounds)	1.0
Ground wire same size as phase wire	4.0
Ground wire one size smaller	4.6
Ground wire two sizes smaller	4.9
Finite earth impedance	3.8–4.2

Source: Anderson, P.M., *Elements of Power System Protection*, Cyclone Copy Center, Ames, IA, 1975.

10.11.4 Components of the Associated Impedance to the Fault

Impedance of the source: If the associated fault duty S given in megavoltamperes at the substation bus is available from transmission system fault studies, the system impedance, that is, "*backup*" impedance, can be calculated as

$$Z_{1,sys} = \frac{V_{L-N}}{I_L}$$

$$= \frac{V_{L-L}}{\sqrt{3} \times I_L} \tag{10.22}$$

but

$$I_L = \frac{S}{\sqrt{3} \times V_{L-L}} \tag{10.23}$$

therefore

$$Z_{1,sys} = \frac{V_{L-L}^2}{S} \tag{10.24}$$

If the system impedance is given at the transmission substation bus rather than at the distribution substation bus, then the subtransmission line impedance has to be involved in the calculations so that the total impedance (i.e., the sum of the system impedance and the subtransmission line impedance) represents the impedance up to the high side of the distribution substation transformer.

If the maximum three-phase fault current on the high-voltage side of the distribution substation transformer is known, then

$$Z_{1,sys} + Z_{1,ST} = \frac{V_{ST,L-L}}{\sqrt{3}(I_{F,3\phi})_{max}} \tag{10.25}$$

where
  $Z_{1,sys}$ is the positive-sequence impedance of system, Ω
  $Z_{1,ST}$ is the positive-sequence impedance of subtransmission line, Ω
  $V_{ST,L-L}$ is the line-to-line subtransmission voltage, V
  $(I_{F,3\phi})_{max}$ is the maximum three-phase fault current referred to subtransmission voltage, A

Note that the impedances found from Equation 10.24 and 10.25 can be referred to the base voltage by using Equation 10.9).

Impedance of the substation transformer: If the percent impedance of the substation transformer is known, the transformer impedance* can be expressed as

$$Z_{1,T} = \frac{(\% Z_T)(V_{L-L}^2)10}{S_{T,3\phi}}$$ (10.26)

where
 $\% Z_T$ is the percent transformer impedance
 V_{L-L} is the line-to-line base voltage, kV
 $S_{T,3\phi}$ is the three-phase transformer rating, kVA

Impedance of the distribution circuits: The impedance values for the distribution circuits depend on the pole-top conductor configurations and can be calculated by means of symmetrical components. For example, Figure 10.25 shows a typical pole-top overhead distribution circuit configuration. The equivalent spacing (i.e., mutual geometric mean distance) of phase wires and the equivalent spacing between phase wires and neutral wire can be determined from

$$dp = D_{eq} = D_m \overset{\Delta}{=} (D_{ab} \times D_{bc} \times D_{ca})^{1/3}$$ (10.27)

and

$$dn \overset{\Delta}{=} (D_{an} \times D_{bn} \times D_{cn})^{1/3}$$ (10.28)

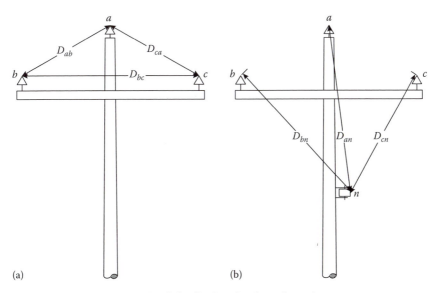

(a)   (b)

FIGURE 10.25 Typical pole-top overhead distribution circuit configuration.

---

* Usually the resistance and reactance values of a substation transformer are approximately equal to 2% and 98% of its impedance, respectively.

Similarly, the mutual reactances (spacing factors) for phase wires, and between phase wires and neutral wire (due to equivalent spacings), can be determined as

$$x_{dp} = 0.05292 \log_{10} dp \quad \Omega/1000\,\text{ft} \tag{10.29}$$

and

$$x_{dn} = 0.05292 \log_{10} dn \quad \Omega/1000\,\text{ft} \tag{10.30}$$

1. If the distribution circuit is a three-phase circuit,
   The positive- and negative-sequence impedances are

$$\bar{z}_1 = \bar{z}_2 = r_{ap} + j(x_{ap} + x_{dp}) \quad \Omega/1000\,\text{ft} \tag{10.31}$$

   and the zero-sequence impedance is

$$\bar{z}_0 = \bar{z}_{0,a} - \frac{\bar{z}_{0,ag}^2}{\bar{z}_{0,g}} \quad \Omega/1000\,\text{ft} \tag{10.32}$$

   with

$$\bar{z}_{0,a} = r_{ap} + r_e + j(x_{ap} + x_e - 2x_{dp}) \quad \Omega/1000\,\text{ft} \tag{10.33}$$

$$\bar{z}_{0,ag} = r_e + j(x_e - 3x_{dn}) \quad \Omega/1000\,\text{ft} \tag{10.34}$$

$$\bar{z}_{0,g} = 3r_{an} + r_e + j(3x_{an} + x_e) \quad \Omega/1000\,\text{ft} \tag{10.35}$$

   where
   r_e is the resistance of earth = 0.0542 Ω/1000 ft
   x_e is the reactance of earth = 0.4676 Ω/1000 ft
   x_{dn} is the spacing factor between phase wires and neutral wire, Ω/1000 ft
   x_{dp} is the spacing factor for phase wires, Ω/1000 ft
   r_{ap} is the resistance of phase wires, Ω/1000 ft
   r_{an} is the resistance of neutral wires, Ω/1000 ft
   x_{ap} is the reactance of phase wire with 1 ft spacing, Ω/1000 ft
   x_{an} is the reactance of neutral wire with 1 ft spacing, Ω/1000 ft
   $z_{0,a}$ is the zero-sequence self-impedance of phase circuit, Ω/1000 ft
   $z_{0,a}$ is the zero-sequence self-impedance of one ground wire, Ω/1000 ft
   $z_{0,ag}$ is the zero-sequence mutual impedance between the phase circuit as one group of conductors and the ground wire as other conductor group, Ω/1000 ft

2. If the distribution circuit is an open-wye and single-phase delta circuit,
   The positive- and negative-sequence impedances are

$$\bar{z}_1 = \bar{z}_2 = r_{ap} + j(x_{ap} + x_{dp}) \quad \Omega/1000\,\text{ft} \tag{10.36}$$

   and the zero-sequence impedance is

$$\bar{z}_0 = \bar{z}_{0,a} - \frac{\bar{z}_{0,ag}^2}{\bar{z}_{0,g}} \quad \Omega/1000\,\text{ft} \tag{10.37}$$

where

$$\bar{z}_{0,a} = r_{ap} + \frac{2r_e}{3} + j\left(x_{ap} + \frac{2x_e}{3} - x_{dp}\right) \quad \Omega/1000\,\text{ft} \tag{10.38}$$

$$\bar{z}_{0,ag} = \frac{2r_e}{3} + j\left(\frac{2x_e}{3} - x_{dn}\right) \quad \Omega/1000\,\text{ft} \tag{10.39}$$

$$\bar{z}_{0,g} = 2r_{an} + \frac{2r_e}{3} + j\left(2x_{an} + \frac{2x_e}{3}\right) \quad \Omega/1000\,\text{ft} \tag{10.40}$$

3. If the distribution circuit is a single-phase multigrounded circuit,
   Its impedance is

$$\bar{z}_{1\phi} = \bar{z}_{0,a} - \frac{\bar{z}_{0,ag}^2}{\bar{z}_{0,g}} \quad \Omega/1000\,\text{ft} \tag{10.41}$$

where

$$\bar{z}_{1\phi} = \bar{z}_{0,a} = r_{ap} + \frac{r_e}{3} + j\left(x_{ap} + \frac{x_e}{3}\right) \quad \Omega/1000\,\text{ft} \tag{10.42}$$

$$\bar{z}_{0,ag} = \frac{r_e}{3} + j\left(\frac{x_e}{3} - x_{dn}\right) \quad \Omega/1000\,\text{ft} \tag{10.43}$$

10.11.5 Sequence-Impedance Tables for the Application of Symmetrical Components

The zero-sequence-impedance equation (given as Equations 10.32, 10.37, or 10.41 in Section 10.11.4), that is,

$$\bar{z}_0 = \bar{z}_{0,a} - \frac{\bar{z}_{0,ag}^2}{\bar{z}_{0,g}} \quad \Omega/1000\,\text{ft}$$

can be expressed as

$$\bar{z}_0 = \bar{z}_{0,a} + \bar{z}_0' \quad \Omega/1000\,\text{ft} \tag{10.44}$$

or

$$\bar{z}_0 = \bar{z}_{0,a} + \bar{z}_0'' \quad \Omega/1000\,\text{ft} \tag{10.45}$$

where
$\bar{Z}_{0,a}$ is the zero-sequence self-impedance of phase circuit, $\Omega/1000$ ft
$\bar{z}_0'$ is the equivalent zero-sequence impedance due to combined effects of zero-sequence self-impedance of one ground wire and zero-sequence mutual impedance between the phase circuit as one group of conductors and the ground wire as another conductor group, assuming a specific vertical distance between the ground wire and phase wires, for example, 38 in.
$\bar{z}_0''$ is the same as $\bar{z}_0'$, except the vertical distance is a different one, for example, 62 in.

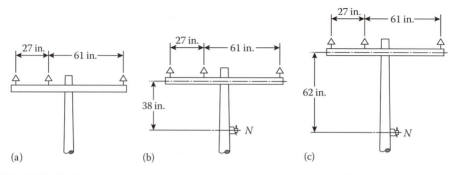

FIGURE 10.26 Various overhead pole-top conductor configurations: (a) without ground wire, $z_0 = z_{0,a}$; (b) with ground wire, $z_0 = z_{0,a} + z_0'$; and (c) with ground wire, $z_0 = z_{0,a} + z_0''$.

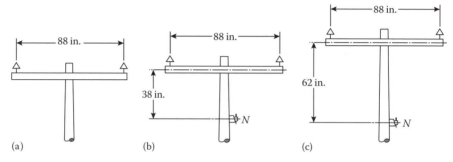

FIGURE 10.27 Various overhead pole-top conductor configurations: (a) without ground wire, $z_0 = z_{0,a}$, (b) with ground wire, $z_0 = z_{0,a} + z_0'$, and (c) with ground wire, $z_0 = z_{0,a} + z_0''$.

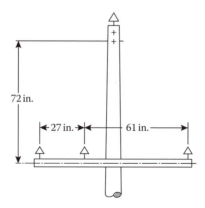

FIGURE 10.28 Various overhead pole-top conductor configurations with ground wire, $z_0 = z_{0,a} + z_0'$.

Therefore, it is possible to develop precalculated sequence-impedance tables for the application of symmetrical components. For example, Figures 10.26 through 10.30 show various overhead pole-top conductor configurations with and without ground wire. The corresponding sequence-impedance values at 60 Hz and 50°C are given in Tables 10.7 through 10.11.

Example 10.2

Assume that a rural substation has a 3750 kVA 69/12.47 kV with LTC transformer feeding a three-phase four-wire 12.47 kV circuit protected by 140 A type L reclosers and 125 A series fuses. It is required to calculate the bolted fault current at point 10, 2 min from the substation on circuit 456319. Assume that the sizes of the phase conductors are 336AS37 (i.e., 336 kcmil

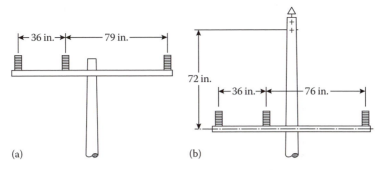

FIGURE 10.29 Various overhead pole-top conductor configurations: (a) without ground wire, $z_0 = z_{0,a}$ and (b) with ground wire, $z_0 = z_{0,a} + z_0'$.

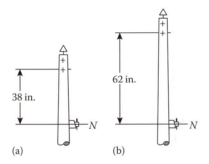

FIGURE 10.30 Single-phase overhead pole-top configurations with ground wires: (a) $z_{1\phi} = z_{1\phi}'$ and (b) $z_{1\phi} = z_{1\phi}''$.

bare-aluminum–steel conductors with 37 strands) and that neutral conductor is 0AS7, spaced 62 in. If the system impedance to the regulated 12.47 kV bus and the system impedance to ground are given as $0.7199 + j3.4619\ \Omega$ and $0.6191 + j3.3397\ \Omega$, respectively, determine the following:

a. The zero- and positive-sequence impedances of the line to point 10
b. The impedance to ground of the line to point 10
c. The total positive-sequence impedance to point 10 including system impedance to the regulated 12.47 kV bus.
d. The total impedance to ground to point 10 including system impedance of the regulated 12.47 kV bus.
e. The three-phase fault current at point 10
f. The L–L fault current at point 10
g. The line-to-ground fault current at point 10

Solution

a. The zero-sequence impedance of the line to point 10 can be found by using Table 10.7 as

$$\overline{Z}_{0,ckt} = 2\left(\overline{z}_{0,a} + \overline{z}_0''\right)5.28$$

$$= 2[(0.1122 + j0.4789) + (-0.0385 - j0.0996)]\,5.28$$

$$= 0.7783 + j4.0054\ \Omega$$

Similarly, the positive-sequence impedance of the line to point 10 can be found as

$$\overline{Z}_{1,ckt} = 2(0.0580 + j0.1208)5.28$$

$$= 0.6125 + j1.2756\ \Omega$$

TABLE 10.7

Sequence-Impedance Values Associated with Figure 10.26, Ω/1000 ft

		Conductor		
Size and Code	$z_1 = z_2$	$z_{0,a}$	z_0'	z''
Bare-aluminum–steel (AS)				
4AS7	$0.4867 + j0.1613$	$0.5409 + j0.5195$	$0.0518 - j0.0543$	$0.0454 - j0.0493$
4AS8	$0.4830 + j0.1605$	$0.5372 + j0.5187$	$0.0520 - j0.0548$	$0.0456 - j0.0497$
3AS7	$0.3920 + j0.1617$	$0.4462 + j0.5198$	$0.0540 - j0.0685$	$0.0472 - j0.0620$
2AS7	$0.3202 + j0.1624$	$0.3743 + j0.5206$	$0.0535 - j0.0827$	$0.0465 - j0.0747$
2AS8	$0.3125 + j0.1581$	$0.3667 + j0.5162$	$0.0543 - j0.0846$	$0.0471 - j0.0764$
1AS8	$0.2614 + j0.1802$	$0.3156 + j0.5384$	$0.0459 - j0.0954$	$0.0395 - j0.0859$
1AS7	$0.2614 + j0.1624$	$0.3156 + j0.5206$	$0.0504 - j0.0970$	$0.0435 - j0.0874$
0AS7	$0.2121 + j0.1607$	$0.2663 + j0.5189$	$0.0451 - j0.1108$	$0.0385 - j0.0996$
000AS7	$0.1377 + j0.1541$	$0.1919 + j0.5123$	$0.0295 - j0.1346$	$0.0242 - j0.1206$
267AS33	$0.0729 + j0.1245$	$0.1271 + j0.4827$	$0.0092 - j0.1663$	$0.0056 - j0.1486$
336AS37	$0.0580 + j0.1208$	$0.1122 + j0.4789$	$0.0008 - j0.1722$	$-0.0020 - j0.1537$
477AS33	$0.0409 + j0.1168$	$0.0951 + j0.4789$	$-0.0101 - j0.1779$	$-0.0119 - j0.1587$
636AS33	$0.0306 + j0.1145$	$0.0848 + j0.4727$	$-0.0175 - j0.1807$	$-0.0184 - j0.1610$
795AS33	$0.0244 + j0.1120$	$0.0786 + j0.4702$	$-0.0221 - j0.1830$	$-0.0226 - j0.1630$
Bare hard-drawn copper (X)				
8X1	$0.7194 + j0.1624$	$0.7739 + j0.5206$	$0.0432 - j0.0340$	$0.0380 - j0.0310$
6X1	$0.4527 + j0.1571$	$0.5069 + j0.5153$	$0.0535 - j0.0588$	$0.0468 - j0.0533$
4X1	$0.2875 + j0.1499$	$0.3417 + j0.5081$	$0.0554 - j0.0910$	$0.0480 - j0.0821$
3X3	$0.2280 + j0.1473$	$0.2822 + j0.5054$	$0.0513 - j0.1080$	$0.0441 - j0.0972$
2X1	$0.1790 + j0.1465$	$0.2332 + j0.5047$	$0.0432 - j0.1237$	$0.0366 - j0.1111$
1X1	$0.1420 + j0.1437$	$0.1962 + j0.5018$	$0.0342 - j0.1366$	$0.0283 - j0.1225$
1X3	$0.1432 + j0.1420$	$0.1976 + j0.5001$	$0.0352 - j0.1367$	$0.0292 - j0.1226$
0X7	$0.1150 + j0.1398$	$0.1692 + j0.4981$	$0.0255 - j0.1466$	$0.0205 - j0.1313$
00X7	$0.0911 + j0.1372$	$0.1453 + j0.4954$	$0.0156 - j0.1547$	$0.0115 - j0.1384$
000X7	$0.0723 + j0.1346$	$0.1265 + j0.4928$	$0.0065 - j0.1608$	$0.0033 - j0.1436$
0000X7	$0.0574 + j0.1317$	$0.1116 + j0.4899$	$0.0016 - j0.1654$	$-0.0041 - j0.1476$
300X12	$0.0407 + j0.1255$	$0.0949 + j0.4837$	$0.0114 - j0.1719$	$-0.0129 - j0.1532$
Bare hard-drawn aluminum (AL)				
0AL7	$0.1843 + j0.1394$	$0.2385 + j0.4976$	$0.0468 - j0.1235$	$0.0398 - j0.1110$
000AL7	$0.1161 + j0.1345$	$0.1703 + j0.4927$	$0.0277 - j0.1485$	$0.0224 - j0.1330$
267AL7	$0.0731 + j0.1292$	$0.1273 + j0.4874$	$0.0082 - j0.1636$	$0.0047 - j0.1462$
477AL19	$0.0413 + j0.1212$	$0.0955 + j0.4794$	$-0.0105 - j0.1747$	$-0.0121 - j0.1558$

b. From Equation 10.17, the impedance to ground of the line to point 10 is

$$\bar{Z}_G = \frac{2\bar{Z}_1 + \bar{Z}_0}{3} = \frac{2(0.6125 + j1.2756) + (0.7783 + j4.0054)}{3}$$

$$= 2.0033 + j2.1855 \ \Omega$$

c. The total positive-sequence impedance is

$$\bar{Z}_1 = \bar{Z}_{1,ckt} + \bar{Z}_{1,ckt} = (0.7199 + j3.4619) + (0.6125 + j1.2756)$$

$$= 1.3324 + j4.7375 \ \Omega$$

TABLE 10.8

Sequence-Impedance Values Associated with Figure 10.27, Ω/1000 ft

Size and Code	$z_1 = z_2$	$z_{0,a}$	z_0'	z_0''
		Conductor		
Bare-aluminum–steel (AS)				
4AS7	$0.4867 + j0.1706$	$0.5229 + j0.3907$	$0.0338 - j0.0357$	$0.0279 - j0.0310$
4AS8	$0.4830 + j0.1698$	$0.5191 + j0.3900$	$0.0340 - j0.0360$	$0.0280 - j0.0313$
3AS7	$0.3920 + j0.1710$	$0.4282 + j0.3911$	$0.0353 - j0.0449$	$0.0289 - j0.0389$
2AS7	$0.3201 + j0.1717$	$0.3562 + j0.3919$	$0.0349 - j0.0542$	$0.0284 - j0.0468$
2AS8	$0.3125 + j0.1674$	$0.3486 + j0.3875$	$0.0354 - j0.0554$	$0.0288 - j0.0479$
1 AS8	$0.2614 + j0.1895$	$0.2975 + j0.4097$	$0.0299 - j0.0625$	$0.0240 - j0.0537$
1AS7	$0.2614 + j0.1717$	$0.2975 + j0.3919$	$0.0328 - j0.0636$	$0.0265 - j0.0547$
0AS7	$0.2121 + j0.1700$	$0.2483 + j0.3902$	$0.0293 - j0.0726$	$0.0233 - j0.0623$
000AS7	$0.1377 + j0.1634$	$0.1738 + j0.3836$	$0.0191 - j0.0882$	$0.0143 - j0.0753$
267AS33	$0.0729 + j0.1339$	$0.1090 + j0.3540$	$0.0057 - j0.1089$	$0.0024 - j0.0926$
336AS33	$0.0580 + j0.1301$	$0.0941 + j0.3502$	$0.0002 - j0.1127$	$-0.0023 - j0.0957$
477AS33	$0.0409 + j0.1261$	$0.0770 + j0.3462$	$-0.0070 - j0.1164$	$-0.0085 - j0.0987$
636AS33	$0.0306 + j0.1238$	$0.0668 + j0.3440$	$-0.0118 - j0.1182$	$-0.0126 - j0.1001$
795AS33	$0.0244 + j0.1214$	$0.0605 + j0.3415$	$-0.0148 - j0.1197$	$-0.0152 - j0.1013$
Bare hard-drawn copper (X)				
8X1	$0.7197 + j0.1717$	$0.7558 + j0.3919$	$0.0282 - j0.0223$	$0.0234 - j0.0196$
6X1	$0.4527 + j0.1664$	$0.4888 + j0.3866$	$0.0349 - j0.0386$	$0.0288 - j0.0335$
4X1	$0.2847 + j0.1611$	$0.3208 + j0.3813$	$0.0357 - j0.0601$	$0.0289 - j0.0518$
4X3	$0.2875 + j0.1592$	$0.3236 + j0.3794$	$0.0361 - j0.0596$	$0.0293 - j0.0514$
3X3	$0.2280 + j0.1566$	$0.2642 + j0.3767$	$0.0334 - j0.0708$	$0.0268 - j0.0608$
2X1	$0.1790 + j0.1558$	$0.2151 + j0.3760$	$0.0281 - j0.0811$	$0.0220 - j0.0695$
1X1	$0.1420 + j0.1530$	$0.1782 + j0.3731$	$0.0221 - j0.0895$	$0.0168 - j0.0765$
1X3	$0.1434 + j0.1513$	$0.1795 + j0.3714$	$0.0228 - j0.0896$	$0.0173 - j0.0766$
0X7	$0.1150 + j0.1492$	$0.1511 + j0.3693$	$0.0164 - j0.0960$	$0.0118 - j0.0819$
00X7	$0.0911 + j0.1466$	$0.1272 + j0.3667$	$0.0099 - j0.1013$	$0.0062 - j0.0862$
000X7	$0.0723 + j0.1439$	$0.1085 + j0.3640$	$0.0040 - j0.1052$	$0.0010 - j0.0894$
0000X7	$0.0574 + j0.1411$	$0.0935 + J0.3612$	$-0.0014 - j0.1083$	$-0.0036 - j0.0919$
300X12	$0.0407 + j0.1348$	$0.0769 + j0.3550$	$-0.0078 - j0.1125$	$-0.0091 - j0.0953$
Bare hard-drawn aluminum (AL)				
0AL7	$0.1843 + j0.1487$	$0.2204 + j0.3689$	$0.0304 - j0.0809$	$0.0240 - j0.0694$
000AL7	$0.1161 + j0.1438$	$0.1522 + j0.3640$	$0.0179 - j0.0972$	$0.0130 - j0.0830$
267AL7	$0.0731 + j0.1385$	$0.1092 + j0.3586$	$0.0050 - j0.1071$	$0.0019 - j0.0911$
477AL19	$0.0413 + j0.1306$	$0.0774 + j0.3507$	$-0.0072 - j0.1143$	$-0.0086 - j0.0969$

d. The total impedance to ground is

$$\bar{Z}_G = \bar{Z}_{G,\text{sys}} + \bar{Z}_{G,\text{ckt}} = (0.6191 + j3.3397) + (2.0033 + j2.1855)$$

$$= 2.6224 + j5.5252\ \Omega$$

e. From Equation 10.7, the three-phase fault at point 10 is

$$I_{f,3\phi} = \left|\frac{\bar{V}_{L-N}}{\bar{Z}_1}\right| = \frac{7200}{4.9213} = 1.463\ \text{A}$$

TABLE 10.9

Sequence-Impedance Values for Bare-Aluminum–Steel (AS) Associated with Figure 10.28, Ω/1000 ft

	Conductor		
Size and Code	$z_1 = z_2$	$z_{0,a}$	z_0'
4AS8	$0.4830 + j0.1605$	$0.5372 + j0.5187$	$0.0439 - j0.0484$
0AS7	$0.2121 + j0.1607$	$0.2663 + j0.5189$	$0.0368 - j0.0967$
000AS7	$0.1377 + j0.1541$	$0.1919 + j0.5123$	$0.0229 - j0.1169$
267AS33	$0.0729 + j0.1208$	$0.1122 + j0.4789$	$-0.0027 - j0.1489$
477AS33	$0.0409 + j0.1168$	$0.0951 + j0.4749$	$-0.0123 - j0.1536$
636AS33	$0.0306 + j0.1145$	$0.0848 + j0.4727$	$-0.0187 - j0.1558$
795AS33	$0.0244 + j0.1120$	$0.0786 + j0.4702$	$-0.0227 - j0.1577$

TABLE 10.10

Sequence-Impedance Values for Bare-Aluminum–Steel (AS) Associated with Figure 10.29, Ω/1000 ft

	Conductor		
Size and Code	$z_1 = z_2$	$z_{0,a}$	z_0'
4AS8	$0.4830 + j0.1660$	$0.5372 + j0.5077$	$0.0406 - j0.0458$
0AS7	$0.2121 + j0.1662$	$0.2663 + j0.5079$	$0.0334 - j0.0909$
000AS7	$0.1377 + j0.1596$	$0.1919 + j0.5012$	$0.0202 - j0.1097$
267AS33	$0.0729 + j0.1301$	$0.1271 + j0.4717$	$0.0029 - j0.1349$
336AS37	$0.0580 + j0.1263$	$0.1122 + j0.4679$	$-0.0040 - j0.1394$
477AS33	$0.0409 + j0.1223$	$0.0951 + j0.4639$	$-0.0131 - j0.1437$
636AS33	$0.0306 + j0.1200$	$0.0808 + j0.4617$	$-0.0191 - j0.1456$
795AS33	$0.0244 + j0.1176$	$0.0786 + j0.4592$	$-0.0229 - j0.1475$

f. From Equation 10.14, the *L–L* fault at point 10 is

$$I_{f,L-L} = 0.866 \times I_{f,3\phi}$$

$$= 0.866 \times 1463$$

$$= 1267\,\text{A}$$

g. From Equation 10.15, the SLG fault at point 10 is

$$I_{f,L-G} = \left| \frac{\bar{V}_{L-N}}{\bar{Z}_G} \right|$$

$$= \frac{7200}{6.1159}$$

$$= 1177.3\,\text{A}$$

TABLE 10.11

Impedance Values Associated with Figure 10.30, Ω/1000 ft

Size and Code	Conductor	
	$z'_{1\phi}$	$z''_{1\phi}$
Bare-aluminum–steel (AS)		
4AS7	0.5230 + j0.2618	0.5202 + j0.2640
4AS8	0.5193 + j0.2609	0.5165 + j0.2631
3AS7	0.4292 + j0.2572	0.4262 + j0.2601
2AS7	0.3570 + j0.2531	0.3540 + j0.2565
2AS8	0.3497 + j0.2481	0.3466 + j0.2516
1AS8	0.2957 + j0.2664	0.2929 + j0.2705
1AS7	0.2973 + j0.2481	0.2942 + j0.2522
0AS7	0.2462 + j0.2415	0.2433 + j0.2464
000AS7	0.1664 +j0.2265	0.1641 + j0.2326
267AS33	0.0946 +j0.1859	0.0930 + j0.1936
336AS37	0.0767 +j0.1800	0.0755 + j0.1880
477AS33	0.0559 +j0.1740	0.0551 + j0.1824
636AS33	0.0431 + j0.1708	0.0426 + j0.1793
795AS33	0.0352 + j0.1675	0.0349 + j0.1762
Bare hard-drawn copper (X)		
8X1	0.7529 + j0.2701	0.7507 + j0.2713
6X1	0.4895 + j0.2561	0.4866 + j0.2585
4X1	0.3221 + j0.2393	0.3189 + j0.2432
4X3	0.3251 + j0.2377	0.3219 + j0.2415
3X3	0.2643 + j0.2291	0.2611 + j0.2338
2X1	0.2124 + j0.2228	0.2096 + j0.2283
1X1	0.1724 + j0.2154	0.1698 + j0.2216
1X3	0.1740 + j0.2137	0.1715 + j0.2198
0X7	0.1423 + j0.2081	0.1401 + j0.2148
00X7	0.1150 + j0.2026	0.1132 + j0.2097
000X7	0.0931 + j0.1978	0.0916 + j0.2053
0000X7	0.0753 + j0.1934	0.0742 + j0.2011
300X 12	0.0552 + j0.1848	0.0546 + j0.1929
Bare hard-drawn aluminum (AL)		
0AL7	0.2190 + j0.2158	0.2159 + j0.2212
000AL7	0.1442 + j0.2021	0.1419 + j0.2089
267AL7	0.0944 + j0.1915	j0.1990
477AL 19	0.0557 + j0.1796	0.0554 + j0.1878

Note that the fault currents are calculated on the basis of a bolted fault. Therefore, they are accurate for faults caused by a low-impedance object making solid contact with the grounds.

However, usually the object causing the fault either has a high impedance or does not make solid contact with the conductors. This introduces an additional impedance into the circuit, which reduces the fault current to some value below the calculated value. Therefore, to be sure that the high-impedance faults will be cleared, it is crucial that all bolted faults clear within 3 s.

10.12 FAULT-CURRENT CALCULATIONS IN PER UNITS

Fault currents can also be determined by using per unit values rather than actual system values, of course. For example, Anderson [3] gives fault-current formulas that use per unit values, as shown in Table 10.12.

Example 10.3

Assume that a distribution substation, shown in Figure 10.31, has a 5000 kVA 69/12.47 kV LTC transformer feeding a three-phase four-wire 12.47 kV distribution system. The transformer has a reactance of 0.065 pu Ω. Assume that the faults are bolted with zero fault impedance and that the maximum and minimum power generations of the system are 600 and 360 MVA, respectively. Use 1 MVA as the three-phase power base:

a. Under the maximum (system) power generation conditions, determine the available three-phase, L–L, and SLG fault currents at buses 1 and 2 in per units, in amperes, and in megavoltamperes.
b. Under the minimum (system) power generation conditions, determine the available three-phase, L–L, and SLG fault currents at buses 1 and 2 in per units, in amperes, and in megavoltamperes.
c. Tabulate the results obtained in parts a and b.

TABLE 10.12
Fault-Current Formulas in Per Units

Fault Type	Fault-Current Formula	Source Connection
3ϕ	$\bar{I}_{f,3\phi} = \dfrac{\bar{V}_f}{\bar{Z}_{\text{sys}} + \bar{Z}_T + \bar{Z}_{\text{ckt}} + \bar{Z}_f}$	Delta or grounded wye
L–L	$\bar{I}_{f,b} = -\bar{I}_{f,c} = -\dfrac{j\sqrt{3} \times \bar{V}_f}{2(\bar{Z}_{\text{sys}} + \bar{Z}_T + \bar{Z}_{\text{ckt}}) + \bar{Z}_f}$	Delta or grounded wye
L–G	$\bar{I}_{f,a} = 0$	Delta or
	$I_{f,a} = \dfrac{3\bar{V}_f}{2\bar{Z}_{\text{sys}} + 3\bar{Z}_T + 6\bar{Z}_{\text{ckt}} + 3\bar{Z}_f}$	Grounded wye[a]

Source: Anderson, P.M., *Elements of Power System Protection*, Cyclone Copy Center, Ames, IA, 1975.
[a] Using $K_0 = 4$, from Table 10.6.

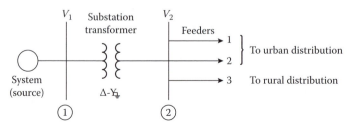

FIGURE 10.31 A distribution substation.

Solution

a. Selecting 69 kV as the voltage base, the impedance base can be determined as

$$Z_B = \frac{V_{B,L-L}^2}{S_{B,3\phi}}$$

$$= \frac{(69 \times 10^3)^2}{1 \times 10^6}$$

$$= 4761\ \Omega$$

Therefore, under the maximum (system) power generation conditions, the system impedance is

$$Z_{sys} = \frac{(69 \times 10^3)^2}{600 \times 10^6}$$

$$= 7.935\ \Omega$$

or

$$Z_{sys} = \frac{7.935}{4761}$$

$$= 0.0017\,\mathrm{pu}\ \Omega$$

Similarly, the three-phase current base can be found as

$$I_B = \frac{S_{B,3\phi}}{\sqrt{3} \times V_{B,L-L}}$$

$$= \frac{1 \times 10^6}{\sqrt{3}(69 \times 10^3)}$$

$$= 8.3674\ \mathrm{A}$$

i. At bus 1, from Table 10.12, the three-phase fault current can be calculated as

$$\bar{I}_{f,3\phi} = \left| \frac{\bar{V}_f}{\bar{Z}_{sys} + \bar{Z}_T + \bar{Z}_{ckt} + \bar{Z}_f} \right|$$

$$= \left| \frac{1.0}{j0.0017 + 0 + 0 + 0} \right|$$

$$\cong 588.2\ \mathrm{pu}\ \mathrm{A}$$

(note that it is assumed that the voltage is 1.0 pu V at the fault point.)

or

$$I_{f,3\phi} = (588.2\ \mathrm{pu}\ \mathrm{A})(8.3674\ \mathrm{A})$$

$$= 4922\ \mathrm{A}$$

or

$$S_{f,3\phi} = \sqrt{3}(69\ \mathrm{kV})(4922) \times 10^{-3}$$

$$\cong 588.2\ \mathrm{MVA}$$

The *L–L* fault current can be calculated by using the appropriate equation from Table 10.12 or from

$$I_{f,L-L} = 0.866 I_{f,3\phi}$$

$$= 0.866(588.2 \text{ pu A})$$

$$\cong 509.38 \text{ pu A}$$

$$= 4262.5 \text{ A}$$

or

$$S_{f,L-L} = (69 \text{ kV})(4267.2 \text{ A})10^{-3}$$

$$= 294.1 \text{ MVA}$$

From Table 10.12, the SLG fault current can be calculated as

$$I_{f,L-G} = \frac{3\overline{V}_f}{2\overline{Z}_{sys} + 3\overline{Z}_T + 6\overline{Z}_{ckt} + 3\overline{Z}_f}$$

$$= \left| \frac{3(1.0)}{2(j0.0017) + 0 + 0 + 0} \right|$$

$$= 882.35 \text{ pu A}$$

$$= 7383 \text{ A}$$

or

$$S_{f,L-G} = \left(\frac{69 \text{ kV}}{\sqrt{3}} \right)(7383 \text{ A}) \, 10^{-3}$$

$$= 294.1 \text{ MVA}$$

ii. At bus 2, since the given transformer reactance of 0.065 pu Ω value is based on 5 MVA, it has to be converted to the new base of 1 MVA. Therefore,

$$Z_{T,new} = Z_{T,old} \left(\frac{V_{B,L-L,old}}{V_{B,L-L,new}} \right)^2 \frac{S_{B,3\phi,new}}{S_{B,3\phi,old}}$$

$$= j0.065 \left(\frac{69 \text{ kV}}{69 \text{ kV}} \right)^2 \frac{1 \text{ MVA}}{5 \text{ MVA}}$$

$$= j0.013 \text{ pu } \Omega$$

$$Z_B = \frac{(12.47 \times 10^3)^2}{1 \times 10^6}$$

$$= 155.5 \, \Omega$$

$$I_B = \frac{1 \times 10^6}{\sqrt{3}(12.47 \times 10^3)}$$

$$= 46.2991 \text{ A}$$

Thus,

$$I_{f,3\phi} = \left| \frac{1.0}{j0.0017 + j0.013 + 0 + 0} \right|$$

$$= 68.0272 \text{ pu A}$$

or

$$I_{f,3\phi} = (68.0272 \text{ pu A})(46.2991 \text{ A})$$

$$\cong 3149.6 \text{ A}$$

or

$$S_{f,3\phi} = \sqrt{3}(12.47 \text{ kV})(3149.6)10^{-3}$$

$$= 68.0272 \text{ MVA}$$

$$I_{f,L-L} = 0.866(I_{f,3\phi})$$

$$= 0.866(68.0272 \text{ pu A})$$

$$= 58.9116 \text{ pu A}$$

$$= 2727.6 \text{ A}$$

or

$$S_{f,L-L} = (12.47 \text{ kV})(2727.6 \text{ A})10^{-3}$$

$$= 34.01 \text{ MVA}$$

$$I_{f,L-G} = \left| \frac{3(1.0)}{2(j0.0017) + 3(j0.013) + 0 + 0} \right|$$

$$= 70.7547 \text{ pu A}$$

$$\cong 3275.9 \text{ A}$$

$$S_{f,L-G} = \left(\frac{12.47 \text{ kV}}{\sqrt{3}} \right)(3275.9 \text{ A})10^{-3}$$

$$\cong 23.58 \text{ MVA}$$

b. Under the minimum (system) power generation conditions, the system impedance becomes

$$Z_{sys} = \frac{(69 \times 10^3)^2}{360 \times 10^6}$$

$$= 13.225 \ \Omega$$

or

$$Z_{sys} = \frac{13.225}{4761}$$

$$= j0.0028 \text{ pu } \Omega$$

i   At bus 1,

$$I_{f,3\phi} = \left| \frac{1.0}{j0.0028 + 0 + 0 + 0} \right|$$

$$= 360 \text{ pu A}$$

$$= 3012.3 \text{ A}$$

or

$$S_{f,3\phi} = \sqrt{3}(69 \text{ kV})(3012.3 \text{ A})10^{-3}$$

$$= 360 \text{ MVA}$$

$$I_{f,L-L} = 0.866 \times I_{f,3\phi}$$

$$= 311.76 \text{ pu A}$$

$$= 2608.7 \text{ A}$$

or

$$S_{f,L-L} = (69 \text{ kV})(2608.7 \text{ A})10^{-3}$$

$$\cong 180.2 \text{ MVA}$$

$$I_{f,L-G} = \left| \frac{3(1.0)}{2(j0.0028) + 0 + 0 + 0} \right|$$

$$= 535.7 \text{ pu A}$$

$$= 4482.5 \text{ A}$$

or

$$S_{f,L-G} = \left(\frac{69 \text{ kV}}{\sqrt{3}} \right)(4487.8 \text{ A})10^{-3}$$

$$\cong 178.6 \text{ MVA}$$

ii.   At bus 2,

$$I_{f,3\phi} = \left| \frac{1.0}{j0.0028 + j0.013 + 0 + 0} \right|$$

$$= 63.2911 \text{ pu A}$$

$$\cong 2930.3 \text{ A}$$

or

$$S_{f,3\phi} = (12.47 \text{ kV})(2933.8 \text{ A})10^{-3}$$

$$\cong 63.29 \text{ MVA}$$

TABLE 10.13
Results of Example 10.3

Bus	Fault	Maximum Generation		Minimum Generation	
		A	MVA	A	MVA
1	3ϕ	4922	588.2	3012.3	360
	L–L	4266.5	294.1	2608.7	180.2
	L–G	7383	294.1	4482.5	178.6
2	3ϕ	3149.6	68.0	2930.3	63.29
	L–L	2727.6	34.0	2537.7	36.6
	L–G	3275.9	23.6	3114.3	23.42

$$I_{f,L-L} = 0.866 \times I_{f,3\phi}$$

$$= 54.81 \text{ pu A}$$

$$\cong 2537.7 \text{ A}$$

or

$$S_{f,L-L} = (12.47 \text{ kV})(2540.7 \text{ A})10^{-3}$$

$$\cong 36.6 \text{ MVA}$$

$$I_{f,L-G} = \left| \frac{3(1.0)}{2(j0.0028) + 3(j0.013) + 0 + 0} \right|$$

$$= 67.2646 \text{ pu A}$$

$$\cong 3114.3 \text{ A}$$

or

$$S_{f,L-G} = \left(\frac{12.47 \text{ kV}}{\sqrt{3}} \right)(3114.3 \text{ A})10^{-3}$$

$$\cong 22.42 \text{ MVA}$$

c. The results are given in Table 10.13.

10.13   SECONDARY-SYSTEM FAULT-CURRENT CALCULATIONS

10.13.1   SINGLE-PHASE 120/240 V THREE-WIRE SECONDARY SERVICE

As shown in Figure 10.32, a line-to-ground fault may involve line l_1 and neutral or line l_2 and neutral. Therefore, the maximum line-to-ground fault current can be calculated as

$$\overline{I}_{f,L-G} = \frac{\overline{V}_{L-N}}{\overline{Z}_{eq}} \tag{10.46}$$

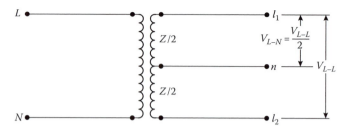

FIGURE 10.32 A line-to-ground fault involving line l_1 and neutral or line l_2 and neutral.

or

$$\overline{I}_{f,L-G} = \frac{0.5\overline{V}_{L-L}}{\overline{Z}_{eq}} \tag{10.47}$$

where

$$\overline{Z}_{eq} = \overline{Z}_T + n^2\overline{Z}_G + \overline{Z}_{G,\,\text{SL}} \tag{10.48}$$

is the equivalent impedance to fault, Ω
$\overline{Z}_T$ is the equivalent impedance of distribution transformer,* Ω

$$= 1.5R_T + j1.2X_T \tag{10.49}$$

$\overline{Z}_G$ is the line-to-ground impedance of primary system, Ω
$\overline{Z}_{G,\text{SL}}$ is the line-to-ground impedance of secondary line, Ω
n is the primary-to-secondary impedance-transfer ratio[†]

$$= \frac{\text{sec } V_{L-N}}{\text{pri } V_{L-N}} \tag{10.50}$$

Also, maximum L–L fault may occur between lines l_1 and l_2. Therefore,

$$\overline{I}_{f,L-L} = \frac{\overline{V}_{L-L}}{\overline{Z}_{eq}} \tag{10.51}$$

where

$$\overline{Z}_{eq} = \overline{Z}_T + n^2\overline{Z}_G + \overline{Z}_{1,\text{SL}} \tag{10.52}$$

$\overline{Z}_T$ is the equivalent impedance of distribution transformer, Ω

$$= Z \tag{10.53}$$

$\overline{Z}_{1,\text{SL}}$ is the positive-sequence impedance of secondary line, Ω

---

* Note that there has been a shift in notation and the symbol Z_T stands for $Z_{1,T}$.
† Note that there has been a shift in notation and the symbol n stands for primary-to-secondary impedance-transfer ratio.
  (It is the inverse of the transformer turns ratio.)

10.13.2 Three-Phase 240/120 or 480/240 V Wye–Delta or Delta–Delta Four-Wire Secondary Service

For a delta-connected secondary, if the primary is connected in wye, its impedance must be converted to its equivalent delta impedance, as indicated in Figure 10.33. Therefore,

$$\bar{Z}_\Delta = \frac{\bar{Z}_1\bar{Z}_1 + \bar{Z}_1\bar{Z}_1 + \bar{Z}_1\bar{Z}_1}{\bar{Z}_1} = 3\bar{Z}_1 \qquad (10.54)$$

where
$\bar{Z}_1$ is the positive-sequence impedance, Ω
$\bar{Z}_\Delta$ is the equivalent delta impedance, Ω

If the primary is already connected in delta, then

$$\bar{Z}_\Delta = \bar{Z}_1 \qquad (10.55)$$

As shown in Figure 10.33, a line-to-ground fault may involve phase a and neutral. The resultant maximum line-to-ground fault current can be expressed as

$$\bar{I}_{f,L-G} = \frac{0.866 \times \bar{V}_{L-L}}{\bar{Z}_{eq}} \qquad (10.56)$$

where

$$\bar{Z}_{eq} = \bar{Z}_T + n^2\bar{Z}_\Delta + \bar{Z}_{G,SL} \qquad (10.57)$$

$$\bar{Z}_T = \frac{(\bar{Z} + \bar{Z}/2)(\bar{Z} + \bar{Z}/2)}{2(\bar{Z} + (\bar{Z}/2))} = \frac{3\bar{Z}}{4} \qquad (10.58)$$

If the line-to-ground fault involves phase b and neutral or phase c and neutral, then the maximum available line-to-ground fault current can be calculated as

$$\bar{I}_{f,L-G} = \frac{0.5 \times \bar{V}_{L-L}}{\bar{Z}_{eq}} \qquad (10.59)$$

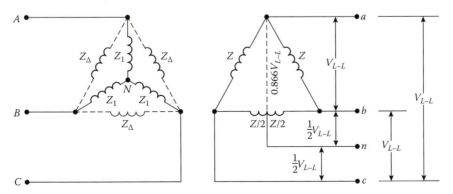

FIGURE 10.33 A wye–delta or delta–delta secondary service. If needed, the wye primary can be replaced with its equivalent delta, as illustrated.

where $\bar{Z}_{eq}$ is found from Equation 10.57 and

$$\bar{Z}_T = \frac{\bar{Z}/2(2\bar{Z}+\bar{Z}/2)}{2(\bar{Z}+(\bar{Z}/2))} = \frac{5\bar{Z}}{12} \tag{10.60}$$

An *L–L* fault may involve phases *a* and *b*, or *b* and *c*, or *c* and *a*. In any case, the maximum *L–L* fault current can be calculated from Equation 10.51 where

$$\bar{Z}_{eq} = \bar{Z}_T + n^2\bar{Z}_\Delta + \bar{Z}_{1,\text{SL}} \tag{10.61}$$

and

$$\bar{Z}_T = \frac{(2\bar{Z})(\bar{Z})}{2\bar{Z}+\bar{Z}} = \frac{2\bar{Z}}{3} \tag{10.62}$$

A three-phase fault, of course, involves all three phases. Therefore,

$$\bar{I}_{f,3\phi} = \frac{\bar{V}_{L-L}}{\sqrt{3} \times \bar{Z}_{eq}} \tag{10.63}$$

where

$$Z_{eq} = Z_T + n^2 Z_\Delta + Z_{1,\text{cable}} \tag{10.64}$$

$$\bar{Z}_T = \bar{Z} \tag{10.65}$$

10.13.3 THREE-PHASE 240/120 OR 480/240 V OPEN-WYE PRIMARY AND FOUR-WIRE OPEN-DELTA SECONDARY SERVICE

Figure 10.34 shows a three-phase open-wye primary and four-wire open-delta secondary connection. If a line-to-ground fault involves phase *b* and neutral or phase *c* and neutral, the maximum available fault current can be calculated as

$$\bar{I}_{f,L-G} = \frac{0.5 \times \bar{V}_{L-L}}{\bar{Z}_{eq}} \tag{10.66}$$

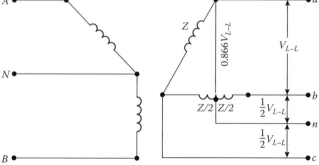

FIGURE 10.34 An open-wye primary and open-delta secondary service.

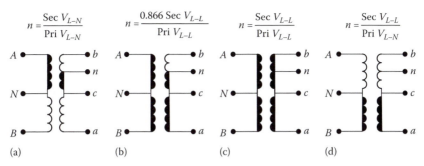

$$n = \frac{\text{Sec } V_{L-N}}{\text{Pri } V_{L-N}} \qquad n = \frac{0.866 \text{ Sec } V_{L-L}}{\text{Pri } V_{L-L}} \qquad n = \frac{\text{Sec } V_{L-L}}{\text{Pri } V_{L-L}} \qquad n = \frac{\text{Sec } V_{L-L}}{\text{Pri } V_{L-N}}$$

(a)   (b)   (c)   (d)

FIGURE 10.35   Various fault-current paths in the transformer and associated impedance-transfer ratios. (The shaded path determines the primary- and secondary-transformer fault impedances.)

where
  $\bar{Z}_{eq}$ is found from Equation 10.48
  n is the transfer ratio found from Figure 10.35a

If the line-to-ground fault involves phase a and neutral, the maximum available fault current can be expressed as

$$\bar{I}_{f,L-G} = \frac{0.886 \times \bar{V}_{L-L}}{\bar{Z}_{eq}} \tag{10.67}$$

where

$$\bar{Z}_{eq} = \bar{Z}_T + n^2 \bar{Z}_1 + \bar{Z}_{G,\text{SL}} \tag{10.68}$$

n is the transfer ratio from Figure 10.35b

$$\bar{Z}_T = \bar{Z} + \frac{\bar{Z}}{2}$$

$$= (R_T + 1.5R_T) + j(X_T + 1.2X_T) \tag{10.69}$$

If an L–L fault involves phases a and b, the available fault current can be calculated from Equation 10.51, where

$$\bar{Z}_{eq} = \bar{Z}_T + n^2 \bar{Z}_1 + \bar{Z}_{1,\text{SL}} \tag{10.70}$$

n is the transfer ratio from Figure 10.35c

$$\bar{Z}_T = 2\bar{Z} \tag{10.71}$$

If an L–L fault involves phases a and c or phases b and c, the available fault current can be calculated by using Equations 10.51 and 10.52, where
  n is the transfer ratio from Figure 10.35d

$$\bar{Z}_T = \bar{Z}$$

Since a three-phase fault involves all three phases, the maximum three-phase fault current can be determined from

$$\overline{I}_{f,3\phi} = \frac{\overline{V}_{L-L}}{\overline{Z}_{eq}} \tag{10.72}$$

where $\overline{Z}_{eq}$ is found from Equation 10.70 and

$$\overline{Z}_T = \frac{(3\overline{Z})(3\overline{Z})}{6\overline{Z}} = \frac{3\overline{Z}}{2} \tag{10.73}$$

10.13.4 THREE-PHASE 208Y/120 V, 480Y/277 V, OR 832Y/480 V FOUR-WIRE WYE–WYE SECONDARY SERVICE

Figure 10.36 shows a three-phase wye–wye-connected four-wire secondary connection. A line-to-ground fault may involve any one of the three phases and neutral. The maximum line-to-ground fault current can be calculated as

$$\overline{I}_{f,L-G} = \frac{\overline{V}_{L-N}}{\overline{Z}_{eq}} \tag{10.74}$$

where $\overline{Z}_{eq}$ is found from Equation 10.48 and

$$\overline{Z}_T = \overline{Z}$$

An L–L fault may involve phases a and b, or b and c, or c and a. The available L–L fault current is

$$\overline{I}_{f,L-L} = \frac{\overline{V}_{L-L}}{2\overline{Z}_{eq}} \tag{10.75}$$

where $\overline{Z}_{eq}$ is determined from Equation 10.70 and

$$\overline{Z}_T = \overline{Z}$$

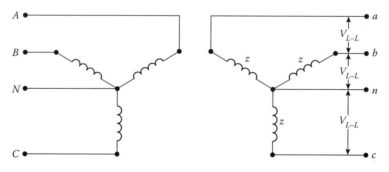

FIGURE 10.36 A three-phase wye–wye-connected four-wire secondary connection.

A three-phase fault may involve all three phases; therefore, the available three-phase fault current can be expressed as

$$\overline{I}_{f,3\phi} = \frac{\overline{V}_{L-L}}{\sqrt{3} \times \overline{Z}_{eq}} = \frac{\overline{V}_{L-N}}{\overline{Z}_{eq}} \tag{10.76}$$

where $\overline{Z}_{eq}$ is determined from Equation 10.70.

Example 10.4

Assume that there is a single-phase L–L secondary fault on a 120/240 V three-wire service, as shown in Figure 10.37, and that the subtransmission system is taken as an infinite bus. The substation transformer is a three-phase 7500 kVA with 7% impedance and 1% resistance. The 12.47 kV primary feeder has three-phase conductors of 336AS37 with a neutral conductor of 0AS7 at 62 in. spacing.

The secondary transformer (i.e., distribution transformer) has 100 kVA capacity with 1.9% impedance and 1% resistance. The 60 ft long self-supporting service cable (SSC) with ACSR neutral has three wires and is given to be 3-4ALSSC. Assume that it has a resistance of 0.4660 Ω/1000 ft and a reactance of 0.0293 Ω/1000 ft. Use Table 10.7 to determine the necessary sequence-impedance values for the primary line and determine the following:

a. The impedance of the substation transformer, Ω
b. The positive- and zero-sequence impedance of the line, Ω
c. The line-to-ground impedance in the primary system, Ω
d. The total impedance through the primary, Ω
e. The total primary impedance referred to secondary, Ω
f. The distribution transformer impedance, Ω
g. The impedance of the secondary cable, Ω
h. The total impedance to the fault, Ω
i. The single-phase L–L fault for the 120/240 V three-wire service, A

Solution

a. Since the impedance of the substation transformer can be expressed as

$$\overline{Z}_T = R_T + jX_T$$

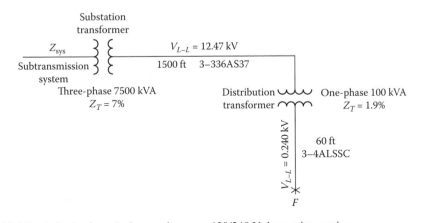

FIGURE 10.37 A single-phase L–L secondary on a 120/240 V three-wire service.

where its reactance is

$$X_T = \left(Z_T^2 - R_T^2\right)^{1/2}$$

$$= (7^2 - 1^2)^{1/2} = 6.9282\% \ \Omega$$

and

$$Z_T = 1 + j6.928\% \ \Omega$$

therefore

$$\bar{Z}_T = \frac{(1 + j6.9282)(12.47)^2(10)}{7500}$$

$$= 0.2073 + j1.4365 \ \Omega$$

b. From Table 10.7, the positive-sequence impedance of the primary line is

$$\bar{Z}_1 = 1.5(0.0580 + j0.1208)$$

$$= 0.0870 + j0.1812 \ \Omega$$

and similarly the zero-sequence impedance is

$$\bar{Z}_0 = 0.1653 + j0.4878 \ \Omega$$

c. From Equation 10.17,

$$\bar{Z}_G = \frac{2\bar{Z}_1 + \bar{Z}_0}{3}$$

$$= \frac{2(0.0870 + j0.1812) + (0.1653 + j0.4878)}{3}$$

$$= 0.1131 + j0.2834 \ \Omega$$

d. Since the subtransmission system is assumed to be an infinite bus,

$$\bar{Z}_{sys} = 0 + j0 \ \Omega$$

therefore the total impedance through the primary is

$$\hat{Z}_{eq} = \bar{Z}_{sys} + \bar{Z}_T + \bar{Z}_G$$

$$= (0 + j0) + (0.2073 + j1.4365) + (0.1131 + j0.2834)$$

$$= 0.3204 + j7199 \ \Omega$$

e. From Equation 10.50, the total primary impedance referred to secondary is

$$n^2\hat{Z}_{eq} = \hat{Z}_{eq}\left(\frac{\sec V_{L-L}}{\text{pri } V_{L-N}}\right)^2$$

$$= (0.3204 + j1.7199)\left(\frac{0.240}{7.2}\right)^2$$

$$= 0.0004 + j0.0019 \ \Omega$$

f. The secondary (i.e., distribution) transformer reactance can be determined as

$$X_T = \left(Z_T^2 - R_T^2\right)^{1/2}$$

$$= (1.9^2 - 1^2)^{1/2} = 1.6155\% \ \Omega$$

Therefore, its impedance can be expressed as

$$\overline{Z}_T = 1 + j1.6155\% \ \Omega$$

or

$$\overline{Z}_T = \frac{(1 + j1.6155)(0.240)^2(10)}{100}$$

$$= 0.0058 + j0.0093 \ \Omega$$

g. Since the impedance of the secondary cable is given in $\Omega/1000$ ft, for a 60 ft length,

$$\overline{Z}_{1,SL} = \frac{60}{1000}(0.4660 + j0.0293)$$

$$= 0.0280 + j0.0018 \ \Omega$$

h. Therefore, the total impedance to the fault can be found as

$$\overline{Z}_{eq} = n^2 \hat{Z}_{eq} + \overline{Z}_T + \overline{Z}_{1,SL}$$

$$= (0.004 + j0.0019) + (0.0058 + j0.0093) + (0.0280 + j0.0018)$$

$$= 0.0342 + j0.0130 \ \Omega$$

i. Thus, from Equation 10.51, the fault current at the secondary fault point F is

$$\overline{I}_{f,L-L} = \frac{\overline{V}_{L-L}}{\overline{Z}_{eq}}$$

$$= \frac{240}{0.0366} = 6559.63 \ A$$

10.14   HIGH-IMPEDANCE FAULTS

The detection of high-impedance faults on electric distribution systems has been one of the most difficult and persistent problems. These faults result from the contact of an energized conductor with surfaces or objects that limit the current to levels below the detection thresholds of conventional protection devices. High-impedance faults often take place when an energized overhead conductor breaks and falls to the ground, creating a serious public hazard.

The typical measured values of primary fault current for a 15 kV class of distribution feeder conductor in contact with various surfaces are 0 A for dry asphalt or sand, 15 A for wet grass, 20 A for dry sod, 25 A for dry grass, 40 A for wet sod, 50 A for wet grass, and 75 A for reinforced concrete. Recent advances in digital technology have provided a means of detecting most of these previously undetectable faults.

However, it is still impossible to detect all high-impedance faults, because of the random and intermittent nature of these low current faults. For example, in one of the methods, an electromechanical time-overcurrent relay, which balanced zero-sequence operating torque against positive- and negative-sequence bias, showed an increased sensitivity but with disappointing security.

Also, several mechanical devices have been developed to catch and ground a broken conductor, but they appear to have questionable reliability and are costly. They are also difficult to install.

The single most important measure that might be taken, in any power system and without product expense, to improve system protection against arcing faults would be to lower circuit-breaker instantaneous trip settings to a level no higher than that required to avoid nuisance tripping under normal conditions. Because of the inadequacies of fuses and arcing-fault currents, recourse to supplementary relaying is necessary to secure adequate protection.

Both grounded and ungrounded systems have proven valuable to arcing-fault burndowns. The *ungrounded* power system tends to present higher probable minimum values of arcing-fault current under certain conditions, when compared to the grounded system. The ungrounded system can have substantial transient overvoltages in the event of faults [16].

Ground fault currents in a system, which is solidly grounded, may have values approaching or exceeding the bolted three-phase short-circuit values, and automatically and prompt interruption of circuits faulted to ground is the intended mode of operation.

The solidly grounded system is the most widely used low-voltage distribution system, in either the industrial or commercial building domain. The ideal solution to the problem would be sensitive to arcing-fault current alone. Since arcing faults in grounded systems almost invariably involve ground, this fat permits a near-perfect approach to the ideal solution.

An excellent method of monitoring the presence of ground fault currents (zero-sequence currents) is provided by the use of a window or doughnut-type CT in combination with an overcurrent relay. All the phase conductors of the circuit to be monitored (plus the neutral conductor, if used) are passed through the window of the CT. With this arrangement (a low-voltage ground-sensor relay combination), only circuit faults involving ground will produce a current in the CT secondary to pick up the relay. By proper matching of the CT and relay, this arrangement may be made quite sensitive, so as to operate on ground currents of 15 A or less.

In the early 1980s, researchers [14] were focused on using an integrated multialgorithm approach in solving this nagging problem. The resultant partial success was due to the following developments in the digital technology: (1) rapid increase in digital processing power, (2) emergence of artificial intelligence methodology, and (3) advances in pattern recognition techniques. The random intermittent nature of these arcing faults required extensive data capture over relatively long-term intervals. Analysis of this expanding database enabled the researchers distinguish high-impedance faults from other power system events. This pattern recognition approach provided the basis for the development of the high-impedance fault detection algorithms.

One of the main algorithms is based on the recognition of sudden and sustained changes in energy in the extracted nonharmonic, as well as the harmonic components, of the feeder CT secondary currents. A second algorithm identifies the distinctive random changes in these extracted signals. In addition to the energy and randomness algorithms, various other algorithms address such things as spectral shape and arc burst patterns to further confirm the presence of arcing high-impedance faults.

While no one of these detection algorithms conclusively distinguishes all types of arcing faults from power system operational events, the integration of all of these algorithms into a knowledge-based system provides sensitive, reliable detection with excellent security. This expert arc-detection system requires a large amount of signal processing in real time during a fault or disturbance to run the various algorithms. Additional intelligence is provided to determine whether a high-impedance fault condition involves a primary conductor on the ground [14–16].

10.15 LIGHTNING PROTECTION

Momentary outages are a main concern for some utilities trying to improve their power quality. Improving protection of overhead distribution circuits from lightning is one way in which some utilities can significantly reduce the number of momentary outages. In most cases, flashovers result in a flash. The fault can be cleared by the substation breaker on the instantaneous setting of the relay, and the system will be back to normal following reclosure. This causes a momentary outage for the entire feeder that, in the past, may not have been a problem. However, modern consumer devices (such as digital clocks, computers, and VCRs) are disrupted by momentary outages, so many utilities are looking for ways to reduce those interruptions.

The voltages by nearby strokes are determined in a similar manner as the method used for direct strokes. For the direct stroke calculations, a current surge is injected into the conductor, and the traveling wave calculations are done. The induced voltage calculations are more complex but are similar in that they use the same traveling wave calculation method. The main difference is that currents are induced at each point along the line instead of a current injected at one point.

The electromagnetic fields created by the lightning stroke induce voltages and currents all along the line, but they can all be computed by considering the fields acting on very small conductor segments. Antenna theory is used to compute the electric and magnetic fields near the line segments due to the current stroke some distance away. Digital analysis becomes necessary due to the complexities involved with multiple phases, arresters, and ground impedances.

10.15.1 A BRIEF REVIEW OF LIGHTNING PHENOMENON

By definition, *lightning* is an electric discharge. It is the high-current discharge of an electrostatic electricity accumulation between the cloud and earth or between the clouds. The mechanism by which a cloud becomes electrically charged is not yet fully understood.

However, it is known that the ice crystals in an active cloud are positively charged, while the water droplets usually carry negative charges. Therefore, a thundercloud has a positive center in its upper section and a negative charge center in its lower section. Electrically speaking, this constitutes dipole. Note that the charge separation is related to the supercooling and occasionally even the freezing of droplets. The disposition of charge concentrations is partially due to the vertical circulation in terms of updrafts and downdrafts.

As a negative charge builds up in the cloud base, a corresponding positive charge is induced on earth, as shown in Figure 10.38a. The voltage gradient in the air between charge centers in cloud (or clouds) or between cloud and earth is not uniform, but it is maximum where the charge concentration is greatest.

When voltage gradient within the cloud builds up to the order of 5–10 kV/cm, the air in the region breaks down and an ionized path called *leader* or *leader stroke* starts to form, moving from the cloud up to the earth, as shown in Figure 10.38b. The tip of the leader has a speed between 10^5 and 2×10^5 m/s (i.e., less than 1000th of the speed of the light of 3×10^8 m/s) and moves in jumps. If photographed by a camera lens of which is moving from left to right, the leader stroke would appear as shown in Figure 10.39. Therefore, the formation of a lightning stroke is a progressive breakdown of the arc path from the cloud to the earth.

As the leader strikes the earth, an extremely bright return streamer, called *return stroke*, propagates upward from the earth to the cloud following the same path, as shown in Figures 10.38c and 10.39. In a sense, the return stroke establishes an electric short circuit between the negative charge deposited along the leader and the electrostatically induced positive charge on the ground.

Therefore, the charge energy from the cloud is related into the ground, neutralizing the charge centers. The initial speed of the return stroke is 10^8 m/s. The current involved in the return stroke has a peak value of from 1 to 200 kA, lasting about 100 μs. About 40 μs later, a second leader, called *dart leader*, may stroke usually following the same path taken by the first leader.

Dart leader is much faster and has no branches and may be produced by discharge between two charge centers in the cloud, as shown in Figure 10.38e. Note the distribution of the negative charge along the stroke path. The process of dart leader and return stroke (Figure 10.38f) can be repeated several times.

The complete process of successive strokes is called *lightning flash*. Thus, a lightning flash may have a single or a sequence of several discrete stokes (as many as 40) separated by about 40 ms, as shown in Figure 10.39.

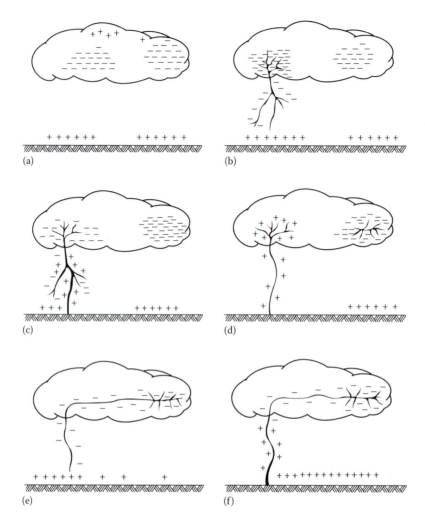

FIGURE 10.38   An illustration of the lightning phenomenon.

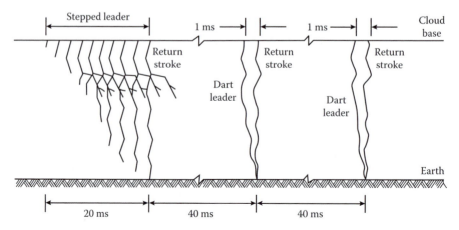

FIGURE 10.39   The complete process of a lightning flash.

10.15.2 Lightning Surges

The voltages produced on overhead lines by lightning may be due to *indirect strokes* or *direct strokes*. In the indirect stroke, induced charges can take place in the lines as a result of close-by lightning strokes to ground. Even though the cloud and earth charges are neutralized through the established cloud-to-earth current path, a charge will be trapped on the line, as shown in Figure 10.40a.

The magnitude of this trapped charge is a function of the initial cloud-to-earth voltage gradient and the closeness of the stroke to the line. Such voltage may also be induced as a result of lightning among clouds, as shown in Figure 10.40. In any case, the voltage induced on the line propagates along the line as a traveling wave until it is dissipated by attenuation, leakage, insulation failure, or arrester operation.

A direct lightning can hit any point on a line. It can hit a pole or somewhere in the span between poles. If lightning hits a pole top, some of the current may flow through the shield wires if there is any, and the remaining current flows through the pole to the earth. The flashover will look like a fireball enveloping the top of the pole. The current will divide at the neutral connection, and a small part of it will travel down the line in both directions to the poles on each side of the struck pole.

When the two traveling lightning current waves reach the two adjacent grounded poles, the current will be traveling to these pole grounds, and a small part will continue down the line to the next pole grounds. In general, all of the current can be assumed to flow into the ground at the pole that is struck and the first two grounded poles adjacent to it. If the earth resistance is high, several more grounded poles may be involved on either side of the struck pole.

However, if lightning hits midspan on a distribution line, the lightning current will divide, and half will travel along the line in each direction at almost the speed of light. If the lightning current flowing through the surge impedance of the line produces a voltage greater than the line insulation can withstand, flashover will take place on the two poles adjacent to the strike point. In addition, strokes that terminate near the line but do not actually hit it can induce voltages high enough to cause flashovers.

To the voltage and current waves, the dead end will appear to be an open circuit. The current wave will be reflected back down the line with a reversed polarity. The voltage wave will double and be reflected back down the line with the same polarity.

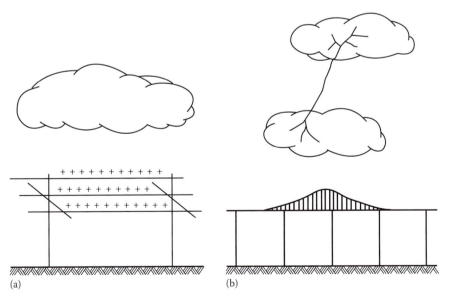

FIGURE 10.40 (a) Induced line charges due to indirect lightning strokes and (b) an occurence of a lightning among clouds.

Because of the low basic lightning impulse level (BIL) of distribution lines and the lightning peak-current magnitudes, voltage doubling at dead ends will produce flashover at all unprotected dead ends, resulting in a *line outage*. Most dead ends have installed surge arresters, which prevent voltage doubling. A surge arrester should be installed on all dead ends.

10.15.3 Lightning Protection

Shunting and *shielding* are two basic methods used to protect lines. With *shunting method*, lightning is permitted to strike the phase conductors, and the lightning current is shunted to ground either by a flashover or by lightning arresters.

With *shielding*, a separate conductor (called *overhead ground wire*) is installed above the phase conductors, and the lightning current is routed to ground without flowing through the phase conductors.

Shielding is used mostly on transmission lines. Shunting is used mainly on distribution lines. The shield wire intercepts the lightning strike. This is accomplished at distribution-line heights by using a 30° protective angle. (This is the angle between the vertical and a straight line between the shield wire and the outside phase conductor.) It is important that there is enough insulation between the phase conductors and the shield system to prevent flashover.

When lightning strikes the shield wire, it travels down the shield to the first structure and down the pole ground to earth. The flow of current in the pole ground results in a voltage between it and the phase conductors. If the insulation strength (i.e., BIL of the line) is exceeded by this voltage, flashover occurs.

Since the lightning current is flowing in the ground circuit instead of the phase conductor, the phenomenon is known as the *backflash*. On the other hand, when lightning strikes a line directly, the raised voltage, at the contact point, propagates in the form of a traveling wave in both directions and raises the potential of the line to the voltage of the downward leader.

If the line is not properly protected against such overvoltage, such voltage may exceed the line-to-ground withstand voltage of the line insulation failure, or preferably the arrester operation establishes a path from the line conductor to ground for the *lightning surge current*.

To achieve reasonable performance with a shield system on distribution lines, the BIL of the path between the insulators and the pole ground is required to be in the 500–600 kV range, and the pole ground impedance has to be less than 10 Ω. Also, *every single* pole is required to have a pole ground installed.

In general, the cost of a *properly designed* shield system will considerably exceed the cost of a lightning arrester-protected system. A lightning arrester-protected line will usually experience fewer outages at less cost. For this reason, shield wires are not recommended on distribution lines.

Lightning protection has the added benefit of reducing equipment damage and line burndowns. Induced flashovers can also be reduced by improved design. Also, building a distribution line on a transmission structure is not a good design option. This is because the number of strikes per mile of the transmission line will be greater than for a distribution line.

Furthermore, the backflash voltage due to strikes to the shield wire will cause flashover on the distribution line, especially if the transmission line pole ground is very close to the distribution-line insulators, with very little wood in the circuit.

10.15.4 Basic Lightning Impulse Level

The voltage level at which flashover will occur on distribution structures is the basic *impulse insulation level* (BIL). BIL is also defined as "*a specific insulation level expressed in kilovolts of the crest value of a standard lightning impulse*." It is determined by testing insulators and equipment using lightning impulse surge generators.

The published voltage impulse for an insulator is defined by the critical flashover voltage for a 1.2 × 50-μs voltage impulse. The voltage across the insulator is increased in steps until flashover occurs. The voltage is adjusted until 50% flashover takes place, which is called the *critical flashover*. Impulse flashover tests are performed for both positive and negative impulses.

BIL is determined statistically from the critical flashover tests and is usually about 10% below critical flashover since the majority of lightning flashes are negative; the essential design value is for negative impulse. For pin insulators, usually used on 7.2/12.47 kV distribution lines, the BIL is approximately 100 kV for negative impulse. However, for the lines, using this insulator on grounded steel crossarms, a 300 kV BIL is needed to prevent flashovers from nearby strikes.

The accurate BIL of a structure can be determined by testing the structure with a surge generator. However, a BIL can also be estimated. One has to remember that the insulator provides the primary insulation for the line. Here, insulation wet flashover values for negative impulses are used.

For example, to estimate BIL for a distribution line, the impulse flashover value for wood or other insulation in the flashover path is then added to the insulator BIL. BIL for wood differs by the type of the wood but usually can be assumed to be about 100 kV/ft dry. Wet value is about 75 kV/ft. Thus, for a structure with a 100 kV BIL insulator and a 3 ft spacing of wood, the BIL would be roughly 325 kV.

The main concern when designing structures is to achieve a 300 kV or greater BIL level to ensure that only direct strikes to the lie will cause a flashover. This is normally achieved by using the wood of the structure itself.

On steel and concrete distribution structures, the only insulation is the conductor insulation. The BI of the structure is the BIL of the insulator. A standard 15 kV insulator has a BIL of about 100 kV. If this insulator is used on a steel pole, the line BIL is 100 kV, and a significant number of flashovers due to nearby lightning strikes can be expected. The insulator BIL requirement is 300 kV. Such an insulator will normally be a 55 kV class insulator.

Fiberglass pins and arms are a good choice from a lightning performance point of view since they normally have a BIL of 200 kV. Adding this to the insulator BIL of 100 gives a 300 kV BIL for the structure. If steel is used, the insulator size should be increased to maintain the 300 kV BIL. Obviously, trade-offs must be made between lightning performance and other considerations such as structural design and economics.

Guy wires, used to help hold poles upright, are generally attached as high on the pole as possible. These guy wires are effectively a grounding point if they do not contain insulating members, and if they are attached high on the pole, the BIL of the configuration will be reduced. The neutral-wire height also affects BIL. On wood poles, the closer the neutral wire to the phase wires, the lower the BIL.

Many of the newer designs have lower BIL than older designs because of tighter phase spacing. The candlestick and spacer cable designs are common in newer construction, and these have lower BIL. The additional spacing and the wood crossarm of the traditional design provide a higher insulation level.

Example 10.5

Surge impedances of overhead distribution lines are in the range of 300–500 Ω. The BIL of distribution lines is in the 100–500 kV range. Determine the following:

  a. The minimum current at which flashover can be expected for distribution lines.
  b. The maximum current at which flashover can be expected for distribution lines.
  c. The minimum total lightning current for the midspan strike.
  d. The maximum total lightning current for a midspan strike.

Solution

a. The minimum current at which flashover can be expected is

$$I_{min} = \frac{min\ BIL}{max\ surge\ impedance} = \frac{100\,kV}{500\,\Omega} = 200\ A \qquad (10.77)$$

b. The maximum current at which flashover can be expected is

$$I_{max} = \frac{max\ BIL}{min\ surge\ impedance} = \frac{500\ kV}{100\ \Omega} = 5\ kA \qquad (10.78)$$

c. For a midspan strike, the current is divided at the strike point, and half is flowed in each direction. The minimum and maximum currents earlier represent half the total current in the lightning flash. Therefore, the minimum total lightning current for a midspan strike is

$$\sum I_{min} = 2 \times I_{min} = 2(200\ A) = 400\ A \qquad (10.79)$$

d. The maximum total lightning current for a midspan strike is

$$\sum I_{max} = 2 \times I_{max} = 2(1.67\,kA) = 3.34\,kA \qquad (10.80)$$

Note that about 99% of lightning first-stroke peak-current magnitudes exceed 3.34 kA.

Except for the small percentage of very-low-magnitude lightning currents, flashover will take place at the two adjacent poles. The process repeats itself, and all of the lightning current will flow to ground at the four poles. However, if the earth resistance is high, current will flow additional adjacent pole grounds.

Example 10.6

Most overhead configurations of distribution lines have a surge impedance between 300 and 500 Ω. Assume that an average lightning stroke current in a stricken phase conductor is 30 kA and determine the following:

a. The voltage level on the stricken phase conductor if the surge impedance is 300 Ω.
b. The voltage level on the stricken phase conductor if the surge impedance is 500 Ω.
c. Will flashovers occur due to direct strikes in parts a and b?

Solution

a. The voltage level on the stricken phase conductor if the surge impedance is 300 Ω is

$$V_{min} = (300\,\Omega)\left(\frac{30\,kA}{2}\right) = 4500\,kV$$

b. The voltage level on the stricken phase conductor if the surge impedance is 500 Ω is

$$V_{max} = (500\,\Omega)\left(\frac{30\,kA}{2}\right) = 7500\,kV$$

c. Yes, since the calculated values earlier are much higher than the BIL of distribution lines, unless some sort of line protection is used, flashovers will occur due to direct strikes.

10.15.5 DETERMINING THE EXPECTED NUMBER OF STRIKES ON A LINE

The unit of measure for ground flash density is strikes per square kilometer per year. The ground flash-density map in Figure 10.41 can be used to determine long-term average ground flash density for any location in the United States. The contours on the map are intervals of two strikes per square kilometer per year.

Any specific location within the country will either be on a contour or between counters. For example, Atlanta is between the 8 and 6 strikes/km² counters but is very close to the 8 contour. Thus, a ground flash density of 8 should be used for Atlanta. Also, notice that the ground flash density for the whole State of California is two. This value can easily be converted into number of strikes/min²/year by multiplying it by 2.59. The resultant number of strikes should be rounded to the nearest whole number. For example, for Atlanta, 8×2.59 results in 21 strikes/min²/year, whereas for California, it is 5 strikes/min²/year. For an engineering design, the low end of the range of ground flash-density values can be taken as 50% of average. The high end of the range can be taken as 200% of average.

As Ben Franklin discovered it, lightning is attracted to tall structures, and this attraction is defined as an area shielded by the structure. This shielded area of the tower S_A in m² is determined from

$$S_A = \frac{N}{N_{gfd}} \tag{10.81}$$

where

N is the number of strikes to the tower

N_{gfd} is the average ground flash density in strikes/min²/year (found from Figure 10.41)

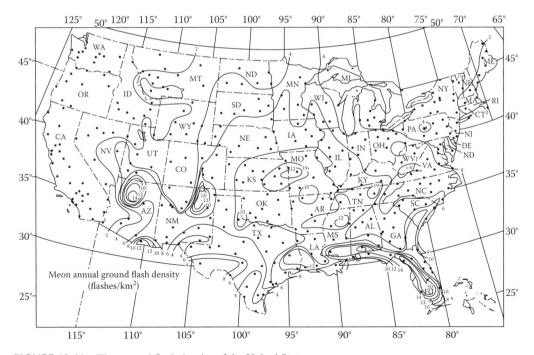

FIGURE 10.41 The ground flash density of the United States.

The number of strikes to a distribution or transmission line in open country (N_{oc}) with a length of 100 km can be found from

$$N_{oc} = N_{gfd}(b + 28h^{0.6}) \times 10^{-1} \text{ strikes/100 km/year} \tag{10.82}$$

where
b is the width between the outside conductors
h is the height of tower in m

or for a length of 1 km, it can be expressed as

$$N_{oc} = N_{gfd}(b + 28h^{0.6}) \times 10^{-3} \text{ strikes/km/year} \tag{10.83}$$

For distribution lines, the width term b can be eliminated by assuming it zero. (It can be shown that the resultant error is less than 2.33%.) In addition by converting to English units, for a length of 1 mile,

$$N_{oc} = N_{gfd}(0.022h^{0.6}) \text{ strikes/min/year} \tag{10.84}$$

For the total line length,

$$\sum N_{oc} = N_{gfd}(0.022h^{0.6})s \text{ strikes/year} \tag{10.85}$$

where
s is the length of the lone in min
N_{gfd} is the average ground flash density in strikes/km²/year

For standard pole lengths (standard setting depths and REA standard construction), the total number of strikes to an open country distribution line can be expressed as

$$\sum N_{oc} = C \times N_{gfd} \times s \tag{10.86a}$$

where C is a constant based on pole length.
For various standard pole lengths, the constant C is given in Table 10.14.

TABLE 10.14
Constant C for REA Standard Pole Lengths

Pole Length, ft	Setting Depth in Soil, ft	Conductor above the Pole Top, ft	Height (h) of Line Aboveground, ft	$C = 0.022h^{0.6}$
30	5.5	0.66	25.16	0.154
35	6.0	0.66	29.16	0.168
40	6.0	0.66	34.66	0.185
45	6.5	0.66	39.16	0.199
50	7.0	0.66	43.66	0.212

Source: From Rural Electrification Administration: *Guide for Making a Sectionalizing Study on Rural Electric Systems*, REA Bulletin 61–2. March, 1958.

The total number of strikes to a distribution line is affected by the nearby trees, other structures, or objects that shield the line from direct strikes. Here, the shielding factor S_f can be defined as

$$S_f = 1 - \frac{N}{\sum N_{oc}} \qquad (10.86b)$$

where

N is the number of strikes to the shielded line
$\sum N_{oc}$ is the number of strikes to the line in open country

The methods used to determine the shielding factor are quite complex and involve the use of electrogeometric models. However, estimates of shielding effects can be made by considering standard shielding cases, and the accuracy of such estimates is sufficient for most engineering decisions.

Here, the value of S_f varies between 0.0 and 1.0:

If there is no shielding of any kind, then $S_f = 0$.
If there are tall trees on both sides of the line and within 100 ft of the line, then $S_f = 0.90$.
If there are tall trees at the edge of the typical 30 ft right of way (15 ft from the center of the line, on both sides), then $S_f = 1.0$.
If the trees present have heights that are 1.5 times the height of the line, then $S_f = 0.70$.
If the height of the one-sided shielding is twice the height of the line and within 50 ft of the line, $S_f = 0.90$.
If the shielding factor is known, the number of predicted direct strikes N to the lie is determined from

$$N = \sum N_{oc}(1 - S_f) \qquad (10.87)$$

Example 10.7

Assume that the NL&NP Company of Kansas has 8000 customers on 2000 miles of line located in central Kansas that can be considered as an open country. The average pole length used on the distribution system is 35 ft. The distribution system has 6000 pole-mounted transformers installed. Determine the following:

a. The average span length, if the system has 35,000 poles
b. The number of strikes per mile per year
c. The number of expected strikes to the adjoining spans of an equipment pole
d. The total expected number of strikes to the adjoining spans of the 6000 transformer poles per year
e. The average time between lightning strikes that can be expected within one span of an "average" equipment pole

Solution

a. For the 35,000 poles, the average span length is

$$S_{avg} = \frac{(2,000\,\text{min})(5,280\,\text{ft/min})}{35,000} \cong 302\,\text{ft}$$

b. From Figure 10.41, the ground flash density in central Kansas is 9 strikes/km²/year. For a 35 ft pole, from Table 10.14, $C = 0.168$. Therefore, the number of strikes is

$$C \times N_{gfd} = 0.168 \times 9 = 1.512 \text{ strikes/min/year}$$

c. Since a direct lightning strike can be expected to cause a flashover on the first pole on either side of the strike point, only concern with the two spans on each side of the equipment location. Therefore, the number of expected strikes is

$$N_{exp} = C \times N_{gfd} \times L = 1.512 \times \frac{2(302\,\text{ft})}{5280\,\text{ft/min}} \cong 0.173 \text{ pole strikes/year}$$

d. The total expected number of strikes to the adjoining spans of the 6000 transformer poles per year is

$$\sum N_{exp} = N_{exp} \times L = 0.173 \times 6000 = 1038 \text{ line strikes/year}$$

e. The average time between lightning strikes that can be expected within one span of an average equipment pole is

$$\frac{1}{N_{exp}} = \frac{1}{0.173} = 5.78 \text{ years}$$

Example 10.8

Consider the distribution system used in Example 10.7 and assume that most transformers are shielded from lightning to some extent by buildings or trees. Therefore, use a *"reasonable estimate"* of 70% for the average shielding factor. It is assumed that 5% of lightning return stokes exceed 100 kA. Determine the following:

a. The total number of strikes per year to transformers if shielding is taken into consideration
b. The total number of expected transformer failures per year due to lightning
c. The annual expected transformer failure rate due to lightning

Solution

a. If the shielding is used, then

$$\sum N_{trf\,strikes} = \sum N_{exp}(1 - S_f) = 1038(1 - 0.70) \cong 311 \text{ transformer strikes/year}$$

b. Since 5% of stoke currents exceed 100 kA, the expected number of transformer failures is

$$\sum N_{trf\,failures} = 0.05 \sum N_{trf\,striks} = 0.05 \times 311 \cong 15$$

c. The annual expected transformer failure rate due to lightning is

$$\lambda_{lightning} = \frac{15}{6000} = 0.0025 \quad \text{or} \quad 25\%$$

that is, 1 out of every 400 installed transformers.

Example 10.9

Consider the distribution system used in Example 10.7 and assume that for the arrester used on the system, a lightning current of 50 kA produces a discharge voltage of 95 kV. The arrester is tank mounted with zero lead length. The NL&NP Company uses 95 kV BIL transformers so that surge voltages in excess of 95 kV can be expected to cause transformer failure. For midspan lightning strikes, the lightning current will divide, and half of the current will flow down the line in each direction. Determine the following:

a. The amount of lightning return stroke current if the transformer is to be subjected to a 50 kA current surge
b. The total annual number of strikes to transformers that can be expected to produce voltages that exceed the transformer BIL and cause failures

Solution

a. For the transformer to be subjected to a 50 kA current surge, it is required that the lightning return stroke current must be

$$2(50 \text{ kA}) = 100 \text{ kA}$$

b. Since 5% of lightning return strokes exceed 100 kA,

$$0.05(1038 \text{ strikes/year}) = 52 \text{ strikes/year}$$

Example 10.10

Assume that the NL&NP Company of California utilizes 30 ft poles in its 100 mile long rural lines. The practical maximum line width occurs with the use of an 8 ft crossarm and a horizontal conductor arrangement. The distance between the outside conductors is 7.4 ft. Assume that the 100 min long line is in open country. Determine the following:

a. The number of strikes to the line
b. The number of strikes to the line, if the line width term b is ignored and/or assumed to be zero
c. Compare the results of parts, and express the difference in percentage

Solution

a. The number of strikes to the 100 min long line can be found from

$$N_{oc} = N_{gfd}(b + 28h^{0.6}) \times 10^{-1} \tag{10.88}$$

where
$N_{gfd} = 2$ strikes/km²/year (from Figure 10.41)
$b = (7.4 \text{ ft})(0.3048 \text{ m/ft}) = 2.256$ m
$h = (25.16 \text{ ft})(0.3048 \text{ m/ft}) = 7.669$ m (25.16 ft is found from Table 10.14)

Thus,

$$N_{oc} = 2(2.256 + 28 \times 7.669^{0.6}) \times 10^{-1} = 19.464 \text{ strikes/year}$$

b. If the line width term b is ignored,

$$N_{oc} = N_{gfd}(28h^{0.6}) \times 10^{-1} = 2(28 \times 7.669^{0.6})10^{-1} = 19.012 \text{ strikes/year}$$

c. Therefore,

$$\Delta N_{oc} = \frac{19.464 - 19.012}{19.464} \times 100 = 2.32\%$$

This 2.32% difference is a maximum value, and for taller or narrower lines, the difference is less than 2.32%.

10.16 INSULATORS

An *insulator* is a material that prevents the flow of an electric current and can be used to support electric conductors. The function of an insulator is to provide for the necessary clearances between the line conductors, between conductors and ground, and between conductors and the pole or tower. Insulators are made up of porcelain, glass, and fiberglass treated with epoxy resins. However, porcelain is still the most common material used for insulators.

The basic types of insulators include (1) *pin-type* insulators, (2) *suspension* insulators, and (3) *strain* insulators. The pin insulator gets its name from the fact that it is supported on a pin. The pin holds the insulator, and the insulator has the conductor tied to it. They may be made in one piece for voltages below 23 kV, in two pieces for voltages from 23 to 46 kV, in three pieces for voltages from 46 to 69 kV, and in four pieces for voltages from 69 to 88 kV. Pin insulators are used in distribution lines and are seldom used on transmission lines having voltages above 44 kV, although some 88 kV lines using pin insulators are in operation. The glass pin insulator is mainly used on low-voltage circuits. The *porcelain pin insulator* is used on secondary mains and services, as well as on primary mains, feeders, and transmission lines.

A modified version of the pin-type insulator is known as the post-type insulator. The *post-type insulators* are used on distribution, subtransmission, and transmission lines and are installed on wood, concrete, and steel poles. The line post insulators are usually made as one-piece solid porcelain units.

Suspension insulators are normally used on subtransmission and transmission lines and consist of a string of interlinking separate disks made of porcelain. A string may consist of many disks depending on the line voltage. (For further information, see Gönen [16, Chapter 4]). For example, as an average, 7 disks are usually used for 115 kV lines and 18 disks are usually used for 345 kV lines.

The *suspension insulator*, as its name implies, is suspended from the crossarm (or a pole or tower) and has the line conductor fastened to the lower end. When there is a dead end of the line or there is corner or a sharp curve, or the line crosses a river, etc., the line will withstand great strain.

The assembly of suspension units arranged to dead-end the conductor of such structure is called a *dead-end*, or *strain*, insulator. In such an arrangement, suspension insulators are used as strain insulators [16,17].

PROBLEMS

10.1 Repeat Example 10.1, assuming that the fault current is 1000 A.
10.2 Repeat Example 10.1, assuming that the fault current is 500 A.
10.3 In Problem 10.2, determine the lacking relay travel that is necessary for the relay to close its contacts and trip its breaker:
  (a) In percent
  (b) In seconds
10.4 Assume that an inverse-time-overcurrent relay is installed at a location on a feeder. It is desired that the substation OCB trip on a sustained current of approximately 400 A and

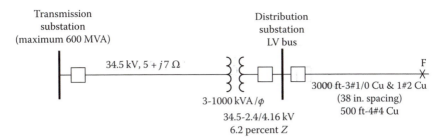

FIGURE P10.1 Distribution circuit of Problem 10.7.

trip in 2 s on a short-circuit current of 4000 A. Assuming that CTs of 60:1 ratio are used, determine the following:

(a) The current-tap setting of the relay
(b) The time setting of the relay

10.5 Repeat Example 10.2, assuming that the transformer is rated 3750 kVA 69/4.16 kV feeding a three-phase four-wire 4.16 kV circuit and that the sizes of the phase conductors and neutral conductor are 267AS33 and OAS7, respectively.

10.6 Repeat Example 10.3, assuming that the faults are bolted and that the fault impedance is 40 Ω.

10.7 Assume that there is a bolted fault at a certain point F on a distribution circuit, as indicated in Figure P10.1. Also assume that the maximum power generation of the system is 600 MVA. Determine the following:

(a) Maximum values of the available three-phase, $L–L$, and SLG fault currents at the fault point F, using actual system values
(b) Minimum value of the available SLG fault, assuming that it is equal to 60 percent of its maximum value found in part a

10.8 Repeat Example 10.4, assuming that the substation transformer's impedance is 7.5 percent and that the distribution transformer has a capacity of 75 kVA with 2 percent impedance. Also assume that the primary line is made of three 477AS33 conductors and a neutral conductor of OAS7 at 62 in spacing and that the lengths of the primary line and secondary cable are 1000 and 50 ft, respectively. Assume that the impedance of the three-wire OALSSC secondary cable is $0.1843 + j0.0273$ Ω/1000 ft.

10.9 Assume that the NL&NP Company of California has 6000 customers on 1000 miles of lines located in central California (which can be considered as an open country). The average pole length used on the distribution system has 4000 pole-mounted transformers installed. Determine the following:

(a) The average span length, if the system has 17,032 poles
(b) The number of strikes per mile per year
(c) The number of expected strikes to the adjoining spans of an equipment pole
(d) The total expected number of strikes to the adjoining spans of the 4000 transformer poles per year
(e) The average time between lightning strikes that can be expected within one span of an "average" equipment pole

10.10 Consider the distribution system given in Problem 10.9 and assume that most transformers are shielded from lightning to some extent by buildings or trees. Therefore, use a "reasonable estimate" of 70% for the average shielding factor. It is assumed that the total expected number of strikes to the transformer poles is 174 strikes/year and that 5% of lightning return strokes exceed 100 kA. Determine the following:

(a) The total number of strikes per year to the transformers if shielding is taken into consideration
(b) The total number of expected transformer failures per year due to lightning
(c) The annual expected transformer failure rate due to lightning

10.11 Consider the distribution system given in Problem 10.10, and assume that for the arrester used on the system, a lightning current of 50 kA produces a discharge voltage of 95 kV. The arrester is tank mounted with zero lead length. The NL&NP Company uses 95 kV BIL transformers so that surge voltages in excess of 95 kV can be expected to cause transformer failure. For midspan lightning strikes, the lightning current will divide, and half of the current will flow down the line in each direction. It is assumed that 5% of lightning return strokes exceed 100 kA. Determine the following:

(a) The amount of lightning return stroke current if the transformer is to be subjected to a 50 kA current surge

(b) The total annual number of strikes to transformers that can be expected to produce voltages that exceed the transformer BIL and cause failures

10.12 Resolve Example 10.3 by using MATLAB®.

10.13 Resolve Example 10.4 by using MATLAB.

REFERENCES

1. Westinghouse Electric Corporation: *Electric Utility Engineering Reference Book-Distribution Systems*, vol. 3, Westinghouse Electric Corporation, East Pittsburgh, PA, 1965.

2. Fink, D. G. and H. W. Beaty: *Standard Handbook for Electrical Engineers*, 11th edn., McGraw-Hill, New York, 1978.

3. Anderson, P. M.: *Elements of Power System Protection*, Cyclone Copy Center, Ames, IA, 1975.

4. General Electric Company: *Overcurrent Protection for Distribution Systems*, Application Manual GET-1751A, Schenectady, NY, 1962.

5. General Electric Company: *Distribution System Feeder Overcurrent Protection*, Application Manual GET-6450, 1979.

6. Westinghouse Electric Corporation: *Westinghouse Transmission and Distribution Reference Book*, Westinghouse Electric Corporation, East Pittsburgh, PA, 1964.

7. IEEE: *Recommended Practice for Protection and Coordination of Industrial and Commercial Power Systems*, IEEE Standard 242-1975, New York, 1975.

8. Anderson, P. M.: *Analysis of Faulted Power Systems*, Iowa State University Press, Ames, IA, 1973.

9. Rural Electrification Administration: *Guide for Making a Sectionalizing Study on Rural Electric Systems*, REA Bulletin 61-2, March, 1958, US Dept. of Agriculture, Washington DC.

10. Wagner, C. F. and R. D. Evans: *Symmetrical Components*, McGraw-Hill, New York, 1933.

11. Stevenson, W. D.: *Elements of Power System Analysis*, 3rd edn., McGraw-Hill, New York, 1975.

12. Gross, C. A.: *Power System Analysis*, Wiley, New York, 1979.

13. Carson, J. R.: Wave propagation in overhead wires with ground return, *Bell Syst. Tech. J.*, 5, October 1926, 539–555.

14. Aucoin, B. M. and B. D. Russell: Distribution high-impedance fault detection utilizing high-frequency current components, *IEEE Trans. Power Appar. Syst.*, PAS-101(6), June 1982, 1596–1606.

15. IEEE: *Detection of Downed Conductors on Utility Distribution Systems*, IEEE Tutorial Course, prod. No. 90EHO310-3-PWR, 1989.

16. Gönen, T.: *Modern Power System Analysis*, Wiley, New York, 1988.

17. Gönen, T.: *Electric Power Transmission System Engineering*, Wiley, New York, 1988.

18. MacGorman, D. R. et al.: *Lightning Strike Density for the Contiguous United States from Thunderstorm Duration Records*, NUREG/CR-3759, National Oceanic and Atmospheric Administration, Prepared for the U. S. Nuclear Regulatory Commission, May 1984.

11 Distribution System Reliability

Mind moves matter.

Virgil

What is mind? No matter. What is matter? Never mind.

Thomas H. Key

If a man said 'all mean are liars', would you believe him?

Author Unknown

11.1 BASIC DEFINITIONS

Most of the following definitions of terms for reporting and analyzing outages of electrical distribution facilities and interruptions are taken from Ref. [1,15] and included here by permission of the Institute of Electrical and Electronics Engineers, Inc.

Outage: Describes the state of a component when it is not available to perform its intended function due to some event directly associated with that component. An outage may or may not cause an interruption of service to consumers depending on system configuration.

Forced outage: An outage caused by emergency conditions directly associated with a component that require the component to be taken out of service immediately, either automatically or as soon as switching operations can be performed, or an outage caused by improper operation of equipment or human error.

Scheduled outage: An outage that results when a component is deliberately taken out of service at a selected time, usually for purposes of construction, preventive maintenance, or repair. The key test to determine if an outage should be classified as forced or scheduled is as follows. If it is possible to defer the outage when such deferment is desirable, the outage is a scheduled outage; otherwise, the outage is a forced outage. Deferring an outage may be desirable, for example, to prevent overload of facilities or an interruption of service to consumers.

Partial outage: "Describes a component state where the capacity of the component to perform its function is reduced but not completely eliminated" [2].

Transient forced outage: A component outage whose cause is immediately self-clearing so that the affected component can be restored to service either automatically or as soon as a switch or circuit breaker can be reclosed or a fuse replaced. An example of a transient forced outage is a lightning flashover that does not permanently disable the flashed component.

Persistent forced outage: A component outage whose cause is not immediately self-clearing but must be corrected by eliminating the hazard or by repairing or replacing the affected component before it can be returned to service. An example of a persistent forced outage is a lightning flashover that shatters an insulator, thereby disabling the component until repair or replacement can be made.

Interruption: The loss of service to one or more consumers or other facilities and is the result of one or more component outages, depending on system configuration.

Forced interruption: An interruption caused by a forced outage.

Scheduled interruption: An interruption caused by a scheduled outage.

Momentary interruption: It has a duration limited to the period required to restore service by automatic or supervisor-controlled switching operations or by manual switching at locations where an operator is immediately available. Such switching operations are typically completed in a few minutes.

Temporary interruption: "It has a duration limited to the period required to restore service by manual switching at locations where an operator is not immediately available. Such switching operations are typically completed within 1–2 h" [2].

Sustained interruption: "It is any interruption not classified as momentary or temporary" [2].

At the present time, there are no industry-wide standard outage reporting procedures. More or less, each electric utility company has its own standards for each type of customer and its own methods of outage reporting and compilation of statistics. A unified scheme for the reporting of outages and the computation of reliability indices would be very useful but is not generally practical due to the differences in service areas, load characteristics, number of customers, and expected service quality.

System interruption frequency index: "The average number of interruptions per customer served per time unit. It is estimated by dividing the accumulated number of customer interruptions in a year by the number of customers served" [3].

Customer interruption frequency index: "The average number of interruptions experienced per customer affected per time unit. It is estimated by dividing the number of customer interruptions observed in a year by the number of customers affected" [3].

Load interruption index: "The average kVA of connected load interrupted per unit time per unit of connected load served. It is formed by dividing the annual load interruption by the connected load" [3].

Customer curtailment index: "The kVA-minutes of connected load interrupted per affected customer per year. It is the ratio of the total annual curtailment to the number of customers affected per year" [3].

Customer interruption duration index: "The interruption duration for customers interrupted during a specific time period. It is determined by dividing the sum of all customer-sustained interruption durations during the specified period by the number of sustained customer interruptions during that period" [3].

Momentary interruption: The complete loss of voltage (<0.1 pu) on one or more phase conductors for a period between 30 cycles and 3 s.

Sustained interruption: The complete loss of voltage (<0.1 pu) on one or more phase conductors for a time greater than 1 min.

According to an IEEE committee report [4], the following basic information should be included in an equipment outage report:

1. Type, design, manufacturer, and other descriptions for classification purposes
2. Date of installation, location on system, length in the case of a line
3. Mode of failure (short-circuit, false operation, etc.)
4. Cause of failure (lightning, tree, etc.)
5. Times (both out of service and back in service, rather than outage duration alone), date, meteorological conditions when the failure occurred
6. Type of outage, forced or scheduled, transient or permanent

Furthermore, the committee has suggested that the total number of similar components in service should also be reported in order to determine outage rate per component per service year. It is also suggested that every component failure, regardless of service interruption, that is, whether it caused a service interruption to a customer or not, should be reported in order to determine component-failure rates properly [4]. Failure reports provide very valuable information for preventive maintenance programs and equipment replacements.

There are various types of probabilistic modeling of components to predict component-failure rates, which include (1) fitting a modified time-varying Weibull distribution to component-failure cases and (2) component survival rate studies. However, in general, there may be some differences between the predicted failure rates and observed failure rates due to the following factors [5]:

1. Definition of failure
2. Actual environment compared with prediction environment
3. Maintainability, support, testing equipment, and special personnel
4. Composition of components and component-failure rates assumed in making the prediction
5. Manufacturing processes including inspection and quality control
6. Distributions of times to failure
7. Independence of component failures

11.2 NATIONAL ELECTRIC RELIABILITY COUNCIL

In 1968, a national organization, the National Electric Reliability Council (NERC), was established to increase the reliability and adequacy of bulk power supply in the electric utility systems of North America. It is a form of nine regional reliability councils and covers all the power systems of the United States and some of the power systems in Canada, including Ontario, British Columbia, Manitoba, New Brunswick, and Alberta, as shown in Figure 11.1.

Here, the terms of reliability and adequacy define two separate but interdependent concepts. The term *reliability* describes the security of the system and the avoidance of power outages, whereas the term *adequacy* refers to having sufficient system capacity to supply the electric energy requirements of the customers.

In general, regional and nationwide annual load forecasts and capability reports are prepared by the NERC. Guidelines to member utilities for system planning and operations are prepared by the regional reliability councils to improve reliability and reduce costs.

Also shown in Figure 11.1 are the total number of bulk power outages reported and the ratio of the number of bulk outages to electric sales for each regional electric reliability council area to provide a meaningful comparison.

Table 11.1 gives the generic and specific causes for outages based upon the *National Electric Reliability Study* [6]. Figure 11.2 shows three different classifications of the reported outage events by (a) types of events, (b) generic subsystems, and (c) generic causes. The cumulative duration to restore customer outages is shown in Figure 11.3, which indicates that 50% of the reported bulk power system customer outages are restored in 60 min or less and 90% of the bulk outages are restored in 7 h or less.

A casual glance at Figure 11.2b may be misleading. Because, in general, utilities do not report their distribution system outages, the 7% figure for the distribution system outages is not realistic. According to *The National Electric Reliability Study* [7], approximately 80% of all interruptions occur due to failures in the distribution system.

The National Electric Reliability Study [7] gives the following conclusions:

1. Although there are adequate methods for evaluating distribution system reliability, there are insufficient data on reliability performance to identify the most cost-effective distribution investments.
2. Most distribution interruptions are initiated by severe weather-related interruptions with a major contributor being inadequate maintenance.
3. Distribution system reliability can be improved by the timely identification and response to failures.

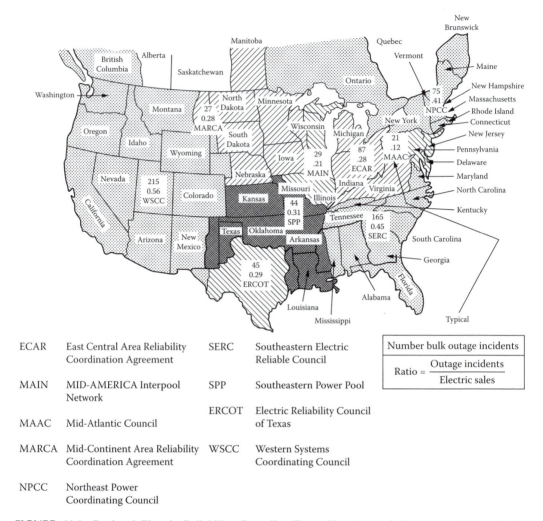

ECAR	East Central Area Reliability Coordination Agreement	SERC	Southeastern Electric Reliable Council
MAIN	MID-AMERICA Interpool Network	SPP	Southeastern Power Pool
MAAC	Mid-Atlantic Council	ERCOT	Electric Reliability Council of Texas
MARCA	Mid-Continent Area Reliability Coordination Agreement	WSCC	Western Systems Coordinating Council
NPCC	Northeast Power Coordinating Council		

$$\text{Number bulk outage incidents}$$
$$\text{Ratio} = \frac{\text{Outage incidents}}{\text{Electric sales}}$$

FIGURE 11.1 Regional Electric Reliability Councils. (From *The National Electric Reliability Study, Technical Study Reports*, U.S. Department of Energy DOE/EP-0005, April 1981.)

11.3 APPROPRIATE LEVELS OF DISTRIBUTION RELIABILITY

The electric utilities are expected to provide continuous and quality electric service to their customers at a reasonable rate by making economical use of available system and apparatus. Here, the term *continuous electric service* has customarily meant meeting the customers' electric energy requirements as demanded, consistent with the safety of personnel and equipment. *Quality electric service* involves meeting the customer demand within specified voltage and frequency limits.

To maintain reliable service to customers, a utility has to have adequate redundancy in its system to prevent a component outage becoming a service interruption to the customers, causing loss of goods, services, or benefits. To calculate the cost of reliability, the cost of an outage must be determined. Table 11.2 gives an example for calculating industrial service interruption cost. Presently, there is at least one public utility commission that requires utilities to pay for damages caused by service interruptions [6].

Reliability costs are used for rate reviews and requests for rate increases. The economic analysis of system reliability can also be a very useful planning tool in determining the capital expenditures required to improve service reliability by providing the real value of additional (and incremental) investments into the system.

TABLE 11.1

Classification of Generic and Specific Causes of Outages

Weather	Miscellaneous	System Components	System Operation
Blizzard/snow	Airplane/helicopter	Electric and mechanical:	System conditions:
Cold	Animal/bird/snake	Fuel supply	Stability
Flood	Vehicle:	Generating unit failure	High/low voltage
Heat	Automobile/truck	Transformer failure	High/low frequency
Hurricane	Crane	Switchgear failure	Line overload/transformer
Ice	Dig-in	Conductor failure	overload
Lightning	Fire/explosion	Tower, pole attachment	Unbalanced load
Rain	Sabotage/vandalism	Insulation failure:	Neighboring power system
Tornado	Tree	Transmission line	Public appeal:
Wind	Unknown	Substation	Commercial and industrial
Other	Other	Surge arrestor	
		Cable failure	All customers
		Voltage control equipment:	Voltage reduction: 0%–2% voltage reduction
		Voltage regulator	Greater than 2–8 voltage reduction
		Automatic tap changer	Rotating blackout
		Capacitor	Utility personnel:
		Reactor	System operator error
		Protection and control:	Powerplant operator error
		Relay failure	Field operator error
		Communication signal error	Maintenance error
		Supervisory control error	Other

Source: The National Electric Reliability Study: Technical Study Reports, U.S. Department of Energy DOE/EP-0005, April 1981.

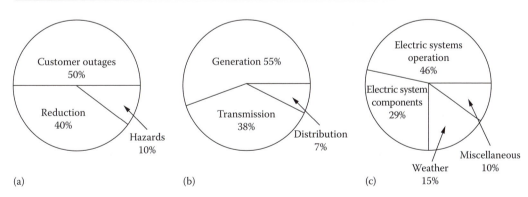

FIGURE 11.2 Classification of reported outage events in the National Electric Reliability Study for the period July 1970–June 1979: (a) types of events, (b) generic subsystems, and (c) generic causes. (From *The National Electric Reliability Study, Technical Study Reports*, U.S. Department of Energy DOE/EP-0005, April 1981.)

As the *National Electric Reliability Study* [6] points out, "it is neither possible nor desirable to avoid all component failures or combinations of component failures that result in service interruptions. The level of reliability can be considered to be *'appropriate'* when the cost of avoiding additional interruptions exceeds the consequences of those interruptions to consumers.

Thus the *appropriate level of reliability* from the consumer perspective may be defined as "*that level of reliability when the sum of the supply costs plus the cost of interruptions which occur are at*

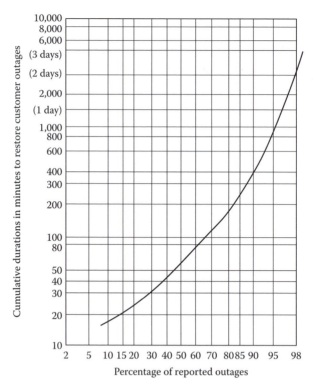

FIGURE 11.3 Cumulative duration in minutes to restore reported customer outages. (From *The National Electric Reliability Study, Technical Study Reports*, U.S. Department of Energy DOE/EP-0005, April 1981.)

a minimum". Figure 11.4 illustrates this theoretical concept. Note that the system's reliability improvement and investment are not linearly related, and that the optimal (or appropriate) reliability level of the system corresponds to the optimal cost, that is, the minimum total cost. However, Billinton [8] points out that "*the most improper parameter is perhaps not the actual level of reliability though this cannot be ignored but the incremental reliability cost. What is the increase in reliability per dollar invested? Where should the next dollar be placed within the system to achieve the maximum reliability benefit?*"

In general, other than "for possible sectionalizing or reconfiguration to minimize either the number of customers affected by an equipment failure or the interruption duration, the only operating option available to the utility to enhance reliability is to minimize the duration of the interruption by the timely repair of the failed equipment(s)" [6].

Experience indicates that most distribution system service interruptions are the result of damage from natural elements, such as lightning, wind, rain, ice, and animals. Other interruptions are attributable to defective materials, equipment failures, and human actions such as vehicles hitting poles, cranes contacting overhead wires, felling of trees, vandalism, and excavation equipment damaging buried cable or apparatus.

Some of the most damaging and extensive service interruptions on distribution systems result from snow or ice storms that cause breaking of overhanging trees, which in turn damage distribution circuits. Hurricanes also cause widespread damage, and tornadoes are even more intensely destructive, though usually very localized. In such severe cases, restoration of service is hindered by the conditions causing the damage, and most utilities do not have a sufficient number of crews with mobile and mechanized equipment to quickly restore all service when a large geographic area is involved.

The coordination of preventive maintenance scheduling with reliability analysis can be very effective. Most utilities design their systems to a specific contingency level, for example, single contingency, so that, due to existing sufficient redundancy and switching alternatives, the failure of

TABLE 11.2

Detailed Industrial Service Interruption Cost Example[a]

Industry	Overlapped Duration (h)	Downtime (h)	Normal Production (h/year)	Fraction of Annual Production Loss	Value-Added Lost[a]	Payroll Lost[a]	Cleanup and Spoil Production[a]	Standby Power Cost[a]	Interruption Cost Lower[a]	Upper[a]	Lower ($/kWh)	Upper ($/kWh)
Food	4	6.00	2,016	0.00298	4,260	1,812	279	0.00	2,091	4,539	2.38	5.17
Tobacco	4	6.00	2,016	0.00298	0	0	0	0.00	0	0	0.00	0.00
Textiles	4	76.00	8,544	0.00890	10,172	5,262	150	0.00	5,413	10,323	10.35	19.74
Apparel	4	6.00	2,016	0.00298	25,309	1,358	83	0.00	1,441	25,391	5.54	97.69
Lumber	4	5.25	2,016	0.00260	1,133	617	248	0.08	865	1,381	3.83	6.12
Furniture	4	6.00	2,016	0.00298	1,074	527	52	0.00	579	1,127	3.51	6.83
Paper	4	14.00	8,544	0.00164	3,006	1,363	144	0.57	1,508	3,151	0.98	2.05
Printing	4	6.00	8,544	0.00070	1,146	569	127	0.00	696	1,273	1.74	3.19
Chemicals	4	24.00	8,544	0.00281	3,899	1,102	27	0.16	1,129	3,925	2.63	9.13
Petroleum refining	4	6.00	8,544	0.00070	888	439	32	0.00	471	919	6.48	6.74
Rubber and plastics	4	6.00	8,544	0.00070	592	325	38	0.00	363	630	2.11	4.12
Leather	4	5.25	2,016	0.00260	1,765	757	563	0.19	1,321	2,328	3.05	5.29
Stone, clay, glass	4	7.75	8,544	0.00091	925	562	380	0.20	942	1,306	2.58	4.54
Primary metal	4	5.25	2,016	0.00061	1,731	818	688	0.23	1,507	2,419	1.71	2.37
Nonelectric machinery	4	5.25	4,864	0.00108	4,851	2,192	944	0.32	3,137	5,795	2.41	3.86
Electric machinery	4	6.00	8,544	0.00070	2,322	1,069	246	0.29	1,315	2,568	3.65	6.75
Transportation equipment	4	5.25	8,544	0.00061	1,739	1,005	858	0.88	1,864	2,598	1.68	3.20
Measuring equipment	4	6.00	4,864	0.00123	2,565	1,112	104	0.00	1,215	2,669	2.34	3.26
Miscellaneous manufacturing	4	6.00	4,864	0.00123	1,817	794	97	0.11	891	1,914	3.72	8.17
Agriculture									21	21	2.90	7.79
					69,293	21,779	5,059	3.07	26,861	74,375	2.81	7.79

Source: The National Electric Reliability Study: Technical Study Reports, U.S. Department of Energy DOE/EP-0005, April 1981.

[a] In thousands of dollars.

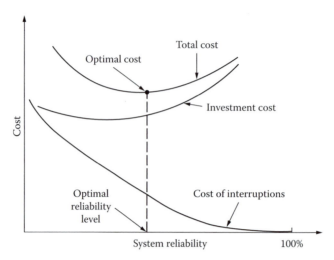

FIGURE 11.4   Cost versus system reliability.

a single component will not cause any customer outages. Therefore, contingency analysis helps to determine the weakest spots of the distribution system.

The special form of contingency analysis in which the probability of a given contingency is clearly and precisely expressed is known as the *risk analysis*. The risk analysis is performed only for important segments of the system and/or customers. The resultant information is used in determining whether to build the system to a specific contingency level or to risk a service interruption. Figure 11.5 shows the flowchart of a reliability planning procedure.

11.4   BASIC RELIABILITY CONCEPTS AND MATHEMATICS

Endrenyi [2] gives the classical definition of reliability as "*the probability of a device or system performing its function adequately, for the period of time intended, under the operating conditions intended.*" In this sense, not only the probability of failure but also its magnitude, duration, and frequency are important.

11.4.1   GENERAL RELIABILITY FUNCTION

It is possible to define the probability of failure of a given component (or system) as a function of time as

$$P(T \leq t) = F(t) \quad t \geq 0 \tag{11.1}$$

where
   T is a random variable representing the failure time
   $F(t)$ is the probability that component will fail by time t

Here, $F(t)$ is the failure distribution function, which is also known as the *unreliability function*. Therefore, the probability that the component will not fail in performing its intended function at a given time t is defined as the *reliability of the component*. Thus, the *reliability function* can be expressed as

$$R(t) = 1 - F(t)$$

$$= P(T > t) \tag{11.2}$$

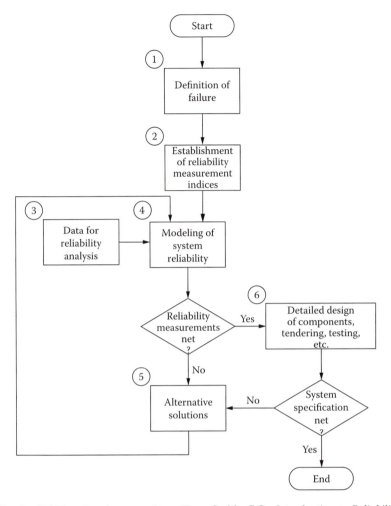

FIGURE 11.5 A reliability planning procedure. (From Smith, C.O., *Introduction to Reliability in Design*, McGraw-Hill, New York, 1976; Albrect, P.F., Overview of power system reliability, *Workshop Proceedings: Power System Reliability-Research Needs and Priorities*, EPRI Report WS-77-60, Palo Alto, CA, October 1978.)

where
  $R(t)$ is the reliability function
  $F(t)$ is the unreliability function

Note that the $R(t)$ reliability function represents the probability that the component will survive at time t.

If the time-to-failure random variable T has a density function $f(t)$, from Equation 11.2,

$$R(t) = 1 - F(t)$$

$$= 1 - \int_0^t f(t)dt$$

$$= \int_t^\infty f(t)dt \qquad (11.3)$$

Therefore, the probability of failure of a given system in a particular time interval (t_1, t_2) can be given either in terms of the unreliability function, as

$$\int_{t_1}^{t_2} f(t)dt = \int_{-\infty}^{t_2} f(t)dt - \int_{-\infty}^{t_1} f(t)dt$$

$$= F(T_2) - F(t_1) \qquad (11.4)$$

or in terms of the reliability function, as

$$\int_{t_1}^{t_2} f(t)dt = \int_{t_1}^{\infty} f(t)dt - \int_{t_2}^{\infty} f(t)dt$$

$$= R(t_1) - R(t_2) \qquad (11.5)$$

Here, the rate at which failures happen in a given time interval (t_1, t_2) is defined as the *hazard rate*, or *failure rate*, during that interval. It is the probability that a failure per unit time happens in the interval, provided that a failure has not happened before the time t_1, that is, at the beginning of the time interval. Therefore,

$$h(t) = \frac{R(t_1) - R(t_2)}{(t_2 - t_1)R(t_1)} \qquad (11.6)$$

If the time interval is redefined so that

$$t_1 = t$$

$$t_2 = t + \Delta t$$

or

$$\Delta t = t_2 - t_1$$

then since the hazard rate is the instantaneous failure rate, it can be defined as

$$h(t) = \lim_{\Delta t \to 0} \frac{P\{\text{a component of age } t \text{ will fail in } \Delta t \mid \text{it has survived up to } t\}}{\Delta t} \qquad (11.7)$$

or

$$h(t) = \lim_{\Delta t \to 0} \frac{R(t) - R(t + \Delta t)}{\Delta t \cdot R(t)}$$

$$= \frac{1}{R(t)}\left[-\frac{d}{dt}R(t)\right]$$

$$= \frac{f(t)}{R(t)} \qquad (11.8)$$

where $f(t)$ is the probability density function

$$= -\frac{dR(t)}{dt}$$

Also, by substituting Equation 11.3 into Equation 11.8,

$$h(t) = \frac{f(t)}{1 - F(t)} \tag{11.9}$$

Therefore,

$$h(t)dt = \frac{dF(t)}{1 - F(t)} \tag{11.10}$$

or

$$\int_0^t h(t)dt = -\text{In}[1 - F(t)] \, |_0^t \tag{11.11}$$

Hence,

$$\text{In}\frac{1 - F(t)}{1 - F(0)} = -\int_0^t h(t)\,dt \tag{11.12}$$

or

$$1 - F(t) = \exp\left[-\int_0^t h(t)\,dt\right] \tag{11.13}$$

Taking derivatives of Equation 11.13 or substituting Equation 11.13 into Equation 11.9,

$$f(t) = h(t)\exp\left[-\int_0^t h(t)dt\right] \tag{11.14}$$

Also, substituting Equation 11.3 into Equation 11.13,

$$R(t) = \exp\left[-\int_0^t h(t)dt\right] \tag{11.15}$$

where

$$\exp[] = e^{[]} \tag{11.16}$$

Let

$$\lambda(t) = h(t) \tag{11.17}$$

hence Equation 11.16 becomes

$$R(t) = \exp\left[-\int_0^t \lambda(t)dt\right] \tag{11.18}$$

Equation 11.18 is known as the *general reliability function*. Note that in Equation 11.18 both the reliability function and the hazard (or failure) rate are functions of time.

Assume that the hazard or failure function is independent of time, that is,

$$h(t) = \lambda \text{ failures/unit time}$$

From Equation 11.14, the failure density function is

$$f(t) = \lambda e^{-\lambda t} \tag{11.19}$$

Therefore, from Equation 11.8, the reliability function can be expressed as

$$R(t) = \frac{f(t)}{h(t)}$$

$$= e^{-\lambda t} \tag{11.20}$$

which is independent of time. Thus, a constant failure rate causes the time-to-failure random variable to be an exponential density function.

Figure 11.6 shows a typical hazard function known as the *bathtub curve*. The curve illustrates that the failure rate is a function of time. The first period represents the *infant mortality* period, which is the period of decreasing failure rate. This initial period is also known as the *debugging period, break-in period, burn-in period, or early life period*. In general, during this period, failures occur due to design or manufacturing errors.

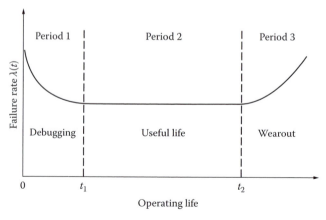

FIGURE 11.6   The *bathtub* hazard function.

The second period is known as the *useful life period*, or *normal operating period*. The failure rates of this period are constant, and the failures are known as *chance failures, random failures*, or *catastrophic failures* since they occur randomly and unpredictably.

The third period is known as the *wear-out period*. Here, the hazard rate increases as equipment deteriorates because of aging or wear as the components approach their "rated lives." If the time t_2 could be predicted with certainty, then equipment could be replaced before this wear-out phase begins.

In summary, since the probability density function is given as

$$f(t) = -\frac{dR(t)}{dt} \tag{11.21}$$

it can be shown that

$$f(t)dt = -dR(t) \tag{11.22}$$

and by integrating Equation 11.22,

$$\int_0^t f(t)dt = -\int_0^{R(t)} R(t)\,dt$$

$$= -[R(t)-1]$$

$$= 1 - R(t) \tag{11.23}$$

However,

$$\int_0^t f(t)\,dt + \int_t^\infty f(t)dt = \int_0^\infty f(t)dt \underline{\underline{\Delta}} 1 \tag{11.24}$$

From Equation 11.24,

$$\int_0^t f(t)\,dt = 1 - \int_t^\infty f(t)\,dt \tag{11.25}$$

Therefore, from Equations 11.23 and 11.25, *reliability* can be expressed as

$$R(t) = \int_t^\infty f(t)\,dt \tag{11.26}$$

However,

$$R(t) + Q(t)\underline{\underline{\Delta}}1 \tag{11.27}$$

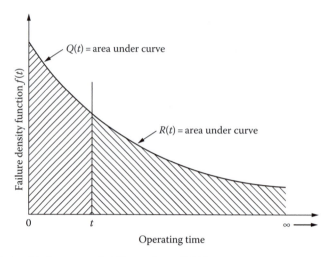

FIGURE 11.7   Relationship between reliability and unreliability.

Thus, the *unreliability* can be expressed as

$$Q(t) = 1 - R(t)$$

$$= 1 - \int_{t}^{\infty} f(t)\,dt$$

or

$$Q(t) = \int_{0}^{t} f(t)\,dt \tag{11.28}$$

Therefore, the relationship between reliability and unreliability can be illustrated graphically, as shown in Figure 11.7.

11.4.2 Basic Single-Component Concepts

Theoretically, the *expected life*, that is, the expected time during which a component will survive and perform successfully, can be expressed as

$$E(T) = \int_{0}^{\infty} t f(t)\,dt \tag{11.29}$$

Substituting Equation 11.21 into Equation 11.29,

$$E(T) = -\int_{0}^{\infty} t \frac{dR(t)}{dt}\,dt \tag{11.30}$$

Integrating by parts,

$$E(T) = -tR(t)\,|_0^\infty + \int_0^\infty R(t)\,dt \qquad (11.31)$$

since

$$R_{(t=0)} = 1 \qquad (11.32)$$

and

$$R_{(t=\infty)} = 0 \qquad (11.33)$$

the first term of Equation 11.31 equals zero, and therefore, the *expected life* can be expressed as

$$E(T) = \int_0^\infty R(t)\,dt \qquad (11.34a)$$

or

$$E(T) = \int_0^\infty \left\{ \exp\left[-\int_0^t \lambda(t)dt \right] \right\} dt \qquad (11.34b)$$

The special case of useful life can be expressed, when there is a constant failure rate, by substituting Equation 11.20 into Equation 11.34a, as

$$E(T) = \int_0^\infty e^{-\lambda t}\,dt = \frac{1}{\lambda} \qquad (11.35)$$

Note that if the system in question is not renewed through maintenance and repairs but simply replaced by a good system, then the $E(T)$ useful life is also defined as the *mean time to failure* and denoted as

$$\mathrm{MTTF} = \bar{m} = \frac{1}{\lambda} \qquad (11.36)$$

where λ is the constant failure rate.

Similarly, if the system in question is renewed through maintenance and repairs, then the $E(T)$ useful life is also defined as the *mean time between failures* and denoted as

$$\mathrm{MTBF} = \bar{T} = \bar{m} = \bar{r} \qquad (11.37)$$

where
$\bar{T}$ is the mean cycle time
$\bar{m}$ is the mean time to failure
$\bar{r}$ is the mean time to repair

Note that the *mean time to repair* is defined as the *reciprocal of the average* (or mean) *repair rate* and denoted as

$$\text{MTTR} = \bar{r} = \frac{1}{\mu} \tag{11.38}$$

where μ is the mean repair rate.

Consider the two-state model shown in Figure 11.8a. Assume that the system is either in the *up* (or in) state or in the *down* (or out) state at a given time, as shown in Figure 11.8b. Therefore, the *mean time to failure* can be reasonably estimated as

$$\text{MTTF} = \bar{m} = \frac{\sum_{i=1}^{n} m_i}{n} \tag{11.39}$$

where

$\bar{m}$ is the mean time to failure
m_i is the observed time to failure for ith cycle
n is the total number of cycles

Similarly, the *mean time to repair* can be reasonably estimated as

$$\text{MTTR} = \bar{r} = \frac{\sum_{i=1}^{n} r_i}{n} \tag{11.40}$$

where

$\bar{r}$ is the mean time to repair
r_i is the time to repair for ith cycle
n is the total number of cycles

Therefore, Equation 11.37 can be reexpressed as

$$\text{MTBF} = \text{MTTF} + \text{MTTR} \tag{11.41}$$

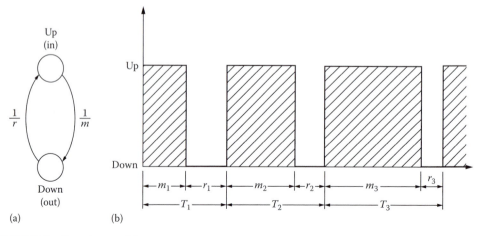

FIGURE 11.8 Two-state model.

The assumption that the behaviors of a repaired system and a new system are identical from a failure standpoint constitutes the base for much of renewal theory. In general, however, perfect renewal is not possible, and in such cases, terms such as the *mean time to the first failure* or the *mean time to the second failure* become appropriate.

Note that the term *mean cycle time* defines the average time that it takes for the component to complete one cycle of operation, that is, failure, repair, and restart. Therefore,

$$\bar{T} = \bar{m} + \bar{r} \tag{11.42}$$

Substituting Equations 11.36 and 11.38 into Equation 11.42,

$$\bar{T} = \frac{1}{\lambda} + \frac{1}{\mu}$$

or (11.43)

$$\bar{T} = \frac{\lambda + \mu}{\lambda \mu}$$

The reciprocal of the mean cycle time is defined as the *mean failure frequency* and denoted as

$$\bar{f} = \frac{1}{\bar{T}}$$

or (11.44)

$$\bar{f} = \frac{\lambda \mu}{\lambda + \mu}$$

When the states of a given component, over a period to time, can be characterized by the two-state model, as shown in Figure 11.8, then it can be assumed that the component is either *up* (i.e., available for service) or *down* (i.e., unavailable for service). Therefore, it can be shown that

$$A + U = 1 \tag{11.45}$$

where
A is the availability of component, that is, the fraction of time component is up
$U = \bar{A}$ is the unavailability of component, that is, the fraction of time component is down

Therefore, on the average, as time t goes to infinity, it can be shown that the *availability* is

$$A \triangleq \frac{\bar{m}}{\bar{T}} = \frac{\text{MTTF}}{\text{MTBF}} \tag{11.46}$$

or

$$A \triangleq \frac{\bar{m}}{\bar{m} + \bar{r}} = \frac{\text{MTTF}}{\text{MTTF} + \text{MTTR}} \tag{11.47}$$

or

$$A = \frac{\mu}{\mu + \lambda} \tag{11.48}$$

Thus, the *unavailability* can be expressed as

$$U \triangleq 1 - A \tag{11.49}$$

Substituting Equation 11.46 into Equation 11.49,

$$U = 1 - \frac{\bar{m}}{\bar{T}}$$

$$= \frac{\bar{T} - \bar{m}}{\bar{T}}$$

$$= \frac{(\bar{m} + \bar{r}) - \bar{m}}{\bar{T}}$$

$$= \frac{\bar{r}}{\bar{T}} = \frac{\text{MTTR}}{\text{MTBF}} \tag{11.50}$$

or

$$U = \frac{\bar{r}}{\bar{r} + \bar{m}} = \frac{\text{MTTR}}{\text{MTTF} + \text{MTTR}} \tag{11.51}$$

or

$$U = \frac{\lambda}{\lambda + \mu} \tag{11.52}$$

Consider Equation 11.47 for a given system's availability, that is,

$$A = \frac{\text{MTTF}}{\text{MTTF} + \text{MTTR}}$$

when the total number of components involved in the system is quite large and

$$\text{MTTF} \gg \text{MTTR}$$

then the division process becomes considerably tedious. However, it is possible to use an approximation form. Therefore, from Equation 11.47,

$$\frac{\text{MTTF}}{\text{MTTF} + \text{MTTR}} = 1 - \frac{\text{MTTR}}{\text{MTTF}} + \cdots (-1)^n \frac{(\text{MTTR})^n}{(\text{MTTF})^n} \tag{11.53}$$

or

$$\frac{\text{MTTF}}{\text{MTTF} + \text{MTTR}} = \sum_{n=0}^{\infty} (-1)^n \frac{(\text{MTTR})^n}{(\text{MTTF})^n} \tag{11.54}$$

or, approximately,

$$\frac{\text{MTTF}}{\text{MTTF} + \text{MTTR}} \cong 1 - \frac{\text{MTTR}}{\text{MTTF}} \tag{11.55}$$

It is somewhat unfortunate, but it has become customary in certain applications, for example, nuclear power plant reliability studies, to employ the MTBF for both unrepairable components and repairable equipment and systems. In any event, however, it represents the same statistical concept of the mean time at which failures occur. Therefore, using this concept, for example, the *availability* is

$$A = \frac{\text{MTBF}}{\text{MTBF} + \text{MTTR}} \tag{11.56}$$

and the *unavailability* is

$$U = \frac{\text{MTTR}}{\text{MTTR} + \text{MTBF}} \tag{11.57}$$

11.5 SERIES SYSTEMS

11.5.1 Unrepairable Components in Series

Figure 11.9 shows a block diagram for a series system that has two components connected in series. Assume that the two components are independent. Therefore, to have the system operate and perform its designated function, both components (or subsystems) must operate successfully. Thus,

$$R_{\text{sys}} = P[E_1 \cap E_2] \tag{11.58}$$

and since it is assumed that the components are independent,

$$R_{\text{sys}} = P(E_1)P(E_2) \tag{11.59}$$

or

$$R_{\text{sys}} = R_1 \times R_2$$

or

$$R_{\text{sys}} = \prod_{i=1}^{2} R_i \tag{11.60}$$

where
   E_i is the event that component i (or subsystem i) operates successfully
   $R_i = P(E_i)$ is the reliability of component i (or subsystem i)
   R_{sys} is the reliability of system (or system reliability index)

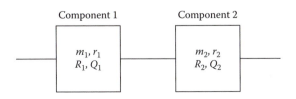

FIGURE 11.9 Block diagram of a series system with two components.

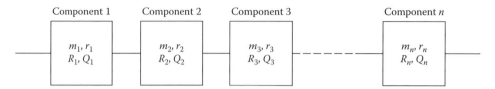

FIGURE 11.10   Block diagram of a series system with n components.

To generalize this concept, consider a series system with n independent components, as shown in Figure 11.10. Therefore, the system reliability can be expressed as

$$R_{sys} = P\left[E_1 \cap E_2 \cap E_3 \cap \cdots \cap E_n\right] \tag{11.61}$$

and since the n components are independent,

$$R_{sys} = P(E_1)P(E_2)P(E_3)\cdots P(E_n) \tag{11.62}$$

or

$$R_{sys} = R_1 \times R_2 \times R_3 \times \cdots \times R_n$$

or

$$R_{sys} = \prod_{i=1}^{n} R_i \tag{11.63}$$

Note that Equation 11.63 is known as the *product rule* or the *chain rule of reliability. System reliability will always be less than or equal to the least-reliable component*, that is,

$$R_{sys} \leq \min_{i}\{R_i\} \tag{11.64}$$

Therefore, the system reliability, due to the characteristic of the series system, is the function of the number of series components and the component reliability level. Thus, the reliability of a series system can be improved by (1) decreasing the number of series components or (2) increasing the component reliabilities. This concept has been illustrated in Figure 11.11.

Assume that the probability that a component will fail is q and it is the same for all components for a given series system. Therefore, the system reliability can be expressed as

$$R_{sys} = (1 - q)^n \tag{11.65}$$

or, according to the binomial theorem,

$$R_{sys} = 1 + n(-q)^1 + \frac{n(n-1)}{2}(-q)^2 + \cdots + (-q)^n \tag{11.66}$$

If the probability of the component failure q is small, an approximate form for the system reliability, from Equation 11.66, can be expressed as

$$R_{sys} \cong 1 - nq \tag{11.67}$$

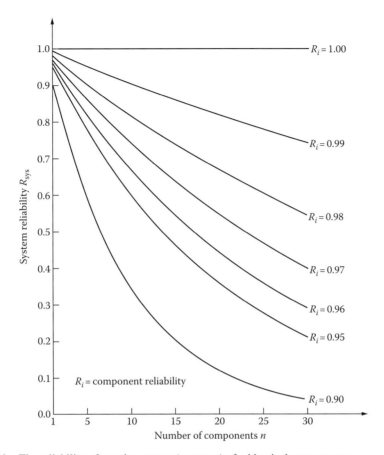

FIGURE 11.11 The reliability of a series system (structure) of n identical components.

where
  n is the total number of components connected in series in the system
  q is the probability of component failure

If the probabilities of component failures, that is, q_i's, are different for each component, then the approximate form of the system reliability can be expressed as

$$R_{sys} \cong 1 - \sum_{i-1}^{n} q_i \qquad (11.68)$$

Example 11.1

Assume that 15 identical components are going to be connected in series in a given system. If the minimum acceptable system reliability is 0.99, determine the approximate value of the component reliability.

Solution

From Equation 11.67,

$$R_{sys} \cong 1 - nq$$

$$0.99 = 1 - 5(q)$$

and

$$q = 0.0007$$

Therefore, the approximate value of the component reliability required to meet the particular system reliability can be found as

$$R_i \cong 0.9993$$

11.5.2 Repairable Components in Series*

Consider a series system with two components, as shown in Figure 11.9. Assume that the components are independent and repairable. Therefore, the *availability* or the *steady-state probability of success* (i.e., operation) of the system can be expressed as

$$A_{sys} = A_1 \times A_2 \tag{11.69}$$

where
    A_{sys} is the availability of system
    A_1 is the availability of component 1
    A_2 is the availability of component 2

Since

$$A_1 = \frac{\bar{m}_1}{\bar{m}_1 + \bar{r}_1} \tag{11.70}$$

and

$$A_2 = \frac{\bar{m}_2}{\bar{m}_2 + \bar{r}_2} \tag{11.71}$$

substituting Equations 11.70 and 11.71 into Equation 11.69 gives

$$A_{sys} = \frac{\bar{m}_1}{\bar{m}_1 + \bar{r}_1} \times \frac{\bar{m}_2}{\bar{m}_2 + \bar{r}_2} \tag{11.72}$$

or

$$A_{sys} = \frac{\bar{m}_{sys}}{\bar{m}_{sys} + \bar{r}_{sys}} \tag{11.73}$$

where
    $\bar{m}_1$ is the mean time to failure of component 1
    $\bar{m}_2$ is the mean time to failure of component 2
    $\bar{m}_{sys}$ is the mean time to failure of system
    $\bar{r}_1$ is the mean time to repair of component 1
    $\bar{r}_2$ is the mean time to repair of component 2
    $\bar{r}_{sys}$ is the mean time to repair of system

---

* The technique presented in this section is primarily based on Ref. [10], by Billinton, Ringlee, and Wood.

The *average frequency of the system failure* is the sum of the average frequency of component 1 failing, given that component 2 is operable, plus the average frequency of component 2 failing while component 1 is operable. Thus,

$$\overline{f}_{sys} = A_2 \times \overline{f}_1 + A_1 \times \overline{f}_2 \tag{11.74}$$

where
$\overline{f}_{sys}$ is the average frequency of system failure
$\overline{f}_i$ is the average frequency of failure of component i
A_i is the availability of component i

Since

$$\overline{f}_i = \frac{1}{\overline{m}_i + \overline{r}_i} \tag{11.75}$$

and

$$A_i = \frac{\overline{m}_i}{\overline{m}_i + \overline{r}_i} \tag{11.76}$$

substituting Equations 11.75 and 11.76 into Equation 11.74 gives

$$\overline{f}_{sys} = \frac{1}{\overline{m}_1 + \overline{r}_1} \times \frac{\overline{m}_2}{\overline{m}_2 + \overline{r}_2} + \frac{1}{\overline{m}_2 + \overline{r}_2} \times \frac{\overline{m}_1}{\overline{m}_1 + \overline{r}_1} \tag{11.77}$$

Note that Equation 11.73 can be expressed as

$$A_{sys} = \overline{m}_{sys} \times \overline{f}_{sys} \tag{11.78}$$

Thus, the *mean time to failure* for a given series system with two components can be expressed as

$$\overline{m}_{sys} = \frac{1}{1/\overline{m}_1 + 1/\overline{m}_2} \tag{11.79}$$

Hence, the *mean time to failure* of a given series system with n components can be expressed as

$$\overline{m}_{sys} = \frac{1}{1/\overline{m}_1 + 1/\overline{m}_2 + \cdots + 1/\overline{m}_n} \tag{11.80}$$

Since the reciprocal of the mean time to failure is defined as the *failure rate*, for the two-component system,

$$\lambda_{sys} = \lambda_1 + \lambda_2 \tag{11.81}$$

and for the *n*-component system,

$$\lambda_{sys} = \lambda_1 + \lambda_2 + \lambda_3 + \cdots + \lambda_n \tag{11.82}$$

Similarly, it can be shown that the *mean time to repair* for the given two-component series system is

$$\bar{r}_{\text{sys}} = \frac{\lambda_1 \bar{r}_1 + \lambda_2 \bar{r}_2 + (\lambda_1 \bar{r}_1)(\lambda_2 \bar{r}_2)}{\lambda_{\text{sys}}} \tag{11.83}$$

or, approximately,*

$$\bar{r}_{\text{sys}} = \frac{\lambda_1 \bar{r}_1 + \lambda_2 \bar{r}_2}{\lambda_{\text{sys}}} \tag{11.84}$$

Therefore, the *mean time to repair* for an *n*-component series system is

$$\bar{r}_{\text{sys}} = \frac{\lambda_1 \bar{r}_1 + \lambda_2 \bar{r}_2 + \lambda_3 \bar{r}_3 + \cdots + \lambda_n \bar{r}_n}{\lambda_{\text{sys}}} \tag{11.85}$$

11.6 PARALLEL SYSTEMS

11.6.1 Unrepairable Components in Parallel

Figure 11.12 shows a block diagram for a system that has two components connected in parallel. Assume that the two components are independent. Therefore, to have the system fail and not be able to perform its designated function, both components must fail simultaneously. Thus, the *system unreliability* is

$$Q_{\text{sys}} = P[\bar{E}_1 \cap \bar{E}_2] \tag{11.86}$$

and since it is assumed that the components are independent,

$$Q_{\text{sys}} = P(\bar{E}_1)P(\bar{E}_2) \tag{11.87}$$

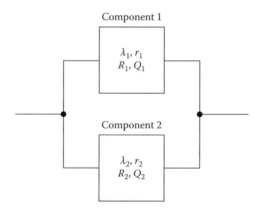

FIGURE 11.12 Block diagram of a parallel system with two components.

---

* Note that Equation 11.84 gives an exact value if there is a dependency between the components; that is, one component must not fail while the other component is on repair.

or

$$Q_{\text{sys}} = \prod_{i=1}^{2}(1 - R_i) \tag{11.88}$$

where
$\bar{E}_i$ is the event that component i fails
$Q_i = P(\bar{E}_i)$ is the unreliability of component i
Q_{sys} is the unreliability of system (or system unreliability index)

Then the *system reliability* is given by the complementary probability as

$$R_{\text{sys}} = 1 - \prod_{i=1}^{2}(1 - R_i) \tag{11.89}$$

for this two-unit redundant system.

To generalize this concept, consider a parallel system with m independent components, as shown in Figure 11.13. Therefore, the *system unreliability* can be expressed as

$$Q_{\text{sys}} = P[\bar{E}_1 \cap \bar{E}_2 \cap \bar{E}_3 \cap \cdots \cap \bar{E}_m] \tag{11.90}$$

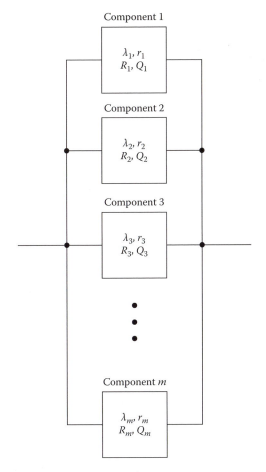

FIGURE 11.13 Block diagram of a parallel system with m components.

and since the m components are independent,

$$Q_{sys} = P(\bar{E}_1)P(\bar{E}_2)P(\bar{E}_3)\cdots P(\bar{E}_m) \tag{11.91}$$

or

$$Q_{sys} = Q_1 \times Q_2 \times Q_3 \times \cdots \times Q_m \tag{11.92}$$

Therefore, the system reliability is

$$
\begin{aligned}
R_{sys} &= 1 - Q_{sys} \\
&= 1 - [Q_1 \times Q_2 \times Q_3 \times \cdots \times Q_m] \\
&= 1 - [(1 - R_1)(1 - R_2)(1 - R_3)\cdots(1 - R_m)] \\
&= 1 - \prod_{i=1}^{m} Q_i
\end{aligned}
$$

or

$$R_{sys} = 1 - \prod_{i=1}^{m}(1 - R_i) \tag{11.93}$$

Note that there is an implied assumption that all units are operating simultaneously and that failures do not influence the reliability of the surviving subsystems.

The instantaneous failure rate of a parallel system is a variable function of the operating time, even though the failure rates and mean times between failures of the particular components are constant. Therefore, the system reliability is the joint function of the mean time between failures of each path and the number of parallel paths. As can be seen in Figure 11.14, for a given component reliability, the marginal gain in the system reliability due to the addition of parallel paths decreases rapidly.

Thus, the greatest gain in system reliability occurs when a second path is added to a single path. The reliability of a parallel system is not a simple exponential but a sum of exponentials. Therefore, the *system reliability* for a two-component parallel system is

$$
\begin{aligned}
R_{sys}(t) &= 1 - (1 - e^{-\lambda_1 t})(1 - e^{-\lambda_2 t}) \\
&= e^{-\lambda_1 t} + e^{-\lambda_2 t} - e^{-(\lambda_1 + \lambda_2)t}
\end{aligned} \tag{11.94}
$$

where
λ_1 is the failure rate of component 1
λ_2 is the failure rate of component 2

11.6.2 REPAIRABLE COMPONENTS IN PARALLEL*

Consider a parallel system with two components as shown in Figure 11.12. Assume that the components are independent and repairable. Therefore, the *unavailability* or the *steady-state probability of failure of the system* can be expressed as

$$U_{sys} = U_1 \times U_2 \tag{11.95}$$

---

* The technique presented in this section is primarily based on Ref. [10], by Billinton, Ringlee, and Wood.

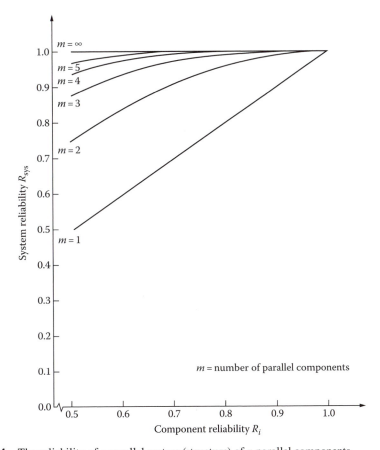

FIGURE 11.14 The reliability of a parallel system (structure) of n parallel components.

where
    U_{sys} is the unavailability of system
    U_1 is the unavailability of component 1
    U_2 is the unavailability of component 2

Since

$$U_1 = 1 - A_1$$

$$= \frac{\lambda_1 \bar{r_1}}{1 + \lambda_1 \bar{r_1}} \tag{11.96}$$

and

$$U_2 = 1 - A_2$$

$$= \frac{\lambda_2 \bar{r_2}}{1 + \lambda_2 \bar{r_2}} \tag{11.97}$$

substituting Equations 11.96 and 11.97 into Equation 11.95 gives

$$U_{sys} = \frac{\lambda_1 \bar{r_1}}{1 + \lambda_1 \bar{r_1}} \times \frac{\lambda_2 \bar{r_2}}{1 + \lambda_2 \bar{r_2}} \tag{11.98}$$

However, the *average frequency of the system failure* is

$$\bar{f}_{sys} = U_2 \bar{f}_1 + U_1 \bar{f}_2 \tag{11.99}$$

where
$\bar{f}_{sys}$ is the average frequency of system failure
$\bar{f}_i$ is the average frequency of failure of component i
U_i is the unavailability of component i

Since

$$\bar{f}_1 = \frac{\lambda_1}{1 + \lambda_1 \bar{r}_1} \tag{11.100}$$

and

$$\bar{f}_2 = \frac{\lambda_2}{1 + \lambda_2 \bar{r}_2} \tag{11.101}$$

substituting equation sets (11.96), (11.97) and (11.100), (11.101) into Equation 11.99 and simplifying gives

$$\bar{f}_{sys} = \frac{\lambda_1 \lambda_2 (\bar{r}_1 + \bar{r}_2)}{(1 + \lambda_1 \bar{r}_1)(1 + \lambda_2 \bar{r}_2)} \tag{11.102}$$

From Equation 11.50, the system *unavailability* can be expressed as

$$U_{sys} \triangleq \frac{\bar{r}_{sys}}{T_{sys}} \tag{11.103}$$

or

$$U_{sys} = \bar{r}_{sys} \times \bar{f}_{sys} \tag{11.104}$$

so that

$$\bar{r}_{sys} = \frac{U_{sys}}{\bar{f}_{sys}} \tag{11.105}$$

Therefore, substituting Equations 11.98 and 11.102 into Equation 11.105, the *average repair time* (or *downtime*) of the two-component parallel system can be expressed as*

$$\bar{r}_{sys} = \frac{\bar{r}_1 \times \bar{r}_2}{\bar{r}_1 + \bar{r}_2} \tag{11.106}$$

---

* Notice the analogy between the total repair time and total (or equivalent) resistance value of a parallel connection of two resistors.

or

$$\frac{1}{\bar{r}_{sys}} = \frac{1}{\bar{r}_1} + \frac{1}{\bar{r}_2}$$

(11.107)

Similarly, from Equation 11.51, the system unavailability can be expressed as

$$U_{sys} \triangleq \frac{\bar{r}_{sys}}{\bar{r}_{sys} + \bar{m}_{sys}}$$

(11.108)

from which

$$\bar{m}_{sys} = \frac{\bar{r}_{sys}(1 - U_{sys})}{U_{sys}}$$

(11.109)

Substituting Equations 11.98 and 11.106 into Equation 11.109, the *average time to failure* (or *operation time*, or *uptime*) of the parallel system can be expressed as

$$\bar{m}_{sys} = \frac{1 + \lambda_1 \bar{r}_1 + \lambda_2 \bar{r}_2}{\lambda_1 \lambda_2 (\bar{r}_1 + \bar{r}_2)}$$

(11.110)

The failure rate of the parallel system is

$$\bar{\lambda}_{sys} \triangleq \frac{1}{\bar{m}_{sys}}$$

(11.111)

or

$$\lambda_{sys} = \frac{\lambda_1 \lambda_2 (\bar{r}_1 + \bar{r}_2)}{1 + \lambda_1 \bar{r}_1 + \lambda_2 \bar{r}_2}$$

(11.112)

When more than two identical units are in parallel and/or when the system is not purely redundant, that is, parallel, the probabilities of the states or modes of the system can be calculated by using the binomial distribution or conditional probabilities.

Example 11.2

Figure 11.15 shows a 4 mi long distribution express feeder that is used to provide electric energy to a load center located in the downtown area of Ghost City from the Ghost River Substation. Approximately 1 mi of the feeder has been built underground due to aesthetic considerations in the vicinity of the downtown area, while the rest of the feeder is overhead. The underground

FIGURE 11.15   A 4 mi long distribution express feeder.

feeder has two termination points. On the average, two faults per circuit-mile for the overhead section and one fault per circuit-mile for the underground section of the feeder have been recorded in the last 10 years.

The annual cable termination fault rate is given as 0.3% per cable termination. Furthermore, based on past experience, it is known that, on the average, the repair times for the overhead section, underground section, and each cable termination are 3, 28, and 3 h, respectively. Using the given information, determine the following:

    a. Total annual fault rate of the feeder
    b. Average annual fault restoration time of the feeder in hours
    c. Unavailability of the feeder
    d. Availability of the feeder

Solution

    a. Total annual fault rate of the feeder is

$$\lambda_{FDR} = \sum_{i=1}^{3} \lambda_i = \lambda_{OH} + \lambda_{UG} + 2\lambda_{CT}$$

where
    λ_{OH} is the total annual fault rate of overhead section of feeder
    λ_{UG} is the total annual fault rate of underground section of feeder
    λ_{CT} is the total annual fault rate of cable terminations

Therefore,

$$\lambda_{FDR} = 3\left(\frac{2}{10}\right) + 1\left(\frac{1}{10}\right) + 2(0.003)$$

$$= 0.706 \text{ faults/year}$$

    b. Average fault restoration time of the feeder per fault is

$$\bar{r}_{FDR} = \sum_{i=1}^{3} \bar{r}_i = \bar{r}_{OH} + \bar{r}_{UG} + 2\bar{r}_{CT}$$

where
    $\bar{r}_{OH}$ is the average repair time for overhead section of feeder, h
    $\bar{r}_{UG}$ is the average repair time for underground section of feeder, h
    $\bar{r}_{CT}$ is the average repair time per cable termination, h

Thus,

$$\bar{r}_{FDR} = 3 + 28 + 2(3)$$

$$= 37 \text{ h}$$

However, the average annual fault restoration time of the feeder is

$$r_{FDR} = \frac{\sum_{i=1}^{3} \lambda_i \times r_i}{\sum_{i=1}^{3} \lambda_i}$$

or

$$\bar{r}_{\text{FDR}} = \frac{(l_{\text{OH}} \times \lambda_{\text{OH}})(\bar{r}_{\text{OH}}) + (l_{\text{UG}} \times \lambda_{\text{UG}})(\bar{r}_{\text{UG}}) + (2\lambda_{\text{CT}})(\bar{r}_{\text{CT}})}{\lambda_{\text{FDR}}}$$

$$= \frac{(3 \times 0.2)(3) + (1 \times 0.1)(28) + (2 \times 0.003)(3)}{0.706}$$

$$= \frac{4.618}{0.706}$$

$$= 6.54\,\text{h}$$

c. Unavailability of the feeder is

$$U_{\text{FDR}} = \frac{\bar{r}_{\text{FDR}}}{\bar{r}_{\text{FDR}} + \bar{m}_{\text{FDR}}}$$

where

$$\bar{m}_{\text{FDR}} = \text{annual mean time to failure}$$

$$= 8760 - \bar{r}_{\text{FDR}}$$

$$= 8760 - 6.54$$

$$= 8753.46\,\text{h/year}$$

Therefore,

$$U_{\text{FDR}} = \frac{6.54}{6.54 + 8753.46}$$

$$= 0.0007\% \quad \text{or} \quad 0.07\%$$

d. Availability of the feeder is

$$\lambda_{\text{FDR}} = 1 - U_{\text{FDR}}$$

$$= 1 - 0.0007$$

$$= 0.9993 \quad \text{or} \quad 99.93\%$$

Example 11.3

Assume that the primary main feeder shown in Figure 11.16 is manually sectionalized and that presently only the first three feeder sections exist and serve customers A, B, and C. The annual average fault rates for primary main and laterals are 0.08 and 0.2 fault/circuit-mile, respectively. The average repair times for each primary main section and for each primary lateral are 3.5 and 1.5 h, respectively. The average time for manual sectionalizing of each feeder section is 0.75 h. Assume that at the time of having one of the feeder sections in fault, the other feeder section(s) are sectionalized manually as long as they are not in the mainstream of the fault current, that is, not in between the faulted section and the circuit breaker. Otherwise, they have to be repaired also.

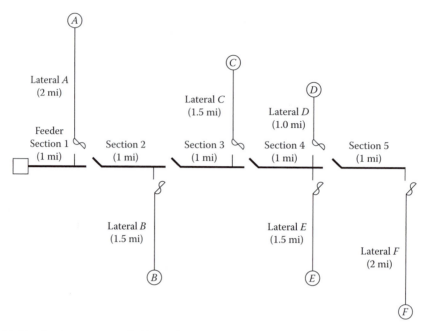

FIGURE 11.16 A primary system for Example 11.3.

Based on the given information, prepare an interruption analysis study for the first contingency only, that is, ignore the possibility of simultaneous outages, and determine the following:

a. The total annual sustained interruption rates for customers A, B, and C.
b. The average annual repair times, that is, downtimes, for customers A, B, and C.
c. Write the necessary codes to solve the problem in MATLAB.

Solution

a. Total annual sustained interruption rates for customers A, B, and C are

$$\lambda_A = \sum_{i=1}^{4} \lambda_i = \lambda_{\text{sec.1}} + \lambda_{\text{sec.2}} + \lambda_{\text{sec.3}} + \lambda_{\text{lat. }A}$$

$$= (1\,\text{mi})(0.08) + (1\,\text{mi})(0.08) + (1\,\text{mi})(0.08) + (2\,\text{mi})(0.2)$$

$$= 0.64\,\text{fault/year}$$

$$\lambda_B = \sum_{i=1}^{4} \lambda_i = \lambda_{\text{sec.1}} + \lambda_{\text{sec.2}} + \lambda_{\text{sec.3}} + \lambda_{\text{lat, }B}$$

$$= (1\,\text{mi})(0.08) + (1\,\text{mi})(0.08) + (1\,\text{mi})(0.08) + (1.5\,\text{mi})(0.2)$$

$$= 0.54\,\text{fault/year}$$

$$\lambda_C = \lambda_B = 0.54\,\text{fault/year}$$

b. Average annual repair time, that is, downtime (or restoration time), for customer A is

$$\bar{r}_A = \frac{\lambda_{\text{sec.1}} \times \bar{r}_{\text{fault}} + \lambda_{\text{sec.2}} \times \bar{r}_{\text{MS}} + \lambda_{\text{sec.3}} \times \bar{r}_{\text{MS}} + \lambda_{\text{lat. }A} \times \bar{r}_{\text{lat fault}}}{\lambda_A}$$

where

$\bar{r}_A$ average repair time for customer A

$\lambda_{sec.\,i}$ is the total fault rate for feeder section i per year

$\bar{r}_{fault}$ is the average repair time for faulted primary main section

$\bar{r}_{lat.\,fault}$ is the average repair time for faulted primary lateral

$\bar{r}_{MS}$ is the average time for manual sectionalizing per section

$\lambda_{lat.\,A}$ is the total fault rate for lateral A per year

Therefore,

$$\bar{r}_A = \frac{(0.08)(3.5 + 0.08)(0.75) + (0.08)(0.75) + (2 \times 0.2)(1.5)1.00}{0.64}$$

$$= \frac{1.00}{0.64}$$

$$= 1.56 \text{ h}$$

Similarly, for customer B,

$$\bar{r}_B = \frac{\lambda_{sec.1} \times \bar{r}_{fault} + \lambda_{sec.2} \times \bar{r}_{MS} + \lambda_{sec.3} \times \bar{r}_{MS} + \lambda_{lat.\,B} \times \bar{r}_{lat,fault}}{\lambda_B}$$

$$= \frac{(0.08)(3.5) + 0.08)(3.5) + (0.08)(0.75) + (1.5 \times 0.2)(1.5)}{0.54}$$

$$= \frac{1.07}{0.54}$$

$$= 1.98 \text{ h}$$

and for customer C,

$$\bar{r}_C = \frac{\lambda_{sec.1} \times \bar{r}_{fault} + \lambda_{sec.2} \times \bar{r}_{MS} + \lambda_{sec.3} \times \bar{r}_{MS} + \lambda_{lat.\,C} \times \bar{r}_{lat,fault}}{\lambda_C}$$

$$= \frac{(0.08)(3.5) + (0.08)(3.5) + (0.08)(3.5) + (1.5 \times 0.2)(1.5)}{0.54}$$

$$= \frac{1.29}{0.54}$$

$$= 2.39 \text{ h}$$

c. Here is the MATLAB script:

~~~~~~

```
clc
clear

% System parameters

% failure rates
lambda_sec1 = 1*0.08;
lambda_sec2 = lambda_sec1;
lambda_sec3 = lambda_sec1;
lambda_sec4 = lambda_sec1;
lambda_latA = 2*0.2;
lambda_latB = 1.5*0.2;
lambda_latC = 1.5*0.2;
```

```
lambda_latD = 1*0.2;
lambda_latE = 1.5*0.2;
lambda_latF = 2*0.2;

% repair times
r_sec = 3.5;
r_lat = 1.5;
r_MS = 0.75;% manual sectionalizing

% Solution to part a

% Total annual sustained interruption rates for customers A, B and C
lambdaA = lambda_sec1 + lambda_sec2 + lambda_sec3 + lambda_latA
lambdaB = lambda_sec1 + lambda_sec2 + lambda_sec3 + lambda_latB
lambdaC = lambdaB

% Solution to part b

% Average annual repair time for customers A, B and C
rA = (lambda_sec1*r_sec + lambda_sec2*r_MS + lambda_sec3*r_MS +
  lambda_latA*r_lat)/lambdaA
rB = (lambda_sec1*r_sec + lambda_sec2*r_sec + lambda_sec3*r_MS +
  lambda_latB*r_lat)/lambdaB
rC = (lambda_sec1*r_sec + lambda_sec2*r_sec + lambda_sec3*r_sec +
  lambda_latC*r_lat)/lambda
```

11.7   SERIES AND PARALLEL COMBINATIONS

Simple combinations of series and parallel subsystems (or components) can be analyzed by successively reducing subsystems into equivalent parallel or series components.

Figure 11.17 shows a parallel–series system that has a *high-level redundancy*. The equivalent reliability of the system with m parallel paths of n components each can be expressed as

$$R_{sys} = 1 - (1 - R^n)^m \tag{11.113}$$

where
   R_{sys} is the equivalent reliability of system
   R^n is the equivalent reliability of a path
   R is the reliability of a component
   n is the total number of components in a path
   m is the total number of paths

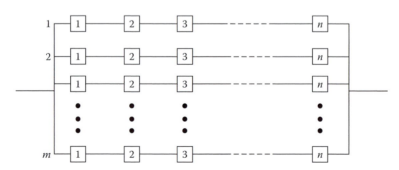

FIGURE 11.17   A parallel–series system.

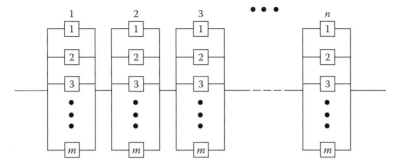

FIGURE 11.18 A series–parallel system.

Figure 11.18 shows a series–parallel system that has a *low-level redundancy*. The equivalent reliability of the system of n series units (or banks) with m parallel components in each unit (or bank) can be expressed as

$$R_{sys} = [1 - (1 - R)^m]^n \tag{11.114}$$

where
R_{sys} is the equivalent reliability of system
$1 - (1 - R)^m$ is the equivalent reliability of a parallel unit (or bank)
R is the reliability of a component
m is the total number of components in a parallel unit (or bank)
n is the total number of units (or banks)

The comparison of the two systems shows that the series–parallel configuration provides higher system reliability than the equivalent parallel–series configuration for a given system. Therefore, it can be concluded that the lower the system level at which redundancy is applied, the larger the effective system reliability. The difference between parallel–series and series–parallel systems is not as pronounced if components have high reliabilities.

Example 11.4

Consider the various combinations of the reliability block diagrams shown in Figure 11.19. Assume that they are based on the logic diagrams of each subsystem and that the reliability of each component is 0.85. Determine the equivalent system reliability of each configuration.

Solution

a. From Equation 11.63, the equivalent system reliability for the series system is

$$R_{eq} = R_{sys} = \prod_{i=1}^{4} R_i$$

$$= (0.85)^4$$

$$= 0.5220$$

b. For the parallel–series system from Equation 11.113,

$$R_{eq} = 1 - (1 - R^4)^2$$

$$= 1 - [1 - (0.85)^4]^2$$

$$= 0.7715$$

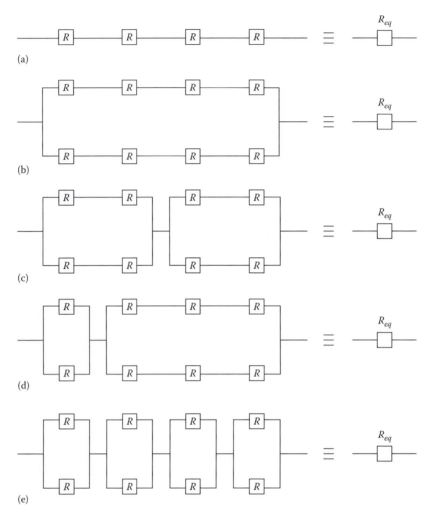

FIGURE 11.19 Various combinations of block diagrams: (a) series, (b) parallel–series, (c) mixed parallel, (d) mixed parallel, and (e) series–parallel.

c. For the mixed-parallel system,

$$R_{eq} = [1-(1-R^2)^2][1-(1-R^2)^2]$$
$$= [1-(1-0.85^2)^2][1-(1-0.85^2)^2]$$
$$= 0.8519$$

d. For the mixed-parallel system,

$$R_{eq} = [1-(1-R)^2][1-(1-R^3)^2]$$
$$= [1-(1-0.85)^2][1-(1-0.85^3)^2]$$
$$= 0.8320$$

e. For the series–parallel system from Equation 11.114,

$$R_{eq} = [1-(1-R)^2]^4$$

$$= [1-(1-0.85)^2]^4$$

$$= 0.9130$$

Example 11.5

Assume that a system has five components, namely, A, B, C, D, and E, as shown in Figure 11.20, and that each component has different reliability as indicated in the figure. Determine the following:

a. The equivalent system reliability.
b. If the equivalent system reliability is desired to beat at least 0.80%, or 80%, design a system configuration to meet this system requirement by using each of the five components at least once.

Solution

a. From Equation 11.63, the equivalent system reliability is

$$R_{eq} = \prod_{i=1}^{5} R_i$$

$$= (0.80)(0.95)(0.99)(0.90)(0.65)$$

$$= 0.4402 \quad \text{or} \quad 44.02\%$$

b. In general, the best way of improving the overall system reliability is to back the less-reliable components by parallel components. Therefore, since the relatively less-reliable components are A and E, they can be backed by parallel redundancy as shown in Figure 11.21. Therefore, the new equivalent system reliability becomes

$$R_{sys} = \prod_{i=1}^{5} R_i$$

$$= [1-(1-0.80)^2](0.95)(0.99)(0.90)[1-(1-0.65)^4]$$

$$= 0.8004 \quad \text{or} \quad 80.04\%$$

Example 11.6

Assume that a three-phase transformer bank consists of three single-phase transformers identified as A, B, and C for the sake of convenience. Assume that (1) transformer A is an old unit and therefore has a reliability of 0.90, (2) transformer B has been in operation for the last 20 years and therefore has been estimated to have a reliability of 0.95, and (3) transformer C is a brand new

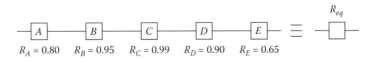

FIGURE 11.20 System configuration.

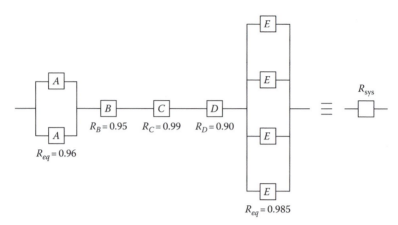

FIGURE 11.21 Imposed system configuration.

one with a reliability of 0.99. Based on the given information and assumption of independence, determine the following:

a. The probability of having no failing transformer at any given time.
b. If one out of the three transformers fails at any given time, what are the probabilities for that unit being the transformer A, or B, or C?
c. If two out of the three transformers fail at any given time, what are the probabilities for those units being the transformers A and B, or B and C, or C and A?
d. What is the probability of having all three transformers out of service at any given time?

Solution

a. The probability of having no failing transformer at any given time is

$$P[A \cap B \cap C] = P(A)P(B)P(C)$$

$$= (0.90)(0.95)(0.99)$$

$$= 0.84645$$

b. If one out of the three transformers fails at any given time, the probabilities for that unit being the transformer A, or B, or C are

$$P[\bar{A} \cap B \cap C] = P(\bar{A})P(B)P(C)$$

$$= (0.10)(0.95)(0.99)$$

$$= 0.09405$$

$$P[A \cap \bar{B} \cap C] = P(A)P(\bar{B})P(C)$$

$$= (0.90)(0.05)(0.99)$$

$$= 0.04455$$

$$P[A \cap B \cap \bar{C}] = P(A)P(B)P(\bar{C})$$

$$= (0.90)(0.95)(0.01)$$

$$= 0.00855$$

c. If two out of the three transformers fail at any given time, the probabilities for those units being the transformers A and B, or B and C, or C and A are

$$P[\bar{A} \cap \bar{B} \cap C] = P(\bar{A})P(\bar{B})P(C)$$

$$= (0.10)(0.05)(0.99)$$

$$= 0.00495$$

$$P[A \cap \bar{B} \cap \bar{C}] = P(A)P(\bar{B})P(\bar{C})$$

$$= (0.90)(0.05)(0.01)$$

$$= 0.00045$$

$$P[\bar{A} \cap B \cap \bar{C}] = P(\bar{A})P(B)P(\bar{C})$$

$$= (0.10)(0.95)(0.01)$$

$$= 0.00095$$

d. The probability of having all three transformers out of service at any given time* is

$$P[\bar{A} \cap \bar{B} \cap \bar{C}] = P(\bar{A})P(\bar{B})P(\bar{C})$$

$$= (0.10)(0.05)(0.01)$$

$$= 0.00005$$

Therefore, the aforementioned reliability calculations can be summarized as given in Table 11.3.

TABLE 11.3
Summary of the Computations

Number of Failed Transformers	System Modes	Probability
0	$A \cap B \cap C$	0.84645
1	$\bar{A} \cap B \cap C$	0.09405
	$A \cap \bar{B} \cap C$	0.04455
	$A \cap B \cap \bar{C}$	0.00855
2	$\bar{A} \cap \bar{B} \cap C$	0.00495
	$A \cap \bar{B} \cap \bar{C}$	0.00045
	$\bar{A} \cap B \cap \bar{C}$	0.00095
3	$\bar{A} \cap \bar{B} \cap \bar{C}$	0.00005
		$\sum = 1.00000$

* Note that as time goes to infinity the reliability goes to zero by definition.

11.8 MARKOV PROCESSES*

A *stochastic process*, $\{X(t); t \in T\}$, is a family of random variables such that for each t contained in the index set T, $X(t)$ is a random variable. Often T is taken to be the set of nonnegative integers.

In reliability studies, the variable t represents time, and $X(t)$ describes the state of the system at time t. The states at a given time t_n actually represent the (exhaustive and mutually exclusive) outcomes of the system at that time. Therefore, the number of possible states may be finite or infinite. For instance, the Poisson distribution

$$P_n(t) = \frac{e^{-\lambda t}(\lambda t)^n}{n!} \quad n = 0, 1, 2,\ldots \tag{11.115}$$

represents a stochastic process with an infinite number of states. If the system starts at time 0, the random variable n represents the number of occurrences between 0 and t. Therefore, the states of the system at any time t are given by $n = 0, 1, 2\ldots$.

A *Markov process* is a stochastic system for which the occurrence of a future state depends on the immediately preceding state and only on it. Because of this reason, the markovian process is characterized by a lack of memory. Therefore, a discrete parameter stochastic process, $\{X(t); t = 0, 1, 2,\ldots\}$, or a continuous parameter stochastic process, $\{X(t); t \geq 0\}$, is a Markov process if it has the following *markovian property*:

$$P\left\{X(t_n) \geq x_n \middle| X(t_1) = x_1,\ X(t_2) = x_2,\ldots,\ X(t_{n-1}) = x_{n-1}\right\}$$

$$= P\left\{X(t_n) = x_n \middle| X(t_{n-1}) = x_{n-1}\right\} \tag{11.116}$$

for any set of n time points, $t_1 < t_2 < \cdots < t_n$ in the index set of the process, and any real numbers $x_1, x_2, \ldots, x_n$. The probability of

$$P_{x_{n-1},x_n} = P\left\{x(t_n) = x_n \middle| X(t_{n-1}) = x_{n-1}\right\} \tag{11.117}$$

is called the *transition probability* and represents the *conditional probability* of the system being in x_n at t_n, given it was x_{n-1} at t_{n-1}. It is also called the *one-step transition probability* due to the fact that it represents the system between t_{n-1}, and t_n. One can define a k-step transition probability as

$$P_{x_n,x_{n+k}} = P\left\{X(t_{n+k}) = x_{n+k} \middle| X(t_n) = x_n\right\} \tag{11.118}$$

or as

$$P_{x_{n-k},x_n} = P\left\{X(t_n) = x_n \middle| X(t_{n-k}) = x_{n-k}\right\} \tag{11.119}$$

A *Markov chain* is defined by a sequence of discrete-valued random variables, $\{X(t_n)\}$, where t_n is discrete-valued or continuous. Therefore, one can also define the Markov chain as the *Markov process with a discrete state space*. Define

$$p_{ij} = P\left\{X(t_n) = j \middle| X(t_{n-1}) = i\right\} \tag{11.120}$$

---

* The fundamental methodology given here was developed by the Russian mathematician A. A. Markov of the University of St. Petersburg around the beginning of the twentieth century.

as the one-step transition probability of going from state i at t_{n-1} to state j at t_n and assume that these probabilities do not change over time. The term used to describe this assumption is *stationarity*. If the transition probability depends only on the time difference, then the Markov chain is defined to be stationary in time. Therefore, a Markov chain is completely defined by its transition probabilities, of going from state i to state j, given in a matrix form:

$$\mathbf{P} = \begin{bmatrix} p_{00} & p_{01} & p_{02} & p_{03} & \cdots & p_{0n} \\ p_{10} & p_{11} & p_{12} & p_{13} & \cdots & p_{1n} \\ p_{20} & p_{21} & p_{22} & p_{23} & \cdots & p_{2n} \\ p_{30} & p_{31} & p_{32} & p_{33} & \cdots & p_{3n} \\ \cdots\cdots\cdots\cdots\cdots\cdots\cdots\cdots\cdots \\ p_{n0} & p_{n1} & p_{n2} & p_{n3} & \cdots & p_{nn} \end{bmatrix} \tag{11.121}$$

The matrix $\mathbf{P}$ is called a *one-step transition matrix* (or *stochastic matrix*) since all the transition probabilities p_{ij}'s are fixed and independent of time. The matrix $\mathbf{P}$ is also called just the *transition matrix* when there is no possibility of confusion. Since the p_{ij}'s are conditional probabilities, they must satisfy the conditions

$$\sum_{j}^{n} p_{ij} = 1 \quad \text{for all } i \tag{11.122}$$

and

$$p_{ij} \geq 0 \quad \text{for all } ij \tag{11.123}$$

where
$i = 0, 1, 2, \ldots, n$
$j = 0, 1, 2, \ldots, n$

Note that when the number of transitions (or states) is not too large, the information in a given transition matrix $\mathbf{P}$ can be represented by a transition diagram. The transition diagram is a pictorial map of the process in which states are represented by nodes and transitions by arrows. Here, the focus is not on time but on the structure of allowable transitions. The arrow from node i to node j is labeled as p_{ij}. Since row i of the matrix $\mathbf{P}$ corresponds to the set of arrows leaving node i, the sum of their probabilities must be equal to unity.

Assume that a given system has two states, namely, state 1 and state 2. For example, here, states 1 and 2 may represent the system being up and down, respectively. Therefore, the associated transition probabilities can be defined as

p_{11} = probability of being in state 1 at time t, given that it was in state 1 at time zero
p_{12} = probability of being in state 2 at time t, given that it was in state 2 at time zero
p_{21} = probability of being in state 2 at time t, given that it was in state 1 at time zero
p_{22} = probability of being in state 1 at time t, given that it was in state 2 at time zero

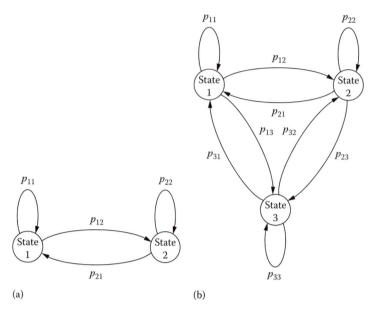

FIGURE 11.22 Transition system (a) for a two-state system and (b) for a three-state system.

Therefore, the associated transition matrix can be expressed as

$$\mathbf{P} = \begin{bmatrix} p_{11} & p_{12} \\ p_{21} & p_{22} \end{bmatrix} \tag{11.124}$$

Figure 11.22a shows the associated transition diagram.

By the same token, if the given system has three states, its transition matrix can be expressed as

$$\mathbf{P} = \begin{bmatrix} p_{11} & p_{12} & p_{13} \\ p_{21} & p_{22} & p_{23} \\ p_{31} & p_{32} & p_{33} \end{bmatrix} \tag{11.125}$$

and its transition diagram can be drawn as shown in Figure 11.22b.

Example 11.7

Based on past history, a distribution engineer of the NL&NP Company has gathered the following information on the operation of the distribution transformers served by the Riverside Substation. The records indicate that only 2% of the transformers that are presently down and therefore being repaired now will be down and therefore will need repair next time. The records also show that 5% of those transformers that are currently up and therefore in service now will be down and therefore will need repair next time. Assuming that the process is discrete, markovian, and has stationary transition probabilities, determine the following:

a. The conditional probabilities
b. The transition matrix
c. The transition diagram

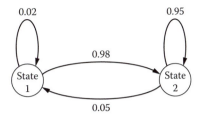

FIGURE 11.23 Transition diagram.

Solution

a. Let t and $t+1$ represent the present time (i.e., *now*) and the next time, respectively. Therefore, the associated conditional probabilities are

$$P\{X_{t+1} = \text{down}|X_t = \text{down}\} = 0.02$$

$$P\{X_{t+1} = \text{up}|X_t = \text{down}\} = 0.98$$

$$P\{X_{t+1} = \text{down}|X_t = \text{up}\} = 0.05$$

$$P\{X_{t+1} = \text{up}|X_t = \text{up}\} = 0.95$$

b. Let numbers 1 and 2 represent the states of down and up, respectively. Therefore, from Equation 11.120 and part a,

$$p_{11} = 0.02 \quad p_{12} = 0.98$$
$$p_{21} = 0.05 \quad p_{22} = 0.95$$

or, from Equation 11.121,

$$\mathbf{P} = \begin{bmatrix} p_{11} & p_{12} \\ p_{21} & p_{22} \end{bmatrix}$$

$$= \begin{bmatrix} 0.02 & 0.98 \\ 0.05 & 0.95 \end{bmatrix}$$

c. Therefore, the transition diagram can be drawn as shown in Figure 11.23.

Example 11.8

Assume that a distribution engineer of the NL&NP Company has studied the feeder outage statistics of the troublesome Riverside Substation and found out (1) that there is a markovian relationship between the feeder outages occurring at the present time and the next time and (2) that the relationship is a stationary one. Assume that the engineer has summarized the findings as shown in Table 11.4. For example, the table shows that if the presently outaged feeder is number 1, then the chances for the next outaged feeder being feeder 1, 2, or 3 are 40%, 30%, and 30%, respectively. Using the given data, determine the following:

a. The conditional outage probabilities
b. The transition matrix
c. The transition diagram

TABLE 11.4

Feeder Outage Data

Presently Outaged Feeder	Chances, in Percent, for the Next Outaged Feeder Being		
	1	2	3
1	40	30	30
2	20	50	30
3	25	25	50

Solution

a. Let t and $t+1$ represent the present time and the next time, respectively. Therefore, the probability of the next outaged feeder being number 1, given it is number 1 now, can be expressed as

$$p_{11} = P\{X_{t+1} = 1 | X_t = 1\} = 0.40$$

where
  X_{t+1} is the outaged feeder at next time
  X_t is the outaged feeder at present time

Similarly,

$$p_{12} = P\{X_{t+1} = 2 | X_t = 1\} = 0.30$$

$$p_{13} = P\{X_{t+1} = 3 | X_t = 1\} = 0.30$$

$$p_{21} = P\{X_{t+1} = 1 | X_t = 2\} = 0.20$$

$$p_{22} = P\{X_{t+1} = 2 | X_t = 2\} = 0.50$$

$$p_{23} = P\{X_{t+1} = 3 | X_t = 2\} = 0.30$$

$$p_{31} = P\{X_{t+1} = 1 | X_t = 3\} = 0.25$$

$$p_{32} = P\{X_{t+1} = 2 | X_t = 3\} = 0.25$$

$$p_{33} = P\{X_{t+1} = 3 | X_t = 3\} = 0.50$$

b. Therefore, the transition matrix is

$$\mathbf{P} = \begin{bmatrix} p_{11} & p_{12} & p_{13} \\ p_{21} & p_{22} & p_{23} \\ p_{31} & p_{32} & p_{33} \end{bmatrix}$$

$$= \begin{bmatrix} 0.40 & 0.30 & 0.30 \\ 0.20 & 0.50 & 0.30 \\ 0.25 & 0.25 & 0.50 \end{bmatrix}$$

c. The associated transition diagram is shown in Figure 11.24.

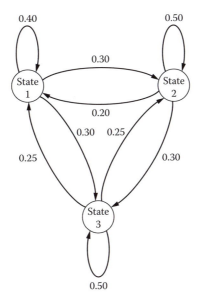

FIGURE 11.24 Transition diagram.

11.8.1 Chapman–Kolmogorov Equations

Assume that S_j represents the exhaustive and mutually exclusive states (outcomes) of a given system at any time, where $j = 0, 1, 9,\dots$ Also assume that the system is markovian and that $p_j^{(0)}$ represents the absolute probability that the system is in state S_j at t_0. Therefore, if $p_j^{(0)}$ and the transition matrix **P** of a given Markov chain are known, one can easily determine the absolute probabilities of the system after n-step transitions. By definition, the one-step transition probabilities are

$$p_{ij} = p_{ij}^{(1)} = P\{X(t_1) = j \,|\, X(t_o) = i\} \tag{11.126}$$

Therefore, the n-step transition probabilities can be defined by induction as

$$p_{ij}^{(n)} = P\{X(t_n) = j \,|\, X(t_o) = i\} \tag{11.127}$$

In other words, $p_{ij}^{(n)}$ is the probability (absolute probability) that the process is in state j at time t_n, given that it was in state i at time t_0. It can be observed from this definition that $p_{ij}^{(0)}$ must be 1 if $i = j$, and 0 otherwise.

The Chapman–Kolmogorov equations provide a method for computing these n-step transition probabilities. In general form, these equations are given as

$$p_{ij}^{(n)} = \sum_k p_{ik}^{(n-m)} \cdot p_{kj}^{(m)} \quad \forall_{ij} \tag{11.128}$$

for any m between zero and n. Note that Equation 11.128 can be represented in matrix form by

$$\mathbf{P}^{(n)} = \mathbf{P}^{(n-m)}\,\mathbf{P}^{(m)} \tag{11.129}$$

Therefore, the elements of a higher-order transition matrix, that is, $\left\|p_{ij}^{(n)}\right\|$, can be obtained directly by matrix multiplication. Hence,

$$\left\|p_{ij}^{(n)}\right\| = \mathbf{P}^{(n-m)}\,\mathbf{P}^{(m)} = \mathbf{P}^{(n)} = \mathbf{P}^n \tag{11.130}$$

Note that a special case of Equation 11.128 is

$$p_{ij}^{(n)} = \sum_k p_{ik}^{(n-1)} \cdot p_{kj} \quad \forall_{ij} \tag{11.131}$$

and therefore, the special cases of Equations 11.129 and 11.130 are

$$\mathbf{P}^{(n)} = \mathbf{P}^{(n-1)}\mathbf{P} \tag{11.132}$$

and

$$\left\|p_{ij}^{(n)}\right\| = \mathbf{P}^{(n-1)}\mathbf{P} = \mathbf{P}^{(n)} = \mathbf{P}^n \tag{11.133}$$

respectively.

The unconditional probabilities such as

$$p_{ij}^{(n)} = P\{X(t_n) = j\} \tag{11.134}$$

are called the *absolute probabilities* or *state probabilities*. To determine the state probabilities, the initial conditions must be known. Therefore,

$$p_{ij}^{(n)} = P\{X(t_n) = j\}$$

$$= P\sum_i \left\{X(t_n) = j \middle| X(t_o) = i\right\} P(t_o) = i\}$$

$$= \sum_i p_i^{(0)} p_{ij}^{(n)} \tag{11.135}$$

Note that Equation 11.135 can be represented in matrix form by

$$\mathbf{p}^{(n)} = \mathbf{p}^{(0)}\mathbf{P}^{(n)} \tag{11.136}$$

where
$\mathbf{p}^{(n)}$ is the vector of state probabilities at time t_n
$\mathbf{p}^{(0)}$ is the vector of initial state probabilities at time t_0
$\mathbf{P}^{(n)}$ is the n-step transition matrix

The state probabilities or absolute probabilities are defined in vector form as

$$\mathbf{p}^{(n)} = \left[p_1^{(n)} p_2^{(n)} p_3^{(n)} \cdots p_k^{(n)} \right] \tag{11.137}$$

and

$$\mathbf{p}^{(0)} = \left[p_1^{(0)} p_2^{(0)} p_3^{(0)} \cdots p_k^{(0)} \right] \tag{11.138}$$

Example 11.9

Consider a Markov chain, with two states, having the one-step transition matrix of

$$\mathbf{P} = \begin{bmatrix} 0.6 & 0.4 \\ 0.3 & 0.7 \end{bmatrix}$$

and the initial state probability vector of

$$\mathbf{p}^{(0)} = [0.8 \quad 0.2]$$

and determine the following:

    a. The vector of state probabilities at time t_1
    b. The vector of state probabilities at time t_4
    c. The vector of state probabilities at time t_8

Solution

    a. From Equation 11.136,

$$\mathbf{p}^{(1)} = \mathbf{p}^{(0)}\mathbf{P}^{(1)}$$

$$= [0.8 \quad 0.2] \begin{bmatrix} 0.6 & 0.4 \\ 0.3 & 0.7 \end{bmatrix}$$

$$= [0.54 \quad 0.46]$$

    b. From Equation 11.136,

$$\mathbf{p}^{(4)} = \mathbf{p}^{(0)}\mathbf{P}^{(4)}$$

       where

$$\mathbf{P}^{(2)} = \mathbf{P}^{(1)}\mathbf{P}^{(1)}$$

$$= \begin{bmatrix} 0.6 & 0.4 \\ 0.3 & 0.7 \end{bmatrix}\begin{bmatrix} 0.6 & 0.4 \\ 0.3 & 0.7 \end{bmatrix}$$

$$= \begin{bmatrix} 0.48 & 0.52 \\ 0.39 & 0.61 \end{bmatrix}$$

       and thus,

$$\mathbf{P}^{(4)} = \mathbf{P}^{(2)}\,\mathbf{P}^{(2)}$$

$$= \begin{bmatrix} 0.48 & 0.52 \\ 0.39 & 0.61 \end{bmatrix}\begin{bmatrix} 0.48 & 0.52 \\ 0.39 & 0.61 \end{bmatrix}$$

$$= \begin{bmatrix} 0.4332 & 0.5668 \\ 0.4251 & 0.5749 \end{bmatrix}$$

Therefore,

$$\mathbf{p}^{(4)} = [0.8 \quad 0.2] \begin{bmatrix} 0.4332 & 0.5668 \\ 0.4251 & 0.5749 \end{bmatrix}$$

$$= [0.4316 \quad 0.5684]$$

c. From Equation 11.136,

$$\mathbf{p}^{(8)} = \mathbf{p}^{(0)} \mathbf{P}^{(8)}$$

where

$$\mathbf{P}^{(8)} = \mathbf{P}^{(4)} \mathbf{P}^{(4)}$$

$$= \begin{bmatrix} 0.4332 & 0.5668 \\ 0.4251 & 0.5749 \end{bmatrix} \begin{bmatrix} 0.4332 & 0.5668 \\ 0.4251 & 0.5749 \end{bmatrix}$$

$$= \begin{bmatrix} 0.4286 & 0.5714 \\ 0.4285 & 0.5715 \end{bmatrix}$$

Therefore,

$$\mathbf{p}^{(8)} = [0.8 \quad 0.2] \begin{bmatrix} 0.4286 & 0.5714 \\ 0.4285 & 0.5715 \end{bmatrix}$$

$$= [0.4286 \quad 0.5714]$$

Here, it is interesting to observe that the rows of the transition matrix $\mathbf{P}^{(8)}$ tend to be the same. Furthermore, the state probability vector $\mathbf{p}^{(8)}$ tends to be the same with the rows of the transition matrix $\mathbf{P}^{(8)}$. These results show that the long-run absolute probabilities are independent of the initial state probabilities, that is, $\mathbf{p}^{(0)}$. Therefore, the resulting probabilities are called the *steady-state probabilities* and defined as the set of π_j, where

$$\pi_j = \lim_{n \to \infty} p_j^{(n)} = \lim_{n \to \infty} P\{X(t_n) = j\} \tag{11.139}$$

In general, the initial state tends to be less important to the n-step transition probability as n increases, such that

$$\lim_{n \to \infty} P\{X(t_n) = j \mid X(t_0) = i\} = \lim_{n \to \infty} P\{X(t_n) = j\} = \Pi_j \tag{11.140}$$

so that one can get the unconditional steady-state probability distribution from the n-step transition probabilities by taking n to infinity without taking the initial states into account. Therefore,

$$\mathbf{P}^{(n)} = \mathbf{P}^{(n-1)} \mathbf{P} \tag{11.141}$$

or

$$\lim_{n \to \infty} \mathbf{P}^{(n)} = \lim_{n \to \infty} \mathbf{P}^{(n-1)} \mathbf{P} \tag{11.142}$$

and thus,

$$\Pi = \Pi \mathbf{P} \tag{11.143}$$

where

$$\Pi = \begin{bmatrix} \pi_1 & \pi_2 & \pi_3 \cdots \pi_k \\ \pi_1 & \pi_2 & \pi_3 \cdots \pi_k \\ \pi_1 & \pi_2 & \pi_3 \cdots \pi_k \\ \cdots\cdots\cdots\cdots\cdots \\ \pi_1 & \pi_2 & \pi_3 \cdots \pi_k \end{bmatrix} \tag{11.144}$$

Note that the matrix Π has identical rows so that each row is a row vector of

$$\Pi = [\pi_1 \quad \pi_2 \quad \pi_3 \cdots \pi_k] \tag{11.145}$$

Since the transpose of a row vector Π is a column vector Π^t, Equation 11.143 can also be expressed as

$$\Pi^t = \mathbf{P}^{(t)}\Pi^{(t)} \tag{11.146}$$

which is a set of linear equations.

To be able to solve equation sets (11.143) or (11.146) for individual π_i's, one additional equation is required. This equation is called the normalizing equation and can be expressed as

$$\sum_{\text{all } i} \pi_i = 1 \tag{11.147}$$

11.8.2 CLASSIFICATION OF STATES IN MARKOV CHAINS

Two states i and j are said to *communicate*, denoted as $i \sim j$, if each is accessible (reachable) from the other, that is, if there exists some sequence of possible transitions that would take the process from state i to state j.

A *closed set* of states is a set such that if the system, once in one of the states of this set, will stay in the set indefinitely; that is, once a closed set is entered, it cannot be left. Therefore, an *ergodic* set of states is a set in which all states communicate and which cannot be left once it is entered. An *ergodic state* is an element of an ergodic set. A state is called *transient* if it is not ergodic. If a single state forms a closed set, the state is called an *absorbing state*. Thus, a state is an absorbing state if and only if $p_{ij} = 1$.

11.9 DEVELOPMENT OF THE STATE-TRANSITION MODEL TO DETERMINE THE STEADY-STATE PROBABILITIES

The Markov technique can be used to determine the steady-state probabilities. The model given in this section is based on the zone-branch technique developed by Koval and Billinton [11,12].

Assuming the process given in a markovian model is irreducible and all states are ergodic, one can derive a set of linear equations to determine the steady-state probabilities as

$$\pi_j = \lim_{t \to \infty} p_{ij}(t) \tag{11.148}$$

Therefore, for example, the system differential equations can be expressed in the matrix form for a single-component state as [13]

$$
\begin{bmatrix} P_0'(t) \\ P_1'(t) \\ P_2'(t) \\ P_3'(t) \end{bmatrix} = \begin{bmatrix} -(\lambda + \hat{n}) & \hat{m} & \mu & 0 \\ \hat{n} & -(\hat{m} + \lambda') & 0 & \mu' \\ \lambda & 0 & -(\mu + \hat{n}) & \hat{m} \\ 0 & \lambda' & \hat{n} & -(\mu + \hat{m}) \end{bmatrix} \begin{bmatrix} P_0(t) \\ P_1(t) \\ P_2(t) \\ P_3(t) \end{bmatrix} \tag{11.149}
$$

where
    λ is the normal weather failure rate of component
    μ is the normal weather repair rate of component
    μ' is the adverse weather failure rate of component
    μ' is the adverse weather repair rate of component

Also,

$$
\hat{n} = \frac{1}{N} \tag{11.150}
$$

and

$$
\hat{m} = \frac{1}{S} \tag{11.151}
$$

where
    N is the expected duration of normal weather period
    S is the expected duration of adverse weather period

Equation 11.148 can be expressed in the matrix form as

$$
\left[\frac{dP(t)}{dt} \right] = \mathbf{P}(t)\mathbf{\Lambda} \tag{11.152}
$$

where
    $[dP(t)/dt]$ is the matrix whose (i, j)th element is $dp_{ij}(t)/dt$
    $\mathbf{P}(t)$ is the matrix whose (i, j)th element is $p_{ij}(t)$
    $\mathbf{\Lambda}$ is the matrix whose (i, j)th element is λ_{ij}

Also, each element in matrix equation (11.152) can be expressed as

$$
\frac{dp_{ij}(t)}{dt} = \sum_k p_{ik}(t)\lambda_{kj} \tag{11.153}
$$

or

$$
\lim_{t \to \infty} \frac{dp_{ij}(t)}{dt} = \lim_{t \to \infty} \sum_k p_{ik}(t)\lambda_{kj} \tag{11.154}
$$

since

$$\lim_{t \to \infty} \frac{dp_{ij}(t)}{dt} = \frac{d}{dt} \lim_{t \to \infty} p_{ij}(t) \tag{11.155}$$

$$\frac{d}{dt} \lim_{t \to \infty} p_{ij}(t) = \sum_k \lim_{t \to \infty} p_{ik}(t) \lambda_{kj} \tag{11.156}$$

or

$$\frac{d\pi_j}{dt} = \sum_k \pi_k \lambda_{kj} \tag{11.157}$$

However, since the differentiation of a constant is zero, that is,

$$\frac{d\pi_j}{dt} = 0 \tag{11.158}$$

Equation 11.157 becomes

$$0 = \sum_k \pi_k \lambda_{kj} \tag{11.159}$$

or, in the matrix form,

$$\mathbf{0} = \mathbf{\Pi}\mathbf{\Lambda} \tag{11.160}$$

where
$\mathbf{0}$ is the row vector of zeros
$\mathbf{\Lambda}$ is the matrix of transition rates
$\mathbf{\Pi}$ is the row vector of steady-state probabilities

Since the equations in the matrix equation (11.160) are dependent, introduction of an additional equation is necessary, that is,

$$\sum \pi_i = 1 \tag{11.161}$$

which is called the *normalizing equation.*
   The matrix $\mathbf{\Lambda}$ can be expressed as

$$\mathbf{\Lambda} = \begin{bmatrix} \lambda_{11} & \lambda_{12} & \cdots & \lambda_{1n} \\ \lambda_{21} & \lambda_{22} & \cdots & \lambda_{2n} \\ \cdots\cdots\cdots\cdots\cdots\cdots \\ \lambda_{n1} & \lambda_{n2} & \cdots & \lambda_{nn} \end{bmatrix} \tag{11.162}$$

where
$\lambda_{ij} = -d_i$ for $i = j$, called the *rate of departure* from state i
$\lambda_{ij} = e_{ij}$ for $i \neq j$, called the *rate of entry* from state i to state j

Therefore, matrix equation (11.162) can be reexpressed as

$$
\Lambda = \begin{bmatrix}
-d_1 & e_{12} & \cdots & e_{1n} \\
e_{21} & -d_2 & \cdots & e_{2n} \\
\hdotsfor{4} \\
e_{n1} & e_{n2} & \cdots & -d_n
\end{bmatrix}
\tag{11.163}
$$

Likewise,

$$
\Pi = [p_1 \quad p_2 \quad \cdots \quad p_n]
\tag{11.164}
$$

Therefore, substituting Equations 11.163 and 11.164 into Equation 11.160,

$$
[0 \quad 0 \quad \cdots \quad 0] = [p_1 \quad p_2 \quad \cdots \quad p_n]
\begin{bmatrix}
-d_1 & e_{12} & \cdots & e_{1n} \\
e_{21} & -d_2 & \cdots & e_{2n} \\
\hdotsfor{4} \\
e_{n1} & e_{n2} & \cdots & -d_n
\end{bmatrix}
\tag{11.165}
$$

or

$$
\begin{aligned}
0 &= -p_1 d_1 + P_2 e_{21} + \cdots + P_n e_{n1} \\
0 &= p_1 e_{12} - p_2 d_2 + \cdots + p_n e_{n2} \\
&\quad \cdots\cdots\cdots\cdots\cdots\cdots\cdots \\
0 &= p_1 e_{1n} + p_2 e_{2n} + \cdots - p_n d_n
\end{aligned}
\tag{11.166}
$$

Therefore,

$$
0 = -p_i \sum d_i + \sum p_j e_{ij}
\tag{11.167}
$$

or

$$
p_i \sum d_i = \sum p_j e_{ij}
\tag{11.168}
$$

Also,

$$
p_1 + p_2 + p_3 + \cdots + p_n = 1
\tag{11.169}
$$

or

$$
p_1 \left(1 + \frac{p_2}{p_1} + \frac{p_3}{p_1} + \cdots + \frac{p_n}{p_1} \right) = 1
\tag{11.170}
$$

As Koval and Billinton [11,12] suggested, once the long-term or steady-state probabilities of each state are computed from Equations 11.168 to 11.170, one can readily calculate the total failure rate and the average repair rate of the particular zone i and branch j. These rates also take into consideration the effects of interruptions on other parts of the system. The total failure rate of zone i branch j is given by Koval and Billinton [11] as

$$\lambda_{ij} = \lambda_s + \sum RIA(ij,k) \times \lambda_i \qquad (11.171)$$

where
λ_{ij} is the total failure rate of zone branch ij
λ_s is the failure rate of supply, that is, feeding substation
$RIA(ij, k)$ is the recognition and isolation array coefficients,
I is the failed zone-branch array coefficient = $FZB(k)$

Likewise, the average downtime, that is, repair time, for each zone i branch j is given as

$$r_{ij} = \frac{\sum DTA(ij,k) \times \lambda_1}{\lambda_{ijT}} \qquad (11.172)$$

or

$$r_{ij} = \frac{\text{total annual outage time of zone branch } i, j}{\text{total failure rate of Zone Branch } i, j} \qquad (11.173)$$

where
r_{ij} is the average repair time for each zone i branch j
$\sum DTA(ij,k)$ is the downtime array coefficients
I is the failed zone-branch array coefficient, = $FZB(k)$

11.10 DISTRIBUTION RELIABILITY INDICES

Since a typical distribution system accounts for 40% of the cost to deliver power and 80% of customer reliability problems, distribution system design and operation is critical for financial success of the utility company and customer satisfaction.

Interruptions and outages can be studied through the use of predictive reliability assessment tools that can predict customer reliability characteristics based on system topology and component reliability data. In order to achieve this, distribution reliability indices are calculated. Such reliability indices should be concerned with both duration and frequency of outage.

They also need to consider overall system conditions as well as specific customer conditions. Using averages all lead to loss of some information such as time until the last customer is returned to service, but averages should give a general trend of conditions for the utility.

Here, it is assumed that as seen as the customer service interrupted, the crews are dispatched and the restoration work starts immediately. Therefore, the duration of interruption is the same as the duration of restoration.

11.11 SUSTAINED INTERRUPTION INDICES

These indices are also known as *customer-based indices.*

11.11.1 SAIFI

System average interruption frequency index (SAIFI) (sustained interruptions). This index is designed to give information about the average frequency of sustained interruptions per customer over a predefined area. Therefore,

$$SAIFI = \frac{\text{total number of customer interruptions}}{\text{total number of customers served}}$$

or

$$SAIFI = \frac{\sum N_i}{N_T} \tag{11.174}$$

where
N_i is the number of interrupted customers for each interruption event during reporting period
N_T is the total number of customers served for the area being indexed

11.11.2 SAIDI

System average interruption duration index (SAIDI). This index is commonly referred to as customer minutes of interruption or customer hours, and is designed to provide information about the average time the customers are interrupted. Thus,

$$SAIDI = \frac{\sum \text{customer interruption durations}}{\text{total number of customers served}}$$

or

$$SAIDI = \frac{\sum r_i N_i}{N_T} \tag{11.175}$$

where r_i is the restoration time for each interruption event.

11.11.3 CAIDI

Customer average interruption duration index (CAIDI). It represents the average time required to restore service to the average customer per sustained interruption. Hence,

$$CAIDI = \frac{\sum \text{customer interruption durations}}{\text{total number of customers interruptions}}$$

or

$$CAIDI = \frac{\sum r_i \times N_i}{N_i} = \frac{SAIDI}{SAIFI} \tag{11.176}$$

11.11.4 CTAIDI

Customer total average interruption duration index (CTAIDI). For customers who actually experienced an interruption, this index represents the total average time in the reporting period they were without power. This index is a hybrid of CAID and is calculated the same except that customers with multiple interruptions are counted only once. Therefore,

$$\text{CTAIDI} = \frac{\sum \text{customer interruption durations}}{\text{total number customers interrupted}}$$

or

$$\text{CTAIDI} = \frac{\sum R_i \times N_i}{CN} \tag{11.177}$$

where *CN* is total number of customers who have experienced a sustained interruption during the reporting period.

In tallying total number of customers interrupted, each individual customer should only be counted once regardless of the number of times interrupted during the reporting period. This applies to both CTAIDI and CAIFI.

11.11.5 CAIFI

Customer average interruption frequency index (CAIFI). This index gives the average frequency of sustained interruptions for those customers experiencing sustained interruptions. The customer is counted only once regardless of the number of times interrupted. Thus,

$$\text{CAIFI} = \frac{\text{total number of customer interruptions}}{\text{total number of customers interrupted}}$$

or

$$\text{CAIFI} = \frac{\sum N_i}{CN} \tag{11.178}$$

11.11.6 ASAI

Average service availability index (ASAI). This index represents the fraction of time (often in percentage) that a customer has power provided during 1 year or the defined reporting period. Hence,

$$\text{ASAI} = \frac{\text{customer hours service availability}}{\text{customer hours service demand}}$$

or

$$\text{ASAI} = \frac{N_T \times (\text{number of hours/year}) - \sum r_i N_i}{N_T \times (\text{number of hours/year})} \tag{11.179}$$

There are 8760 h in a regular year, 8784 in a leap year.

11.11.7 ASIFI

Average system interruption frequency index (ASIFI). This index was specifically designed to calculate reliability based on load rather than number of customers. It is an important index for areas that serve mainly industrial/commercial customers.

It is also used by utilities that do not have elaborate customer tracking systems. Similar to SAIFI, it gives information on the system average frequency of interruption. Therefore,

$$\text{ASIFI} = \frac{\text{connected kVA interrupted}}{\text{total connected kVA served}}$$

or

$$\text{ASIFI} = \frac{\sum L_i}{L_T} \tag{11.180}$$

where
$\sum L_i$ is the total connected kVA load interrupted for each interruption event
L_T is the total connected kVA load served

11.11.8 ASIDI

Average system interruption duration index (ASIDI). This index was designed with the same philosophy as ASIFI, but it provides information on system average duration of interruptions. Thus,

$$\text{ASIDI} = \frac{\text{connected kVA duration interrupted}}{\text{total connected kVA served}}$$

or

$$\text{ASIDI} = \frac{\sum r_i \times L_i}{L_T} \tag{11.181}$$

11.11.9 CEMI$_n$

Customers experiencing multiple interruptions (CEMI$_n$). This index is designed to track the number n of sustained interruptions to a specific customer. Its purpose is to help identify customer trouble that cannot be seen by using averages. Hence,

$$\text{CEMI}_n = \frac{\text{total number of customers that experienced more sustained interruptions}}{\text{total number of customers served}}$$

or

$$\text{CEMI}_n = \frac{CN_{(k>n)}}{N_T} \tag{11.182}$$

where $CN_{(k>n)}$ is the total number of customers who have experienced more than n sustained interruptions during the reporting period.

11.12 OTHER INDICES (MOMENTARY)

11.12.1 MAIFI

Momentary average interruption frequency index (MAIFI). This index is very similar to SAIFI, but it tracks the average frequency of momentary interruptions. Therefore,

$$MAIFI = \frac{\text{total number of customer momentary interruptions}}{\text{total number of customers served}}$$

or

$$MAIFI = \frac{\sum ID_i \times N_i}{N_T} \tag{11.183}$$

where ID_i is the number of interrupting device operations.

MAIFI is the same as SAIFI, but it is for short-duration rather than long-duration interruptions.

11.12.2 MAIFI$_E$

Momentary average interruption event frequency index (MAIFI$_E$). This index is very similar to SAIFI, but it tracks the average frequency of momentary interruption events. Thus,

$$MAIFI_E = \frac{\text{total number of customer momentary interruption events}}{\text{total number of customers served}}$$

or

$$MAIFI_E = \frac{\sum ID_E \times N_i}{N_T} \tag{11.184}$$

where ID_E is the interrupting device events during reporting period.

Here, N_i is the number of customers experiencing momentary interrupting events. This index does not include the events immediately preceding a lockout.

Momentary interruptions are most commonly tracked by using breaker and recloser counts, which implies that most counts of momentaries are based on MAIFI and MAIFI$_E$. To accurately count MAIFI$_E$, a utility must have a supervisory control and data acquisition (SCADA) system or other time-tagging recording equipment.

11.12.3 CEMSMI$_n$

Customers experiencing multiple sustained interruption and momentary interruption events (CEMSMI$_n$). This index is designed to track the number n of both sustained interruption and momentary interruption events to a set of specific customers.

Its purpose is to help identify customer trouble that cannot be seen by using averages. Hence,

$$CEMSMI_n = \frac{\text{total number of customers that experienced more than } n \text{ interruptions}}{\text{total number of customers served}}$$

or

$$\text{CEMSMI}_n = \frac{CNT_{(k>n)}}{N_T}$$ (11.185)

where $CNT_{(k>n)}$ is the total number of customers who have experienced more than n sustained interruption and momentary interruption events during the reporting period.

11.13 LOAD- AND ENERGY-BASED INDICES

There are also load- and energy-based indices. In determination of such indices, one has to know the average load at each load bus. This average load L_{avg} at a bus is found from

$$L_{avg} = L_{peak} \times F_{LD}$$ (11.186)

where
L_{avg} is the peak load (demand)
F_{LD} is the load factor

The average load can also be found from

$$L_{avg} = \frac{\text{total energy demanded in period of interest}}{\text{period of interest}}$$

If the period of interest is a year,

$$L_{avg} = \frac{\text{total annual energy demanded}}{8760}$$ (11.187)

11.13.1 ENS

Energy not supplied index (ENS). This index represents the total energy *not supplied* by the system and is expressed as

$$\text{ENS} = \sum L_{avg,i} \times r_i$$ (11.188)

where $L_{avg,i}$ is the average load connected to load point i.

11.13.2 AENS

Average energy not supplied (AENS). This index represents the average energy not supplied by the system.

$$\text{AENS} = \frac{\text{total energy not supplied}}{\text{total number of customers served}}$$ (11.189)

or

$$\text{AENS} = \frac{\sum L_{avg,i} \times r_i}{N_T}$$ (11.190)

This index is the same as the average system curtailment index (ASCI).

11.13.3 ACCI

Average customer curtailment index (ACCI). This index represents the total energy not supplied per affected customer by the system.

$$ACCI = \frac{\text{total energy not supplied}}{\text{total number of customers affected}}$$

or

$$ACCI = \frac{\sum L_{avg,i} \times r_i}{CN} \tag{11.191}$$

It is a useful index for monitoring the changes of average energy *not supplied* between one calendar year and another.

Example 11.10

The *Ghost Town Municipal Electric Utility Company* (GMEU) has a small distribution system for which the information is given in Tables 11.5 and 11.6. Assume that the duration of interruption is the same as the restoration time. Determine the following reliability indices:

a. SAIFI
b. CAIFI
c. SAIDI
d. CAIDI
e. ASAI
f. ASIDI
g. ENS
h. AENS
i. ACCI

Solution

a. $SAIFI = \dfrac{\sum N_i}{N_T} = \dfrac{950}{1,000} = 0.95$ interruptions/customer served

b. $CAIFI = \dfrac{\sum N_i}{CN} = \dfrac{950}{700} = 1.357$ interruptions/customer affected

TABLE 11.5
Distribution System Data of GMEU Company

Load Point	Number of Customers (N_i)	Average Load Connected (kW) ($L_{avg,i}$)
1	250	2,300
2	300	3,700
3	200	2,500
4	250	1,600
	$N_T = 1,000$	$L_T = 10,100$

TABLE 11.6

Annual Interruption Effects

Load Point Affected	Number of Customers Interrupted (N_i)	Load Interrupted (kW) (L_i)
1	250	2,300
2	200	2,500
3	250	1,600
4	250	1,600
	950	8,000

Load Point Affected	Duration of Interruptions (h) ($d_i = r_i$)	Customer Hours Curtailed ($r_i \times N_i$)	Energy Not Supplied (kWh) ($r_i \times L_i$)
1	2	500	4,600
2	3	600	7,500
3	1	250	1,600
4	1	250	1,600
		1,600	15,300

CN, number of customers affected = 250 + 200 + 250 = 700.

c. $\text{SAIDI} = \dfrac{\sum r_i \times N_i}{N_T} = \dfrac{1,600}{1,000} = 1.6 \text{ h/customer served} = 96 \text{ min/customer served}$

d. $\text{CAIDI} = \dfrac{\sum r_i \times N_i}{N_i} = \dfrac{1,600}{950} = 1.684 \text{ h/customer interrupted}$

$= 101.05 \text{ min/customer interrupted}$

e. $\text{ASAI} = \dfrac{N_T \times 8,760 - \sum r_i \times N_i}{N_T \times 8,760} = \dfrac{1,000 \times 8,760 - 1,600}{1,000 \times 8,760} = 0.999817$

f. $\text{ASIDI} = \dfrac{\sum r_i \times L_i}{L_T} = \dfrac{15,300}{10,100} = 1.515$

g. $\text{ENS} = \sum L_{avg,i} \times r_i = 15,300 \text{ kWh}$

h. $\text{AENS} = \dfrac{\text{ENS}}{N_T} = \dfrac{15,300}{1,000} = 15.3 \text{ kWh/customer affected}$

i. $\text{ACCI} = \dfrac{\text{ENS}}{\text{CN}} = \dfrac{15,300}{700} = 21.857 \text{ kWh/customer affected}$

11.14 USAGE OF RELIABILITY INDICES

Based on the two industry-wide surveys, the *Working group on System Design of IEEE Power Engineering Society's T&D Subcommittee* has determined that the most commonly used indices are SAIDI, SAIFI, CAIDI, and ASAI in the descending popularity order of 70%, 80%, 66.7%, and

63.3%, respectively. Most utilities track one or more of the reliability indices to help them understand how the distribution system is performing.

For example, removing the instantaneous trip from the substation recloser has an effect on the whole circuit. The first area to look at is the effect on the reliability indices. With the advent of the digital clock and electronic equipment, a newer index (i.e., MAIFI, which tracks momentary outages) is gaining in popularity.

With the substation recloser instantaneous trip on, the SAIDI and CAIDI indices should be low, due to the "fuse saving" effect when clearing momentary faults. The MAIFI, however, will be high due to the blinks on the whole circuit. By removing the instantaneous trip, the MAIFI should be reduced but the SAIDI will increase.

11.15 BENEFITS OF RELIABILITY MODELING IN SYSTEM PERFORMANCE

A *reliability assessment model* quantifies reliability characteristics based on system topology and component reliability data. The aforementioned reliability indices can be used to assess the past performance of a distribution system. Assessment of system performance is valuable for various reasons.

For example, it establishes the changes in system performance and thus helps to identify weak areas and the need for reinforcement. It also identifies overloaded and undersized equipment that degrades system reliability. Also, it establishes existing indices that can be used in the future reliability assessments. It enables previous predictions to be compared with actual operating experience. Such results can benefit many aspects of distribution planning, engineering, and operations. Reliability problems associated with expansion plans can be predicted.

However, a *reliability assessment study* can help to quantify the impact of design improvement options. Adding a recloser to a circuit will improve reliability, but by how much? Reliability models answer this question. Typical improvement options that can be studied based on a *predictive reliability model* include the following:

1. New feeders and feeder expansions
2. Load transfers between feeders
3. New substation and substation expansions
4. New feeder tie points
5. Line reclosers
6. Sectionalizing switches
7. Feeder automation
8. Replacement of aging equipment
9. Replacing circuits by underground cables

According to Brown [22], reliability studies can help to identify the number of sectionalizing switches that should be placed on a feeder, the optimal location of devices, and the optimal ratings of new equipment. Adding a tie switch may reduce index by 10 min, and reconductoring for contingencies may reduce SAIDI by 5 min. Since reconductoring permits the tie switch to be effective, doing both projects may result in a SAIDI reduction of 30 min, doubling the cost-effectiveness of each project.

Cost-effectiveness is determined by computing the cost of each reliability improvement option and computing a benefit/cost ratio. This is a measure of how much reliability is purchased with each dollar being spent. Once all projects are ranked in order of cost-effectiveness, projects and project combinations can be approved in order of descending cost-effectiveness until reliability targets are met or budget constraints become binding. This process is referred to as *value-based planning and engineering.* In a given distribution system, reliability improvements can be achieved by various means, which include the following [22]:

1. *Increased line sectionalizing*: It is accomplished by placing normally closed switching devices on a feeder. Adding fault interrupting devices (fuses and reclosers) improves devices by reducing the number of customers interrupted by downstream faults. Adding switches without fault interrupting capability improves reliability by permitting more flexibility during post-fault system reconfiguration.
2. *New tie points*: A tie point is a normally open switch that permits a feeder to be connected to an adjacent feeder. Adding new tie points increases the number of possible transfer paths and may be a cost-effective way to improve reliability on feeders with low transfer capability.
3. *Capacity constrained load transfers*: Following a fault, operators and crews can reconfigurate a distribution system to restore power to as many customers as possible. Reconfiguration is only permitted if it does not load a piece of equipment above its emergency rating. If a load transfer is not permitted because it will overload a component, the component is charged with a capacity constraint. System reliability is reduced, because the equipment does not have sufficient capacity for reconfiguration to take place.
4. *Transfer path upgrades*: A transfer path is an alternate path to serve load after a fault takes place. If a transfer path is capacity constrained due to small conductor sizes, reconductoring may be a cost-effective way to improve reliability.
5. *Feeder automation*: SCADA-controlled switches on feeders permit post-fault system reconfiguration to take place much more quickly than with manual switches, permitting certain customers to experience a momentary interruption rather than a sustained interruption.

In summary, distribution system reliability assessment is crucial in providing customers more with less cost. Today, computer softwares are commercially available, and the time has come for utilities to treat reliability issues with the same analytical rigor as capacity issues.

11.16 ECONOMICS OF RELIABILITY ASSESSMENT

Typically, as investment in system reliability increases, the reliability improves, but it is not a linear relationship. By calculating the cost of each proposed improvement and finding a ration of the increased benefit to the increased cost, the cost-effectiveness can be quantified.

Once the cost-effectiveness of the improvement options has been quantified, they can be prioritized for implementation. This incremental analysis of how reliability improves and affects the various indices versus the additional cost is necessary in order to help ensure that scarce resources are used most effectively.

Quantifying the additional cost of improved reliability is important, but additional considerations are needed for a more complete analysis. The costs associated with an outage are placed side by side against the investment costs for comparison in helping to find the true optimal reliability solution. Outage costs are generally divided between utility outage costs and customer outage costs.

Utility outage costs include the loss of revenue for energy not supplied, and the increased maintenance and repair costs to restore power to the customers affected. According to Billinton and Wang [23], the maintenance and repair costs can be quantified as

$$C_{m\&r} = \sum_{i}^{n} C_l + C_{\text{comp}} \ \$$$

(11.192)

where
    C_l is the labor cost for each repair and maintenance action, in dollars
    C_{comp} is the component replacement or repair costs, in dollars

Therefore, the total utility cost for an outage is

$$C_{\text{out}} = (\text{ENS}) \times (\text{cost/kWh}) + C_{m\&r} \ \$ \tag{11.193}$$

While the outage costs to the utility can be significant, often the costs to the customer are far greater. These costs vary greatly by customer sector type and geographical location. Industrial customers have costs associated with loss of manufacture, damaged equipment, extra maintenance, loss of products and/or supplies to spoilage, restarting costs, and greatly reduced worker productivity effectiveness.

Commercial customers may lose business during the outage and experience many of the same losses as industrial customers, but on a possibly smaller scale.

Residential customers typically have costs during a given outage that are far less than the previous two, but food spoilage, loss of heat during winter, or air conditioning during a heat wave can be disproportionately large for some individual customers. In general, customer outage costs are more difficult to quantify. Through collection of data from industry and customer surveys, a formulation of sector damage functions is derived, which lead to composite damage functions.

According to Lawton et al. [24], the *sector customer damage function (SCDF)* is a cost function of each customer sector. The *composite customer damage function (CCDF)* is an aggregation of the SCDF at specified load points and is weighted proportionally to the load at the load points. For n customers,

$$\text{CCDF} = \sum_{i=1}^{n} C_i + \text{SCDF}_i \ \$/\text{kW} \tag{11.194}$$

where C_i is the energy demand of customer type i.

Therefore, the customer outage cost by sector is

$$\text{COST}_i = \sum_{i=1}^{n} \text{SCDF}_i \times L_i \ \$ \tag{11.195}$$

where L_i is the average load at load point i.

Since the CCDF is a function of outage attributes, customer characteristics, and geographical characteristics, it is important to have accurate information about these variables. Although outage attributes include duration, season, time of day, advance notice, and day of the week, the most heavily weighted factor is outage duration.

The total customer cost for all applicable sectors can be found for a particular load point from

$$\text{COST} = \sum_{i=1}^{n} \text{CCDF}_i \times L_i \ \$ \tag{11.196}$$

or

$$\text{COST} = \sum_{i=1}^{n} C_i \times \text{SCDF}_i \times L_i \ \$ \tag{11.197}$$

However, using the CCDF marks the outage cost that is borne disproportionately by the different sectors. For a reliability planning, in addition to the load point indices of λ, r, and U, one has to determine the following reliability cost/worth indices [23]:

1. *Expected energy not supplied (EENS) index*: It is defined as

$$\text{EENS}_i = \sum_{j=1}^{N_e} L_i \times r_i \times \lambda_{ij} \text{ energy per customer unit time} \tag{11.198}$$

where
N_e is the total number of elements in the distribution system
L_i is the average load at load point i
r_{ij} is the failure duration at load point i due to component j
λ_{ij} is the failure rate at load point i due to component j

2. *Expected customer outage cost (ECOST) index*: It is defined as

$$\text{ECOST}_i = \sum_{i=1}^{n} \text{SCDF}_{ij} \times L_i \times \lambda_{ij} \ \$ \tag{11.199}$$

where SCDF_{ij} is the sector customer damage function at load point i due to component j
3. *Interrupted energy assessment rate (IEAR) indices*: It is defined as

$$\text{IEAR}_i = \frac{\text{ECOST}_i}{\text{EENS}_i} \ \$ \tag{11.200}$$

This index provides a quantitative worth of the reliability for a particular load point in terms of cost for each unit of energy not supplied.

The reliability cost/worth analysis provides a more comprehensive analysis of the time reliability cost of the system. In addition to the incentives for improving the system indices and keeping system costs under control, costs help to ensure that the reliability investment costs are apportioned judiciously for maximum benefit to both the utility and the end user.

Reliability is terribly important for the customer. In one study performed in the Eastern United States in 2002, the *"average"* residential customer cost for an outage duration of 1 h was approximately \$3, for a small-to-medium commercial customer the cost was \$1,200, and for a large industrial customer the cost was \$82,000 [24]. Providing a comprehensive reliability cost/worth assessment is a tool in order to help ensure a reliable electricity supply is available and that the system costs of the utility company are well justified.

PROBLEMS

11.1   Assume that the given experiment is tossing a coin three times and that a single outcome is defined as a certain succession of heads (H) and tails (T), for example, (HHT).
   a.   How many possible outcomes are there? Name them.
   b.   What is the probability of tossing three heads, that is, (HHH)?
   c.   What is the probability of getting heads on the first two tosses?
   d.   What is the probability of getting heads on any two tosses?
11.2   Two cards are drawn from a shuffled deck. What is the probability that both cards will be aces?

11.3 Two cards are drawn from a shuffled deck.
    a.   What is the probability that two cards will be the same suit?
    b.   What is the probability if the first card is replaced in the deck before the second one is drawn?

11.4 Assume that a substation transformer has a constant hazard rate of 0.005 per day.
    a.   What is the probability that it will fail during the next 5 years?
    b.   What is the probability that it will not fail?

11.5 Consider the substation transformer in Problem 11.4 and determine the probability that it will fail during year 6, given that it survives 5 years without any failure.

11.6 What is the MTTF for the substation transformer of Problem 11.4?

11.7 Determine the following for a parallel connection of three components:
    a.   The reliability
    b.   The availability
    c.   The MTTF
    d.   The frequency
    e.   The hazard rate

11.8 A large factory of the International Zubits Company has 10 identical loads that switch on and off intermittently and independently with a probability p of being "on." Testing of the loads over a long period has shown that, on the average, each load is on for a period of 12 min/h. Suppose that when switched on, each load draws some X kVA from the Ghost River Substation that is rated $7X$ kVA. Find the probability that the substation will experience an overload. (*Hint*: Apply the binomial expansion.)

11.9 Verify Equation 11.79.

11.10 Verify Equation 11.83.

11.11 Using Equation 11.78, derive and prove that the mean time to repair a two-component system is

$$\bar{r}_{sys} = \frac{(\bar{m}_1 + \bar{r}_1)(\bar{m}_2 + \bar{r}_2) - \bar{m}_1\bar{m}_2}{\bar{m}_1 + \bar{m}_2}$$

11.12 Calculate the equivalent reliability of each of the system configurations in Figure P11.12, assuming that each component has the indicated reliability.

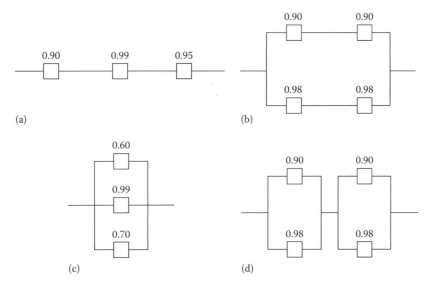

FIGURE P11.12 Various system configurations: (a) in series, (b) in series and parallel, (c) in parallel, and (d) in parallel and series, connections.

11.13 Calculate the equivalent reliability of each of the system configurations in Figure P11.13, assuming that each component has the indicated reliability.

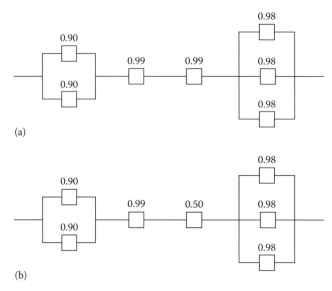

(a)

(b)

FIGURE P11.13   Various system configurations: (a) series connections of number of combinations, and (b) the same as (a) but with different reliabilities.

11.14 Determine the equivalent reliability of the system in Figure P11.14.

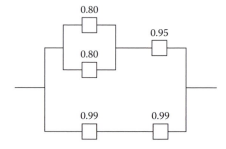

FIGURE P11.14   System configuration for Problem 11.14.

11.15 Using the results of Example 11.6, determine the following:
   a.   The probability of having any one of the three transformers out of service at any given time.
   b.   The probability of having any two of the three transformers out of service at any given time.
11.16 Using the results of Example 11.6, determine the following:
   a.   The probability of having at least one of the three transformers out of service at any given time.
   b.   The probability of having at least two of the three transformers out of service at any given time.
11.17 Repeat Example 11.2, assuming that the underground section of the feeder has been increased another mile due to growth in the downtown area and that on the average, the annual fault rate of the underground section has increased to 0.3 due to the growth and aging.

11.18 Repeat Example 11.3 for customers $D–F$, assuming that they all exist as shown in Figure 11.16.

11.19 Repeat Problem 11.18 but assume that during emergency the end of the existing feeder can be connected to and supplied by a second feeder over a normally open tie breaker.

11.20 Verify Equation 11.172 for a two-component system.

11.21 Verify Equation 11.172 for an n-component system.

11.22 Derive Equation 11.131 based upon the definition of n-step transition probabilities of a Markov chain.

11.23 Use the data given in Example 11.8 and assume that feeder 1 has just had an outage. Using the joint probability concept of the classical probability theory techniques and the system's probability tree diagram, determine the probability that there will be an outage on feeder 2 at the time after the next outage.

11.24 Repeat Problem 11.23 by using the Markov chains concept rather than the classical probability theory techniques.

11.25 Use the data given in Example 11.8 and the Markov chains concept. Assuming that there is an outage on feeder 3 at the present time, determine the following:
a. The probabilities of being in each of the respective states at time t_1.
b. The probabilities of being in each of the respective states at time t_2.

11.26 Use the data given in Example 11.8 and the Markov chains concept. Assume that there is an outage on feeder 2 at the present time and determine the probabilities associated with this outage at time t_4.

11.27 Use the data given in Example 11.8 and the Markov chains concept. Determine the complete outage probabilities at time t_4.

11.28 Derive Equation 11.187 from Equation 11.186.

11.29 Consider a radial feeder supplying three laterals and assume that the distribution system data and annual interruption effects of a utility company are given in Tables P11.29A and B, respectively. Assume that the duration of interruption is the same as the restoration time. Determine the following reliability indices:
a. SAIFI
b. CAIFI
c. SAIDI
d. CAIDI
e. ASAI
f. ASIDI
g. ENS
h. AENS
i. ACCI

TABLE P11.29A
Distribution System Data

Load Point	Number of Customers (N_i)	Average Load Connected in kW ($L_{avg,i}$)
1	1800	8400
2	1300	6000
3	900	4600
	$N_T = 4000$	$L_T = 1900$

TABLE P11.29B

Annual Interruption Effects

Load Point Affected	Number of Customers Interrupted (N_i)	Load Interrupted in kW (L_i)
2	800	3,600
3	600	2,800
3	300	1,800
3	600	2,800
2	500	2,400
3	300	1,800
	—	—
	3100	15,200

Load Point Affected	Duration of Interruptions (h) ($d_i = r_i$)	Customer Hours Curtailed ($r_i \times N_i$)	Energy Not Supplied (kWh) ($r_i \times L_i$)
2	3	2400	10,800
3	3	1800	8,400
3	2	600	3,600
3	1	600	2,800
2	1.5	750	3,600
3	1.5	450	2,700
		—	—
		6600	31,900

CN, number of customers affected = 800 + 600 + 300 + 500 = 2,200.

11.30 Assume that a radial feeder is made up of three sections (i.e., sections A, B, and C) and that a load is connected at the end of each section. Therefore, there are three loads, that is, L_1, L_2, and L_3. Table P11.30A gives the component data for the radial feeder. Table P11.30B gives the load point indices for the radial feeder. Finally, Table P11.30C gives the distribution system data. Determine the following reliability indices:
   a. SAIFI
   b. SAIDI
   c. CAIDI
   d. ASAI
   e. ENS
   f. AENS

TABLE P11.30A

Component Data for the Radial Feeder

Line	λ (Faults/Year)	r (h)
A	0.20	6.0
B	0.10	5.0
C	0.15	8.0

TABLE P11.30B
Distribution System Data

Load Point	λ_L (Faults/Year)	r_L (h)	U_L (h/Year)
L_1	0.20	6.0	1.2
L_2	0.30	5.7	1.7
L_3	0.45	6.4	2.9

TABLE P11.30C
Additional Distribution System Data

Load Point	Number of Customers	Average Load Demand (kW)
L_1	200	1000
L_2	150	700
L_3	100	400
	—	—
	450	2100

11.31 Resolve Example 11.3 by using MATLAB. Assume that all the quantities remain the same.
11.32 Resolve Example 11.9 by using MATLAB.

REFERENCES

1. IEEE Committee Reports: Proposed definitions of terms for reporting and analyzing outages of electrical transmission and distribution facilities and interruptions, *IEEE Trans. Power Appar. Syst.*, 87, May 5, 1968, 1318–1323.
2. Endrenyi, J.: *Reliability Modeling in Electric Power Systems*, Wiley, New York, 1978.
3. IEEE Committee Report: Definitions of customer and load reliability indices for evaluating electric power performance, Paper A75 588-4, presented at the *IEEE PES Summer Meeting*, San Francisco, CA, July 20–25, 1975.
4. IEEE Committee Report: List of transmission and distribution components for use in outage reporting and reliability calculations, *IEEE Trans. Power Appar. Syst.*, PAS-95(4), July/August 1976, 1210–1215.
5. Smith, C. O.: *Introduction to Reliability in Design*, McGraw-Hill, New York, 1976.
6. The National Electric Reliability Study: Technical Study Reports, U.S. Department of Energy DOE/EP-0005, April 1981.
7. The National Electric Reliability Study: Executive Summary, U.S. Department of Energy, DOE/EP-0003, April 1981.
8. Billinton, R.: *Power System Reliability Evaluation*, Gordon and Breach, New York, 1978.
9. Albrect, P. F.: Overview of power system reliability, *Workshop Proceedings: Power System Reliability-Research Needs and Priorities*, EPRI Report WS-77-60, Palo Alto, CA, October 1978.
10. Billinton, R., R. J. Ringlee, and A. J. Wood: *Power-System Reliability Calculations*, M.I.T., Cambridge, MA, 1973.
11. Koval, D. O. and R. Billinton: Evaluation of distribution circuit reliability, Paper F77 067-2, *IEEE PES Winter Meeting*, New York, NY, January–February 1977.
12. Koval, D. O. and R. Billinton: Evaluation of elements of distribution circuit outage durations, Paper A77 685-1, *IEEE PES Summer Meeting*, Mexico City, Mexico, July 17–22, 1977.
13. Billinton, R. and M. S. Grover: Quantitative evaluation of permanent outages in distribution systems, *IEEE Trans. Power Appar. Syst.*, PAS-94, May/June 1975, 733–741.

14. Gönen, T. and M. Tahani: Distribution system reliability analysis, *Proceedings of the IEEE MEXICON-80 International Conference*, Mexico City, Mexico, October 22–25, 1980.

15. *Standard Definitions in Power Operations Terminology Including Terms for Reporting and Analyzing Outages of Electrical Transmission and Distribution Facilities and Interruptions to Customer Service*, IEEE Standard 346–1973, 1973.

16. Heising, C. R.: Reliability of electrical power transmission and distribution equipment, *Proceedings of the Reliability Engineering Conference for the Electrical Power Industry*, Seattle, WA, February 1974.

17. Electric Power Research Institute: *Analysis of Distribution R&D Planning*, EPRI Report 329, Palo Alto, CA, October 1975.

18. Howard, R. A.: *Dynamic Probabilistic Systems, Vol. I: Markov Models*, Wiley, New York, 1971.

19. Markov, A.: Extension of the limit theorems of probability theory to a sum of variables connected in a chain, *Izv. Akad. Nauk St. Petersburg* (translated as Notes of the Imperial Academy of Sciences of St. Petersburg), December 5, 1907.

20. Gönen, T. and M. Tahani: Distribution system reliability performance, *IEEE Midwest Power Symposium*, Purdue University, West Lafayette, IN, October 27–28, 1980.

21. Gönen, T. et al.: Development of advanced methods for planning electric energy distribution systems, U.S. Department of Energy, October 1979. Available from the National Technical Information Service, U.S. Department of Commerce, Springfield, VA.

22. Brown, E. R. et al.: Assessing the reliability of distribution systems, *IEEE Comput. Appl. Power*, 14, January 1, 2001, 33–49.

23. Billinton, R. and P. Wang: Distribution system reliability cost/worth analysis using analytical and sequential simulation techniques, *IEEE Trans. Power Syst.*, 13, November 1998, 1245–1250.

24. Lawton, L. et al.: A framework and review of customer outage costs: Integration and analysis of electric utility outage cost surveys, Environmental Energy Technologies Division, Lawrance Berkley National Laboratory, LBNL-54365, Berkley, CA, November 2003.

12 Electric Power Quality

Only one thing is certain—that is, nothing is certain,
If this statement is true, it is also false.

Ancient Paradox

12.1 BASIC DEFINITIONS

Harmonics: Sinusoidal voltages or currents having frequencies that are an integer multiples of the fundamental frequency at which the supply system is designed to operate.

Total harmonic distortion (THD): The ratio of the root-mean-square (rms) of the harmonic content to the rms value of the fundamental quantity, expressed as a percent of the fundamental.

Displacement factor (DPF): The ratio of active power (watts) to apparent power (voltamperes).

True power factor (TPF): The ratio of the active power of the fundamental wave, in watts, to the apparent power of the fundamental wave, in rms voltamperes (including the harmonic components).

Triplen harmonics: A term frequency used to refer to the odd multiples of the third harmonic, which deserve special attention because of their natural tendency to be zero sequence.

Total demand distortion (TDD): The ratio of the rms of the harmonic current to the rms value of the rated or maximum demand fundamental current, expressed as a percent.

Harmonic distortion: Periodic distortion of the sign wave.

Harmonic resonance: A condition in which the power system is resonating near one of the major harmonics being produced by nonlinear elements in the system, hence increasing the harmonic distortion.

Nonlinear load: An electric load that draws current discontinuously or whose impedances varies throughout the cycle of the input ac voltage waveform.

Notch: A switching (or other) disturbance of the normal power voltage waveform, lasting less than a half cycle; which is initially of opposite polarity than the waveform. It includes complete loss of voltage for up to a 0.5 cycle.

Notching: A periodic disturbance caused by normal operation of a power electronic device, when its current is commutated from one phase to another.

K-factor: A factor used to quantify the load impact of electric arc furnaces on the power system.

Swell: An increase to between 1.1 and 1.8 pu in rms voltage or current at the power frequency for durations *from 0.5 cycle to 1 min.*

Overvoltage: A voltage that has a value at least 10% above the nominal voltage for a period of time greater than 1 min.

Undervoltage: A voltage that has a value at least 10% below the nominal voltage for a period of time greater than 1 min.

Sag: A decrease to between 0.1 and 0.9 pu in rms voltage and current at the power frequency for a duration of 0.5 cycles to 1 min.

Cress factor: A value that is displayed on many power quality monitoring instruments representing the ratio of the crest value of the measured waveform to the rms value of the waveform. For example, the cress factor of a sinusoidal wave is 1.414.

Isolated ground: It originates at an isolated ground-type receptacle or equipment input terminal block and terminates at the point where neutral and ground are bonded at the power source. Its conductor is insulated from the metallic raceway and all ground points throughout its length.

Waveform distortion: A steady-state deviation from an ideal sine wave of power frequency principally characterized by the special content of the deviation.

Voltage fluctuation: A series of voltage changes or a cyclical variation of the voltage envelope.

Voltage magnification: The magnification of capacitor switching oscillatory transient voltage on the primary side by capacitors on the secondary side of a transformer.

Voltage interruption: Disappearance of the supply voltage on one or more phases. It can be momentary, temporary, or sustained.

Recovery voltage: The voltage that occurs across the terminals of a pole of a circuit interrupting device upon interruption of the current.

Oscillatory transient: A sudden and nonpower frequency change in the steady-state condition of voltage or current that includes both positive and negative polarity values.

Noise: An unwanted electric signal with a less than 200 kHz superimposed upon the power system voltage or current in phase conductors or found on neutral conductors or signal lines. It is not a harmonic distortion or transient. It disturbs microcomputers and programmable controllers.

Voltage imbalance (or unbalance): The maximum deviation from the average of the three-phase voltages or currents, divided by the average of the three-phase voltages or currents, expressed in percent.

Impulsive transient: A sudden (nonpower) frequency change in the steady-state condition of the voltage or current that is unidirectional in polarity.

Flicker: Impression of unsteadiness of visual sensation induced by a light stimulus whose luminance or spectral distribution fluctuates with time.

Frequency deviation: An increase or decrease in the power frequency. Its duration varies from a few cycles to several hours.

Momentary interruption: The complete loss of voltage (<0.1 pu) on one or more phase conductors for a period between 30 cycles and 3 s.

Sustained interruption: The complete loss of voltage (<0.1 pu) on one or more phase conductors for a time greater than 1 min.

Phase shift: The displacement in time of one voltage waveform relative to other voltage waveform(s).

Low-side surges: The current surge that appears to be injected into the transformer secondary terminals upon a lighting strike to grounded conductors in the vicinity.

Passive filter: A combination of inductors, capacitors, and resistors designed to eliminate one or more harmonics. The most common variety is simply an inductor in series with a shunt capacitor, which short-circuits the major distorting harmonic component from the system.

Active filter: Any of a number of sophisticated power electronic devices for eliminating harmonic distortion.

12.2 DEFINITION OF ELECTRIC POWER QUALITY

In general, there is no single definition of the term electric power quality that is acceptable by everyone. According to Heydt [3], the electric power quality can be defined "as the goodness of the electric power quality supply in terms of its voltage wave shape, its current wave shape, its frequency, its voltage regulation, as well as level of impulses, and noise, and the absence of momentary outages."

Occasionally, some additional considerations are included in the definition of electric power quality. These concerns include reliability, electromagnetic compatibility, and even generation supply concerns. Distribution engineers usually focus on the load bus voltage in terms of maintaining its rated sinusoid voltage and frequency, in addition to other concerns, including spikes, notches, and outages. The growing utilization of electronic equipment has increased the interest in power quality in recent year.

The more specific definitions of the electric power quality depend on the points of view. For example, some utility companies may define power quality as reliability and point out to statistics demonstrating that the power system is 99.98% reliable. The equipment manufacturers may define it as those characteristics of the power supply that enable their equipment to work properly.

However, customers may define the electric power quality in terms of the absence of any power quality problems. From the customer point of view, the power quality problem is defined as "any power problem manifested in voltage, current, or frequency deviations that result in failure or unsatisfactory operation of customer's equipment" [4].

In general, *electric power quality issues* cover the entire electric power system, but their main emphases are in the primary and secondary distribution systems. Since usually the loads cause the distortion in bus voltage wave shape and are generally connected to the secondary system, the secondary system receives more attention than the primary system. However, occasionally, transmission and generation system are also included in some power quality analysis and evaluations.

12.3 CLASSIFICATION OF POWER QUALITY

The electric power quality disturbances can be classified in terms of the *steady-state disturbance* that is often periodic and lasts for a long period of time and the *transient disturbance* that generally lasts for a few milliseconds and then decays down to zero.

The first one is usually less obvious, less harmful, and lasts for a long time, but the cost involved may be very high. The second one is usually more obvious in its harmful effects and the costs involved may be extremely high. In the United States, it is estimated that the annual cost of transient power quality problems is anywhere between 100 million and 3 billion dollars, depending on the year [3].

The electric power quality issues include a wide variety of electromagnetic phenomena on the power systems. The International Electrotechnical Commission (IEC) classifies electromagnetic phenomena into various groups, as given in Table 12.1. Note that the definition of *waveform distortion* includes *harmonics, interharmonics, dc in ac networks*, and *notching* phenomena. The categories and characteristics of power system electromagnetic phenomena are given in Table 12.2.

Note that long-duration voltage variations can be either *overvoltages* or *undervoltages*. They generally are not the result of system faults, but are caused by load variations on the system and system switching operations.

A *sag*, or *dip*, is a decrease to between 0.1 and 0.9 pu in rms voltage or current at the power frequency for durations from 0.5 cycles to 1 min. A *swell* is an increase to between 1.1 and 1.8 pu in rms voltage or current at the power frequency for durations from 0.5 cycle to 1 min. As with sags, swells are usually associated with system fault conditions, but they are not a common a voltage sags.

Typical examples of swell-producing events include the temporary voltage rise on the unfaulted phases during an SLG fault. Swells can also be caused by switching off a large load or energizing a large capacitor bank. *Waveform distortion* is defined as a steady-state deviation from an ideal sine wave.

TABLE 12.1
Classification of Electromagnetic
Disturbances according to IEC

Conducted Low-Frequency Phenomena
- Harmonics, interharmonics
- Signaling voltages
- Voltage fluctuations
- Voltage dips and interruptions
- Voltage unbalance
- Power frequency variations
- Induced low frequency voltages
- dc in ac networks

Radiated Low-Frequency Phenomena
- Magnetic fields
- Electric fields

Conducted High-Frequency Phenomena
- Induced continuous wave (CW) voltages or currents
- Unidirectional transients
- Oscillatory transients

Radiated High-Frequency Phenomena
- Magnetic fields
- Electric fields
- Electromagnetic fields
- Continuous waves
- Transients

Electrostatic Discharge Phenomena (ESD)
Nuclear Electromagnetic Pulse (NEMP)

The main types of waveform distortions include *dc offset, harmonics, interharmonics, notching,* and *noise.* Figure 12.1 shows various types of disturbances. In the United States, most of residential, commercial, and industrial systems use line-to-neutral voltages that are equal or less than 277 V. The basic sources and characteristics of surges and transients in primary and secondary distribution networks are given in Table 12.3.

12.4 TYPES OF DISTURBANCES

Switching of reactive loads, for example, transformers and capacitors, create transients in the kilohertz range. Figure 12.2a shows *phase-neutral* transients resulting from addition of *capacitive load.* Figure 12.2b shows *neutral-ground* transient resulting from addition of *inductive load.* Electromechanical switching device interacts with the distributed inductance and capacitance in the ac distribution and loads to create *electric fast transients* (EFTs). For example, Figure 12.2c shows *phase-neutral transients* resulting from *arching and bouncing contactor.*

12.4.1 HARMONIC DISTORTION

Harmonics is blamed for many power quality disturbances that are actually transients. Even though transient disturbances may also have high-frequency components (not associated with the system fundamental frequency), transients and harmonics are distinctly different phenomena and are analyzed differently. Transients are usually dissipated within a few cycles, for example, transients that result from switching a capacitor bank.

TABLE 12.2

Categories and Characteristics of Power System Electromagnetic Phenomena

Categories	Typical Spectral Content	Typical Duration	Typical Voltage Magnitude
1.0 Transients			
1.1 Impulsive			
• Nanosecond	5 ns rise	<50 ns	
• Microsecond	1 μs rise	50 ns–1 ms	
• Millisecond	0.1 ms rise	>1 ms	
1.2 Oscillatory			
• Low frequency	<5 kHz	0.3–50 ms	0–4 pu
• Medium frequency	5–500 kHz	20 μs	0–8 pu
• High frequency	0.5–5 MHz	5 μs	0–4 pu
2.0 Short-duration variations			
2.1 Instantaneous			
• Interruption		0.5–30 cycles	<0.1 pu
• Sag (dip)		0.5–30 cycles	0.1–0.9 pu
• Swell		0.5–30 cycles	1.1–1.8 pu
2.2 Momentary			
• Interruption		30 cycles–3 s	<0.1 pu
• Sag (dip)		30 cycles–3 s	0.1–0.9 pu
• Swell		30 cycles–3 s	1.1–1.4 pu
2.3 Temporary			
• Interruption		3 s–1 min	<0.1 pu
• Sag (dip)		3 s–1 min	0.1–0.9 pu
• Swell		3 s–1 min	1.1–1.2 pu
3.0 Long-duration variations			
3.1 Interruption, sustained		>1 min	0.0 pu
3.2 Undervoltages		>1 min	0.8–0.9 pu
3.3 Overvoltages		>1 min	1.1–1.2 pu
4.0 Voltage distortion		Steady state	0.5%–2%
5.0 Waveform distortion			
5.1 DC offset		Steady state	0%–0.1%
5.2 Harmonics	0–100th harmonic	Steady state	0%–20%
5.3 Interharmonics	0–6 KHz	Steady state	0%–2%
5.4 Notching		Steady state	
5.5 Noise	Broadband	Steady state	0–1%
6.0 Voltage fluctuations	<25 Hz	Intermittent	0.1%–7%
7.0 Power frequency variations		<10 s	

In contrast, harmonics take place in steady state and are integer multiples of the fundamental frequency. Also, the waveform distortion that produces the harmonics is continuously present or at least for several seconds. Usually, harmonics are associated with the continuous operation of a load. However, transformer energization is a transient case but can result in a significant waveform distortion for many seconds. Furthermore, this is known to cause system resonance, especially when an underground cabled system is being fed by the transformer.

Harmonic distortion is caused by *nonlinear* devices in the distribution system. Here, a nonlinear device is defined as the one in which the current is not proportional to the applied voltage, that is, while the applied voltage is perfectly sinusoidal, the resulting current is distorted. Increasing the voltage by a small amount may cause the current to double and take on a different wave shape.

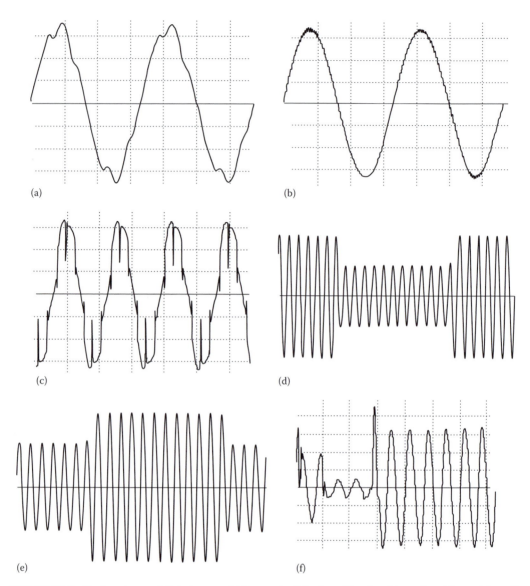

FIGURE 12.1 Various types of disturbances: (a) harmonic distortion, (b) noise, (c) notches, (d) sag, (e) swell, and (f) surge.

Any periodic and distorted waveform can be expressed as a sum of sinusoids with different frequencies. When the waveform is identical from one cycle to the next, it can be represented by the sum of pure sine waves in which the frequency of each sinusoid is an integer multiple of the fundamental frequency of the distorted wave. This multiple is called a *harmonic* of the fundamental. The sum of sinusoids is referred to as a *Fourier series*. In this way, it is much easier to determine the system resonance to an input that is sinusoidal.

For example, the system is analyzed separately at each harmonic using the conventional steady-state analysis techniques. The outputs at each frequency are then combined to form a new Fourier series, from which the output waveform may be determined, if necessary. Usually, only the magnitudes of the harmonics are needed. When both the positive and negative half cycles of a waveform have identical shapes, the Fourier series has only odd harmonics.

TABLE 12.3

Sources and Characteristics of Surge Voltages in Primary and Secondary Distribution Circuits

Type	Source	Characteristics
System switching transients	Line switching, capacitor switching	Propagates in secondary circuits with attenuation at distribution transformers
	Minor load switching	Switching of large commercial or residential loads
	Transients resulting from circuit breaker and fuse operations due to faults	Fast breakers (e.g., vacuum) may cause high-current interruption in the µs range.
Lightning	Direct stroke to primary	Worst case can be in 100 kA range. Typical impulse in primary in the range of 1–6 kA.
	Stroke near primary	Induced in adjacent circuit by magnetic induction. Amplitude of impulse dependent on proximity and intensity of stroke.
	Direct stroke to secondary	Worst case can be in 100 kA range. Typical impulses in 0.5–6 kA range.
	Stoke near secondary	Induced in adjacent circuits by magnetic induction. Overhead circuits with considerable exposure are most likely to experience near-stoke phenomenon (e.g., rural electric circuits).
	Common ground current	Distribution of lightning stokes currents in the earth and in metallic ground circuits cause common coupling with power system ground circuits.

Source: Heydt, G.T., *Electric Power Quality*, 1st edn., Stars in a Circle Publications, West LaFayatte, IN, 1991.

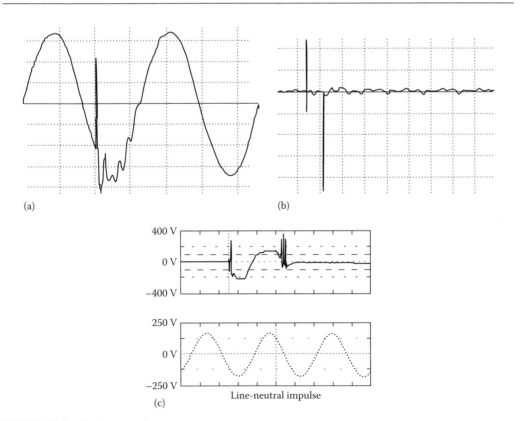

FIGURE 12.2 Various transients.

The presence of even harmonics is often an indication that there is something wrong either with the load equipment or with the transducer used to make the measurement. However, there are exceptions, for example, half-wave rectifiers and arc furnaces when the arc is random.

In a distribution system, most *nonlinearities* can be found in its *shunt* elements, that is, loads. Its series impedance, that is, short-circuit impedance between the sources and the load, is sufficiently nonlinear. Nonlinear loads appear to be sources of harmonic current in shunt with and injecting harmonic currents *into* the power system. For most of harmonics study, it is customary to treat these harmonic-generating loads simply as harmonic current sources, that is, harmonic current generators.

Harmonics, which do little or no useful work, cause extra power losses in distribution transformers, feeders, and some conventional loads such as motors. Harmonics also cause interference in communication circuits, resonance in power systems, and abnormal operations of protection and control equipment.

In the past, most harmonic problems were caused by large single-phase harmonic sources, and they were handled effectively on a case-by-case basis. However, because of the growing use of harmonic-generating power electronic loads, the background distortion levels are gradually increasing. Dealing with such problems is more difficult than dealing with those caused by single-harmonic sources.

12.4.2 CBEMA AND ITI CURVES

Protection of the equipment against the hostile environment is the goal of the technology of electromagnetic compatibility. The Computer Business Equipment Manufacturer Association (CBEMA) has developed the CBEMA curve, shown in Figure 12.3, which can be used to evaluate the voltage quality of a power system with respect to voltage interruptions, dips or undervoltages, and swells or overvoltages.

It was developed as a guideline to help CBEMA members in the design of the power supply for their computer and electronic equipment. A portion of the curve was adopted IEEE Standard (Std.) 446 that is typically used in the analysis of power quality monitoring results.

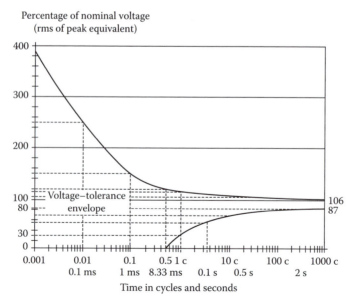

FIGURE 12.3 CBEMA curve.

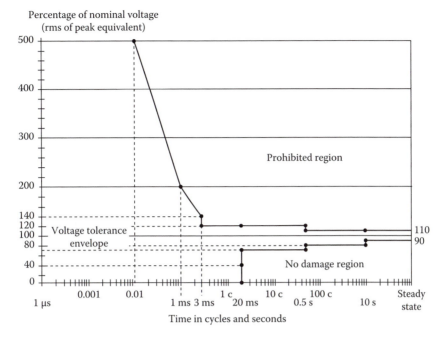

FIGURE 12.4   ITI curve.

The curve shows the magnitude and duration of voltage variations on the power system. The region between the two sides of the curve is the tolerance envelope within which electronic equipment is expected to operate reliably.

In power systems, the only portion of the curve that is used is from 0.1 cycles and higher due to limitations in power quality monitoring instruments. The CBEMA has been replaced by ITI curve, shown in Figure 12.4. It is similar to the CBEMA curve and specifically applies to common 120 V computer equipment. Although developed for 120 V computer equipment, the curve has been applied to general power quality equipment. This curve is also being used in power quality studies.

12.5   MEASUREMENTS OF ELECTRIC POWER QUALITY

12.5.1   RMS Voltage and Current

The expressions for the rms voltage and current are

$$V_{\text{rms}} = \sum_{h=1}^{\infty} \left(\frac{V_h}{\sqrt{2}} \right)^2 \tag{12.1}$$

and

$$I_{\text{rms}} = \sum_{h=1}^{\infty} \left(\frac{I_h}{\sqrt{2}} \right)^2 \tag{12.2}$$

Here, it is assumed that V_h and I_h are also given in rms.

12.5.2 DISTRIBUTION FACTORS

There are several indices that have been used to measure electric power quality. The most widely used one is the *total harmonic distortion* (THD). The total harmonic distortion THD_V, also known as the *voltage distortion factor* (VDF), is defined as

$$\text{THD}_V = \frac{\sqrt{\sum_{h=2}^{\infty} V_h^2}}{V_1} \tag{12.3a}$$

or

$$\text{THD}_V = \sqrt{\left(\frac{V_{\text{rms}}}{V_1}\right)^2 - 1} \tag{12.3b}$$

where
  V_h is the harmonic voltage at harmonic frequency "h" in rms
  V_1 is the rated fundamental voltage in rms
  h is the harmonic order ($h = 1$ corresponds to the fundamental)

Similarly, the total harmonic distortion THD_I, also known as the current distortion factor (CDF), is defined as

$$\text{THD}_I = \frac{\sqrt{\sum_{h=2}^{\infty} I_h^2}}{I_1} \tag{12.4a}$$

or

$$\text{THD}_I = \sqrt{\left(\frac{I_{\text{rms}}}{I_1}\right)^2 - 1} \tag{12.4b}$$

where
  I_h is the harmonic current at harmonic frequency "h" in rms
  I_1 is the rated fundamental current in rms

The rms voltage and current can now be expressed in terms of THD as

$$V_{\text{rms}} = \sqrt{\sum_{h=1}^{\infty} V_h^2} = V_1 \sqrt{1 + \text{THD}_V^2} \tag{12.5}$$

and

$$I_{\text{rms}} = \sqrt{\sum_{h=1}^{\infty} I_h^2} = I_1 \sqrt{1 + \text{THD}_I^2} \tag{12.6}$$

For balanced three-phase voltages, the line-to-line neutral voltage is used in Equation 12.3. But, in the unbalanced case, it is necessary to calculate a different THD for each phase. The voltage THD is almost always a meaningful number. However, this is not the case for the current.

The current THD definition causes some confusion because there is a nonlinear relationship between the magnitude of the harmonic components and percent THD. With the definition of THD, one losses an intuitive feeling for how distorted a particular wave form may be. Distortions greater than 100% are possible, and a waveform with 120% does not contain twice the harmonic components of a waveform with 60% distortion. For the lower levels (less than 10%) of THD, the THD definition is fairly linear. But, for higher levels of THD, which are possible for real-world current distortion, the THD definition is very nonlinear. Also, a small current may have a high THD but not be a significant threat to the system.

This difficulty may be avoided by referring THD to the fundamental of the peak demand current rather than the fundamental of the present sample. This is called *total demand distortion* (TDD) and serves as the basis for the guidelines in IEEE Std. 519-1992. Therefore,

$$\text{TDD} = \frac{\sqrt{\sum_{h=2}^{\infty} I_h^2}}{I_L} \times 100 \tag{12.7}$$

where I_L is the maximum demand load current in rms amps.

When discussing distortion in power distribution systems, it is important to be specific as to the quantity being measured and the conditions of measurements. For example, an equipment may have an "output distortion of 5%." Is this voltage or current distortion? Under what load conditions is it taking place? Transformers often have a specification such as "1% maximum output voltage distortion."

What is not stated is that this voltage distortion specification applies only to a linear load and that the transformer-generated voltage distortion (i.e., 1%) is additive to any voltage distortion that may be present on the input voltage source. When supplying nonlinear loads, the transformer voltage distortion will be higher.

Finally, the *distortion index* (DIN) is commonly used in countries outside of North America. It is defined as

$$\text{DIN} = \sqrt{\frac{\sum_{h=2}^{\infty} V_h^2}{\sum_{h=1}^{\infty} V_h^2}} \tag{12.8}$$

from which

$$\text{DIN} = \frac{\text{THD}}{\sqrt{1 + \text{THD}^2}} \tag{12.9}$$

12.5.3 Active (Real) and Reactive Power

The following relationships are for active (real) and reactive power apply. Active power is

$$p(t) = v(t) \times i(t) \tag{12.10}$$

which has the average

$$P = \frac{1}{T} \int_0^T p(t)dt \tag{12.11a}$$

or

$$P = \frac{1}{2}\sum_{h=1}^{\infty} v_h i_h \cos(\theta_h - \phi_h) \qquad (12.11b)$$

or

$$P = \sum_{h=1}^{\infty} V_h I_h \cos(\theta_h - \phi_h) \qquad (12.11c)$$

The real power is defined as

$$Q = \frac{1}{2}\sum_{h=1}^{\infty} v_h i_h \sin(\theta_h - \phi_h) \qquad (12.12a)$$

or

$$Q = \sum_{h=1}^{\infty} V_h I_h \sin(\theta_h - \phi_h) \qquad (12.12b)$$

12.5.4 Apparent Power

Based on the aforementioned formulas for voltage and current, the apparent power is

$$S = V_{rms} I_{rms} \qquad (12.13a)$$

or

$$S = \sqrt{\sum_{h=1}^{\infty} V_h^2 I_h^2} \qquad (12.13b)$$

or

$$S = V_1 I_1 \sqrt{1 + THD_V^2}\, \sqrt{1 + THD_I^2} \qquad (12.13c)$$

or

$$S = S_1 \sqrt{1 + THD_V^2}\, \sqrt{1 + THD_I^2} \qquad (12.13d)$$

where S_1 is the apparent power at the fundamental frequency.

12.5.5 Power Factor

For purely sinusoidal voltage and current, the average power (or true average active power)

$$P_{avg} = \frac{1}{2} V_m I_m \cos\theta \qquad (12.14)$$

or

$$P_{avg} = V_{rms} I_{rms} \cos \theta \tag{12.15}$$

where

$$V_{rms} = \frac{1}{\sqrt{2}} V_m$$

$$I_{rms} = \frac{1}{\sqrt{2}} I_m$$

$\cos \theta$ is the power factor (PF)

For the sake of simplicity in notation, Equation 12.15 can be expressed as

$$P = VI \cos \theta \tag{12.16}$$

The $\cos \theta$ factor is called the PF. It is said that the PF *leads* when the current leads the voltage and *lags* when the current lags the voltage:

$$PF = \cos \theta = \frac{P}{S} \tag{12.17}$$

This PF is now called the *displacement power factor* (DF). Also,

$$S^2 = P^2 + Q^2 \tag{12.18}$$

But, for the nonsinusoidal case,

$$P^2 = \frac{1}{4} \sum_{h=1}^{\infty} V_h^2 I_h^2 \cos^2 \theta_h \tag{12.19a}$$

or

$$P^2 = \frac{1}{4} \left[V_1^2 I_1^2 \cos^2 \theta_1 + V_2^2 I_2^2 \cos^2 \theta_2 + V_3^2 I_3^2 \cos^2 \theta_3 + \cdots \right] \tag{12.19b}$$

Note that here, the V_h and I_h quantities are the peak quantities:

$$Q^2 = \frac{1}{4} \sum_{h=1}^{\infty} V_h^2 I_h^2 \sin^2 \theta_h \tag{12.20a}$$

or

$$Q^2 = \frac{1}{4} \left[V_1^2 I_1^2 \sin^2 \theta_1 + V_2^2 I_2^2 \sin^2 \theta_2 + V_3^2 I_3^2 \sin^2 \theta_3 + \cdots \right] \tag{12.20b}$$

Therefore, because of harmonic distortion,

$$S^2 > P^2 + Q^2$$

but

$$S^2 = P^2 + Q^2 + D^2 \tag{12.21}$$

where

$$D^2 = \frac{1}{4}\sum_{h=1}^{\infty} V_h^2 I_h^2 \tag{12.22a}$$

or

$$D^2 = \frac{1}{4}\left[V_1^2 I_1^2 + V_2^2 I_2^2 + V_3^2 I_3^2 + \cdots \right] \tag{12.22b}$$

or

$$D^2 = S^2 - (P^2 + Q^2) > 0 \tag{12.23}$$

or

$$D = [S^2 - (P^2 + Q^2)]^{1/2} \tag{12.24}$$

Here, D represents *distortion power* and is called *distortion voltamperes*. It represents all cross products of voltage and current at different frequencies, which yield no average power. Since the PF is a measure of the power utilization efficiency of the load,

$$PF = \frac{\mathrm{Re\,al\;power}\;(power\;consumed)}{\mathrm{Apparent\;power}\;(power\;delivered)}$$

Thus,

$$TPF = \frac{P}{S_{\mathrm{rms}}} = \frac{P}{V_{\mathrm{rms}} I_{\mathrm{rms}}} \tag{12.25a}$$

or

$$TPF = \frac{P}{\sqrt{P^2 + Q^2 + D^2}} \tag{12.25b}$$

The PFs that can be found from Equation 12.26a and b are called the *true power factor* (TPF). The PF that can be found from Equation 12.10 is redefined as the *displacement power factor*. Also, the TPF can be expressed as

$$TPF = \frac{P}{S_{\mathrm{rms}}} = \frac{P}{S_1} \times \frac{1}{\sqrt{1 + \mathrm{THD}_V^2}\sqrt{1 + \mathrm{THD}_I^2}} \tag{12.26a}$$

or

$$TPF = PF \times DPF \tag{12.26b}$$

where
   $DF = P/S_1$ is the displacement power factor
   DPF is the distortion power factor

or

$$DPF = \frac{1}{\sqrt{1+THD_V^2}\sqrt{1+THD_I^2}}$$ (12.27a)

or

$$DPF = \frac{V_1}{V_{rms}} \times \frac{I_1}{I_{rms}} = \frac{S_1}{S_{rms}}$$ (12.27b)

Note that when harmonics are involved, from Equation 12.26, the TPF is

$$TPF_h = \min\left(\frac{1}{\sqrt{1+THD_V^2}}, \frac{1}{\sqrt{1+THD_I^2}}\right) < PF = \frac{P}{S_1}$$

Furthermore, one should not be mislead by a nameplate PF of unity. The unity PF is attainable only with pure sinusoids. What is actually provided is the displacement PF.

Power quality monitoring instruments now commonly report both the displacement factors as well as the TPFs. The displacement factor is typically used in determining PF adjustments on a utility bill since it is related to the displacement of the fundamental voltage and current.

However, sizing capacitors for PF correction is no longer simple. It is not possible to get unity PF due to the distortion power presence. In fact, if resonance effects are significant after installing the capacitors, D can become large and PF would decrease. (In most cases, D is less than 5%.) This results from the fact that the power is proportional to the product of voltage and current.

Capacitors basically compensate only for the fundamental frequency reactive power and cannot completely correct the TPF to unity when there are harmonics present. In fact, capacitors can make the PF worse by creating resonance conditions that magnify the harmonic distortion.

The maximum to which the TPF that can be corrected can approximately be found from

$$TPF \cong \sqrt{\frac{1}{1+THD_I^2}}$$ (12.28)

where
  THD_I is in per units
  THD_V is zero

12.5.6 CURRENT AND VOLTAGE CREST FACTORS

The current crest factor (CCF) is defined as

$$CCF = \frac{\sum_{h=2} I_h}{I_1}$$ (12.29)

and the voltage crest factor (VCF) is defined as

$$VCF = \frac{\sum_{h=2} V_h}{I_1}$$ (12.30)

Neglecting phase angles, the total peak current or voltage would be

$$I_{peak} = \sum_{h=1} I_h = I_1(1 + CCF)$$ (12.31)

or

$$V_{peak} = \sum_{h=1} V_h = V_1(1 + VCF)$$ (12.32)

The corresponding pu increase in total peak current or voltage is then

$$\Delta I_{peak\,pu} = \frac{\Delta I_{peak}}{I_1} = \frac{I_{peak}I_1}{I_1} = \frac{I_{peak}}{I_1} - 1 = CCF$$ (12.33)

or

$$\Delta V_{peak\,pu} = \frac{\Delta V_{peak}}{V_1} = \frac{V_{peak}V_1}{V_1} = \frac{V_{peak}}{V_1} - 1 = CCF$$ (12.34)

Note that $I_{peak}/I_{rms} = \sqrt{2}$ is only true for the case of a pure sinusoid, and the same applies for voltage.

Example 12.1

Based on the output of a harmonic analyzer, it has been determined that a nonlinear load has a total rms current of 75 A. It also has 38, 21, 4.6, and 3.5 A for the third, fifth, seventh, and ninth harmonic currents, respectively. The instrument used in has been programmed to present the resulting data in amps rather than in percentages. Based on the given information, determine the following:

  a. The fundamental current in amps
  b. The amounts of the third, fifth, seventh, and ninth harmonic currents in percentages
  c. The amount of the THD

Solution

  a. Since $I_{rms} = \left(I_1^2 + I_3^2 + I_5^2 + I_7^2 + I_9^2 \right)^{1/2}$,

$$I_1 = \left[I_{rms}^2 - \left(I_3^2 + I_5^2 + I_7^2 + I_9^2 \right) \right]^{1/2} = \left[75^2 - (38^2 + 21^2 + 4.6^2 + 3.5^2) \right]^{1/2} = 60.88 \text{ A}$$

  b. Hence,

$$I_3 = \frac{I_3}{I_1} = \frac{38\,\text{A}}{60.88\,\text{A}} = 0.6242 \quad \text{or} \quad 62.42\%$$

$$I_5 = \frac{I_5}{I_1} = \frac{21\,\text{A}}{60.88\,\text{A}} = 0.3449 \quad \text{or} \quad 34.49\%$$

$$I_7 = \frac{I_7}{I_1} = \frac{4.6\,\text{A}}{60.88\,\text{A}} = 0.0756 \quad \text{or} \quad 7.56\%$$

$$I_9 = \frac{I_9}{I_1} = \frac{3.5\,\text{A}}{60.88\,\text{A}} = 0.0575 \quad \text{or} \quad 5.75\%$$

c. Since $I_1 = I_{\text{rms}} / \sqrt{1 + \text{THD}^2}$,

$$\sqrt{1 + \text{THD}^2} = \frac{I_{\text{rms}}}{I_1} = \frac{75\,\text{A}}{60.88\,\text{A}} \cong 1.232$$

Thus,

$$1 + \text{THD}^2 = 1.232^2 = 1.5178$$

or

$$\text{THD} \cong 0.72 \quad \text{or} \quad 72\%$$

or

$$\text{THD} = \left(I_3^2 + I_5^2 + I_7^2 + I_9^2 \right)^{1/2}$$

$$= (0.6242^2 + 0.3449^2 + 0.0756^2 + 0.0575^2)^{1/2} \cong 0.72 \quad \text{or} \quad 72\%$$

12.5.7 TELEPHONE INTERFERENCE AND THE $I \cdot T$ PRODUCT

Harmonics generate telephone interference through inductive coupling. The $I \cdot T$ product, used to measure telephone interference, is defined as

$$I \cdot T = \sqrt{\sum_{h=1}^{\infty} (I_h T_h)^2}$$

where T_h is the telephone interference weighting factor at the hth harmonic. (It includes the audio effects as well as inductive coupling effects.)

The telephone interference factor (TIF) is defined as

$$\text{TIF} = \frac{\sqrt{\sum_{h=2}^{\infty} (I_h T_h)^2}}{I_1}$$

Table 12.4 gives the telephone interference weighting factors for various harmonics based on Table 12.2 of IEEE Std. 519-1992.

TABLE 12.4
Standard Telephone Interference Weighting Factors

h	1	3	5	7	9	11	13	15	17	19	21	23
T_h	0.5	30	225	650	1320	2260	3360	4350	5100	5630	6050	6370

TABLE 12.5
IEEE Std. 519-1992 Limits for Harmonic Voltage Distortion in Percent at PCC

	2.3–69 kV	69–161 kV	>161 kV
Maximum individual voltage division	3.0	1.5	1.0
Total voltage distortion, THD_V	5.0	2.5	1.5

Example 12.2

A 4.16 kV three-phase feeder is supplying a purely resistive load of 5400 kVA. It has been determined that there are 175 V of zero-sequence third harmonic and 75 V of negative-sequence fifth harmonic. Determine the following:

 a. The total voltage distortion.
 b. Is the THD below the IEEE Std. 519-1992 for the 4.16 kV distribution system?

Solution

 a.

$$\text{THD} = \frac{\sqrt{V_3^2 + V_5^2}}{V_1} \times 100 = \frac{\sqrt{175^2 + 75^2}}{4160} \times 100 = 4.58\%$$

 b. From Table 12.5, the THD_V limit for 4.16 kV is 5%. Since the THD calculated is 4.58%, it is less than the limit of 5% recommended by IEEE Std. 519-1992 for 4.16 kV distribution systems.

Example 12.3

According to ANSI 368 Std., telephone interference from a 4.16 kV distribution system is unlikely to occur when the $I \cdot T$ index is below 10,000. Consider the load given in Example 12.2 and assume that the TIF weightings for the fundamental, the third, and the fifth harmonics are 0.5, 30, and 225, respectively. Determine the following:

 a. The I_1, I_3, and I_5 currents in amps.
 b. The $I \cdot T_1$, $I \cdot T_2$, and $I \cdot T_5$ indices.
 c. The total $I \cdot T$ index.
 d. Is the total $I \cdot T$ index is less the ANSI 368 Std. limit?
 e. The total TIF index.

Solution

 a. $I_1 = I_L = \dfrac{S_{3\phi}/3}{V_{L-L}/\sqrt{3}} = \dfrac{5400\,\text{kVA/3}}{4.16\,\text{kV}/\sqrt{3}} = 748.56\,\text{A}$

 and the resistance is

$$R = \frac{V_{L-L}/\sqrt{3}}{I_L} = \frac{V_1/\sqrt{3}}{I_1} = \frac{4160/\sqrt{3}}{748.56} = 3.2123\,\Omega$$

The harmonic currents are

$$I_3 = \frac{V_3/\sqrt{3}}{R} = \frac{175/\sqrt{3}}{3.2123} = 31.4902\,\text{A}$$

$$I_5 = \frac{V_5/\sqrt{3}}{R} = \frac{75/\sqrt{3}}{3.2123} = 13.4956\,\text{A}$$

b. The $I \cdot T$ indices are

$$I \cdot T_1 = (748.56) \times 0.5 = 374.28$$

$$I \cdot T_3 = (31.4902) \times 30 = 404.868$$

$$I \cdot T_5 = (13.4956) \times 225 = 7085.302$$

c. $$I \cdot T = \sqrt{\sum_{h=1} (I_h T_h)^2} = \sqrt{374.28^2 + 404.868^2 + 7085.302^2} = 7106.72$$

d. Since 7106.72 < 10,000 limit, it is well below the ANSI Std. limit.
e. The total *TIF* index for this case is

$$\text{TIF} = \frac{\sqrt{(0.5 \times 4160)^2 + (30 \times 175)^2 + (225 \times 75)^2}}{\sqrt{4160^2 + 175^2 + 75^2}} = \frac{17794.79}{4164.35} = 4.27$$

Typical requirements of TIF are between 15 and 50.

12.6 POWER IN PASSIVE ELEMENTS

12.6.1 POWER IN A PURE RESISTANCE

Real (or active) power dissipated in a resistor is given by

$$P = \frac{1}{2}\sum_{h=1} V_h I_h = \frac{1}{2}\sum_{h=1} I_h^2 R_h = \frac{1}{2}\sum_{h=1} \frac{V_h^2}{R_h}$$

where R_h is the resistance at the hth harmonic.
    If the resistance is assumed to be constant, that is, ignoring the skin effect, then

$$P = \frac{1}{2R}\sum_{h=1} V_h^2$$

$$= \frac{V_1^2}{2R}\left(1 + \text{THD}_V^2\right)$$

$$= P_1\left(1 + \text{THD}_V^2\right)$$

$$= P_1 \sum_{h=1} V_{h(\text{pu})}^2 \tag{12.35}$$

Alternatively, expressed in terms of current,

$$P = \frac{R}{2} \sum_{h=1} I_h^2$$

$$= \frac{I_1^2 R}{2} \left(1 + \text{THD}_I^2\right)$$

$$= P_1 \left(1 + \text{THD}_I^2\right)$$

$$= P_1 \sum_{h=1} I_{h(\text{pu})}^2 \tag{12.36}$$

Note that the aforementioned equations can be reexpressed in pu as

$$P_{\text{pu}} = \frac{P}{P_1} = 1 + \text{THD}_V^2 = \sum_{h=1} V_{h(\text{pu})}^2$$

$$= 1 + \text{THD}_I^2 = \sum_{h=1} I_{h(\text{pu})}^2 \tag{12.37}$$

where
   P is the total power loss in the resistance
   P_1 is the power loss in the resistance at the fundamental frequency

$$V_{h(\text{pu})} = \frac{V_h}{V_1}$$

$$I_{h(\text{pu})} = \frac{I_h}{I_1}$$

For a purely resistive element, it can be observed from Equation 12.36 that

$$\text{THD}_V = \text{THD}_I.$$

12.6.2 Power in a Pure Inductance

Power in a pure inductance can be expressed as

$$Q_L = \frac{1}{2} \sum_{h=1} V_h I_h = \sum_{h=1} V_{h(\text{rms})} I_{h(\text{rms})} \tag{12.38}$$

where

$$V_1 = j2\pi f_1 L I_1$$

$$V_h = j2\pi f_1 L I_h$$

f_1 is the fundamental frequency.

Thus,

$$\frac{V_h}{V_1} = h \times \frac{I_h}{I_1} \tag{12.39}$$

so that

$$\frac{Q_L}{Q_{L1}} = \frac{\frac{1}{2}\sum_{h=1} V_h I_h}{\frac{1}{2} V_1 I_1}$$

$$= \sum_{h=1} h \left(\frac{I_h}{I_1}\right)^2 = \sum_{h=1} \frac{1}{h} \left(\frac{V_h}{V_1}\right)^2 \tag{12.40}$$

or

$$Q_{L(\mathrm{pu})} = \sum_{h=1} h \times I_{h(\mathrm{pu})}^2 = \sum_{h=1} \frac{V_{h(\mathrm{pu})}^2}{h} \tag{12.41}$$

12.6.3 POWER IN A PURE CAPACITANCE

Power in a pure capacitance can be expressed as

$$Q_c = -\frac{1}{2}\sum_{h=1} V_h I_h = -\sum_{h=1} V_{h(\mathrm{rms})} I_{h(\mathrm{rms})} \tag{12.42}$$

The *negative sign* indicates that the reactive power is delivered to the load:

$$V_1 = \frac{I_1}{j2\pi f_1 C}$$

and

$$V_h = \frac{I_h}{j2\pi h f_1 C}$$

Thus,

$$\frac{V_h}{V_1} = \frac{I_h}{h \times I_1} \tag{12.43}$$

Hence,

$$\frac{Q_c}{Q_{c1}} = \frac{-\frac{1}{2}\sum_{h=1} V_h I_h}{-\frac{1}{2} V_1 I_1}$$

$$= \sum_{h=1} h \left(\frac{V_h}{V_1}\right)^2 = \sum_{h=1} \frac{1}{h} \left(\frac{I_h}{I_1}\right)^2 \tag{12.44}$$

or

$$Q_{c(\mathrm{pu})} = \sum_{h=1} h \times V_{h(\mathrm{pu})}^2 = \sum_{h=1} \frac{I_{h(\mathrm{pu})}^2}{h} \tag{12.45}$$

12.7 HARMONIC DISTORTION LIMITS

IEEE Std. 519-1992 is entitled *Recommended Practices and Requirements for Harmonic Control in Electric Power Systems* [7]. It gives the recommended practice for electric power system designers to control the harmonic distortion that might otherwise determine electric power quality.

The recommended practice is to be used as a guideline in the design of power system with nonlinear loads. The limits set are for steady-state operation and are recommended for "worse-case" conditions. The underlying philosophy is that the customer should limit harmonic currents and the electric utility should limit harmonic voltages.

It does not specify the highest-order harmonics to be limited. Also, it does not differentiate between single-phase and three-phase systems. Thus, the recommended harmonic limits equally apply to both. It does also address direct current that is not a harmonic.

12.7.1 Voltage Distortion Limits

The current edition of IEEE 519-1992 establishes limits on voltage distortion that a utility may supply a user. This assumes almost unlimited ability for the utility to absorb harmonic currents from user. It is obvious that in order for the utility to meet the voltage distortion limits, some limits must be placed on the amount of harmonic current that users can inject the power system.

Table 12.5 gives the new IEEE 519 voltage distortion limits at the point of common coupling (PCC) to the utility and other users. The concept of PCC is illustrated in Figure 12.5. It is the location where another customer can be served from the system. It can be located at either the primary or the secondary of a supply transformer depending on whether or not multiple customers are supplied from the transformer.

12.7.2 Current Distortion Limits

The harmonic currents from an individual customer are evaluated at the PCC where the utility can supply other customers. The limits are dependent on the customer load in relation to the system short-circuit capacity at the PCC. Note that all current limits are expressed as a percentage of the customer's average maximum demand load current.

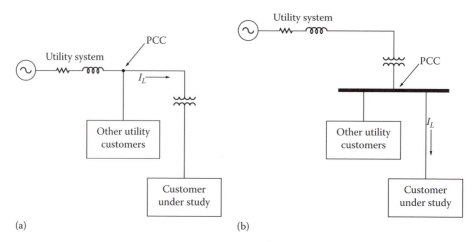

(a)   (b)

FIGURE 12.5 Selection of PCC.

Table 12.6 gives the new IEEE 519 harmonic current distortion limits at the PCC to the customer. The current distortion limits vary by the size of the user relative to the utility system capacity. The limits attempt to prevent users for a disproportionately using the utility's harmonic current absorption capacity as well as reducing the possibility of harmonic distortion problems. According to the changes that are suggested in reference [8] include the expansion of voltage levels up to and beyond 161 kV, are included in Tables 12.7 and 12.8, respectively.

TABLE 12.6

IEEE Std. 519-1992 Limits Imposed on Customers (120 V–69 kV) for Harmonic Current Distortion in Percent of I_L for Odd Harmonic h at the PCC

I_{sc}/I_L	$h < 11$	$11 \leq h < 17$	$17 \leq h < 23$	$23 \leq h < 35$	$35 \leq h$	TDD
<20[a,b]	4.0	2.0	1.5	0.6	0.3	5.0
20–50	7.0	3.5	2.5	1.0	0.5	8.0
50–100	10.0	4.5	4.0	1.5	0.7	12.0
100–1000	12.0	5.5	5.0	2.0	1.0	15.0
>1000	15.0	7.0	6.0	2.5	1.4	20.0

[a] I_{sc} is the short-circuit current at the PCC. I_L is the maximum demand load current (fundamental frequency component) also at the PCC. It can be calculated as the average of the maximum monthly demand currents for the previous 12 months or it may have to be estimated.

[b] All power generation equipment applications are limited to these values of current distortion regardless of the actual short-circuit ratio I_{sc}/I_L. The individual harmonic component limits apply to the odd harmonic components. Even harmonic components are limited to 25% of the limits in the table. Current distortions that result in a dc offset, for example, half-wave converters, are not allowed.

TABLE 12.7

IEEE Std. 519-1992 Limits Imposed on Customers (69–161 kV) for Harmonic Current Distortion in Percent of I_L for Odd Harmonic h at the PCC

I_{sc}/I_L	$h < 11$	$11 \leq h < 17$	$17 \leq h < 23$	$23 \leq h < 35$	$35 \leq h$	TDD
<20[a]	2.0	1.0	0.75	0.3	0.15	2.5
20–50	3.5	1.75	1.25	0.5	0.25	4.0
50–100	5.0	2.25	2.0	1.25	0.35	6.0
100–1000	6.0	2.75	2.5	1.0	0.5	7.5
>1000	7.5	3.5	3.0	1.25	0.7	10.0

[a] See the footnotes of Table 12.6.

TABLE 12.8

IEEE Std. 519-1992 Limits Imposed on Customers (above 161 kV) for Harmonic Current Distortion in Percent of I_L for Odd Harmonic h at the PCC

I_{sc}/I_L	$h < 11$	$11 \leq h < 17$	$17 \leq h < 23$	$23 \leq h < 35$	$35 \leq h$	TDD
<50[a]	2.0	1.0	0.75	0.3	0.15	2.5
≥50	3.5	1.75	1.25	0.5	0.25	4.0

[a] See the footnotes of Table 12.6.

Since harmonic effects differ substantially depending on the equipment affected, the severity of the harmonic effects imposed on all types of equipment cannot completely be connected to a few simple harmonic indices. Also, the harmonic characteristics of the utility circuit seen from the PCC are often not known accurately. Therefore, good engineering judgment often dictated to review a case-by-case basis. However, through a judicious application of the recommended practice, the interferences between different loads and the system can be minimized. According to IEEE 519-1992, the evaluation procedure for newly installed nonlinear loads includes the following:

1. Definition of the PCC
2. Determination of the I_{sc}, I_L, and I_{sc}/I_L at the PCC
3. Finding the harmonic current and current distortion of the nonlinear load
4. Determination of whether or not the harmonic current and current distortions in step 3 satisfy IEEE 519-1992 recommendation limits
5. Taking necessary remedies to meet the guidelines

Preventive solutions, such as IEEE 519-1992 guidelines for dealing with harmonics, are the best course of action. However, if these guidelines are not satisfied, the remedial solution, such as passive or active filtering, should be included at the time of installation to avoid potential problems. Meanwhile, the I_{sc}/I_L ratio may vary due to different PCC choices. The risk should be reevaluated whenever the I_{sc}/I_L ratio is unchanged.

Harmonic controls can be exercised at the utility and end-user sides. IEEE Std.519 attempts to establish reasonable harmonic goals for electric systems that contain nonlinear loads. The objectives are the following: (1) customers should limit harmonic currents, since they have control over their loads; (2) electric utilities should limit harmonic voltages, since they have control over the system impedances; and (3) both parties share the responsibility for holding harmonic levels in check.

12.8 EFFECTS OF HARMONICS

Harmonics adversely affect virtually every component in the power system with additional dielectric, thermal, and/or mechanical stresses. Harmonics cause increased losses and equipment loss of life. For example, when magnetic devices, such as motors, transformers, and relay coils, are operated from a distorted voltage source, they experience increased heating due to higher iron and copper losses.

Harmonics typically also cause additional audible noise. In motors and generators, severe harmonic distortion can also cause oscillating torques that may excite mechanical resonances. In general, unless specifically designed to accommodate harmonics, magnetic devices should not be operated from voltage sources having more than 5% THD.

Wiring is also affected by harmonic currents. In the case of parallel resonance, the associated wiring may be subjected to abnormally high harmonic current flow. The conductors also experience additional heating beyond the normal I^2R_{dc} losses, due to skin effect and proximity effect that vary by frequency and wiring construction. The ac resistance that includes these effects needs to be calculated at each harmonic current frequency and the wiring ampacity must be derated. Normally, the derating required for harmonic is minimal and can be ignored if conservative wire sizing methods are used.

In general, when capacitors are applied to a power system having significant nonlinear loads, some necessary precautions must be taken to prevent parallel resonance. Even without resonance, additional harmonic current will flow in the capacitors causing additional losses and reduced life. With resonance, the high harmonic voltages and currents can cause capacitor fuses or capacitor failures.

TABLE 12.9

Comparison of Sensing Techniques for Various Waveforms

Waveform	Average Sensing RMS Calibrated (A)	Peak Sensing RMS Calibrated (A)	True-RMS Sensing (A)
Sine wave	100	100	100
Square wave	111	71	100
Triangle wave	97	122	100
Rectifier-capacitor power supply current	50	201	100
Personal computer load	833	168	100

Metering and overcurrent protection can also be affected by harmonics unless they are designed with *true-rms* sensing. Many devices such as meters and overcurrent protection can be peak or average sensing devices that are rms calibrated assuming sinusoidal waveforms. When nonsinusoidal waveforms are present, significant sensing errors can result. PF meters that are based on phase angle measurements cannot be relied upon with nonlinear loads. Solid state circuit breakers without *true-rms* sensing may experience nuisance tripping or, worse yet, fail to trip when used with nonlinear loads. Table 12.9 shows various sensing errors under different sensing techniques for various waveforms.

Other loads may also be affected by harmonics. The voltage and/or current harmonics can be coupled into sensitive loads and appears as noise or interference, causing degradation of performance or misoperation. Examples of these are voltage harmonics causing picture quality problems in television monitors and current harmonics causing interference in telephone circuits. Some computer systems are also known to be sensitive to voltage distortion, as evidenced by computer vendor specifications that include limits on voltage distortion, typically 5% THD.

Furthermore, harmonics may affect capacitor banks in many ways that include the following:

1. Capacitors can be overloaded by harmonic currents. This is due to the fact that their reactances decrease with frequency makes them the sinks for harmonics.
2. Harmonics tend to increase the dielectric losses of capacitors, causing additional heating and loss of life.
3. Capacitors combined with source inductance may develop a parallel resonant circuits, causing harmonics to be amplified. The resulting voltages may greatly over exceed the voltage ratings of the capacitors, causing capacitor damage and/or blown out fuses. For a remedy, relocating capacitors changes the source-to-capacitor inductive reactance hence prevents the occurrence of the parallel resonance with the supply. Also, varying the reactive power output of a capacitor bank will change the resonant frequency.

12.9 SOURCES OF HARMONICS

As explained in Section 8.10, voltage sources, that is, generators, inverters, and transformers, produce voltage distortion. Good generators produce minimal voltage distortions that are usually less than 0.5% THD. Inverter voltage distortion depends on the inverter design.

Most online uninterruptible power supply (UPS) inverters produce less than 4%–5% THD with linear load. Many off-line, that is, standby, UPS inverters produce distorted voltages with greater than 25% of THD.

Transformers add voltage distortion to the input voltage waveform. Most good power transformers will add less than 0.5%–1.0% THD with linear loads.

Today, the major sources of distortion in a power system are the harmonic currents of nonlinear loads. The magnetizing currents of transformers and other magnetic devices are usually quite insignificant. Arc or discharge loads, such as arc furnaces or gas discharge lighting, represent very nonlinear loads. Static power converters represent one of the more widely used nonlinear loads. Static power converters include motor drives, battery charges, UPS, and omnipresent electronic power supply. The levels of current distortion of static power converters are dependent upon their designs. For example, electric power supplies have been observed to produce up to 140% current distortion, while newer designs can have as low as 3% THD.

Succinctly put, harmonic distortion has many causes. Distortion voltage sources will cause harmonic currents to flow in linear loads. Nonlinear loads will draw distorted currents from otherwise sinusoidal voltage source. Distorted currents flowing through the power system impedance will cause voltage distortion.

In power systems, the nonlinear load can be modeled as a load for the fundamental current and as a current source for the harmonic currents. The harmonic currents flow from the nonlinear load toward the power source, following the paths of least impedance, as shown in Figure 12.6. The voltage drops in the power system components at each harmonic current frequency will add to, or subtract from, the generated voltage to produce the distorted voltage system.

On radial primary feeders and industrial plant power systems, generally, the harmonic currents flow from the harmonic-producing load toward the power system source, as illustrated in Figure 12.7a. This tendency can be used to locate sources of harmonics. For example, one can use a power quality monitoring device, which is capable of showing harmonic contents of the current, and measure the harmonic currents in each branch, starting at the beginning of the circuit, and trace the harmonics to the source.

However, if there are PF capacitors, this flow pattern can be altered for at least one of the harmonics. For example, adding a capacitor may draw a large amount of harmonic current into that portion of the circuit, as illustrated in Figure 12.7b. Because of this, it is usually required that all capacitors are temporarily disconnected to accurately locate the sources of harmonics.

In the presence of resonance involving a capacitor bank, it is very easy to differentiate harmonic currents due to actual harmonic sources from harmonic currents that are strictly due to resonance. The resonance currents have one dominant harmonic riding on top of the fundamental sinusoidal sine wave. Thus, a large single harmonic almost always indicates resonance.

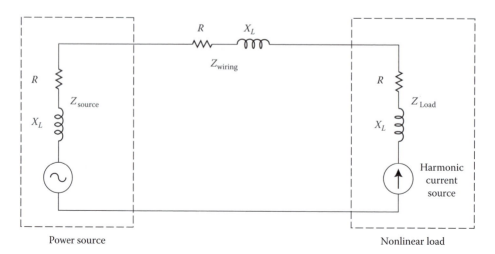

FIGURE 12.6 Representation of a nonlinear load.

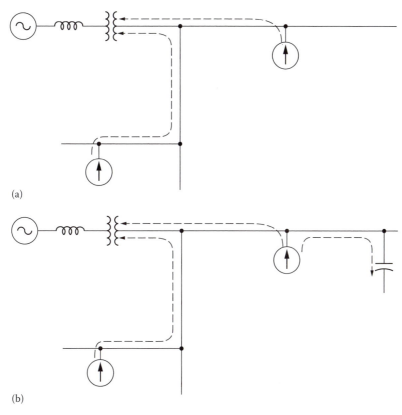

(a)

(b)

FIGURE 12.7 General flow of harmonic currents in a radial power system: (a) without power capacitors and (b) with power capacitors.

12.10 DERATING TRANSFORMERS

Transformers serving nonlinear loads exhibit increased eddy current losses due to harmonic currents generated by those loads. Because of this, the transformer rating is derated using a K-factor.

12.10.1 *K*-FACTOR

Both the Underwriters Laboratories (UL) and transformer manufacturers established a rating method called *K-factor* to indicate their suitability for nonsinusoidal load currents.

This K-factor relates transformer capability to serve varying degrees of nonlinear load without exceeding the rated temperature rise limits. It is based on the predicted losses of a transformer. In per unit, the K-factor is

$$K = \frac{\sum_{h=1}^{\infty} I_h^2 \times h^2}{\sum_{h=1}^{\infty} I_h^2} \tag{12.46}$$

where I_h is the rms current at harmonic h, in per unit of rated rms load current

According to UL specification, the rms current of any single harmonic that is greater than the 10th harmonic be considered as no greater than $1/h$ of the fundamental rms current.

Today, manufacturers build special K-factor transformers. Standard K-factor ratings are 4, 9, 13, 20, 30, 40, and 50. For linear loads, the K-factor is always one. For nonlinear loads, if harmonic currents are known, the *K*-factor is calculated and compared against the transformer's nameplate *K*-factor. As long as the load *K*-factor is equal to, or less than, the transformer *K*-factor, the transformer does not need to be derated.

12.10.2 Transformer Derating

For transformers, ANSI/IEEE Std. C75.110 [5] provides a method to derate the transformer capacity when supplying nonlinear loads. The transformer derating is based on additional eddy current losses due to the harmonic current and that these losses are proportional to the square of the frequency. Thus,

$$\text{Transformer derating} = \sqrt{\frac{1 + P_{\text{ec-r}}}{1 + \dfrac{\sum\limits_{h=1}^{\infty} I_h^2 h^2}{\sum\limits_{h=1}^{\infty} I_h^2} \times P_{\text{ec-r}}}} \tag{12.47}$$

or

$$\text{Transformer derating} = \sqrt{\frac{1 + P_{\text{ec-r}}}{1 + K \times P_{\text{ec-r}}}} \tag{12.48}$$

where
$P_{\text{ec-r}}$ is the maximum transformer per unit eddy current loss factor (typically, between 0.05 and 0.10 per units for dry-type transformers)
I_h is the harmonic current, normalized by dividing it by the fundamental current
h is the harmonic order

Table 12.10 gives some of the typical values of $P_{\text{ec-r}}$ based on the transformer type and size.

Example 12.4

Assume that the per unit harmonic currents are 1.000, 0.016, 0.261, 0.050, 0.003, 0.089, 0.031, 0.002, 0.048, 0.026, 0.001, 0.033, and 0.021 pu A for the harmonic order of 1, 3, 5, 7, 9, 11, 13, 15, 17, 19, 21, 23, and 25, respectively. Also assume that the eddy current loss factor is 8%. Based on ANSI/IEEE Std. C75.110, determine the following:

    a. The *K*-factor of the transformer
    b. The transformer derating based on the standard

TABLE 12.10
Typical Values of $P_{\text{ec-r}}$

Type	MVA	Voltage	$P_{\text{ec-r}}$ (%)
Dry	≤1	—	3.8
	≥1.5	5 kV (HV)	12–20
	≤1.5	15 kV (HV)	9–15
Oil filled	≤2.5	480 V (LV)	1
	2.5–5	480 V (LV)	1–5
	>5	480 V (LV)	9–15

TABLE 12.11

The Results of Example 12.4, Part (a)

Harmonic (h)	Currents (pu)	I^2	$I^2 \times h^2$
1	1.000	1.000	1.000
3	0.016	0.000	0.002
5	0.261	0.068	1.703
7	0.050	0.003	0.123
9	0.003	0.000	0.001
11	0.089	0.008	0.958
13	0.031	0.001	0.162
15	0.002	0.000	0.001
17	0.048	0.002	0.666
19	0.026	0.001	0.244
21	0.001	0.000	0.000
23	0.033	0.001	0.576
25	0.021	0.000	0.276
	Totals	1.084	5.712

Solution

a. The results are given in Table 12.11.
   Thus, the K-factor is

$$K = \frac{\sum_{h=1}^{\infty} I_h^2 h^2}{\sum_{h=1}^{\infty} I_h^2} = \frac{5.712}{1.084} \cong 5.3$$

b. According to the standard, the transformer derating is

$$\text{Transformer derating} = \sqrt{\frac{1+P_{\text{ec-r}}}{1+K \times P_{\text{ec-r}}}} = \sqrt{\frac{1+0.08}{1+5.3 \times 0.08}} \cong 0.87\,\text{pu} \quad \text{or} \quad 87\%$$

12.11 NEUTRAL CONDUCTOR OVERLOADING

When single-phase electronic loads are supplied with a three-phase four-wire circuit, there is a concern for the current magnitudes in the neutral conductor. Neutral current loading in the three-phase circuits with linear loads is simply a function of the load balance among the three phases. With relatively balanced circuits, the neutral current magnitude is quite small.

In the past, this has resulted in a practice of undersizing the neutral conductor in a relation to the phase conductors. Power system engineers are accustomed to the traditional rule that *balanced three-phase systems have no neutral currents.* However, this rule is not true when power electronic loads are present.

With electronic loads supplied by switch-made power supplies and fluorescent lighting with electronic ballasts, the harmonic components in the load currents can result in much higher neutral current magnitudes. This is because the odd triplen harmonics (3, 9, 15, etc.) produced by these loads show up as zero-sequence components for balanced circuits.

Instead of canceling in the neutral (as is the case with positive- and negative-sequence components), zero-sequence components add directly in the neutral. The third harmonic is usually the largest single-harmonic component in single-phase power supplies or electronic ballasts. As shown in the next example, the neutral current in such cases will approximately be 173% of the rms phase current magnitude.

The conclusion from this calculation is that neutral conductors in circuits supplying electronic loads should not be undersized. In fact, they should have almost twice the ampacity of the phase conductors. An alternative method to wire these circuits is to provide a neutral conductor with each phase conductor.

Also, many PCs have third harmonic currents greater than 80%. In such cases, the neutral current will be at least 3(80%) = 240% of the fundamental phase current. Therefore, when PC loads dominate a building circuit, it is good engineering practice for each phase to have its own neutral wire or for the shared neutral wire to have at least twice the current rating of each phase wire. Overloaded neutral current are usually only a local problem inside a building, for example, at a service panel.

However, the neutral current concern is not as significant on the 480 V system. The zero-sequence components from the power supply loads are trapped in the delta winding of the step-down transformers to the 120 V circuits. Therefore, the only circuits with any neutral current concern are those supplying fluorescent lighting loads connected to line to neutral, that is, 277 V. In this case, the third harmonic components are much lower.

Typical electronic ballast should not have a third harmonic component exceeding 30% of the fundamental. This means that the neutral current magnitude should always be less than the phase current magnitude in circuits supplying fluorescent lighting load. In these circuits, it is sufficient to make the sizes of neutral conductors the same as the phase conductors.

Office areas and computer rooms with high concentrations of single-phase line-to-neutral power supplies are particularly vulnerable to overheated neutral conductors and distribution transformers. Trends in computer systems over the last several years have increased the likelihood of high neutral currents. Computer systems have shifted from three-phase to single-phase power supplies. Development of switched-mode power supplies allows connection directly to the line-to-neutral voltage without a step-down transformer.

Additionally, there were of buildings not specifically designed to accommodate computer systems. The CBEMA has recognized this concern and alerted the industry to problems caused by harmonics from computer power supplies.

The *possible solutions to neutral conductor overloading* include the following:

1. A separate neutral conductor is provided with each phase conductor in a three-phase circuit that serves single-phase nonlinear loads.
2. When a shared neutral is used in a three-phase circuit with single-phase nonlinear loads, the neutral conductor capacity should approximately double the phase-conductor capacity.
3. In order to limit the neutral currents, delta–wye transformers specifically designed for nonlinear loads can be used. They should be located as close as possible to the nonlinear loads, for example, computer rooms, to minimize neutral conductor length and cancel triplen harmonics.
4. The transformer can be derated, or oversized, in accordance with ANSI/IEEE C57.110 to compensate for the additional losses due to the harmonics.
5. The transformers should be provided with supplemental transformer overcurrent protection, for example, winding temperature sensors.
6. The third harmonic currents can be controlled by placing filters at the individual loads, if rewiring is an expensive solution.

Example 12.5

In an office building, measurement of a line current of branch circuit serving exclusively computer load has been made using a harmonic analyzer. The outputs of the harmonic analyzer are phase current waveform and the spectrum of current supplying such electronic power loads. For a

60 Hz, 58.5 A rms fundamental current, it is observed from the spectrum that there is 100% fundamental and odd triplen harmonics of 63.3%, 4.4%, 1.9%, 0.6%, 0.2%, and 0.2% for 3rd, 9th, 15th, 21st, 27th, and 33rd order, respectively. If it is assumed that loads on the three phases are balanced and all have this same characteristic, determine the following:

a. The approximate rms value of the phase current in per units
b. The approximate rms value of the neutral current in per units
c. The ratio of the neutral current to the phase current

Solution

a. The approximate rms value of the phase current is

$$I_{phase} = \left(I_1^2 + I_3^2\right)^{1/2} = (1.0^2 + 0.706^2)^{1/2} = 1.2241\,pu$$

where

$$I_3 = (63.3 + 4.4 + 1.9 + 0.6 + 0.2 + 0.2)\% = 70.6\% = 0.706\,pu$$

b. The approximate rms value of the neutral current is

$$I_{neutral} = (I_3 + I_3 + I_3) = 0.706 + 0.706 + 0.706 = 2.118\,pu$$

c. Hence, the ratio of the neutral current to the phase current is

$$\frac{I_{neutral}}{I_{phase}} = \frac{2.118\,pu}{1.2241\,pu} = 1.73$$

or

$$I_{neutral} = 1.73 \times I_{phase}$$

Example 12.6

A commercial building is being served by 480 V so that its fluorescent lighting loads can be supplied by a line-to-neutral voltage of 277 V. It is observed that the third harmonic components are much lower. For instance, typical electronic ballast used with the fluorescent lighting should not have a third harmonic component exceeding 30% of the fundamental. For this worse-case analysis, determine the following:

a. The approximate rms value of the phase current in per units
b. The approximate rms value of the neutral current in per units
c. The ratio of the neutral current to the phase current

Solution

a. The approximate rms value of the phase current is

$$I_{phase} = \left(I_1^2 + I_3^2\right)^{1/2} = (1.0^2 + 0.3^2)^{1/2} = 1.04\,pu$$

b. The approximate rms value of the neutral current is

$$I_{neutral} = (I_3 + I_3 + I_3) = 0.3 + 0.3 + 0.3 = 0.9\,pu$$

c. Hence, the ratio of the neutral current to the phase current is

$$\frac{I_{\text{neutral}}}{I_{\text{phase}}} = \frac{0.9 \text{ pu}}{1.04 \text{ pu}} = 0.87$$

or

$$I_{\text{neutral}} = 0.87 \times I_{\text{phase}}$$

This means that the neutral current magnitude should always be less than the phase current magnitude.

12.12 CAPACITOR BANKS AND POWER FACTOR CORRECTION

As discussed in Chapter 8, capacitor banks used in parallel with an inductive load provide this load with reactive power. They reduce the system's reactive and apparent power and, therefore, cause its PF to increase.

Furthermore, capacitor current causes voltage rise that results in lower line losses and voltage drops loading to an improved efficiency and voltage regulation. Based on the power triangle shown in Figure 12.8, the reactive power delivered by the capacitor bank Q_c is

$$Q_c = Q_1 - Q_2$$
$$= P(\tan\theta_1 - \tan\theta_2)$$
$$= P\left[\tan\left(\cos^{-1}\text{PF}_1\right) - \tan\left(\cos^{-1}\text{PF}_2\right)\right] \qquad (12.49)$$

where
P is the real power delivered by the system and absorbed by the load
Q_1 is the load's reactive power
Q_2 is the system reactive power after the capacitor bank connection

As it can be observed from the following equation, since a low PF means a high current,

$$I = \frac{P_{3\phi}}{\sqrt{3}V_{L-L}\cos\theta}$$

the disadvantages of a low PF include the following: (1) increased line losses, (2) increased generator and transformer ratings, and (3) extra regulation equipment for the case of low lagging PF.

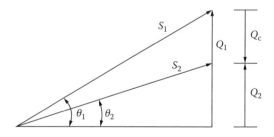

FIGURE 12.8 Power triangle for a PF correction capacitor bank.

12.13 SHORT-CIRCUIT CAPACITY OR MVA

Where a new circuit to be added to an existing bus in a complex power system, short-circuit capacity or MVA (or kVA) data provide the equivalent impedance of the power system up to that bus. The three-phase short-circuit MVA is determined from

$$MVA_{sc(3\phi)} = \frac{\sqrt{3}I_{3\phi}kV_{L-L}}{1000} \tag{12.50}$$

where
$I_{3\phi}$ is the total three-phase fault current in A
kV_{L-L} is the system phase-to-phase voltage in kV

$$I_{3\phi} = \frac{1000\,MVA_{sc(3\phi)}}{\sqrt{3}kV_{L-L}} \tag{12.51}$$

Alternatively,

$$MVA_{sc(3\phi)} = \frac{(kV_{L-L})^2}{Z_{sc}} \tag{12.52}$$

from which

$$Z_{sc} = \frac{(kV_{L-L})^2}{MVA_{sc(3\phi)}} = \frac{1000\,kV_{L-L}}{\sqrt{3}I_{3\phi}} = \frac{V_{L-N}}{I_{3\phi}} \tag{12.53}$$

It is often in power systems, the short-circuit impedance is equal to the short-circuit reactance, ignoring the resistance and shunt capacitance involved. Hence, the three-phase short-circuit MVA is found from

$$MVA_{sc(3\phi)} = \frac{(kV_{L-L})^2}{X_{sc}} \tag{12.54}$$

from which

$$X_{sc} = \frac{(kV_{L-L})^2}{MVA_{sc(3\phi)}} = \frac{1000\,kV_{L-L}}{\sqrt{3}I_{3\phi}} = \frac{V_{L-N}}{I_{3\phi}} \tag{12.55}$$

12.14 SYSTEM RESPONSE CHARACTERISTICS

All circuits containing both capacitance and inductance have one or more natural resonant frequencies. When one of these frequencies corresponds to an exciting frequency being produced by nonlinear loads, harmonic resonance can occur. Voltage and current will be dominated by the resonant frequency and can be highly distorted. The response of the power system at each harmonic frequency determines the true impact of the nonlinear load on harmonic voltage distortion.

Somewhat surprisingly, power systems are quite tolerant of the currents injected by harmonic-producing loads unless there is some adverse interaction with the system impedance. The response of the power system at each harmonic frequency determines the true impact of the nonlinear load on harmonic voltage distortion.

12.14.1 System Impedance

Since at the fundamental frequency power systems are mainly inductive, their equivalent impedances are also called the short-circuit reactance. In utility distribution systems as well as industrial power systems, capacitive effects are frequently ignored.

The short-circuit impedance Z_{sc} (to the point on a power network at which a capacitor located) can be calculated from fault study results as

$$\mathbf{Z}_{sc} = R_{sc} + jX_{sc} = \frac{kV_{L-L}^2}{MVA_{sc(3\phi)}} \tag{12.56}$$

where
R_{sc} is the short-circuit resistance
X_{sc} is the short-circuit reactance
kV_{L-L} is the phase-to-phase voltage, kV
$MVA_{sc(3\phi)}$ is the three-phase short-circuit MVA

The inductive reactance portion of the impedance changes linearly with frequency. The reactance at the hth harmonic is found from the fundamental-impedance reactance X_1 by

$$X_h = h \times X_1 \tag{12.57}$$

In general, the resistance of most power system components does not change significantly for the harmonics less than the ninth. However, this is not the case for the lines and cables as well as transformers.

For the lines and cables, the resistance changes roughly by the square root of the frequency once the skin effect becomes significant in the conductor at a higher frequency.

For larger transformers, their resistances may vary almost proportionately with the frequency because of the eddy current losses. At utilization voltages, such as industrial power systems, using the transformer impedance X_T as X_{sc} may be a good approximation so that

$$X_{sc} \cong X_T \tag{12.58}$$

Generally, this X_{sc} is about 90% of the total impedance. It usually suffices for the assessment of whether or not there will be a harmonic resonance problem. If the transformer impedance is given in percent, from its per unit value, its impedance value in ohms can be found from

$$X_T = \frac{kV_{L-L}^2}{MVA_{sc(3\phi)}} \times Z_{T,pu} \tag{12.59}$$

Here, it is assumed that the transformer's resistance is negligibly small.

12.14.2 Capacitor Impedance

Shunt capacitors substantially change the system impedance variation with frequency. They do not create harmonics. However, severe harmonic distortion can sometimes be attributed to their presence. While the reactance of inductive components increases proportionately to frequency, capacitive reactance X_c decreases proportionately:

$$X_c = \frac{1}{2\pi f C} \tag{12.60}$$

where C is the capacitance in farads.

The equivalent line-to-neutral capacitive reactance at the fundamental frequency of a capacitor bank is found from

$$X_c = \frac{V^2}{Q} \tag{12.61a}$$

or

$$X_c = \frac{kV^2}{M \, \text{var}} = \frac{1000 \times kV^2}{k \, \text{var}} \tag{12.61b}$$

12.15 BUS VOLTAGE RISE AND RESONANCE

Assume that a switched capacitor bank is connected to a bus that has an impedance load, as shown in Figure 12.9, and that the short-circuit capacity of the bus is MVA_{sc}. With the resistance ignored, after the switch is closed, the equivalent (short-circuit) impedance of the system (or the source) is

$$X_s = X_{sc} = \omega_1 L_s = \frac{kV_{rated}^2}{MVA_{sc}} \tag{12.62}$$

or in per units,

$$X_{s,pu} = \frac{X_s}{Z_B} \tag{12.63a}$$

or

$$X_{s,pu} = X_{sys,pu} = \frac{S_B}{MVA_{sc}} = \frac{1}{MVA_{sc,pu}} \tag{12.63b}$$

where S_B is in MVA and $kV_B = kV_{rated}$.

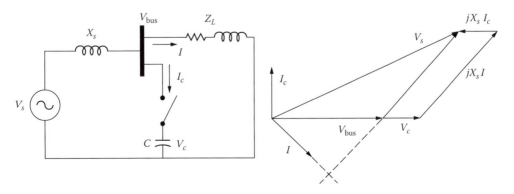

FIGURE 12.9 Power system with shunt switched capacitor.

Also,

$$X_{c,\text{pu}} = \frac{X_c}{Z_B} \tag{12.64a}$$

or

$$X_{c,\text{pu}} = \frac{1}{Q_{c,\text{pu}}} = \frac{1}{M\,\text{var}_{c,\text{pu}}} \tag{12.64b}$$

and the resonant frequency of the system is

$$f_r = \frac{1}{2\pi\sqrt{L_s C}} \tag{12.65a}$$

or

$$f_r = f_1 \sqrt{\frac{X_{c,\text{pu}}}{X_{s,\text{pu}}}} = f_1 \sqrt{\frac{MVA_{\text{sc,pu}}}{M\,\text{var}_{c,\text{pu}}}} \tag{12.65b}$$

Since at the resonance,

$$h_r = \frac{f_r}{f_1} \tag{12.66a}$$

or

$$h_r = \sqrt{\frac{X_{c,\text{pu}}}{X_{s,\text{pu}}}} = \sqrt{\frac{MVA_{\text{sc,pu}}}{M\,\text{var}_{c,\text{pu}}}} \tag{12.66b}$$

Before the connection of the capacitor bank,

$$V_s = V_{\text{bus}} + jX_s I_c \tag{12.67}$$

and after the capacitor bank is switched,

$$V_s' = V_{\text{bus}} + jX_s(I + I_c) \tag{12.68}$$

where

$$I_c = j\frac{V_c}{X_c} = j\frac{V_{\text{bus}}}{X_c} \tag{12.69}$$

Assuming that $\bar{V}_s$ remains constant, the phase voltage rise at the bus due to the capacitor bank connection is

$$\Delta V_{\text{bus}} = \left| V_{\text{bus}}' \right| - \left| V_{\text{bus}} \right| \tag{12.70a}$$

or

$$\Delta V_{\text{bus}} = \left| -jX_s \bar{I}_c \right| = \frac{X_s}{X_c} \left| V'_{\text{bus}} \right| \tag{12.70b}$$

Thus,

$$\Delta V_{\text{bus, pu}} = \frac{\Delta V_{\text{bus}}}{\left| V'_{\text{bus}} \right|} = \frac{X_s}{X_c} \tag{12.71a}$$

or

$$\Delta V_{\text{bus, pu}} = \omega_1^2 L_s C = (2\pi)^2 f_1^2 L_s C \tag{12.71b}$$

so that

$$f_r = \frac{1}{2\pi \sqrt{L_s C}} = \frac{f_1}{\sqrt{\Delta V_{\text{bus, pu}}}} \tag{12.72}$$

Since

$$h_r = \frac{f_r}{f_1} \tag{12.73}$$

or

$$h_r = \frac{1}{\sqrt{\Delta V_{\text{bus, pu}}}} \tag{12.74}$$

or

$$\Delta V_{\text{bus, pu}} = \frac{1}{h_r^2} \tag{12.75}$$

From Equation 12.74, one can observe that a 0.04 per unit rise in bus voltage due to the switching on a capacitor bank results in a resonance at

$$h_r = \frac{1}{\sqrt{0.04}} = 5\text{th harmonic}$$

Similarly, a 0.02 pu bus voltage rise results in

$$h_r = \frac{1}{\sqrt{0.02}} = 7.07\text{th harmonic}$$

It can also be shown that

$$\frac{V_c}{\left| V_{\text{bus}} \right|} = \frac{\left| V'_{\text{bus}} \right|}{\left| V_{\text{bus}} \right|} = \frac{h_r^2}{h_r^2 - 1} \tag{12.76}$$

Example 12.7

A three-phase 12.47 kV, 5 MVA capacitor bank is causing a bus voltage increase of 500 V when switched on. Determine the following:

a. The per unit increase in bus voltage
b. The resonant harmonic order
c. The harmonic frequency at the resonance

Solution

a. The per unit increase in bus voltage is

$$\Delta V_{bus,\,pu} = \frac{500\,V}{12{,}470\,V} \cong 0.04\,pu$$

b. The resonant harmonic order is

$$h_r = \frac{1}{\sqrt{\Delta V_{bus,\,pu}}} = \frac{1}{\sqrt{0.04}} = 5$$

c. The harmonic frequency at the resonance is

$$f_r = f_1 \times h_r = 60 \times 5 = 300\,Hz$$

12.16 HARMONIC AMPLIFICATION

Consider the capacitor switching that is illustrated in Figure 12.10. When the capacitor is switched on, the bus voltage can be expressed as

$$V_c = V'_{bus} = \frac{-jX_c}{Z_s - jX_c}V_s \tag{12.77a}$$

or

$$V_c = V'_{bus} = \frac{V_s}{1 - \omega_1^2 L_s C + j\omega_1^2 CR_s} \tag{12.77b}$$

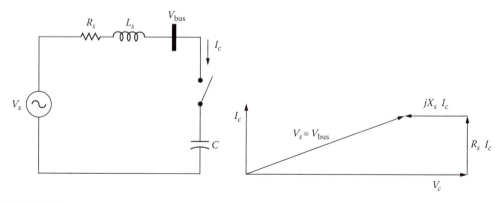

FIGURE 12.10 Capacitor switching.

At the *resonance*,

$$\omega_r = \omega_1 h_r = \frac{1}{\sqrt{L_s C}} \qquad (12.78)$$

or

$$h_r = \frac{\omega_r}{\omega_1} = \frac{1}{\omega_1 \sqrt{L_s C}} = \sqrt{\frac{X_c}{X_s}} = \sqrt{\frac{MVA_{sc}}{M\,var_c}} \qquad (12.79)$$

Hence, the hth harmonic capacitor voltage (or the *capacitor voltage at resonance*) can be expressed as

$$V_c(h) = \frac{V_s}{j\omega C R_s} = -j\frac{V_s}{R_s}\sqrt{\frac{L_s}{C}} \qquad (12.80a)$$

or

$$V_c(h) = -j\frac{Z_s}{R_s}V_s = -jA_f V_s \qquad (12.80b)$$

where
   Z_s is the characteristic impedance

$$Z_s = \sqrt{\frac{L_s}{C}} = \sqrt{X_s X_c} \qquad (12.81)$$

A_f is the amplification factor

$$A_f = \frac{Z_s}{R_s} \qquad (12.82)$$

From Equation 12.80a and b, one can observe that harmonics corresponding or close to the resonant frequency are amplified. The resulting voltages highly exceed the standard voltage rating, causing capacitor damage or fuse blowouts. The *amplification factor* can also be expressed as

$$A_f = \frac{Z_s}{R_s} = \frac{\sqrt{L_s/C}}{R_s} = \frac{\sqrt{X_s X_c}}{R_s} \qquad (12.83a)$$

or

$$A_f = \frac{X_s}{R_s} \times h_r \qquad (12.83b)$$

According to ANSI/IEEE Std.18-1992, shunt capacitors can be continuously operated in a harmonic environment provided that [15]

1. Reactive power does not exceed 135% of rating

$$\frac{Q_c}{Q_{c1}} = \sum_{h=1} h \left(\frac{V_h}{V_1}\right)^2 = \sum_{h=1} \frac{1}{h}\left(\frac{I_h}{I}\right)^2 \leq 1.35$$

2. Peak current does not exceed 180% of rated peak

$$\frac{I_{peak}}{I_1} = 1 + \text{CCF} \leq 1.8$$

3. Peak voltage does not exceed 120% of rated

$$\frac{V_{peak}}{V_1} = 1 + \text{VCF} \leq 1.2$$

4. RMS voltage does not exceed 110% of rated

$$\frac{V_{rms}}{V_1} = \sqrt{1 + \text{THD}_V^2} \leq 1.1 \quad \text{or} \quad \text{THD}_V \leq \sqrt{0.21} = 45.8\%$$

Example 12.8

A three-phase wye–wye connected 138/13.8 kV 50 MVA transformer with an impedance of 0.25% + j12% is connected between high- and low-voltage buses. Assume that a wye-connected switched capacitor bank is connected to the low-voltage bus of 13.8 kV and that the capacitor bank is made up of three 4 Mvar capacitors. Assume that at the 138 kV bus, the short-circuit MVA of the external system is 4000 MVA and its X/R ratio is 7. Use a MVA base of 100 MVA and determine the following:

   a. The impedance bases for the HV and LV sides
   b. The short-circuit impedance of the power system at the 138 kV bus
   c. The transformer impedance in per units
   d. The short-circuit impedance at the 13.8 kV bus in per units
   e. The X/R ratio and the short-circuit MVA at the 13.8 kV bus in per units
   f. The reactance of the capacitor per phase in ohms and per units
   g. The resonant harmonic order
   h. The characteristic impedance in per units
   i. The amplification factor

Solution

   a. Since $MVA_{B(HV)} = MVA_{B(LV)} = 100$ MVA and $kV_{B(HV)} = 138$ kV, $kV_{B(LV)} = 13.8$ kV,

$$Z_{B(HV)} = \frac{kV_{B(HV)}^2}{MVA_{B(HV)}} = \frac{138^2}{100} = 190.44 \ \Omega$$

and

$$Z_{B(LV)} = \frac{kV_{B(LV)}^2}{MVA_{B(LV)}} = \frac{13.8^2}{100} = 1.9044 \ \Omega$$

b. Since $MVA_{sc(sys)} = 4000$ MVA $= 40$ pu,

$$Z_{sc(sys)} = \frac{1}{40} \angle \tan^{-1} 7 = 0.025 \angle \tan^{-1} 7$$

$$= 0.003536 + j0.024749 \text{ pu} = 0.6734 + j4.7132 \, \Omega$$

c. The transformer impedance is

$$Z_T = (0.0025 + j0.12) \frac{100 \text{ MVA}}{50 \text{ MVA}} = 0.005 + j0.24 \text{ pu}$$

d. Looking from the 13.8 kV bus,

$$Z_{sc} = Z_{sc(sys)} + Z_T = 0.008536 + j0.26475 \text{ pu} = 0.26489 \angle 88.1533° \text{ pu}$$

e. The short-circuit MVA at the 13.8 kV bus is

$$MVA_{sc} = \frac{1}{Z_{sc, \text{pu}}} = \frac{1}{0.26489} = 3.775 \text{ pu}$$

and the X/R ratio is

$$\left(\frac{X}{R} \right)_{13.8} = \tan 88.1533° = 31.0153$$

f. Since the capacitor bank size is 4 Mvar per phase,

$$Q_c = M \, var_c = 4 \text{ Mvar} = 0.04 \text{ pu}$$

so that

$$X_c = \frac{kV^2}{Q_c} = \frac{13.8^2}{4} = 47.61 \, \Omega \text{ per phase} = 25 \text{ pu}$$

g. The resonant harmonic order of the resonance between the capacitor bank and system inductance is

$$h_r = \frac{f_r}{f_1} = \sqrt{\frac{MVA_{sc}}{M \, var_c}} = \sqrt{\frac{3.775 \text{ pu}}{0.04 \text{ pu}}} \cong 9.715$$

h. The characteristic impedance is

$$Z_c = \sqrt{X_{sc} X_c} = \sqrt{0.26475 \times 25} \cong 2.573 \text{ pu}$$

i. The amplification factor is

$$A_f = h_r \left(\frac{X}{R} \right)_{13.8} = 9.715 \times 31.0153 \cong 301.3$$

12.17   RESONANCE

The *resonance* is defined as an operating condition such that the magnitude of the impedance of the circuit passes through an extremum, that is, maximum or minimum. *Series resonance* occurs in a series RLC circuit that has equal inductive and capacitive reactances, so that the circuit imped-ance is low and a small exciting voltage results in a huge current. Similarly, parallel RLC circuit has equal inductive and capacitive reactances, so that circuit impedance is low and a small exciting current develops a large voltage.

The *resonance phenomenon*, or *near-resonance condition*, is the cause of the most of the har-monic distortion problems in power systems. Therefore, *at the resonance,*

$$X_{L_r} = \omega_r L = X_{C_r} = \frac{1}{\omega_r C}$$

where its *resonant frequency* is

$$f_r = \frac{1}{2\pi\sqrt{LC}} = \frac{f_1}{\omega_1\sqrt{LC}} = f_1\sqrt{\frac{X_C}{X_L}} \text{ Hz}$$

where
f_1 is the fundamental frequency
X_C is the capacitor's reactance at the fundamental frequency
X_L is the inductor's reactance at the fundamental frequency

Notice that f_r is independent of the circuit resistance. The *harmonic order of resonant frequency* is

$$h_r = \frac{f_r}{f_1} = \frac{1}{\omega_1\sqrt{LC}} = \sqrt{\frac{X_C}{X_L}} \qquad (12.84)$$

The resonance can cause *nuisance tripping* of sensitive electronic loads and high harmonic cur-rents in feeder capacitor banks. In severe cases, capacitors produce audible noise, and they some-times bulge. *Parallel resonance* occurs when the power system presents a parallel combination of power system inductance and PF correction capacitors at the nonlinear load. The product of the harmonic impedance and injection current produces high harmonic voltages. *Series resonance* occurs when the system inductance and capacitors are in series, or nearly in series, with respect to the nonlinear load point. *For parallel resonance, the highest voltage distortion is at the nonlinear load.* However, *for series resonance, the highest distortion is at a remote point, perhaps miles away or on an adjacent feeder served by the same substation transformer.*

12.17.1   SERIES RESONANCE

Consider the series RLC circuit of Figure 12.11a that is made up of R_L, X_L, and X_C at the frequency f. Its equivalent impedance is

$$Z = R + j(X_L - X_C) = R + \left(\omega L - \frac{1}{\omega C}\right) \qquad (12.85)$$

For any harmonic h,

$$\mathbf{Z}(h) = R + j\left(h \times X_L - \frac{X_C}{h}\right) \qquad (12.86)$$

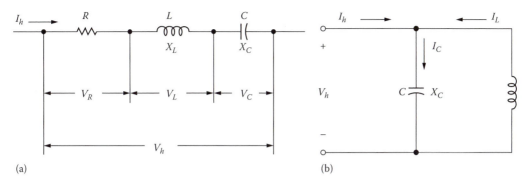

FIGURE 12.11   Resonance circuits for (a) series resonance and (b) parallel resonance.

so that

$$|\mathbf{Z}(h)| = \left[R^2 + \left(h \times X_L - \frac{X_C}{h} \right)^2 \right]^{1/2}$$ (12.87)

At resonance, $h = h_r$ and accordingly,

$$h_r X_L = \frac{X_C}{h_r} = X_r$$ (12.88)

from which

$$h_r = \sqrt{\frac{X_C}{X_L}}$$ (12.89)

and

$$X_r^2 = X_L X_C = \frac{L}{C}$$ (12.90)

or

$$X_r = \sqrt{X_L X_C} = \sqrt{\frac{L}{C}}$$ (12.91)

As a result, the impedance of the circuit at the resonance is then purely resistive and is only equal to R. That is,

$$\mathbf{Z}(h_r) = R$$ (12.92)

The quality factor Q is

$$Q = \frac{X_r}{R}$$ (12.93)

Example 12.9

A series RLC circuit has $X_L = 0.2 \ \Omega$, $X_C = 1.8 \ \Omega$, and $Q = 100$. Determine the following:

    a. The harmonic order of the series resonance
    b. The reactance of the circuit at the resonance
    c. The value of R

Solution

    a. Its harmonic order is

$$h_r = \sqrt{\frac{X_C}{X_L}} = \sqrt{\frac{1.8}{0.2}} = 3$$

    b. Its circuit reactance at the resonance is

$$X_r = \sqrt{X_L X_C} = \sqrt{0.2 \times 1.8} = 0.6 \ \Omega$$

    c. The value of the resistance is

$$R = \frac{X_r}{Q} = \frac{0.6}{100} = 0.006 \ \Omega$$

12.17.2 PARALLEL RESONANCE

Consider a parallel RCC circuit of Figure 12.11b that is made up of R, X_L, and X_C at a frequency f. Its equivalent impedance is

$$\mathbf{Z} = \frac{j((RX_L X_C)/(X_L - X_C))}{R - j((X_L X_C)/(X_L - X_C))} = \frac{-jRX_L X_C}{R(X_L - X_C) - jX_L X_C} \tag{12.94}$$

For any harmonic h,

$$X_L(h) = h \times X_L \quad \text{and} \quad X_C(h) = \frac{X_C}{h}$$

so that

$$X_L(h)X_C(h) = X_L X_C \tag{12.95}$$

and the impedance is

$$\mathbf{Z}(h) = \frac{-jRX_L X_C}{R\left(h \times X_L - \dfrac{X_C}{h}\right) - jX_L X_C} \tag{12.96}$$

or

$$|\mathbf{Z}(h)| = \frac{RX_L X_C}{\left\{\left[R\left(h \times X_L - \dfrac{X_C}{h}\right)\right]^2 + [X_L X_C]^2\right\}^{1/2}} \tag{12.97}$$

At resonance, $h = h_r$ and accordingly,

$$h_r X_L = \frac{X_C}{h_r} = X_r \tag{12.98}$$

from which

$$h_r = \sqrt{\frac{X_C}{X_L}} \tag{12.99}$$

and

$$X_r^2 = X_L X_C = \frac{L}{C} \tag{12.100}$$

or

$$X_r = \sqrt{X_L X_C} = \sqrt{\frac{L}{C}} \tag{12.101}$$

Again, the impedance of the circuit is equal to R. That is,

$$\mathbf{Z}(h_r) = R$$

The quality factor Q is

$$Q = \frac{R}{X_r} \tag{12.102}$$

Here, the critical damping takes place at $Q = 0.5$ or $R = 0.5X_r$. Quality factor determines the sharpness of the frequency response. Q varies considerably by location on the power system. It might be less than 5 on a distribution feeder and more than 30 on the secondary bus of a large step-down transformer.

Example 12.10

For a given parallel RLC circuit having $X_L = 0.926\ \Omega$, $X_C = 75\ \Omega$, and $Q = 5$, determine the following:

   a. Its harmonic order
   b. Its circuit reactance at the resonance
   c. The value of R

Solution
   a. Its harmonic order is

$$h_r = \sqrt{\frac{X_C}{X_L}} = \sqrt{\frac{75}{0.926}} \cong 9$$

   b. Its circuit reactance at the resonance is

$$X_r = \sqrt{X_L X_C} = \sqrt{0.926 \times 75} = 8.333\ \Omega$$

c. The value of R is

$$R = Q \times X_r = 5 \times 8.333 = 41.665 \ \Omega$$

Note that the resistance of the circuit varies with different quality factors.

12.17.3 Effects of Harmonics on the Resonance

In the presence of harmonics, the resonance takes place when the source (or system) reactance X_{s_r} is equal to the reactance of the capacitor X_{C_r} at the tuned frequency, as follows:

$$X_{C_r} = \frac{X_{C_1}}{h_r} = X_{s_r} = h_r \times X_{s_1} \tag{12.103}$$

and at an angular resonant frequency of

$$\omega_r = h_r \times \omega_1 = \frac{1}{\sqrt{L_{s_1} C_1}} \text{ rad/s} \tag{12.104}$$

or

$$f_r = h_r \times f_1 = \frac{1}{2\pi \sqrt{L_{s_1} C_1}} \text{ Hz} \tag{12.105}$$

where
  X_{C_1} is the reactance of the capacitor at the fundamental frequency
  X_{s_1} is the inductive reactance of the source at the fundamental frequency
  $L_{s1} = L_s$ is the inductance of the source at the fundamental frequency
  $C_1 = C$ is the capacitance of the capacitor at the fundamental frequency

from which the harmonic order h_r to cause resonance can be found as

$$h_r = \frac{f_r}{f_1} = \frac{1}{\omega_1 \sqrt{L_{s_1} C_1}} \tag{12.106a}$$

or

$$h_r = \sqrt{\frac{X_{c1}}{X_{s1}}} = \sqrt{\frac{MVA_{s,\,pu}}{M \, var_{c,\,pu}}} \tag{12.106b}$$

Let $X_{sc} = X_s = X_{s1}$, $X_c = X_{c1}$ and $MVA_{sc} = MVA_s$, then

$$h_r = \sqrt{\frac{X_c}{X_{sc}}} = \sqrt{\frac{MVA_{sc}}{M \, var}} \tag{12.107}$$

so that a capacitor with a reactance of $X_{c1} = h_r^2 \times X_{s1}$ or $X_c = h_1^2 \times X_s$ excites resonance at the h_r th harmonic order.

In order to tune a capacitor to a certain harmonic (or designing a capacitor to trap, i.e., to filter a certain harmonic) requires the addition of a reactor. At the tuned harmonic,

$$X_{L_{\text{tuned}}} = X_{C_{\text{tuned}}} = X_{\text{tuned}}$$

or

$$h_{\text{tuned}} X_L = \frac{X_C}{X_{\text{tuned}}} \tag{12.108}$$

where its characteristic reactance can be expressed as

$$X_{\text{tuned}} = \sqrt{X_L X_C} = \sqrt{\frac{L}{C}} \tag{12.109}$$

The tuned frequency is then

$$f_{\text{tuned}} = h_{\text{tuned}} f_1 \tag{12.110a}$$

or

$$f_{\text{tuned}} = \frac{1}{2\pi\sqrt{LC}} \text{ Hz} \tag{12.110b}$$

Hence, the inductive reactance of the reactor is

$$X_L = \frac{X_C}{h_{\text{tuned}}^2} \tag{12.111a}$$

or

$$X_L = \frac{h_r^2}{h_{\text{tuned}}^2} X_s \tag{12.111b}$$

If $f_{\text{tuned}} = f_r$ (or $h_{\text{tuned}} = h_r$), then Equation 12.111b becomes $X_L = X_s$. Also, Equation 12.110a and b becomes

$$h_{\text{tuned}} = \frac{f_{\text{tuned}}}{f_1} = \frac{1}{\omega_1 \sqrt{LC}} \tag{12.112a}$$

or

$$h_{\text{tuned}} = \sqrt{\frac{X_C}{X_L}} = h_r \sqrt{\frac{X_s}{X_L}} = h_r \sqrt{\frac{X_{\text{sc}}}{X_L}} \tag{12.112b}$$

Example 12.11

A 34.5 kV three-phase 5.325 Mvar capacitor bank is to be installed at a bus that has a short-circuit MVA of 900 MVA. Investigate the possibility of having a resonance and eliminate it. Determine the following:

a. The harmonic order of the resonance.
b. The capacitive reactance of the capacitor bank in ohms.
c. Design the capacitor bank that will trap the resultant harmonic by adding a reactor in series with the capacitor. Find the required reactor size X_L.
d. The characteristic reactance.

    e. Select the filter quality factor as 50 and find the resistance of the reactor.
    f. The impedance of this resultant series-tuned filter at any harmonic order h.
    g. The rated filter size.

Solution

    a. The harmonic order of the resonance due to the interaction between the capacitor bank and the system is

$$h_r = \frac{f_r}{f_1} = \sqrt{\frac{X_C}{X_{sc}}} = \sqrt{\frac{MVA_{sc}}{Q_C}} = \sqrt{\frac{MVA_{sc}}{M\,var_C}} = \sqrt{\frac{900}{5.325}} = 13$$

    b. The capacitive reactance of the capacitor bank is

$$X_C = \frac{kV_{L-L}^2}{Q_{c,3\phi}} = \frac{34.52}{5.325} \cong 223.521\ \Omega \text{ per phase}$$

    c. The required reactor size is

$$X_L = \frac{X_C}{h_{tuned}^2} = \frac{223.521}{13^2} \cong 1.323\ \Omega$$

    d. The characteristic reactance is

$$X_{tuned} = \sqrt{X_L X_C} = \sqrt{1.323 \times 223.521} \cong 17.196\ \Omega$$

    e. Since $Q = 50$,

$$R = \frac{X_{tuned}}{Q} = \frac{\sqrt{X_L X_C}}{Q} = \frac{17.196\ \Omega}{50} \cong 0.344\ \Omega$$

    f. The impedance function of the filter is

$$\mathbf{Z}_{filter}(h) = R + j\left(hX_L - \frac{X_C}{h}\right)$$

$$= 0.344 + j\left(1.323h - \frac{223.521}{h}\right)\Omega$$

    g. The rated filter size is

$$Q_{filter} = \frac{kV^2}{X_C - X_L} = \frac{h_{tuned}^2}{h_{tuned}^2 - 1} \times Q_C$$

$$= \frac{13^2}{13^2 - 1} \times 5.325 \cong 5.357\ \text{Mvar}$$

12.17.4 PRACTICAL EXAMPLES OF RESONANCE CIRCUITS

Figure 12.12 shows practical examples of possible series and parallel resonant conditions. Figure 12.12a shoes a step-down transformer supplying loads including PF correction capacitors from a bus that has a considerable nonlinear load. Its equivalent circuit is shown in Figure 12.12b. Normally, the harmonic currents generated by the nonlinear load would flow to the utility.

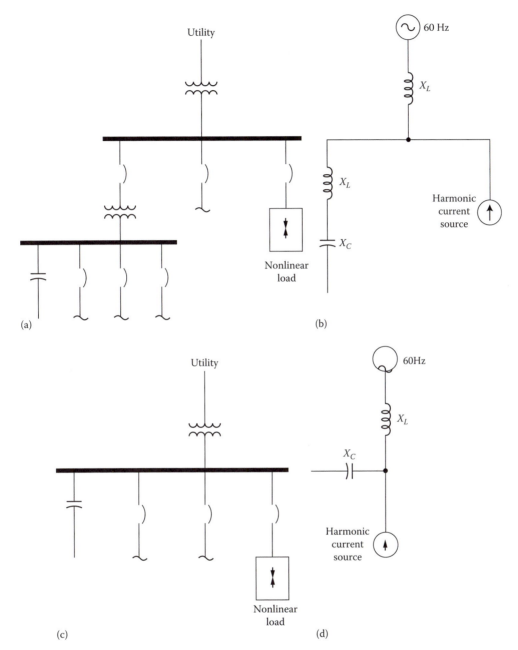

FIGURE 12.12   Practical examples of resonance circuits: (a) series resonance circuit, (b) its equivalent circuit, (c) parallel resonance circuit, and (d) its equivalent circuit.

However, if at one of the nonlinear load's significant harmonic current frequencies (typically, the 5th, 7th, 11th, or 13th harmonic) the step-down transformer's inductive reactance equals the power-factor-correction capacitor's reactance, then the resulting series resonant circuit will attract the harmonic current from the nonlinear load. The additional unexpected harmonic current flow through the transformer and capacitors will cause additional heating and possibly overload.

Figure 12.12c depicts a potentially more troublesome problem, that is, parallel resonance. Its equivalent circuit is shown in Figure 12.12d. In this case, PF correction capacitors are applied to the same voltage bus that feeds significant nonlinear loads.

If the inductive reactance of the upstream transformer equals the capacitive reactance at one of the nonlinear load's harmonic current frequencies, then parallel resonance takes place. With parallel resonance, high currents can oscillate in the resonance circuit and the voltage bus waveform can be severely distorted.

As discussed before, from the harmonic source's point of view, at harmonic frequencies, shunt capacitors appear to be in parallel with the equivalent system inductance, as shown in Figure 12.13a and b.

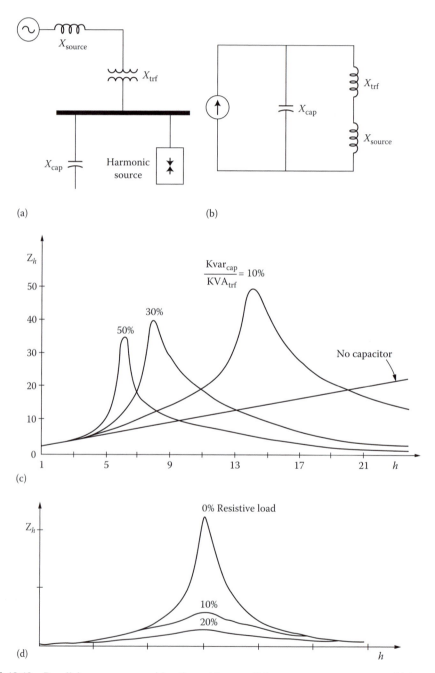

FIGURE 12.13 Parallel resonance considerations: (a) a parallel resonance prone system, (b) its equivalent circuit, (c) effects of capacitor sizes, and (d) effects of resistive loads.

At frequencies other than fundamental, the power system generation appears to be short circuit. When there is a parallel resonance situation, that is, at certain frequency where X_c and the total system reactance are equal, the apparent impedance seen by the source harmonic currents becomes very large. Figure 12.13c shows the system frequency response as capacitor size is varied in relation to transformer as well as in the case of having no capacitor.

If one of the peaks lines up with a common harmonic current produced by the load, there will be a much greater voltage drop across the apparent impedance than the case of no capacitors.

However, the alignment of the resonant harmonic with the common source harmonic is not always problematic. Often, the damping provided by resistance of the system is sufficient to prevent any catastrophic voltages or currents, as shown in Figure 12.13d.

As one can see, even a 10% resistance loading has a considerable effect on the peak impedance. Because of this fact, if there is a considerable length of lines or cables between the capacitor bus and the nearest upstream transformer, the resonance will be suppressed.

Since the resistances of lines and cables are significantly large, catastrophic harmonic problems due to capacitors do not appear often on distribution feeders. Therefore, resistive loads will damp resonance and cause a significant reduction in the harmonic distortion.

However, very little damping is achieved if any from motor loads, since they are basically inductive. On the contrary, they may increase distortion by shifting system resonant frequency closer to a significant harmonic. But small fractional-horsepower motors may contribute considerably to damping because of their lower X/R ratios.

The worst resonant conditions take place when capacitors are installed on substation buses where the transformer dominates the system impedance and has a high X/R ratio, the relative resistance is low, and associated parallel resonant impedance peak is very high and sharp. This phenomenon is known to be *the cause of the failure in capacitors, transformers, or load equipment.*

Example 12.12

A three-phase wye–wye-connected transformer with $X = 10\%$ is supplying a 40 MVA load at a lagging PF of 0.9. At the low-voltage bus of 12.47 kV, three-phase wye-connected capacitor bank is to be connected to correct the PF to 0.95. A distribution engineer is asked to investigate the problem, knowing that the short-circuit MVA at the 345 kV bus is 2000 MVA. Use a MVA base of 100 MVA and determine the following:

a. The current bases for the HV and LV sides of the transformer in amps
b. The impedance bases for the HV and LV sides in ohms
c. The short-circuit reactance of the system at the 345 kV bus in per units and ohms
d. The short-circuit reactance of the system at the 12.47 kV bus in per units and ohms
e. The short-circuit MVA of the system at the 12.47 kV bus in per units and MVA
f. The real power of the load at the lagging PF of 0.9 in per units and MW
g. The size of the capacitor bank needed to correct the PF to 0.95 lagging in per units and Mvar
h. The resonant harmonic order at which the interaction between the capacitor bank and system inductance initiates resonance
i. The reactance of each capacitor per phase in per units and ohms

Solution

a. Since $MVA_{B(HV)} = MVA_{B(LV)} = 100$ MVA and $kV_{B(HV)} = 345$ KV, $kV_{B(LV)} = 12.47$ kV. The current bases for the HV and LV sides are

$$I_{B(HV)} = \frac{MVA_{B(HV)}}{\sqrt{3}kV_{B(HV)}} = \frac{100,000 \text{ kVA}}{\sqrt{3}(345 \text{ kV})} = 167.55 \text{ A}$$

and

$$I_{B(LV)} = \frac{MVA_{B(LV)}}{\sqrt{3}kV_{B(LV)}} = \frac{100,000 \text{ kVA}}{\sqrt{3}(12.47 \text{ kV})} = 4635.4 \text{ A}$$

b. The impedance bases for the HV and LV sides are

$$Z_{B(HV)} = \frac{kV_{B(HV)}^2}{MVA_{B(HV)}} = \frac{345^2}{100} = 1190.25 \text{ } \Omega$$

and

$$Z_{B(LV)} = \frac{kV_{B(LV)}^2}{MVA_{B(LV)}} = \frac{12.47^2}{100} = 1.555 \text{ } \Omega$$

c. Since $MVA_{sc(sys)} = MVA_{sc(source)} = 2000 \text{ MVA} = 20 \text{ pu}$,

$$X_{sc(sys)} = \frac{1}{MVA_{sc(sys)pu}} = \frac{1}{20 \text{ pu}} = 0.05 \text{ pu}$$

or

$$X_{sc(sys)} = \frac{kV_{L-L}^2}{MVA_{sc(sys)}} = \frac{345^2}{2000} = 59.513 \text{ } \Omega$$

d. Since

$$X_T = 0.10 \times \frac{100 \text{ MVA}}{60 \text{ MVA}} = 0.117 \text{ pu}$$

Looking from the LV bus of 12.47 kV,

$$X_{sc} = X_{sc(sys)} + X_T = 0.05 + 0.1667 = 0.2167 \text{ pu}$$

or

$$X_{sc} = (0.2167 \text{ pu}) \times Z_{B(LV)} = 0.2176 \times 1.555 = 0.3367 \text{ } \Omega$$

e. The MVA_{sc} at the 12.47 kV bus in per units and MVA are

$$MVA_{sc} = \frac{1}{X_{sc(pu)}} = \frac{1}{0.2167 \text{ pu}} = 4.6147 \text{ pu}$$

or

$$MVA_{sc} = (4.6147 \text{ pu}) \times MVA_{B(LV)} = (4.6147 \text{ pu}) \times 100 = 461.47 \text{ MVA}$$

f. The real power of load is

$$P = S \times \cos\theta = (40 \text{ MVA}) \times 0.9 = 36 \text{ MVA} \quad \text{or} \quad 0.36 \text{ pu}$$

g. The real size of the three-phase capacitor bank needed to correct the PF is

$$Q_c = P(\tan\theta_1 - \tan\theta_2)$$

$$= (36\ \text{MVA})[\tan(\cos^{-1}0.9) - \tan(\cos^{-1}0.95)] = 5.603\ \text{Mvar} \quad \text{or} \quad 0.05603\ \text{pu}$$

h. Since the capacitor bank is wye connected,

$$I_c = I_L = \frac{5603\ \text{kvar}}{\sqrt{3}(12.47\ \text{kV})} = 259.72\ \text{A}$$

Thus,

$$X_c = \frac{V_{L-N}}{I_c} = \frac{12{,}470/\sqrt{3}}{259.72} = 27.75\ \Omega\ \text{per phase}$$

or

$$X_c = \frac{27.75\ \Omega}{Z_{B(LV)}} = \frac{27.75\ \Omega}{1.555\ \Omega} = 17.845\ \text{pu}$$

i. The interaction between the capacitance bank and system inductance initiates resonance at

$$h_r = \frac{f_r}{f_1} = \sqrt{\frac{X_C}{X_{sc}}} = \sqrt{\frac{17.845\ \text{pu}}{0.2167\ \text{pu}}} = 9.075 \cong 9.08$$

or

$$h_r = \sqrt{\frac{MVA_{sc}}{Q_c}} = \sqrt{\frac{4.6147\ \text{pu}}{0.05603\ \text{pu}}} = 9.075 \cong 9.08$$

12.18 HARMONIC CONTROL SOLUTIONS

In general, harmonics become a problem if (1) the source of harmonic currents is to large, (2) the system response intensifies one or more harmonics, and (3) the currents' path is electrically too long, causing either high-voltage distortion or telephone interference.

When these types of problems happen, the following options are the main ones to control the harmonics: (1) decrease the harmonic currents generated by the nonlinear loads; (2) add filters to either get rid of the harmonic currents from the system, supply the harmonic currents locally, or block the currents locally from entering the system; and (3) modify the system frequency response to avoid adverse interaction with harmonic currents.

This can be done by feeder sectionalizing, adding or removing capacitor banks, changing the size of the capacitor banks, adding shunt filters, or adding reactors to detune system away from harmful resonances.

Usually, not much can be done with existing load equipment to substantially reduce its harmonic currents. One exception to these devices is pulse-width modulated (PWM) adjustable-speed drives (ASDs) that change the dc bus capacitor directly from the line. Here, adding a line reactor in series will considerably decrease harmonics as well as provide transient protection benefits.

Transformer connections can also be used to reduce harmonic currents in three-phase systems. For example, delta-connected transformers can block the flow of the zero-sequence triplen harmonics from the line. Also, zigzag and grounding transformers can shunt the triplens off the line.

The filter used can be shunt or series filters. The shunt filter application works by short-circuiting the harmonic currents as close to the source of distortion as practical. It keeps the harmonic currents out of the supply system. It is the most common type of filtering used due to economics and its tendency to smooth the load voltage as well as its elimination of the harmonic current.

The series filter blocks the harmonic currents. It has a parallel-tuned circuit that presents high impedance to the harmonic current. It is not often used since it is difficult to insulate and has very distorted load voltage. It is commonly used in the neutral of a grounded-wye capacitor to block the flow of triplen harmonics while still having a good ground at fundamental frequency.

In addition, it is possible to use active filters. Active filters work by electronically supplying the harmonic component of the current into a nonlinear load.

Furthermore, adverse system responses to harmonics can be modified by using one of the following methods: (1) adding a shunt filter, (2) adding a reactor to detune the system, (3) changing the capacitor size, (4) moving a capacitor to a point on the system with a different short-circuit impedance or higher losses (when adding a capacitor bank results in telephone interference, moving the bank to another branch of the feeder may solve the problem), and (5) removing the capacitor and accepting its consequences may be the best economic choice.

12.18.1 PASSIVE FILTERS

Passive (or passive tuned) filters are relatively inexpensive, but they have potential for adverse interactions with the power system. They are used either to shunt the harmonic currents off the line or to block their flow between parts of the system by tuning the elements to create a resonance at a selected harmonic frequency. As shown in Figure 12.14, passive filters are made up of inductance, capacitance, and resistance elements. A single-tuned "notch" filter is the most common type of filter since it is often sufficient for the application and inexpensive.

Figure 12.15 shows typical 480 V single-tuned wye- or delta-connected filters. Such notch filter is series tuned to present low impedance to a specific harmonic current and is connected in shunt with power system. As a result, harmonic currents are diverted from their normal flow path on the line into filter.

Notch filters provide PF correction in addition to harmonic suppression. As shown in the figure, a typical delta-connected low-voltage capacitor bank converted into a filter by adding an inductance (reactor) in series. *The tuned frequency for such combination is selected somewhere below the fifth harmonic (e.g., 4.7) to prevent a parallel resonance at any characteristic harmonic. This is in order*

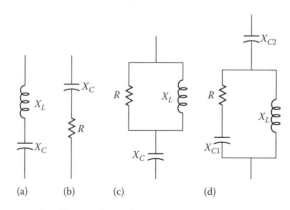

FIGURE 12.14 Common passive filter configurations: (a) type I, (b) type II, (c) type III, and (d) type IV.

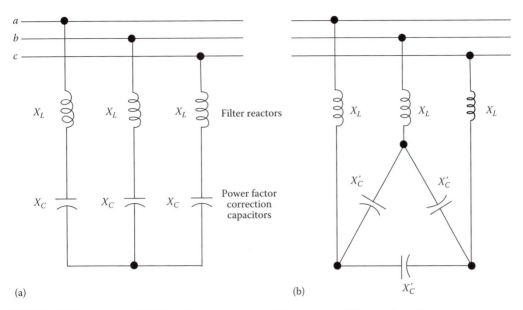

(a)   (b)

FIGURE 12.15 A typical 480 V single-tuned wye- or delta-connected filter configurations.

to provide a margin of safety in case there is some change in system parameters later. This point represents the notch harmonic, h_{notch}, and is related to the fundamental frequency reactance X_1 by

$$h_{\text{notch}} = \sqrt{\frac{X_c}{3X_1}} \tag{12.113}$$

Here, X_c is the reactance of one leg of the delta rather than the equivalent line-to-neutral capacitive reactance. If line-to-line voltage and three-phase capacitive reactive power are used to calculate X_c, then it should not be divided by 3 in Equation 12.113.

Note that if such filters were tuned exactly to the harmonic, changes in inductance or capacitance with failure or due to changes in temperature might push the parallel resonance higher into the harmonic. As a result, the situation becomes much worse than having no filter.

Because of this, filters are added to the system beginning with the lowest problematic harmonics. Hence, installing a seventh-order harmonic filter usually dictates the installation of a fifth-order harmonic filter.

Also, it is usually a good idea to use capacitors with a higher voltage rating in filter applications because of the voltage rise across the reactor at the fundamental frequency and due to the harmonic loading. In this case, 600 V capacitors are used for a 480 V application.

In general, capacitors on utility distribution systems are connected in wye. It provides a path for the zero-sequence triplen harmonics by changing the neutral connection.

Also, placing a reactor in the neutral of a capacitor is a common way to force the bank to filter only zero-sequence harmonics. It is often used to get rid of telephone interference. Usually, a tapped reactor is inserted into the neutral, and the tap is adjusted according to the harmonic causing the interference to minimize the problem.

Passive filters should always be placed on a bus where X_{sc} is constant. The parallel resonance will be much lower with standby generation than utility system. Because of this, filters are often *removed for standby operation.* Furthermore, filters should be designed according to the bus capacity not only for the load.

Note that tuned capacitor banks act as a harmonic filter for the fifth harmonic. They will have to absorb some percentage of the fifth harmonic current from loads within the facility and also will have to absorb fifth harmonic current due to fifth harmonic voltage distortion on the utility supply system. IEEE 519-1992 allows the voltage distortion on the supply system to be as high as 3% at an individual harmonic on medium voltage systems. Thus, this level of fifth harmonic distortion should be assumed for filter design purposes. The general methodology for applying filters is explained in the following steps:

1. Only a single-tuned shunt filter designed for the lowest produce frequency is applied at first.
2. The voltage distortion level at the low-voltage bus is found out.
3. The effectiveness of the filter designed is checked by changing the elements of the filter in conformity with the specified tolerances.
4. It is assured that the resulting parallel resonance is not close to a harmonic frequency by reviewing the frequency response characteristics.
5. The requirement for having several filters, for example, fifth and seventh or third, fifth, and seventh, is considered in the application.

Consider the single-tuned 480 V notch filter shown in Figure 12.15. Such filter should be tuned slightly below the harmonic frequency of concern. This permits for tolerances in the filter components and prevents the filter from acting as a short circuit for the offending harmonic current. It minimizes the possibility of having dangerous harmonic resonance if the system parameters change and cause the tuning frequently to shift slightly higher.

The actual fundamental frequency compensation provided by a derated capacitor bank is found from

$$Q_{actual} = Q_{rated} \left(\frac{V_{actual}}{V_{rated}} \right)^2 \tag{12.114}$$

The fundamental frequency current of the capacitor bank is

$$I_{c(FL)} = \frac{Q_{actual}}{\sqrt{3}V_{actual}} \tag{12.115}$$

The equivalent single-phase reactance of the capacitor bank is

$$X_{c(wye)} = \frac{V^2}{Q_c} \tag{12.116}$$

The reactance of the filter reactor is found from

$$X_L = X_{reactor} = \frac{X_c}{h_{tuned}^2} \tag{12.117}$$

where h_t is the tuned harmonic. The fundamental frequency current of the filter becomes

$$I_{filter(FL)} = \frac{V_{bus}}{\sqrt{3}(X_c + X_{reactor})} \tag{12.118}$$

Since the filter draws more fundamental current than the capacitor alone, the supplied var compensation is larger than the capacitor rating and is found from

$$Q_{supplied} = \sqrt{3}V_{bus}I_{filter(FL)} \tag{12.119}$$

The tuning characteristic of the filter earlier is defined by its quality factor, Q. It is a measure of sharpness of tuning. For such series filter, it is given by

$$Q = \frac{X_h}{R} = \frac{X_{L_h}}{R} = \frac{h \times X_{reactor}}{R} \tag{12.120}$$

where

h is the tuned harmonic
$X_L = X_{reactor}$ is the reactance of filter reactor at fundamental frequency
R is the series resistance of filter

Usually, the value of R is only the resistance of the inductor that results in a very large value of Q and a very strong filtering. Normally, this is satisfactory for a typical single-filter usage. It is a very economical filter operation due to its small energy consumption.

However, occasionally, it might be required to have some losses to be able to dampen the system response. To achieve this, a resistor is added in parallel with the reactor to create a high-pass filter. In such a case, the quality factor is given by

$$Q = \frac{R}{h \times X_L} \tag{12.121}$$

Here, the larger the Q, the sharper the tuning. It is not economical to operate such filter at the fifth and seventh harmonics because of the amount of losses. However, they are used at the 11th and 13th or higher order of harmonics.

In special cases where tuned capacitor banks are not sufficient to control harmonic current levels, a more complicated filter design may be required. This is often difficult and a more detailed harmonic study will normally be required. Figure 12.16 gives the general procedure for designing these filters.

Significant derating of the filters may be required to handle harmonics from the power system. Including the contribution from the power system is part of the process of selecting a minimum-size filter at each tuned frequency. The filter size must be large enough to absorb the power system harmonics.

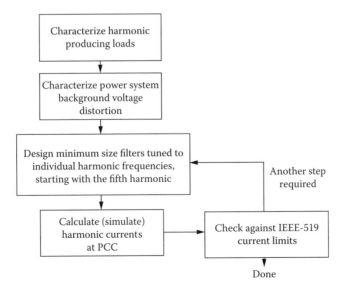

FIGURE 12.16   General procedure for designing individually tuned filter steps for harmonic control.

The design may result in excessive kvar due to the number of filter steps and filter sizes needed for harmonic control. This would result in leading PF and possible overvoltages. In some rare cases, even three or four steps (e.g., 5, 7, 11 or 5, 7, 13) may not be sufficient to control the higher-order harmonic components to the levels specified in IEEE Std. 519-1992.

If the aforementioned concerns result in some unacceptable filter designs, it may be possible to control the harmonics with modifications to nonlinear loads, for example, multiphase configurations or active front ends, or electronically with active filters.

Example 12.13

A 60 Hz 600 V three-phase delta-connected 600 kvar capacitor bank will be used as a part of a single-tuned 480 V filter. The filter will be used for the fifth harmonic of nonlinear loads of an industrial plant. Set the resonant at 4.7 harmonic for a margin of safety. The facility has 500 hp of ASDs connected at 480 V. Design a single-tuned filter and determine the following:

a. The actual fundamental frequency compensation provided by a derated capacitor bank.
b. The full-load fundamental frequency current of the capacitor bank.
c. The wye equivalent single-phase reactance of the capacitor bank.
d. The reactance of the serially connected filter reactor.
e. The full-load current of the filter.
f. The reactive power supplied to the filter.
g. Compare the capacitor ratings with the standard capacitor limits that are given in IEEE Std. 18-1980. Are they within the limits?

Solution

a. The full-load fundamental frequency current of the capacitor bank is

$$Q_{actual} = Q_{rated} \left(\frac{V_{actual}}{V_{rated}} \right)^2 = (600 \text{ kvar}) \left(\frac{480 \text{ V}}{600 \text{ V}} \right)^2 \cong 384 \text{ kvar}$$

b. The full-load fundamental frequency current of the capacitor bank is

$$I_{c(FL)} = \left(\frac{Q_{actual}}{\sqrt{3} V_{actual}} \right) = \frac{384 \text{ kvar}}{\sqrt{3}(0.480 \text{ kV})} \cong 461.0 \text{ A}$$

c. The wye equivalent single-phase reactance of the capacitor bank is

$$X_{c(wye)} = \frac{kV^2}{Q_c} = \frac{kV_{rated}^2}{M \text{ var}_{rated}} = \frac{(0.600 \text{ kV})^2}{0.600 \text{ Mvar}} = 0.6 \text{ } \Omega$$

d. The reactance of the serially connected filter reactor is

$$X_L = X_{reactor} = \frac{X_{c(wye)}}{h^2} = \frac{0.6 \text{ } \Omega}{4.7^2} \cong 0.0272 \text{ } \Omega$$

e. The full-load current of the filter is

$$I_{filter(FL)} = \frac{V_{L-L}}{\sqrt{3}(X_c + X_{reactor})} = \frac{480 \text{ V}}{\sqrt{3}(-0.6 + 0.0272)} \cong 483.8 \text{ A}$$

TABLE 12.12

Harmonic Filter Design Spreadsheet for Example 12.13

System Information

Filter specification: 5th	Power system frequency: 60 Hz
Capacitor bank rating: 600 kvar	Capacitor rating: 600 V
Rated bank current: 577 A	60 Hz
Nominal bus voltage: 480 V	Derated capacitor: 384 kvar
Capacitor current (actual): 461.9 A	Total harmonic load: 500 kVA
Filter tuning harmonic: 4.7th	Filter tuning frequency: 282 Hz
Cap impedance (wye equivalent): 0.6000 Ω	Cap value (wye equivalent): 4421.0 μF
Reactor impedance: 0.0272 Ω	Reactor rating: 0.0272 mH
Filter full-load current (actual): 483.8 A	Supplied compensation: 402 kvar
Filter full-load current (rated): 604.7 A	Utility side V_h: 3.00% V_h
Transformer nameplate: 1500 kVA	
(Rating and impedance): 6.00%	(Utility harmonic voltage source)
Load harmonic current: 35.00% fund	
Utility harmonic current: 134.5 A	Load harmonic current: 210.5 A
	Max total harmonic current: 345.0 A

Capacitor Duty Calculations

Filter rms current: 594.2 A	Fundamental cap voltage: 502.8 V
Harmonic cap voltage: 71.7 V	Maximum peak voltage: 574.5 V
RMS capacitor voltage: 507.8 V	Maximum peak current: 828.8 A

Capacitor Limits (IEEE Std. 18-1980) **Filter Configuration**

	Limit (%)	Actual (%)
Peak Voltage	120	96
Current	180	103
Kvar	135	87
RMS voltage	110	85

Three delta-connected 600 kvar and 600 V rated capacitors connected over three $X_L = 0.0272$ Ω reactors to a 480 V bus

Filter Reactor Design Specifications

Reactor impedance: 0.0272 Ω	Reactor rating: 0.0720 mH
Fundamental current: 483.8 A	Harmonic current: 345.0 A

---

f. The reactive power supplied to the filter is

$$Q_{supplied} = Q_{filter(FL)} = \sqrt{3}V_{L-L}I_{filter(FL)} = \sqrt{3}(480 \text{ V})(483.8 \text{ A}) \cong 402 \text{ kvar}$$

g. Table 12.12 shows the design spreadsheet of the filter. The standard capacitor limits that are given in IEEE Std. 18-1080 are shown at the bottom of the table. As one can see, the capacitor ratings are within the limits of the standard.

12.18.2 ACTIVE FILTERS

Active filtering is a new technology that uses intelligent circuits to measure harmonics and take corrective actions. Active filters use either the phase-cancellation principle by injecting equal but opposite harmonics, or they inject/absorb current bursts to hold the voltage waveform within an acceptable tolerance of sinusoidal.

They are much more expensive than passive filters, but they have some great advantages. For example, they do not resonate with the system. Because of this advantage, they can be used in very difficult parallel resonance spots where passive filters cannot operate successfully.

They are very useful for large distorting loads fed from somewhat weak points on the power system. Also, they can be used for more than one harmonics at a time and are useful against other power quality problems such as flickers.

The main idea is to replace the missing sine wave portion in a nonlinear load. In an active filter, an electronic control monitors the line voltage and/or current, switching the power electronics very precisely to track the load current or voltage and force it to be sinusoidal. Either an inductor is used to store up current to be injected into the system at the appropriate instant or a capacitor is used instead. As a result, the load current is distorted as demanded by the nonlinear load but the current seen by the system is much more sinusoidal. Active filters correct both harmonics and PF of the load.

12.19 HARMONIC FILTER DESIGN

As previously discussed, in order to tune a capacitor to a certain harmonic (or designing a capacitor to trap, i.e., to filter a certain harmonic), it requires the addition of a reactor. At the tuned harmonic of h_{tuned},

$$X_{L_{tuned}} = X_{C_{tuned}}$$

or

$$X_{tuned} = X_{L_{tuned}} = h_{tuned} \times X_{L_1} = X_{C_{tuned}} = \frac{X_{C_1}}{h_{tuned}}$$

so that

$$X_{tuned} = X_{L_{tuned}} = X_{C_{tuned}} = \sqrt{X_{L_1} X_{C_1}} = \sqrt{\frac{L_1}{C_1}} \tag{12.122}$$

Thus, the tuned frequency is

$$f_{tuned} = h_{tuned} \times f_1 = \frac{1}{2\pi \sqrt{L_1 C_1}} \tag{12.123}$$

and the tuning order is

$$h_{tuned} = \frac{f_{tuned}}{f_1} = \frac{1}{\omega_1 \sqrt{L_1 C_1}} = \sqrt{\frac{X_{C_1}}{X_{L_1}}} \tag{12.124}$$

The inductive reactance of the reactor is

$$X_{L_1} = \frac{X_{C_1}}{h_{tuned}^2} \tag{12.125}$$

Capacitors are sensitive to peak voltages. Because of this, they need to be able to withstand the total peak voltage across it. Thus, a capacitor has to have a voltage rating that is equal to the algebraic sum of the fundamental and tuned harmonic voltages. That is,

$$V_C = V_{C_1} + V_{C_{tuned}} \tag{12.126a}$$

or

$$V_C = X_{C_1} I_{C_1} + X_{C_{\text{tuned}}} I_{C_{\text{tuned}}} \tag{12.126b}$$

But, a capacitor tuned to a particular harmonic may absorb other harmonics as well. Accordingly, a capacitor should have a voltage rating of

$$V_{C(L-L)} = \sum_{h=1} V_{C_{h(L-L)}} = \sum_{h=1} \sqrt{3} X_{C_h} I_{C_h} = \sum_{h=1} \sqrt{3} \frac{X_C}{h} \times I_{C_h} \tag{12.127}$$

even though its rms voltage is

$$V_{C(\text{rms})} = \sqrt{\sum_{h=1} V_{C_{h(L-L)}}^2} = \sqrt{3 \sum_{h=1} \left(\frac{X_C}{h} \times I_{C_h} \right)^2} \tag{12.128}$$

The reactive power absorbed by the capacitor bank can be expressed as

$$Q_L = \sum_{h=1} V_{L_h} I_{L_h} = \sum_{h=1} h \times X_L I_{L_h}^2 = \sum_{h=1} \frac{V_{L_h}^2}{h \times X_L} \tag{12.129}$$

and the reactive power delivered by the capacitor bank is

$$Q_C = \sum_{h=1} V_{C_h} I_{C_h} = \sum_{h=1} \frac{X_C}{h} \times I_{C_h}^2 = \sum_{h=1} \frac{h}{X_C} \times V_{C_h}^2 \tag{12.130}$$

12.19.1 Series-Tuned Filters

A series-tuned filter is basically a capacitor designed to trap a certain harmonic by the addition of a reactor having $X_L = X_C$ at the tuned frequency f_{tuned}. Steps of designing a series-tuned filter to the h_{tuned} harmonic include the following:

1. Estimate the capacitor size Q_C in Mvar to be equal to the reactive power requirement of the harmonic source.
2. Determine the reactance of the capacitor from

$$X_C = \frac{kV^2}{Q_C} \tag{12.131}$$

3. Find the size of the reactor that is necessary to trap the h_t harmonic from

$$X_L = \frac{X_C}{h_{\text{tuned}}^2} \tag{12.132}$$

4. Find out the resistance of the reactor from

$$R = \frac{X_t}{Q} \tag{12.133}$$

where Q is the quality factor of the filter, $30 < Q < 100$.

5. Find out the characteristic reactance of the filter from

$$X_{\text{tuned}} = X_{L_{\text{tuned}}} = X_{\text{tuned}} = \sqrt{X_L X_C} = \sqrt{\frac{L}{C}} \tag{12.134}$$

6. Determine the filter size from

$$Q_{\text{filter}} = \frac{kV^2}{X_C - X_L}$$

$$= \frac{kV^2}{\left(X_C - \left(X_C/h_{\text{tuned}}^2\right)\right)}$$

$$= \frac{h_{\text{tuned}}^2}{h_{\text{tuned}}^2 - 1} \times Q_C \tag{12.135}$$

7. Give the impedance function of the filter at any harmonic h:

$$\mathbf{Z}_{\text{filter}}(h) = R + j\left(h \times X_L - \frac{X_C}{h}\right) \tag{12.136}$$

so that

$$\left|\mathbf{Z}_{\text{filter}}(h)\right| = \left[R^2 + \left(h \times X_L - \frac{X_C}{h}\right)^2\right]^{1/2} \tag{12.137}$$

8. Calculate the ratio of the fundamental component of the voltage across the capacitor to the fundamental component of the voltage at the bus from

$$\frac{\mathbf{V}_{C_1}}{\mathbf{V}_{\text{bus}_1}} = \frac{-jX_{C_1}}{j\left(X_{L_1} - X_{C_1}\right)} = \frac{h_{\text{tuned}}^2}{h_{\text{tuned}}^2 - 1} \tag{12.138}$$

9. Calculate the ratio of the capacitor voltage at the tuned frequency to the bus voltage at the tuned frequency from

$$\frac{\mathbf{V}_{C_{\text{tuned}}}}{\mathbf{V}_{\text{bus}_{\text{tuned}}}} = \frac{-jX_{C_{\text{tuned}}}}{R + j\left(X_{L_{\text{tuned}}} - X_{C_{\text{tuned}}}\right)} = -j\frac{X_{\text{tuned}}}{R} = -jQ \tag{12.139}$$

where

$$X_{\text{tuned}} = X_{L_{\text{tuned}}} = X_{C_1} = \sqrt{X_{L_1} X_{C_1}} = \sqrt{\frac{L_1}{C_1}} \tag{12.140}$$

and

$$Q = \text{the filter's quality factor} = \frac{X_{\text{tuned}}}{R} \tag{12.141}$$

10. Determine the bus voltage from

$$V_{bus_1} = \frac{h_{tuned}^2 - 1}{h_{tuned}^2} \times V_{C_1} = V_{C_1} - \frac{V_{C_1}}{h_{tuned}^2} = V_{C_1} - V_{L_1} \tag{12.142}$$

Example 12.14

Assume that a series-tuned filter is tuned to the ninth harmonic. If $X_C = 324\ \Omega$, determine the following:

    a. The reactor size of the filter
    b. The characteristic reactance of the filter
    c. The size of the reactor resistance, if the quality factor is 100

Solution
    a. The reactor size is

$$X_L = \frac{X_C}{h_{tuned}^2} = \frac{324\ \Omega}{9^2} = 4\ \Omega$$

    b. The characteristic reactance of the filter is

$$X_{tuned} = \sqrt{X_L X_C} = \sqrt{4 \times 324} = 36\ \Omega$$

    c. The size of the reactor resistance is

$$R = \frac{X_{tuned}}{Q} = \frac{36}{100} = 0.36\ \Omega$$

Example 12.15

Suppose that for a 34.5 kV series-tuned filter $X_C = 676\ \Omega$, $X_L = 4\ \Omega$, and $R = 1.3\ \Omega$, determine the following:

    a. The tuning order of the filter.
    b. The quality factor of the filter.
    c. The reactive power delivered by the capacitor bank.
    d. The rated size of the filter.
    e. If the filter is used to suppress the resonance at the 13th harmonic, find the short-circuit MVA at the filter's location.

Solution
    a. The tuning order of the filter is

$$h_{tuned} = \sqrt{\frac{X_C}{X_L}} = \sqrt{\frac{676}{4}} = 13$$

    b. The quality factor of the filter is

$$Q = \frac{X_{tuned}}{R} = \frac{\sqrt{X_L X_C}}{R} = \frac{\sqrt{4 \times 676}}{1.3} = 40$$

c. The reactive power delivered by the capacitor bank is

$$Q_C = \frac{kV^2}{X_C} = \frac{34.5^2}{676} \cong 1.761\,\text{Mvar}$$

d. The rated size of the filter is

$$Q_{filter} = \frac{kV^2}{X_C - X_L} = \frac{h_{tuned}^2}{h_{tuned}^2 - 1} \times Q_C = \frac{13^2}{13^2 - 1} \times 1.761 = 1.771\,\text{Mvar}$$

e. The short-circuit MVA is

$$MVA_{sc} = h_r^2 \times Q_C = 13^2 \times 1.761 \cong 297.61\,\text{MVA}$$

12.19.2 Second-Order Damped Filters

The steps of designing a second-order damped filter tuned to the h_{tuned} harmonic include the following:

1. Decide the capacitor size Q_C in Mvar for the reactive power requirement of a harmonic source.
2. Calculate the reactance of the capacitor from

$$X_C = \frac{kV^2}{Q_C} \qquad\qquad (12.143)$$

3. Find the size of the reactor that is necessary to trap the h_{tuned} harmonic from

$$X_L = \frac{X_C}{h_{tuned}^2} \qquad\qquad (12.144)$$

4. Determine the size of the resistor bank from

$$R = X_{tuned}Q \qquad\qquad (12.145)$$

   where Q is the quality factor of the filter, $0.5 < Q < 5$.
5. Find the characteristic reactance of the filter from

$$X_{tuned} = X_{L_{tuned}} = X_{C_{tuned}} = \sqrt{X_L X_C} \qquad\qquad (12.146)$$

6. Determine the rated filter size from

$$Q_{filter} = \frac{kV^2}{X_c - X_L} = \frac{h_{tuned}^2}{h_{tuned}^2 - 1} \times Q_c \qquad\qquad (12.147)$$

7. Give the impedance function of the filter at any harmonic h:

$$\mathbf{Z}_{filter}(h) = \frac{jRhX_L}{R + jhX_L} - j\frac{X_C}{h} \qquad\qquad (12.148)$$

or

$$Z_{\text{filter}}(h) = \frac{R(hX_L)^2}{R^2 + (hX_L)^2} + j\left(\frac{R^2 hX_L}{R^2 + (hX_L)^2} - \frac{X_C}{h}\right) \qquad (12.149)$$

8. Calculate the current of the reactor from

$$I_{L_h} = \frac{R}{\sqrt{R^2 + X_{L_h}^2}} \times I_{\text{filter}_h} \qquad (12.150)$$

or

$$I_{L_h} = \frac{Q}{\sqrt{Q^2 + (h/h_{\text{tuned}})^2}} \times I_{\text{filter}_h} \qquad (12.151)$$

9. Determine the current of the resistor from

$$I_{R_h} = \frac{X_{L_h}}{\sqrt{R^2 + X_{L_h}^2}} \times I_{\text{filter}_h} = \frac{(h/h_{\text{tuned}})}{\sqrt{Q^2 + (h/h_{\text{tuned}})^2}} \times I_{\text{filter}_h} \qquad (12.152)$$

or

$$I_{R_h} = \frac{h}{h_{\text{tuned}}} \times \frac{I_{L_h}}{Q} = \frac{hX_L}{R} \times I_{L_h} \qquad (12.153)$$

10. Find the power loss in the resistor from

$$P_R = \sum_{h=1} RI_{Rh}^2 \qquad (12.154a)$$

or

$$P_R = \frac{X_L^2}{R} \sum_{h=1} (h \times I_{L_h})^2 \qquad (12.154b)$$

Example 12.16

Assume that a second-order damped filter is to be tuned to $h_{\text{tuned}} \geq 13$. If $X_C = 2.5\ \Omega$, determine the following:

    a. The size of the reactor
    b. The characteristic reactance
    c. The sizes of the resistor bank for the quality factors of 0.5 and 5

Solution

    a. The size of the reactor is

$$X_L = \frac{X_C}{h_{\text{tuned}}^2} = \frac{2.5\ \Omega}{13^2} \cong 0.0148\ \Omega$$

b. The characteristic reactance is

$$X_{\text{tuned}} = \sqrt{X_L X_C} = \sqrt{0.0148 \times 2.5} \cong 0.192 \, \Omega$$

c. The sizes of the reactor bank are

$$\text{For } Q = 0.5: \quad R = X_{\text{tuned}}Q = 0.192 \times 0.5 \cong 0.096 \, \Omega$$

$$\text{For } Q = 5: \quad R = 0.196 \times 5 = 0.96 \, \Omega$$

Example 12.17

A 34.5 kV 6 Mvar capacitor bank is being used as a second-order damped filter tuned to $h_{\text{tuned}} \geq 5$. Determine the following:

a. The size of the capacitor reactance of the filter
b. The size of the filter
c. The characteristic reactance of the filter
d. The size of the resistor bank for the quality factors of 0.5, 2, 3, 5
e. The rated filter size

Solution

a. The size of the capacitor reactance of the filter is

$$X_C = \frac{34.5^2}{6} = 198.375 \, \Omega$$

b. The reactor size of the filter is

$$X_L = \frac{X_C}{h_{\text{tuned}}^2} = 7.935 \, \Omega$$

c. The sizes of the resistor bank are

$$\text{For } Q = 0.5: \quad R = X_{\text{tuned}} \, Q = 39.675 \times 0.5 \cong 19.838 \, \Omega$$

$$\text{For } Q = 2: \quad R = 39.675 \times 2 = 79.35 \, \Omega$$

$$\text{For } Q = 3: \quad R = 39.675 \times 3 = 119.025 \, \Omega$$

$$\text{For } Q = 5: \quad R = 39.675 \times 5 = 198.375 \, \Omega$$

d. The reactor size is

$$Q_{\text{filter}} = \frac{kV^2}{X_C - X_L} = \frac{h_{\text{tuned}}^2}{h_{\text{tuned}}^2 - 1} \times Q_C = \frac{5^2}{5^2 - 1} \times 198.375 = 206.64 \text{ Mvar}$$

where

$$Q_C = \frac{kV^2}{X_C} = \frac{34.5^2}{6} = 198.375$$

12.20 LOAD MODELING IN THE PRESENCE OF HARMONICS

12.20.1 IMPEDANCE IN THE PRESENCE OF HARMONICS

The impedance of an inductive element, which has resistance of R and reactance of $X_L = 2\pi f L$, is normally expressed as

$$\mathbf{Z} = R + jX_L$$

at the fundamental frequency. However, in the presence of harmonics, the impedance of such element becomes

$$\mathbf{Z}(h) = R + jh \times X_L \tag{12.155}$$

where h is the harmonic order.

Similarly, a capacitive element has a reactance of $X_C = 1/(2\pi f C)$ at the fundamental frequency. In the presence of harmonics, the reactance becomes

$$X_C(h) = \frac{X_C}{h} \tag{12.156}$$

12.20.2 SKIN EFFECT

As the frequency increases, conductor current concentrates toward the surface, so that the ac resistance increases and the internal inductance decreases. Therefore, in modeling the power system components for a harmonics study, the impact of skin effects must be taken into account in determining the impedances of individual system components. Some researches represent passive loads at a harmonic order of h as

$$\mathbf{Z}(h) = \sqrt{h} \times R + jh \times X_L \tag{12.157}$$

where
R is the load resistance at the fundamental frequency
X is the load reactance at the fundamental frequency
h is the harmonic order

Note that some other researches use a factor of $0.6\sqrt{h}$ instead of $\sqrt{h}$ as the weighting coefficient for frequency dependence of the resistive component. Taking *skin effect* into account in the presence of harmonics, the *impedance of a transformer* is given as

$$\mathbf{Z}(h) = h(R + jX) \tag{12.158}$$

Similarly, the *impedance of a generator* is given as

$$\mathbf{Z}(h) = \sqrt{h} \times R + jh \times X \tag{12.159}$$

The *impedance of a transmission line* is represented by

$$\mathbf{Z}(h) = \sqrt{h}(R + jX_L) = \sqrt{h}\mathbf{Z}_L \tag{12.160}$$

12.20.3 LOAD MODELS

In harmonics studies involving mainly a transmission network, the loads are usually made up of equivalent parts of the distribution network, specified by the consumption of active and reactive power. Normally, a *parallel model* is used and the *equivalent load impedance* is represented by

$$\mathbf{Z}_p = j \frac{R_p \times X_p}{R_p + jX_p} \tag{12.161}$$

where

$$R_p \text{ is the load resistance in ohms} = \frac{V^2}{P}$$

$$X_p \text{ is the load resistance in ohms} = \frac{V^2}{Q}$$

There are many variations of this parallel form of load representation. For example, some researches suggest to use

$$R_p = \frac{V^2}{(0.1 \times h + 0.9)P} \tag{12.162}$$

and

$$X_p = \frac{V^2}{(0.1 \times h + 0.9)Q} \tag{12.163}$$

where P and Q are fundamental frequency active and reactive powers, respectively.

Due to difficulties involved, the power electronic loads are often left open-circuited when calculating harmonic impedances. However, their effective harmonic impedances need to be considered when the power ratings are relatively high, such as arc furnaces and aluminum smelters. An alternative approach to explicit load representation is the use of empirical models derived from measurements [14].

Example 12.18

A three-phase purely resistive load of 50 kW is being supplied directly from a 60 Hz three-phase 480 V bus. At the time of measuring, the load was using 48 kW and the voltage waveform had 12 V of negative-sequence fifth harmonic and 9 V of positive-sequence seventh harmonic. Assuming that the load resistance varies with the square root of the harmonic order h, determine the following:

    a. The values of the load resistance
    b. The components of the load current
    c. The THD index for the voltage
    d. The THD index for the current
    e. The TDD index for current

Solution

a. The values of the load resistance are

$$R_1 = \frac{V_1^2}{P_{1\phi}} = \frac{(480/\sqrt{3})2}{(48,000/3)} \cong 4.81\,\Omega$$

$$R_5 = R_1 \times \sqrt{h} = 4.81\sqrt{5} \cong 2.15\,\Omega$$

$$R_7 = R_1 \times \sqrt{h} = 4.81\sqrt{7} \cong 12.73\,\Omega$$

b. The components of the load are

$$I_{rms} = \frac{50,000}{\sqrt{3} \times 480} \cong 60.212\ A$$

$$I_1 = \frac{(V_1/\sqrt{3})}{R_1} = \frac{(480/\sqrt{3})}{4.81} = 57.683\ A$$

$$I_5 = \frac{(V_5/\sqrt{3})}{R_5} = \frac{(12/\sqrt{3})}{2.15} = 3.226\ A$$

$$I_7 = \frac{(V_7/\sqrt{3})}{R_7} = \frac{(9/\sqrt{3})}{12.73} = 0.409\ A$$

c. The THD index for the voltage is

$$THD_V = \frac{\sqrt{\left(V_5^2 + V_7^2\right)}}{V_1} = \frac{\sqrt{\left(11^2 + 8^2\right)}}{480} \cong 0.02834$$

d. The THD index for the current is

$$THD_I = \frac{\sqrt{\left(I_5^2 + I_7^2\right)}}{I_1} = \frac{\sqrt{(3.226^2 + 0.409^2)}}{57.683} \cong 0.0564$$

e. The TDD index for the current is

$$TDD_I = \frac{\sqrt{\left(I_5^2 + I_7^2\right)}}{I_{rms}} = \frac{\sqrt{(3.226^2 + 0.409^2)}}{60.212} \cong 0.054$$

PROBLEMS

12.1 The harmonic currents of a transformer are given as 1.00, 0.33, 0.20, 0.14, 0.11, 0.09, 0.08, 0.07, 0.06, 0.05, and 0.05 in pu A for the harmonic order of 1, 3, 5, 7, 9, 11, 13, 15, 17, 19, and 21, respectively. Also assume that the eddy current loss factor is 10%. Based on ANSI/IEEE Std.C75.110, determine the following:

    a.   The K-factor of the transformer

    b.   The transformer derating based on the standard

12.2 Consider an industrial load bus where the transformer impedance is dominant. If a parallel resonance condition is created by its 1800 kVA transformer, with 5% impedance, and 400 kvar PF correction capacitor bank, determine:

    a.   The resonant harmonic

    b.   The approximate or parallel resonant frequency

12.3 A 60 HZ 480 V three-phase delta-connected 500 kvar capacitor bank will be used as a part of a single-tuned 480 V filter. The filter will be used for the fifth harmonic of nonlinear loads of an industrial plant. Set the resonant frequency at 4.7 harmonic for a margin of safety. The facility has 500 hp of ASDs connected to 480 V. Design a single-tuned filter and determine the following:

a. The actual fundamental frequency compensation provided by a derated capacitor bank

b. The full-load fundamental frequency current of the capacitor bank

12.4 An electric car battery charger is 5 kW and that is supplied by a 5 kVA, 2400/240 V 60 Hz single-phase transformer with an impedance of $0.021 + j0.008$ pu ohms. Assume that everything else is the same as before. Determine the following:

a. The low-voltage side base impedance of the transformer.

b. The per unit impedance of the service drop line.

c. The value of the ratio I_{sc}/I_L.

d. Based on IEEE Std. 519-1992, find the maximum limits of odd current harmonics at the meter in steady-state operation, expressed in percent of the fundamental load current I_L.

12.5 Based on the output of a harmonic analyzer, a nonlinear load current has rms total of 10.5 A, total odd harmonics of 115.2%, total even harmonics of 13.8%, and a THD of 128.3%. Its total odd harmonic distribution is given in Table P12.5. Consider the current waveform and spectrum of a distorted current and determine the following:

a. The fundamental current in amps

b. The 300 Hz harmonic current in amps

c. The 660 Hz harmonic current in amps

d. The crest factor

TABLE P12.5
The Output of the Harmonic Analyzer

	Percentage (%)
Fundamental	100.0
3rd	70.4
5th	28.8
7th	0.7
9th	3.8
11th	1.5
13th	3.0
15th	1.2
17th	2.1
19th	0.9
21st	1.1
23rd	0.4
25th	0.3
27th	0.3
29th	0.4
31st	0.3
33rd	0.5

12.6 Based on the output of harmonic analyzer, a nonlinear load has a total rms current of 43.3 A. It also has 22.8, 12, 2.20, and 2.48 A for the third, fifth, seventh, and ninth harmonic currents, respectively. Here, the instrument used has been programmed to present the resulting data in amps rather than in percentages. Based on the given information, determine the following:

a. The fundamental current in amps

b. The amounts of the third, fifth, seventh, and ninth harmonic currents in percentages

c. The amount of the THD

12.7 The illumination of a large office building is being provided by fluorescent lighting with electronic ballasts. A line current measurement of a branch circuit serving exclusively such fluorescent lighting has been made by using a harmonic analyzer. The output of the harmonic analyzer is a phase current waveform and spectrum of current supplying such electronic power loads. For a 60 Hz 15.2 A fundamental rms current, it is observed from the spectrum that there is 100% fundamental odd triplen harmonics of 19.9%, 2.4%, 0.4%, 0.1%, and 0.1% for the 3rd, 9th, 15th, 21st, and 27th order, respectively. It is assumed that loads on the three phases are balanced and all have the same characteristic, determine the following:

a. The approximate value of the rms phase current in per units

b. The approximate value of the rms neutral current in per units

c. The ratio of the neutral current to the phase current

12.8 In an office building, a line current measurement of a branch circuit serving some nonlinear loads has been made by using a harmonic analyzer. The output of the analyzer is phase current waveform and spectrum of current supplying such electronic loads. For a 60 Hz 105 A fundamental rms current, it is observed from the spectrum that there is 100% fundamental and odd triplen harmonics of 70.4%, 3.8%, 1.2%, 1.1%, 0.3%, and 0.5% for 3rd, 9th, 15th, 21st, 27th, and 33rd order, respectively. Assume that loads on the three phases are balanced and all have the same characteristic, determine the following:

a. The approximate value of the rms phase current in per units

b. The approximate value of the rms neutral current in per units

c. The ratio of the neutral current to the phase current

12.9 In a large office building, there are 500 combinations of personal computers and printers. The harmonic spectrum of the total current shows the third harmonic (70%), followed by the fifth (60%), seventh (40%), and ninth (22%). Assume that each PC's fundamental current is 1 A. If a 500 kVA, 12.47 kV/480 V transformer supplies the building at 0.95 lagging PF, determine the following:

a. The total rms load current

b. The total fundamental load current

c. The third harmonic load current

d. The fifth harmonic load current

e. The seventh harmonic load current

f. The ninth harmonic load current

g. The TDD index of the load

h. The transformer neutral current

12.10 A 4.16 kV three-phase feeder is supplying a purely resistive load of 4500 kVA. It has been determined that there are 80 V of zero-sequence third harmonic and 180 V of negative-sequence fifth harmonic. Determine the following:

a. The total voltage distortion.

b. Is the THD below the IEEE Std. 519-1992 for the 4.16-kV distribution system?

12.11 According to ANSI 368 Std., telephone interference from a 4.16 kV distribution system is unlikely to occur when $I \cdot T$ index is below 10,000. Consider the load given in Problem 12.10 and assume that the TIF weightings for the fundamental, the third, and the fifth harmonics are 0.5, 30, and 225, respectively. Determine the following:

a. The I_1, I_2, I_3, and I_5 currents in amps.
b. The indices of $I \cdot T_1$, $I \cdot T_2$, $I \cdot T_3$, and $I \cdot T_5$.
c. The total $I \cdot T$ index.
d. Is the total $I \cdot T \cdot T$ index below the ANSI 368 Std. limit?
e. The total TIF index.

12.12 Repeat Example 12.3 if the load is 6,300 kVA.

12.13 A three-phase wye–wye-connected 230/13.8 kV 80 MVA transformer with $X = 19\%$ is supplying a 50 MVA load at a lagging PF of 0.9. At the low-voltage bus of 13.8 kV, a three-phase wye-connected capacitor bank is to be connected to correct the PF to 0.95. A distribution engineer is asked to investigate the problem, knowing that the short-circuit MVA at the 230 kV bus is 1600 MVA. Use a MVA base of 100 MVA and determine the following:

a. The current bases for the HV and LV sides in amps
b. The impedance bases for the HV and LV sides in ohms
c. The short-circuit reactance of the system at the 230 kV bus in per units and ohms
d. The short-circuit reactance of the system at the 13.8 kV bus in per units and ohms
e. The short-circuit MVA of the system at the 13.8 kV bus in per units and MVA
f. The real power of the load at the lagging PF of 0.9 in per units and MW
g. The size of the capacitor bank needed to correct the PF to 0.95 lagging in per units and Mvar
h. The reactance of each capacitor per phase in per units and ohms
i. The resonant harmonic at which the interaction between the capacitor bank and system inductance initiates resonance
j. The reactance of each capacitor in per units and ohms, if the capacitor bank is connected in delta

12.14 Verify Equation 12.82 by derivation.

12.15 A three-phase 13.8 kV 10 MVA capacitor bank is causing a bus voltage increase of 800 V when switched on. Determine the following:

a. The increase in bus voltage in per units
b. The resonant harmonic
c. The harmonic frequency at the resonance

12.16 A three-phase wye–wye-connected 115/12.47 kV 60 MVA transformer with an impedance of $0.3\% + j13\%$ is connected between high- and low-voltage buses. Assume that a wye-connected switched capacitor bank is connected to the low-voltage bus of 12.47 kV and that the capacitor bank is made up of three 3 Mvar. At the 115 kV bus, the short-circuit MVA of the external system is 2000 MVA and its X/R ratio is 6.5. Use MVA base of 100 MVA and determine the following:

a. The impedance bases for the HV and LV sides
b. The short-circuit impedance of the power system at the 115 kV bus
c. The transformer impedance in per units
d. The short-circuit impedance at the 12.47 kV bus in per units
e. The X/R ratio and the short-circuit MVA at the 12.47 kV bus in per units
f. The reactance of the capacitor per phase in ohms and per units
g. The resonant harmonic order
h. The characteristic impedance in per units
i. The amplification factor

12.17 A series-tuned filter is tuned to the 11th harmonic. If $X_C = 605\ \Omega$, determine the following:

a. The reactor size of the filter
b. The characteristic reactance of the filter
c. The size of the reactor resistance, if the filter quality factor is 90

12.18 Consider a 34.5 kV series-tuned filter that has $X_C = 423.5\ \Omega$, $X_L = 3.5\ \Omega$, and, determine the following:

a. The tuning order of the filter.
b. The quality factor of the filter.
c. The reactive power delivered by the capacitor bank.
d. The rated size of the filter.
e. If the filter is used to suppress the resonance at the 11th harmonic, determine the short-circuit MVA at the filter's location.

12.19 Assume that a second-order damped filter is to be tuned to $h_{tuned} \geq 15$. If $X_C = 1.8\ \Omega$, determine the following:

a. The size of the reactor
b. The characteristic reactance
c. The size of the resistor bank for the quality factors of 0.5 and 5

12.20 A 12.47 kV 3 Mvar capacitor bank is being tuned to $h_{tuned} \geq 9$. Determine the following:

a. The capacitor reactance size
b. The reactor size
c. The characteristic reactance
d. The resistor bank sizes for the quality factors of 0.5, 2, 3, 5
e. The rated filter size

12.21 Consider a single-phase power line, with an impedance of $1 + j4$ ohms, connected to a 7.2 kV power source. Assume that a fifth harmonic current source of 100 A is connected to the line and that the line resistance is constant at the fifth harmonic current level. Determine the following:

a. The equivalent circuit of the system
b. The magnitude of the line impedance
c. The voltage drop of the line
d. The percent voltage drop of the line

12.22 Consider Problem 12.21 and assume that there is a capacitor connected just before the harmonic current source. Its capacitive reactance is 260 Ω. Determine the following:

a. The reactive power of the capacitor
b. The capacitive reactance of the capacitor at the fifth harmonic
c. The resonant harmonic

REFERENCES

1. Gönen, T. and A. A. Mahmoud: Bibliography of power system harmonics, Part I. *IEEE Trans. Power Appar. Syst.*, PAS-103(9), September 1984, 2460–2469.
2. Gönen, T. and A. A. Mahmoud: Bibliography of power system harmonics, Part II. *IEEE Trans. Power Appar. Syst.*, PAS-103(9), September 1984, 2470–2479.
3. Heydt, G. T.: *Electric Power Quality*, 1st Ed., Stars in a Circle Publications, West LaFayatte, IN, 1991.
4. Dugan, C. R., M. F. McGranaghan, and H. W. Beaty: *Electric Power Quality*, McGraw-Hill, New York, 1996.
5. *Recommended Practice for Establishing Transformer Capability When Supplying Nonsinusoidal Load Currents*, ANSI/IEEE C57.110-1986, New York, 1986.
6. *IEEE Tutorial Course: Power System Harmonics*, 84 EHO221-2-PWR, IEEE Power Engineering Society, New York, 1984.
7. *IEEE Recommended Practices and Requirements for Harmonic Control in Electric Power Systems*, IEEE Std. 519-1992, IEEE, New York, 1993.
8. *IEEE Guide for Applying Harmonic Limits on Power Systems*, Power System Harmonics Committee Report, IEEE Power Engineering Society, New York, 1994.
9. Arrilaga, J.: *Power System Harmonics*, Wiley, New York, 1985.
10. Bollen, M. H. J.: *Understanding Power Quality Problems: Voltage Sags and Interruptions*, IEEE Press, New York, 2000.
11. Kennedy, B. W.: *Power Quality Primer*, McGraw-Hill, New York, 2000.

12. Porter, G. and J. A. Van Sciver: *Power Quality Solutions: Case Studies for Trouble Shooters*, Fairmont Press, Lilburn, GA, 1998.
13. Shepherd, W. and P. Zand: *Energy Flow and Power Factor in Nonsinusoidal Circuits*, Cambridge University Press, Cambridge, U.K., 1979.
14. Arrilaga, J., N. R. Watson, and S. Chen: *Power System Quality Assessment*, Wiley, New York, 2000.
15. Wakileh, G. J.: *Power Systems Harmonics*, Springer-Verlag, Berlin, Germany, 2001.
16. National Technical Information Service: Federal Information Processing Standards Publication 94: *Guidelines on Electric Power for ADP Installations*.
17. Information Technology Industry Council: *ITI curve Application Note*, available at http://www.itic.org/iss-pol/techdocs/curve.Pdf.

13 Distributed Generation and Renewable Energy

Simplicity is the most deceitful mistress that ever betrayed man.

Henry Brooks Adams

It was involuntary. They sank my boat.

John F. Kennedy. 1965 (Remark when asked how he become a hero.)

13.1 INTRODUCTION

Renewable energy is of many types, including wind, solar, hydro, geothermal (earth heat), and biomass (waste material). All renewable energy with the exception of tidal and geothermal power, and even the energy of fossil fuels, ultimately comes from the sun. About 1%–2% of energy coming from the sun is converted into wind energy. Today, the most prevalent renewable energy resources are wind and solar that will be reviewed briefly in this chapter.

13.2 RENEWABLE ENERGY

Renewable energy* is also a naturally *distributed resource*, that is, it can provide energy to remote areas without the requirement for elaborate energy transportation systems. However, it needs to be pointed out that it is not always a requirement that the renewable energy has to be converted into electricity. Solar water heating and wind-powered water pumping are good examples for it.

Presently, the largest renewable energy technology application (with the exception of hydro) has taken place is wind power, with 95 GW, worldwide by the end of 2007. In 2003, renewable energy contributed 13.5% of the world's total primary energy (2.2% hydro, 10.8% combustible renewables and waste, and 0.5% geothermal, solar, and wind). Even though the combustible renewables are used for heat, the contributions for electric generation were somewhat different: hydro contributed 15.9% and geothermal, solar, wind, and combustibles contributed 1.9%.

The capacity of the world's hydro plant is over 800 GW, and the capacity of fast-developing wind energy has sustained a 25% compound growth for well over a decade and was about 60 GW by the end of 2005. Also, in 2003, the world's electricity production from hydro was about 2654 TWh and all other renewables provided 310 TWh.

It is known that the world's primary energy demand almost doubled between 1971 and 2003. Furthermore, it is projected to increase by another 40% by 2020. It is also known that in the last 30 years, there has been a considerable shift away from oil and toward the natural gas. The natural

---

* It is often that in a classroom, some student who is indeed curious or perhaps just to be smart, Alex asks: "Can we harness lightning as an energy source?" The answer is that lightning is very powerful and very dangerous. But lightning strikes are very brief and infrequent, and therefore, the amount of energy that could be gained (and theoretically stored) would be small in comparison to overall electrical needs. One lightning strike has enough energy (about 1500 MJ) to power a 100 W light bulb for almost half of a year. However, you would need to harness over 58,000 lightning strikes each day to equal the electricity production capability of a large (1 GW) power plant.

gas accounted for 21% of primary energy and 19% of electricity generation, worldwide, in 2003. The impetus for this change has primarily been the increasing concerns over global warming due to carbon dioxide emissions.

13.3 IMPACT OF DISPERSED STORAGE AND GENERATION

Following the oil embargo and the rising prices of oil, the efforts toward the development of alternative energy sources (*preferably renewable resources*) for generating electric energy have been increased. Furthermore, opportunities for small power producers and cogenerators have been enhanced by recent legislative initiatives, for example, the *Public Utility Regulatory Policies Act* (PURPA) of 1978, and by the subsequent interpretations by the *Federal Energy Regulatory Commission* (FERC) in 1980.

The following definitions of the criteria affecting facilities under PURPA are given in Section 201 of PURPA:

- *A small power production facility* is one that produces electric energy solely by the use of primary fuels of biomass, waste, renewable resources, or any combination thereof. Furthermore, the capacity of such production sources together with other facilities located at the same site must not exceed 80 MW.
- *A cogeneration facility* is one that produces electricity and steam or forms of useful energy for industrial, commercial, heating, or cooling applications.
- *A qualified facility* is any small power production or cogeneration facility that conforms to the previous definitions and is owned by an entity not primarily engaged in generation or sale of electric power.

In general, these generators are small (typically ranging in size from 100 kW to 10 MW and connectable to either side of the meter) and can be economically connected only to the distribution system, as shown in Figure 13.1. They are defined as *dispersed storage and generation (DSG)* devices. If properly planned and operated, DSG may provide benefits to distribution systems by reducing capacity requirements, improving reliability, and reducing losses. Examples of DSG technologies include hydroelectric and diesel generators, wind electric systems, solar electric systems, batteries, storage space and water heaters, storage air conditioners, hydroelectric pumped storage, photovoltaics (PVs), and fuel cells.

13.4 INTEGRATING RENEWABLES INTO POWER SYSTEMS

Before going any further, it might be appropriate to define some terminologies including the terms *grid*, *grid connected*, and *national grid*. The term *grid* is usually used to describe the totality of the electric power network. The term *grid connected* means connected to any part of the power network. On the other hand, the term *national grid* usually means the extra-high voltage (EHV) transmission network.

The physical connection of a generator to the network is defined as *integrated*. But it is required that the necessary attention must be given to the secure and safe operation of the system and the control of the generator to achieve optimality in terms of the energy resource usage.

In general, the integration of the generators powered by the renewable energy sources essentially is the same as fossil fuel-powered generators and is based on the same methodology. However, renewable energy sources are very often variable and geographically dispersed.

A renewable energy generator can be defined as stand-alone or grid-connected. A *stand-alone renewable energy generator* provides for the greater part of the demand with or without other generators or storage. On the other hand, in *a grid-connected system*, the renewable energy generator supplies power to a large interconnected network that is also supplied power by other generators. Here, the power supplied by the renewable energy generator is only a small portion of the power

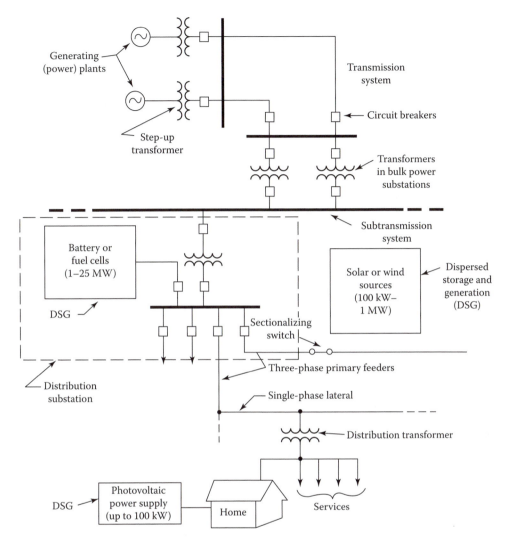

FIGURE 13.1 Connecting DSGs into utility system.

supplied to the grid with respect to power supplied by other connected generators. The connection point is called the *point of common coupling* (PCC)*.

13.5 DISTRIBUTED GENERATION

It appears that there is a general consensus that by the end of this century the most of our electric energy will be provided by renewable energy sources. As said previously, small generators cannot be connected to the transmission system due to high cost of high-voltage transformers and switchgear.

Thus, small generators must be connected to the distribution system network. Such generation is known as *distributed generation (DG)* or *dispersed generation*. It is also called *embedded generation* since it is embedded in the distribution network.

Power in such power systems may flow from point to point within the distribution network. As a result, such unusual flow pattern may create additional challenges in the effective operation and protection of the distribution network.

---

* See Gönen [12], Chapter 12.

Due to decreasing fossil fuel resources, poor energy efficiency, and environmental pollution concerns, the new approach for generating power locally at distribution voltage level by employing nonconventional/renewable energy sources such as natural gas, wind power, solar PV cells, biogas, cogeneration systems [which are *combined heat and power* (CHP) systems, Stirling engines, and microturbines.]

These new energy sources are connected to (or integrated into) the utility distribution network. As aforementioned, such power generation is called DG and its energy resources are known as *distributed energy resources* (*DERs*). Furthermore, the distribution network becomes *active* with the integration of DG and thus is known as *active distribution network*. The properties of DG include the following:

1. It is normally less than 50 MW.
2. It is neither centrally dispatched nor centrally planned by the power utility.
3. The distributed generators or power sources are generally connected to the distribution systems, which typically have voltages of 240 V up to 34.5 kV.

The *development and integration of the DG* were based on the technical, economic, and environmental benefits that include the following:

1. Reduction of environmental pollution and global warming concerns, as dictated by the Kyoto Protocol, and use of the nonconventional/renewable energy resources as a viable solution.
2. As a result of rapid load growth, the fossil fuel reserves are increasingly depleted. Therefore, the use of nonconventional/renewable energy resources is increasingly becoming a requirement. Also, the use of DERs is to produce clean power without the associated pollution of the environment.
3. DERs are usually modular units of small capacity because of their lower energy density and dependence on geographical conditions of a region.
4. The overall power quality and reliability improves due to contributions of the stand-alone and grid-connected operations of DERs in generation augmentation. Such DG integration further increases due to deregulated environment and open access to the distribution network.
5. The overall plant energy efficiency increases and also associated thermal pollution of the environment decreases because of the use of the DG such as cogeneration or CHP plants.

Furthermore, it is possible to connect a DER separately to the utility distribution network, or it may be connected as a microgrid due to the fact that the power is produced at low voltage. Thus, the microgrid can be connected to the utility's network as a separate semiautonomous entity.

13.6   RENEWABLE ENERGY PENETRATION

The proportion of electric energy or power being supplied from wind turbines or from other renewable energy sources is usually referred to as the penetration. It is usually given in percentage. The average penetration is defined by the following equation;

$$\text{Average penetration} = \frac{\left(\begin{array}{c}\text{Annual energy from renewable}\\ \text{energy powered generators (kWh)}\end{array}\right)}{\left(\begin{array}{c}\text{Total annual energy}\\ \text{delivered to loads (kWh)}\end{array}\right)} \tag{13.1}$$

The term *average penetration* is used when fuel or CO_2-emission savings are being considered.

However, for other purposes, including system control, use the following definition:

$$\text{Instantaneous penetration} = \frac{\left(\begin{array}{c}\text{Power from renewable energy}\\\text{powered generators (kW)}\end{array}\right)}{\left(\dfrac{\begin{array}{c}\text{Power from renewable energy}\\\text{powered generators (kW)}\end{array}}{\begin{array}{c}\text{Total power delivered}\\\text{to loads (kW)}\end{array}}\right)} \qquad (13.2)$$

In general, the maximum instantaneous penetration is much greater than the average penetration.

13.7 ACTIVE DISTRIBUTION NETWORK

It is also called "*generation embedded distribution network.*" In the past, distribution networks had a unidirectional electric power transportation. That is, distribution networks were stable passive networks.

Today, the distribution networks are becoming active by the addition of DG that causes bidirectional power flows in the networks. Today's distribution networks started to involve not only demand-side management but also integration of DG.

In order to have good active distribution networks that have flexible and intelligent operation and control, the following should be provided:

1. Adaptive protection and control
2. Wide-area active control
3. Advanced sensors and measurements
4. Network management apparatus
5. Real-time network simulation
6. Distributive penetrating communication network
7. Knowledge and data extraction by intelligent methods
8. New and modern design of transmission and distribution systems

13.8 CONCEPT OF MICROGRID

A microgrid is basically an active distribution network and is made up of a collection of DG systems and various loads at distribution voltage level. They are generally small low-voltage combined heat loads of a small community. The examples of such small community include university or school campuses, a commercial area, an industrial site, a municipal region or a trade center, a housing estate, or a suburban locality.

The generators or microsources used in a microgrid are generally based on renewable/nonconventional distribution energy resources. They are integrated together to provide power at distribution voltage level. In order to introduce the microgrid to the utility power system as a single controlled unit that meets local energy demand for reliability and security, the microsources must have power electronic interfaces (PEIs) and controls to provide the necessary flexibility to the semiautonomous entity so that it can maintain the dictated power quality and energy output.

A *microgrid* is different than a conventional power plant. The differences include the following:

1. Power generated at distribution voltage level and can thus be directly provided to the utility's distribution system.
2. They are of much smaller capacity with respect to the large generators in conventional power plants.

3. They are usually installed closer to the customers' locations so that the electric/heat loads can be efficiently served with proper voltage level and frequency and ignorable line losses.
4. They are ideal for providing electric power to remote locations.
5. The fundamental advantage of microgrids to a power grid is that they can be treated as a controlled entity within the power system.
6. The fundamental advantage of microgrids to customers is that they meet the electric/heat requirements locally. This means that they can receive uninterruptable power, reduced feeder losses, improved local reliability, and local voltage support.
7. The fundamental advantage to the environment is that they reduce environmental pollution and global warming by utilizing low-carbon technology.

However, before microgrids can be extensively established to provide a stable and secure operation, there are a number of technical, regulatory, and environmental issues that need to be addressed that include the establishment of standards and regulations for operating the microgrids in synchronism with the power utility, low energy content of the fuels involved, and the climate-dependent nature of the production of the DERs.

Figure 13.2 shows a microgrid connection scheme. Microgrid is connected to the medium voltage (MV) utility "main grid" through the PPC circuit breaker. Microsource and storage devices are connected to the feeders *B* and *C* through *microsource controllers* (MCs). Some loads on feeders *B* and *C* are considered to be priority loads (i.e., needing uninterruptable power supply), while the rest are non-priority loads. On the other hand, feeder *B* had only non-priority electric loads.

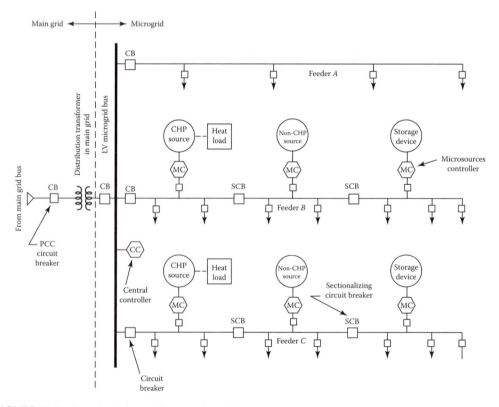

FIGURE 13.2   A typical microgrid connection scheme.

The microgrid has two modes of operations: (1) grid-connected and (2) stand-alone. In the first mode, the microgrid imports or exports power from or to the main grid. In the event of any disturbance in the main grid, the microgrid switches aver to stand-alone mode but still supply power to the priority loads. This is achieved by opening the necessary circuit breakers. But feeder *A* will be left alone so that it can ride through the disturbance.

The main functions of central controller (CC) include *energy management module* (EMM) and *protection coordination module* (PCM). The EMM supplies the set points for active and reactive power output, voltage, and frequency to each microgrid controller (MC). This is done by advanced communication and artificial intelligent techniques, whereas the PCM answers to microgrid and main grid faults and loss of grid situations so that proper protection coordination of the microgrid is achieved.

Chowdhuri et al. [1] define the functions of the CC in the grid-connected mode and in the stand-alone mode. The functions of the CC in the grid-connected mode include the following:

1. Monitoring system diagnostics by gathering information from the microsources and loads
2. Performing state estimation and security assessment evaluation, economic generation scheduling and active and reactive power control of the microsources, and demand-side management functions by employing collected information
3. Ensuring synchronized operation with the main grid maintaining the power exchange at priori contract points

The functions of the CC in the stand-alone mode are as follows:

1. Performing active and reactive power control of the microsources to keep stable voltage and frequency at load ends
2. Adapting load interruption/load-shedding strategies using demand-side management with storage device support for maintaining power balance and bus voltage
3. Beginning a local "cold start" to ensure improved reliability and continuity of service
4. Switching over the microgrid to grid-connected mode after main grid supply is restored without hindering the stability of either grid

Chowdhuri et al. [1] list the following technical and economic advantages of microgrid for the electric power industry:

1. Reducing environmental problems and issues
2. Reducing some operational and investment issues
3. Improving power utility and reliability
4. Increasing cost savings
5. Solving market issues

13.9 WIND ENERGY AND WIND ENERGY CONVERSION SYSTEM

Besides home wind electric generation, a number of electric utilities around the world have built larger wind turbines to supply electric power to their customers. In 2009, worldwide more than 1,000,000 windmills of about 120 GW installed power generation capacity were in operation, as given in Table 13.1. This was based on the understanding that ultimately, additional energy sources causing less pollution are necessary. Due to favorable tax regulations in the 1980s, about 12,000 wind turbines providing power ranging from 20 kW to about 200 kW were installed in California.

TABLE 13.1
Installed Wind Power Capacity Worldwide, as of 2009

	Rated Capacity (MW)	Share Worldwide (%)
United States	25,200	21
Germany	23,900	20
Spain	16,800	14
China	12,200	10
India	9,600	8

FIGURE 13.3 Solar and wind applications in the city of Kassel in the state of Hessen, Germany. (SMA Solar Technology AG.)

Germany had the leadership in wind turbine applications in the past. But, since then, the United States has taken over the leadership. (Figure 13.3 shows solar and wind applications in the city of Kassel in the state of Hessen in Germany. Figure 13.4 shows solar and wind turbine applications in the state of Rheinland-Pfalz in Germany. Figure 13.9 shows solar and wind turbine applications in the state of Rheinland-Pfalz in Germany.) The average commercial size of wind energy conversion system (WECS) was 300 kW until the mid-1990s. Today, there are wind turbines with a capacity of up to 6 MW that have been developed and installed. Since 1973, prices have dropped as performance has improved. Today, the cost of a wind turbine is below $2/W of installed capacity, and large wind farms with several hundred megawatt capacities are being developed over several months. For example, it is now quite common for wind power plants (wind farms) with collections of utility-scale turbines to be able to sell electricity for fewer than four cents per kWh. Early developments in California were basically in the form of wind farms, with tens of wind turbines, even up to 100 or more in some cases. The reasons for this development include the economies of scale that can be achieved by building wind farms, especially in construction and grid connection costs, and even possibly by getting quantity discounts from the turbine manufacturers. It is interesting to point out that the market introduction of wind energy is being done.

The European accessible onshore wind resource has been estimated at* 4800 TWh/year taking into account typical wind turbine efficiencies, with the European offshore resource in the

---

* Note that a terawatt is denoted as TW so that

    1 TWh = 1 × 10¹² Wh = 1000 GWh

    T = tera = 10¹²

    1 MWh = 1000 kWh

    1 MW wind power produces 2 GWh/year on land and 3 GWh/year offshore.

FIGURE 13.4 Solar and wind turbine applications in the state of Rheinland-Pfalz in Germany. (SMA Solar Technology AG.)

region of 3000 TWh/year although this is very dependent on the assumed allowable distance from the shore. According to a recent report [1], by 2030 the EU could be generating 965 TWh from onshore and offshore wind, amounting to 22.6% of electricity requirements. The world onshore resource is approximately 53,000 TWh/year, considering siting constraints. Note the annual electricity demand for the United Kingdom and the United States are 350 and 3500 TWh, respectively.

13.9.1 Advantages and Disadvantages of Wind Energy Conversion Systems

The wind energy is the fastest growing energy source in the world due to many advantages that it offers. Continuous research efforts are being made even further to increase the use of wind energy.

13.9.2 Advantages of a Wind Energy Conversion System

a. It is one of the lowest-cost renewable energy technologies that exist today.
b. It is available as a domestic source of energy in many countries worldwide and not restricted to only few countries, as in case of oil.
c. It is energized by naturally flowing wind; thus, it is a clean source of energy. It does not pollute the air and cause acid rain or greenhouse gases.
d. It can also be built on farms or ranches and hence can provide the economy in rural areas using only a small fraction of the land. Thus, it still provides opportunity to the landowners to use their land. Also, it provides rent income to the landowners for the use of the land.

13.9.3 Disadvantages of a Wind Energy Conversion System

a. The main challenge to using wind as a source of power is that the wind is intermittent and it does not always blow when electricity is needed. It cannot be stored; not all winds can be harnessed to meet the timing of electricity demands. At the present time, the use of energy storage in battery banks is not economical for large wind turbines.
b. Despite the fact that the cost of wind power has come down substantially in the past 10 years, the technology requires a higher initial investment than the solutions using fossil fuels. Hence, depending on the wind profile at the site, the wind farm may or may not be as cost competitive as a fossil fuel-based power plant.
c. It may have to compete with other uses for the land, and those alternative uses may be more highly valued than electricity generation.
d. It is often that good sites are located in remote locations, far from cities where the electricity is needed. Thus, the cost of connecting remote wind farms to the supply grid* may be prohibitive.
e. There may be some concerns over the noise generated by the rotor blades and esthetic problems that can be minimized through technological developments or by correctly siting wind plants [2].

13.9.4 Categories of Wind Turbines

Wind turbines turn the kinetic energy of the moving air into electric power or mechanical work. There are various WECSs. They can be classified as (1) horizontal-axis converters, (2) vertical-axis converters, and (3) upstream power stations.

Figure 13.5 shows three-blade wind energy converter that is the most common type of horizontal-axis converter for generating electricity worldwide. It shows the front and side views of a three-blade horizontal-axis wind energy converter. It has only a few rotor blades. Another conventional (older) type of horizontal-axis rotor is the multiblade wind converter. The horizontal-axis converters are of two types: with fast rotation or slow rotation.

The vertical axis converters are of two types: (1) Darrieus and (2) Savonius. The Darrieus converter has a vertical axis construction. They do not depend on the direction of the wind. But they have a low starting torque. Because of this, they need the help of a generator working as a motor or the help of Savonius rotor installed on top of the vertical axis.

The wind velocity increases substantially with height; as a result, the horizontal-axis wheels on towers are more economical. In the 1980s, a large number of Darrieus converters were installed in California, but a further expansion into a higher power range and their application worldwide has not happened. The Savonius rotor is used as a measurement device especially for wind velocity. However, it is used for power production for very small capacities under 100 W. The last technique mentioned previously is also known as *"upstream power station"* or *thermal tower.* It is a mix between a wind converter and a solar collector, poor efficiency, only about 1%.

Note that the terms "wind energy converters," "windmills," or "wind turbines" represent the same thing. The first one is the technical name of the system, whereas the other two are popularly used terms. Today, there are various types of wind energy converters that are in operation, as shown in Figure 13.6. Figure 10.7 shows eight different classes of wind turbines used in the Altamont pass in California.

Over the last 25 years, the size of the largest commercial wind turbines has increased from approximately 50 kW to 2 MW, with machines up to 6 MW under design. Figure 13.7 shows the

---

* The term *grid* is often used loosely to describe the totality of the network. For example, *grid connected* means connected to any part of the electric network. The term *national grid* usually means the EHV transmission network. Similarly, *integration* means the physical connection of the generator to the network for secure, safe, and optimal operation of the electrical system.

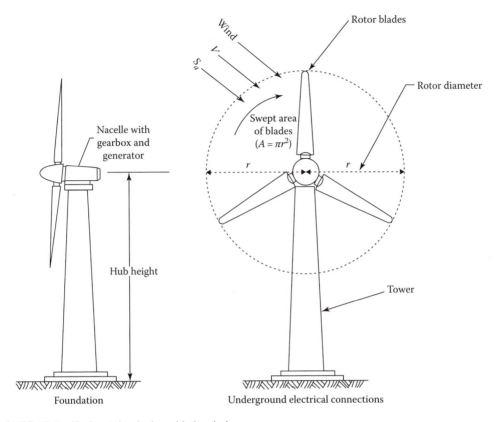

FIGURE 13.5 Horizontal-axis three-blade wind energy.

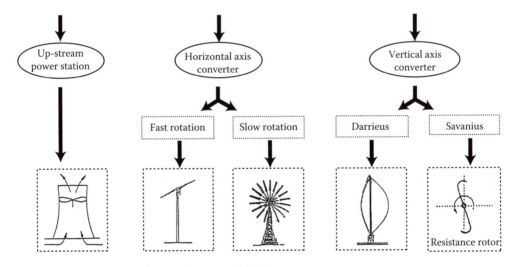

FIGURE 13.6 Overview of differential types of wind energy converters.

main subsystems of a typical horizontal design. Figure 13.8 shows the main subsystems of a typical horizontal-axis wind turbine. These include the rotor, including the blades and supporting hub; the drive train, which includes the rotating parts of the wind turbine (except the rotor), including shafts gearbox, coupling, a mechanical brake, and the generator; the nacelle and main frame, including wind turbine housing, bedplate, and the yaw system; the tower and the foundation; and the machine controls, the switchgear, transformers, and possibly electronic power converters.

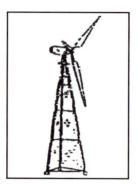

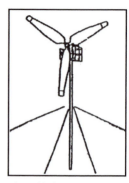

Turbine type:	Three-blade lattice (downwind)	Three-blade lattice (upwind)	Three-blade Guyed-pipe tower
Tower height:	60–80 ft	45–80 ft	40–60–80 ft
Rotor diameter:	59 ft	50–56 ft	33–80 ft
Description:	Downwind free yaw	Upwind	Downwind
Number:	3350 (1989)	248	1559
	3640 (1990)		

Turbine type:	Two-blade lattice (downwind)	Medium tubular	Large tubular
Tower height:	80 ft	100–150 ft	82 ft
Rotor diameter:	54 ft	50–82 ft	102 ft
Description:	Downwind free yaw	Upwind	Upwind
Number:	346	1421	135

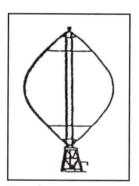
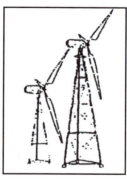

Turbine type:	Vertical axis	Windwall
Tower height:	90–106 ft	140 ft
Rotor diameter:	56–62 ft	59 ft
Description:	–	Downwind, free yaw
Number:	169	103

FIGURE 13.7 Eight categories of wind turbines used in the Altamont Pass in California. (From Orloff, S. and Flannery, A., Wind turbine effects on avian activity, habitat use, and mortality in altamont pass and Solano County wind resource areas: 1989–1991, California Energy Commission Report, No. P700-92,002; Stanon, C., Wind farm visual impact and its assessment, *Wind Directions*, BWEA, August 1995, pp. 8–9.)

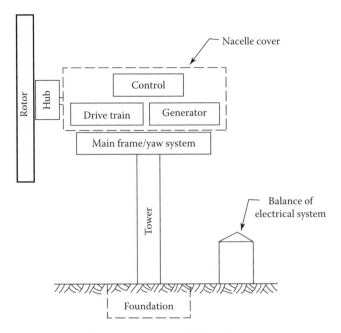

FIGURE 13.8 Major components of a horizontal-axis wind turbine.

There are a number of options in wind machine design and construction. These options include the number of blades (normally two or three); the blade material, construction method, and profile; the rotor orientation, downward or upward of tower; hub design, rigid, teetering, or hinged; fixed or variable rotor speed; orientation by self-aligning action (free yaw) or direct control (active yaw); power control via aerodynamic control (stall control) or variable pitch blades (pitch control); synchronous or induction generator; and gearbox or direct-drive generator.

Almost all wind turbines use either induction or synchronous generators. Both of these designs entail a constant or nonconstant rotational speed of the generator when the generator is directly connected to a utility network. The majority of wind turbines installed in grid-connected applications use induction generators. An induction generator operates within a narrow range of speeds slightly higher than a narrow range of speeds slightly higher than its synchronous speed. The main advantage of induction generators is that they are rugged, inexpensive, and easy to connect to an electric network. An induction generator is much simpler to connect to the grid than a synchronous generator.

The *nacelle* of horizontal-axis turbine contains a bedplate on which the components are mounted. There is a main shaft with main bearings, a generator, and a yaw motor that turns the nacelle and rotor into the wind. The nacelle cover protects the contents from the weather. Nacelle and yaw system include the wind turbine housing, the machine bedplate or main frame, and the yaw orientation system. The main frame provides for the mounting and proper alignment of the drive train components.

A yaw orientation system is needed to keep the rotor shaft properly aligned with the wind. The main component is a large bearing that connects the main frame to the tower. An active yaw drive, generally used with an upwind turbine, has one or more yaw motors, each of which drives a pinion gear against a bull gear attached to the yaw bearing. This mechanism is controlled by an automatic yaw control system with its wind direction sensor usually mounted on the nacelle of the wind turbine. Sometimes yaw brakes are used with this type of design to hold the nacelle of the wind turbine. Free yaw systems are normally used on downwind wind machines. They can self-align with the wind. The control system of a wind turbine includes sensors, controllers, power amplifiers, and actuators.

13.9.5 Types of Generators Used in Wind Turbines

There are three types of electrical machines that can convert mechanical power into electric power, which are the direct current (dc) generator, the synchronous alternator, and the induction generator. In the past, the shunt-wound dc generators were commonly used in small battery charging wind turbines. In these generators, the field is on the stator and the armature is on the rotor. A commutator on the rotor rectifies the generated power to dc.

By regulating the speed of the generator (i.e., wind turbine) and/or its field, the dc voltage can be maintained with a specified range. Speed regulation is usually performed by changing the pitch of the propeller blades. If the dc voltage is sensed, the field strength can be varied according to the control of the generated voltage. As illustrated in Figure 13.9a, a transmission that increases the rotating blade speed to that required for the generator has to be included.

The field current and thus magnetic field increase with operating speed. The armature voltage and electrical torque also increases with speed. The actual speed of the turbine is determined by a balance between the torque from the turbine rotor and the electrical torque. Since the wind speed is variable over a wide range, some regulation method must be used, as shown in Figure 13.10.

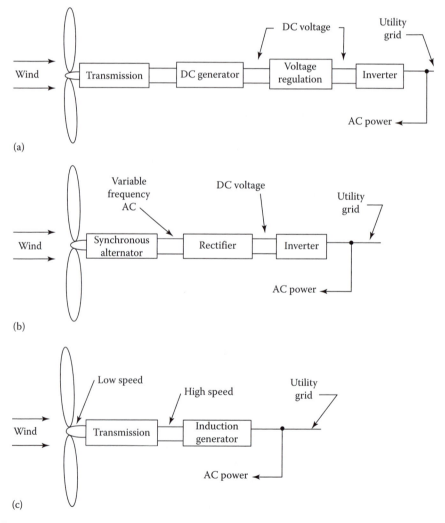

FIGURE 13.9 Block diagram of a WECS: (a) using a dc generator, (b) using a synchronous alternator, and (c) using induction generator.

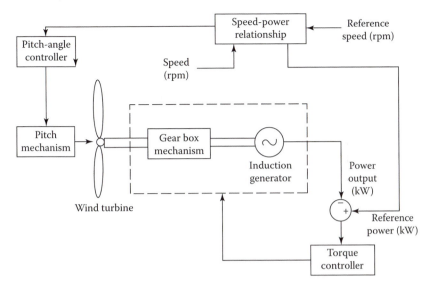

FIGURE 13.10 Variable-speed pitch-regulated wind turbine.

A wind machine typically rotates at a speed in the range of 50–100 rpm (i.e., about 5–10 rad/s). Depending on the generator, this has to be geared up to 1000–2000 rpm (i.e., about 100–200 rad/s). The net efficiency of the energy conversion system is a function of the efficiency of the blades, transmission, generator, regulating circuitry, and inverter. However, dc generators of this type are seldom used today because of high costs and maintenance requirements (due to the commutators).

Permanent magnet generators are used in most small wind turbine generators, up to at least 10 kW. Here, permanent magnets provide the magnetic field. Hence, there is no need for field windings, or supply current to the field, nor there is any need for commutators, slip rings, or brushes. The permanent magnet generator is quite rugged since the machine construction is so simple. Their operating principles are similar to that of synchronous machines, with the exception that they are run asynchronously. In other words, they are not generally connected directly to the alternating current (ac) network. The power produced by the generator is initially variable voltage and frequency ac. This ac variable voltage is often rectified immediately to dc. The resultant dc power then either directed to dc loads or battery storage, or else it is inverted to ac with a fixed frequency and voltage.

Synchronous machines operate at constant speed, with only the power angle changing as the torque varies. Synchronous machines hence have a very "stiff" response to fluctuating conditions. An alternator produces an ac voltage whose frequency is proportional to shaft speed. Even with speed regulation, there will still be enough of a variation in frequency and phase to prevent connection of the alternator directly to the utility grid.

Therefore, the alternator is permitted to turn at different speeds, producing a variable-frequency output. The alternator output is then rectified, converting it to dc, as shown in Figure 13.9b. The magnitude will be constant since the alternator field is constant. It is usually a permanent magnet alternator. The dc is now fed to a synchronous inverter, whose line frequency output can be connected directly to the utility grid. Here, the need for transmission is eliminated and the alternator can be connected directly to the wind wheel.

The induction generator is well suited for a wind energy system provided that utility power is available. But in order for the induction machine to operate as a generator, a separate source of reactive power is necessary to excite the machine. Also, the induction generator must be driven slightly faster than synchronous speed. However, it is not necessary for the speed to be constant, merely to maintain a negative slip. Rated power and peak efficiency are generally achieved at about –3% slip, not the speed of its rotor.

The only components required for this WECS are a transmission to gear the speed of the blades up to that necessary for the negative slip and the induction generator, as represented in Figure 12.9c. However, in the event of a loss of utility, power automatically disables the WECS since the field excitation no longer exists. The net system efficiency depends on the efficiency of the blades, transmission, and generator. But some means of speed regulation is required to maintain the required slip.

Note that when a constant torque is applied to the rotor of an induction machine, it will operate at a constant slip. If the applied torque is varying, then the speed of the rotor will vary as well. This relationship can be described by the following equation:

$$J \frac{d\omega_r}{dt} = Q_e - Q_r \tag{13.3}$$

where
  J is the moment of inertia of the generator rotor
  ω_r is the angular speed of the generator rotor (rad/s)
  Q_e is the applied electrical torque
  Q_r is the torque applied to the generator rotor.

Induction machines are somewhat "softer" in their dynamic response to changing conditions than are synchronous machines. This is due to the fact that induction machines undergo a small but significant speed change (slip) as the torque in or out changes.

Induction machines are designed to operate at a specific operating point. This operating point is usually defined as the rated power at a specific frequency and voltage. However, in wind turbine applications, there may be a number of cases when the machine may run at off-design conditions. These conditions include starting, operation below rated power, variable-speed operation, and operation in the presence of harmonics. The operation below rated power, but at rated frequency and voltage, is a common occurrence. It normally presents few problems. But efficiency and power factor are generally both lower under such conditions.

In general, there are a number of benefits of running a wind turbine rotor at variable speed. A wind turbine with an induction generator can be run at variable speed if the electronic power converter of approximate design is included in the system between the generator and the rest of the electric network.

Such converters operate by changing the frequency of the ac supply at the terminals of the generator. These converters also have to vary the applied voltage. It is due to the fact that an induction machine performs best when the ratio between frequency and voltage, that is, "*volts to hertz ratio*," of the supply is constant or almost constant. When that ratio departs from the design value, a number of problems can take place. For example, currents may be higher, causing higher losses and possible damage to the generator windings.

Finally, operation in the presence of harmonics can take place, if there is a power electronic converter of significant size on the system to which the induction machine is connected. Also, harmonics may cause bearing and electrical insulation damage and may interfere with electrical control or data signal as well.

13.9.6 Wind Turbine Operating Systems

Depending on controllability, wind turbine operating systems are categorized as (1) constant-speed wind turbines and (2) variable-speed wind turbines.

13.9.6.1 Constant-Speed Wind Turbines

They operate at almost constant speed as predetermined by the generator design of gearbox ratio. The control schemes are always aimed at maximizing either energy capture by controlling the rotor torque or the power output at high winds by regulating the pitch angle. Based on the control strategy, constant-speed wind turbines are again subdivided into (1) stall-regulated turbines and (2) pitch-regulated turbines.

Constant-speed stall-regulated turbines have no options for any control input. Its turbine blades are designed with a fixed pitch to operate near the original tip speed ratio (TSR) for a given wind speed. When wind speed increases, it causes a reduced rotor efficiency and limitation of the power output. The same result can be achieved by operating the wind turbine at two distinct constant operating speeds by either changing the number of poles of the induction generator or changing the gear ratio.

The stall regulation has the advantage of simplicity. But it has the disadvantage of not being able to capture wind energy in an efficient manner at wind speeds other than the design speed. They use pitch regulation for staring up. They have the following advantages:

1. They have a simple, robust construction and electrically efficient design.
2. They are highly reliable since they have fewer parts.
3. No current harmonics are produced since there is no frequency conversion.
4. They have a lower capital cost in comparison to variable-speed wind turbines.

On the other hand, their disadvantages include the following:

1. They are aerodynamically less efficient.
2. They are prone to mechanical stress and are noisier.

13.9.6.2 Variable-Speed Wind Turbines

Figure 13.10 shows a typical variable-speed pitch-regulated wind turbine system. It has two methods for controlling the turbine operation in terms of speed changes and blade pitch changes. The control strategies that are usually used are power optimization strategy and power limitation strategy.

Power optimization strategy is used when the wind speed is below the rated value. It optimizes the energy capture by keeping the speed constant based on the optimum TSR. However, if speed is changed because of load variation, the generator may be overloaded for wind speeds above nominal value. In order to prevent this, methods like generator torque control are employed to control the speed.

On the other hand, the power limitation strategy is used for wind speeds above the rated value by changing the blade pitch to reduce the aerodynamic efficiency. The advantages of the variable-speed wind turbine systems include the following:

1. They are subjected to less mechanical stress and they have high energy capture capacity.
2. They are aerodynamically efficient and have low transient torque.
3. They require no mechanical damping systems since the electric system can effectively provide the damping.
4. They do not suffer from synchronization problems or voltage sags because they have stiff electrical controls.

The disadvantages of the variable-speed wind turbine systems include the following:

1. They are more expensive.
2. They may require complex control strategies.
3. They have lower electrical efficiency.

In general, in order to indicate how much wind power there is in a country, the total installed capacity is used as a measure. Every wind turbine has a rated power (maximum power) that can vary from a few hundred watts to 5000 kW (5 MW). The number of turbines does not give any information on how much of wind power they can produce.

How much wind a wind turbine can produce depends not only on its rated power but also on the wind conditions. In order to get an indication of how much a certain amount of installed (rated) power will produce per year, use the following rule of thumb: "*1 MW wind power produces 2* GWh *per year on land and 3* GWh *per year offshore.*"

13.9.7 METEOROLOGY OF WIND

The fundamental driving force of air is a difference in air pressure between two regions. This air pressure is governed by various physical laws. One of them is known as *Boyle's law*. It states that *the product of pressure and volume of a gas at a constant temperature must be constant.* Thus,

$$p_1 v_1 = p_2 v_2 \tag{13.4}$$

Another law is *Charles' law*. It states that *for a constant pressure, the volume of a gas varies directly with absolute temperature.* Hence,

$$\frac{v_1}{T_1} = \frac{v_2}{T_2} \tag{13.5}$$

Therefore, at −273.15°C or 0 K, the volume of a gas becomes zero.

The laws of Charles and Boyle can be combined into the ideal gas law. That is,

$$pv = nRT \tag{13.6}$$

where
p is the pressure in pascal (N/m²)
v is the volume of gas in cubic meters
n is the number of kilomoles of gas
R is the universal gas constant
T is the temperature in kelvin

At standstill conditions (i.e., 0°C and 1 atm), 1 kmol of gas occupies 22.414 m³ and the universal gas constant is 8314.5 J/(kmol·K), where J represents a joule or newton meter of energy. The pressure of 1 atm at 0°C is then

$$p = \frac{[8314.5 \text{ J/(kmol} \cdot \text{K)}](273.15 \text{ K})}{22.414 \text{ m}^3}$$

$$= 101{,}325 \text{ Pa}$$

$$= 101.325 \text{ kPa} \tag{13.7}$$

The mass of 1 kmol of dry air is 28.97 kg. For all ordinary purposes, dry air behaves like an ideal gas.

The density ρ of a gas is the mass m of 1 kmol divided by the volume v of that kilomole:

$$\rho = \frac{m}{v} \tag{13.8}$$

The volume of 1 kmol varies with pressure and temperature as defined by Equation 13.6. By inserting Equation 13.8 into Equation 13.9, the density can be expressed by the following equation:

$$\rho = \frac{mp}{RT}$$

$$= \frac{3.484\, p}{RT}\ \text{kg/m}^3 \tag{13.9}$$

where
p is in kilopascal (kPa)
T is in kelvin (K)

This expression yields a density for dry air at standard conditions of 1.293 kg/m³.

The common unit of pressure used in the past for meteorological work has been the bar (i.e., 100 kPa) and the millibar (100 Pa). A standard atmosphere is 1.01325 bar or 1013.25 millibar.

Atmospheric pressure has also been given by the height of mercury in an evacuated tube. This height is 29.92 in. or 760 mm of mercury for a standard atmosphere. Also note that the chemist uses 0°C as standard temperature, whereas engineers have often used 68°F (20°C) or 77°F (25°C) as standard temperature. Therefore, here standard conditions are always defined to be 0°C and 101.3 kPa pressure.

Most wind-speed measurements are made about 10 m above the ground. Typically, small wind turbines are mounted 20–30 m above ground level, while the propeller tip may read a height of more than 100 m on the large turbines. Thus, an estimate of wind-speed variation with height is needed. Here, let us examine a property that is known as *atmospheric stability* in the atmosphere.

Pressure decreases quickly with height at low attitudes, where density is high, and slowly at high altitudes where density is low. At sea level and a temperature of 273 K, the average pressure is 101.3 kPa. A pressure of half this value is reached at about 5500 m.

A temperature decrease of 30°C will often be related to a pressure increase of 2–3 kPa. The atmospheric pressure tends to be a little higher in the early morning than in the middle of the afternoon. Winter pressure tends to be higher than summer pressures.

The power output of a wind turbine is proportional to air density, which in turn is proportional to air pressure. Hence, a wind speed produces loss power from a given wind turbine at higher elevations, due to the fact that the air pressure is less. A wind turbine located at an elevation of 1000 m above sea level will produce only about 90% of the power it would produce at sea level, for the same wind speed and air temperature.

However, there are many good wind sites in the United States at elevations above 1000 m. The air density at a proposed wind turbine site is estimated by determining the average pressure at that elevation from Figure 13.11 and then using Equation 12.7 to find density. The ambient temperature must be used in the equation.

Example 13.1

Consider a wind turbine that is rated at 100 kW in a 10 m/s wind speed in air at standard conditions. If power output is directly proportional to air density, determine the power output of the wind turbine in a 10 m/s wind speed at a temperature of 20°C at a site that has the elevation of

a. 1000 m above sea level
b. 2000 m above sea level

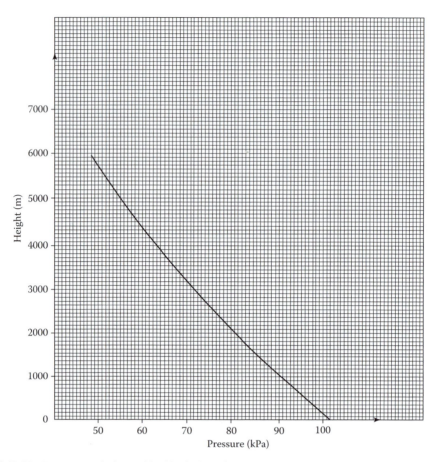

FIGURE 13.11   Pressure variations with altitude for US standard atmosphere.

Solution

a. From Figure 13.11, the average pressure at the 1000 m elevation is 90 kPa, and from Equation 13.9, the density at 20°C = 293 K is

$$\rho = \frac{3.484p}{T}$$

$$= \frac{3.484(90)}{293}$$

$$= 1.070$$

Thus, the power output at the conditions is just the ratio of this density to the density at standard conditions times the power at standard conditions:

$$P_{\text{new}} = P_{\text{old}}\left(\frac{\rho_{\text{new}}}{\rho_{\text{old}}}\right)$$

$$= 100\left(\frac{1.070}{1.293}\right)$$

$$= 82.75 \text{ kW}$$

b. From Figure 13.11, the average pressure of the 2000 m elevation is 80 kPa, and since the temperature is still 20°C = 293 K, the density is

$$\rho = \frac{3.484p}{T}$$

$$= \frac{3.484(80)}{293}$$

$$= 0.951$$

Hence, the power at the 2000 m elevation is

$$P_{new} = P_{old}\left(\frac{\rho_{new}}{\rho_{old}}\right)$$

$$= 100\left(\frac{0.951}{1.293}\right)$$

$$= 73.55\,kW$$

Note that the power output has dropped from 100 to 82.75 kW at the same wind speed at the 1000 m elevation and to 73.55 kW at the 2000 m elevation due to the fact that there are lesser air densities at the higher elevations.

13.9.7.1 Power in the Wind

The wind speed is always fluctuating, and thus, the energy content of the wind is always changing. The variation depends on the weather and on local surface conditions and obstacles to the wind flow. Power output from a wind turbine will vary as the wind varies, even though the most rapid variations will to some extent be compensated for by the inertia of the wind turbine rotor.

It is common knowledge around the globe that it is windier during the daytime than at night. This variation is mostly as a result of temperature differences that tend to be larger during the day than at night.

Furthermore, the wind is also more turbulent and tends to change direction more frequently during the day than at night. Therefore, forecasting the amount of electric energy that can be harnessed over a period of time is extremely difficult.

Consider the wind turbine shown in Figure 13.5 and assume that the wind blows perpendicularly through a circular cross-sectional area. A wind generator will capture only the wind power caught by the given swept area A that can be expressed in watts in SI system as

$$P = \frac{1}{2}\rho_a A v^3 \tag{13.10}$$

where
ρ_a is the mass density of air (and is relatively constant)
A is the circular cross-sectional area in m² (i.e., $A = \pi r^2$)
r is the radius of the circular cross-sectional area in m
v is the wind velocity in m/s

For the average mass density of air, $\rho_a = 1.24$ kg/m³
or in British system in ft.lb/s as

$$P = \frac{1}{2}\rho_a A v^3\left(\frac{746}{550}\right) \tag{13.11a}$$

or

$$P = 0.678\, \rho_a A v^3 \tag{13.11b}$$

where
   $= 0.0024\ \text{lb} \cdot \text{s}^2/\text{ft}^4$
   A is in ft^2
   v is in ft/s

Note that since the wind speed is usually given in miles per hour (mph), it needs to be converted into ft/s by using

$$v_{\text{ft/s}} = 1.47 v_{\text{mph}} \tag{13.12}$$

The following equation gives an improved version of the previous equation to determine the power in the wind in watts,

$$P = \frac{1}{2}\rho_a A v^3 C_p \tag{13.13}$$

where C_p is the *turbine power coefficient*, which represents the power conversion efficiency of wind turbine. It gives a measure of the amount of power extracted by the turbine rotor. Its value varies with *rotor design* and the *TSR*.

TSR is the *relative speed of the rotor* and the wind and has a maximum practical value of about 0.4. The ratio of the tip speed of the machine turbine blades to wind speed is found from

$$\lambda = \frac{r \times \Omega}{v} \tag{13.14}$$

where
   r is the radius of the circular cross-sectional area (i.e., turbine radius)
   Ω is the tip speed of the machine turbine blades
   v is the wind speed

Figure 13.12 shows various tip speed diagrams for various types of wind energy converters.

Here, the TSR (λ) is the relation between the speed v_{tip} and undisturbed wind speed v_0 and is signified by λ. Thus,

$$\lambda = \frac{\text{tangential velocity of blade tip}}{\text{wind speed}} \tag{13.15a}$$

or

$$\lambda = \frac{v_{\text{tip}}}{v_0} \tag{13.15b}$$

Previously, the power present in a wind for a given velocity and swept area was given by Equation 13.10 or 13.11b. However, all of this power cannot be collected by wind turbine. The theoretical maximum fraction of available wind power that can be collected by a wind turbine is given by the *Betz coefficient*.

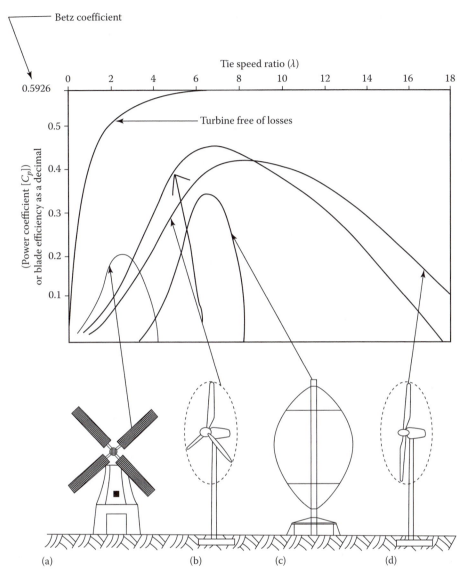

FIGURE 13.12 TSR diagrams for various types of wind energy converters. (The power coefficient gives a measure of how large a share of the wind's power a turbine can utilize. The theoretical maximum of the value is $16/27 = 0.5926$. The diagram shows the relation between TSR and power coefficient for different types of wind turbines: (a) windmill, (b) modern turbine with three blades, (c) vertical-axis Darrieus turbine, and (d) modern turbine with two blades.)

The energy in the wind is kinetic energy. In order to capture this energy, the blades of a wind turbine have to slow down as it passes through them. Hence, after the wind has passed through the wind turbine, its velocity (thus, its kinetic energy) is less than it originally had. Here, the energy it lost has been converted to the kinetic energy of the rotating blades. If after passing through the blades, the wind speed has decreased to one-third of its initial value, the blades will have theoretically captured a maximum fraction of the available wind energy. This maximum energy is given by

$$\text{Beta coefficient} = 0.5926$$

This means that the actual power input for a wind turbine will be (at best) 59% of the power provided by Equation 13.10 or 13.11b. The actual blade efficiency is somewhat less than the Betz coefficient. It is a function of a quantity called the TSR λ, as explained in Equation 13.14. The power coefficient C_p gives a measure of how large portion of the wind's power a turbine can utilize. The theoretical maximum value of C_p is $16/27 = 0.5926$. The curves in Figure 13.12 show the relation between TSR and power coefficient for different types of wind turbines: (1) windmill, (2) modern turbine with three blades, (3) vertical-axis Darrieus turbines, and (4) modern turbine with two blades. Note that the turbine power coefficient C_p is maximum at the $\lambda_{optimal}$. Also note that the wind turbine system uses induction generators that are independent of torque variation while speed varies between 1% and 2%. In general, there is a great amount of power in the wind. However, this mechanical power when it is converted to electric power is reduced substantially. A typical WECS has an efficiency of 20%–30%.

Example 13.2

Determine the amount of power that is present in a 10 m/s wind striking a windmill whose blades have a radius of 5 m.

Solution

The area swept by the blades of the wind turbine is

$$A = \pi r^2$$

$$= \pi (5 \text{ m})^2$$

$$\cong 78.54 \text{ m}^2$$

Thus, the power that is present in the wind is

$$P = \frac{1}{2}\rho_a A v^3$$

$$= \frac{1}{2}(1.24)(78.54)(10)^3$$

$$\cong 48,695 \text{ W}$$

If the turbine power coefficient (C_p) is 0.20, then the amount that will be converted to usable electric power is

$$P = 48,695 C_p$$

$$= 48,695(0.20)$$

$$\cong 9.739 \text{ W}$$

which is considerably lesser than the power that is preset in the wind.

13.9.8 EFFECTS OF A WIND FORCE

In any WECS, the support of the tower on which the wind generator is mounted must be considered; when a wind blows on a wind turbine, it applies a force on the blades. This wind force applied to the blades is determined in SI or British system from

$$F_w = 0.44\rho_a A v^2 \tag{13.16}$$

Additionally, the wind force applied on the tower (F_t) carrying the wind turbine has to be considered. The resultant effect of these forces is to develop a moment about the tower base in the clockwise direction. This overturning moment is a function of the wind speed, size of the blades, and the height of the wind turbine.

Because of this, large wind turbines mounted on high towers must be properly supported. Also, many wind turbines have an automatic high-wind shutdown feature. This feature automatically turns the blades so that they become parallel to the wind and it can escape any damage to the WECS system.

13.9.9 Impact of Tower Height on Wind Power

As general rule, a taller tower is expected to result in higher-speed winds to the wind turbine. However, surface winds can also be affected by the irregularities or roughness of the earth's surface or by the existing forest and/or buildings in the vicinity. The relationship between the wind speed and the height of the wind turbine can be expressed as

$$\frac{v}{v_0} = \left(\frac{H}{H_0}\right)^\alpha$$

(13.17)

where

v is the wind speed at height H
v_0 is the reference (or known) wind speed at reference height of H_0
α is the roughness (friction) sufficient

In Europe, the relationship in Equation 13.17 is modified as

$$\frac{v}{v_0} = \frac{\ell n(H/2)}{\ell n(H_0/2)}$$

(13.18)

There are many factors that affect wind, for example, elevation, contour of the ground in the surrounding areas, tall buildings, and trees. The average wind speed will be probably different at different tower heights. In the event that the average wind speed at different heights is the same, the location with shorter height should be considered since such application results in less expensive tower.

Furthermore, at a higher elevation having greater wind, it is possible to use a smaller wind turbine with shorter blade diameter, rather than using a large turbine with larger blade diameter at a lower elevation for obtaining the same amount of power.

The value of the exponent α in Equation 13.17 depends on the roughness of the terrain given in Table 13.2

TABLE 13.2
Roughness Coefficient for Various Class Types of Terrain

Roughness Class	Terrain Description	Roughness Coefficient (α)
Class 0	(Open water)	$\alpha = 0.1$
Class 1	(Open plain)	$\alpha = 0.15$
Class 2	(Countryside with farms)	$\alpha = 0.2$
Class 3	(Villages and low forest)	$\alpha = 0.3$

Example 13.3

If the average wind speeds on an open plain (roughness class 1) is known to be 6 m/s at 10 m height, determine the wind speed at 50 m height.

Solution

From Table 13.2, $\alpha = 0.15$ and using Equation 13.15,

$$\frac{v}{v_0} = \left(\frac{H}{H_0}\right)^{\alpha} = \left(\frac{50}{10}\right)^{0.15}$$

or

$$\frac{v_{50}}{6} = \left(\frac{50}{10}\right)^{0.15}$$

Thus, at 50 m height,

$$v_{50} = 6\left(\frac{50}{10}\right)^{0.15} = 7.6 \text{ m/s}$$

Example 13.4

Assume that the average wind speed at a point A is 6 m/s, while at point B 7 m/s. In order to capture 2 kW, determine the blade diameter d for a wind turbine operating

    a. At point A
    b. At point B

Solution

    a. Using Equation 13.12, the given wind speed needs to be converted to ft/s as

$$v = 1.47 \times v_{(mph)}$$

$$= 1.47 \times 6$$

$$= 8.82 \text{ ft/s} \quad (\text{at point } A)$$

and

$$v = 1.47 \times v_{(mph)}$$

$$= 1.47 \times 7$$

$$= 10.29 \text{ ft/s} \quad (\text{at point } B)$$

From Equation 13.11b, at point A,

$$A = \frac{P}{0.678\rho_a v^3}$$

$$= \frac{2000 \text{ W}}{0.678 \times 0.0024 \times 8.82^3}$$

$$\cong 1791.4 \text{ ft}^3$$

Since

$$A = \pi \frac{d^2}{4}$$

then

$$d = \sqrt{\frac{4A}{\pi}}$$

$$= \sqrt{\frac{4(1791.4)}{\pi}}$$

$$\cong 47.76 \text{ ft}$$

b. At point B,

$$A = \frac{P}{0.678 \rho_a v^3}$$

$$= \frac{2000 \text{ W}}{0.678 \times 0.0024 \times 10.29^3}$$

$$\cong 1128.09 \text{ ft}^3$$

Thus,

$$d = \sqrt{\frac{4A}{\pi}}$$

$$= \sqrt{\frac{4(1128.09)}{\pi}}$$

$$\cong 37.9 \text{ ft}$$

Therefore, a smaller (cheaper) wind turbine could be employed at point A and provide the same power as a larger wind turbine at point B.

13.9.10 WIND MEASUREMENTS

Wind measurement equipment usually consists of an anemometer, which measures wind speed, and a wind vane, which measures wind direction. In most countries, a national meteorological institute has measured and collected data on the winds since the nineteenth century. They register wind speed, wind direction, temperature, and other kinds of meteorological data several times a day (every 4 h, day and night) all year around. These data are reported daily to a central institution.

Nowadays, wind data are registered automatically. These observations make up the basis for the *wind statistics* that are used to describe wind climate in different regions and to create so-called wind atlas data that are used to calculate how much wind turbines can be expected to produce at different sites.

However, in the past, weather observers read the anemometer every 4 h, day and night. They observed the anemometer for a couple of minutes and recorded the average wind speed for that period. However, wind-speed data are affected by the anemometer height, the human factor in reading the wind speed, and the quality and maintenance of the anemometer.

A typical wind-cup anemometer works with a diametric flow of air. As the wind blows, the anemometer rotates at a speed proportional to the wind speed. Typically, a permanent magnet dc generator is connected to the rotating shaft. A voltage is thus produced that is proportional to the wind speed at every instant of time. The second instrument that is required is a *wind data compilator*. It is an electronic instrument that is connected to the anemometer and records the wind speed continuously.

Example 13.5

Consider a wind turbine that has blades with 8 ft radius. At its location where it is mounted, data were taken and it was discovered that the wind speed was 3 mph for 3 h and 12 mph for another 3 h time period. Determine the amount of energy that can be intercepted by the wind turbine.

Solution

The energy needs to be determined independently for each 3 h period.

During the first 3 h time period,
the average wind speed is

$$V_{avg} = \left(1.47\,\frac{ft/s}{mph}\right)(3\,mph) = 4.41\,ft/s$$

By using Equation 13.11b,

$$P = 0.678 \times 0.0024 \times 314.16 \times 4.41^3$$

$$= 43.84\,W$$

where

$$A = \pi(10\,ft)^2$$

$$\cong 314.16\,ft^2$$

$$Energy = (43.84\,W)(3\,h)$$

$$= 131.52\,Wh$$

$$= 0.13152\,kWh$$

During the second 3 h time period,
the average wind speed is

$$V_{avg} = \left(1.47\,\frac{ft/s}{mph}\right)(12\,mph) = 17.64\,ft/s$$

By using Equation 13.11b,

$$P = 0.678 \times 0.0024 \times 314.16 \times 17.64^3$$

$$= 2806\,W$$

Therefore, the total energy generated during the total period is

$$\text{Energy} = (2.806 \text{ kW})(3 \text{ h})$$

$$= 8.418 \text{ kWh}$$

Hence, total energy is

$$\text{Total energy} = 0.13152 + 8.418$$

$$\cong 8.56 \text{ kWh}$$

13.9.11 CHARACTERISTICS OF A WIND GENERATOR

The most important characteristic of a wind generator is its power curve. Normally, it is a graph provided by the manufacturer of a particular wind turbine. It shows the approximate power output as a function of wind speed. Figure 13.13 shows a typical power curve for a wind turbine rated 3 kW/25 mph. The power curve of a wind generator provides important information. In addition, to provide information for the obtainable power output at any given wind speed, it provides information about the cut-in speed, the rated power, the rated speed, and the shutdown speed.

Here, the minimum wind speed required to start the blades turning and producing a useful output is defined as the *cut-in speed*. The maximum power output that the wind turbine will produce is called the *rated power*.

The minimum wind speed needed for the wind turbine to produce rated power is known as the *rated speed*. The *shutdown speed* is also called the *furling speed*. It is the maximum operational speed of the wind turbine. Beyond this speed, in order to prevent damage to the system from high winds, the blades are either folded back or turned to a high-pitch position.

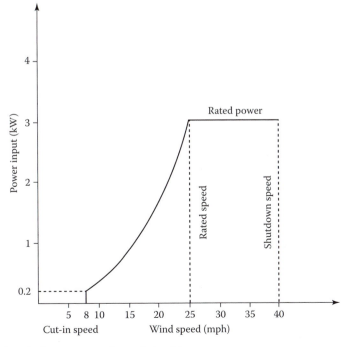

FIGURE 13.13 A typical power curve for a wind turbine.

Example 13.6

Consider the wind turbine whose power curve of its generator is shown in Figure 13.13. It is rated 3 kW/25 mph, as indicated in the figure. Assume that during an 8 h period, the wind had the following average speeds: 6 mph for 2 h duration, 10 mph for 3 h duration, 15 mph for 2 h duration, and 20 mph for 1 h duration. Determine the resultant electric output for the 8 h period.

Solution

The energy is calculated for each of the four wind speeds and time intervals:
At 6 mph, it is below the cut-in speed in Figure 13.13; thus, the output is zero.

At 10 mph, the output from the curve is 0.35 kW:

$$Energy = (0.35 \text{ kW})(3 \text{ h})$$

$$= 1.05 \text{ kWh}$$

At 15 mph, the output from the curve is 0.85 kW:

$$Energy = (0.85 \text{ kW})(2 \text{ h})$$

$$= 1.7 \text{ kWh}$$

At 20 mph, the output from the curve is 1.65 kW:

$$Energy = (1.65 \text{ kW})(1 \text{ h})$$

$$= 1.65 \text{ kWh}$$

Thus, the total energy for 8 h duration is

$$Total \ energy = 0 \text{ kWh} + 1.05 \text{ kWh} + 1.7 \text{ kWh} + 1.65 \text{ kWh}$$

$$= 3.855 \text{ kWh}$$

13.9.12 EFFICIENCY AND PERFORMANCE

How much energy a wind turbine can produce is a function of a number of factors: the rotor swept area, the hub height, and how efficiently the wind turbine can convert the kinetic energy of the wind. Also the additional factors include the mean wind speed and the frequency distribution at the site where the wind turbine is installed.

The power of the wind that is available to a turbine is proportional to the rotor swept area A and the cube of the wind speed v. Over the years, the rotor swept area of wind turbines has increased steadily and thus so has the rated power of the wind turbines, as given in Table 13.3. Note that the production figures given in the table are based on a site with average wind resources. It appears that since the 1980s, the power of wind turbines has doubled every 4–5 years on the average.

Example 13.7

Assume that a wind generator whose power curve is shown in Figure 13.13 has a blade diameter of 16 ft. If its power output is at 120 V at 60 Hz, determine the net efficiency of this WECS at a wind speed of 20 mph.

TABLE 13.3

Development of Wind Turbine Size, 1980–2005

Year	1980	1985	1990	1995	2000	2005
Power (kW)	50	100	250	600	1000	2500
Diameter (m)	15	20	30	40	55	80
Swept area (m²)	177	314	706	1256	2375	5024
Production (MWh/year)	90	150	450	1200	2000	5000

Source: From Wizelius, T., *Developing Wind Power Projects: Theory and Practice*, Earthscan, London, U.K., 2007.

Solution

First, it is necessary to convert the wind speed from mph to ft/s by using Equation 13.12:

$$V_{(ft/s)} = 1.47 \times V_{(mph)}$$

$$= 1.47 \times 20$$

$$= 29.4 \text{ ft/s}$$

The input power is found from Equation 13.11b as

$$P_{in} = 0.678 \rho_a A v^3$$

$$= 0.678 \rho_a \times \pi r^2 \times v^3$$

$$= 0.678 \times 0.0024 \times \pi \times 8^2 \times 29.4^3$$

$$= 8314 \text{ W}$$

From Figure 13.13, the output power at 20 mph is

$$P_{out} = 1.65 \text{ kW} = 1650 \text{ W}$$

Thus, the efficiency of the system is

$$\eta = \frac{P_{out}}{P_{in}} \times 100$$

$$= \frac{1650 \text{ W}}{8314 \text{ W}} \times 100$$

$$= 19.84\%$$

Example 13.8

Assuming that the wind turbine with three blades in Example 13.7 is rotating at 100 rpm, find the blade efficiency at a wind speed of 20 mph.

Solution

$$V_0 = \text{wind speed}$$

$$= 20 \times 1.47$$

$$= 29.4 \text{ ft/s}$$

The circumference that the blade tip traces out is

$$2\pi r = 2\pi \times (8 \text{ ft})$$

$$= 50.27 \text{ ft}$$

The blade tip speed is

$$v_{tip} = (50.27 \text{ ft/rev})100 \text{ rpm}\left(\frac{1}{60 \text{ s/min}}\right)$$

$$= 83.78 \text{ ft/s}$$

From Equation 13.15b,

$$\lambda = \frac{v_{tip}}{v_0}$$

$$= \frac{83.78 \text{ ft/s}}{29.4 \text{ ft/s}}$$

$$\cong 2.85$$

From Figure 13.12, for $\lambda = 2.85$, the blade efficiency is about 13%. Note that the share of power in the wind that can be utilized by the rotor is called the *power coefficient, C_p.*

Example 13.9

Assume that a WECS shown in Figure 13.9c uses a three-phase six-pole induction machine. The line frequency is 60 Hz and the average wind speed is 12 mph. The blades have a 30 mph diameter and peak efficiency when the TSR is 8.3. If the generator efficiency is a maximum at a negative lip of 3.3%, determine the transmission gear ratio for the peak system efficiency.

Solution

At first, the speeds required for the blades and generator have to be found. Then the transmission will be selected to match the two speeds. The required generator speed is

$$n_g = [1-(-s)]\frac{120f}{p}$$

$$= [1-(-0.033)]\frac{120 \times 60}{6}$$

$$= 1239.6 \text{ rpm}$$

From Equation 13.12, the average wind speed is

$$v_0 = 1.47(12 \text{ mph})$$

$$= 17.64 \text{ ft/s}$$

From Equation 13.15b, the blade tip speed is

$$v_{tip} = \lambda \times v_0$$

$$= 8.3 \times 17.64$$

$$= 146.412 \text{ ft/s}$$

The circumference traced out by the blade tip is

$$2\pi(15 \text{ ft}) = 94.248 \text{ ft/rev}$$

Hence, the blade tip speed must be

$$\frac{146.412 \text{ ft/s}}{94.248 \text{ ft/rev}} = 1.5535 \text{ rev/s}$$

Converting this to rpm,

$$(1.5535 \text{ rev/s})(60 \text{ s/min}) = 93.21 \text{ rpm}$$

Therefore, the transmission must gear up from 93.21 rpm to 1239.6 rpm. Hence, the required gear ratio is

$$\text{Gear ratio} = \frac{1239.6 \text{ rpm}}{93.21 \text{ rpm}}$$

$$\cong 13.3$$

13.9.13 EFFICIENCY OF A WIND TURBINE

In order to calculate the efficiency of a wind turbine, the efficiency of its components has to be calculated at first.

13.9.13.1 Generator Efficiency

A wind turbine can never utilize all the power in the wind. The amount of power that can be utilized by a wind turbine is given by the power coefficient C_p. It is known that (*based on Bets' law*) the maximum value of this coefficient is 0.59. It varies with the wind speed. For most wind turbines, the maximum value varies between 0.45 and 0.50 at a wind speed of 8–10 m/s for most wind turbines.

In order to convert the power in the wind from the revolving rotor to electric power, it is passed through a gearbox and a generator or, for direct-drive turbines, through a generator and an inverter. In this conversion process, some power will be lost. Also, the efficiency of the individual components will vary with the wind speed.

It is known that a generator is most efficient when it is running at its nominal power. On a wind turbine, most of the time the generator is operating on *partial load*, that is, it runs on lower power when the wind speed is lower than the nominal wind speed. As a result, the standard generator efficiency will then be reduced, as given in Table 13.4.

There is also a relationship between the physical size of a generator and efficiency. That is, efficiency increases with the size of the generator, since losses to heat are reduced, as given in Table 13.5.

TABLE 13.4
Generator Efficiency

% of Full Load	5	10	20	50	100
Efficiency	0.4	0.8	0.90	0.97	1.00

TABLE 13.5
Relationship between Size and Efficiency

Nominal Power (kW)	5	50	500	1000
Efficiency		0.84	0.89	0.94	0.95

For example, a 1 MW wind turbine running at 20% of its nominal power (200 kW) has an efficiency of $0.95 \times 0.90 = 85\%$. Note that the relationship between efficiency, size, and partial load can also differ between different models and manufacturers.

13.9.13.2 Gearbox

Typically on a large modern wind turbine, the rotor has a rotational speed of 20–30 rpm, while the generator will need to rotate at 1520 rpm. In order to increase the speed, a gearbox is used. If the turbine rotor runs at 30 rpm, a gear change of 30:1520 = 1:50.7 is required. That is, 1 rev of the main shaft has to be increased to 50.7 rev on the secondary shaft that is connected to the generator.

Generally, a gearbox has several steps; thus, the rotational speed is increased stepwise. Losses can be estimated at 1% per step. In wind turbines, three-step gearboxes are usually used and the efficiency of the gearbox will then be about 97%.

However, wind turbines with a direct-drive generators and variable speed do not need any gearbox. Instead the frequency and voltage of the electric current will vary with the rotational speed. Thus, the current has to be rectified to dc and then converted by an inverter to ac with the same frequency and voltage as the grid. The efficiency of such an inverter is also about 97%.

13.9.13.3 Overall Efficiency

In summary, the overall efficiency η_{total} l of a wind turbine is the product of the turbine rotor's power coefficient C_p and the efficiency of the gearbox (or inverter) and generator

$$\eta_{\text{total}} = C_p \times \mu_{\text{gear}} \times \mu_{\text{generator}} \tag{13.19}$$

Often C_p is set to 0.59 and μ_{rotor} (or μ_r) is used to show how large a share of the theoretically available power the rotor can utilize. For example, if the power coefficient $C_p = 0.49$, the rotor turbine charges is then

$$\mu_r = \frac{0.49}{0.59} = 0.83$$

The efficiency of a wind turbine changes with the wind speed. When the wind speed is below the nominal wind speed, the efficiency of the generator will decrease, and if the turbine has a fixed rotational speed, the TSR will change, that is, the ever smaller share of the power in the wind will be utilized and C_p will decrease successfully. Since the wind turbines are used to convert wind power to electric power, and thus another coefficient is used, C_p, which indicates the turbine rotor's power coefficient.

13.9.13.4 Other Factors to Define the Efficiency

In order to estimate efficiency, the following factors are also often used:

$$\text{Power/swept area} = \frac{\text{Production per year}}{\text{Rotor swept area}} \text{ kWh/m}^2 \tag{13.20}$$

$$\frac{\text{Power production}}{\text{nominal power}} = \frac{\text{Production per year}}{\text{Rotor swept area}} \text{ kWh/kW} \qquad (13.21)$$

$$\text{Capacity factor} = \left(\frac{\text{Production per year}}{(\text{Nominal power}) \times 8760}\right) \times 100\% \qquad (13.22)$$

$$\text{Full load hours} = \left(\frac{\text{Production per year}}{\text{Nominal power}}\right) \times 100\% \qquad (13.23)$$

$$\text{Cost efficiency} = \frac{\text{Investment cost}}{\text{Production per year}} \text{ \$/kWh/year} \qquad (13.24)$$

$$\text{Availability} = \left(\frac{8760\,\text{h} - \text{stop hours}}{8760\,\text{h}}\right) \times 100\% \qquad (13.25)$$

Availability is the technical reliability of a wind turbine. If the wind turbine is out of operation due to faults or scheduled service and maintenance for 5 days a year, the technical availability is 98.6%. (A year is normally taken as 360 days for such calculations.)

It means that the turbine could produce power for 98.6% of the time, if there was always enough wind to make the run. The technical lifetime for a turbine is estimated at 20–25 years. However, its economic lifetime can be shorter due to increased maintenance costs as the turbine gets old.

There is another factor that is used to indicate the capacity factor. It is called annual load factor and defined as

$$\text{Annual load duration factor} = (\text{Capacity factor}) \times 8760\% \qquad (13.26)$$

Here, the significance of load duration is that it expresses that number of hours for which the wind turbine can be considered to be virtually operating at its rated capacity in 1 year.

In general, in order to indicate how much wind power there is in a country, the total installed capacity is used as a measure. Every wind turbine has a rated power (maximum power) that can vary from a few hundred watts to 5000 kW (5 MW). The number of turbines does not give any information on how much of wind power they can produce.

How much wind a wind turbine can produce depends not only on its rated power but also on the wind conditions. In order to get an indication of how much a certain amount of installed (rated) power will produce per year, use the following rule of thumb: "*1 MW wind power produces 2 GWh per year on land and 3 GWh per year offshore.*"

Example 13.10

Consider a 4 MW wind turbine that is under maintenance for 400 h in 1 year; out of 8760 h of 1 year. If it actually produced 8000 MWh due to fluctuations in wind availability, determine the following:

    a. The availability factor of the wind turbine
    b. The capacity factor of the wind turbine
    c. The annual load duration of the wind turbine

Solution

a. The availability factor of the wind turbine is

$$\text{Availability factor} = \left(\frac{8760\,\text{h} - \text{stop hours}}{8760\,\text{h}} \right) \times 100\%$$

$$= \frac{8760 - 400}{8760} \times 100$$

$$= 0.9543 \quad \text{or} \quad 95.43\%$$

b. The capacity factor of the wind turbine is

$$\text{Capacity factor} = \left(\frac{\text{Production per year}}{(\text{Nominal power}) \times 8760} \right) \times 100\%$$

$$= \frac{8000\,\text{MWh}}{(2\,\text{MW}) \times 8760} \times 100$$

$$= 0.4566 \quad \text{or} \quad 45.66\%$$

c. The annual load duration for the wind turbine is

$$\text{Annual load duration factor} = (\text{Capacity factor}) \times 8760\%$$

$$= (0.4566) \times 8760$$

$$= 3999.8\,\text{h}$$

However, the capacity factor of 0.4566 (or 45.66%) does not mean that the wind turbine is only running less than half of the time. Rather, a wind turbine at a typical location would normally run for about 65%–90% of the time. But, much of the tie, it will be generating at less than full capacity, causing its capacity factor lower.

13.9.14 GRID CONNECTION

The term *grid* is often used loosely to describe the totality of the network. Specifically, *grid connected* means connected to any part of the network. The term *national grid* usually means the EHV transmission network.

Integration particularly means the physical connection of the generator to the network with due regard to the secure and safe operation of the system and the control of the generator so that the energy resource is utilized optimally. The integration of generator power from wind turbine (or any other renewable energy sources) is basically similar to that of fossil fuel-powered generator and is based on the same principles. However, renewable energy sources are often variable and geographically dispersed. The connection point is referred to as the PCC.

Wind power can be classified as small and non-grid connected, small and grid connected, large and non-grid connected, and large and grid connected. The small and non-grid-connected type of wind turbine can be used in a location that is not served by a utility. It can be improved by adding batteries to level out supply and demand. The cost will be high about $0.50/kWh. The small and grid-connected wind turbine is usually not economically feasible.

The economic feasibility can be improved, if the local utility is willing to provide an arrangement that is called *net metering*. In such system, the meter runs backward when the turbine is generating more than the owner is consuming at the moment. The owner pays a monthly charge for the wires to his home.

In general, utilities want to buy at wholesale and sell at retail. It is often that the owner might pay $0.08–$0.15/kWh and get paid $0.02/kWh for the wind-generated electricity that is far from enough to economically justify a wind turbine.

Wind speed is the main factor in determining electricity cost, in terms of influencing the energy yield, and approximately, at the locations with wind speeds of 8 m/s, it will yield electricity at one-third of the cost for a 5 m/s site. Wind speeds of approximately 5 m/s can typically be found at the locations away from the coastal areas. However, wind energy developers usually intend to find higher wind speeds. Levels at about 7 m/s can be found in many coastal regions.

The large and non-grid-connected wind turbines are installed on islands or in some native villages where it is virtually impossible to connect to a large grid. In such places, one or more wind turbines can be installed in parallel with the diesel generators so that the wind turbines can act as fuel savers when the wind is blowing. This system can operate easily. In general, the justification for having the small or the large wind turbines must be based on whether or not it will result in *a lower net cost to society*, including the environmental benefits of wind generation. Today, wind turbines with ratings near 1 MW or more are now common.

However, this is still small compared to the needs of a utility, so clusters of turbines are placed together to form *wind farms* or *wind plants* with total ratings of 10–100 MW, or even more. Presently, Southern California Edison (SCE) Company is working on Tehachapi Renewable Transmission Project (TRTP) of 500 kV. The purpose of the proposed TRTP project is to provide the electrical facilities necessary to integrate levels of new wind generation in excess of 700 MW and up to approximately 4500 MW in the future in the Tehachapi Wind Resource Area (TWRA) in Southern California.

The voltage level of large wind turbines, in general, is 600 V, so-called industrial voltage. Therefore, they can be connected to a factory without a transformer. Smaller wind turbines, up to 300 kW, which were common in the near past, have a voltage of 480 V and can be connected directly via a feeder cable to a farm or a house. However, usually wind turbines are connected to the power grid through a transformer that increases the voltage level from 480 or 600 V to the higher voltage, normally 10 or 20 kV, in the distribution grid. A suitable transformer is installed on the ground next to the tower for smaller- and medium-sized wind turbines. But in large wind turbines, the transformer is often a component of the turbine itself.

In modern wind turbines, the power that is supplied into the power grid can be converted by power electronics to achieve the phase angle and reactive power that the grid needs at the point where the wind turbine is connected to improve power quality in the grid. However, the power electronic equipment can cause a main problem, namely, *harmonics*, that is, currents with frequencies that are multiples of 60 Hz, and has a negative effect on power quality. Such "*dirt*" can, to some extent, be "*cleaned of*" by different kinds of filters. Unfortunately, such equipment is expensive and seldom takes care of all the "*dirt*."

13.9.15 Some Further Issues Related to Wind Energy

In general, integration of wind power plants into the electric power system presents challenges to power system planners and operators. Wind plants naturally operate when the wind blows, and their power levels vary with the strength of the wind. Thus, they are not dispatchable in the traditional sense. Wind is primarily an energy source. Its main function is displacement of fossil fuel combustion in existing generating units.

These units maintain system balance and reliability, so no new conventional generation is required as "*backup*" for wind plants. Wind also provides some effective load-carrying capability and therefore contributes to planning reserves but not day-to-day operating reserves. Wind's variability and uncertainty do increase the operating costs of the non-wind portion of the power system, but generally by modest amounts.

Nowadays, wind studies in the United States employ sophisticated atmospheric (*mesoscale numerical weather prediction*) models to develop credible wind power time series for use in the integration analysis. Today, it is in general accepted that integration studies should use this type of data, synchronized with load data, when actual wind data are not available [5].

According to Smith et al. [5], wind-integration studies performed in recent years have provided important new insights into the *impact wind's variability* and uncertainty will have on system operation and operating costs. Their conclusions include the following:

1. Several studies of very high penetrations of wind (up to 25% energy and 35% capacity) have concluded that the power system can handle these high penetrations without compromising system operation.
2. The importance of detailed wind resource modeling has been clearly demonstrated.
3. The importance of increased flexibility in the non-wind portion of the generating mix has been clearly demonstrated.
4. The value of good wind forecasting has been clearly demonstrated to reduce unit commitment costs in the day-ahead time frame.
5. The difficulties of maintaining system balance under light-load conditions with significant wind variability constitute a serious problem.
6. Even though wind is mainly an energy resource, it does provide modest amounts of additional installed capacity for planning-reserve purposes.
7. There is a great value sharing balancing functions over large regions with a diversity of loads, generators, and wind resources.

13.9.16 Development of Transmission System for Wind Energy in the United States

In the United States, existing wind farms are in remote areas with respect to load centers. Transmission system owners have been unable to build new high-voltage transmission lines to remote areas where there may be a high-potential wind energy source but little existing generation or load.

Also, it is uneconomic to build transmission capacity to the peak power capacity of wind farms. But if transmission capacity is built to a number lower than the peak, it can lead to congestion when wind production is greater than the transmission capacity. That is, wind developers may find it economical to build wind capacity even though they know that congestion may develop and remains for a period of time.

When it comes to building new transmission lines, it appears that, due to limited funds, the emphasis is on the *eliminating bottlenecks in high-load corridors*. Also, in the past, new transmission lines have been approved only if there is *a proven need for improved system reliability*. Because of these concerns, the utility companies that are interested in building wind farms have not been able to build new power plants in remote but wind-rich areas if there is no transmission line that has the capacity to transfer the plant output to major load centers. As a result, this chicken-and-egg dilemma delays the development of new wind plants and transmission lines to deliver the wind energy to load centers.

However, there has been some progress in California, Texas, and Colorado. For example, in California, the Tehachapi region has the potential for more than 7000 MW (7 GW) of new wind generation, but the opportunity to develop it was stalled because there was no way to fund the necessary expansion of the bulk 500 kV transmission system. SCE received the California Independent System Operator's (CALISO) approval for the $1.4 billion Tehachapi Transmission project in 2007. Some transmission segments are now under construction, and a few more are in the proposal stage. The project completion date is given as 2013.

13.9.17 Energy Storage

When wind production exceeds the transmission system capacity or congestions takes place on the system, storage can capture the "lost" energy and then discharges back to the grid when the congestion eases. Here, the main idea is to use storage to increase the effective capacity of the wind farm.

The unconstrained wind farm output is known as *potential capacity factor*. It is the total capacity if transmission capacity is built to wind peak. Whereas the *actual capacity factor* defines the real capacity in case the transmission constraint restricts the output, the *effective capacity factor* is defined as the capacity that can be achieved through the use of storage.

If there is unlimited storage capacity, the effective capacity factor would equal the potential full capacity factor of the wind farm. But other factors such as cost and size force wind developers to establish a balance between the maximum (*unconstrained*) and minimum (take no action) capacity factors. The economics of such application is a function of the wind farm power duration, the transmission congestion duration, and the ratio of the storage capacity to wind farm capacity (both in terms of power and duration) [6].

Succinctly put, wind is *not a constant resource*. Wind velocities follow *regular diurnal patterns*; that is, wind does not blow consistently throughout the day, but rather reaches peaks at specific and typical times and declines in the same manner. In the plains, wind velocities might be greatest at night time.

In mountain regions, on the other hand, wind velocity might be greatest in the early morning and late afternoon as well as early evening and lowest during the daytime or in the middle of the night. Offshore wind is typically more reliable; still, the pattern throughout the day varies.

According to Fioravanti et al. [6], although emerging storage technologies are making great progress, the megawatt capacity needed to shift the potential generation of wind farms tends to outsize the capabilities of the storage technologies.

There are basically two main technologies that have the capacity to perform in this application: *pumped hydro* and *compressed air energy storage* (CAES). There are other storage technologies that include batteries, flywheels, ultra-capacitors, and to some extent PVs.

Most of these technologies are best suited for power quality and reliability enhancement applications, due to their relative energy storage capabilities and power density characteristics, even though some large battery installations could be used for peak shaving.

All of the storage technologies have a power electronic converter interface and can be used together with other distributed utility (DU) technologies to provide *"seamless"* transitions when power quality is a requirement.

The earliest known use of pumped hydro technology was in Zurich, Switzerland, in 1982. The relative, low-efficiency, and low-cost pumped hydro is more often than compensated by the ability to avoid expensive peaking power. When available, pumped hydro plants are excellent solutions to solve the diurnal problems. But there are presently limited siting possibilities for new pumped hydro resources.

An alternative is the *CAES*. In such system, the off-peak electric energy is used to run meters and compressors to pump air into a limestone cavern. At times of the need that is during peak, the pressured air is let through a recuperator–turbine–generator system. The produced electric energy is returned to the power system at the time of peak. This system of CAES has been developed by the Electric Power Research Institute (EPRI) over the years.

However, CAES is not a pure storage system since natural gas is added to the compressed air going through the turbine to boost power production and overall efficiency. Without this injection, the overall efficiencies of the cycle would be low—about 70%–80%.

CAES is a peaking gas turbine power plant that consumes less than 403 of the gas used in a combined-cycle gas turbine to produce the same amount of electric output power.

This is accomplished by blending compressed air to the input fuel to the turbine. By compressing air during off-peak periods when energy prices are very low, the plant's output can produce electricity during peak periods at lower costs than conventional stand-alone gas turbines can achieve.

Today, the EPRI has an advanced CAES system designed around a simpler system using advanced turbine technology. It is developed for plants in the 150–400 MW range with underground storage reservoirs of up to 10 h of compressed air at 1500 lb/in.^{2}

Depending on the reservoir size, multiple units can be deployed. The largest plant under construction in the United State would have an initial rating of 800 MW. EPRI is also studying an aboveground CAES alternative with high-pressure air stored in a series of large pipes. These smaller systems are targeted at ratings of up to 15 MW for 2 h. CAES has the potential to be very large scale when underground storage is used. The first commercial CAES was a 290 MW unit built in Huntorf, Germany, in 1978 [6].

13.9.18 WIND POWER FORECASTING

Wind power forecasting plays a key role in dealing with the challenge of balancing the system supply and demand, given the uncertainty involved with the wind plant output. According to Smith et al. [5], wind forecasting is a prerequisite for the integration of a large share of wind power in an electricity system, as it links the weather-dependent production with the scheduled production of conventional power plants and the forecast of the electricity demand, the latter being predictable with reasonable accuracy.

The essential application of wind power forecasting is to reduce the need for balancing energy and reserve power, which are needed to integrate wind power into the balancing of supply system and demand in the electricity supply system (i.e., to optimize the power plant scheduling). This leads to lower integration costs for wind power, lower emissions from the power plants used for balancing, and subsequently a higher value of wind power.

A second application is to provide forecasts of wind power feed-in for grid operation and grid security evaluation, as wind farms are often connected to remote areas of the transmission grid. In order to forecast congestion as well as losses due to high physical glows, the grid operator required to know the current and future wind power feed-in at each grid connection point [5].

Therefore, as wind power capacity rapidly increases, forecast accuracy becomes increasingly important. This is especially true for large onshore or offshore wind farms, where an accurate forecast is crucial due to the high concentration of capacity in a small area.

Luckily, in recent years, the forecasting accuracy has improved steadily and will be more likely even better in the future. It has been discovered that if many wind farms are forecasted together, the forecast error decreases. The larger the involved regions, the more accurate the resultant forecast since the forecast errors of different regions will partially cancel each other out.

Today, utility companies study *not only wind-integration costs* but also *operational savings due to disputed fuel and emissions.* Since 2006, wind-integration studies have evolved to consider higher wind penetration and larger regions, which leads to a greater focus in new transmission needs. According to Corbus et al. [7], such a regional study approach dictates that *additional questions to be answered* include the following:

1. How do local wind resources compare with higher capacity factor wind that requires more transmission?
2. How does the geographic diversity of wind power reduce wind-integration costs (i.e., by spreading the wind over a larger region and thereby *"smoothing out"* some of the variability)?
3. How does offshore wind compare with on shore wind?
4. How does balancing area consolidation or cooperation affect wind power integration costs?
5. How much new transmission is needed to facilitate high penetration of wind power?
6. What is the role and value of wind forecasting?
7. What role do shorter scheduling intervals have to play?
8. What are the wind power integration costs spread over large market footprints and regions?
9. What additional operating reserves are needed for large wind power developments?

According to Ackermann et al. [8], experience with the integration of high amounts of wind generation into power systems around the world has shown no incidents in which wind generation has directly or indirectly caused unmanageable operational problems. The key elements *for the successful integration of high penetration* levels of wind power are as follows:

1. There must be well-functioning markets over large geographic areas—combining a number of balancing areas—that enable an economical way of sharing balancing resources. This situation also enables aggregation of a more diverse portfolio of wind plants, which reduces the output variability. *Well-functioning markets* must also offer a range of scheduling periods, (i.e., day ahead, hour ahead, and real time) to accommodate the uncertainty in wind plant forecasts. The basic requirement for such a well-functioning market over large geographic areas is an appropriately designed transmission system to interconnect the different network areas.
2. Advanced wind-forecasting systems based on a variety of weather input and their active integration into power system operation are needed.
3. New simulation tools are necessary to evaluate the impact of wind power on the security of supply and load balancing in near real time.
4. The corresponding "*right to curtail*" wind power, when necessary from the system security point of view.

13.10 SOLAR ENERGY

13.10.1 SOLAR ENERGY SYSTEMS

Even though solar energy is a very small portion of the energy system today, the size of the resource is enormous. The average intensity of light outside the atmosphere (known as the solar constant) is about 1353 W/m^2. In order to produce a gigawatt of power, an area of about 5 km^2 would be needed, assuming a conversion of 20%. The earth receives more energy from the sun in 1 h than the global population uses in an entire year. Figure 13.14 shows wind and solar applications in the city of Huleka in South Africa.

Furthermore, the *solar* PV industry is growing very fast, sustaining an annual growth rate of more than 40% for the last decade. Because of this fast growth, decreasing costs, and a vast

FIGURE 13.14 Wind and solar application in the city of Huleka in South Africa. (SMA Solar Technology AG.)

technical potential, solar energy is becoming an important alternative for the future energy needs. With increasing applications of distributed and utility-scale PV as well as *concentrating solar power* (CSP), solar technologies start to play an important role in meeting the world's energy demand.

The term *photovoltaic* describes the conversion process to convert light energy directly to electric energy. The developments in semiconductor technology cause the invention of the PV cell (also known as solar cell) in the early 1950s.

PV and CSP technologies both use the sun to generate electricity. However, they do it in different ways. The earth's surface receives sunlight in either a direct or diffuse form. Direct sunlight is solar radiation whose path is directly from the sun's disk and shines perpendicular to the plane of a solar device. This is the form used by CSP systems and concentrating PV systems, in which the reflection or focusing of the sun's light is essential to the electricity-generating process. Flat-plate or non-concentrating, PV systems can also use direct sunlight. Figure 13.15 shows solar applications on the roof of Oregon State Capital in Salem, Oregon. Figure 13.16 shows solar installations at Montalto di Castro in Italy. Figure 13.17 shows solar applications on the rooftop of a barn in Bayern, Germany. Figure 13.18 shows solar module used in the city of Kassel in the state of Hessen, Germany. Figure 13.19 shows solar and wind turbine applications in the state of Rheinland-Pfalz in Germany.

FIGURE 13.15   Solar applications on the roof of Oregon State Capital, Salem, Oregon, United States. (SMA Solar Technology AG.)

FIGURE 13.16   Solar installations at Montalto di Castro in Italy. (SMA Solar Technology AG.)

FIGURE 13.17 Solar applications on the rooftop of a barn in Bayern, Germany. (SMA Solar Technology AG.)

FIGURE 13.18 Solar module used in the city of Kassel in the state of Hessen, Germany. (SMA Solar Technology AG.)

PV (or *solar electric*) systems use semiconductor solar cells to convert sunlight directly into electricity. In contrast, CSP (or *solar thermal electric*) systems use mirrors to concentrate sunlight and exploit the sun's thermal energy. This energy heats a fluid that can be used to drive a turbine or piston, hence producing electricity. Succinctly put, PV uses the sun's light to produce electricity directly, whereas CSP uses the sun's heat to produce electricity indirectly. Again, PV and CSP both use the sun to produce electricity. However, they use different forms of the sun's radiation.

Other solar radiation is diffuse, meaning the sunlight reaches the earth's surface after passing through thin cloud cover or reflecting off of particles or surfaces. *Global radiation* is the sum of the direct and diffuse components of sunlight. This global radiation, as well as direct or diffuse radiation alone, can be used by *flat-plate* PV systems to generate electricity.

As said before, PV and CSP are different forms of solar technologies. Similarly, there are different PV materials and designs for generating electricity, which include crystalline silicon, thin films, concentrating PV, and future-generation PV, in addition to associated balance of systems components.

FIGURE 13.19   Solar and wind turbine applications in the state of Rheinland-Pfalz in Germany. (SMA Solar Technology AG.)

The new technologies, such as depositing solar modules onto a flexible plastic substrate or using solar "*inks*" (e.g., copper indium gallium selenide) and a "printing" process to produce film solar panels, are ready to drastically reduce the cost of solar power plants to less than $1/W.

Solar power plants can be built where they are most needed in the grid because siting PV arrays is usually much easier than siting a conventional power plant. Also, unlike conventional power plants, modular PV plants can be expanded incrementally as demand increases. It is expected that municipal solar power plants with few megawatt capacity built close to load centers will become common during the next decade.

13.10.2   CRYSTALLINE SILICON

Silicon was one of the very first materials that were used in early PV devices, and it still has more than 90% of the market share in today's commercial solar cell market. The silicon-based cells are known as *first-generation PV*. Pure silicon is mixed with very small amounts of other elements such as boron and phosphorous, which become positive- and negative-type semiconductor materials, respectively. Putting the two materials in contact with one another creates a built-in potential field.

Thus, when this semiconductor device is subject to the sunlight, the energy of the sunlight frees electrons that then move out of the cell, because of the potential field, into wires that form an electric circuit. Such "*PV*" effect needs no moving parts and does not use up any of the material in the process of generating electricity. Figure 13.20 shows solar applications on a building in the city of Laatzen in the state of Niedersachsen, Germany. Figure 13.21 shows solar applications in a sports stadium in the city of Mainz in the state of Rheinland-Pfalz in Germany. Figure 13.22 shows solar application in a German school in San Salvador (SMA Solar Technology AG). Figure 13.23 shows solar rooftop applications in the state of Baden-Wurttemberg, Germany.

A typical solar cell has a glass or plastic cover or other encapsulated cover, an antireflective surface layer, a front contact to permit electrons to enter a circuit, and a black contact to permit the semiconductor layers where the electrons start and finish their flow. The thickness of a crystalline silicon (c-Si) cell may be 170–200 μm (10^{-6}).

Figure 13.24 shows a typical 3 in. diameter cell. It will produce a voltage of 0.57 V when sunlight shines upon it under open-circuit conditions. This voltage would be the same regardless of how big

FIGURE 13.20 Solar applications on a building in the city of Laatzen in the state of Niedersachsen, Germany. (SMA Solar Technology AG.)

FIGURE 13.21 Solar applications in a sports stadium in the city of Mainz in the state of Rheinland-Pfalz in Germany. (SMA Solar Technology AG.)

the size of the cell is. But its current supply is directly proportional to its surface area. In that sense, a solar cell can be considered a constant current source as well as a voltage source. The solar cell is a nonlinear device and its performance is subject to its characteristic curve.

Figure 13.25 shows a typical I–V characteristic for a PV cell. Note that when the current is zero (i.e., no load), the voltage (i.e., the *open-circuit voltage* V_{oc}) is about 0.6 V. As the load resistance increases, causing the voltage output of the cell to increase, the current remains relatively constant until the "knee" of the curve is reached. The current then drops off quickly, with only a small increase in voltage, until the open-circuit condition is reached. At this point, the open-circuit voltage is obtained and no current is drawn from the device.

FIGURE 13.22   Solar application in a German school in San Salvador. (SMA Solar Technology AG.)

FIGURE 13.23   Solar rooftop applications in the state of Baden-Wurttemberg, Germany. (SMA Solar Technology AG.)

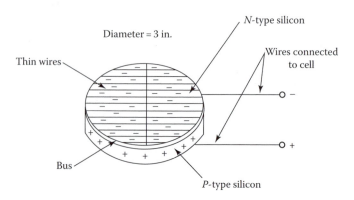

FIGURE 13.24   Typical 3 in. diameter cell.

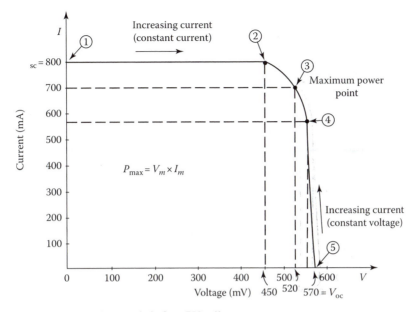

FIGURE 13.25 Typical *I–V* characteristic for a PV cell.

A solar cell will produce a voltage of 0.57 V when the sunlight shines upon it under open-circuit conditions. This voltage would be the same regardless of how big the size of the cell is. But its current supply is directly proportional to its surface area. In that sense, a solar cell can be considered a constant current source as well as a voltage source. The solar cell is a nonlinear device and its performance is subject to its characteristic curve.

The power output of any electrical device, including a solar cell, is the output voltage times the output current under the same conditions. The open-circuit voltage is a point of no power, that is, the current is zero. Similarly, the short-circuit condition produces no power because the voltage is zero. The maximum power point is the best combination of voltage and current. This is the point at which the load resistance matches the solar cell internal resistance.

The power into the cell is a function of the cell area and the power density of light. Once these are fixed, the peak efficiency takes place when the power output is maximum. Thus, the maximum power point should be selected as the operating point of the cell. The maximum power out takes place somewhere around the center of the knee of the curve.

The power into the solar cell is a function of the cell area and the power density of the light. Hence, for a given cell, the peak efficiency takes place when the power output is a maximum. The voltage and current can then be changed electrically to their desired values. In general, the solar cell responds well to all forms of visible light. Thus, they can operate indoors from incandescent or florescent lamps.

The peak power current changes proportionally to the amount of sunlight, but the voltage drops only slightly with large changes in the light intensity. Hence, a solar cell system can be designed to extract enough usable power to trickle-charge a storage battery even on a cloudy day.

It is estimated that the sun is constantly emitting 1.7×10^{23} kW of power. A very small portion of this (about 8.5×10^{23} kW) reaches the earth. About 30% of this is lost and 70% (about 6×10^{13} kW) penetrates our atmosphere. The amount of power per unit area that is received from the sum is defined as *power density*. When the sun is directly overhead on a clear day, the power density of sunlight is about 100 mW/cm². The power density of sunlight is also defined with a unit called the sun. Hence,

$$1 \text{ sun} = 100 \text{ mW/cm}^2 = 1 \text{ kW/m}^2 \qquad (13.27)$$

On a cloudy day, the *power density* of sunlight might be

$$30 \text{ mW/cm}^2 = 0.3 \text{ sun}$$

Energy density is another quantity that is also used to measure sunlight. Its unit is the Langley.

$$1 \text{ Ly} = 11.62 \text{ Wh/m}^2 \tag{13.28}$$

Example 13.11

Determine the energy density in Langleys, if the strength of sunshine is 1 sun for a period of 2 min.

Solution

$$\text{Power density} = 1 \text{ sun} = 1 \text{ kW/m}^2 = 1000 \text{ W/m}^2$$

$$\text{Energy} = \text{power} \times \text{time}$$

$$= (1000 \text{ W/m}^2)\left(\frac{1 \text{ min}}{60 \text{ min/h}}\right)$$

$$= 16.67 \text{ Wh/m}^2$$

To determine the energy density, use Equation 13.28:

$$\text{Energy density} = \frac{16.7 \text{ Wh/m}^2}{11.62 \text{ Wh/m}^2}$$

$$= 1.424 \text{ Ly}$$

Example 13.12

Assume that the solar cell whose characteristic is shown in Figure 13.25 is connected to a resistive load. Determine the required value of load resistance R_L to obtain at each of the operating points of 1, 2, 3, 4, and 5 on the following characteristics:

Solution

At point 1:	$V = 0$,	$I = I_{sc} = 800$ mA
At point 2:	$V = 450$ mV,	$I = 800$ mA
At point 3:	$V = 700$ mV,	$I = 520$ mA
At point 4:	$V = 570$ mV,	$I = 550$ mA
At point 5:	$V = 570$ mV,	$I = 0$ mA

$$R_L = \frac{570 \text{ mV}}{0} = \infty \Rightarrow \text{open circuit}$$

Example 13.13

Determine the electric power (output power) that can be obtained from the solar cell at each of the points in Example 13.12.

Solution

At point 1:	$P = (0 \text{ mV})(800 \text{ mA}) = 0 \text{ W}$
At point 2:	$P = (450 \text{ mV})(800 \text{ mA}) = 0.36 \text{ W}$
At point 3:	$P = (700 \text{ mV})(520 \text{ mA}) = 0.364 \text{ W}$
At point 4:	$P = (570 \text{ mV})(550 \text{ mA}) = 0.313 \text{ W}$
At point 5:	$P = (570 \text{ mV})(0 \text{ mA}) = 0 \text{ W}$

Notice that a good power output is obtained between points 2 and 3. In fact, maximum power output is obtained at the point 3, that is, the *maximum power point*.

Example 13.14

Determine the energy density in langley, if the strength of sunlight is 1/2 sun for a period of 2 min.

Solution

$$\text{Power density} = \frac{1}{2}\text{sun}$$

$$= 0.5 \text{ kW/m}^2$$

$$= 500 \text{ W/m}^2$$

$$\text{Energy} = \text{power} \times \text{time}$$

By using Equation 13.28, the energy density in Langley is

$$\text{Energy density} = \frac{8.333 \text{ Wh/m}^2}{11.62 \text{ Wh/m}^2}$$

$$= 0.717 \text{ Ly}$$

It can be seen that still the same amount of energy density is obtained, but it would take twice as long at 1/2 sun.

Example 13.15

Consider the rooftop of a home measuring 12 m × 15 m that is all covered with PV cells to provide the electric energy requirement by that home. Assume that the sun is at its peak (i.e., having strength of 1 sun) for 3 h every day and the efficiency of a solar cell is 10%. (i.e., 10% of the sunlight power that falls on the cell is converted to electric power.) Determine the average daily electric energy converted by the rooftop.

Solution

The area of the roof is

$$A = 12 \text{ m} \times 15 \text{ m} = 180 \text{ m}^2$$

The amount of power collected by the rooftop is

$$P = (1 \text{ kW/m}^2)(180 \text{ m}^2) = 180 \text{ kW}$$

Since the cells are 10% efficient, the electric power converted by the cells will be

$$P_{\text{elec}} = 0.1 \times 180 \text{ kW} = 18 \text{ kW}$$

Assuming that a peak sun exists for 2 h every day, the average daily energy is

$$18 \text{ kW} \times 3 \text{ h} = 54 \text{ kWh}$$

Example 13.16

Consider the results of Example 13.15 and assume that the cost of electric energy to the homeowner is $0.10/kWh charged by the utility company. If the cost of solar roof is about $5000, determine the following:

     a. The amount of electric energy produced by solar cells per year
     b. The amount of savings on electric energy to the homeowner per year
     c. The break-even period for the solar panels to pay for themselves in months

Solution

     a. The amount of electric energy produced by the solar panels per year is

$$\text{Annual electric energy produced} = (54 \text{ kWh/day})(365 \text{ days/year})$$

$$= 19,710 \text{ kWh/year}$$

     b. The amount of savings to the homeowner due to electric energy produced by the solar panels per year is

$$\text{Annual savings} = (19,710 \text{ kWh/year})(\$0.10/\text{kWh})$$

$$= \$1871/\text{year}$$

     c. The break-even time for the solar panels to pay for themselves is

$$n = \frac{\$5000}{\$1871/\text{year}}$$

$$= 2.67 \text{ years}$$

13.10.3   Effect of Sunlight on Solar Cell's Performance

The *I–V* characteristic given in Figure 13.22 is based on the assumption that the solar cell is operating under a bright noontime sun. In the event that the power density of the sunlight decreases, the output of the cell decreases accordingly. The reasons for such decrease may include the following: The sun is not shining directly on the cell due to the fact that it is just rising or setting; the sun is not shining on the cell because it is winter. (In the northern hemisphere, the sun has a southern exposure. On the other hand, in the southern hemisphere, the opposite is true, that is, in winter, the sun has a northern exposure.) There may exist tall trees or structures that cast a shadow on the cell, during certain times of the day; it is a cloudy or overcast day.

     The angular position on the earth's surface north or south of the equator is defined as the *latitude*. The equator itself has 0° latitude and divides the earth into two equal hemispheres. For every 69 miles north or south of the equator, the latitude increases by 1°. For instance, the north pole has a latitude of 90° north and the south pole has a latitude of 90° south. Table 13.6 gives the latitudes of selected cities around the world.

     Sun's position varies at a given time. For example, at noontime, the sun's position varies about 40° from its highest position in June to its lowest position in December. In March and September (equinox), the sun's center is directly over the equator. Its apparent position in the sky then is approximately equal to your latitude. Thus, in New York City, the noontime sun will appear to be

TABLE 13.6

Latitudes of Selected Cities around the World

Location	Latitude	Location	Latitude
Athens, Greece	38° N	Madrid, Spain	40° N
Berlin, Germany	53° N	Miami, Florida	26° N
Bogota, Columbia	2° N	Montreal, Canada	46° N
Bombay, India	20° N	Moscow, Russia	55° N
Buenos Aires, Argentina	20° N	Munich, Germany	48° N
Cairo, Egypt	30° N	Oslo, Norway	60° N
Edinburgh, Scotland	56° N	Paris, France	49° N
Entebbe, Uganda	0°	Quito, Ecuador	0°
Honolulu, Hawaii	20° N	Rio de Janeiro, Brazil	23° S
Houston, Texas	30° N	Rome, Italy	42° N
Kansas City, Missouri	39° N	Seattle, Washington	47° N
Las Vegas, Nevada	36° N	Sydney, Australia	35° S
Lima, Peru	12° S	Thule, Greenland	77° N
London, England	52° N	Tokyo, Japan	36° N
Los Angeles, California	34° N	Valparaiso, Chile	36° N

40° south of vertical during the equinox. In June, it will be 20° (40°–20°) south of vertical, and in December, 60° (40° + 20°) south of vertical.

For maximum energy absorption from the sun, a solar cell should be tilted south (in the northern hemisphere) by the angle of the latitude of the location on the earth.

Example 13.17

At what angle a solar cell should be tilted to get the most energy from the sun if it is located in

a. Quito, Ecuador?
b. Athens, Greece?
c. Valparaiso, Chile?

Solution

It is illustrated in Figure 13.26 below.

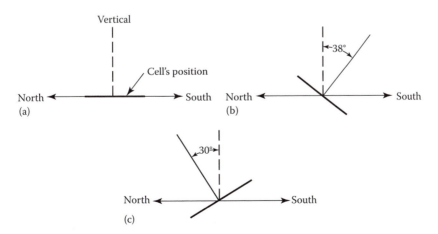

FIGURE 13.26 Solution for Example 13.17.

13.10.4 Effects of Changing Strength of the Sun on a Solar Cell

It is often that characteristics of a solar cell include the effects of a variation in power density. The current output of a cell is directly proportional to the sunlight.

Thus, under a very weak sun, the cell puts out very little current. However, its voltage is still quite high. Thus, one can conclude that under open-circuit conditions, the voltage is relatively independent of sunlight. But the effect of the load on the operating point as the power density varies is more important.

Example 13.18

Assume that a fixed resistive load selected to force the solar cell shown in Figure 13.27 operates at the point where maximum power conversion takes place.

a. If that point has full sunlight (1 sun), determine the value of the load resistance R_L at the maximum power point a in Figure 13.27.
b. Determine the value of the power that can be obtained at such point a.
c. Let's say due to passing by cloud now, the power density drops to 0.5 sun, and as a result, the operating point of the cell changes. At the corresponding lower sun curve (with the current output of 300 mA and the fixed load resistance), determine the value of the voltage.
d. Correspondingly, the new operating point is at b in Figure 13.27. Determine the power at this point.
e. Find the new load resistance that is necessary to achieve this.

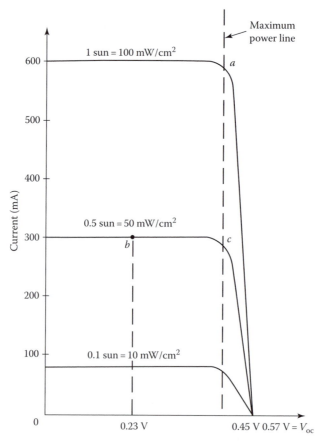

FIGURE 13.27 Variation of I–V characteristic of a solar cell due to changing power density.

f. As it can be observed from Figure 13.27 that the solar cell is no longer operating at the knee of the curve, consequently, the maximum power is not being converted. To achieve the maximum power conversion, the operating point has to be moved to the point c in Figure 13.27. As a result, there has to be a new load resistance. Determine its value.

g. Determine the corresponding power at the point c. (Note that the power at the point c is one-half the power at the point a.)

Solution

a. The value of the load resistance is

$$R_L = \frac{0.45 \text{ V}}{0.58 \text{ A}} \cong 0.78 \text{ } \Omega$$

b. The value of the power at such point a is

$$P_a = (0.45 \text{ V})(0.58 \text{ A}) = 0.26 \text{ W}$$

c. The value of the voltage at point a is

$$V = R_L I$$

$$= (0.78 \text{ } \Omega)(0.3 \text{ A})$$

$$= 0.23 \text{ V}$$

d. Thus, the power at the new point b is

$$P_b = (0.23 \text{ V})(0.3 \text{ A})$$

$$= 0.069 \text{ W}$$

Consequently, the maximum power is not being converted.

e. In order to find the new load resistance that is necessary to achieve the maximum power, the operating point has to be moved to the point c in Figure 13.27.

f. As a result, there has to be a new load resistance. Hence,

$$\text{new } R_L = \frac{0.45 \text{ V}}{0.29 \text{ A}} \cong 1.55 \text{ } \Omega$$

g. Hence, the corresponding power at point c becomes

$$P_c = (0.45 \text{ V})(0.29 \text{ A})$$

$$\cong 0.13 \text{ W}$$

Note that the power at the point c is one-half the power at the point a.

Therefore, it can be concluded that the current output is directly proportional to the sunlight (i.e., power density). However, the power will not be proportional unless the load is changed as the intensity of the light varies. Accordingly, the maximum power output of solar cell is a function of not only sunlight but also the load. Because of this, inverters employed for solar energy conversion systems have special tracking circuitry that continuously adjusts the loading on the solar cells while monitoring power output.

TABLE 13.7

Data for Example 13.19

Power Density (Suns)	Time (h)	Energy Density (Sun-Hours)
0.2	1	0.2
1.0	1	1.0
0.9	2	1.8
1.0	3	3.0
0.9	2	1.8
0.3	1	0.3
0.2	1	0.2
0.1	1	0.1
		Total = 8.4 sun-h

The predictability of how much power can be obtained from the sun from hour to hour or day to day is a very serious problem from the scheduling point of view. This problem is also caused by the variation in sunlight. Because of this, it is often that the average data on sunlight are used.

Example 13.19

In order to determine how much energy can be received from the sun by one solar cell in a city in California, data were collected on a sunny day and the data are given in Table 13.7. During the day, the sun rises at 7 AM and sets at 7 PM.

Solution

Since from Table 13.7, the total energy density for the day is 8.4 sun-h, it is equivalent to having a full or peak sun (i.e., 1 sun) for 8.4 h. Keeping this in mind and using the power calculated in part (c) for 1 sun power density in Example 13.17,

$$P = 0.26 \text{ W} \quad \text{(for 1 sun)}$$

Thus, the total energy* that can be produced by one solar cell can be found as

$$\text{Total energy} = (0.26 \text{ W/sun-h})(8.4 \text{ sun-h})$$

$$= 2.184 \text{ W}$$

13.10.5 TEMPERATURE'S EFFECT ON CELL CHARACTERISTICS

The ratings of solar cells are based on the minimum current they supply at 0.45 V under a full sun at 25°C (77°F). Its output is a function of its cell temperature. As temperature increases, the current will increase, while the voltage will decrease by about 2.1 mV/°C. As a result, its power output and consequently its cell efficiency decrease.

---

* An alternative method would be to use a set of curves to the ones in Figure 13.4. Hence, for each power density of the sunlight in Table 13.6, calculate the power at the new of the curve, assuming that the load is adjusted for maximum power. To determine the energy for each time interval, the calculated maximum powers are multiplied by the time of duration. The total energy is found by adding energies for the day.

The opposite takes place as the temperature decreases; that is, cell operates more efficiently when they are cooler. Because of this, commercially manufactured solar panels, that is, groups of interconnected cells, have a metal (usually, aluminum) that plays the role of a heat sink. Otherwise, in areas of low latitude, the temperatures of a solar cell can reach 80°C (i.e., 176°F) without a heat sink. At other temperatures, the voltage and the current of the cell can be determined from

$$E_0 = E_R - 0.0021(T - 25) \qquad (13.29)$$

and

$$I_0 = I_R + 0.025A(T - 25) \qquad (13.30)$$

where
E_R and I_R are the cell ratings in volts and milliamperes, respectively, at 25°C
E_0 and I_0 will be the cell voltage and current at the new temperature T in degrees Celsius
A is the cell area in square centimeter

Example 13.20

Assume that a solar cell is rated 600 mA, 0.45 V, at 25°C, and that the cell area is 30 cm². If as the cell is under a full sun and providing maximum power, the temperature increases to about 50°C, determine the following:

a. Its power output at 25°C
b. Its voltage, current, and power output at 50°C
c. The amount of percentage drop in power output due to the increased temperature

Solution

a. Its power output at 25°C can be found from its rated voltage and current values as

$$P = (0.45 \text{ V})(600 \text{ mA})$$

$$= 270 \text{ mW}$$

b. From Equation 13.29, its new voltage is

$$E_0 = E_R - 0.0021(T - 25)$$

$$= 0.45 - 0.0021(50 - 25)$$

$$\cong 0.40 \text{ V}$$

And from Equation 13.30, the new current is

$$I_0 = I_R + 0.0025 \, A(T - 25)$$

$$= 600 \text{ mA} + \left(0.025 \frac{\text{mA}}{\text{deg-cm}^2}\right)(30 \text{ cm}^2)[(50 - 25) \text{ deg}]$$

$$= 618.75 \text{ mA} \cong 619 \text{ mA}$$

Hence, the new power output of the cell is

$$P = (0.40 \text{ V})(618.75 \text{ mA})$$

$$= 247.5 \text{ mW}$$

c. The amount of drop power output in percentage is

$$\%P_{\text{drop}} = \frac{270 - 247.5}{270} \times 100$$

$$\cong 8.33\%$$

Temperature effects have to be considered when a PV system is designed. In general, the power estimate is increased by about 10% to take into account the loss due to increased cell temperature.

13.10.6 EFFICIENCY OF SOLAR CELLS

In general, *solar cell efficiency* is defined as the ratio of the electric power output to the sunlight power it receives. The maximum theoretical efficiency of a silicon solar cell is about 25%. Today's cells have rated efficiencies of 10%–16%.

The efficiency of a cell is a function of number of things, including the number and thickness of the wires connected to the top of the cell and light reflected from the surface of the cell. Also, when cells are connected together to form panels, the panel efficiency will be based on the cell's shape. Figure 13.20 shows solar rooftop applications in the state of Baden-Wurttemberg, Germany.

Example 13.21

Consider a circular cell that has a diameter of 2.5 in. If it has a rating at 25°C of 1200 mA and 0.45 V in a full sun, determine the cell's efficiency.

Solution

In order to determine the cell's efficiency, first, let us find the cell area. Since the radius of the cell is

$$r = \frac{d}{2} = \frac{2.5 \text{ in.}}{2} = 1.25 \text{ in.}$$

or

$$r = (1.25 \text{ in.})(2.54 \text{ cm/in.}) = 3.175 \text{ cm}$$

then

$$A = \pi r^2 = \pi(3.175 \text{ cm})^2 = 31.67 \text{ cm}^2$$

Since 1 sun = 100 mW/cm²,

$$P_{\text{in}} = (100 \text{ mW/cm}^2)(31.67 \text{ cm}^2) = 3167 \text{ mW}$$

$$P_{\text{out}} = (0.45 \text{ V})(1200 \text{ mA})$$

$$= 540 \text{ mW}$$

Thus, the efficiency is

$$\eta = \frac{P_{out}}{P_{in}} \times 100$$

$$= \frac{540\ mW}{3167\ mW} \times 100$$

$$= 17\%$$

13.10.7 INTERCONNECTION OF SOLAR CELLS

Typical solar cell output is 800 mA at 0.45 V. But most everyday applications dictate more than 800 mA at 0.45 V. However, solar cells can be treated just like batteries. The net voltage can be increased by connecting them in series, and the net current can be increased by connecting them in parallel. For instance, if 12 identical cells each rated 1 A and 0.45 V in a full sun are connected, as shown in Figure 13.28, the net output of the system will be 3 A at 1.8 V. For each parallel path, the current increases by 1 A, and for each cell in series, the voltage increases by 0.45 V.

It may be of an interest that solar cells are more flexible than batteries in that they can be broken into pieces to get odd ratings. The voltage output from a piece of a cell will still be the rated voltage of the whole cell. But the current will be proportional to the area of the piece of cell. It is a common practice to cut the circular cells in halves and quadrant cells. If many solar cells are connected in series and parallel to form a permanent unit, it is called a *solar panel*.

However, the current produced by a series connection of multiple cells will be the minimum of all the cells. For example, if three cells each rated 1 A are connected in series with a half of the same cell (rated 0.5 A), the current will be 0.5 A. In practice, solar panels are manufactured in different sizes and ratings. Such panels can be interconnected to form *solar arrays* or *modules*.

Example 13.22

Consider a solar panel that is rated 20 W at 1.0 V. Determine the following:

    a. How many of these panels are needed to supply 1 A of current at 120 V?
    b. How should they be connected?
    c. The value of the necessary load resistance.
    d. The total power that can be obtained from this array of solar cells.

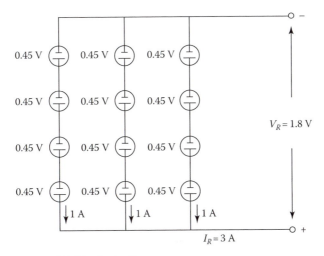

FIGURE 13.28 Connection of 12 identical cells.

Solution

a. Since each panel produces 12 V, to get 120 V, the number of panels required in series is

$$\text{\# of panels in series} = \frac{\text{Total voltage}}{\text{volts/panel}}$$

$$= \frac{120 \text{ V}}{10 \text{ V/panel}}$$

$$= 12 \text{ panels}$$

b. The current rating of each panel can be found from panel's power and voltage ratings. Hence,

$$I = \frac{20 \text{ W}}{10 \text{ V}} = 2 \text{ A}$$

Thus, each path, of seriously connected panels, will provide 2 A. Therefore,

$$\text{\# of paths} = \frac{\text{Total current}}{\text{Current/path}}$$

$$= \frac{10 \text{ A}}{2 \text{ A/path}}$$

$$= 5 \text{ paths}$$

Hence, five parallel paths each having 12 panels in series are required to form an array to meet the power requirement. The resultant panel arrangement is shown in Figure 13.29.

c. The value of the required load resistance is found as

$$R_L = \frac{120 \text{ V}}{10 \text{ A}} = 12 \ \Omega$$

This input resistance facilitates the maximum power conversion in a full sun. However, as the lighting changes, this input impedance needs to be changed in order to get the maximum power conversion.

d. The total power that can be obtained from this solar array is

$$\text{Total power} = (120 \text{ V})(10 \text{ A})$$

$$= 1.2 \text{ kW}$$

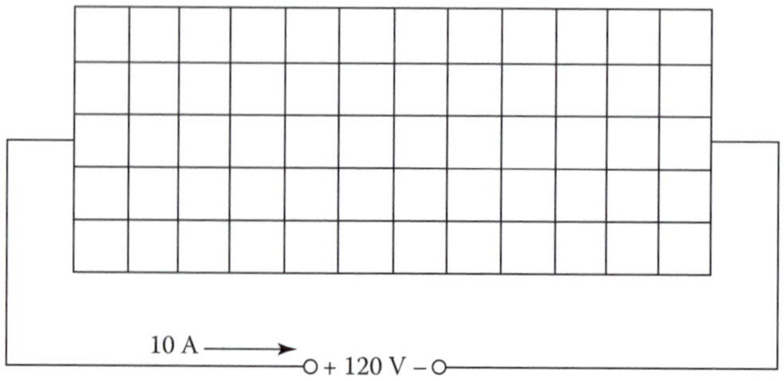

FIGURE 13.29 Panel arrangement for Example 13.22.

Alternatively,

$$\text{Total power} = (60 \text{ panels})\left(\frac{20 \text{ W}}{\text{panel}}\right)$$

$$= 1.2 \text{ kW}$$

13.10.8 OVERALL SYSTEM CONFIGURATION

PV cells and modules are configurable from 1 to 5 MW. A solar generator has basically two possible fundamental configurations. The first one is a stand-alone system, as shown in Figure 13.30a. It can be used in the location where there is no utility power available. During peak sun-hours, the solar array supplies all the ac power needs and keeps the storage batteries fully charged. The storage batteries have to have the capability of storing enough energy to supply the power required when the sun goes down.

In the second system, shown in Figure 13.30b, the solar generator is connected to the utility grid. Here, there is no need for the storage batteries. The dc power from the solar array gets inverted to ac power. The inverter output provided the ac power needs, during peak sun-hours. In the event that these needs are low, the power will be fed back into the utility grid for credit. On the other hand, during nighttime hours, the ac power requirements are met by the power supplied by the utility grid. Figure 13.30 only shows the overall system configuration, without the necessary circuit breakers and/or meters.

The overall system frequency must take into account of the effectiveness of all other components (such as inverter, batteries, and any additional circuitry) in addition to the efficiency of the solar array. Figure 13.31 shows a solar module used in the city of Kassel in the state of Hessen, Germany.

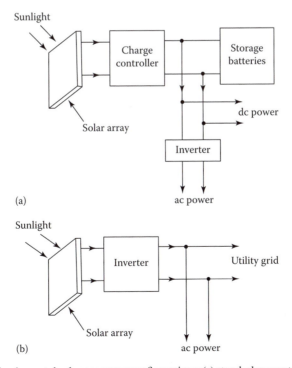

FIGURE 13.30 Two fundamental solar generator configurations: (a) stand-alone system and (b) supplemental or cogeneration system.

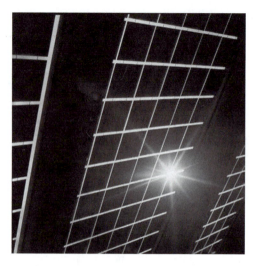

FIGURE 13.31   Solar module used in the city of Kassel in the state of Hessen, Germany. (SMA Solar Technology.)

A given solar energy system is designed based on the size of the solar array requirement (for a given specific location and electric energy needs for that location) and the amount of electric energy that the array can supply. After the location and solar array size and its rating are known, the energy output from the system can be easily determined.

Example 13.23

Smith lives in Miami, Florida (with yearly average of 4.7 peak sun-hours per day), and is considering placing a PV array on his south-facing roof. The roof is unshaded and has a size of 25 by 44 ft. The electric power produced is to be converted to ac and used in his home. Any excess power produced will be back into the local utility company's grid system. Analyze the system configuration and determine how much energy Smith can expect to get from the solar installation.

Solution

The system configuration will be like the one shown in Figure 13.30b. But, in addition, there will be a circuit breaker and an ac watt-hour meter will be placed between the inverter and the utility grid. The amount of energy that will be sold to the utility company through the grid connection will be measured by the watt-hour meter. The local utility company by law (PURPA Act of 1078) is obliged to pay for this energy.

The solar panel chosen for this application is rated 16.2 V, 2.4 A, and 39 W and has 10% efficiency at 25°C in full sun. Its dimensions are 1 by 4 ft. At 50°C, the ratings of the solar panel become 14.7 V, 2.27 A, and 55 W. If about 4 ft is permitted at the edges of the roof for a work area, 63 solar panels can easily be placed on the roof. They could be placed end to end with seven in a row ($9 \times 4 = 36$ ft) and nine rows total ($9 \times 1 = 9$ ft), as illustrated in Figure 13.32. Hence, the maximum power output will be

$$(81\,\text{panels})\left(\frac{39\,\text{W}}{\text{panel}}\right) = 3159\,\text{W}$$

A 3.2 kW single-phase inverter will be employed. Its maximum input current is specified as 25 A, and the input voltage can vary from 60 to 120 V dc. The output voltage is 120 V ac. The seven panels in each row will be connected in series. This will provide a range of voltage (from 25°C to 50°C) of

$$7 \times 16.2 = 113.4\,\text{V} \quad (25°C)$$

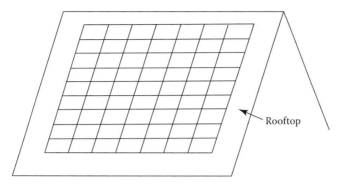

FIGURE 13.32 Installation of solar panels in Example 13.23.

to

$$7 \times 14.7 = 102.9 \text{ V} \quad (50°C)$$

which is within the range of the inverter. The nine rows will be connected in parallel. Hence, the peak current from the solar array (at 50°C) will be

$$9 \times 2.27 = 20.43 \text{ A}$$

which is within the inverter limit. The panel should be tilted 26° toward south. Since in Miami, Florida, one can expect an average of 4.7 peak sun-hours per day, multiplying it by the peak power output at 50°C will provide the average daily supplied by the solar array. Hence,

$$\text{At } 50°C: \quad P = (102.9 \text{ V})(20.43 \text{ A}) \cong 2102 = 2.102 \text{ kW}$$

$$\text{Average daily energy} = (2.1 \text{ kW})(4.7 \text{ h}) \cong 9.87 \text{ kWh}$$

If the inverter efficiency is 95%, the average energy produced by the system per day will be

$$(9.87 \text{ kWh/day})(0.95) = 9.4 \text{ kWh/day}$$

By multiplying this by 30 days,

$$(9.4 \text{ kWh/day})(30 \text{ days}) = 282 \text{ kWh/month}$$

or

$$(9.4 \text{ kWh/day})(365 \text{ days/year}) = 3431 \text{kWh/year}$$

At 10 cents/kWh, Smith will save about \$343/year in electric bills. The solar array would cost

$$(3159 \text{ W})(\$1/\text{W}) = \$3159$$

Thus, the break-even point for this investment is

$$n = \frac{\$3159}{\$343}$$

$$= 9.2 \text{ years}$$

Smart will recover his initial investment in 9.2 years. But if the cost of solar array drops down to $0.50/peak watt (as predicted previously), the investment cost would be

$$(3159 \text{ W})(\$0.50/\text{W}) = \$1579.50$$

Thus,

$$n = \frac{\$1579.50}{\$343 / \text{year}}$$

$$\cong 4.6 \text{ years}$$

At this price, Smart will recover his investment in less than 5 years.

13.10.9 THIN-FILM PV

Second-generation PV devices are a more recent development. They are made of layers of semiconductor materials that are much thinner than those in silicon cells. The thickness of a cell is on the order of only 2–3 μm thick,

In the event that silicon is used, it is typically in the form of *amorphous* (i.e., not crystallized) *silicon* (a-Si), which has no discernible crystal structure. Also, microcrystalline silicon thin-film devices are being developed. In addition, other thin-film materials have also been developed and commercialized, including *cadmium telluride* (CdTe) and *copper indium gallium diselenide* (CIGS). These PV devices need much less material than traditional c-Si devices.

According to Kroposki et al. [9], thin films generally have lower solar conversion efficiency than the c-Si cells. Here, the conversion efficiency is defined as the percentage of the sun's power shining on the cell. For instance, if 1000 W of solar power illuminates a cell and 150 W of electricity is generated, then the cell has a solar conversion efficiency of 15%.

A commercial silicon cell may have an efficiency of about 20%, whereas a commercial CdTe cell's efficiency is about 11%. The thin-film cell uses less material and can be deposited with a method that is much less energy intensive than silicon. Less material also causes lighter weight.

Also, some thin-film technologies do not use rigid wafers; instead, they can be deposited on flexible layers of stainless steel or plastic. Depending on the application, such flexibility might be highly desirable. In general, thin-film PV is less expensive to manufacture and easier to implement [8].

13.10.10 CONCENTRATING PV

There is another type of second-generation PV device that is based on high-efficiency multifunction cell that uses compounds from the group III and group V elements of the periodic table of elements. For example, such multifunction solar cell design has three layers, each of which absorbs a different portion of the solar spectrum to use in generating electricity.

The top layer may be made of gallium indium phosphide and the bottom layer of germanium. This type of design may have a high efficiency of about 40%. This is due to the fact that each layer in this design is designed to absorb and use a different portion of the solar spectrum.

But such design is expensive since the group III and V materials are costly to produce. However, the cost can go down substantially, if a relatively inexpensive lens or mirror can be employed to focus sunlight on just a small area of cells [9]. For instance, if a 10 × 10 in. lens focuses this area of incident sun onto a 0.5 × 0.5 in. cell, the concentration factor becomes 400×, that is, 100 m²/0.25 in.² Thus, such cell with the lens can produce as much power as a 10 × 10 in. cell with lens, but at about 1/400 of the cell cost. Even though, they are not suitable for small projects, concentration systems could be very effective in large power generation for several homes.

13.10.11 PV Balance of Systems

Balance of systems compromises all of the components of a PV system beyond the actual PV module that produces the power. A frame structure is also required toward the sun, to stabilize it in the outer elements, including wind and snow.

PV systems produce dc electricity. Hence, if ac is needed, the balance of systems has to include an inverter. But the inverter decreases the overall system efficiency by an additional 5%–10%. However, the system efficiency can be improved by connecting a tracking system to the solar modules. The trackers can be of single axis or dual axis.

The single-axis trackers aligned with the axis in a north–south direction permit the module to trade the sun's progress across the sky from east to west during the day. Dual-axis trackers further improve the module's orientation, permitting the sun to always illuminate the cells perpendicular to the plane of the module. The result is the maximum energy output from the system.

In general, residential and commercial PV systems are directly connected to the grid without energy storage. Including battery energy storage would increase the reliability of the system. Hence, the batteries store excess power that is generated from the PV array to be used later. Figure 13.33 shows solar applications in a sports stadium in the city of Mainz in the state of Rheinland-Pfalz in Germany. Figure 13.34 shows solar application in the Ineco airport in the city of Valencia, Spain.

13.10.12 Types of Conversion Technologies

There are two primary technologies for the conversion of sunlight into electricity. PV cells depend on the use of semiconductor devices for the direct conversion of the solar radiation into electric energy. The typical efficiencies of such commercial crystalline PV cells are in the range of 12%–18%, even though there have been experimental cells built that are capable of over 30%.

The second type of technology is based on *solar thermal systems*. It is known as CSP. It involves intermediate conversion of solar energy into thermal energy in the form of steam, which in turn is employed to drive a turbogenerator. To have high temperatures, thermal systems invariably use concentrators by the use of mirrors either in the form of parabolic troughs or thermal towers.

FIGURE 13.33 Solar applications in a sports stadium in the city of Mainz in the state of Rheinland-Pfalz in Germany. (SMA Solar Technology AG.)

FIGURE 13.34 Solar application in the INECO airport in the city of Valencia, Spain. (SMA Solar Technology AG.)

But, presently, generation of electricity by either technology is considerably more expensive than traditional means. The CSP systems are essentially categorized based on how the systems collect solar energy. The three basic systems are the linear, tower, and disk systems.

13.10.13 Linear CSP Systems

In such systems, CSP collectors capture the sun's energy with large mirrors that reflect and focus the sunlight onto a linear receiver tube. Inside the receiver, there is a fluid that is heated by the sunlight and then employed to create superheated steam that causes a turbine to rotate in order to drive a generator to produce electricity.

It is also possible to produce the steam directly in the solar field. Here, no heat exchanger is employed, but the system uses expensive high-pressure piping system in the entire solar field. It has a lower operating temperature.

Essentially, concentrating collector fields in such systems are made of a large number of collectors in parallel rows that are usually aligned in north–south orientation to increase both summertime and annual energy collection. Using its single-axis sun-tracking system, the system facilitates the mirrors to track the sun from east to west during the day, causing the sun to reflect continuously onto the receiver tubes.

Such trough designs can use thermal storage. In that case, the collector field is built oversized in order to heat a storage system during the day that in the evening can be used to produce additional steam to generate electricity.

It is also possible to design the parabolic trough plants as hybrid systems that fuel to supplement the solar output during periods of low solar radiation. In such applications, usually a natural gas-fired heater or gas–steam boiler/reheater is used [9].

13.10.14 Power Tower CSP Systems

In such system, several large, flat, sun-tracking mirrors, which are called heliostats, focus sunlight onto a receiver at the top of a tower. The receiver has a heat-transfer fluid that is heated to produce steam. The heated steam in turn is employed in a typical turbine generator to generate electricity. The heat-transfer fluid is usually water/steam although in advanced designs replaced by molten nitrate salt due to its better heat-transfer and energy storage capabilities. Presently, such systems have been developed to produce up to 200 MW of electricity.

13.10.15   DISH/ENGINE CSP SYSTEMS

According to Kroposki et al. [9], these systems generate relatively small amounts (3–25 kW) of electricity with respect to other CSP technologies. Here, a solar concentrator (or dish) collects the solar energy radiating directly from the sun. The resultant beam of concentrated sunlight is reflected onto a thermal receiver that collects the solar heat. The dish is attached on a structure that tracks the sun continuously throughout the day to reflect the highest percentage of sunlight that is possibly onto the thermal receiver.

The power conversion unit is made of the *thermal receiver* and the *engine/generator.* The thermal receiver is the interface between the dish and the engine/generator. Its function is to absorb the concentrated beams of solar energy and, after converting them to heat, to transfer this heat to the engine/generator.

The *thermal receiver* can be made of a bank of tubes with a cooling fluid (hydrogen or helium) that is used as a transfer medium. Other thermal receivers are made of heat pipes, where the boiling and condensing of an intermediate fluid transfers the heat to the engine.

The *engine/generator system* is the subsystem of the dish/engine CSP system. It takes the heat from the thermal receiver and uses it to produce electricity. Most commonly, a *Stirling engine* is used as the heat engine. It uses a heated fluid to move pistons and create mechanical power to rotate the shaft of the generator to produce electric power.

The last subsystem of the CSP system is *thermal energy storage system.* It provides a solution for the curtailed energy production when the sun sets or is blocked by the clouds.

13.10.16   PV APPLICATIONS

PV cells were first used in power satellites. By the end of the 1990s, PV electric generation was cost competitive with the marginal cost of production with gas turbines. As a result, a number of utilities have introduced utility-interactive PV systems to supply portion of their total customer demand. Figure 13.24 shows solar application in the INECO airport in the city of Valencia, Spain.

Some of these systems have been residential and commercial rooftop systems and other systems have been larger ground-mounted systems. PV systems are classified as utility-interactive (grid-connected) or stand-alone systems.

13.10.16.1   Utility-Interactive PV Systems

Utility-interactive PV systems are categorized by IEEE Standard 929 as small, medium, or large PV systems. Small systems are less than 10 kW, medium systems range from 10 to 500 kW, and large systems are larger than 500 kW. Each size dictates different consideration for the utility interaction.

Since the output of PV modules is dc, it is, of course, necessary to convert this output to ac before connecting it to the grid; it is done by an inverter. It is also called a power conditioning unit (PCU). A typical small utility system of a few kilowatts consists of an array of modules selected by either a total cost or an available roof area. The modules are connected to produce an output voltage ranging from 48 to 300 V, as a function of the dc input requirements of the PCU. One or two PCUs are used to interface the PV output to the utility at 120 V or 120/240 V.

The point of utility connection is the load side of the circuit breaker in the distribution panel of the building of the PV system that is connected on the customer side of the revenue meter. However, medium- and large-scale utility-interactive systems differ from small-scale systems only in the possibility that the utility may dictate different interfacing conditions with respect to disconnect and/or power quality.

13.10.16.2   Stand-Alone PV Systems

They are used when it is not possible to connect to the utility grid. Examples include water-pumping systems, PV-powered fans, power systems for remote installations, and portable highway signs. Some of them include battery storage to operate the system under sun or no-sun situations.

PROBLEMS

13.1 Consider a wind turbine that is rated at 100 kW in a 10 m/s wind speed in air at standard conditions. If power output is directly proportional to air density, what is the power output of the wind turbine in a 10 m/s wind speed at an elevation of 2000 m above sea level at a temperature of 20°C?

13.2 Consider Example 13.4 and assume that the average wind speed at point *a* is 10 mph, while at point *b* is 8 mph. In order to capture 2 kW, determine the blade diameter (*d*) for a wind turbine operating:
   a.   At point *a*
   b.   At point *b*

13.3 Consider a wind turbine with blades of 10 ft radius. At the location, wind speed was 5 mph for 3 h and 15 mph for another 3 h time period. Determine the amount of energy that can be intercepted by the wind turbine.

13.4 Consider the wind turbine given in Example 13.10 and assume that the wind turbine is 2 MW and is under maintenance 200 h in 1 year out of a total of 8760 h of 1 year. If it is actually produced 4000 MWh due to fluctuations in wind availability, determine the following:
   a.   The availability factor of the wind turbine
   b.   The capacity factor of the wind turbine
   c.   The annual load duration of the wind turbine

13.5 Consider Example 13.6 and assume that during an 8 h period, the wind had the following average speeds: 4 mph for 2 h duration, 12 mph for 2 h duration, 17 mph for 1 h duration, and 23 mph for 3 h duration. Determine the resultant electric output for the 8 h period.

13.6 Assume that a wind generator whose power curve is shown in Figure 13.9 has a blade diameter of 18 ft. If its power output is at 120 V and 60 Hz, determine the net efficiency of the WECS at a wind speed of 15 mph.

13.7 Assume that the wind turbine with three blades in Problem 13.6 has been replaced by one with two blades that is rotating at 90 rpm; find the blade efficiency at a wind speed of 16 mph.

13.8 A WECS has a eight-pole 60 Hz three-phase synchronous alternator driven at synchronous speed. The blades have an 8 m diameter and a peak efficiency when the TSR = 5. Determine the transmission gear ratio for peak system efficiency at a wind speed of 6 m/s?

13.9 A WECS uses a 6-pole 60 Hz three-phase induction generator. It is excited by a three-phase 60 Hz power line. The blades have 11 m diameter and peak efficiency when the TSR = 6. If the generator efficiency is a maximum at a slip of −3.3%, what should the transmission gear ratio be for peak system efficiency at a wind speed of 5 m/s?

13.10 Assume that the WECS shown in Figure 12.9c uses a four-pole three-phase induction machine. The line frequency is 60 Hz and the average wind speed is 15 mph. The blades have a 32 ft diameter and peak efficiency when the TSR = 6. If the generator efficiency is a maximum at a negative slip of 3.3%, determine the transmission gear ratio that is necessary for the peak system efficiency.

13.11 Determine the energy density in langleys, if the strength of sunlight is 1 sun for a period of 15 min.

13.12 Consider a type of rooftop of a home measuring 15 m × 20 m that is all covered with PV cells to provide the electric energy requirement by that home. If the peak sun per day is 4 h and the efficiency of solar cell is 10%, find the average daily electric energy converted by the rooftop.

13.13 Consider the results of Problem 13.12, and assume that the cost of electric energy to the homeowner is $0.19/kWh that is changed by the utility company. If the cost of solar roof is about $5000, determine the following:
   a.   The amount of electric energy produced by the solar cells per year
   b.   The amount of savings on electric energy produced by the solar cells per year
   c.   The break-even period for the solar panels to pay for themselves in months

TABLE P17.1

Necessary Energy Data for the Cottage

Appliance	Current (A)	Time (h)	Battery Drain (A·A)	Power (W)	Energy (W·h)
Refrigerator	2	22	44	25	550
Miscellaneous (radio, TV, lighting, toaster, etc.)	5	4	20	75	300
Total			64	100	850

13.14 Consider a circular solar cell that has a 4 in. diameter that is rated 2.0 A at 0.45 V. If a certain application dictates 1.8 V and draws 0.5 A, how can the cell be modified to satisfy the requirements of such application?

13.15 A circular cell has a diameter of 3 in. If it has a rating at 25°C of 800 mA and 0.45 V in a full sun, determine its efficiency.

13.16 A circular cell has a diameter of 3 in. If it has a rating at 25°C of 1200 mA and 0.45 V in a full sun, determine its efficiency.

13.17 Assume that Smart owns a vacation cottage in Napa Valley. He is considering installing a solar system to meet the electricity needs at the cottage. Table P17.1 gives the energy demands at the cottage. Assume that there is five peak sun-hours per day at the location of the cottage.

All the electric equipment requires 12 V dc. Design the system and determine the following:

a. Design the solar system
b. The changing current of the battery for every hour of peak sun
c. The size of the battery that is needed
d. The peak power of the panels
e. If the cost is $1/peak watt, find the cost of panels alone

13.18 Consider a solar module rated 5.3 V, 38 W. If it is used to provide power to an application that needs 30 V and 25 A current, determine the following:

a. The total number of solar modules required
b. The type of connection that is needed in terms of series and parallel connected modules
c. The necessary load resistance value to get the rated power

13.19 Assume that a roof measures 12 m × 18 m and that the solar array under consideration is 10% efficient and that is tilted 30° toward the south and the yearly average peak sun-hours per day in Dallas, Texas, and Sacramento, California, is 4.7 peak sun-hours and 5 peak sun-hours, respectively. Determine the average daily electric energy converted by the rooftop array, if the house is located at

a. Dallas, Texas
b. Sacramento, California

REFERENCES

1. Manwell, J.F., J.G. McGowan, and A.L. Rogers: *Wind Energy Explained: Theory, Design, and Application*, Wiley, West Sussex, England, 2003.
2. Orloff, S. and A. Flannery: Wind turbine effects on avian activity, habitat use, and mortality in altamont pass and Solano County wind resource areas: 1989–1991, California Energy Commission Report, No. P700–92,002.
3. Stanon, C.: Wind farm visual impact and its assessment, *Wind Directions*, BWEA, August 1995, pp. 8–9.
4. Wizelius, T.: *Developing Wind Power Projects: Theory and Practice*, Earthscan, London, U.K., 2007.
5. Smith, J.C. et al.: A mighty wind, *IEEE Power Energy Mag.*, 7(2), March/April 2009, 41–57.
6. Fioravanti, R., V. Khoi, and W. Stadlin: Large-scale solutions, *IEEE Power Energy Mag.*, 7(4), July/August 2009, 48–57.

7. Grant, W. et al.: Change in the air, *IEEE Power Energy Mag.*, 7(6), November/December 2009, 36–46.
8. Key, T.: Finding a bright spot, *IEEE Power Energy Mag.*, May/June 2009, 34–44.
9. Kroposki, B., R. Margolis, and D. Ton: Harnessing the sun, *IEEE Power Energy Mag.*, May/June 2009, 22–33.
10. *Wind Energy: A Vision for Europe* in 2030, Report from TPWind Advisory Council, European Wind Energy technology Platform, 2007.
11. Wagner, H.J. and J. Mathur: *Introduction to Wind Energy Systems*, Springer, Berlin, Germany, 2009.
12. Stanon, C.: Wind farm visual impact and design of wind farms in the landscape, *Wind Energy Conversion*, 1994, BWEA, pp. 249–255.
13. Gönen, T.: *Electric Power Distribution System Engineering*, 2nd edn., CRC Press, Boca Raton, FL, 2008, pp. 47–64.

GENERAL REFERENCES

Ackermann, T. et al.: Where the wind blows, *IEEE Power Energy Mag.*, 7(5), November/December 2009, 65–75.

ANSI/IEEE P929, *IEEE* recommended practice for utility interface of residential and intermediate photovoltaic (PV) systems, IEEE Standards Coordinating Committee 21, Photovoltaics, Draft 10, February 1999.

Babic, J., Walling, R., O'Brien, K., and B. Kroposki: The sun also rises, *IEEE Power Energy Mag.*, May/June 2009, 45–54.

Buresch, M.: *Photovoltaic Energy Systems: Design and Installations*, McGraw-Hill, New York, 1953.

Denholm, P. and R.M. Margolis: Evaluating the limits of solar photovoltaics (PV) in electric power systems utilizing energy storage and other enabling technology, *Energy Policy*, 35(9), 2007a, 4424–4433.

Denholm, P. and R.M. Margolis: Evaluating the limits of solar photovoltaics (PV) in traditional electric power systems, *Energy Policy*, 35(5), 2007b, 3852–2861.

Jha, A.R.: *Solar Cell Technology and Applications*, CRC Press, Boca Raton, FL, 2010.

Key, T.: Finding a bright spot, *IEEE Power Energy Mag.*, May/June 3009, 34–44.

Komp, R.: *Practical Photovoltaics: Electricity from Solar Cells*, 3rd edn., Aatec Publications, Ann Arbor, MI, 2002.

Komp, R.J.: *Practical Photovoltaics-Electricity from Solar Cells*, 2nd edn., Aatce Publications, Ann Arbor, MI, 1984.

Manwell, J.F., J.G. McGowan, and A.L. Rogers: *Wind Energy Explained: Theory, Design, and Application*, Wiley, West Sussex, England, 2003.

Maycook, P.D. and E.N. Stirewalt: *Photovoltaics: Sunlight to Electricity in One Step*, Brick House Publishing Company, Andover, MA, 1981.

Mehos, M., Kabel, D., and P. Smithers: Planting the seed, *IEEE Power Energy Mag.*, May/June 2009, 55–62.

Nelson, J.: *The Physics of Solar Cells*, Imperial College Press, London, U.K., 2003.

Pagliaro, M., Palmisano, G., and R. Criminna: *Flexible Solar Cells*, Wiley Vch, Weinheim, Germany, 2008.

Stand-Alone Photovoltaic Systems: A Handbook of Recommended Design Practices, Sandia National Laboratories, Albuquerque, NM, 1996.

Stanon, C.: Wind farm visual impact and design of wind farms in the landscape, *Wind Energy Conversion*, 1994, BWEA, pp. 249–255.

Wind Energy: A Vision for Europe in 2030, Report from TPWind Advisory Council, European Wind Energy technology Platform, 2007.

Wizelius, T.: *Developing Wind Power Projects: Theory and Practice*, Earthscan, London, U.K., 2007.

Zweibel, K.: *Harnessing Solar Power*, Plenum Press, New York, 1990.

14 Energy Storage Systems for Electric Power Utility Systems

Nothing comes from nothing.
The darkness comes from darkness.
Pain comes from the darkness.
And we call it wisdom. It is pain!

Randell Jarrell

Power when wielded by abnormal energy is the most serious of facts.

Henry Brooks Adams

14.1 INTRODUCTION*

In the past, batteries of all sizes have been part of everyday applications for decades. For example, besides automobiles, battery rooms were also found in electric power plants and substations where reliable power was required for operations of switchgear, critical standby systems, and possibly black start of the station. Often batteries for large switchgear lineups were 125 or 250 V nominal systems and feature redundant battery chargers with independent power sources. Separate battery rooms might have been provided to protect against loss of the station due to a fire in a battery bank. For stations that were capable of black start, power from battery system might have been required for many purposes including switchgear operations.

However, for a long time, large utility-scale batteries have been slow to develop, due to technological limitations and the demands and costs of delivering utility-scale power, when necessary, for use on the grid. Recent developments indicate that utility-scale batteries finally reached a point of technological development in which they can be integrated into grid in select applications to ensure a constant power supply. This development will affect the entire alternative energy field, as large amounts of electricity generated by alternative means from solar to wind to any alternative power source can be stored in large quantities and used when necessary. This development requires huge wind installations or solar applications that could someday provide reliable electricity throughout the world. In summary, the use of intermittent or variable sources of energy, such as solar and wind energies, and some of the forms derived from moving water, often requires some means of energy storage.

Energy storage can not only potentially benefit solar energy systems as well as other renewable energy resources but also benefit the transmission and distribution systems because storage applications can be used to mitigate diurnal or other congestion patterns and, in effect, store energy until the transmission system is capable of delivering it where needed. By storing energy from variable resources, such as wind and solar power, energy storage could provide firm generation from these units, permit the energy produced to be used more efficiently, and provide supplementary transmission benefits.

Therefore, the adverse impacts of large-scale photovoltaic (PV) power generation systems connected to the power grid and developed output control technologies with integrated battery storage

---

* This chapter is the reprint of chapter 14 of Electrical Machines by T. Gönen, 2nd ed., CRC Press, 2012. Reprinted with the permission of the CRC Press.

are still under the study. The sodium–sulfur (Na–S) battery is designed to absorb fluctuations in the PV output within its limit of kW and kWh capacities. For more efficient and effective operation of the Na–S battery, several control algorithms of a battery system for smoothing PV output are being developed by the industry [1].

14.2 STORAGE SYSTEMS

At the present time, there is a great interest in the possible applications of energy storage in power systems. The interested parties include the electric utilities, energy service companies, and automobile manufacturers (for electric vehicle applications). For example, the ability to store large amounts of energy would permit electric utility companies to have greater flexibility in their operation because with this option, the supply and demand do not have to be matched instantaneously. Hopefully, the availability of the proper battery at the proper price will finally make the electric vehicle a reality.

The battery technologies are diverse and at different stages of development. They include a variety of batteries: high-speed flywheels, supercapacitors, and regenerative fuel cells. Local energy storage would assist interms of embedded generation from renewable energy by providing a buffer between the variability of supply and demand. Potential benefits include capacity reduction, frequency support, standing reserve provision, and cold start capability. Depending on technical requirements, and geographical settings, a given utility may avail one or more of these technologies.

Power applications, such as uninterruptible power supply backup for data centers and automotive starting batteries, represent the largest market for lead–acid batteries, whereas laptop batteries and power tools have caused incredible growth for lithium-ion. For bulk energy storage in utility grids, pumped hydropower plants dominate, with approximately 100 GW in service around the world.

Even though many utilities possess pumped storage plant, little focus has been placed on their proper potential roles in meeting power demand or shave off demand peaks, and this way partially decouple energy production from energy consumption. Energy storage can perform the same roles, but may also be used as a generation source, either replacing expensive, low-efficiency storage capability or load scheduling, and the generation capacity would be required to meet the average electrical demand only rather than the peak demand. Expensive network upgrades can be deferred.

By enabling thermal generating units to operate closer to rated capacity, higher thermal efficiencies are achieved, and both system fuel costs and CO_2 emissions are reduced. Even further benefits also come from reducing demand variability, and thus the requirement for load cycling of generating units and the requirements for additional regulating reserves. As a result, the balancing costs that may be associated with wind variability can be reduced. Also, expensive standing reserve, in the form of open-cycle gas turbines, diesel engines, etc., can also be reduced, since both energy storage and load management can provide a similar role. In general, power applications would be storage systems rated for 1 h or less, and energy applications would be for longer periods. Figure 14.1 presents a comparison of storage technologies in terms of power-level applications and storage time.

Today, power applications for each of these technologies are being found in electric grid. For example, in the transmission system for bulk power storage as well as in the residential feeder circuit for smaller systems. The following abbreviations are used in Figure 14.1: Li-ion for lithium-ion battery, NiCd for nickel–cadmium, NiMH for nickel–metal hydride battery, CAES for compressed air energy storage, SMES for superconducting magnetic energy storage, VRB for vanadium redox battery, ZnBr for zinc–bromine battery, Na–S for sodium–sulfur battery, Zebra battery for high-temperature battery (used at substations), and super caps for supercapacitors.

14.3 STORAGE DEVICES

The list for conventional technologies includes the large hydro, compressed air energy storage (CAES), and pumped hydro.

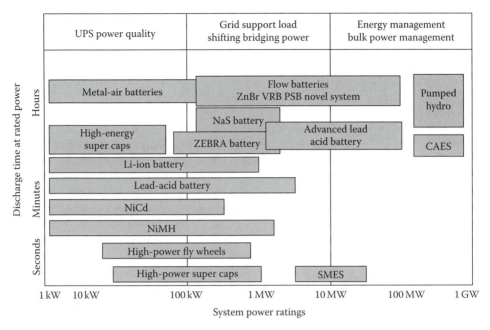

FIGURE 14.1 Comparison of storage technologies.

14.3.1 Large Hydro

It is an oldest renewable source of power/energy. Small hydro systems vary from 100 kW to 30 MW, while micro hydropower plants are smaller than 100 kW. Small hydropower generators work at variable speeds because the water upon which they depend flows at variable speeds. Induction generators are normally used with turbine system. The turbine converts the water's (kinetic) energy to mechanical rotational energy. The available power (P) from the water flow is expressed as

$$P_{avail} = Q \times H \tag{14.1}$$

where
Q is the discharged water in m^3/s
H is the net head in m

Hydroelectric plants typically have fast ramp-up and ramp-down rates, proving strong regulating capabilities, and their marginal generation cost is close to zero. In many countries, a natural synergy exists between hydroelectric generation/pumped storage and wind power. Clearly, if hydro generation is being replaced by wind energy, then emission levels will not be affected, but the hydro energy can be transformed into potential energy stored for later use. Existing hydroelectric plant can reduce the output, using reservoirs as storage, to avoid wind energy curtailment.

14.3.2 Compressed Air Storage

It involves the storage of compressed air in disused underground cavities, for example, exhausted salt mines. Alternatively, an underground storage complex can be created using a network of large diameter pipes. Later, the compressed air can be released as part of the generation cycle, providing a cycle efficiency of approximately 75%. In an open-cycle gas turbine or combined-cycle gas turbine plant, incoming air is compressed by the gas turbine compressor before being ignited with the incoming fuel supply. The exhaust gases are then expanded within the turbine, driving both an electrical generator and the compressor.

A modern CAES is a peaking gas turbine power plant that consumes less than 40% of the gas used in a combined-cycle gas turbine (and less than 60% gas is used by a single-cycle gas turbine) to produce the same amount of electric output power. It is accomplished by blending compressed air to the input fuel to the turbine by compressing air during peak periods at lower costs than conventional stand-alone gas turbines.

It is required that plants are near proper underground geological formations, such as salt caverns, mines, or depleted gas wells. The first commercial CAES plant was a 290-MW unit built in Handorf, Germany, in 1978. The second one was a 110-MW unit built in McIntosh, Alabama, in 1991. They are fast-acting units and typically can be put into service in 15 min when it is required. The Electric Power Research Institute has developed an advanced CAES system designed around a simpler way using advanced-turbine technology. The largest plant under consideration in the United States has a rating of 800 MW.

14.3.3 Pumped Hydro

The most widely established large-scale form of energy storage is hydroelectric pumped storage. It is an excellent energy storage technique, but unfortunately, few attractive sites exist and initial investment costs are very high. Typically, such plant operates on a diurnal basis—charging at night during periods of low demand (and low-priced energy) and discharging at day during times of high or peak demand. A pumped storage plant may have the capacity for 4–8 h of peak generation with 1–2 h of reserve, although in some cases, the discharge time can extend to a few days.

A typical pumped hydro plant consists of two interconnected reservoirs (lakes), tunnels that convey water from one reservoir to another, valves, hydro machinery (a water pump—turbine), a motor-generator, transformers, a transmission switchyard, and a transmission connection. The amount of stored electricity is proportional to the product of the total volume of water and the differential height between reservoirs. For example, storing 1000 MWh (deliverable in a system with an elevation change of 300 m) dictates a water volume of about 1.4 million m^3. The earliest application of pumped hydro technology was in Zurich, Switzerland, in 1882. It was realized early that a Francis turbine could also be used as a pump, such as the one used in the Hiwassee Dam Unit 2, in 1956, and has a rating of 59.5 MW.

Today, the global capacity of pumped hydro storage plants totals more than 95 GW, with approximately 20 GW operating in the United States. The original intent of these plants was to provide off-peak base loading for large coal and nuclear plants to optimize their overall performance and provide peaking energy each day. Since then their duties also include frequency regulation in the generation mode.

There are also less conventional technologies, including hydrogen, flywheels, high-power fuel cells, high-power supercapacitors, superconducting magnetic energy storage (SMES), heat or cold storage systems, and high-power batteries.

14.3.4 Hydrogen

Hydrogen has been proposed as the energy store (carrier) for the future and the basis for a new transport economy. The reasons for this are simple: hydrogen is the lightest chemical element, thus offering the best energy/mass ratio of any fuel and in a fuel cell can generate electricity efficiently and cleanly. Indeed, the waste product (water) can be electrolyzed to make more fuel (hydrogen).

Hydrogen can be transported conveniently over long distances using pipelines or tankers, so that generation and utilization take place in distinct locations, while a variety of storage forms are possible (gaseous, liquid, metal hydrating, etc.). It can be produced by the electrolysis of water using energy from a renewable resource. It can then be burned as a fuel to generate electricity.

Alternatively, it can be piped as a gas or liquid to consumers to be used locally providing both electricity and heating in a total energy scheme, or it can be used for transport. For transport needs, fuel cells in vehicles combine multi-fuel capability, high efficiency with zero (or low) exhaust emissions, and low noise.

The combustion of hydrogen provides energy plus water with no harmful emissions or by-products. If electricity is the final product, this process may not be attractive since the overall efficiency is usually below 50%. Because of this, the interest in hydrogen is usually for transportation purposes, which also depends on having proper storage systems.

In the future, hydrogen pipeline infrastructures are likely to be developed around the world. Excess hydrogen (i.e., energy) could be stored by temporarily increasing the gas pressure. Large wind farms could be used to power hydrogen-processing facilities, and pipelines (in lieu of large electric transmission lines) could carry bulk hydrogen, as the energy source, to major population centers.

Thus, hydrogen (similar to transporting and storing natural gas) would be stored as necessary to match the demand for fuel cells for electricity and hydrogen-powered cars. This scheme has the further benefit of reducing wind power variability, since the wind energy is not directly used for electrical generation. For distances greater than 1000 km, energy transportation by hydrogen carrier should be more economical than high-voltage electrical transmission. However, there is a question on the overall efficiencies of creating large quantities of hydrogen to power fuel cells to create electricity.

14.3.5 High-Power Flywheels

It is a kinetic-energy-storage device. In this method, energy is stored in very fast (approaching 75,000 rotations/min) rotating mass of flywheels. In the past, the flywheels had severe problems with maintenance, losses associated with bearings, material strength, and related severe failure management problems at high speeds.

Modern flywheels are made of fiber-reinforced composites. The flywheel motor/generator is interfaced to the main through a power electronic converter. At the present time, this technology is expensive and only used for select applications.

14.3.6 High-Power Flow Batteries

They operate similar to that of car batteries but without electrodes. Instead, when the flow cell is used as a "sink," the electric energy is converted into chemical energy by "charging" two liquid electrolyte solutions. The stored energy can be released on discharge. In common with all dc systems connected to the ac network, a bidirectional power electronic converter is needed.

Succinctly put, they use electrolyte liquids flowing through a cell stack with ion exchange through a microporous membrane to generate an electrical charge. Several different chemistries have been developed for use in utility power applications. Their advantage is their ability to scale systems independently in terms of power and energy. More cell stacks means increased power rating. Also, a greater volume of electrolytes means an increased runtime. Furthermore, flow batteries operate at ambient temperatures rather than high temperatures.

Zinc–bromine flow batteries are being used for utility applications. The battery operates with a solution of zinc bromide salt dissolved in water and stored in two tanks. The battery is charged or discharged by pumping the electrolytes through a reactor cell.

14.3.7 High-Power Supercapacitors

They are also called *ultracapacitors*. They consist of a pair of metal foil electrodes, each of which has an activated carbon material deposed on one side. These sides are separated by a proper membrane and then rolled into a package. Its operation is based on an electrostatic effect whereby charging and discharging take place with the totally physical (not chemical) reversible movement of ions.

Therefore, there are some fundamental differences between ultracapacitors and battery technologies including long shelf and operating life as well as large charge–discharge cycles of up to 500,000.

Supercapacitors are electrochemical capacitors. They look and perform similar to lithium-ion batteries. They store energy in the two series capacitors and the electric double layer that is formed between each of the electrodes and the electrolyte ions. The distance over which the charge separation takes place is just a few angstroms (a unit of length equal to 10^{-10} m). The extremely large surface area makes the capacitance and energy density of these devices thousands times larger than those of conventional electrolytic capacitors.

The electrodes are often made with porous carbon material. The electrolyte is either aqueous or organic. The aqueous capacitors have a lower energy density due to a lower cell voltage, but are less expensive and work in a wider temperature range. The asymmetrical capacitors that use metal for one of the electrodes have a significantly larger energy density than the symmetric ones do and also have a lower leakage current.

In comparison to lead–acid batteries, electrochemical capacitors have lower energy density, but they can be cycled hundreds of thousands of times and are much more powerful than batteries. They have fast charge and discharge capability. They have been applied for blade-pitch control devices for individual wind turbine generators to control the rate at which power increases and decreases with changes in wind velocity. This is highly necessary if wind turbines are connected to weak utility grids [5].

In California, Palmdale Water District uses a 450 kW supercapacitor to regulate the output of a 950 kW wind turbine attached to the treatment plant microgrid. This arrangement helps to reduce network congestion in the area, while providing reliable supply to critical loads in the microgrid.

14.3.8 Super Conducting Magnetic Energy Storage

As a result of recent developments in power electronics and superconductivity, the interest in using SMES units to store energy and/or damp power system oscillations has increased. It stores energy within a magnetic field created by the flow of direct current in a coil of superconducting material.

In a sense, SMES can be seen as a controllable current source whose magnitude and phase can be changed within one cycle. The upper limit of this source is imposed by the dc current in the superconducting coil. Typically, the coil is maintained in its superconducting state through immersion in liquid helium at 4.2 K within a vacuum-insulated cryostat. A power electronic converter interfaces the SMES to the grid and controls the energy flow bidirectionally. With the recent development of materials that exhibit superconductivity closer to room temperature, this technology may become economically viable [3].

Figure 14.2 shows a typical configuration of an SMES unit with a double gate-turn-off (GTO) thyristor bridge. In the configuration, the superconducting coil (L) is coupled to the transmission system via two converters and transformers. The converter firing angles, α_1 and α_2, are determined by the PQI controller in order to control the real and reactive power outputs and the dc current (I) in the coil.

The control strategy is determined by the modulation controller of SMES to damp out power swings in the network. The active and reactive power available from SMES depends on the type of ac/dc tool for transient stability enhancement and can be used to support primary frequency regulation [4].

14.3.9 Heat or Cold Storage

There has been a long tradition of using thermal storage to assist in power system operation, especially in the United Kingdom. This technology involves modulation of the energy absorbed by individual consumer electric heating elements and refrigeration systems for the benefit of overall system power balance. An aggregation of a large number of dynamically controlled loads has the potential of providing added frequency stability and smoothing to power networks, both at times of sudden increase in demand (or less of generation) and during times of fluctuating wind or other renewable power.

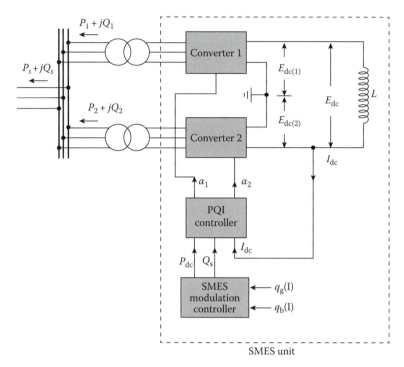

FIGURE 14.2 SMES unit with double GTO thyristor bridge. (From Gönen, T., *Electric Power Transmission System Engineering: Analysis and Design*, CRC Press, Boca Raton, FL, 2009. With permission.)

Such devices could displace some reserve and may cause a substantial reduction in the governor activity of remaining generators. The potential demand that could be operated under dynamic control is considerable. Deep-freeze units, industrial and commercial refrigeration, air–conditioning, as well as water heating systems could provide dynamic demand control. The potential available in a developed country could be several GW. This concept is not limited for small applications. In Europe, a very large thermal storage system (up to 10,000 MWh) is being proposed [5].

14.4 BATTERY TYPES

Battery systems are quiet and nonpolluting. They can be installed near load centers and existing suburban substations. Their efficiencies are in the range of 85% and can respond to load changes within 20 ms. Lead–acid batteries as large as 10 MW with 4 h of storage have been used in several U.S., European, and Japanese utilities.

Although the input and output energies of a battery are electrical, the storage is in chemical form. Chemical batteries are individual cells filed with a conducting medium—electrolyte that, when connected together, form a *battery*. Multiple batteries connected together form a *battery bank*. Essentially, there are two basic types of batteries: *primary battery* (nonrechargeable) and *secondary batteries* (rechargeable).

14.4.1 SECONDARY BATTERIES

Secondary batteries are rechargeable batteries. They are further divided into two categories based on the operating temperature of the electrolytes. *Ambient operating temperature batteries* have either *aqueous* (flooded) or *nonaqueous* electrolytes. *High operating temperature batteries* (molten electrodes) have either solid or molten electrolytes. Rechargeable lead–acid and nickel–cadmium batteries have been used widely by utilities for small-scale backup, load leveling, etc.

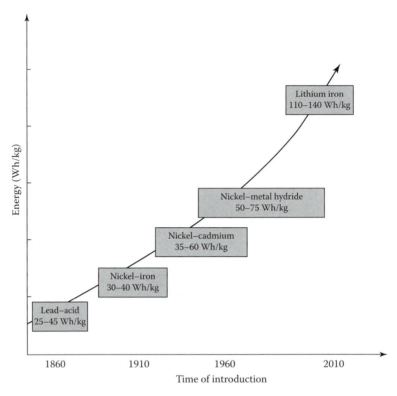

FIGURE 14.3 The trend of exponential improvement in battery performance.

The largest (nickel–cadmium) battery installation is a 45 MW, 10 MWh installation in Fairbanks, Alaska, built in 2003 and designed to provide a guaranteed 27 MW for at least 15 min following local power outages. For similar reasons, the largest (20 MW, 14 MWh) lead–acid system was installed by the Puerto Rico Electric Power Authority in 1994 and later repowered in 2004. But given the fairly toxic nature of materials involved, low efficiency (70%–80%), and the limited life and energy density, secondary batteries based on other designs are being sought for utility applications. Batteries in electric vehicles are the secondary rechargeable type and are in either of the two subcategories.

A battery for an electric vehicle has to satisfy certain performance goals that include quick discharge and charge capability, long-cycle life, low cost, recyclability, *high specific energy* (i.e., the amount of usable energy, measured in watt-hours per kilogram), *high energy density* (amount of energy stored per unit volume), *specific power* (defines the potential for acceleration), and the ability to work in extreme heat or cold.

However, at the present time, there is no battery that is available that meets all these criteria. Figure 14.3 shows the trend of exponential improvement in battery performance over the years. Today, a large variety of battery types are being used in electric power systems for grid support applications.

14.4.2 Sodium–Sulfur Batteries

This battery is a high-performance battery, with the electrolyte operating at temperatures of 572°F (300°C). It consists of a liquid (molten) sulfur positive electrode and a molten sodium negative electrode separated by a solid beta alumina ceramic electrode. The electrolyte permits only positive sodium ions to pass through it and combine with sulfur to form sodium polysulfides.

The sodium component of this battery explodes on contact with water, which raises certain safety questions. The materials of the battery have to be capable of withstanding the high internal temperatures they create, as well as freezing and thawing cycles. This battery has a very high specific energy of 110 Wh/kg. During discharge, positive sodium ions flow through the electrolyte and electrons flow in the external circuit of the battery, providing about 2 V. This process is reversible since charging causes sodium polysulfides to release the positive ions back through the electrolyte to recombine as elemental sodium. The Na–S battery cells are efficient (about 89%). This battery system is capable of 6 h of discharge time on a daily basis.

This technology for large-scale applications was perfected in Japan. Presently, there are 190 battery systems in service in Japan, totaling more than 270 MW of capacity with stored energy suitable for 6 h of daily peak shaving. The largest single Na–S battery installation is a 34-MW, 245-MWh system for wind power stabilization in northern Japan. The battery will permit the output of the 51-MW wind farm to be 100% dispatchable during on-peak periods.

According to Roberts [5], in the United States, utilities have applied 9 MW of Na–S batteries for peak shaving, backup power, firming wind capacity, and other applications. Zebra battery is another high-temperature battery and is based on sodium nickel chloride chemistry. It is used for electric transportation applications in Europe. Recently, it is being considered for utility applications as well [5].

14.4.3 FLOW BATTERY TECHNOLOGY

The performance of flow batteries is similar to a hydrogen fuel cell. They use electrolyte liquids flowing through a microporous membrane to generate an electrical charge. They store and release electrical energy through a reversible electrochemical reaction between two liquid electrolytes.

The liquids are separated by an ion-exchange membrane, allowing the electrolytes to flow into and out from the cell through separate manifolds and to be transformed electrochemically within the cell. For their utility applications, various chemistries have been developed. In standby mode, the batteries have a response time of the order of milliseconds to seconds, making them suitable for frequency and voltage support. One of the advantages of such flow battery design is the ability to scale systems independently in terms of power and energy. For example, more cell stacks permit for an increase in power rating, and a greater volume of electrolytes provides for more runtime. Plus, flow batteries operate at ambient (instead of high) temperature levels.

14.4.3.1 Zinc–Bromine Flow Battery

In utility applications, zinc–bromine flow batteries are being used. This battery operates with a solution of zinc bromide salt dissolved in water and stored in tow tanks. The battery is charged or discharged by pumping the electrolytes through a reactor cell.

During the charging cycle, metallic zinc from the electrolyte solution is plated onto the negative electrode surface of the reactor cell. The bromide is converted to bromine at the positive surface of the electrode in the reactor cell and then is stored in the other electrolyte tank as a safe chemically complex oily liquid. During the discharge of the battery, the process is reversed, and the metallic zinc plated on the negative electrode is dissolved in the electrolyte solution and available for the next charge cycle.

In order to create different system ratings and duration times, flow battery manufacturers use modular construction. For example, a zinc bromide flow battery package with a rating of 500 kW for 2 h. Other packages are being applied at utilities with ratings of up to 2.8 MWh packaged in a 53-ft trailer.

14.4.3.2 Vanadium Redox Flow Battery

Another type of flow battery is the vanadium redox battery (VRB). During its charge and discharge cycles, positive hydrogen ions are exchanged between the two electrolyte tanks through a hydrogen-ion permeable polymer membrane. Similar to the zinc–bromine battery, the VRB system's power and energy ratings are independent of each other [5].

14.4.4 LITHIUM-ION BATTERIES

Among the available battery technologies today, the lithium-ion battery has the greatest applications. It can be applicable in a large variety of shapes and sizes, permitting the battery to efficiently fill the available space, such as a cell phone or a laptop computer. They are also lighter in weight in comparison to other aqueous battery technologies, such as lead–acid batteries. They have the highest power density (110–140 Wh/kg) of all batteries on the commercial market on a per-unit-of-volume basis.

The leading lithium-ion cell design is a combination of lithiated nickel, cobalt, and aluminum oxides, referred to as an NCA cell. There are two lithium-ion designs that are starting to be employed in higher-power utility grid applications: lithium titanate and lithium iron phosphate.

14.4.4.1 Lithium–Titanate Batteries

This battery uses manganese in the cathodes and titanate in the anodes. This chemistry provides for a very stable design with fast charge capability and good performance at low temperatures.

The batteries can be at lower temperatures. They can be discharged to 0% and have a relatively long life. They are used in utility power ancillary service applications (e.g., frequency regulation).

14.4.4.2 Lithium Ion Phosphate Batteries

It is a newer and safer technology in which it is more difficult to release oxygen from the electrode, which reduces the risk of fire in the battery cells. It is more resistant to overcharge when operated in a range of up to 100% state of charge. They are also used in utility power ancillary service applications.

14.4.5 LEAD–ACID BATTERIES

They are the oldest and most mature among all the battery technologies. Because of their large applications, lead–acid batteries have the lowest cost of all battery technologies. This battery operates at an ambient temperature and has an aqueous electrolyte. Even though the lead–acid battery is relatively inexpensive, it is very heavy, with a limited usable energy by weight (specific energy).

A cousin of this battery is the deep-cycle lead–acid battery, now widely used in golf carts and forklifts. The first electric cars built also employed this technology. Lead–acid batteries should not be discharged by more than 80% of their rated capacity or depth of discharge. Exceeding the 80% of the depth discharge shortens the life of the battery. They are inexpensive, readily available, and are highly recyclable, using the elaborate recycling system already in place.

For utility application, a 40-MWh lead–acid battery was installed in the Southern California grid in 1988 to demonstrate the peak shaving capabilities of batteries in a grid application. The application of the battery demonstrated the value of stored energy in the grid; however, the limited cycling capability of lead–acid made the overall economics of the system unacceptable. However, for backup power sources in large power plants, lead–acid batteries are still used as "black start" sources in case of emergencies [5]. Their long life and lower costs make them ideal for applications with low-duty cycles.

Research continues to try to improve these batteries. For example, a lead–acid nonaqueous (gelled lead acid) battery uses an electrolyte paste instead of liquid. These batteries do not have to be mounted in an upright position. There is no electrolyte to spill in the accident. But nonaqueous lead–acid batteries typically do not have a high life cycle and are more expensive than flooded deep-cycle lead–acid batteries.

14.4.5.1 Advanced Lead–Acid Batteries

In order to significantly extend the life of lead–acid batteries, carbon is added to the negative electrode. As a result, their life is significantly extended in cycling applications. But lead–acid batteries fail due to sulfation in the negative plate that increases as they are cycled more.

Adding as much as 40% of activated carbon to the negative electrode composition increases the battery's life up to 2000 cycles. This represents a three-to-four times improvement over the current lead–acid designs. This extended life coupled with lower costs will lead storage developers to revisit lead–acid technology for grid applications.

14.4.6 NICKEL–CADMIUM BATTERIES

Nickel–iron (Edison cells) and nickel–cadmium (Ni–Cad) pocket and sintered plate batteries have been in use for many years. Both of these batteries have a specific energy of approximately 25 Wh/lb (55 Wh/kg), which is higher than advanced lead–acid batteries. Both are nontoxic, while Ni–cads are toxic. They can be discharged to 100% of depth of discharge without damage. The biggest obstacle to the utilization of these batteries is their cost. In the past, the Ni–Cad batteries represented a substantial increase in battery power. They are rugged, durable with good cycling capability and a broad discharge range.

In power systems, Ni–Cad batteries have been used in a variety of backup power applications and were chosen to provide "spinning reserve" for a transmission project in Alaska [5]. It involved a 26-MW Ni–Cad battery rated for 15 min, which represents the largest battery in a utility application in North America. Today, they are still being used for utility applications. For example, it is used for the power ramp rate control for smoothing with weak power grids (such as island power systems).

14.5 OPERATIONAL PROBLEMS IN BATTERY USAGE

The storage-battery-integrated PV system recovers the energy that would have been lost when voltage is over the limitation value. Since the risk of overvoltage is higher when the reverse power flow is greater, the state of charge of the storage battery should not be full at around noon. Thus, for efficient operation of the storage battery, only part of the surplus power that is greater than the load demand should be charged into the storage battery.

According to Hara et al. [1], the following problems must be considered when operating the storage battery:

1. The storage battery must be at a discharge state in the morning to prepare for charging around noon.
2. If the lead–acid battery is left in a discharge state, it may deteriorate and shorten the life time.
3. The frequency of use of the storage batteries may be varied by the impedance of the distribution line and by a power flow condition.
4. There are round-trip energy losses of the storage battery and power conditioning system increases when charging and discharging larger amounts of energy.

14.6 FUEL CELLS

They were first developed in 1839 and put to practical use in the 1960s by NASA to generate fuel for electricity needed by the spacecrafts Apollo and Gemini. The stored hydrogen can be converted back to electricity using an open-cycle gas turbine. However, in that case, electrical efficiency tends

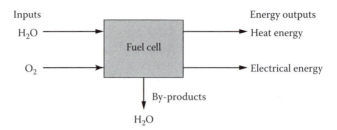

FIGURE 14.4 A block diagram of a fuel cell system.

to be low, even ignoring transportation losses and those associated with converting the electricity to hydrogen in the first place. Fuel cells are quiet, clean, and highly efficient on-site generators of electricity that use the electrochemical process to convert fuel into electricity. This is the reverse electrolysis. It has few moving parts and produces very little waste heat or gas. In addition to generating electricity, fuel cells can also serve as a thermal energy source for water and space heating or for cooling absorption.

Fuel cells offer an alternative approach and essentially consist of an electrolyte (liquid or solid) membrane sandwiched between two electrodes. A block diagram of a fuel cell is shown in Figure 14.4.

A single fuel cell produces output voltage less than 1 V. Thus, in order to produce higher voltages, fuel cells are stacked on top of each other and are serially connected forming a full cell system. Electrical efficiencies of fuel cells lie between 36% and 60%, according to the type and system configuration. By using conventional heat recovery apparatus, the overall efficiency can be improved to about 85%.

System reforming of liquid hydrocarbons (C_nH_m) is a potential way of providing hydrogen-rich fuel for fuel cells. This is a preferred method since storage of hydrogen is quite hazardous and expensive. Reformers facilitate a continuous supply of hydrogen without having to use bulky pressurized hydrogen tanks or hydrogen vehicles for distribution. The endothermic reaction that takes place in the reforming process in the presence of a catalyst is

$$C_nH_m + nH_2O \rightarrow nCO + \left(\frac{m}{2} + n\right)H_2 \tag{14.2}$$

and

$$CO + H_2O \rightarrow CO_2 + H_2 \tag{14.3}$$

Carbon monoxide combines steam to produce more hydrogen through the water gas shift reaction. Figure 14.5 shows the flows and reactions in a fuel cell.

Fuel cells are classified according to the nature of the electrolyte used and the operating temperature, with each type requiring particular materials and fuels. The electrochemical efficiency tends to increase with fuel cell temperature. It is often the nature of the membrane that dictates the operating temperature, and expensive catalysts, such as platinum, may be required to step up the rate of electrochemical reactions. Fuel cells can run using hydrogen, natural gas, methanol, coal, or gasoline.

In addition to this raw fuel of hydrogen, more environmentally friendly fuels, such as biogas, and biomass, may be used. For most fuel cells, such fuels must be transformed into hydrogen using a reformer or coal gasifier. However, high-temperature fuel cells can generally use a fossil fuel (natural gas, coal gas, etc.) directly. Polluting emissions are produced, but since hydrogen is passed over

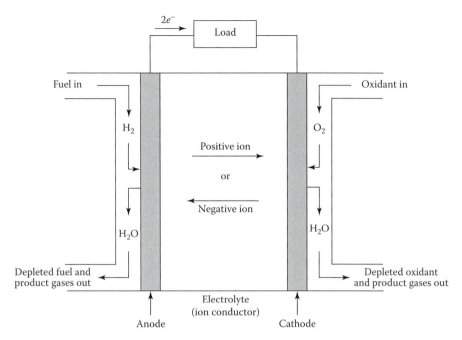

FIGURE 14.5 Flows and reactions in a fuel cell.

one electrode (anode), hydrogen molecules separate to the cathode where they combine with oxygen to form water. The oxygen supply may be derived from air or as a stored by-product from the water electrolysis (forming hydrogen).

For large-scale utility storage applications, the choice of technology will depend on the ability to use pure hydrogen (electrolyzed from water) as the fuel, the electrical efficiency of conversion, and the load-following capability of the fuel cell, thus providing a degree of regulation from fluctuating wind or other renewable sources. Of the various options available, SO and PEM seem most likely to succeed. The efficiency for conversion of fuel to electricity can be as high as 65%, which is nearly twice as efficient as conventional power plants. Also, small-scale fuel cell plants are just as efficient as the large ones, whether they operate at the full load or not. Because of their modular nature, they can be placed at or near load centers, resulting in savings of transmission network expansion.

A fuel cell power plant is essentially made of three subsystems or sections. In the fuel-processing section, the natural gas or other hydrocarbon fuel is converted to hydrogen-rich fuel. This process is known as a steam catalytic reforming process. This fuel is then fed to the power section, where it reacts with oxygen from the air in a large number of individual fuel cells to produce dc electricity and by-product heat in the form of usable steam or hot water. For a power plant, the number of fuel cells can vary from several hundred (for a 40-kW plant) to several thousand (for a multi-megawatt plant). In the third stage, the dc electricity is converted in the power conditioning subsystem to electric utility-grade ac electricity.

In the power section of fuel cell, which has the electrodes and the electrolyte, two separate electrochemical reactions happen: an oxidation half-reaction, taking place at the anode, and a reduction half-reaction occurring at the cathode. The anode and the cathode are separated from each other by the electrolyte. During the oxidation half-reaction at the anode, gaseous hydrogen produces hydrogen ions, which travel through the ionically conducting membrane to the cathode. At the same time, electrons travel through an external circuit to the cathode. In the reduction half-reaction at the cathode, oxygen supplied from air combines with the hydrogen ions and electrons to form water and excess heat. Hence, the fuel products of the overall reaction are electricity, water, and excess heat.

14.6.1 Types of Fuel Cells

Since the electrolyte defines the key properties, specifically the operating temperature, of the fuel cell, fuel cells are categorized based on their electrolyte type, as described later:

1. Polymer electrolyte membrane (PEM)
2. Alkaline fuel cell (AFC)
3. Phosphoric acid fuel cell (PAFC)
4. Molten carbonate fuel cell (MCFC)
5. Solid oxide fuel cell (SOFC)

These fuel cells operate at different temperatures, and each of them is best suited to specific applications. Table 14.1 gives a brief comparison of the five cell technologies introduced earlier.

14.6.1.1 Polymer Electrolyte Membrane

It is one of a family of fuel cells that are in various stages of development. The electrolyte in a PEM cell is a type of polymer and is usually referred to as a membrane, thus is the name. PEMs are somewhat unusual electrolytes, that is, in the presence of water, which the membrane readily absorbs, the negative ions are rigidly held within their structure. Only the positive (H) ions contained within the membrane are mobile and are free to carry positive charges through the membrane in one direction only, from anode to cathode. At the same time, the organic nature of the PEM structure makes it an electron insulator, forcing it to travel through the outside circuit providing electric power to the load. Each of the two electrodes is made of porous carbon to which

TABLE 14.1
Brief Comparison of Five Fuel Cell Technologies

Type	Electrolyte	Operating Temperature (°C)	Applications	Advantages
PEM	Solid organic polymer	60–90	Electric utility, transportation, portable power	H_2
	Metal oxide (Y_2O_3/ZrO_2)	700–1000		H_2, CH_4, biogas, etc. Solid electrolyte reduces corrosion, low temperature, quick start-up. Efficiency is 35%–55%
Direct alcohol	Polymer membrane/ liquid alkaline	60–120	Transportation, portable power	H_2, CH_4, biogas, coal gas etc. Its efficiency is 35%–40%
Alkaline (AFC)	Aqueous solution of potassium hydroxide soaked in a matrix	50–90	Military, space	Cathode reaction faster in alkaline electrolyte; thus high performance. It uses H_2 as fuel. Efficiency is 50%–60%
Phosphoric acid (PAFC)	Liquid phosphoric acid soaked in a matrix	150–220	Electric utility, transportation and heat	Its efficiency is 45%–55%. Up to 85% efficiency in regeneration of electricity
Molten carbonate (MCFC)	Liquid solution of lithium sodium, and/or potassium carbonates soaked in a matrix	600–750	Electric utility	Higher efficiency, fuel flexibility. inexpensive catalysts. It uses H_2, CH_4, biogas, coal gas, etc.
Solid oxide (SOFC)	Solid zirconium oxide to which a small amount of yttria is added	600–1000	Electric utility	Higher efficiency, fuel flexibility, inexpensive catalysts. Solid electrolyte advantage like PEM

very small platinum particles are bonded. The electrodes are slightly porous so that the gases can diffuse through them to reach the catalyst. Also as both platinum and carbon conduct electrons well, they are able to move freely through the electrodes [7]. Chemical reactions that take place inside a PEM fuel cell are the following:

At anode

$$2H_2 \rightarrow 4H^+ + 4e^- \tag{14.4}$$

At cathode

$$O_2 + 4H^+ + 4e^- \rightarrow H_2O \tag{14.5}$$

Net reaction

$$2H_2 + O_2 = 2H_2O \tag{14.6}$$

Here, hydrogen gas diffuses through the polymer electrolyte until it meets a platinum particle in the anode. The platinum catalyzes dissociation of the hydrogen molecule into two hydrogen atoms (H) bonded to two neighboring platinum atoms. Only then can each H atom release an electron to form a hydrogen ion (H+), which travels to the same time as the free electron through the other circuit. At the cathode, the oxygen molecule interacts with the hydrogen ion and the electron from the outside circuit to form water. The performance of the PEM fuel cell is limited mainly by the slow rate of the oxygen reduction half-reaction at the cathode, which is 100 times slower than the hydrogen oxidation half-reaction at the anode [7].

14.6.1.2 Phosphoric Acid Fuel Cell

This technology has moved from the laboratory R&D to the first stages of the commercial application. Today, 200-kW plants are available and have been built at more than 70 sites in the United States, Japan, and Europe. Operating at approximately 200°C, the PAFC plant also produces heat for domestic hot water and space heating, and its electrical efficiency is close to 40%. Its high cost is the only thing that stops it from its wide commercial acceptance. At the present time, capital costs of PAFC plant is about \$2500–\$4000/kW. According to Rahman [7], if it is reduced down to \$1000–\$1500/kW, this technology may be accepted by the power industry. The chemical reactions that take place at two electrodes are as follows:

At anode

$$2H_2 \rightarrow 4H^+ + 4e^- \tag{14.7}$$

At cathode

$$O_2 + 4H^+ + 4e^- \rightarrow 2H_2O \tag{14.8}$$

14.6.1.3 Molten Carbonate Fuel Cell

This technology is attractive because it offers several potential advantages over PAFC. Carbon monoxide, which positions the PAFC, is indirectly used as a fuel in the MCFC. The higher operating temperature of about 650°C makes the MCFC a better candidate for combined cycle applications whereby the fuel cell exhaust can be used as input to the intake of a gas turbine or the boiler of a steam turbine. The total efficiency can approach 85%. It is just about to enter the commercial market. Capital costs involved are expected to be lower than PAFC. MCFCs are now being tested in full-scale demonstration plants [7]. The chemical reactions that take place inside the cell are the following:

At anode

$$2H_2 + 2CO_3^2 \rightarrow 2H_2O + 2CO_2 + 4e^-$$ (14.9)

and

$$2CO + 2CO_3^2 \rightarrow 4CO_2 + 4e^-$$ (14.10)

At cathode

$$O_2 + 2CO_2 + 4e^- \rightarrow 2O_3^2$$ (14.11)

14.6.1.4 Solid Oxide Fuel Cell

According to Rahman [7], an SOFC is currently being demonstrated at a 100-kW plant. This technology dictates very significant changes in the structure of the cell. It uses a solid electrolyte, a ceramic material, so the electrolyte does not need to be replenished during the operational life of the cell.

The results of this are simplification in design, operation, and maintenance, as well as having the potential to reduce costs. This offers the potential to reduce costs. This offers the stability and reliability of all solid-state construction and permits higher-temperature operation.

The ceramic makeup of the cell lends itself to cost-effective fabrication techniques. Its tolerance to impure fuel streams makes SOFC systems especially attractive for utilizing H_2 and CO from natural gas steam-reforming and coal gasification plants [7]. The chemical reactions that take place inside the cell are as follows:

At anode

$$2H_2 + 2O^{2-} \rightarrow 2H_2O + 4e^-$$ (14.12)

and

$$2CO + 2O^{2-} \rightarrow 2CO_2 + 4e^-$$ (14.13)

At cathode

$$O_2 + 4e^- \rightarrow 2O^{2-}$$ (14.14)

REFERENCES

1. Hara, R. et al.: Testing the technologies, *IEEE Power & Energy Magazine*, May/June 2009, 77–85.
2. Gönen, T.: *Electric Power Transmission System Engineering: Analysis and Design*, CRC Press, Boca Raton, FL, 2009.
3. Gönen, T.: High-temperature superconductors a technical article in *McGraw-Hill Encyclopedia of Science & Technology*, 7th edn., Vol. 7, 1992, pp. 127–129.
4. Gönen, T, P. M. Anderson, and D. Bowen: Energy and the future, *Proceedings of the 1st World Hydrogen Energy Conference*, 3(2c), 1977, 55–78.

5. Roberts, B.: Capital grid power, *IEEE Power & Energy Magazine*, July/August 2009, 32–41.
6. Jasinski, R.: *High-Energy Batteries*, Plenium Press, New York, 1967.
7. Rahman, S.: Advanced energy technologies in *Electric Power Generation, Transmission, and Distribution*, L.L. Grigsby, ed., CRC Press, Boca Raton, FL, 2007.
8. Béguin, F. and E. Frackowiak: *Carbons for Electrochemical Energy Storage and Conversion Systems*, CRC Press, Boca Raton, FL, 2010.
9. Barak, M.: *Electrochemical Power Sources*, Peter Peregrinus Ltd., Stevenage, U.K., 1980.
10. Zimmerman, A. H.: *Nickel-Hydrogen Batteries*, The Aerospace Press, Reston, VA, 2009.
11. Schalkwijk, W. A. and B. Scrosati: *Advances in Lithium-Ion Batteries*, Kluwer Academic/Plenum Publishers, New York, 2002.
12. Kiehne, H. A.: *Battery Technology Handbook*, Marcel Dekker, Inc., New York, 1987.
13. Sutton, G. W.: *Direct Energy Conversion*, McGraw-Hill Book Company, New York, 1966.
14. Gasik, M.: *Materials for Fuel Cells*, CRC Press, Boca Raton, FL, 2008.
15. Rajalakshmi, N. and K. S. Dhathathreyan: *Present Trends in Fuel Cell Technology Development*, Nova Science Publishers, Inc., New York, 2008.
16. Barclay, F.: *Fuel Cells, Engines and Hydrogen*, Wiley, New York, 2006.
17. Ozawa, K.: *Lithium Ion Rechargeable Batteries*, Wiley-VCH, Weinhein, Germany, 2009.
18. Soo, S. L.: *Direct Energy Conversion*, Prentice Hall, Englewoods Cliffs, NJ, 1968.

15 Concept of Smart Grid and Its Applications

To dream the impossible dream,
To reach the unreachable star.

Joe Darion, in *The Impossible Dream*

An active field of science is like an immense anthills; the individual almost vanishes into the mass of minds tumbling over each other, carrying information from place to place, passing it around at the speed of light.

Lewis Thomas, in *Natural Science*

15.1 BASIC DEFINITIONS

Grid: The transmission system. It is the interconnected group of power lines and associated equipment for moving electric energy of high voltage between points of supply and points at which it is delivered to other electric systems or transformed to a lower voltage for delivery to customers.

Smart grid: The modernization of the grid by installing intelligent electronic devices in terms of sensor electronic switches, smart meters, and also advanced communication, and data acquisition and interactive software with real-time control that optimize the operation of the whole electric system and make more efficient utilization of the grid assets. It is such grid that is called the *smart grid*.

Legacy power systems: The presently existing power systems that have no smart grid applications yet.

Intelligent electronic device (IED): Any device incorporating one or more processors with the capability to receive or send data and control from or to an external source (e.g., electronic meters, digital relays, and controllers).

IED integration: Integration of protection, control, and data acquisition function into a minimal number of platforms to reduce capital and operating costs, reduce panel and control room space, and eliminate redundant equipment and databases.

International electrotechnical commission (IEC): An international organization whose mission is to prepare and publish standards for all electric, electronic, and related technologies.

Remote terminal unit (RTU): The entire complement of devices, functional modules, and assemblies that are electrically interconnected to affect the remote station supervisory functions. The equipment includes the interface with the communication channel but does not include the interconnecting channel.

Conventional RTU: Designed primarily for hardwired input/output (I/O) and has little or no capability to talk to downstream IEDs.

Remote access: Access to a control system or IED by a user whose operation terminal is not directly connected to the control systems or IED. Transport mechanisms typical of remote access include dial-up modem, frame relay, ISDN, Internet, and wireless technologies.

Protocol: A formal set of conventions governing the formal and relative timing of message exchange between two communication terminals; a strict procedure required to initiate and maintain communication. A communication protocol allows communication between two devices. The devices must have the same protocol (and version) implemented, otherwise the differences will result in communication errors. The substation integration and automation architecture must allow devices from different suppliers to communicate (interoperate) using an industry standard protocol.

Substation automation (SA): Deployment of substation and feeder operating functions and applications ranging from supervisory control and data acquisition (SCADA) and alarm processing to integrated volt/var control (IVVC) in order to optimize the management of capital assets and enhance operation and maintenance efficiencies with minimal human intervention.

Data concentrator: Designed primarily for IED integration and may also have limited capability for hardwired I/O.

Operational data: Also called SCADA, data and are instantaneous values of power system analog and status points (e.g., amps, volts, MW, Mvar, circuit breaker status, switch position). The operational data is conveyed to the SCADA master station at the scan rate of the SCADA system using the SCADA system's communication protocol, for example, DNP3.

Nonoperational data: Consists of files and waveforms (e.g., event summaries, oscillographic event reports, or sequential event records) in addition to SADA-like points (e.g., status and analog points) that have logical state or numerical value.

Home area network (*HAN*): The HAN stands for "home area network"; it is used to identify the network of communicating loads, appliances, and sensors beyond the smart meter and within the customer's property.

Security: The protection of computer hardware and software from accidental or malicious access, use, modification, destruction, or disclosure.

National Institute of Standards and Technology (*NIST*): Under the federal law of Energy Independence and Security Act of 2007, the National Institute of Standards and Technology (NIST) has been given the key role of coordinating development of a framework for smart grid standards.

Distributed network protocol (*DNP3*): A non-propriety communication protocol that is designed to optimize the transmission of data acquisition information and control commands from one computer to another.

CIGRE: An International Conference on Large High-Voltage Electric Systems. It is recognized as a permanent nongovernmental and nonprofit-making international association based in France. It focuses on issues related to the planning and operation of power systems, as well as the design, construction, maintenance, and disposal of high-voltage equipment and plants.

Cyber security: Security from threats conveyed by computer or computer terminals; also, the protection of other physical assets from modification or damage from accidental or malicious misuse of computer-based control facilities.

Intrusion detection system (*IDS*): A device that monitors the traffic on a communication line with the aim of detecting and reporting unauthorized users of the facilities. It is programmed to identify and track specific patterns of activities.

Port: A communication pathway into or out of a computer or networked device such as a server. Ports are often numbered and associated with specific application programs. Well-known applications have standard port numbers; for example, port 80 is used for HTTP traffic (web traffic).

Demand response (DR): *It defines the consumers' behavioral change to the changing rates.* It enables demand-side resources and improves the economic operation of electric power markets by aligning prices more closely with the value customers place on electric power.

Demand response management (DRM): It is similar to the old term known as "load management." It can provide competitive pressure to reduce wholesale power prices, increase awareness of energy usage, provide for more efficient operation of markets, mitigate market power, enhance reliability, and, in combination with certain new technologies, support the use of renewable energy resources, distributed generation, and advanced metering.

Distributed energy resources (DER): It is a phrase that refers to the new energy resources that are connected into a utility distribution network.

Plug-in hybrid electric vehicles (PHEVs): The vehicles that are designed and built to operate using either gas or electricity, by charging their batteries at night by plugging them into receptacles.

Wide area network (WAN): WAN is used to identify the network of upstream utility assets, including power plants, distribution storage, and substations. The interface between WAN and LAN and LAN and HAN is provided by smart meters.

Wide area monitoring/measurement system (WAMS): A monitoring/measurement system that is dedicated to monitor a WAN. It is based on a low latency networking.

Local area network (LAN): A local area network that facilitates for the enabled IEDs can be directly connected to the substation automation LAN. It is used to identify the network of integrated smart meters, field components, and gateways that constitute the logical network between distribution substations and customer's premises. The non-LAN-enabled IEDs require a network interface module (NIM) for protocol and physical interface conversion. A substation LAN is typically high speed and extends into the switchyard, which speeds the transfer of measurements, indications, control commands, and configuration and historical data between intelligent devices at the site.

Neighborhood area networks (NAN): The physical connections to the meter nodes change from WAN to NAN technologies. It is achieved by introducing one or more routers at the borders of the NAN that is connected to WAN, enabling bidirectional data streams between WAN and NAN.

Short-circuit analysis (SCA): It is used to calculate the short-circuit current to evaluate the possible impact of a fault on the network.

Active network: It is a passive network that has been converted to an active one by the connection of distributed generation in terms of, for example, CHP cogens and/or other renewable-based energy-producing units such as wind turbines or solar units.

Distribution management system (DMS): A system that uses voltage regulators or transformers with load tap changers (LTCs) to automatically raise or lower the voltage in response to changes in load. It also uses capacitor banks to supply some of the reactive power that would otherwise be drawn from the supply substations.

Substation automation (SA): Deployment of substation and feeder operating functions and applications ranging from SCADA and alarm processing to IVVC in order to optimize the management of capital assets and enhance operation and maintenance efficiencies with minimal human intervention.

Open systems: A computer system that embodies supplier-independent standards so that software can be applied on many different platforms and can interoperate with other applications on local and remote systems.

Data processing applications: The software applications that provide various users access to the data of the substation controller in order to provide instructions and programming to the substation controller, collect data from the substation controller, and perform the necessary functions.

Data collection applications: The software applications that provide the access to other systems and components that have data elements necessary for the substation controller to perform its functions.

Control database: All data reside in a single location, whether from a data processing application, data collection application, or derived from the substation controller itself.

Distributed network protocol (*DNP*): A comprehensive protocol to achieve open, standard-based interoperability between substation computers, RTUs, IEDs, and master stations (with the exception of inter-master communications) for the electric utility industry. The present version of DNP is version 3, that is, DNP3. Level 1 has the least functionality (used for simple IEDs), and level 3 has the most functionality (used for SCADA master station-communications front-end processors). Its advantages include the following: interoperability between multi-supplier devices; fewer protocols to support in the field; not needing protocol translators; reduced software costs; shorter delivery schedules; less need for testing, maintenance, and training; improved documentation; support for independent user group and third-party sources; easier system expansion; faster adoption of new technology; longer product life; and huge operational savings.

Distributed generation (*DG*): It refers to small generators that are connected to the distribution system network. Such generation is known as *distributed generation* (DG) or *dispersed generation*, or *embedded generation*.

Automated metering infrastructure (*AMI*): The network of communications between the substation and substation regulators, switches, capacitor banks, distributed regulators, real-time metering, alternative points of supplies, as well as active networks, if they exist.

15.2 INTRODUCTION

Today, there are many who do not like the usage of the term *smart grid*, but would have preferred to use *smarter grid*, because they feel that the existing grid is already plenty smart. What is needed is a "more efficient or advanced grid" in terms of the usage of advanced communication technology and information technology (IT) and other advanced technologies, and improved efficiencies. Arguably, because of that, it would have been the best to call the new perceived grid the "intelligent grid." Alas, the name is already coined for it. So, there it is.

Today's (legacy) electric grid is a one-way flow of electricity and according to the NIST [3]:

- Responsible for 405 of human-caused CO_2 production
- Centralized, bulk generation, mainly coal and natural gas
- Has controllable generation and predictable loads
- Limited automation and situational awareness
- Lots of customized proprietary systems
- Lack of customer-side data to manage and reduce energy uses

In that sense, the smart grid can simply be defined as "the grid with brain." It is a modernized grid that enables bidirectional flows of energy and uses two-way communication and control capabilities that will lead to an array of new functionalities and applications. The smart grid will permit the two-way flow of both electricity and information.

The move toward the smart grid is fueled by a number of needs. For example, there is the need for improved grid reliability while dealing with an aging infrastructure, and there is the need for environmental compliance and energy conservation. Also, there is the need for improved operational

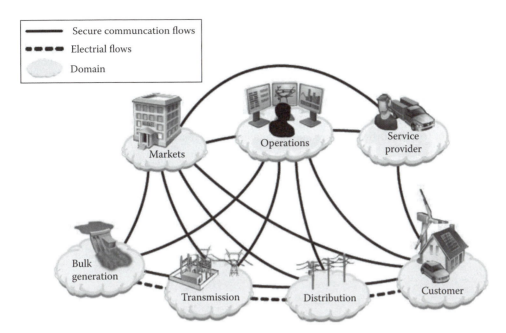

FIGURE 15.1   The conceptual representation of the smart grid network framework of NIST. (From *U.S. National Institute of Standards and Technology*: NIST framework and roadmap for smart grid interoperability standards, Release 1.0, September 2009, Washington, DC.)

efficiencies and customer service. Figure 15.1 shows the conceptual representation of the smart grid network framework of NIST. It represents the fundamental domains of the electric power system, namely, bulk power generation, transmission, distribution, customer operations, operation provider, customer, and markets. It also shows the electric flows and secure communication flows within the system.

Under the federal law of Energy Independence and Security Act of 2007, the National Institute of Standards and Technology (NIST) has been given the key role of coordinating development of a framework for smart grid standards. Thus, NIST's national coordinator for smart grid interoperability launched a three-phase plan to jump-start development and promote widespread adoption of smart grid interoperability standards [3]:

- Engage stakeholders in a participatory public process to identify applicable standards, gaps in currently available standards, and priorities for new standardization activities
- Establish a formal private–public partnership to drive longer-term progress
- Develop and implement a framework for testing and certification

Standardized architectural concepts, data models, and protocols are essential to achieve interoperability, reliability, security, and evolvability. New measurement methods and models are needed to sense, control, and optimize the grid's new operational paradigm. Today, the industry can benefit from similar large-scale experience-developing architecture and protocols for modernization of the telecom network and the Internet.

Furthermore, with an increase in regulating influence and the focus on smart grid advanced technologies, there is a renewed interest in increasing the investment in distribution networks to defer infrastructure build-out and to reduce operating and maintenance costs through improving grid efficiency, network reliability, and asset management (AM) programs. Thus, since the roots of power system issues are usually found in the electric distribution system, the point of departure for the grid overhaul is the distribution system [3].

The electric power grid is now focusing on a large number of technological innovations. Today, utility companies around the world are incorporating new technologies in their various operations and infrastructures. At the bottom of this transformation is the requirement to make more efficient use of present assets. Several utility companies have developed their own vision of future smart distribution systems to reach the smart grid objectives. Figure 15.2 shows such representation in terms of a smart grid tree. Notice that the trunk of the tree is the AM, which is the base of smart grid development. Based on this foundation, the utility company builds its smart grid system by a careful overhaul of their IT infrastructure, communication, and network infrastructure. Well-designed

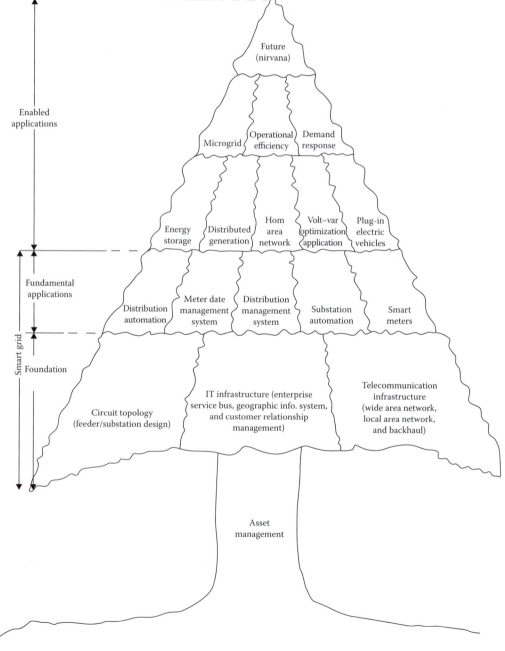

FIGURE 15.2   The representation of a smart grid as a tree.

layer of intelligence over utility assets that enables the emergence of the smart grid capabilities will be built on vertical integration of the upper-layer applications. For example, an important capability such as microgrid may not be possible without the integration of distributed generation (DG) and home area networks (HANs).

Thus, the emergence of the truly smart grid will require a drastic overhaul of the existing system. It will require the establishment of distributed control and monitoring systems within and alongside the present (legacy) electric power grid. Most likely, this change will be gradual but continuous.

The conventional methods of load management (LM) and load estimation (LE) in the traditional distribution management system (DMS) are no longer effective, causing other DMS applications ineffective or altogether useless. However, the impact of demand response management (DRM) and consumer behaviors may be mandated and predicted, from the utility pricing rules and rewarding policies for specific time periods.

As said before, most of the utilities believe that the biggest return on investment will be investing in distribution automation (DA) that will provide them with fast increasing capability over time. Thus, "blind" and manual operations, along with electromechanical components in the electric distribution grid, will need to be transformed into a "smart grid."

Such transformation is necessary to meet environmental targets, to accommodate a greater emphasis on demand response (DR), and to support plug-in hybrid electric vehicles (PHEVs) as well as DG and storage capabilities. Also, as succinctly put by Gellings [4], the attributions of the good smart grid are

1. Absolute reliability of supply
2. Optimal use of bulk power generation and storage in combination with distributed resources and controllable/dispatchable consumer loads to assure lowest cost
3. Minimal environmental impact of electricity production and delivery
4. Reduction in electricity used in the generation of electricity and an increase in the efficiency of the power delivery system and in the efficiency and effectiveness of end use
5. Resiliency of supply and delivery from physical and cyber-attacks and major natural phenomena (e.g., hurricanes, earthquakes, and tsunamis)
6. Assuring optimal power quality for all consumers who require it
7. Monitoring of all critical components of the power system to enable automated maintenance and outage prevention

Furthermore, the recommended renewable portfolio standard (RPS) mechanism generally places an obligation on the utility companies to provide a minimum percentage of their electricity from approved renewable energy sources. According to the US Environmental Protection Agency, as of August 2008, 32 states plus the District of Columbia had established RPS targets.

Together, these states represent for almost half of the electricity sales in the United States. The RPS targets presently range from a low 2% to a high 25% of electricity generation, with California leading the pact that requires 20% of the energy supply coming from renewable resources by 2010 and 33% by 2020. RPS noncompliance penalties imposed by states range from $10 to $25 per MWh.

On the average, a typical household in the United States uses 920 kWh of electricity per month with appliances accounting for 64.7% of electricity consumption.

The people who run the grid are generator owners and transmission owners. Also, from the system point of view, they are the independent system organizations (ISOs and RTOs). They monitor system loads and voltage profiles, operate transmission facilities and direct generation, define operating limits and develop contingency plans, and implement emergency procedures. Also, reliability coordinators play essential role. The North American Reliability Cooperation (NERC) develops and enforces reliability standards; monitors the bulk power systems; assesses future adequacy; audits owners, operators, and uses for preparedness; and educates and trains industry personnel.

Many states regulatory commissions have initiated proceedings or adopted policies for the implementation of automated metering infrastructure (AMI) to enable distributed resources. In this ruling on October 17, 2008, the Federal Energy Regulatory Commission (FERC) established a policy aimed at eliminating barriers to the participation of DR in the organized power markets (independent service operators [ISOs] and regional transmission organizations [RTOs]) by ensuring the comparable treatment of resources. In this ruling,

According to McDonald [5], FERC states that "the demand response can provide competitive pressure to reduce wholesale power prices: increases awareness of energy usage; provides for more efficient operation of markets; mitigates market power; enhances reliability; and in combination with certain new technologies, can support the use of renewable energy resources, DG, and advanced metering. Thus, enabling demand-side resources, improves the economic operation of electric power markets by aligning prices more closely with the value customers place on electric power."

Among other things, the order directs RTOs and ISOs to accept bids from distributed resources (DR) for energy and ancillary services, eliminate penalties for taking less energy than scheduled, and permit aggregators to bid DR on behalf of retail customers.

It is well known that the reliable supply of electric power plays a critical role in the economy. The new operating strategies for environmental compliance, together with our aging transmission and distribution infrastructure, constitute a great challenge to the security, reliability, and quality of the electric power supply.

When implemented throughout the system, intermittent energy resources, such as wind, will greatly stress transmission grid operation. The distribution grid will be also stressed with the introduction and, perhaps, rapid adaptation of on-site solar generation as well as PHEVs and plug-in electric vehicles (PEVs). Such plug-in vehicles could considerably increase the circuit loading if the charging times and schedules are not properly managed and controlled.

Major upgrades to distribution system infrastructure may flow patterns due to the integration of the DG and microgrids. Therefore, the existing power delivery infrastructure can be substantially improved through automation and information management.

As succinctly put by Farhangi [2], the convergence of communication technology and IT, with power system engineering helped by a number of new approaches, technologies, and applications, permits the existing grid to penetrate condition monitoring and AM, especially on the distribution system.

It is obvious that smart distribution system applications are at the core of the energy delivery systems between the transmission system and customers; all smart distribution applications target three main objectives:

1. Improving distribution network performance
2. Improving distribution system energy efficiency by increasing distribution network transit and capacity
3. Empowering the customer by providing choices

The smart distribution applications will have to integrate power system technologies such as DA, volt/var control (VVC), advanced distribution line monitoring with telecommunications, and data management technologies. The primary purpose of VVC is to maintain acceptable voltage at all points along the distribution feeder under all loading conditions. The purpose of using voltage regulators or transformers with *load tap changers* (LTCs) is to automatically raise or lower the voltage in response to changes in load.

It is often that there is a need that requires the use of *capacitor banks* to supply some reactive power that would otherwise be drawn from the supply substations. As penetration of intermittent renewable resource-based generating units increase in the future, high-speed dynamic load/var control will play a significant role in *maintaining power quality* and *voltage stability* on the distribution feeders.

Hence, as said before, electric utility companies believe that investing in DA will provide them with increasing capabilities over time. Thus, the first step in the evolution of the smart grid starts at the distribution side, enabling new applications and operational efficiencies to be introduced into the system.

15.3 NEED FOR ESTABLISHMENT OF SMART GRID

The electric power delivery system has often been referred to as the greatest and most complex machine ever built. It is made of wires, electric machines, electric towers, transformers, and circuit breakers—all put together in some fashion. The existing electricity grid is unidirectional in nature. Its overall conversion efficiency is very low, about 33% or 34%. In the United States, there are 140 million customers of electricity. They can be divided into three categories: *residential* (with 122 million customers and 37% of electricity sales), *commercial* (with 17 million customers and 35% of electricity sales), and *industrial* (with less than 1 million customers and 28% of electricity sales).

That means it converts only one-third of fuel energy into electricity, without recovering even its waste heat. Approximately 8% of its output is lost along its transmission lines, while 20% of its generation capacity exists just to meet its peak demand (i.e., it is being used for only 5% of the time). Furthermore, the existing electric system is vulnerable for domino-effect failures due to hierarchical topology of its assets.

The electric grid of today is one-way flow of electricity, as shown in Figure 15.3. According to NIST, its properties can be categorized as follows:

- Centralized, bulk generators, mainly coal and natural gas
- Responsible for 40% of human-caused CO_2 production
- Controllable generation and predictable loads
- Limited automation and situational awareness
- Lots of customized proprietary systems
- Lack of customer-side data to manage and reduce energy losses

The electric power grid infrastructure of the United States was built more than 50 years ago. It has become a complex spider web of power lines and aging networks and systems with obsolescent technology and outdated communications. In addition, there is a growing demand for lower carbon emissions, renewable energy sources, and improved system reliability and security. Figure 15.4 shows 2007 electric generation by source.

Increasing efficiency is a key priority. In that process, the following facts have to be taken into account [3]:

- Half of US coal plants are greater than 40 years old.
- Average substation transformer age is greater than 40 years.

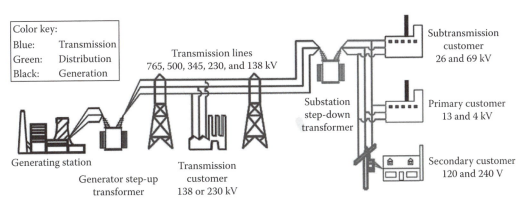

FIGURE 15.3 Legacy systems (today's electric grid). (From *U.S. National Institute of Standards and Technology*: NIST framework and roadmap for smart grid interoperability standards, Release 1.0, September 2009, Washington, DC.)

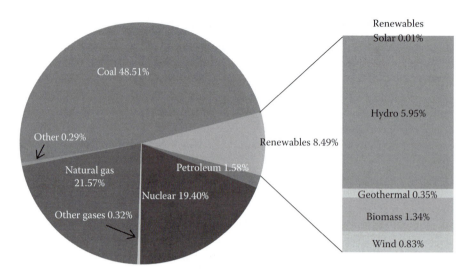

FIGURE 15.4 2007 electric generation by source. (From *U.S. National Institute of Standards and Technology*: NIST framework and roadmap for smart grid interoperability standards, Release 1.0, September 2009, Washington, DC.)

- Projected investment and expansion expenditure is predicted to be between $1.5 and $2 trillion by 2030.
- The US per capita annual electricity consumption is about 13,000 kWh that is largest among the industrialized countries. For example, Japan's per capita usage is 7,900 kWh.
- Smart grid helps utilities reduce delivery losses and average consumption and therefore reduce investment otherwise would be required.

The smart grid integrates IT and advanced communications into the power system in order to

- Increase system efficiency and cost effectiveness
- Provide customer tools to manage energy use
- Improve reliability, resiliency, and power quality
- Enable use of innovative technologies, including renewables, storage, and electric vehicles

All these requirements dictate the modernization of the grid by installing intelligent electronic devices (IEDs) in terms of sensor electronic switches, smart meters, and also advanced communication, and data acquisition and interactive software with real-time control that optimize the operation of the whole electric system and make more efficient utilization of the grid assets. It is such grid that is called the *smart grid*.

It is envisioned that such smart grid would integrate the renewable energy sources, especially wind and solar, with conventional power plants in a coordinated and intelligent way that would not only improve reliability and service continuity but would effectively reduce energy consumption and significantly reduce the carbon emissions. Based on a July 2009 Smart Grid Report of the United States Department of Energy [11], a smart grid has to have the following functions:

1. *Optimize asset utilization and operating efficiency*: The smart grid optimizes the utilization of the existing and new assets, improves load factors, and lowers system losses in order to maximize the operational efficiency and reduce the cost. Advanced sensing and robust communications will allow early problem detection, preventive maintenance, and correction action.
2. *Provide the power quality for the range of needs*: The smart grid will enable utilities to balance load sensitivities with power quality, and consumers will have the option of

purchasing varying grades of power quality at different prices. Also, irregularities caused by certain consumer loads will be buffered to prevent propagation.

3. *Accommodate all generation and storage options*: The smart grid will integrate all types of electric generation and storage systems, including small-scale power plants that serve their loads, known as *distributed generation*, with a simplified interconnection process analogous to "*plug-and-play.*"

4. *Enable informed participation by customers*: The smart grid will give consumers information, control, and options that enable them to become active participants in the grid. *Well-informed* customers will modify consumption based on balancing their demands and resources with the electric system's capability to meet those demands.

5. *Enable new products, services, and markets*: The smart grid will enable market participation, allowing buyers and sellers to bid on their energy resources through the supply and demand interactions of markets and real-time price quotes.

6. *Operate resiliently to disturbances, attacks, and natural disasters*: The smart grid operates resiliently, that is, it has the ability to withstand and recover from disturbances in a self-healing manner to prevent or mitigate power outages, and to maintain reliability, stability, and service continuity. The smart grid will operate resiliently against attack and natural disaster. It incorporates new technology and higher cyber security, covering the entire electric system, reducing physical and cyber vulnerabilities, and enabling a rapid recovery from disruptions.

Therefore, the next-generation electricity grid, known as the "smart grid" or "intelligent grid," is designed to address the major shortcomings of the existing grid. Basically, the smart grid is required to provide the electric power utility industry with full visibility and penetrative control and monitoring over its assets and services. It is required to be self-healing and resilient to system abnormalities.

Furthermore, the smart grid needs to provide an improved platform for utility companies to engage with each other and do energy transactions across the system. The smart grids are expected to provide tremendous operational benefits to power utilities around the world because they provide a platform for enterprise-wide solutions that deliver far-reaching benefits to both utilities and their end customers. In 2009, "smart grid interoperability panel" (SGIP) was created. It has 1900 people, 750 member organizations; it is open to the public as well, with international participation. It coordinates standards developed by standards development organizations (SDOs). It identifies requirements and prioritizes standards development programs.

However, the development of the smart grid is not easy due to many reasons. First of all, the present power grid of the United States is a very large, fragmented, and complex system. For example, 22% of the world electricity is consumed by the United States. Accordingly, this country has 3,200 electric utility companies, 17,000 power plants, 800 GW peak demand, 165,000 miles of high-voltage lines, and 6 million miles of distribution lines; covers 140 million meters; has 41 trillion in assets; and creates $350 billion annual revenues.

Today, the utilities and their suppliers are already integrating IT and advanced communications into the power system in order to

- Increase system efficiency and cost effectiveness
- Provide customers tools to manage energy use
- Improve reliability, resiliency, and power quality
- Enable use of innovative technologies including renewables, storage, and electric vehicles

The development of new technologies and applications in distribution management can help drive optimization of smart grid and assets. Hence, the smart grid is the result of convergence of communication technology and communication technology with power system engineering.

The following are the NIST SG research and development (R&D) vision [3]:

- The smart grid is a complex system of systems that incorporates many new technologies and operating paradigms in an end-to-end system that functions very differently than the legacy grid in order to deliver power more efficiently, reliably, and cleanly.
- NIST develops the measurement science and standards, including interoperability and cyber-security standards, necessary to ensure that the performance of the smart grid—at the system, subsystem, and end-user levels—can be measured, controlled, and optimized to meet performance requirements, especially for safety and security, reliability and resilience, agility and stability, and energy efficiency.

Table 15.1 provides a comparison of features of the smart power grid with the existing power grid. In summary, a smart grid is the use of sensors, communications, computational ability, and control in some form to enhance the overall functionality of the electric power delivery system. In other words, a dumb system becomes smart by sensing, communicating, applying intelligence, exercising control, and, through feedback, continually adjusting. This allows several functions that permit optimization, in combination, of the use of bulk generation, and storage, transmission, distribution, distributed resources, and consumer end uses toward goals that ensure reliability and optimize or minimize the use of energy, mitigate environmental impact, manage assets, and minimize costs. In other words, the philosophy of the smart grid is a brand new way of looking at the electric power delivery system and its operation to achieve the optimality and maximum efficiency and effectiveness. Presently, it is hoped and expected that a smart grid will provide the following:

1. Higher penetration of renewable resources.
2. Extensive and effective communication overlay from generation to consumers.
3. The use of advanced sensors and high-speed control to make the grid more robust.
4. It will provide higher operating efficiency.
5. It will provide a greater resiliency against attack and natural disasters.
6. It will provide effective automated metering and rapid service restoration after storms.
7. It will facilitate real-time or time-of-use pricing of the electric energy.
8. It will provide greater customer participation in generation and selling of the energy generated by using renewable resources, such as the ones shown in Figures 15.5 through 15.9. Figures show various solar and wind applications as examples of DG.

TABLE 15.1

Comparison of the Features of the Smart Grid with the Existing Grid

Smart Grid	Existing Grid
Digital	Electromechanical
Two-way communication	One-way communication
DG	Centralized generation
Network	Hierarchical
Sensors throughout	Few sensors
Self-monitoring	Blind
Self-healing	Manual restoration
Adaptive and islanding	Failures and blackouts
Old-fashion customer metering	Intelligent customer metering
Remote checking/testing	Manual checking/testing
Pervasive control	Limited control
Many customer choices	Few customer choices

FIGURE 15.5   Solar and wind applications in the city of Kassel in the state of Hessen, Germany. (SMA Solar Technology AG.)

FIGURE 15.6   Solar and wind turbine applications in the state of Rheinland-Pfalz in Germany. (SMA Solar Technology AG.)

FIGURE 15.7   Solar installations in Germany. (SMA Solar Technology AG.)

FIGURE 15.8   Solar applications on the roof of Munich Temple in Germany. (SMA Solar Technology AG.)

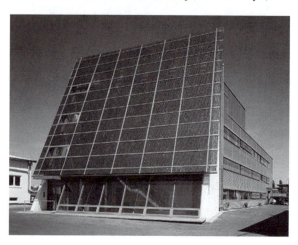

FIGURE 15.9   Solar application in the city of Kassel in the state of Hessen in Germany. (SMA Solar Technology AG.)

In summary, a smart grid is an electricity network that can intelligently integrate the actions of all users connected to it, that is, generators, consumers, and those that do both, in order to efficiently deliver sustainable, economic, and secure electricity supplies.

A smart grid uses innovative products and services together with intelligent monitoring, control, communication, and self-healing technologies in order to

1. Better facilitate the connection and operation of generators of all sizes and technologies
2. Permit consumers to play a part in optimizing the operation of the system
3. Provide consumers with greater information and choice of supply
4. Deliver increased levels of reliability and security of supply
5. Significantly reduce the environmental impact of the whole electricity supply system

15.4 SMART GRID APPLICATIONS VERSUS BUSINESS OBJECTIVES

As in any successful organization, the smart distribution applications of a successful utility company have to match its business objectives. Otherwise, the result becomes a catastrophic failure.

Example 15.1

Consider the typical business objective of *improving distribution network performance* of a utility company, and match it with the following specific *smart distribution applications*:

    a. Reliability
    b. Power quality

Solution

    1. The smart grid applications *for improving distribution system reliability* are as follows:
        A. Fault detection at distribution feeder devices
        B. DA
            a. Automatic reconfiguration of feeders
            b. Remote controlling of feeders, switches, and breakers
        C. Advanced distribution line monitoring
            a. Accurate fault location based on wave-shape analysis
    2. The smart grid applications *for improving power quality* are
        A. Power quality measurements
        B. Advanced distribution monitoring

Example 15.2

Consider the typical business objective of *increasing distribution network transit and capacity* (i.e., *distribution system energy efficiency*) of a utility company, and match it with the typical and specific *smart distribution applications*.

Solution

The smart grid applications for *increasing distribution network transit and capacity* are as follows:

    1. Distributed resource integration
        A. Dispersed generation and distributed energy resources (DERs)
        B. Storage
    2. DR

Example 15.3

Consider the typical business objective of *empowering the customer* of a utility company, and match it with the specific *smart distribution applications*.

Solution

The smart grid applications for the typical business objective of *empowering the customer* of a utility company are as follows:

    1. Integrated voltage and var control (IVVC)
        A. Voltage control from sensors on the distribution system
        B. Control of capacitors
    2. Advanced distribution line monitoring and control
        A. Feeder protection settings

15.5 ROOTS OF THE MOTIVATION FOR THE SMART GRID

The electric power grid is now focusing on a large number of technological innovations. Utility companies around the world are incorporating new technologies in their various operations and infrastructures.

At the bottom of this transformation is the requirement to make more efficient use of present assets. Several utility companies have developed their own vision of future smart distribution systems to reach the smart grid objectives. Table 15.2 provides a list of enabled applications. Table 15.3 provides a list of smart grid applications.

Based on this foundation given in these tables, the utility company builds its smart grid system by a careful overhaul of their IT infrastructure and communication, and network infrastructure well-designed layer of intelligence over utility assets enables the emergence of the smart grid capabilities to be built on vertical integration of the upper-layer applications. For example, an important capability such as microgrid may not be possible without the integration of DG and HANs.

TABLE 15.2
Enabled Applications

- Microgrid
- Operational efficiency
- DR
- Energy storage
- DG
- HAN
- VVO application
- Plug-in electric vehicles

TABLE 15.3
Smart Grid Applications

1. Fundamental Applications of Smart Grid
   A. DA
   B. Meter data management system
   C. DMS
   D. SA
   E. Smart meters
2. Smart Grid Foundation
   A. Circuit topology
      a. Feeder design
      b. Substation design
   B. IT infrastructure
      a. Enterprise service bus
      b. GIS
      c. Customer relationship management
   C. Telecommunication infrastructure
      a. WAN
      b. LAN
      c. Backhaul

Thus, the emergence of the truly smart grid will require a drastic overhaul of the existing system. It will require the establishment of distributed control and monitoring systems within and alongside of the present electric power grid. Most likely this change will be gradual but continuous.

Such smart grids will facilitate the DG and cogeneration of energy as well as the integration of alternative sources of energy and the management and control of a power system's emissions and carbon footprint. Furthermore, they will help the utilities to make more efficient use of their present assets through a peak shaving DR and service quality control [2].

Here, the dilemma of a utility company is how to establish the smart grid to achieve the highest possible rate of return on the needed investments for such fundamental overhauls, the new architecture, protocols, and standards toward the smart grid.

However, as supply constraints continue, there will be more focus on the operational effectiveness of the distribution network in terms of cost reduction and capacity relief. Monitoring and control requirements for the distribution system will increase, and the integrated smart grid architecture will benefit from data exchange between the DMS and other project applications. The appearance of widespread distribution generation and consumer-demand response programs also introduces considerable impact to the DMS operation.

It is important to point out that smart grid technologies will add a tremendous amount of real-time and operational data with the increase in sensors and the need for more information on the operations of the system. In addition, utility customers will be able to generate and deliver electricity to the grid or consume the electricity from the grid based on predetermined rules and schedules. Using Ethernet TCP/IP sensors, transducers, and communication protocol, as illustrated in Figure 15.10, customers, through the use of smart meters, can control loads, such as washers and dryers, space heaters, air conditioners, electric stoves, refrigerators, and hot water tanks.

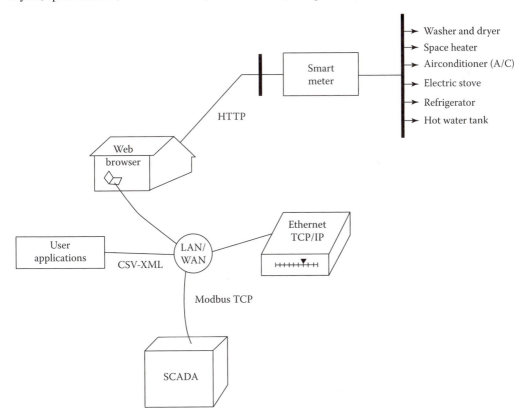

FIGURE 15.10 Application of Ethernet TCP/IP sensors, transducers, and communication protocol for load control.

Thus, the roles of the consumers have changed; they are no longer only buyers but are sellers and/or buyers, switching back and forth from time to time. This results in two-way power flows in the grid and the need for monitoring and controlling the generation and consumption points on the distribution network.

As a result of these changes, the distribution generation will be from dissimilar sources and subject to great uncertainty. At the same time, the electricity consumption of the individual customers is also subject to a great uncertainty when they respond to the real-time pricing and rewarding policies of power utilities for economic benefits.

The conventional methods of LM and LE in the traditional DMS are no longer effective, causing other DMS applications ineffective or altogether useless.

However, the impact of DRM and consumer behaviors may be mandated and predicted, from the utility pricing rules and rewarding policies for specified time periods. The fundamental benefits of automation applied to a distribution system are as follows:

1. Released capacity
2. Reduced losses
3. Increased service reliability
4. Extension of the lives of equipment
5. Effective utilization of assets

The drivers for advanced DA include the following:

1. Worldwide energy consumption is increasing due to population growth and increased energy use per capita in most developing countries.
2. Increased emphasis on system efficiency, reliability, and quality.
3. The need to serve increasing amounts of sensitive loads.
4. Need to do more with less capital expenditure.
5. Performance-based rates.
6. Increasing focus on renewable energy due to the increasing costs to extract and utilize the fossil fuels in an environmentally benign manner is becoming increasingly expensive.
7. Availability of real-time analysis tools for faster decision making.
8. Worldwide energy consumption.

In general, DA and control functions include the following [1]:

1. Discretionary load switching
2. Peak load pricing
3. Load shedding
4. Cold load pickup
5. Load reconfiguration
6. Voltage regulation
7. Transformer load management (TLM)
8. Feeder load management (FLM)
9. Capacitor control
10. Dispersed storage and generation
11. Fault detection, control, and isolation
12. Load studies
13. Condition and state monitoring
14. Automatic customer meter reading
15. Remote service connection or disconnection
16. Switching operations

15.6 DISTRIBUTION AUTOMATION

In a typical DA application, the included tasks are shown in Figure 15.11, which includes medium voltage (MV) regulator automation, MV post-type monitoring stations, MV switch automation, and MV recloser automation. However, in a smart grid development, there might be additional tasks that can be automated at the distribution level. For example, IEEE definition of the *DA* is "a system that enables an electric utility to remotely monitor, coordinate, and operate distribution companies in real-time mode from remote locations." The automation also includes the distribution system monitoring and AMI. Figure 15.12 shows the tasks involved in the distribution system monitoring. Figure 15.13 shows the tasks of an AMI. The figure illustrates the relationships between the communication network called the AMI and the components of a substation (i.e., substation regulators,

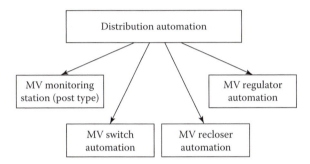

FIGURE 15.11 Tasks involved in the distribution level automation (at the MV level).

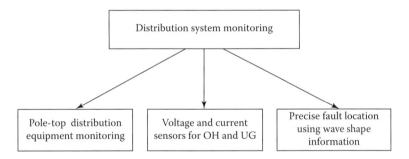

FIGURE 15.12 The tasks involved in the distribution system monitoring.

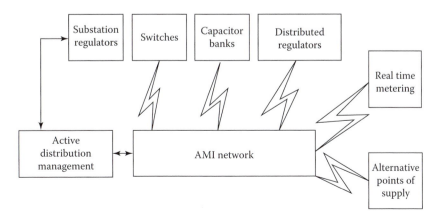

FIGURE 15.13 The tasks of an AMI.

switches, capacitor banks, distributed regulators, real-time metering, alternative points of supplies, as well as active networks [if they exists]) that have to be established.

For example, various distribution management system (MIS) applications that are commonly used today include the following:

1. *Fault detection, isolation, and service restoration (FDIR)*: It is designed to improve system reliability. It detects a fault on a feeder section based on the remote measurements from the feeder terminal units (FTUs), rapidly isolates the faulted feeder section, and then restores service to the *unfaulted* feeder sections.
2. *The topology processor (TP)*: It is a background, offline processor that accurately determines the distribution network topology and connectivity for display colorization and to provide accurate network data for other DMS control applications.
3. *Optimal network reconfiguration (ONR)*: It is a module that recommends switching operations to reconfigure the distribution network to minimize network energy losses, maintain optimum voltage profile, and balance the loading conditions among the substation transformers, the distribution feeders, and the network phases.
4. *IVVC*: It has three basic objectives: reducing feeder network losses by energizing or de-energizing the feeder capacitor banks, ensuring that an optimum voltage profile is maintained along the feeder during normal operating conditions, and reducing peak load through feeder voltage reduction by controlling the transformer tap positions in substations and voltage regulators on feeder sections, as illustrated in Figure 15.14.
5. *Switch order management (SOM)*: It is very useful for system operators in real-time operation. It provides advanced analysis and execution features to better manage all switch operations in the system.
6. *Dynamic load modeling/load estimation (LM/LE)*: It is the base module in DMS. It uses all the available information from the distribution network to accurately estimate individual loads and aggregate bulk loads.
7. *The dispatcher training simulator (DTS)*: It is used to simulate the effects of normal and abnormal operating conditions and switching scenarios before they are applied to the real system.

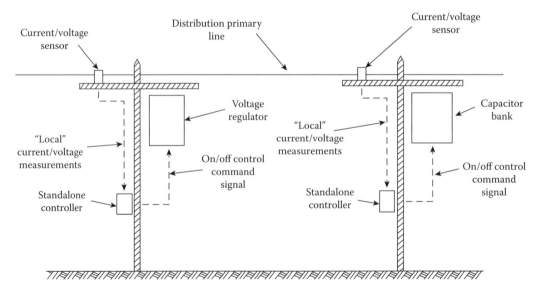

FIGURE 15.14 An illustration for how to control volt/var flows related to a distribution primary line using the individual, independent, stand-alone volt/var regulating equipment under the *traditional* VVC approach.

8. *Short-circuit analysis* (*SCA*): It is used to calculate the short-circuit current to evaluate the possible impact of a fault on the network.

9. *Relay protection coordination* (RPC): It manages and verifies the relay settings of the distribution feeders under various operating conditions and network configurations.

10. *Optimal capacitor placement/optimal voltage regulator placement* (OCP/OVP): It is used to determine the optimal locations for capacitor banks and voltage regulators in the distribution networks for the most effective control of the feeder volt/var profile.

15.7 ACTIVE DISTRIBUTION NETWORKS

An *active network* is a passive network that has been converted to an active one by the connection of DG in terms of, for example, combined heat and power (CHP) cogens and/or other renewable-based energy-producing units such as wind turbines or solar units, as shown in Figures 15.15 through 15.17.

The necessary monitoring, communications, and control in terms of both preventive and corrective actions are provided for such networks. They are flexible, adaptable (most likely autonomous

FIGURE 15.15 Solar applications in Bruchweg stadium- FSV Mainz 05, Germany. (SMA Solar Technology AG.)

FIGURE 15.16 Solar applications in the state of Crevillente in Spain. (SMA Solar Technology AG.)

FIGURE 15.17 Solar applications in Montalto di Castro in Italy. (SMA Solar Technology AG.)

such as the case with microgrids), and most likely intelligent. Hence, *active distribution networks* are the distribution networks to which renewable energy sources are connected. An active distribution network can be considered as an active network. Microgrids are *autonomous active networks*. The active network management is receiving considerable attention in the development of smart grid.

Also, in active distribution networks, the older style voltage regulators were often designed for hardly a pure radial situation, that is, power flow is always from the same direction (from the substation). They may not work correctly if power flow is from the opposite direction. For example, they could raise voltage during light load, creating higher voltage situation, or they could lower voltage during heavy load, creating low voltage situation.

Therefore, it is necessary to use *bidirectional* voltage regulator controller to handle feeder reconfiguration. Feeder reconfiguration may become a more frequent occurrence due to load transferred to another feeder during service restoration or due to having an ONR to reduce losses. In general, a distribution generation of sufficient size can reverse power flow.

To this day, there is no generally accepted definition of active network management yet. However, there is an acceptable definition of active network. By that definition, an active network includes the DG, renewables, monitoring, communications and control, and preventive and corrective actions; it is defined as flexible, adoptable, autonomous, and intelligent. Active network management includes dispatch, network reconfiguration, dynamic constraints, fault level management, demand-side management (DSM), and active voltage control.

15.8 INTEGRATION OF SMART GRID WITH DISTRIBUTION MANAGEMENT SYSTEM

Figure 15.18 shows the integration of the existing DMS with smart grid. The model involves planning and forecasting, outage management system (OMS), collection of outage data, AM, involving geographic information system (GIS), AMI, meter distribution management system (MDMS), customer information systems (CIS), as well as distribution system model. It also shows the DMS in terms of DA, the supervisory control and data acquisition (SCADA), VVC, efficiency, DSM, DG, and PHEV.

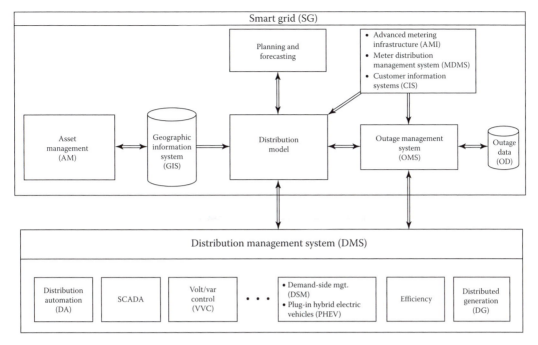

FIGURE 15.18 Integration of the existing DMS with smart grid.

15.9 VOLT/VAR CONTROL IN DISTRIBUTION NETWORKS

In general, it is agreed that the following are the three main approaches to VVC in the distribution systems:

1. Traditional approach
2. SCADA volt/var approach
3. IVVC optimization approach

15.9.1 TRADITIONAL APPROACH TO VOLT/VAR CONTROL IN THE DISTRIBUTION NETWORKS

In the traditional approach to VVC in the distribution networks, the process is controlled by using individual, independent, and stand-alone volt/var regulating equipment such as substation transformer's LTC, by using line voltage regulators, and by using fixed and switched capacitor banks. Figure 15.14 shows an illustration for how to control volt/var flows related to a distribution primary line using the individual, independent, stand-alone volt/var regulating equipment under the *traditional* VVC approach.

Note that VVC is a fundamental operating requirement of all electric distribution systems. Its primary purpose is to maintain acceptable voltage at all points along the distribution feeder under all loading conditions, that is, under the full-load or light-load conditions.

Example 15.4

Conceptually illustrate how to manage a typical distribution primary line (i.e., feeder) volt/var flows by using the traditional VVC approach. Use the individual, independent, stand-alone volt/var regulating equipment such as substation transformer's LTC; line voltage regulators; and fixed and switched capacitor banks.

Solution

Figure 15.14 illustrates how to manage volt/var flows related to a distribution primary line using the individual, independent, stand-alone volt/var regulating equipment under the *traditional VVC approach*. Note that the current/voltage sensors located on the pole tops send the "local" current/voltage measurements to stand-alone controllers.

They, in turn, send out individual on/off control/command signal to the associated capacitor bank and/or voltage regulators, after using the substation transformer's LTCs. The process continues until appropriate voltage profile for the feeder is obtained. However, such traditional approach has the following limitations:

1. *Power factor correction/loss reduction*: Many traditional capacitor bank controllers have voltage control, that is, they switch on when voltage is low. This approach is good at maintaining acceptable voltage and has reactive power controllers, but it is expensive since it requires the addition of CTs. Also, the approach is good at power factor correction during peak-load times, but it may not come on at all during off-peak times. As a result, power factor is nearly unity during the peak-load times, but it is low during the off-peak times, causing higher electric losses.
2. *Monitoring of switched capacitor bank performance*: It is well known that switch capacitor banks are often out of service due to blown out fuses, etc. With the traditional approach, the switched capacitor bank could be out of service for extended periods without the operator knowing it. This results in higher losses due to the capacitor bank being out of service. As a result, it requires routine inspections that are costly.
3. *Voltage regulation problem when large DG unit is connected*: As a result of connecting a large DG unit, load current through voltage regulator will be reduced since the voltage regulator *thinks* the load is light on the feeder. Thus, the voltage regulator lowers its tap settings in order to avoid *light-load, high-voltage* condition. This, in turn, makes the actual *heavy-load, low-voltage* condition even worse.

These problems can be remedied by implementing DMS. Such system uses voltage regulators or transformers with LTCs that automatically raise or lower the voltage in response to changes in load. Also, capacitor banks are used to supply some of the reactive power that would otherwise be drawn from the supply substations.

However, today, utilities are seeking to do more with VVC than just keeping voltage within the allowable limits. System optimization is an important part of the normal operating strategy under smart grid.

Especially in the future, as penetration of intermittent renewable generating resources increases, high-speed dynamic VVC will be essential in sustaining power quality and voltage stability on the distribution feeders. But, the traditional approach has the previously mentioned limitations.

15.9.2 SCADA Approach to Control Volt/VAR in the Distribution Networks

Volt/var power apparatus is monitored and controlled by SCADA system. Such VVC is typically handled by two separate (independent) systems, that is, by var dispatch system that controls capacitor banks to improve power factor, reduce electric losses, etc., or by *voltage control* system that controls LTCs and/or voltage regulators to reduce demand and/or energy consumption, which is also known as *conservation voltage reduction** (CVR).

Operation of these systems is primarily *based on a stored set of predetermined rules*. For example, *if power factor is less than 0.95, then switch capacitor bank #1 off*. The overall objective of

---

* CVR is the practice of calibrating substation voltage regulating equipment so that system voltages are maintained. The US utilities average CVR factor is about 0.8. Here, CVR factor = $\Delta P/\Delta V$. The CVR performed during peak load period can be viewed as demand (capacity) reduction. The resultant annual energy savings due to CVR can be found from

$$\text{Annual energy savings} = \begin{pmatrix} \text{Average} \\ \text{load} \end{pmatrix}\begin{pmatrix} \text{Number of} \\ \text{hours per year} \end{pmatrix}\begin{pmatrix} \% \text{ voltage} \\ \text{reduction} \end{pmatrix}\begin{pmatrix} \text{CVR} \\ \text{factor} \end{pmatrix}\begin{pmatrix} \text{Dollar value} \\ \text{from kWh sales} \end{pmatrix} - \begin{pmatrix} \text{Loss of revenue} \\ \text{from kWh sales} \end{pmatrix}$$

var dispatch is to maintain power factor as close as to unity at the beginning of the feeder without causing leading factor.

The *objectives of SCADA voltage control* are the following:

1. Maintain acceptable voltage at all locations under all loading conditions
2. Operate at a low voltage as possible to reduce power consumption through CVR

Example 15.5

Using the SCADA volt/var approach, conceptually illustrate how to manage a typical distribution primary line volt/var flows using var dispatch components. Use switched and fixed capacitor banks, capacitor bank control interface, communication facility, means of monitoring three-phase var flow at the substation, and master station running var dispatch software.

Solution

Figure 15.19 shows var dispatch components. Note that var dispatch processor contains rules for capacitor switching. There is a one-way communication link for capacitor bank control between the var dispatch processor and the capacitor bank controller. (Note that at the present time, the capacitor bank is de-energized by having the capacitor switch at off position.)

Substation remote terminal unit (RTU) measures the real and reactive power at substation end of the feeder; accordingly var dispatch processor sends commands to the capacitor bank controller. VAR dispatch processor applies the rules to determine if the capacitor bank switching is needed. If the reactive power is below the threshold, there is no need for any action, as illustrated in Figure 15.20. (*Note that there is no communication between the radios.*)

On the other hand, if the reactive power is above the threshold, action is required. Thus, var dispatch processor applies the rules to determine whether capacitor bank switching is needed.

The necessary communication is established between the radios, and the capacitor bank controller sends a command to capacitor switch, and it switches to the "on" position, and the capacitor bank is energized, as illustrated in Figure 15.21.

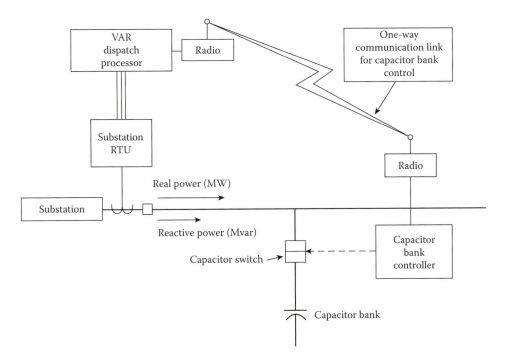

FIGURE 15.19 Var dispatch components of a SCADA system.

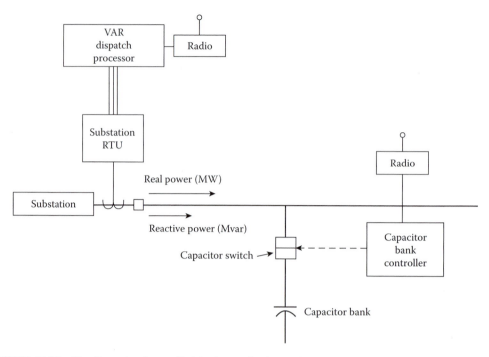

FIGURE 15.20   Var dispatch rules applied (and no action is required).

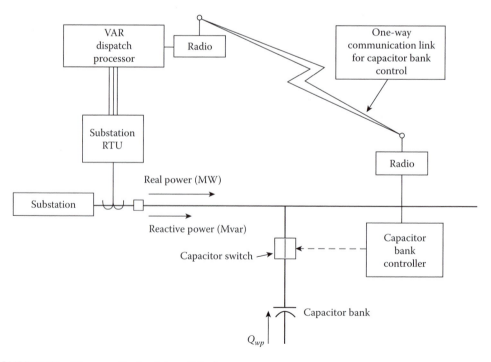

FIGURE 15.21   The capacitor bank is switched *on*.

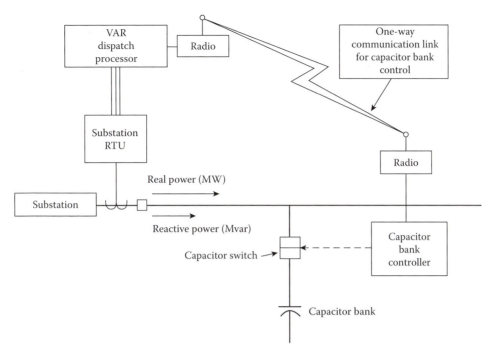

FIGURE 15.22 Change in reactive power is detected, and the capacitor bank is switched *off.*

Thus, the reactive power coming from the supplier is reduced. This change is detected by the substation RTUs. They measure real and reactive power at the substation end of the feeder. The var dispatch processor applies the rules for capacitor switching and sends a signal to the capacitor bank controller, which, in turn, de-energizes the capacitor bank by switching the capacitor switch to the "off" position, as illustrated in Figure 15.22.

At the substation end of the feeder, the var dispatch processor applies the rules for capacitor switching and sends a signal to the capacitor bank controller, which, in turn, de-energized the capacitor bank by switching the capacitor switch to the "off" position, as illustrated in Figure 15.22.

Note that such application of SCADA system provides (1) self-monitoring, (2) operator overrode capability, and (3) some improvement inefficiency. Here, the objectives of SCADA voltage control are (1) maintaining acceptable voltage at all locations under all loading conditions and (2) operating at as low voltage as possible to reduce power consumption, that is, CVR. However, SCADA cannot accomplish the following:

1. It does not adapt to changing feeder configurations (i.e., the rules are fixed in advance).
2. It does not adapt to varying operating needs (i.e., the rules are fixed in advance).
3. The overall efficiency with SCADA is improved with respect to the traditional approach, but it is not necessarily optimal under all conditions.
4. The operations of var/volt devices are not coordinated. It does not adapt well to the presence of modern grid devices such as DG.

15.9.3 Integrated Volt/VAR Control Optimization

What is needed is an IVVC optimization approach (i.e., centralized approach) that develops and executes a coordinated "optimal" switching plan for all voltage control devices based on an optimal power flow program to decide the plans for action.

In the process, it achieves utility-specific objectives as well, which include (1) minimizing power demand in terms of total customer demand and distribution system power losses, (2) minimizing "wear and tear" on the control equipment, and (3) maximizing revenue that is the difference between energy sales and energy prime cost.

Example 15.6

Consider the system given in Example 15.5, and conceptually design an IVVC optimization system configuration that develops a coordinated "optimal" switching plan for all voltage control equipment and execute the plan.

Solution

Figure 15.23 shows the integrated volt/var optimization (IVVO) system configuration. All the inputs from essential devices are fed to the VVC regulation coordination algorithm. They are provided through the communication links between the devices and the coordination algorithm.

Here, the volt/var regulation coordination algorithm manages tap changer settings, inverter and rotating machine var levels, and capacitors to regulate voltage, reduce losses, and conserve energy and system resources. The AMI, MDMS, and line switch provide inputs to distribution SCADA.

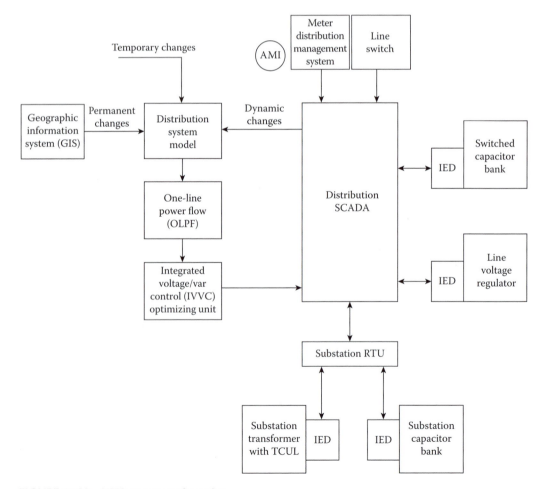

FIGURE 15.23 VVO system configuration.

The inputs of bank voltages and status, switch control are also provided by the switched capacitor banks and line voltage regulators. IVVC-optimizing engine develops a coordinated "optimal" switching plan for *all* voltage control devices and executes the plan. IVVC regulates an accurate, up-to-date electric model based on control of substation and feeder devices.

The AMI provides information on voltage feedback and accurate load data. The line switch provides information on switch status. The line voltage regulator provides inputs on monitor and control tap position and measure load voltage and load. The bank voltage and status, switch control of substation capacitor bank, as well as substation transformer with tap-charger under load (TCUL) are also used as inputs to the SCADA system.

Also, it is required that the substation transformer (with TCUL) to monitor and control tap position and measure load voltage and load. Temporary changes (i.e., cuts, jumpers, manual switching) as well as permanent asset changes (i.e., line extension and/or reconductoring) are inputted into distribution system model. Also provided are the real-time updates of dynamic changes. The necessary power flow results are provided by OLPF that calculates losses, voltage profile, etc. The IVVC determines optimal set of control actions to achieve a desired objective and provides the optimal switching plan to SCADA. In general, the IVVC has the following benefits:

1. When the network reconfiguration takes place, the dynamic model upgrades automatically.
2. The VVC actions are coordinated.
3. The system can model the effects of DG and other modern grid elements.
4. It produces the "optimal" results.
5. It accommodates varying operating objectives depending on present need.

15.10 EXISTING ELECTRIC POWER GRID

The present electricity grid is the result of fast urbanization and infrastructure development. However, the growth of the electric power system has been influenced by economic, political, and geographic factors that are utility specific. Nevertheless, the basic topology of the present power system has remained the same. As it can be easily observed, the basic topology is a vertical one.

Due to lack of proper communications and real data, the system is over-engineered to meet maximum expected peak demand of its aggregated customer load. Since the peak demand is infrequent due to its nature, the system is intrinsically vulnerable due to increase in demand for electric energy and decreased investments in plant and equipment, resulting in extensive blackouts.

In order to prevent it and maintain any expensive upstream plant and equipment without any damage, the utilities have established various command-and-control functions, such as SCADA systems that will be discussed in detail in the next sections. However, the application of the SCADA has remained not totally effective and has covered about 15%–20% of the distribution system. Primarily, the SCADA has been implemented into the transmission system.

15.11 SUPERVISORY CONTROL AND DATA ACQUISITION

SCADA is the equipment and procedures for controlling one or more remote stations from a master control station. It includes the digital control equipment, sensing and telemetry equipment, and two-way communications to and from the master stations and the remotely controlled stations.

The SCADA digital control equipment includes the control computers and terminals for data display and entry. The sensing and telemetry equipment includes the sensors, digital to analog and analog to digital converters, actuators, and relays used at the remote station to sense operating and alarm conditions and to remotely activate equipment such as circuit breakers.

The communication equipment includes the modems (modulator/demodulator) for transmitting the digital data and the communication link (radio, phone line, and microwave link, or power line).

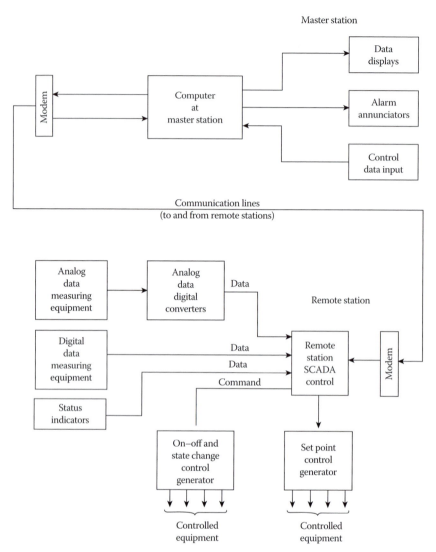

FIGURE 15.24   SCADA.

Figure 15.24 shows a block diagram of a SCADA system. Typical functions that can be performed by SCADA are the following:

1. Control and indication of the position of a two- or three-position device, for example, a motor-driven switch or a circuit breaker
2. State indication without control, for example, transformer fans on or off
3. Control without indication, for example, capacitors switched in or out
4. Set point control of remote control station, for example, nominal voltage for an automatic tap changer
5. Alarm sensing, for example, fire or the performance of a non-commanded function
6. Permit operators to initiate operations at remote stations from a central control station
7. Initiation and recognition of sequences of events, for example, routing power around a bad transformer by opening and closing circuit breakers or sectionalizing a bus with a fault on it
8. Data acquisition from metering equipment, usually via analog/digital converter and digital communication link

Today, in this country, all routine substation functions are remotely controlled. For example, a complete SCADA system can perform the following substation functions:

1. Automatic bus sectionalizing
2. Automatic reclosing after a fault
3. Synchronous check
4. Protection of equipment in a substation
5. Fault reporting
6. Transformer load balancing
7. Voltage and reactive power control
8. Equipment condition monitoring
9. Data acquisition
10. Status monitoring
11. Data logging

All SCADA systems have two-way data and voice communication between the master and the remote stations. Modems at the sending and receiving ends modulate, that is, put information on the carrier frequency, and demodulate, that is, remove information from the carrier, respectively.

Here, digital codes are utilized for such information exchange with various error detection schemes to assure that all data are received correctly. The RTU properly codes remote station information into the proper digital form for the modem to transmit and to convert the signals received from the master into the proper form for each piece of remote equipment.

When a SCADA system is in operation, it scans all routine alarm and monitoring functions periodically by sending the proper digital code to interrogate, or poll, each device. The polled device sends its data and status to the master station. The total scan time for a substation might be 30 s to several minutes subject to the speed of the SCADA system and the substation size. If an alarm condition takes place, it interrupts a normal scan. Upon an alarm, the computer polls the device at the substation that indicated the alarm.

It is possible for an alarm to trigger a computer-initiated sequence of events, for example, breaker action to sectionalize a faulted bus. Each of the activated equipment has a code to activate it, that is, to make it listen, and another code to cause the controlled action to take place.

Also, some alarm conditions may sound an alarm at the control station that indicates action is required by an operator. In that case, the operator initiates the action via a keyboard or a CRT. Of course, the computers used in SCADA systems must have considerable memory to store all the data, codes for the controlled devices, and the programs for automatic response to abnormal events.

15.12 ADVANCED SCADA CONCEPTS

The increasing competitive business environment of utilities, due to deregulation, is causing a reexamination of SCADA as a part of the process of utility operations, not as a process unto itself. The present business environment dictates the incorporation of hardware and software of the modern SCADA system into the corporation-wide, management information system strategy to maximize the benefits to the utility.

Today, the dedicated islands of automation gave way to the corporate information system. Tomorrow, in advanced systems, SCADA will be a function performed by workstation-based applications, interconnected through a *wide area network* (WAN) to create a virtual system, as shown in Figure 15.25.

This arrangement will provide the SCADA applications access to a host of other applications, for example, substation controllers, automated mapping/facility management system, trouble call analysis, crew dispatching, and demand-side LM. The WAN will also provide the traditional link between the utility's energy management system (EMS) and SCADA processors. The workstation-based applications will also provide for flexible expansion and economic system reconfiguration.

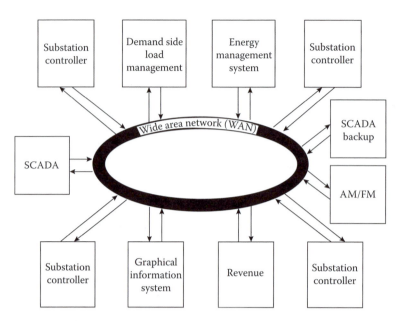

FIGURE 15.25 SCADA in a virtual system established by a WAN. (From Gönen, T., *Electric Power Transmission System Engineering*, 2nd ed., CRC Press, Boca Raton, FL, 2009.)

Also, unlike the centralized database of most existing SCADA systems, the advanced SCADA system database will exist in dynamic pieces that are distributed throughout the network. Modifications to any of the interconnected elements will be immediately available to all users, including the SCADA system. SCADA will have to become a more involved partner in the process of economic delivery and maintained quality of service to the end user.

In most applications today, SCADA and the EMS operate only on the transmission and generation sides of the system. In the future, economic dispatch algorithms will include demand-side (load) management and voltage control/reduction solutions. The control and its hardware and software resources will cease to exist.

15.12.1 SUBSTATION CONTROLLERS

In the future, RTUs will not only provide station telemetry and control to the master station but will also provide other primary functions such as system protection, local operation, *graphical user interface* (GUI), and data gathering/concentration from other subsystems.

Therefore, the future's RTUs will evolve into a class of devices that performs multiple substation control, protection, and operation functions. Besides these functions, the substation controller also develops and processes data required by the SCADA master, and it processes control commands and messages received from the SCADA master.

The substation controller will provide a gateway function to process and transmit data from the substation to the WAN. The substation controller is basically a computer system designed to operate in a substation environment. As shown in Figure 15.27, it has hardware modules and software in terms of the following:

  1. *Data processing applications*: These software applications provide various users access to the data of the substation controller in order to provide instructions and programming to the substation controller, collect data from the substation controller, and perform the necessary functions.

2. *Data collection applications*: These software applications provide the access to other systems and components that have data elements necessary for the substation controller to perform its functions.

3. *Control database*: All data reside in a single location, whether from a data processing application, data collection application, or derived from the substation controller itself.

Therefore, the substation controller is a system that is made up of many different types of hardware and software components and may not even be in a single location. Here, RTU may exist only as a software application within the substation controller system. Substation controllers will make all data available on WANS. They will eliminate separate stand-alone systems and thus provide greater cost savings to the utility company.

According to Sciacca and Block [7], the SCADA planner must look beyond the traditional roles of SCADA. For example, the planner must consider the following issues:

1. Reduction of substation design and construction costs
2. Reduction of substation operating costs
3. Overall lowering of power system operating costs
4. Development of information for non-SCADA functions
5. Utilization of existing resources and company standard for hardware, software, and database generation
6. Expansion of automated operations at the subtransmission and distribution levels
7. Improved customer relations

To accomplish these, the SCADA planner must join forces with the substation engineer to become an integrated team. Each must ask the other, "How can your requirements be met in a manner that provides positive benefits for my business?"

15.13   ADVANCED DEVELOPMENTS FOR INTEGRATED SUBSTATION AUTOMATION

Since the substation integration and automation technology is fairly new, there are no industry standard definitions with the exception of the following definitions:

IED: Any device incorporating one or more processors with the capability to receive or send data/control from or to an external source, for example, digital relays, controllers, and electronic multifunction meters.

IED integration: Integration of protection, control, and data acquisition functions into a minimal number of platforms to reduce capital and operating costs, reduce panel and control room space, and eliminate redundant equipment and databases.

Substation automation (SA): Deployment of substation and feeder operating functions and applications ranging from SCADA and alarm processing to IVVC in order to optimize the management of capital assets and enhance operation and maintenance efficiencies with minimal human intervention.

Open systems: A computer system that embodies supplier-independent standards so that software can be applied on many different platforms and can interoperate with other applications on local and remote systems.

An SA project prior to the 1990s typically involved three major functional areas: SCADA; plus station control, metering, and display; and plus protection.

In recent years, the utility industry has started using IEDs in their systems. These IEDs provided additional functions and features, including self-check and diagnostics, communication interfaces, the ability to store historical data, and integrated RTU input/output (I/O). The IED also enabled redundant equipment to be eliminated, as multiple functions were integrated into a single piece of equipment. For example, when interfaced to the potential transformers and current transformers of an individual circuit, the IED could simultaneously handle protection, metering, and remote control.

As more and more traditional SA functions become integrated into single piece of equipment, the definition of IED began to expand. The term is now applied to any microprocessor-based device with a communication port and therefore includes protection relays, meters, RTUs, *programmable logic controllers* (PLCs), load survey and operator-indicating meters, digital fault recorders, revenue meters, and power equipment controllers of various types.

The IED can thus be considered as the first level of automation integration. Additional economies of scale can be obtained by connecting all of the IEDs into a single integrated substation control system. The use of a fully integrated control system can lead to further streamlining of redundant equipment, as well as reduced costs for wiring, communications, maintenance, and operation, and improved power quality and reliability.

However, the process of implementation has been slow, largely because hardware interfaces and protocols for IEDs are not standardized. Protocols are as numerous as the vendors, and in fact more so, since products even from same end or often have different protocols. Figure 15.26 shows the configuration of an SA system.

The electric utility SA system uses a variety of devices integrated into a functional package by a communication technology for the purpose of monitoring and controlling the substation. Common communication connections include utility operations centers, finance offices, and engineering centers.

Communications for other users is usually through a bridge, gateway, or processor. A library of standard symbols should be used to represent the substation power apparatus on graphical displays. In fact, this library should be established and used in all substations and coordinated with other systems in the utility, such as distribution SCADA system, the EMS, the GIS, and the trouble call management system.

According to McDonald [5], the *global positioning system* (GPS) satellite clock time reference shown in Figure 15.27 provides a time reference for the SA system and IEDs in the substation.

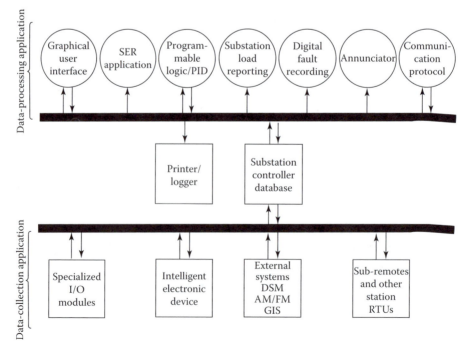

FIGURE 15.26 Substation controller. (From Gönen, T., *Electric Power Transmission System Engineering*, 2nd ed., CRC Press, Boca Raton, FL, 2009.)

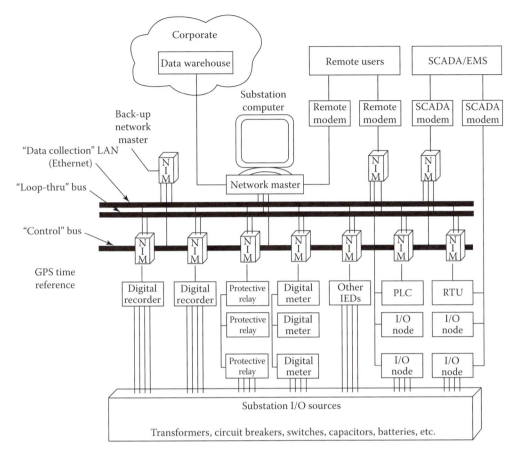

FIGURE 15.27 Configuration of SA system. (From Gönen, T., *Electric Power Distribution System Engineering*, 2nd ed., CRC Press, Boca Raton, FL, 2008.)

The host processor provides the GUI and the historical information system for achieving operational and nonoperational data.

The SCADA interface knows which SA system points are sent to the SCADA system, as well as the SCADA system protocol. The *local area network* (LAN)-enabled IEDs can be directly connected to the SA LAN. The non-LAN-enabled IEDs require a network interface module (NIM) for protocol and physical interface conversion.

A substation LAN has typically high speed and extends into the switchyard, which speeds the transfer of measurements, indications, control commands, and configuration and historical data between intelligent devices at the site.

This architecture reduces the amount and complexity of cabling currently required between intelligent devices. Also, it increases the communications bandwidth available to support faster updates and more advanced functions. Other benefits of an open LAN architecture can include creation of a foundation for future upgrades, access to third-party equipment, and increased interoperability.

In the United States, there are two major LAN standards, namely, Ethernet and PROFIBUS. Ethernet's great strength is the availability of its hardware and options from a myriad of vendors, not to mention industry-standard network-protocol support, multiple application-layer support and quality, and sheer quantity of test equipment. Because of these qualifications, Ethernet is more popular in this country, whereas PROFIBUS is widely used in Europe.

There are interfaces to substation IEDs to acquire data, determine the operating status of each IED, support all communication protocols used by the IEDs, and support standard protocols being developed. Besides SCADA, there may be an interface to the EMS that allows system operators to monitor and control each substation and the EMS to receive data from the substation integration and automation system at different time intervals.

The data warehouse enables users to access substation data while maintaining a firewall to protect substation control and operation functions. The utility has to decide who will use the SA system data, the type of data required, the nature of their application, and the frequency of the data, or update, required for each user.

A communication protocol permits communication between two devices. The devices must have the same protocol and its version implemented. Any protocol differences will result in communication errors. The substation integration and automation architecture must permit devices from different supplies to communicate employing an industry-standard protocol. The primary capability of an IED is its stand-alone capability, for example, protecting the power system for a relay IED. Its secondary capability is its integration capabilities, such as its physical interface, for example, RS-232, RS-485, Ethernet, and its communication protocol, for example, Modbus, Modbus Plus, DNP3, UCA2, and MMS.

To get all IEDs and their heterogeneous protocols onto a common substation LAN and platform, the gateway approach is best. The gateway will act not only as an interface between the local network physical layer and the RS-232/RS-485 ports found on the IEDs but also as a protocol converter, translating the IED's native protocol (like SEL, DNP3, or Modbus) into the protocol standard found on the substation's local network.

Two approaches can be used when using gateways to interface to the substation network. In one, a single low-cost gateway is used for each IED, and in the other, a multi-ported gateway interfaces with multiple IEDs. Which approach is more economical will depend on where the intelligent devices are located. If the IEDs are clustered in a central location, then the multi-ported gateway is certainly better.

The design of the substation integration and automation for new substations is easier than the one for existing substations. The new substation will typically have many IEDs for different functions, and the majority of operational data for the SCADA system will come from these IEDs. The IEDs will be integrated with digital two-way communications.

Typically, there are no conventional RTU in new substations. The RTU functionality is addressed using IEDs and PLCs and an integration network, using digital communications. In existing substations, there are several alternative approaches, depending on whether the substation has a conventional RTU installed.

The utility has three choices for their conventional substation RTUs: (1) integrate RTU with IEDs, (2) integrate RTU as another substation IED, and (3) retire RTU and use IEDs and PLCs, as with a new substation.

The environment of a substation is challenging for SA equipment. Substation control buildings are seldom heated or air-conditioned. Ambient temperatures can range from well below freezing to above 100°F (40°C). Metal-clad switchyard substations can reach ambient temperatures in excess of 140°F (50°C). Temperature changes stress the stability of measuring components in IEDs, RTUs, and transducers. In many environments, self-contained heating or air-conditioning may be recommended.

In summary, the integrated substation control system architecture (which is made up of IEDs, LANs, protocols, GUIs, and substation computers) is the foundation of the automated substation. However, the application building blocks consisting of operating and maintenance software are what produce the really substantial savings that can justify investment in an integrated substation control system.

15.14 EVOLUTION OF SMART GRID

As illustrated in Figure 15.28, the metering side of the distribution system has received most of the attention in terms of the recent infrastructure investments. The earlier applications in the distribution system included the automated meter reading (AMR). The AMR technology has facilitated utilities with the ability to read the customer consumption meters, alarms, and status remotely.

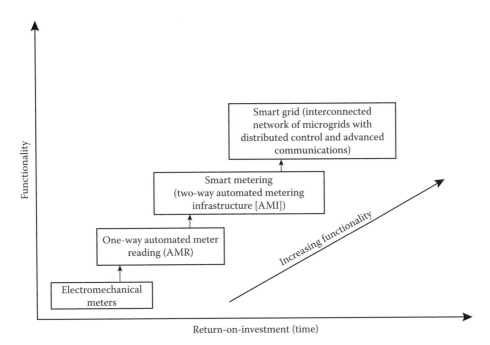

FIGURE 15.28   The evolution of smart grid as a function of return-on-investment versus time.

As shown in Figure 15.29, even though AMR technology has received a substantial attention initially, in time it became clear that AMR is not the answer for the DSM, primarily due to its one-way communication nature. It simply reads the customers' meter data.

It does not permit the transition to the smart grid where extensive control at all levels is essential [2]. Thus, AMR technology applications become extinct. It was replaced by AMI. This system provides utility companies with a two-way communication system to the customers' meters and the ability to modify customers' service-level parameters.

Hence, by using AMI, utilities can reach to their goals in load management and increased revenues. With AMI technology, power companies not only can collect instantaneous information about individual and aggregated demand but can also modify the energy consumption, as well as implement their cost-cutting measures. As said by Farhangi [2], the emergence of AMI started a concerted move by stakeholders to further refine the ever-changing concepts around the smart grid.

As a next step, according to Farhangi [2], the smart grid requires to leverage the AMI infrastructure and implement its distribution command-and-control strategies over the AMI backbone. The penetrating control and intelligent that are properties of the smart grid have to be located across all geographic areas, as well as components, and functions of the power system. The distinguished three elements mentioned in the preceding text, that is, geographic areas, components, and functions, determine the topology of the smart grid and its components.

Again, it is important to point out that smart distribution systems are essential part of the smart grid; here are the necessary steps to establish a smart distribution system:

Step 1: Design information models based on overall requirements of a smart distribution system.
Step 2: Establish substation data integration and associated applications, such as traditional SCADA and fault location.
Step 3: Add feeder automation for selected applications. For example, automatic reconfiguration and VVC.
Step 4: Add advanced metering integration. For example, state estimation and outage management.

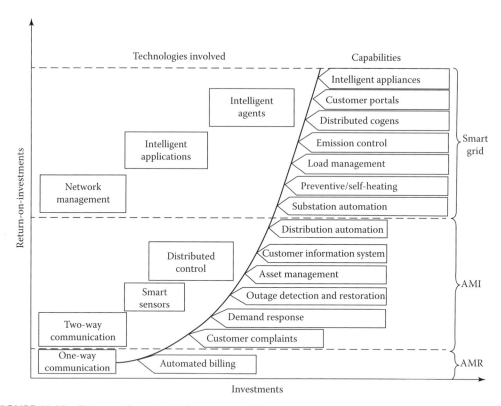

FIGURE 15.29 Return on investments for a smart grid.

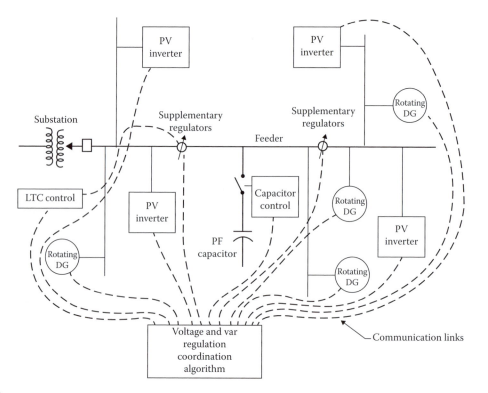

FIGURE 15.30 The additional steps that are necessary to achieve the VVO.

Step 5: Add integrated applications, such as energy management optimization, risk assessment, and advanced equipment diagnostics.

Step 6: Add interface to DERs as well as islanding, advanced energy management and optimization, and real and reactive power (PQ) management capabilities.

The necessary next steps to achieve the volt/var optimization (VVO) have been illustrated in Figure 15.30. The volt/var regulation coordination algorithm manages tap changer settings, inverter and rotating machine var levels, and capacitors to regulate voltage, reduce losses, and conserve energy and system resources. The necessary links between the algorithm and the individual components have been indicated in the figure.

15.15 SMART MICROGRIDS

As succinctly put by Farhangi [2], the smart grid is the collection of all technologies, concepts, topologies, and approaches that permit to maintain hierarchies of generation, transmission, and distribution to be replaced with an end-to-end, organically intelligent, fully integrated environment where the business processes, objectives, and needs of all stakeholders are supported by the efficient exchange of data, services, and transactions.

A smart grid is hence defined as a grid that accommodates a wide variety of generation options, for example, central, distributed, intermittent, and mobile. It provides customers with the ability to interact with the EMS to adjust their energy use and reduce their energy costs.

A smart grid also has to be a self-healing system. It foresees the forthcoming failures and takes the necessary corrective actions to avoid or mitigate system problems. A smart grid uses the IT to continuously optimize the employment of its capital assets while minimizing operational and maintenance costs [2].

However, the smart grid should not be seen as a replacement for the present electric power grid but a complement to it. Thus, the smart grid can coexist with the present electric power grid, adding to its capabilities, functionalities, and capacities by means of evolutionary path. This dictates a topology for the smart grid that permits for organic growth, the inclusion of forward-looking technologies, and full backward compatibility with the present systems [2].

The smart grid can also be defined as the ad hoc integration of complementary components, subsystems, and functions under the extensive control of a highly intelligent and distributed command-and-control system.

Furthermore, the organic growth and evolution of the smart grid is achieved by the inclusion of intelligent microgrids. Here, the *microgrid* is defined as interconnected networks of distributed energy systems, including loads and resources. So, it can function whether they are connected to or separated from the electric power grid [2].

As succinctly put by Keyhani [19], smart microgrid systems consist of renewable green energy sources with their associated power converters, efficient transformers, and storage systems.

According to Keyhani [19], a microgrid renewable green (MRG) energy DG system has to be also designed to provide an intelligent to act as grid optimization manager that would facilitate the control of various customer loads according to pricing trends and grid stress by altering customer's use of power. This is accomplished by smart meters in terms of shedding customer loads and permitting distribution generation to come online, if the price of power is above a set limit. The MRG system's EMS is in communication with all individual smart meters located at residential, commercial, and industrial customer sites. Here, the EMS is given information from the power grid and the open access same-time information system that is known as OASIS. Thus, the EMS has a two-way communication with the smart meters.

In addition, according to Keyhani [19], EMS can control power flow in microgrid according to load forecasts, weather forecasts, unit availability, and price fluctuations. Such MRG systems can facilitate the operation of clusters of load system and renewable DGs as a single controllable load

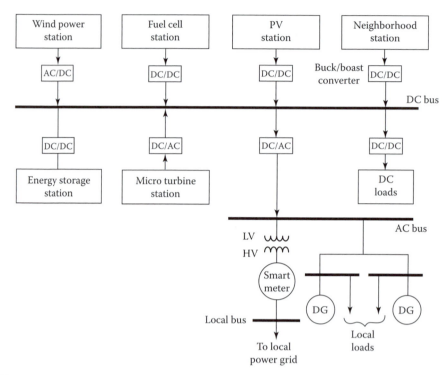

FIGURE 15.31 The dc and ac schematics of an MRG energy DG system.

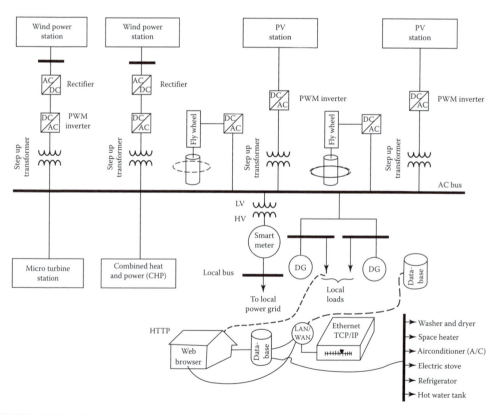

FIGURE 15.32 The ac schematics of an MRG energy DG system.

system and as a single dispatchable generation source, respectively. The interconnection point of the smart microgrid to the local power grid is represented by a node that is called locational marginal pricing (LMP). This cost represents the cost of energy at the location. Figure 15.31 shows the dc and ac schematics of an MRG energy DG system.

Figure 15.32 shows the ac schematics of an MRG energy DG system. Here, green energy sources of microturbines, fuel cells, or other renewable sources, for example, wind farms and solar generating stations, can be connected to a dc or an ac bus, employing standard interchangeable converters. The MRG systems have to be able to operate both in *synchronized operation* with the local power grid and in the *island mode* of operation. The MRG system inverter must be able to control active end reactive power at lagging, leading, or unit power factors. But, the voltage control, that is, the reactive power (*vars*) control, is left to the EMS of the local power grid.

15.16 TOPOLOGY OF A MICROGRID

As discussed in Section 11.13, a small microgrid network can operate in both grid-connected and islanded modes. The topology of a smart microgrid is illustrated in Figure 15.33. As said by Farhangi [2], a small grid integrates the following components:

1. It includes power plants capable of meeting local demand as well as feeding the unused energy back to the electric power grid. They are known as cogenerators and often use renewable resources, such as sun, wind, and biomass. Some microgrids have CHP thermal power plants that are capable of recovering the waste heat in terms of district cooling or heating in the vicinity of the power plant.
2. It employs local and distributed power-storage capability to smooth out the intermittent performance of renewable energy sources.
3. It services a variety of loads, including residential, commercial, and industrial loads.

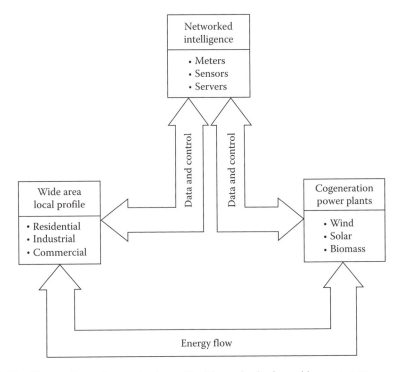

FIGURE 15.33   The topology of a smart microgrid with required microgrid components.

4. It has communication infrastructure that facilitates system components to exchange information and commands reliably and securely.
5. It employs smart meters and sensors capable of measuring a number of consumption parameters (e.g., real and reactive powers, voltage, current, and demand) with acceptable accuracy.
6. It includes an intelligent core, made of integrated networking, computing, and communication infrastructure elements, which appear to users in terms of energy management applications that permit command and control on all network nodes.
7. It includes smart terminations, loads, and appliances capable of communicating their status and accepting commands to adjust and control their performance and service levels according to consumer and/or utility requirements.

15.17   FUTURE OF A SMART GRID

Farhangi [2] predicts that the smart grid of the future will be interconnected through dedicated highways for power exchange, and data and commands, as shown in Figure 15.34. But, it is expected that not all microgrids will have the same capabilities and needs. It will be subject to the load diversity, geography, economics, and the mix of the primary energy resources.

The necessary AMI systems now being established will facilitate the evolution of the smart grid. However, due to the high costs involved, it is foreseeable that the new and the old grids may coexist for some time. Eventually though, it is expected that the system will replace the old grid.

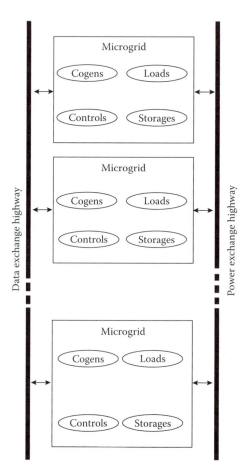

FIGURE 15.34   The envisioned smart grid of the future.

Thus, during the transition period, there will be a hybrid system. The new power grid will appear as a system of organically integrated collection of smart grids with extensive command-and-control functions implemented at all levels.

15.18 STANDARDS OF SMART GRIDS

It is very possible that some substantial problems emerge when distinctively different systems, components, and functions begin to be integrated as part of a distributed command-and-control system of a smart grid.

A part of the problem is the fact that at the present time, there are no commonly accepted interfaces, messaging and control protocols, and standards that would be abided by to ensure a common communication vocabulary among system components of a smart grid.

In order to assist the development of the required standards, the power industry is slowly adopting different terminologies for the purpose of segmentation of the command-and-control layers of the smart grid. The examples of this include HAN, LAN, and WAN.

The HAN stands for "home area network"; it is used to identify the network of communicating loads, appliances, and sensors beyond the smart meter and within the customer's property.

The LAN denotes the local area network. It is used to identify the network of integrated smart meters, field components, and gateways that constitute the logical network between distribution substations and customer's premises.

Finally, WAN is used to identify the network of upstream utility assets, including power plants, distribution storage, and substations. As shown in Figure 15.35, the interface between WAN and LAN and LAN and HAN is provided by smart meters.

In the United States, the US NIST is leading the effort for the standardization for smart grid [3]. In the United States, the most common standard that is used for substations is IEC 61850. It operates in real-time environment.

In Europe and other places, similar efforts indicate the need for the development of common information model (CIM) to enable vertical and lateral integration of applications and functions within the smart grid. CIM is a unified modeling language (UML) based on information model representing real-world objects and information entities exchanged within the value chain of the electric power industry. CIM is based on IEC 61970/61968 applications. It uses IBM's "rational rose" modeling tool. It is available in many forms, for example, cat, mch, html, xml, and owl. It enables

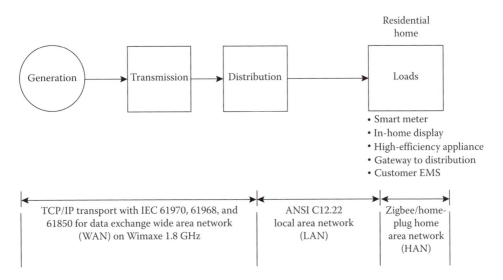

FIGURE 15.35 Development of standards for the smart grid.

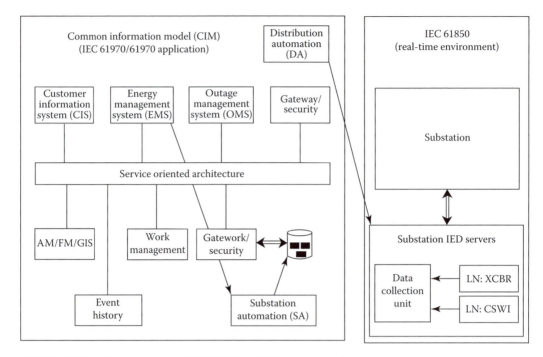

FIGURE 15.36   The application of IEC 61850 and CIM to a substation environment.

data access in a standard way. It also enables integration of applications and systems. It uses common language to navigate and access complex data structures in any database. It provides a common model behind all messages exchanged between systems. It is not tied to a particular application's viewpoint of the world.

Finally, it is the basis for defining information exchange models. It is being developed and standardized by IEC. Figure 15.36 shows the application of IEC 61850 and CIM to a substation environment. Figure 15.37 shows an example how to develop the CIM for distribution applications as well as its application to the field operations.

Figure 15.38 shows the application of the CIM in the interface reference model (IRM). It provides the framework for identifying information exchange requirements among utility business functions. The left-hand side of the figure represents the distribution management business functions, such as electric distribution network planning, constructing, maintaining, and operating. On the other hand, the right-hand side of the figure represents the business functions that are external to distribution management, such as generation and transmission management, enterprise resource planning, supply chain, and general corporate services. All the IEC 61968 activity diagrams and sequence diagrams are organized by the IRM.

Out of the proposed standards, IEC 61850 and its related standards appear to be favorites for WAN data communication, supporting TCP/IP, among other protocols, over fiber or a 1.8 GHz WiMax.

In North America, ANSI C12.12, and its related standards, is considered as the favorite LAN standard, facilitating a new generation of smart meters capable of communicating substation gateways over numerous wireless technologies.

Also, the European community is pushing for the development of the AMI standard for Europe, replacing the aging DLMS/COSEM standard. Thus, it is pushing for efforts to develop a European counterpart for ANSI-C12.22.

It appears that ZigBee with Smart Energy Profile is the favorite for HAN, partially due to the lack of initiatives by the home appliance manufacturers.

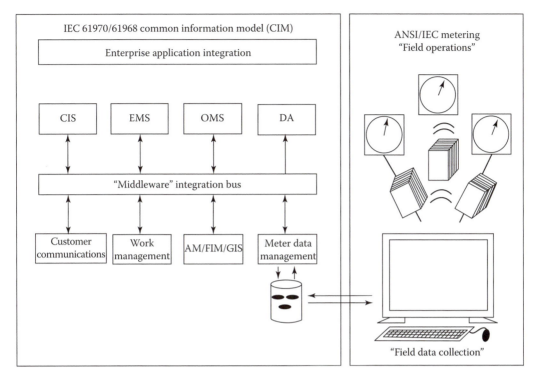

FIGURE 15.37   Developing the CIM for distribution applications as well as its application to the field operations.

15.19   ASSET MANAGEMENT

One of the major objectives of all utilities is to generate sufficient cash flow to cover their operating costs. As part of AM activities, a careful analysis of the substation equipment and operation can identify critical equipment that creates a significant fire hazard. Assets especially in substation control facilities are very critical and should be diligently reviewed to determine the adequacy of the planned fire protection.

As said before, with an increase in regulating influence and the focus on smart grid advanced technologies, there is a renewed interest in increasing the investment in distribution networks to defer infrastructure build-out and to reduce operating and maintenance costs through improving grid efficiency, network reliability, and AM programs.

Thus, since the roots of power system issues are usually found in the electric power distribution system, the point of departure for the grid overhaul is the distribution system. For example, proper loading, as well as overloading, issues of transformers should be considered as the careful application of AM principles. This will also address the risk of major outages in substations. As part of the application of AM principles, it will benefit the lives of substation transformers by establishing proper overloading principles to increase the useful life of the existing transformers.

In general, the smart grid optimizes the utilization of the existing and new assets, improves load factors, and lowers losses in order to maximize the operational efficiency and reduce the cost. Advanced sensing and robust communications will allow early problem detection, preventive maintenance, and corrective action.

Therefore, AM is very important for the utility companies in many respects. It will advance the applications of advanced DA. It will promote the constant monitoring of the conditions in the substations and other existing facilities. It will manage real-time loading of the plant and equipment. It will encourage establishing scheduled maintenance and will facilitate the increase in plant and equipment utilization.

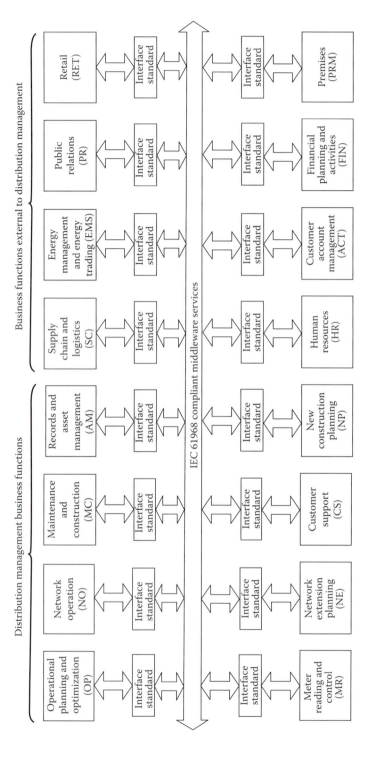

FIGURE 15.38   The IEC 61968 IRM showing activity diagrams and sequence diagrams that are organized by the IRM.

In summary, the proper application of the AM will increase the useful lives of the plant and equipment, delay the timing of the replacement of the plant and equipment, cause the increase in the net profit by decreasing the expenditures involved, and decrease the operational hazards to the plant and equipment as well as to the personnel involved.

15.20 EXISTING CHALLENGES TO THE APPLICATION OF THE CONCEPT OF SMART GRIDS

It is estimated that the electric power grid will make a transition from an electromechanically controlled system to an electronically controlled network within the next two decades. According to Amin and Wollenberg [5], there are some fundamental challenges to achieve this transition, namely,

1. The lack of transmission capacity to meet the substantially increasing loads
2. The difficulties of grid operation in a competitive market environment
3. The redefinition of power system planning and operation in the competitive era
4. The determination of the optimum type, mix, and placement of sensing, communication, and control hardware
5. The coordination of centralized and decentralized control

Smart grids are not really about doing things a lot differently than the way they are being done today. Instead, they are about doing more of what is being done, that is, sharing communication and infrastructures, filling in product gaps, and leveraging existing technologies to a greater extent while driving a higher level of integration to realize the synergies across enterprise integration.

A smart grid is not an off-the-shelf product or something that can be installed and turn on the next day. Rather, it is an integrated solution of technologies driving incremental benefits in capital expenditures, operation and maintenance expenditures, and customer and societal benefits.

A well-designed smart grid imitative build on the existing infrastructure provides a greater level of integration at the enterprise level and has a long-term focus. It is definitely a one-time solution but a change in how utilities look at a set of technologies that can enable both strategic and operational processes. It is the means to leverage benefits across applications and remove the barriers that are created by the past company practices.

15.21 EVOLUTION OF SMART GRID

Figure 15.39 illustrates the possible future application of the smart grid concept at the substation level as well as between substations. The figure shows the AMI infrastructure and implementation of distribution command-and-control strategies over the AMI backbone. Note that the penetrating control and intelligent that are the properties of smart grid have to be located across all geographic areas, as well as components, and functions of the power system. As said before, the distinguished three elements mentioned earlier, that is, geographic areas, components, and functions, determine the topology of the smart grid and its components.

Figure 15.40 shows the present and future research areas in the application of the smart grid concept into distribution systems. The areas can essentially be categorized into two areas, namely, the applications and technology, and the infrastructure. The area of applications and technology includes the new technologies; new applications and systems; technology transfer, industry coordination, technology watch, and application guides; PV integration; distribution efficiency; and PHEV integration. The area of infrastructure includes communication infrastructure, information system integration, and security.

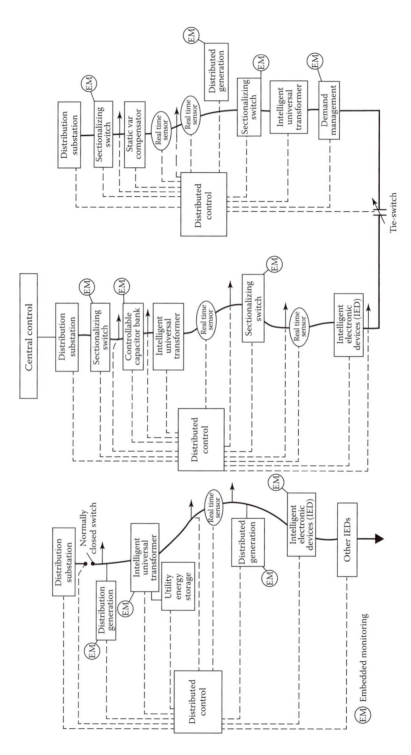

FIGURE 15.39   Illustrates the possible future application of the smart grid concept at the substation level as well as between substations.

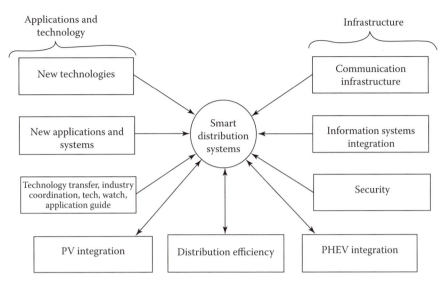

FIGURE 15.40   Present and future research areas in smart grid applications.

REFERENCES

1. Gönen, T.: *Electric Power Distribution System Engineering*, 2nd ed., CRC Press, Boca Raton, FL, 2008.
2. Farhangi, H.: The path of the smart grid, *IEEE Power Energy Mag.*, 8(1), January/February 2010, 18–28.
3. *U.S. National Institute of Standards and Technology*: NIST framework and roadmap for smart grid interoperability standards, Release 1.0, September 2009, Washington, DC.
4. Gellings, C. W.: *The Smart Grid*, CRC Press, Boca Raton, FL, 2009.
5. McDonald, D. J.: Substation integration and automation, *Electric Power Substation Engineering*, Chapter 7, CRC Press, Boca Raton, FL, 2003.
7. Sciacca, S. C. and W. R. Block: Advanced SCADA concepts, *IEEE Comput. Appl. Power*, 8(1), January 1995, 23–28.
8. Gönen, T.: *Engineering Economy for Engineering Managers: With Computer Applications*, Wiley, New York, 1990.
9. Gönen, T.: *Electric Power Transmission System Engineering*, 2nd ed., CRC Press, Boca Raton, FL, 2009.
10. Amin, M. and B. F. Wollenberg: Toward a smart grid-power delivery for the 21st century, *IEEE Power Energy Mag.*, 3(5), September/October 2005, 34–41.
11. http://www.oe.energy.gov/DocumentsandMedia/SGSRMain_lowres.pdf
12. Farhangi, H.: Intelligent microgrid research at BCIT, *Proceedings of IEEE Electric Power System Conference*, (EPEC'08), Vancouver, British Columbia, Canada, October 2008, pp. 1–7.
13. FERC: *Assessment of Demand Response and Advanced Metering*, 2007 Staff Report, September 2007 [on line]. Available: http://www.ferc.gov/legal/staff-reports/09-07-demand-response.pdf
14. Kintner-Meyer, M., K. Schneider, and R. Pratt: Impacts assessment of plug-in hybrid vehicles on electric utilities and regional U.S. power grids. [on line]. Available: http://www.pnl.gov/energy/eed/etd/pdfs/phev_feasibility_analysis_combined.pdf
15. IEEE PES Power & Energy Society, Smart distribution systems tutorial, *IEEE PES General Meeting*, Minneapolis, MN, July 2110.
16. Gnadt, P. A. and J. S. Lawler: *Automating Electric Utility Distribution Systems*, Prentice Hall, Englewood Cliffs, NJ, 1990.
17. Gönen, T.: *Modern Power System Analysis*, 2nd ed., CRC Press, Boca Raton, FL, 2013.
18. More, D. and D. McDonnell: Smart grid meets distribution utility reality, *Electric Light & Power*, March 2007, 85(2)1–6.
19. Keyhani, A.: *Design of Smart Power Grid Renewable Energy Systems*, Wiley, Hoboken, NJ, 2011.
20. Bricker, S., L. Rubin, and T. Gönen: Substation automation techniques and advantages, *IEEE Comput. Appl. Power*, 14(3), July 2001, 31–37.

Appendix A: Impedance Tables for Lines, Transformers, and Underground Cables

TABLE A.1

Characteristics of Copper Conductors, Hard-Drawn, 97.3% Conductivity

Size of Conductor Circular Mils	AWG or B & S	Number of Strands	Diameter of Individual Strands (in.)	Outside Diameter (in.)	Breaking Strength (lb)	Weight (lb/mi)	Approx. Current Carrying Capacity[a] (amps)	Geometric Mean Radius at 60 Cycles (ft)
1,000,000	—	37	0.1644	1.151	43,830	16,300	1300	0.0368
900,000	—	37	0.1560	1.092	39,610	14,670	1220	0.0349
800,000	—	37	0.1470	1.029	35,120	13,040	1130	0.0329
750,000	—	37	0.1424	0.997	33,400	12,230	1090	0.0319
700,000	—	37	0.1375	0.963	31,170	11,410	1040	0.0306
500,000	—	37	0.1273	0.891	27,020	9,781	940	0.0285
500,000	—	37	0.1162	0.814	22,610	8,161	840	0.0260
500,000	—	19	0.1622	0.811	21,590	8,161	840	0.0256
450,000	—	19	0.1539	0.770	19,750	7,336	780	0.0243
400,000	—	19	0.1451	0.726	17,560	6,521	730	0.0229
350,000	—	19	0.1357	0.679	16,890	5,706	670	0.0214
350,000	—	12	0.1708	0.710	16,140	5,706	670	0.0225
300,000	—	19	0.1257	0.629	13,510	4,891	610	0.01987
300,000	—	12	0.1581	0.657	13,170	4,891	610	0.0208
250,000	—	19	0.1147	0.574	11,360	4,076	540	0.01813
250,000	—	12	0.1443	0.600	11,130	4,076	540	0.01902
211,600	4/0	19	0.1055	0.528	9,617	3,450	480	0.01668
211,600	4/0	12	0.1328	0.552	9,483	3,450	490	0.01750
211,600	4/0	7	0.1739	0.522	9,154	3,450	480	0.01579
167,800	3/0	12	0.1183	0.492	7,556	2,736	420	0.01569
167,800	3/0	7	0.1548	0.464	7,366	2,736	420	0.01404
133,100	2/0	7	0.1379	0.414	5,926	2,170	360	0.01252
106,600	1/0	7	0.1228	0.368	4,752	1,720	310	0.01113
83,690	1	7	0.1093	0.328	3,804	1,364	270	0.00992
63,690	1	3	0.1670	0.360	3,620	1,351	270	0.01016
66,370	2	7	0.0974	0.292	3,045	1,082	230	0.00883
66,370	2	3	0.1487	0.320	2,913	1,071	240	0.00903
66,370	2	1	—	0.258	3,003	1,061	220	0.00836
52,630	3	7	0.0867	0.260	2,433	858	200	0.00787
52,630	3	3	0.1325	0.286	2,359	850	200	0.00805
52,630	3	1	—	0.229	2,439	841	190	0.00745
41,740	4	3	0.1180	0.254	1,879	674	180	0.00717
41,740	4	1	—	0.204	1,970	667	170	0.00663
33,100	5	3	0.1050	0.226	1,605	534	180	0.00638
33,100	5	1	—	0.1819	1,591	529	140	0.00590
26,250	6	3	0.0935	0.201	1,205	424	130	0.00568
26,250	6	1	—	0.1620	1,280	420	120	0.00526
20,820	7	1	—	0.1443	1,030	333	110	0.00468
16,510	8	1	—	0.1286	826	264	90	0.00417

Source: Westinghouse Electric Corporation, *Electric Utility Engineering Reference Book—Distribution Systems*, East Pittsburgh, PA, 1965.

[a] For conductor at 75°C, air at 25°C, wind 1.4 mi/h (2 ft/s), frequency = 60 cycles.

	r_a Resistance (Ω/Conductor/mi)							X_a Inductive Reactance (Ω/Conductor/mi) at 1 ft Spacing			X'_a Shunt Capacitive Reactance (MΩ·mi/Conductor) at 1 ft Spacing		
	25°C (77°F)			50°C (122°F)									
DC	25 Cycles	50 Cycles	60 Cycles	DC	25 Cycles	50 Cycles	60 Cycles	25 Cycles	50 Cycles	60 Cycles	25 Cycles	50 Cycles	60 Cycles
0.0585	0.0594	0.0620	0.0634	0.0640	0.0648	0.0672	0.0685	0.1666	0.333	0.400	0.216	0.1081	0.0901
0.0650	0.0658	0.0682	0.0695	0.0711	0.0718	0.0740	0.0752	0.1693	0.339	0.406	0.220	0.1100	0.0916
0.0731	0.0739	0.0760	0.0772	0.0800	0.0808	0.0826	0.0837	0.1722	0.344	0.413	0.224	0.1121	0.0934
0.0780	0.0787	0.0807	0.6818	0.0853	0.0859	0.0878	0.0888	0.1739	0.348	0.417	10.225	0.1132	0.0943
0.0836	0.0842	0.0661	0.0671	0.0914	0.0920	0.0937	0.0947	0.1789	0.352	0.422	0.229	0.1145	0.0954
0.0975	0.0981	0.0997	0.1006	0.1066	0.1071	0.1086	0.1095	0.1799	0.360	0.432	0.235	0.1173	0.0977
0.1170	0.1175	0.1188	0.1196	0.1280	0.1283	0.1296	0.1303	0.1845	0.369	0.443	0.241	0.1206	0.1004
0.1170	0.1175	0.1188	0.1196	0.1280	0.1283	0.1296	0.1303	0.1853	0.371	0.445	0.241	0.1206	0.1006
0.1300	0.1304	0.1316	0.1323	0.1422	0.1426	0.1437	0.1443	0.1879	0.376	0.451	0.245	0.1224	0.1020
0.1462	0.1466	0.1477	0.1484	0.1600	0.1603	0.1613	0.1519	0.1909	0.382	0.458	0.249	0.1245	0.1038
0.1671	0.1675	0.1684	0.1690	0.1828	0.1831	0.1840	0.1845	0.1943	0.389	0.466	0.254	0.1269	0.1058
0.1671	0.1675	0.1684	0.1690	0.1828	0.1831	0.1840	0.1845	0.1918	0.384	0.460	0.251	0.1253	0.1044
0.1950	0.1953	0.1961	0.1966	0.213	0.214	0.214	0.215	0.1982	0.396	0.476	0.259	0.1296	0.1060
0.1950	0.1953	0.1961	0.1966	0.213	0.214	0.214	0.215	0.1957	0.392	0.470	0.256	0.1281	0.1068
0.234	0.234	0.235	0.236	0.256	0.256	0.257	0.257	0.203	0.406	0.487	0.266	0.1329	0.1108
0.234	0.234	0.235	0.236	0.256	0.256	0.257	0.257	0.200	0.401	0.481	0.263	0.1313	0.1094
0.276	0.277	0.277	0.278	0.302	0.303	0.303	0.303	0.207	0.414	0.497	0.272	0.1359	0.1132
0.276	0.277	0.277	0.278	0.302	0.303	0.303	0.303	0.208	0.409	0.491	0.269	0.1343	0.1119
0.276	0.277	0.277	0.278	0.302	0.303	0.303	0.303	0.210	0.420	0.603	0.273	0.1363	0.1136
0.349	0.349	0.349	0.350	0.381	0.381	0.382	0.382	0.210	0.421	0.606	0.277	0.1384	0.1153
0.349	0.349	0.349	0.350	0.381	0.381	0.382	0.382	0.216	0.431	0.518	0.281	0.1405	0.1171
0.440	0.440	0.440	0.440	0.481	0.481	0.481	0.481	0.222	0.443	0.532	0.289	0.1445	0.1205
0.555	0.555	0.555	0.555	0.606	0.607	0.607	0.607	0.227	0.455	0.546	0.298	0.1488	0.1240
0.599	0.699	0.699	0.699	0.766				0.233	0.467	0.560	0.306	0.1528	0.1274
0.692	0.692	0.692	0.692	0.757				0.232	0.464	0.557	0.299	0.1495	0.1246
0.881	0.882	0.882	0.882	0.964				0.239	0.478	0.574	0.314	0.1570	0.1308
0.873				0.956				0.238	0.476	0.571	0.307	0.1637	0.1281
0.884				0.946				0.242	0.484	0.581	0.323	0.1614	0.1346
1.112				1.216				0.245	0.490	0.588	0.322	0.1611	0.1343
1.101				1.204				0.244	0.488	0.585	0.316	0.1578	0.1315
1.090	Same as DC			1.192	Same as DC			0.248	0.496	0.595	0.331	0.1656	0.1380
1.388				1.518				0.250	0.499	0.599	0.324	0.1619	0.1349
1.374				1.503				0.264	0.507	0.609	0.339	0.1697	0.1416
1.750				1.914				0.256	0.511	0.613	0.332	0.1661	0.1384
1.733				1.895				0.260	0.519	0.623	0.348	0.1738	0.1449
2.21				2.41				0.262	0.523	0.628	0.341	0.1703	0.1419
2.18				2.39				0.265	0.531	0.637	0.356	0.1779	0.1483
2.75				3.01				0.271	0.542	0.651	0.364	0.1821	0.1517
3.47				3.80				0.277	0.564	0.665	0.372	0.1862	0.1652

TABLE A.2
Characteristics of Anaconda Hollow Copper Conductors

Design Number	Size of Conductor Circular Mils or AWG	Wires Number	Diameter (in.)	Outside Diameter (in.)	Breaking Strain (lb)	Weight (lb/mi)	Geometric Mean Radius at 60 Cycles (ft)	Approx. Current Carrying Capacity (amps)[a]
966	890,500	28	0.1610	1.650	36,000	15,085	0.0612	1395
96R 1	750,000	42	0.1296	1.155	34,200	12,345	0.0408	1160
939	650,000	50	0.1097	1.126	29,500	10,761	0.0406	1060
360R 1	600,000	50	0.1053	1.007	27,500	9,905	0.0387	1020
938	550,000	50	0.1009	1.036	25,200	9,103	0.0373	960
4R 5	510,000	50	0.0970	1.000	22,700	8,485	0.0360	910
892R 3	500,000	18	0.1558	1.080	21,400	8,263	0.0394	900
933	450,000	21	0.1353	1.074	19,300	7,476	0.0398	850
924	400,000	21	0.1227	1.014	17,200	6,642	0.0376	810
925R 1	380,500	22	0.1211	1.003	16,300	6,331	0.0373	780
565R 1	350,000	21	0.1196	0.950	15,100	5,813	0.0353	750
936	350,000	15	0.1444	0.860	15,400	5,776	0.0311	740
378R 1	350,000	30	0.1059	0.736	16,100	5,739	0.0253	700
954	321,000	22	0.1113	0.920	13,850	5,343	0.0340	700
935	300,000	18	0.1205	0.839	13,100	4,984	0.0307	670
903R 1	300,000	15	0.1338	0.797	13,200	4,953	0.0289	660
178R 2	300,000	12	0.1507	0.750	13,050	4,937	0.0266	650
926	250,000	18	0.1100	0.766	10,950	4,155	0.0279	600
915R 1	250,000	15	0.1214	0.725	11,000	4,148	0.0266	590
24R 1	250,000	12	0.1368	0.683	11,000	4,133	0.0245	580
923	4/0	18	0.1005	0.700	9,300	3,521	0.0255	530
922	4/0	15	0.1109	0.663	9,300	3,510	0.0238	520
50R 2	4/0	14	0.1152	0.650	9,300	3,510	0.0234	520
158R 1	3/0	16	0.0961	0.606	7,500	2,785	0.0221	460
495R 1	3/0	15	0.0996	0.595	7,600	2,785	0.0214	460
570R 2	3/0	12	0.1123	0.560	7,600	2,772	0.0201	450
909R 2	2/0	15	0.0880	0.530	5,950	2,213	0.0191	370
412R 2	2/0	14	0.0913	0.515	6,000	2,207	0.0184	370
937	2/0	13	0.0950	0.505	6,000	2,203	0.0181	370
930	125,600	14	0.0885	0.500	5,650	2,083	0.0180	360
934	121,300	15	0.0836	0.500	5,400	2,015	0.0179	350
901	119,400	12	0.0936	0.470	5,300	1,979	0.0165	340

Source: Westinghouse Electric Corporation, *Electric Utility Engineering Reference Book—Distribution Systems*, East Pittsburgh, PA, 1965.

[a] For conductor at 75°C, air at 25°C, wind 1.4 mi/h (2 ft/s), frequency = 60 cycles, average tarnished surface.

| r_a Resistance (Ω/Conductor/mi) | | | | X_a Inductive Reactance (Ω/Conductor/mi) at 1 ft Spacing | | | X'_a Shunt Capacitive Reactance (MΩ·mi/Conductor) at 1 ft Spacing | | |
| 25°C (77°F) | | 50°C (122°F) | | | | | | | |
DC 25 Cycles	50 Cycles 60 Cycles	DC 25 Cycles	50 Cycles 60 Cycles	25 Cycles	50 Cycles	60 Cycles	25 Cycles	50 Cycles	60 Cycles
0.0671	0.0676	0.0734	0.0739	0.1412	0.282	0.339	0.1907	0.0953	0.0794
0.0786	0.0791	0.0860	0.0865	0.1617	0.323	0.388	0.216	0.1080	0.0900
0.0909	0.0915	0.0994	0.1001	0.1621	0.324	0.389	0.218	0.1089	0.0908
0.0984	0.0991	0.1077	0.1084	0.1644	0.329	0.395	0.221	0.1105	0.0921
0.1076	0.1081	0.1177	0.1183	0.1663	0.333	0.399	0.224	0.1119	0.0932
0.1173	0.1178	0.1283	0.1289	0.1681	0.336	0.404	0.226	0.1131	0.0943
0.1178	0.1184	0.1289	0.1296	0.1630	0.326	0.391	0.221	0.1164	0.0920
0.1319	0.1324	0.1443	0.1448	0.1630	0.326	0.391	0.221	0.1106	0.0922
0.1485	0.1491	0.1624	0.1631	0.1658	0.332	0.398	0.225	0.1126	0.0939
0.1565	0.1572	0.1712	0.1719	0.1663	0.333	0.399	0.226	0.1130	0.0942
0.1695	0.1700	0.1854	0.1860	0.1691	0.338	0.406	0.230	0.1150	0.0958
0.1690	0.1695	0.1849	0.1854	0.1754	0.351	0.421	0.237	0.1185	0.0988
0.1685	0.1690	0.1843	0.1849	0.1860	0.372	0.446	0.248	0.1241	0.1034
0.1851	0.1856	0.202	0.203	0.1710	0.342	0.410	0.232	0.1161	0.0968
0.1980	0.1985	0.216	0.217	0.1761	0.352	0.423	0.239	0.1194	0.0995
0.1969	0.1975	0.215	0.216	0.1793	0.359	0.430	0.242	0.1212	0.1010
0.1964	0.1969	0.215	0.216	0.1833	0.367	0.440	0.247	0.1234	0.1028
0.238	0.239	0.260	0.261	0.1810	0.362	0.434	0.245	0.1226	0.1022
0.237	0.238	0.259	0.260	0.1834	0.367	0.440	0.249	0.1246	0.1038
0.237	0.238	0.259	0.260	0.1876	0.375	0.450	0.253	0.1267	0.1066
0.281	0.282	0.307	0.308	0.1855	0.371	0.445	0.252	0.1258	0.1049
0.281	0.282	0.307	0.308	0.1889	0.378	0.453	0.256	0.1278	0.1065
0.280	0.281	0.306	0.307	0.1898	0.380	0.455	0.257	0.1285	0.1071
0.354	0.355	0.387	0.388	0.1928	0.386	0.463	0.262	0.1310	0.1091
0.353	0.354	0.386	0.387	0.1943	0.389	0.466	0.263	0.1316	0.1097
0.352	0.353	0.385	0.386	0.1976	0.395	0.474	0.268	0.1338	0.1115
0.446	0.446	0.487	0.487	0.200	0.400	0.481	0.271	0.1357	0.1131
0.446	0.446	0.487	0.487	0.202	0.404	0.485	0.274	0.1368	0.1140
0.446	0.446	0.487	0.487	0.203	0.406	0.487	0.275	0.1375	0.1146
0.473	0.473	0.517	0.517	0.203	0.406	0.487	0.276	0.1378	0.1149
0.491	0.491	0.537	0.537	0.203	0.407	0.488	0.276	0.1378	0.1149
0.507	0.507	0.555	0.555	0.207	0.415	0.498	0.280	0.1400	0.1167

TABLE A.3

Characteristics of General Cable Type HH Hollow Copper Conductors

Conductor Size Circular Mils or AWG	Outside[a] Diameter (in.)	Wall Thickness (in.)	Weight (lb/mi)	Breaking Strength (lb)	Geometric Mean Radius (ft)	Approx. Current Carrying Capacity[b] (amps)
1,000,000	2.103	0.150[c]	16,160	43,190	0.0833	1620
950,000	2.035	0.147[c]	15,350	41,030	0.0805	1565
900,000	1.966	0.144[c]	14,540	38,870	0.0778	1505
850,000	1.901	0.140[c]	13,730	36,710	0.0751	1450
800,000	1.820	0.137[c]	12,920	34,550	0.0722	1390
790,000	1.650	0.131[d]	12,760	34,120	0.0646	1335
750,000	1.750	0.133[c]	12,120	32,390	0.0691	1325
700,000	1.686	0.130[c]	11,310	30,230	0.0665	1265
650,000	1.610	0.126[c]	10,500	28,070	0.0635	1200
600,000	1.558	0.123[c]	9,692	25,910	0.0615	1140
550,000	1.478	0.119[c]	8,884	23,750	0.0583	1075
512,000	1.400	0.115[c]	8,270	22,110	0.0551	1020
500,000	1.390	0.115[c]	8,076	21,590	0.0547	1005
500,000	1.268	0.109[d]	8,074	21,590	0.0494	978
500,000	1.100	0.130[d]	8,068	21,590	0.0420	937
500,000	1.020	0.144[d]	8,063	21,590	0.0384	915
450,000	1.317	0.111[c]	7,268	19,430	0.0518	939
450,000	1.188	0.105[d]	7,266	19,430	0.0462	910
400,000	1.218	0.106[c]	6,460	17,270	0.0478	864
400,000	1.103	0.100[d]	6,458	17,270	0.0428	838
350,000	1.128	0.102[c]	5,653	15,110	0.0443	790
350,000	1.014	0.096[d]	5,650	15,110	0.0393	764
300,000	1.020	0.096[c]	4,845	12,950	0.0399	709
300,000	0.919	0.091[d]	4,843	12,950	0.0355	687
250,000	0.914	0.091[c]	4,037	10,790	0.0357	626
250,000	0.818	0.086[d]	4,036	10,790	0.0315	606
250,000	0.766	0.094[d]	4,034	10,790	0.0292	594
214,500	0.650	0.098[d]	3,459	9,265	0.0243	524
4/0	0.733	0.082[d]	3,415	9,140	0.0281	539
3/0	0.608	0.080[d]	2,707	7,240	0.0230	454
2/0	0.500	0.080[d]	2,146	5,750	0.0180	382

Source: Westinghouse Electric Corporation, *Electric Utility Engineering Reference Book—Distribution Systems*, East Pittsburgh, PA, 1965.

[a] Conductors of smaller diameter for given cross-sectional are also available; in the naught sizes, some additional diameter expansion is possible.

[b] For conductor at 75°C, air at 25°C, wind 1.4 mi/h (2 ft/s), frequency = 60 cycles.

[c] Thickness at edges of interlocked segments.

[d] Thickness uniform throughout.

| r_a Resistance (Ω/Conductor/mi) | | | | | | | | X_a Inductive Reactance (Ω/Conductor/mi) at 1 ft Spacing | | | X'_a Shunt Capacitive Reactance (MΩ·mi/Conductor) at 1 ft Spacing | | |
| 25°C (77°F) | | | | 50°C (122°F) | | | | | | | | | |
DC	25 Cycles	50 Cycles	60 Cycles	DC	25 Cycles	50 Cycles	60 Cycles	25 Cycles	50 Cycles	60 Cycles	25 Cycles	50 Cycles	60 Cycles
0.0576	0.0576	0.0577	0.0577	0.0630	0.0630	0.0631	0.0631	0.1257	0.251	0.302	0.1734	0.0867	0.0722
0.0606	0.0606	0.0607	0.0607	0.0663	0.0664	0.0664	0.0664	0.1274	0.255	0.306	0.1757	0.0879	0.0732
0.0640	0.0640	0.0641	0.0641	0.0700	0.0701	0.0701	0.0701	0.1291	0.258	0.310	0.1782	0.0891	0.0742
0.0677	0.0678	0.0678	0.0678	0.0741	0.0742	0.0742	0.0742	0.1309	0.262	0.314	0.1805	0.0903	0.0752
0.0720	0.0720	0.0720	0.0721	0.0788	0.0788	0.0788	0.0788	0.1329	0.266	0.319	0.1833	0.0917	0.0764
0.0729	0.0729	0.0730	0.0730	0.0797	0.0798	0.0799	0.0799	0.1385	0.277	0.332	0.1906	0.0953	0.0794
0.0768	0.0768	0.0768	0.0769	0.0840	0.0840	0.0841	0.0841	0.1351	0.270	0.324	0.1864	0.0932	0.0777
0.0822	0.0823	0.0823	0.0823	0.0900	0.0900	0.0901	0.0901	0.1370	0.274	0.329	0.1891	0.0945	0.0788
0.0886	0.0886	0.0886	0.0887	0.0969	0.0970	0.0970	0.0970	0.1394	0.279	0.335	0.1924	0.0962	0.0802
0.0959	0.0960	0.0960	0.0960	0.1050	0.1051	0.1051	0.1051	0.1410	0.282	0.338	0.1947	0.0974	0.0811
0.1047	0.1048	0.1048	0.1048	0.1146	0.1146	0.1147	0.1147	0.1437	0.287	0.345	0.1985	0.0992	0.0827
0.1124	0.1125	0.1125	0.1125	0.1230	0.1230	0.1231	0.1231	0.1466	0.293	0.352	0.202	0.1012	0.0843
0.1151	0.1151	0.1152	0.1152	0.1259	0.1260	0.1260	0.1260	0.1469	0.294	0.353	0.203	0.1014	0.0845
0.1151	0.1152	0.1152	0.1152	0.1259	0.1260	0.1260	0.1261	0.1521	0.304	0.365	0.209	0.1047	0.0872
0.1150	0.1151	0.1152	0.1153	0.1258	0.1259	0.1260	0.1260	0.1603	0.321	0.385	0.219	0.1098	0.0915
0.1150	0.1150	0.1152	0.1152	0.1258	0.1259	0.1260	0.1261	0.1648	0.330	0.396	0.225	0.1124	0.0937
0.1279	0.1280	0.1280	0.1280	0.1400	0.1401	0.1401	0.1401	0.1496	0.299	0.359	0.207	0.1033	0.0861
0.1278	0.1279	0.1279	0.1280	0.1399	0.1400	0.1400	0.1401	0.1554	0.311	0.373	0.214	0.1070	0.0892
0.1439	0.1440	0.1440	0.1440	0.1575	0.1576	0.1576	0.1576	0.1537	0.307	0.369	0.212	0.1061	0.0884
0.1438	0.1439	0.1439	0.1440	0.1574	0.1575	0.1575	0.1576	0.1593	0.319	0.382	0.219	0.1097	0.0914
0.1644	0.1645	0.1645	0.1645	0.1799	0.1800	0.1800	0.1800	0.1576	0.315	0.378	0.218	0.1089	0.0907
0.1644	0.1645	0.1645	0.1646	0.1799	0.1800	0.1800	0.1801	0.1637	0.328	0.393	0.225	0.1127	0.0939
0.1918	0.1919	0.1919	0.1919	0.210	0.210	0.210	0.210	0.1628	0.326	0.391	0.225	0.1124	0.0937
0.1917	0.1918	0.1918	0.1919	0.210	0.210	0.210	0.210	0.1688	0.338	0.405	0.232	0.1162	0.0968
0.230	0.230	0.230	0.230	0.252	0.252	0.252	0.252	0.1685	0.337	0.404	0.233	0.1163	0.0970
0.230	0.230	0.230	0.230	0.252	0.252	0.252	0.252	0.1748	0.350	0.420	0.241	0.1203	0.1002
0.230	0.230	0.230	0.230	0.252	0.252	0.252	0.252	0.1787	0.357	0.429	0.245	0.1226	0.1022
0.268	0.268	0.268	0.268	0.293	0.293	0.293	0.294	0.1879	0.376	0.451	0.257	0.1285	0.1071
0.272	0.272	0.272	0.272	0.297	0.297	0.298	0.298	0.1806	0.361	0.433	0.248	0.1242	0.1035
0.343	0.343	0.343	0.343	0.375	0.375	0.375	0.375	0.1907	0.381	0.458	0.262	0.1309	0.1091
0.432	0.432	0.432	0.432	0.472	0.473	0.473	0.473	0.201	0.403	0.483	0.276	0.1378	0.1149

TABLE A.4
Characteristics of Alcoa Aluminum Conductors, Hard-Drawn, 61% Conductivity

Size of Conductor Circular Mils or AWG	No. of Strands	Diameter of Individual Strands (in.)	Outside Diameter (in.)	Ultimate Strength (lb)	Weight (lb/mi)	Geometric Mean Radius at 60 Cycles (ft)	Approx. Current Carrying Capacity[a] (amps)
6	7	0.0612	0.184	528	130	0.00556	100
4	7	0.0772	0.232	826	207	0.00700	134
3	7	0.0867	0.260	1,022	261	0.00787	155
2	7	0.0974	0.292	1,266	329	0.00883	180
1	7	0.1094	0.328	1,537	414	0.00992	209
1/0	7	0.1228	0.368	1,865	523	0.01113	242
1/0	19	0.0745	0.373	2,090	523	0.01177	244
2/0	7	0.1379	0.414	2,350	659	0.01251	282
2/0	19	0.0837	0.419	2,586	659	0.01321	283
3/0	7	0.1548	0.464	2,845	832	0.01404	327
3/0	19	0.0940	0.470	3,200	832	0.01483	328
4/0	7	0.1739	0.522	3,590	1049	0.01577	380
4/0	19	0.1055	0.528	3,890	1049	0.01666	381
250,000	37	0.0822	0.575	4,860	1239	0.01841	425
266,800	7	0.1953	0.586	4,525	1322	0.01771	441
266,800	37	0.0849	0.594	5,180	1322	0.01902	443
300,000	19	0.1257	0.629	5,300	1487	0.01983	478
300,000	37	0.0900	0.630	5,830	1487	0.02017	478
336,400	19	0.1331	0.666	5,940	1667	0.02100	514
336,400	37	0.0954	0.668	6,400	1667	0.02135	514
350,000	37	0.0973	0.681	6,680	1735	0.02178	528
397,500	19	0.1447	0.724	6,880	1967	0.02283	575
477,000	19	0.1585	0.793	8,090	2364	0.02501	646
500,000	19	0.1623	0.812	8,475	2478	0.02560	664
500,000	37	0.1162	0.813	9,010	2478	0.02603	664
556,500	19	0.1711	0.856	9,440	2758	0.02701	710
636,000	37	0.1311	0.918	11,240	3152	0.02936	776
715,500	37	0.1391	0.974	12,640	3546	0.03114	817
750,000	37	0.1424	0.997	12,980	3717	0.03188	864
750,000	61	0.1109	0.998	13,510	3717	0.03211	864
795,000	37	0.1466	1.026	13,770	3940	0.03283	897
874,500	37	0.1538	1.077	14,830	4334	0.03443	949
954,000	37	0.1606	1.024	16,180	4728	0.03596	1000
1,000,000	61	0.1280	1.152	17,670	4956	0.03707	1030
1,000,000	91	0.1048	1.153	18,380	4956	0.03720	1030
1,033,500	37	0.1672	1.170	18,260	5122	0.03743	1050
1,113,000	61	0.1351	1.216	19,660	5517	0.03910	1110
1,192,500	61	0.1398	1.258	21,000	5908	0.04048	1160
1,192,500	91	0.1145	1.259	21,400	5908	0.04062	1160
1,272,000	61	0.1444	1.300	22,000	6299	0.04180	1210
1,351,500	61	0.1489	1.340	23,400	6700	0.04309	1250
1,431,000	61	0.1532	1.379	24,300	7091	0.04434	1300
1,510,500	61	0.1574	1.417	25,600	7487	0.04556	1320
1,590,000	61	0.1615	1.454	27,000	7883	0.04674	1380
1,590,000	91	0.1322	1.454	28,100	7883	0.04691	1380

Source: Westinghouse Electric Corporation, *Electric Utility Engineering Reference Book—Distribution Systems*, East Pittsburgh, PA, 1965.

[a] For conductor at 75°C, wind 1.4 mi/h (2 ft/s), frequency = 60 cycles.

r_a Resistance (Ω/Conductor/mi)								X_a Inductive Reactance (Ω/Conductor/mi) at 1 ft Spacing			X_a Shunt Capacitive Reactance (MΩ·mi/ Conductor) at 1 ft Spacing		
25°C (77°F)				50°C (122°F)									
DC	25 Cycles	50 Cycles	60 Cycles	DC	25 Cycles	50 Cycles	60 Cycles	25 Cycles	50 Cycles	60 Cycles	25 Cycles	50 Cycles	60 Cycles
3.56	3.56	3.56	3.56	3.91	3.91	3.91	3.91	0.2626	0.5251	0.6301	0.3468	0.1734	0.1445
2.24	2.24	2.24	2.24	2.46	2.46	2.46	2.46	0.2509	0.5017	0.6201	0.3302	0.1651	0.1376
1.77	1.77	1.77	1.77	1.95	1.95	1.95	1.95	0.2450	0.4899	0.5879	0.3221	0.1610	0.1342
1.41	1.41	1.41	1.41	1.55	1.55	1.55	1.55	0.2391	0.4782	0.5739	0.3139	0.1570	0.1308
1.12	1.12	1.12	1.12	1.23	1.23	1.23	1.23	0.2333	0.4665	0.5598	0.3055	0.1528	0.1273
0.885	0.8851	0.8853	0.885	0.973	0.9731	0.9732	0.973	0.2264	0.4528	0.5434	0.2976	0.1488	0.1240
0.885	0.8851	0.8853	0.885	0.973	0.9731	0.9732	0.973	0.2246	0.4492	0.5391	0.2964	0.1482	0.1235
0.702	0.7021	0.7024	0.702	0.771	0.7711	0.7713	0.771	0.2216	0.4431	0.5317	0.2890	0.1445	0.1204
0.702	0.7021	0.7024	0.702	0.771	0.7711	0.7713	0.771	0.2188	0.4376	0.5251	0.2882	0.1441	0.1201
0.557	0.5571	0.5574	0.558	0.612	0.6121	0.6124	0.613	0.2157	0.4314	0.5177	0.2810	0.1405	0.1171
0.557	0.5571	0.5574	0.558	0.612	0.6121	0.6124	0.613	0.2129	0.4258	0.5110	0.2801	0.1400	0.1167
0.441	0.4411	0.4415	0.442	0.485	0.4851	0.4855	0.486	0.2099	0.4196	0.5036	0.2726	0.1363	0.1136
0.441	0.4411	0.4415	0.442	0.485	0.4851	0.4855	0.486	0.2071	0.4141	0.4969	0.2717	0.1358	0.1132
0.374	0.3741	0.3746	0.375	0.411	0.4111	0.4115	0.412	0.2020	0.4040	0.4848	0.2657	0.1328	0.1107
0.350	0.3502	0.3506	0.351	0.385	0.3852	0.3855	0.386	0.2040	0.4079	0.4895	0.2642	0.1321	0.1101
0.350	0.3502	0.3506	0.351	0.385	0.3852	0.3855	0.386	0.2004	0.4007	0.4809	0.2633	0.1316	0.1097
0.311	0.3112	0.3117	0.312	0.342	0.3422	0.3426	0.343	0.1983	0.3965	0.4758	0.2592	0.1296	0.1080
0.311	0.3112	0.3117	0.312	0.342	0.3422	0.3426	0.343	0.1974	0.3947	0.4737	0.2592	0.1296	0.1080
0.278	0.2782	0.2788	0.279	0.306	0.3062	0.3067	0.307	0.1953	0.3907	0.4688	0.2551	0.1276	0.1063
0.278	0.2782	0.2788	0.279	0.306	0.3062	0.3067	0.307	0.1945	0.3890	0.4668	0.2549	0.1274	0.1062
0.267	0.2672	0.2678	0.268	0.294	0.2942	0.2947	0.295	0.1935	0.3870	0.4644	0.2537	0.1268	0.1057
0.235	0.2352	0.2359	0.236	0.258	0.2582	0.2589	0.259	0.1911	0.3822	0.4587	0.2491	0.1246	0.1038
0.196	0.1963	0.1971	0.198	0.215	0.2153	0.2160	0.216	0.1865	0.3730	0.4476	0.2429	0.1214	0.1012
0.187	0.1873	0.1882	0.189	0.206	0.2062	0.2070	0.208	0.1853	0.3707	0.4448	0.2412	0.1206	0.1005
0.187	0.1873	0.1882	0.189	0.206	0.2062	0.2070	0.208	0.1845	0.3689	0.4427	0.2410	0.1205	0.1004
0.168	0.1683	0.1693	0.170	0.185	0.1853	0.1862	0.187	0.1826	0.3652	0.4383	0.2374	0.1187	0.0989
0.147	0.1474	0.1484	0.149	0.162	0.1623	0.1633	0.164	0.1785	0.3569	0.4283	0.2323	0.1162	0.0968
0.137	0.1314	0.1326	0.133	0.144	0.1444	0.1455	0.146	0.1754	0.3508	0.4210	0.2282	0.1141	0.0951
0.125	0.1254	0.1267	0.127	0.137	0.1374	0.1385	0.139	0.1743	0.3485	0.4182	0.2266	0.1133	0.0944
0.125	0.1254	0.1267	0.127	0.137	0.1374	0.1385	0.139	0.1739	0.3477	0.4173	0.2263	0.1132	0.0943
0.117	0.1175	0.1188	0.120	0.129	0.1294	0.1306	0.131	0.1728	0.3455	0.4146	0.2244	0.1122	0.0935
0.107	0.1075	0.1089	0.110	0.118	0.1185	0.1198	0.121	0.1703	0.3407	0.4088	0.2210	0.1105	0.0921
0.0979	0.0985	0.1002	0.100	0.108	0.1085	0.1100	0.111	0.1682	0.3363	0.4036	0.2179	0.1090	0.0908
0.0934	0.0940	0.0956	0.0966	0.103	0.1035	0.1050	0.106	0.1666	0.3332	0.3998	0.2162	0.1081	0.0901
0.0934	0.0940	0.0956	0.0966	0.103	0.1035	0.1050	0.106	0.1664	0.3328	0.3994	0.2160	0.1080	0.0900
0.0904	0.0910	0.0927	0.0936	0.0994	0.0999	0.1015	0.102	0.1661	0.3322	0.3987	0.2150	0.1075	0.0895
0.0839	0.0845	0.0864	0.0874	0.0922	0.0928	0.0945	0.0954	0.1639	0.3278	0.3934	0.2124	0.1062	0.0885
0.0783	0.0790	0.0810	0.0821	0.0860	0.0866	0.0884	0.0895	0.1622	0.3243	0.3892	0.2100	0.1050	0.0875
0.0783	0.0790	0.0810	0.0821	0.0860	0.0866	0.0884	0.0895	0.1620	0.3240	0.3888	0.2098	0.1049	0.0874
0.0734	0.0741	0.0762	0.0774	0.0806	0.0813	0.0832	0.0843	0.1606	0.3211	0.3853	0.2076	0.1038	0.0865
0.0691	0.0699	0.0721	0.0733	0.0760	0.0767	0.0787	0.0798	0.1590	0.3180	0.3816	0.2054	0.1027	0.0856
0.0653	0.0661	0.0685	0.0697	0.0718	0.0725	0.0747	0.0759	0.1576	0.3152	0.3782	0.2033	0.1016	0.0847
0.0618	0.0627	0.0651	0.0665	0.0679	0.0687	0.0710	0.0722	0.1562	0.3123	0.3748	0.2014	0.1007	0.0839
0.0597	0.0596	0.0622	0.0636	0.0645	0.0653	0.0677	0.0690	0.1549	0.3098	0.3718	0.1997	0.0998	0.0832
0.0587	0.0596	0.0622	0.0636	0.0645	0.0653	0.0677	0.0690	0.1547	0.3094	0.3713	0.1997	0.0998	0.0832

TABLE A.5
Characteristics of Aluminum Cable, Steel Reinforced (Aluminum Company of America)

Circular Mils or AWG Aluminum	Aluminum Strands	Layers	Strand Diameter (in.)	Steel Strands	Strand Diameter (in.)	Outside Diameter (in.)	Copper Equivalent[a] Circular Miles or AWG	Ultimate Strength (lb)	Weight (lb/mi)	Geometric Mean Radius at 60 Cycles (ft)	Approx. Current Carrying Capacity[b] (amps)	r_a Resistance 25°C (77°F) 25 DC	Cycles
1,590,000	54	3	0.1716	19	0.1030	1.545	1,000,000	56,000	10,777	0.0520	1380	0.0587	0.0588
1,510,500	54	3	0.1673	19	0.1004	1.506	950,000	53,200	10,237	0.0507	1340	0.0618	0.0619
1,431,000	54	3	0.1628	19	0.0977	1.465	900,000	50,400	9,699	0.0493	1300	0.0652	0.0653
1,351,000	54	3	0.1582	19	0.0949	1.424	850,000	47,600	9,160	0.0479	1250	0.0691	0.0692
1,272,000	54	3	0.1535	19	0.0921	1.382	800,000	44,800	8,621	0.0465	1200	0.0734	0.0735
1,192,500	54	3	0.1486	19	0.0892	1.338	750,000	43,100	8,082	0.0450	1160	0.0783	0.0784
1,113,000	54	3	0.1436	19	0.0862	1.293	700,000	40,200	7,544	0.0435	1110	0.0839	0.0840
1,033,500	54	3	0.1384	7	0.1384	1.246	650,000	37,100	7,019	0.0420	1060	0.0903	0.0905
954,000	54	3	0.1329	7	0.1329	1.196	600,000	34,200	6,479	0.0403	1010	0.0979	0.0980
900,000	54	3	0.1291	7	0.1291	1.162	566,000	32,300	6,112	0.0391	970	0.104	0.104
874,500	54	3	0.1273	7	0.1273	1.146	550,000	31,400	5,940	0.0386	950	0.107	0.107
795,000	54	3	0.1214	7	0.1214	1.093	500,000	28,500	5,399	0.0368	900	0.117	0.118
795,000	26	2	0.1749	7	0.1360	1.108	500,000	31,200	5,770	0.0375	900	0.117	0.117
795,000	30	2	0.1628	19	0.0977	1.140	500,000	38,400	6,517	0.0393	910	0.117	0.117
715,500	54	3	0.1151	7	0.1151	1.036	450,000	26,300	4,859	0.0349	830	0.131	0.131
715,500	26	2	0.1659	7	0.1290	1.051	450,000	28,100	5,193	0.0355	840	0.131	0.131
715,500	30	2	0.1544	19	0.0926	1.081	450,000	34,600	5,865	0.0372	840	0.131	0.131
666,000	54	3	0.1111	7	0.1111	1.000	419,000	24,500	4,527	0.0337	800	0.140	0.140
636,000	54	3	0.1085	7	0.1085	0.977	400,000	23,600	4,319	0.0329	770	0.147	0.147
636,000	26	2	0.1564	7	0.1216	0.990	400,000	25,000	4,616	0.0335	780	0.147	0.147
636,000	30	2	0.1456	19	0.0874	1.019	400,000	31,500	5,213	0.0351	780	0.147	0.147
605,000	54	3	0.1059	7	0.1059	0.953	380,500	22,500	4,109	0.0321	750	0.154	0.155
605,000	26	2	0.1525	7	0.1186	0.966	380,500	24,100	4,391	0.0327	760	0.154	0.154
556,500	26	2	0.1463	7	0.1138	0.927	350,000	22,400	4,039	0.0313	730	0.168	0.168
556,500	30	2	0.1362	7	0.1362	0.953	350,000	27,200	4,588	0.0328	730	0.168	0.168
500,000	30	2	0.1291	7	0.1291	0.904	314,500	24,400	4,122	0.0311	690	0.187	0.187
477,000	26	2	0.1355	7	0.1054	0.858	300,000	19,430	3,462	0.0290	670	0.196	0.196
477,000	30	2	0.1261	7	0.1261	0.883	300,000	23,300	3,933	0.0304	670	0.196	0.196
397,500	26	2	0.1236	7	0.0961	0.783	250,000	16,190	2,885	0.0265	590	0.235	
397,500	30	2	0.1151	7	0.1151	0.806	250,000	19,980	3,277	0.0278	600	0.235	Same
336,400	26	2	0.1138	7	0.0885	0.721	4/0	14,050	2,442	0.0244	530	0.278	
336,400	30	2	0.1059	7	0.1059	0.741	4/0	17,040	2,774	0.0255	530	0.278	
300,000	26	2	0.1074	7	0.0835	0.680	188,700	12,650	2,178	0.0230	490	0.311	
300,000	30	2	0.1000	7	0.1000	0.700	188,700	15,430	2,473	0.0241	500	0.311	
266,800	26	2	0.1013	7	0.0788	0.642	3/0	11,250	1,936	0.0217	460	0.350	

For Current Approx. 75% Capacity[c]

Circular Mils or AWG Aluminum	Aluminum Strands	Layers	Strand Diameter (in.)	Steel Strands	Strand Diameter (in.)	Outside Diameter (in.)	Copper Equivalent[a] Circular Miles or AWG	Ultimate Strength (lb)	Weight (lb/mi)	Geometric Mean Radius at 60 Cycles (ft)	Approx. Current Carrying Capacity (amps)	r_a DC	Cycles
266,800	6	1	0.2109	7	0.0703	0.633	3/0	9,645	1,802	0.00684	460	0.351	0.351
4/0	6	1	0.1878	1	0.1878	0.563	2/0	8,420	1,542	0.00814	340	0.441	0.442
3/0	6	1	0.1672	1	0.1672	0.502	1/0	6,675	1,223	0.00600	300	0.556	0.557
2/0	6	1	0.1490	1	0.1490	0.447	1	5,345	970	0.00510	270	0.702	0.702
1/0	6	1	0.1327	1	0.1327	0.398	2	4,280	769	0.00446	230	0.885	0.885
1	6	1	0.1182	1	0.1182	0.355	3	3,480	610	0.00418	200	1.12	1.12
2	6	1	0.1052	1	0.1052	0.316	4	2,790	484	0.00418	180	1.41	1.41
2	7	1	0.0974	1	0.1299	0.325	4	3,525	566	0.00504	180	1.41	1.41
3	6	1	0.0937	1	0.0937	0.281	5	2,250	384	0.00430	160	1.78	1.78
4	6	1	0.0834	1	0.0834	0.250	6	1,830	304	0.00437	140	2.24	2.24
4	7	1	0.0772	1	0.1029	0.257	6	2,288	356	0.00452	140	2.24	2.24
5	6	1	0.0743	1	0.0743	0.223	7	1,460	241	0.00416	120	2.82	2.82
6	6	1	0.0661	1	0.0661	0.198	8	1,170	191	0.00394	100	3.56	3.56

[a] Based on copper 97%, aluminum 61% conductivity.

[b] For conductor at 75°C, air at 25°C, wind 1.4 mi/h (2 ft/s), frequency = 60 cycles.

[c] "Current Approx. 75% Capacity" is 75% of the "Approx. Current Carrying Capacity in Amps." and is approximately the current which will produce 50°C conductor temperature (25°C rise) with 25°C air temperature, wind 1.4 mi/h.

(Ω/Conductor/mi)						X_a'			X_a'		
Small Currents		50°C (122°F) Current Approx. 75% Capacity^c				x_a Inductive Reactance (Ω/Conductor/mi) at 1 ft Spacing All Currents			X_a' Shunt Capacitive Reactance (MΩ·mi/Conductor) at 1 ft Spacing		
50 Cycles	60 Cycles	DC	25 Cycles	50 Cycles	60 Cycles	25 Cycles	50 Cycles	60 Cycles	25 Cycles	50 Cycles	60 Cycles
0.0590	0.0591	0.0646	0.0656	0.0675	0.0684	0.1495	0.299	0.359	0.1953	0.0977	0.0814
0.0621	0.0622	0.0680	0.0690	0.0710	0.0720	0.1508	0.302	0.362	0.1971	0.0986	0.0821
0.0655	0.0656	0.0718	0.0729	0.0749	0.0760	0.1522	0.304	0.365	0.1991	0.0996	0.0830
0.0694	0.0695	0.0761	0.0771	0.0792	0.0803	0.1536	0.307	0.369	0.201	0.1006	0.0838
0.0737	0.0738	0.0808	0.0819	0.0840	0.0851	0.1551	0.310	0.372	0.203	0.1016	0.0847
0.0786	0.0788	0.0862	0.0872	0.0894	0.0906	0.1568	0.314	0.376	0.206	0.1028	0.0857
0.0842	0.0844	0.0924	0.0935	0.0957	0.0969	0.1585	0.317	0.380	0.208	0.1040	0.0867
0.0907	0.0908	0.0994	0.1005	0.1025	0.1035	0.1603	0.321	0.385	0.211	0.1053	0.0878
0.0981	0.0982	0.1078	0.1088	0.1118	0.1128	0.1624	0.325	0.390	0.214	0.1068	0.0890
0.104	0.104	0.1145	0.1155	0.1175	0.1185	0.1639	0.328	0.393	0.216	0.1078	0.0898
0.107	0.108	0.1178	0.1188	0.1218	0.1228	0.1646	0.329	0.395	0.217	0.1083	0.0903
0.118	0.119	0.1288	0.1308	0.1358	0.1378	0.1670	0.334	0.401	0.220	0.1100	0.0917
0.117	0.117	0.1288	0.1288	0.1288	0.1288	0.1660	0.332	0.399	0.219	0.1095	0.0912
0.117	0.117	0.1288	0.1288	0.1288	0.1288	0.1637	0.327	0.393	0.217	0.1085	0.0904
0.131	0.132	0.1442	0.1452	0.1472	0.1482	0.1697	0.339	0.407	0.224	0.1119	0.0932
0.131	0.131	0.1442	0.1442	0.1442	0.1442	0.1687	0.337	0.405	0.223	0.1114	0.0928
0.131	0.131	0.1442	0.1442	0.1442	0.1442	0.1664	0.333	0.399	0.221	0.1104	0.0920
0.141	0.141	0.1541	0.1571	0.1591	0.1601	0.1715	0.343	0.412	0.226	0.1132	0.0943
0.148	0.148	0.1618	0.1638	0.1678	0.1688	0.1726	0.345	0.414	0.228	0.1140	0.0950
0.147	0.147	0.1618	0.1618	0.1618	0.1618	0.1718	0.344	0.412	0.227	0.1135	0.0946
0.147	0.147	0.1618	0.1618	0.1618	0.1618	0.1693	0.339	0.406	0.225	0.1125	0.0937
0.155	0.155	0.1695	0.1715	0.1755	0.1775	0.1739	0.348	0.417	0.230	0.1149	0.0957
0.154	0.154	0.1700	0.1720	0.1720	0.1720	0.1730	0.346	0.415	0.229	0.1144	0.0953
0.168	0.168	0.1849	0.1859	0.1859	0.1859	0.1751	0.350	0.420	0.232	0.1159	0.0965
0.168	0.168	0.1849	0.1859	0.1859	0.1859	0.1728	0.346	0.415	0.230	0.1149	0.0957
0.187	0.187	0.206				0.1754	0.351	0.421	0.234	0.1167	0.0973
0.196	0.196	0.216				0.1790	0.358	0.430	0.237	0.1186	0.0988
0.196	0.196	0.216				0.1766	0.353	0.424	0.235	0.1176	0.0980
		0.259				0.1836	0.367	0.441	0.244	0.1219	0.1015
as DC		0.259	Same as DC			0.1812	0.362	0.435	0.242	0.1208	0.1006
		0.306				0.1872	0.376	0.451	0.250	0.1248	0.1039
		0.306				0.1855	0.371	0.445	0.248	0.1238	0.1032
		0.342				0.1908	0.382	0.458	0.254	0.1269	0.1057
		0.342				0.1883	0.377	0.452	0.252	0.1258	0.1049
		0.385				0.1936	0.387	0.465	0.258	0.1289	0.1074

Single Layer Conductors

50 Cycles	60 Cycles	DC	25 Cycles	50 Cycles	60 Cycles	Small Currents			Current Approx. 75% Capacity^c			25 Cycles	50 Cycles	60 Cycles
						25 Cycles	50 Cycles	60 Cycles	25 Cycles	50 Cycles	60 Cycles			
0.351	0.352	0.386	0.430	0.510	0.552	0.194	0.388	0.466	0.252	0.504	0.605	0.259	0.1294	0.1079
0.444	0.445	0.485	0.514	0.567	0.592	0.218	0.437	0.524	0.242	0.484	0.581	0.267	0.1336	0.1113
0.559	0.560	0.612	0.642	0.697	0.723	0.225	0.450	0.540	0.259	0.517	0.621	0.275	0.1377	0.1147
0.704	0.706	0.773	0.806	0.866	0.895	0.231	0.462	0.554	0.267	0.534	0.641	0.284	0.1418	0.1182
0.887	0.888	0.974	1.01	1.08	1.12	0.237	0.473	0.568	0.273	0.547	0.656	0.292	0.1460	0.1216
1.12	1.12	1.23	1.27	1.34	1.38	0.242	0.483	0.580	0.277	0.554	0.665	0.300	0.1500	0.1250
1.41	1.41	1.55	1.59	1.66	1.69	0.247	0.493	0.592	0.277	0.554	0.665	0.308	0.1542	0.1285
1.41	1.41	1.55	1.59	1.62	1.65	0.247	0.493	0.592	0.267	0.535	0.642	0.306	0.1532	0.1276
1.78	1.78	1.95	1.95	2.04	2.07	0.252	0.503	0.604	0.275	0.551	0.661	0.317	0.1583	0.1320
2.24	2.24	2.47	2.50	2.54	2.57	0.257	0.514	0.611	0.274	0.549	0.659	0.325	0.1627	0.1355
2.24	2.24	2.47	2.50	2.53	2.55	0.257	0.515	0.618	0.273	0.545	0.655	0.323	0.1615	0.1346
2.82	2.82	3.10	3.12	3.16	3.18	0.262	0.525	0.630	0.279	0.557	0.665	0.333	0.1666	0.1388
3.56	3.56	3.92	3.94	3.97	3.98	0.268	0.536	0.643	0.281	0.561	0.673	0.342	0.1708	0.1423

TABLE A.6

Characteristics of "Expanded" Aluminum Cable, Steel Reinforced (Aluminum Company of America)

Circular Mils AWG Aluminum	Aluminum			Steel		Filler Section				Outside Diameter (in.)	Copper Equivalent Circular Miles or AWG	Ultimate Strength (lb)	Weight (lb/mi)	Geometric Mean Radius at 60 Cycles (ft)	Approx. Current Carrying Capacity (amps)
						Aluminum		Paper							
	Strands	Layers	Strand Diameter (in.)	Strands	Strand Diameter (in.)	Strand	Strand Diameter (in.)	Strands	Layers						
850,000	54	2	0.1255	19	0.0834	4	0.1182	23	2	1.38	534,000	35,371	7,200		
1,150,000	54	2	0.1409	19	0.0921	4	0.1353	24	2	1.55	724,000	41,900	9,070	(a)	(a)
1,338,000	66	2	0.1350	19	0.100	4	0.184	18	2	1.75	840,000	49,278	11,340		

(a) Electrical characteristics not available until laboratory measurements are completed.

r_a Resistance (Ω/Conductor/mi)								x_a			X_a'		
25°C (77°F) Small Currents				50°C (122°F) Current Approx. 75% Capacity‡				Inductive Reactance (Ω/Conductor/mi) at 1 ft Spacing All Currents			Shunt Capacitive Reactance (MΩ·mi/Conductor) at 1 ft Spacing		
DC	25 Cycles	50 Cycles	60 Cycles	DC	25 Cycles	50 Cycles	60 Cycles	25 Cycles	50 Cycles	60 Cycles	25 Cycles	50 Cycles	60 Cycles
	(a)				(a)			(a)	(a)	(a)	(a)	(a)	(a)

TABLE A.7
Characteristics of Copperweld Copper Conductors

Nominal Designation	(Number and Diameter of Wires) Copperweld	Copper	Outside Diameter (in.)	Copper Equivalent Circular Mile or AWG	Rated Breaking Load (lb)	Weight (lb/mi)	Geometric Mean Radius at 60 Cycles (ft)	Approx. Current Carrying Capacity at 60 Cycles (amps)[a]
350 E	7x. 1576″	12x. 1576″	0.788	350,000	32,420	7409	0.0220	660
350 EK	4x. 1470″	15x. 1470″	0.735	350,000	23,850	6536	0.0245	680
350 V	3x. 1751″	9x. 1893″	0.754	350,000	23,480	6578	0.0226	650
300 E	7x. 1459″	12x. 1459″	0.729	300,000	27,770	6351	−0.0204	500
300 EK	4x. 1361″	15x. 1361″	0.680	300,000	20,960	5602	0.0227	610
300 V	3x. 1621″	9x.1752″	0.698	300,000	20,730	5639	0.0208	590
250 E	7x. 1332″	12x. 1332″	0.666	250,000	23,920	5292	0.01859	540
250 EK	4x. 1242″	15x. 1242″	0.621	250,000	17,840	4669	0.0207	540
250 V	3x. 1480″	9x.1600″	0.637	250,000	17,420	4699	0.01911	530
4/0 E	7x. 1225″	12x. 1225″	0.613	4/0	20,730	4479	0.01711	480
4/0 G	2x. 1944″	5x. 1944″	0.583	4/0	1,540	4168	0.01409	460
4/0 EK	4x. 1143″	15x. 1143″	0.571	4/0	15,370	3951	0.01903	490
4/0 V	3x. 1361″	9x. 1472″	0.586	4/0	15,000	3977	0.01758	470
4/0 F	1x. 1833″	6x. 1833″	0.550	4/0	12,290	3750	0.01558″	470
3/0 E	7x. 1091″	12x. 1091″	0.545	3/0	16,800	3522	0.01521	420
3/0 J	3x. 1851″	4x.1851″	0.555	3/0	16,170	3732	0.01158	410
310 G	2x. 1731″	2x. 1731″	0.519	3/0	12,860	3305	0.01254	400
3/0 EK	4x. 1018″	4x. 1018″	0.509	3/0	12,370	3134	0.01697	420
3/0 V	3x. 1311″	9x. 1311″	0.522	3/0	12,220	3154	0.01566	410
3/0 F	1x. 1632″	6x. 1632″	0.490	3/0	9980	2974	0.01388	410
2/0 K	4x. 1780″	3x. 1780″	0.534	2/0	17,600	3411	0.00912	360
2/0 J	3x. 1648″	4x. 1648″	0.494	2/0	13,430	2960	0.01029	350
2/0 G	2x. 1542″	6x. 1542″	0.463	2/0	10,510	2622	0.01119	350
2/0 V	3x. 1080″	9x. 1167″	0.465	2/0	9,846	2502	0.01395	360
2/0 F	1x. 1454″	6x. 1454″	0.436	2/0	8,094	2359	0.01235	350
1/0 K	4x. 1585″	3x. 1585″	0.475	1/0	14,490	2703	0.00812	310
1/0 J	3x. 1467″	4x. 1467″	0.440	1/0	10,970	2346	0.00917	310
1/0 G	2x. 1373″	5x. 1373″	0.412	1/0	8,563	2078	0.00995	310
1/0 F	1x. 1294″	6x. 1294″	0.388	1/0	6,536	1870	0.01099	310
1 N	5x. 1546″	2x. 1546″	0.464	1	15,410	2541	0.00638	280
1 K	4x. 1412″	3x. 1412″	0.423	1	11,900	2144	0.00723	270
1 J	3x. 1307″	4x. 1307″	0.392	1	9,000	1881	0.00817	270
1 G	2x. 1222″	5x. 1222″	0.367	1	6,956	1649	0.00887	260
1 F	1x. 1153″	6x. 1153″	0.346	1	5,266	1483	0.00980	270
2 P	6x. 1540″	1x. 1540″	0.452	2	16,870	2487	0.00501	250
2 N	5x. 1377″	2x. 1377″	0.413	2	12,880	2015	0.00568	240
2 K	4x. 1257″	3x. 1257″	0.377	2	9,730	1701	0.00644	240
2 J	3x. 1164″	4x.1164″	0.349	2	7,322	1476	0.00727	230
2 A	1x. 1699″	2x.1699″	0.366	2	5,876	1356	0.00763	240
2 G	2x. 1089″	5x.1089″	0.327	2	5,626	1307	0.00790	230

r_a Resistance (Ω/Conductor/mi) at 25°C (77°F) Small Currents				r_a Resistance (Ω/Conductor/mi) at 50°C (122°F) Current Approx. 75% of Capacity[b]				X_a Inductive Reactance (Ω/Conductor/mi) 1 ft Spacing Average Currents			X'_a Capacitive Reactance (MΩ·mi/Conductor) 1 ft Spacing		
DC	25 Cycles	50 Cycles	60 Cycles	DC	25 Cycles	50 Cycles	60 Cycles	25 Cycles	50 Cycles	60 Cycles	25 Cycles	50 Cycles	60 Cycles
0.1658	0.1728	0.1789	0.1812	0.1812	0.1915	0.201	0.204	0.1929	0.386	0.463	0.243	0.1216	0.1014
0.1658	0.1682	0.1700	0.1705	0.1812	0.1845	0.1873	0.1882	0.1875	0.375	0.450	0.248	0.1241	0.1034
0.1655	0.1725	0.1800	0.1828	0.1809	0.1910	0.202	0.206	0.1915	0.383	0.460	0.246	0.1232	0.1027
0.1934	0.200	0.207	0.209	0.211	0.222	0.232	0.235	0.1969	0.394	0.473	0.249	0.1244	0.1037
0.1934	0.1958	0.1976	0.198	0.211	0.215	0.218	0.219	0.1914	0.383	0.460	0.254	0.1269	0.1057
0.1930	0.200	0.208	0.210	0.211	0.222	0.233	0.237	0.1954	0.391	0.469	0.252	0.1259	0.1050
0.232	0.239	0.245	0.248	0.254	0.265	0.275	0.279	0.202	0.403	0.484	0.255	0.1276	0.1604
0.232	0.235	0.236	0.237	0.254	0.258	0.261	0.261	0.1960	0.392	0.471	0.260	0.1301	0.1084
0.232	0.239	0.246	0.249	0.253	0.264	0.276	0.281	0.200	0.400	0.480	0.258	0.1292	0.1077
0.274	0.281	0.287	0.290	0.300	0.312	0.323	0.326	0.206	0.411	0.493	0.261	0.1306	0.1088
0.273	0.284	0.294	0.298	0.299	0.318	0.336	0.342	0.215	0.431	0.517	0.265	0.1324	0.1103
0.274	0.277	0.278	0.279	0.300	0.304	0.307	0.308	0.200	0.401	0.481	0.266	0.1331	0.1109
0.274	0.281	0.288	0.291	0.299	0.311	0.323	0.328	0.204	0.409	0.490	0.264	0.1322	0.1101
0.273	0.280	0.285	0.287	0.299	0.309	0.318	0.322	0.210	0.421	0.505	0.269	0.1344	0.1220
0.346	0.353	0.359	0.361	0.378	0.391	0.402	0.407	0.212	0.423	0.608	0.270	0.1348	0.1123
0.344	0.356	0.367	0.372	0.377	0.398	0.419	0.428	0.225	0.451	0.541	0.268	0.1341	0.1118
0.344	0.355	0.365	0.369	0.377	0.397	0.416	0.423	0.221	0.443	0.531	0.273	0.1365	0.1137
0.346	0.348	0.350	0.351	0.378	0.382	0.386	0.386	0.206	0.412	0.495	0.274	0.1372	0.1143
0.345	0.352	0.360	0.362	0.377	0.390	0.403	0.408	0.210	0.420	0.504	0.273	0.1363	0.1136
0.344	0.351	0.366	0.358	0.377	0.388	0.397	0.401	0.216	0.432	0.519	0.277	0.1385	0.1155
0.434	0.447	0.459	0.466	0.475	0.499	0.524	0.535	0.237	0.476	0.570	0.271	0.1355	0.1129
0.434	0.446	0.457	0.462	0.475	0.498	0.520	0.530	0.231	0.463	0.555	0.277	0.1383	0.1152
0.434	0.445	0.456	0.459	0.475	0.497	0.518	0.526	0.227	0.454	0.545	0.281	0.1406	0.1171
0.435	0.442	0.450	0.452	0.476	0.489	0.504	0.509	0.216	0.432	0.518	0.281	0.1404	0.1170
0.434	0.441	0.446	0.448	0.475	0.487	0.497	0.501	0.222	0.444	0.533	0.285	0.1427	0.1189
0.548	0.560	0.573	0.579	0.599	0.625	0.652	0.664	0.243	0.487	0.584	0.279	0.1397	0.1164
0.548	0.559	0.570	0.576	0.699	0.624	0.648	0.659	0.237	0.474	0.589	0.285	0.1423	0.1188
0.548	0.559	0.568	0.573	0.699	0.623	0.645	0.654	0.233	0.466	0.559	0.289	0.1447	0.1206
0.548	0.554	0.559	0.562	0.599	0.612	0.622	0.627	0.228	0.456	0.547	0.294	0.1469	0.1224
0.691	0.705	0.719	0.726	0.755	0.787	0.818	0.832	0.256	0.512	0.614	0.281	0.1405	0.1171
0.691	0.704	0.716	0.722	0.755	0.784	0.813	0.825	0.249	0.498	0.598	0.288	0.1438	0.1198
0.691	0.703	0.714	0.719	0.755	0.783	0.808	0.820	0.243	0.486	0.583	0.293	0.1465	0.1221
0.691	0.702	0.712	0.716	0.755	0.781	0.805	0.815	0.239	0.478	0.573	0.298	0.1488	0.1240
0.691	0.698	0.704	0.705	0.755	0.769	0.781	0.786	0.234	0.468	0.561	0.302	0.1509	0.1258
0.871	0.886	0.901	0.909	0.952	0.988	1.024	1.040	0.268	0.536	0.643	0.281	0.1406	0.1172
0.871	0.885	0.899	0.906	0.952	0.986	1.020	1.035	0.261	0.523	0.627	0.289	0.1445	0.1208
0.871	0.884	0.896	0.902	0.952	0.983	1.014	1.028	0.255	0.510	0.612	0.296	0.1479	0.1232
0.871	0.883	0.894	0.899	0.952	0.982	1.010	1.022	0.249	0.498	0.598	0.301	0.1506	0.1255
0.869	0.875	0.880	0.882	0.950	0.962	0.973	0.979	0.247	0.493	0.592	0.298	0.1489	0.1241
0.871	0.882	0.892	0.896	0.952	0.980	1.006	1.016	0.246	0.489	0.587	0.306	0.1529	0.1276

(continued)

TABLE A.7 (continued)
Characteristics of Copperweld Copper Conductors

Nominal Designation	Size of Conductor (Number and Diameter of Wires) Copperweld	Copper	Outside Diameter (in.)	Copper Equivalent Circular Mile or AWG	Rated Breaking Load (lb)	Weight (lb/mi)	Geometric Mean Radius at 60 Cycles (ft)	Approx. Current Carrying Capacity at 60 Cycles (amps)[a]
2 F	1x. 1026″	6x. 1026″	0.308	2	4,233	1176	0.00873	230
3 P	6x. 1371″	1x. 1371″	0.411	3	13,910	1973	0.00445	220
3 N	5x. 1226″	2x. 1226″	0.368	3	10,390	1598	0.00506	210
3 K	4x. 1120″	3x. 1120″	0.336	3	7,910	1349	0.00674	210
3 J	3x. 1036″	4x. 1036″	0.311	3	5,956	1171	0.00648	200
3 A	1x. 1513″	2x. 1513″	0.326	3	4,810	1075	0.00679	210
4 P	6x. 1221″	1x. 1221″	0.366	4	11,420	1584	0.00397	190
4 N	5x. 1092″	2x. 1092″	0.328	4	8,460	1267	0.00.451	180
4 D	2x. 1615″	1x. 1615″	0.348	4	7,340	1191	0.00586	190
4 A	1x. 1347″	2x. 1347″	0.290	4	3,938	853	0.00604	180
5 P	6x. 1087″	1x. 1087″	0.326	5	9,311	1240	0.00353	160
5 D	2x. 1438″	1x. 1438″	0.310	5	6,035	944	0.00504	160
5 A	1x. 1200″	2x. 1200″	0.258	5	3,193	675	0.00538	160
6 D	2x. 1281″	1x. 1281″	0.276	6	4,942	749	0.00449	140
6 A	1x. 1068″	2x. 1068″	0.230	6	2,585	536	0.00479	140
6 C	1x. 1046″	2x. 1046″	0.225	6	2,143	514	0.00469	130
7 D	2x. 1141″	1x. 1141″	0.246	7	4,022	594	0.00400	120
7 A	1x. 1266″	2x. 0895″	0.223	7	2,754	495	0.00441	120
8 D	2x. 1016″	1x. 1016″	0.219	8	3,256	471	0.00356	110
8 A	1x. 1127″	2x. 0797″	0.199	8	2,233	392	0.00394	100
8 C	1x. 0808″	2x. 0834″	0.179	8	1,362	320	0.00373	100
9½ D	2x. 0808″	1x. 0808″	0.174	9½	1,743	298	0.00283	85

Source: Westinghouse Electric Corporation, *Electric Utility Engineering Reference Book—Distribution Systems*, East Pittsburgh, PA, 1965.

[a] Based on a conductor temperature of 75°C and an ambient of 25°C wind 1.4 mi/h (2 ft/s), (frequency = 60 cycles, average tarnished surface).

[b] Resistances at 50°C total temperature, based on an ambient of 25°C plus 25°C rise due to heating effect of current. The of approximate magnitude of the current necessary to produce the 25°C rise is 75% of the "approximate current-carrying capacity 60 cycles."

r_a Resistance (Ω/Conductor/mi) at 25°C (77°F) Small Currents				r_a Resistance (Ω/Conductor/mi) at 50°C (122°F) Current Approx. 75% of Capacity[b]				X_a Inductive Reactance (Ω/Conductor/mi) 1 ft Spacing Average Currents			X'_a Capacitive Reactance (MΩ·mi/ Conductor) 1 ft Spacing		
DC	25 Cycles	50 Cycles	60 Cycles	DC	25 Cycles	50 Cycles	60 Cycles	25 Cycles	50 Cycles	60 Cycles	25 Cycles	50 Cycles	60 Cycles
0.871	0.878	0.884	0.885	0.952	0.967	0.979	0.986	0.230	0.479	0.576	0.310	0.1551	0.1292
1.098	1.113	1.127	1.136	1.200	1.239	1.273	1.296	0.274	0.647	0.657	0.290	0.1448	0.1207
1.098	1.112	1.126	1.133	1.200	1.237	1.273	1.289	0.267	0.634	0.641	0.298	0.1487	0.1239
1.098	1.111	1.123	1.129	1.200	1.233	1.267	1.281	0.261	0.622	0.626	0.304	0.1520	0.1266
1.098	1.110	1.121	1.126	1.200	1.232	1.262	1.275	0.255	0.609	0.611	0.309	0.1547	0.1289
1.096	1.102	1.107	1.109	1.198	1.211	1.226	1.229	0.252	0.606	0.606	0.306	0.1531	0.1275
1.385	1.400	1.414	1.423	1.514	1.555	1.598	1.616	0.280	0.559	0.671	0.298	0.1489	0.1241
1.385	1.399	1.413	1.420	1.514	1.554	1.593	1.610	0.273	0.546	0.655	0.306	0.1528	0.1274
1.382	1.389	1.396	1.399	1.511	1.529	1.544	1.542	0.262	0.523	0.628	0.301	0.1507	0.1256
1.382	1.388	1.393	1.395	1.511	1.525	1.540	1.545	0.258	0.517	0.620	0.316	0.1572	0.1310
1.747	1.762	1.776	1.785	1.909	1.954	2.00	2.02	0.285	0.571	0.685	0.306	0.1531	0.1275
1.742	1.749	1.756	1.759	1.905	1.924	1.941	1.939	0.268	0.535	0.642	0.310	0.1548	0.1290
1.742	1.748	1.753	1.755	1.905	1.920	1.938	1.941	0.264	0.528	0.634	0.323	0.1514	0.1245
2.20	2.21	2.21	2.22	2.40	2.42	2.44	2.44	0.273	0.547	0.555	0.318	0.1590	0.1325
2.20	2.20	2.21	2.21	2.40	2.42	2.44	2.44	0.270	0.540	0.648	0.331	0.1655	0.1379
2.20	2.20	2.21	2.21	2.40	2.42	2.44	2.44	0.271	0.542	0.651	0.333	0.1663	0.1384
2.77	2.78	2.79	2.79	3.03	3.06	3.07	3.07	0.279	0.558	0.670	0.326	0.1831	0.1359
2.77	2.78	2.78	2.78	3.03	3.06	3.07	3.07	0.274	0.548	0.658	0.333	0.1665	0.1388
3.49	3.50	3.51	3.51	3.82	3.84	3.86	3.86	0.285	0.570	0.684	0.334	0.1872	0.1392
3.49	3.50	3.51	3.51	3.82	3.84	3.86	3.87	0.280	0.560	0.672	0.341	0.1706	0.1422
3.49	3.50	3.51	3.51	3.82	3.84	3.86	3.86	0.283	0.565	0.679	0.349	0.1744	0.1453
4.91	4.92	4.92	4.93	5.37	5.39	5.42	5.42	0.297	0.593	0.712	0.351	0.1754	0.1462

TABLE A.8
Characteristics of Copperweld Conductors

Nominal Conductor Size	Number and Size of Wires	Outside Diameter (in.)	Area of Conductor Circular Mile	Rated Breaking Load (lb) Strength		Weight (lb/mi)	Geometric Mean Radius at 60 Cycles and Average Currents (ft)	Approx. Current Carrying Capacity[a] (amps) at 60 Cycles
				High	Extra High			
30% conductivity								
7/8[a]	19 No. 5	0.910	628,900	55,570	66,910	9344	0.00758	620
18/16[a]	19 No. 6	0.810	498,800	45,830	55,530	7410	0.00675	540
23/32[a]	19 No. 7	0.721	395,500	37,740	45,850	5877	0.00501	470
21/32[a]	19 No. 8	0.642	313,700	31,040	37,690	4560	0.00535	410
9/16[a]	19 No. 9	0.572	248,800	25,500	30,610	3698	0.00477	350
5/8[a]	7 No. 4	0.613	292,200	24,780	29,430	4324	0.00511	410
9/16[a]	7 No. 5	0.546	231,700	20,470	24,650	3429	0.00455	350
1/2[a]	7 No. 6	0.485	183,800	16,890	20,460	2719	0.00405	310
7/16[a]	7 No. 7	0.433	145,700	13,910	15,890	2157	0.00351	270
3/8[a]	7 No. 8	0.385	115,600	11,440	13,890	1710	0.00321	230
11/32[a]	7 No. 9	0.343	91,650	9,393	11,280	1356	0.00286	200
9/16[a]	7 No. 10	0.306	72,680	7,758	9,196	1076	0.00255	170
3 No. 5	3 No. 5	0.392	99,310	9,262	11,860	1467	0.00457	220
3 No. 6	3 No. 6	0.349	78,750	7,639	9,754	1163	0.00407	190
3 No. 7	3 No. 7	0.311	62,450	6,291	7,922	922.4	0.00363	160
3 No. 8	3 No. 8	0.277	49,530	5,174	6,282	731.5	0.00323	140
3 No. 9	3 No. 9	0.247	39,280	4,250	6,129	580.1	0.00288	120
3 No. 10	3 No. 10	0.220	31,150	3,509	4,160	460.0	0.00257	110
40% conductivity								
7/6[a]	19 No. 5	0.910	628,900	50,240	—	9344	0.01175	690
18/16[a]	19 No. 6	0.810	498,800	41,600	—	7410	0.01046	610
23/32[a]	19 No. 7	0.721	395,500	34,390	—	5877	0.00931	530
21/32[a]	19 No. 8	0.642	313,700	28,380	—	4660	0.00829	470
9/16[a]	19 No. 9	0.572	248,800	23,390	—	3696	0.00739	410
5/8[a]	7 No. 4	0.613	292,200	22,310	—	4324	0.00792	470
9/16[a]	7 No. 5	0.546	231,700	18,510	—	3429	0.00705	410
1/2[a]	7 No. 6	0.486	183,800	15,330	—	2719	0.00628	350
7/16[a]	7 No. 7	0.433	145,700	12,670	—	2157	0.00559	310
3/8[a]	7 No. 8	0.385	115,600	10,460	—	1710	0.00497	270
11/32[a]	7 No. 9	0.343	91,650	8,616	—	1356	0.00443	230
8/16[a]	7 No. 10	0.306	72,680	7,121	—	1076	0.00395	200
3 No. 5	3 No. 5	0.392	99,310	8,373	—	1467	0.00621	250
3 No. 6	3 No. 6	0.349	78,750	6,934	—	1163	0.00553	220
3 No. 7	3 No. 7	0.311	62,450	5,732	—	922.4	0.00492	190
3 No. 8	3 No. 8	0.277	49,530	4,730	—	731.5	0.00439	160
3 No. 9	3 No. 9	0.247	39,280	3,898	—	580.1	0.00391	140
3 No. 10	3 No. 10	1.220	31,150	3,221	—	460.0	0.00348	120
3 No. 12	3 No. 12	0.174	19,590	2,236	—	289.3	0.00276	90

Source: Westinghouse Electric Corporation, *Electric Utility Engineering Reference Book—Distribution Systems*, East Pittsburgh, PA, 1965.

[a] Based on a conductor temperature of 125°C and an ambient of 25°C.

[b] Resistances at 75°C total temperature, based on an ambient of 25°C plus 50°C rise due to heating effect of current. The 60 cycles of approximate magnitude of the current necessary to produce the 50°C rise is 75% of the "approximate current-carrying capacity 60 cycles."

r_a Resistance (Ω/Conductor/mi) at 25°C (77°F) Small Currents				r_a Resistance (Ω/Conductor/mi) at 75°C (157°F) Current Approx. 75% of Capacity[b]				X_a Inductive Reactance (Ω/Conductor/mi) 1 ft Spacing Average Currents			X'_a Capacity Reactance (MΩ·mi/Conductor) 1 ft Spacing		
DC	25 Cycles	50 Cycles	60 Cycles	DC	25 Cycles	50 Cycles	60 Cycles	25 Cycles	50 Cycles	60 Cycles	25 Cycles	50 Cycles	60 Cycles
0.306	0.316	0.328	0.331	0.363	0.419	0.476	0.499	0.261	0.493	0.592	0.233	0.1165	0.0971
0.386	0.396	0.406	0.411	0.458	0.518	0.580	0.605	0.267	0.505	0.605	0.241	0.1206	0.1006
0.486	0.495	0.506	0.511	0.577	0.643	0.710	0.737	0.273	0.517	0.621	0.250	0.1248	0.1040
0.613	0.623	0.633	0.638	0.728	0.799	0.872	0.902	0.279	0.529	0.635	0.258	0.1289	0.1074
0.773	0.783	0.793	0.798	0.917	0.995	1.076	1.106	0.285	0.541	0.649	0.266	0.1330	0.1109
0.656	0.664	0.672	0.676	0.778	0.824	0.870	0.887	0.281	0.533	0.640	0.261	0.1306	0.1088
0.827	0.836	0.843	0.847	0.981	1.030	1.080	1.090	0.287	0.545	0.654	0.269	0.1347	0.1122
1.042	1.050	1.058	1.062	1.237	1.290	1.343	1.354	0.293	0.557	0.668	0.278	0.1388	0.1157
1.315	1.323	1.331	1.335	1.550	1.617	1.675	1.897	0.299	0.569	0.683	0.286	0.1420	0.1191
1.658	1.656	1.574	1.578	1.957	2.03	2.09	2.12	0.305	0.581	0.597	0.294	0.1471	0.1226
2.09	2.10	2.11	2.11	2.48	2.55	2.81	2.64	0.311	0.592	0.711	0.303	0.1512	0.1260
2.64	2.64	2.65	2.66	3.13	3.20	3.27	3.30	0.316	0.804	0.725	0.311	0.1553	0.1294
1.926	1.931	1.936	1.938	2.29	2.31	2.34	2.35	0.289	0.545	0.654	0.293	0.1465	0.1221
2.43	2.43	2.44	2.44	2.88	2.91	2.94	2.95	0.295	0.556	0.688	0.301	0.1506	0.1255
3.06	3.07	3.07	3.07	3.63	3.66	3.70	3.71	0.301	0.568	0.682	0.310	0.1547	0.1289
3.86	3.87	3.87	3.87	4.58	4.61	4.65	4.66	0.307	0.580	0.695	0.318	0.1589	0.1324
4.87	4.87	4.88	4.88	5.78	5.81	5.85	5.86	0.313	0.591	0.710	0.326	0.1629	0.1358
6.14	6.14	6.15	6.15	7.28	7.32	7.36	7.38	0.319	0.603	0.724	0.334	0.1671	0.1392
0.229	0.239	0.249	0.254	0.272	0.321	0.371	0.391	0.236	0.449	0.539	0.233	0.1165	0.0971
0.289	0.299	0.309	0.314	0.343	0.395	0.450	0.472	0.241	0.461	0.553	0.241	0.1206	0.1005
0.365	0.375	0.385	0.390	0.433	0.490	0.549	0.573	0.247	0.473	0.567	0.250	0.1248	0.1040
0.460	0.470	0.480	0.485	0.546	0.608	0.672	0.698	0.253	0.485	0.582	0.258	0.1289	0.1074
0.580	0.590	0.800	0.605	0.688	0.756	0.826	0.753	0.259	0.496	0.595	0.266	0.1330	0.1109
0.492	0.500	0.508	0.512	0.584	0.824	0.664	0.680	0.255	0.489	0.587	0.261	0.1306	0.1088
0.620	0.628	0.636	0.640	0.736	0.780	0.843	0.840	0.261	0.501	0.601	0.269	0.1347	0.1122
0.782	0.790	0.798	0.802	0.928	0.975	1.021	1.040	0.267	0.513	0.615	0.278	0.1388	0.1167
0.986	0.994	1.002	1.006	1.170	1.220	1.271	1.291	0.273	0.524	0.629	0.286	0.1429	0.1191
1.244	1.252	1.260	1.264	1.476	1.530	1.584	1.606	0.279	0.536	0.644	0.294	0.1471	0.1226
1.568	1.576	1.584	1.588	1.851	1.919	1.978	2.00	0.285	0.548	0.658	0.303	0.1512	0.1260
1.978	1.986	1.994	1.998	2.35	2.41	2.47	2.50	0.291	0.559	0.671	0.311	0.1553	0.1294
1.445	1.450	1.455	1.457	1.714	1.738	1.762	1.772	0.269	0.514	0.617	0.293	0.1485	0.1221
1.821	1.826	1.831	1.833	2.16	2.19	2.21	2.22	0.275	0.526	0.631	0.301	0.1506	0.1255
2.30	2.30	2.31	2.31	2.73	2.75	2.78	2.79	0.281	0.537	0.645	0.310	0.1547	0.1289
2.90	2.90	2.91	2.91	3.44	3.47	3.50	3.51	0.286	0.549	0.659	0.318	0.1589	0.1324
3.65	3.66	3.66	3.66	4.33	4.37	4.40	4.41	0.292	0.561	0.673	0.326	0.1629	0.1358
4.61	4.61	4.62	4.62	5.46	5.50	5.53	5.55	0.297	0.572	0.687	0.334	0.1671	0.1392
7.32	7.33	7.33	7.34	8.69	8.73	8.77	8.78	0.310	0.596	0.715	0.361	0.1754	0.1462

TABLE A.9
Electrical Characteristics of Overhead Ground Wires

Part A: Alumoweld Strand

| Strand (AWG) | Resistance (Ω/mi) | | | | 60 Hz Reactance for 1 ft Radius | | 60 Hz Geometric Mean Radius (ft) |
| | Small Currents | | 75% of Capacitive | | | | |
	25°C OC	25°C 60 Hz	75°C OC	75°C 60 Hz	Inductive (Ω/mi)	Capacitive (MΩ·mi)	
7 NO. 5	1.217	1.240	1.432	1.669	0.707	0.1122	0.002958
7 NO. 6	1.507	1.536	1.773	2.010	0.721	0.1157	0.002633
7 NO. 7	1.900	1.937	2.240	2.470	0.735	0.1191	0.002345
7 NO. 8	2.400	2.440	2.820	3.060	0.749	0.1226	0.002085
7 NO. 9	3.020	3.080	3.560	3.800	0.763	0.1260	0.001858
7 NO. 10	3.810	3.880	4.480	4.730	0.777	0.1294	0.001658
3 NO. 5	2.780	2.780	3.270	3.560	0.707	0.1221	0.002940
3 NO. 6	3.510	3.510	4.130	4.410	0.721	0.1255	0.002618
3 NO. 7	4.420	4.420	5.210	5.470	0.735	0.1289	0.002333
3 NO. 8	5.580	5.580	6.570	6.820	0.749	0.1324	0.002078
3 NO. 9	7.040	7.040	8.280	8.520	0.763	0.1358	0.001853
3 NO. 10	8.870	8.870	10.440	10.670	0.777	0.1392	0.001650

Part B: Single-Layer ACSR

| Code | 25°C DC | Resistance (Ω/mi) 60 Hz, 75°C | | | 60 Hz Reactance for 1 ft Radius Inductive (Ω/mi) at 75°C | | | Capacitive (MΩ·mi) |
		$I = 0$ A	$I = 100$ A	$I = 200$ A	$I = 0$ A	$I = 100$ A	$I = 200$ A	
Brahma	0.394	0.470	0.510	0.565	0.500	0.520	0.545	0.1043
Cochin	0.400	0.480	0.520	0.590	0.505	0.515	0.550	0.1065
Dorking	0.443	0.535	0.575	0.650	0.515	0.530	0.565	0.1079
Dotterel	0.479	0.565	0.620	0.705	0.515	0.530	0.575	0.1091
Guinea	0.531	0.630	0.685	0.780	0.520	0.545	0.590	0.1106
Leghorn	0.630	0.760	0.810	0.930	0.530	0.550	0.605	0.1131
Minorca	0.765	0.915	0.980	1.130	0.540	0.570	0.640	0.1160
Petrel	0.830	1.000	1.065	1.220	0.550	0.580	0.655	0.1172
Grouse	1.080	1.295	1.420	1.520	0.570	0.640	0.675	0.1240

Part C: Steel Conductors

| Grade (7-Strand) | Diameter (in.) | Resistance (Ω/mi) at 60 Hz | | | 60 Hz Reactance for 1 ft Radius Inductive (Ω/mi) | | | Capacitive (MΩ·mi) |
		$I = 0$ A	$I = 30$ A	$I = 60$ A	$I = 0$ A	$I = 30$ A	$I = 60$ A	
Ordinary	1/4	9.5	11.4	11.3	1.3970	3.7431	3.4379	0.1354
Ordinary	9/32	7.1	9.2	9.0	1.2027	3.0734	2.5146	0.1319
Ordinary	5/16	5.4	7.5	7.8	0.8382	2.5146	2.0409	0.1288
Ordinary	3/8	4.3	6.5	6.6	0.8382	2.2352	1.9687	0.1234
Ordinary	1/2	2.3	4.3	5.0	0.7049	1.6893	1.4236	0.1148
E.B.	1/4	8.0	12.0	10.1	1.2027	4.4704	3.1565	0.1354
E.B.	9/32	6.0	10.0	8.7	1.1305	3.7783	2.6255	0.1319
E.B.	5/16	4.9	8.0	7.0	0.9843	2.9401	2.5146	0.1288
E.B.	3/8	3.7	7.0	6.3	0.8382	2.5997	2.4303	0.1234
E.B.	1/2	2.1	4.9	5.0	0.7049	1.8715	1.7616	0.1148
E.B.B.	1/4	7.0	12.8	10.9	1.6764	5.1401	3.9482	0.1354
E.B.B.	9/32	5.4	10.9	8.7	1.1305	4.4833	3.7783	0.1319
E.B.B.	5/16	4.0	9.0	6.8	0.9843	3.6322	3.0734	0.1288
E.B.B.	3/8	3.5	7.9	6.0	0.8382	3.1168	2.7940	0.1234
E.B.B.	1/2	2.0	5.7	4.7	0.7049	2.3461	2.2352	0.1148

Source: Reprinted from Anderson, P.M., *Analysis of Faulted Power Systems*, The Iowa State University Press, Ames, IA, Copyright 1973. With permission.

TABLE A.10

Inductive Reactance Spacing Factor X_d, Ω/(Conductor mi), at 60 Hz

Ft	0.0	0.1	0.2	0.3	0.4	0.5	0.6	0.7	0.8	0.9
0		−0.2794	−0.1953	−0.1461	−0.1112	−0.0841	−0.0620	−0.0433	−0.0271	−0.0128
1	0.0	0.0116	0.0221	0.0318	0.0408	0.0492	0.0570	0.0644	0.0713	0.0779
2	0.0841	0.0900	0.0957	0.1011	0.1062	0.1112	0.1159	0.1205	0.1249	0.1292
3	0.1333	0.1373	0.1411	0.1449	0.1485	0.1520	0.1554	0.1588	0.1620	0.1651
4	0.1682	0.1712	0.1741	0.1770	0.1798	0.1825	0.1852	0.1878	0.1903	0.1928
5	0.1953	0.1977	0.2001	0.2024	0.2046	0.2069	0.2090	0.2112	0.2133	0.2154
6	0.2174	0.2194	0.2214	0.2233	0.2252	0.2271	0.2290	0.2308	0.2326	0.2344
7	0.2361	0.2378	0.2395	0.2412	0.2429	0.2445	0.2461	0.2477	0.2493	0.2508
8	0.2523	0.2538	0.2553	0.2568	0.2582	0.2597	0.2611	0.2625	0.2639	0.2653
9	0.2666	0.2680	0.2693	0.2706	0.2719	0.2732	0.2744	0.2757	0.2769	0.2782
10	0.2794	0.2806	0.2818	0.2830	0.2842	0.2853	0.2865	0.2876	0.2887	0.2899
11	0.2910	0.2921	0.2932	0.2942	0.2953	0.2964	0.2974	0.2985	0.2995	0.3005
12	0.3015	0.3025	0.3035	0.3045	0.3055	0.3065	0.3074	0.3084	0.3094	0.3103
13	0.3112	0.3122	0.3131	0.3140	0.3149	0.3158	0.3167	0.3176	0.3185	0.3194
14	0.3202	0.3211	0.3219	0.3228	0.3236	0.3245	0.3253	0.3261	0.3270	0.3278
15	0.3286	0.3294	0.3302	0.3310	0.3318	0.3326	0.3334	0.3341	0.3349	0.3357
16	0.3364	0.3372	0.3379	0.3387	0.3394	0.3402	0.3409	0.3416	0.3424	0.3431
17	0.3438	0.3445	0.3452	0.3459	0.3466	0.3473	0.3480	0.3487	0.3494	0.3500
18	0.3507	0.3514	0.3521	0.3527	0.3534	0.3540	0.3547	0.3554	0.3560	0.3566
19	0.3573	0.3579	0.3586	0.3592	0.3598	0.3604	0.3611	0.3617	0.3623	0.3629
20	0.3635	0.3641	0.3647	0.3653	0.3659	0.3665	0.3671	0.3677	0.3683	0.3688
21	0.3694	0.3700	0.3706	0.3711	0.3717	0.3723	0.3728	0.3734	0.3740	0.3745
22	0.3751	0.3756	0.3762	0.3767	0.3773	0.3778	0.3783	0.3789	0.3794	0.3799
23	0.3805	0.3810	0.3815	0.3820	0.3826	0.3831	0.3836	0.3841	0.3846	0.3851
24	0.3856	0.3861	0.3866	0.3871	0.3876	0.3881	0.3886	0.3891	0.3896	0.3901
25	0.3906	0.3911	0.3916	0.3920	0.3925	0.3930	0.3935	0.3939	0.3944	0.3949
26	0.3953	0.3958	0.3963	0.3967	0.3972	0.3977	0.3981	0.3986	0.3990	0.3995
27	0.3999	0.4004	0.4008	0.4013	0.4017	0.4021	0.4026	0.4030	0.4035	0.4039
28	0.4043	0.4048	0.4052	0.4056	0.4061	0.4065	0.4069	0.4073	0.4078	0.4082
29	0.4086	0.4090	0.4094	0.4098	0.4103	0.4107	0.4111	0.4115	0.4119	0.4123
30	0.4127	0.4131	0.4135	0.4139	0.4143	0.4147	0.4151	0.4155	0.4159	0.4163
31	0.4167	0.4171	0.4175	0.4179	0.4182	0.4186	0.4190	0.4194	0.4198	0.4202
32	0.4205	0.4209	0.4213	0.4217	0.4220	0.4224	0.4228	0.4232	0.4235	0.4239
33	0.4243	0.4246	0.4250	0.4254	0.4257	0.4261	0.4265	0.4268	0.4272	0.4275
34	0.4279	0.4283	0.4286	0.4290	0.4293	0.4297	0.4300	0.4304	0.4307	0.4311
35	0.4314	0.4318	0.4321	0.4324	0.4328	0.4331	0.4335	0.4338	0.4342	0.4345
36	0.4348	0.4352	0.4355	0.4358	0.4362	0.4365	0.4368	0.4372	0.4375	0.4378
37	0.4382	0.4385	0.4388	0.4391	0.4395	0.4398	0.4401	0.4404	0.4408	0.4411
38	0.4414	0.4417	0.4420	0.4423	0.4427	0.4430	0.4433	0.4436	0.4439	0.4442
39	0.4445	0.4449	0.4452	0.4455	0.4458	0.4461	0.4464	0.4467	0.4470	0.4473
40	0.4476	0.4479	0.4492	0.4485	0.4488	0.4491	0.4494	0.4497	0.4500	0.4503
41	0.4506	0.4509	0.4512	0.4515	0.4518	0.4521	0.4524	0.4527	0.4530	0.4532
42	0.4535	0.4538	0.4541	0.4544	0.4547	0.4550	0.4553	0.4555	0.4558	0.4561
43	0.4564	0.4567	0.4570	0.4572	0.4575	0.4578	0.4581	0.4584	0.4586	0.4589
44	0.4592	0.4595	0.4597	0.4600	0.4603	0.4606	0.4608	0.4611	0.4614	0.4616
45	0.4619	0.4622	0.4624	0.4627	0.4630	0.4632	0.4635	0.4638	0.4640	0.4643
46	0.4646	0.4648	0.4651	0.4654	0.4656	0.4659	0.4661	0.4664	0.4667	0.4669
47	0.4672	0.4674	0.4677	0.4680	0.4682	0.4685	0.4687	0.4690	0.4692	0.4695
48	0.4697	0.4700	0.4702	0.4705	0.4707	0.4710	0.4712	0.4715	0.4717	0.4720
49	0.4722	0.4725	0.4727	0.4730	0.4732	0.4735	0.4737	0.4740	0.4742	0.4744
50	0.4747	0.4749	0.4752	0.4754	0.4757	0.4759	0.4761	0.4764	0.4766	0.4769

(continued)

TABLE A.10 (continued)
Inductive Reactance Spacing Factor X_d, Ω/(Conductor mi), at 60 Hz

Ft	0.0	0.1	0.2	0.3	0.4	0.5	0.6	0.7	0.8	0.9
51	0.4771	0.4773	0.4776	0.4778	0.4780	0.4783	0.4785	0.4787	0.4790	0.4792
52	0.4795	0.4797	0.4799	0.4801	0.4804	0.4806	0.4808	0.4811	0.4813	0.4815
53	0.4818	0.4820	0.4822	0.4824	0.4827	0.4829	0.4831	0.4834	0.4836	0.4838
54	0.4840	0.4843	0.4845	0.4847	0.4849	0.4851	0.4854	0.4856	0.4858	0.4860
55	0.4863	0.4865	0.4867	0.4869	0.4871	0.4874	0.4876	0.4878	0.4880	0.4882
56	0.4884	0.4887	0.4889	0.4891	0.4893	0.4895	0.4897	0.4900	0.4902	0.4904
57	0.4906	0.4908	0.4910	0.4912	0.4914	0.4917	0.4919	0.4921	0.4923	0.4925
58	0.4927	0.4929	0.4931	0.4933	0.4935	0.4937	0.4940	0.4942	0.4944	0.4946
59	0.4948	0.4950	0.4952	0.4954	0.4956	0.4958	0.4960	0.4962	0.4964	0.4966
60	0.4968	0.4970	0.4972	0.4974	0.4976	0.4978	0.4980	0.4982	0.4984	0.4986
61	0.4988	0.4990	0.4992	0.4994	0.4996	0.4998	0.5000	0.5002	0.5004	0.5006
62	0.5008	0.5010	0.5012	0.5014	0.5016	0.5018	0.5020	0.5022	0.5023	0.5025
63	0.5027	0.5029	0.5031	0.5033	0.5035	0.5037	0.5039	0.5041	0.5043	0.5045
64	0.5046	0.5048	0.5050	0.5052	0.5054	0.5056	0.5058	0.5060	0.5062	0.5063
65	0.5065	0.5067	0.5069	0.5071	0.5073	0.5075	0.5076	0.5078	0.5080	0.5082
66	0.5084	0.5086	0.5087	0.5089	0.5091	0.5093	0.5095	0.5097	0.5098	0.5100
67	0.5102	0.5104	0.5106	0.5107	0.5109	0.5111	0.5113	0.5115	0.5116	0.5118
68	0.5120	0.5122	0.5124	0.5125	0.5127	0.5129	0.5131	0.5132	0.5134	0.5136
69	0.5138	0.5139	0.5141	0.5143	0.5145	0.5147	0.5148	0.5150	0.5152	0.5153
70	0.5155	0.5157	0.5159	0.5160	0.5162	0.5164	0.5166	0.5167	0.5169	0.5171
71	0.5172	0.5174	0.5176	0.5178	0.5179	0.5181	0.5183	0.5184	0.5186	0.5188
72	0.5189	0.5191	0.5193	0.5194	0.5196	0.5198	0.5199	0.5201	0.5203	0.5204
73	0.5206	0.5208	0.5209	0.5211	0.5213	0.5214	0.5216	0.5218	0.5219	0.5221
74	0.5223	0.5224	0.5226	0.5228	0.5229	0.5231	0.5232	0.5234	0.5236	0.5237
75	0.5239	0.5241	0.5242	0.5244	0.5245	0.5247	0.5249	0.5250	0.5252	0.5253
76	0.5255	0.5257	0.5258	0.5260	0.5261	0.5263	0.5265	0.5266	0.5268	0.5269
77	0.5271	0.5272	0.5274	0.5276	0.5277	0.5279	0.5280	0.5282	0.5283	0.5285
78	0.5287	0.5288	0.5290	0.5291	0.5293	0.5294	0.5296	0.5297	0.5299	0.5300
79	0.5302	0.5304	0.5305	0.5307	0.5308	0.5310	0.5311	0.5313	0.5314	0.5316
80	0.5317	0.5319	0.5320	0.5322	0.5323	0.5325	0.5326	0.5328	0.5329	0.5331
81	0.5332	0.5334	0.5335	0.5337	0.5338	0.5340	0.5341	0.5343	0.5344	0.5346
82	0.5347	0.5349	0.5350	0.5352	0.5353	0.5355	0.5356	0.5358	0.5359	0.5360
83	0.5362	0.5363	0.5365	0.5366	0.5368	0.5369	0.5371	0.5372	0.5374	0.5375
84	0.5376	0.5378	0.5379	0.5381	0.5382	0.5384	0.5385	0.5387	0.5388	0.5389
85	0.5391	0.5392	0.5394	0.5395	0.5396	0.5398	0.5399	0.5401	0.5402	0.5404
86	0.5405	0.5406	0.5408	0.5409	0.5411	0.5412	0.5413	0.5415	0.5416	0.5418
87	0.5419	0.5420	0.5422	0.5423	0.5425	0.5426	0.5427	0.5429	0.5430	0.5432
88	0.5433	0.5434	0.5436	0.5437	0.5438	0.5440	0.5441	0.5442	0.5444	0.5445
89	0.5447	0.5448	0.5449	0.5451	0.5452	0.5453	0.5455	0.5456	0.5457	0.5459
90	0.5460	0.5461	0.5463	0.5464	0.5466	0.5467	0.5468	0.5470	0.5471	0.5472
91	0.5474	0.5475	0.5476	0.5478	0.5479	0.5480	0.5482	0.5483	0.5484	0.5486
92	0.5487	0.5488	0.5489	0.5491	0.5492	0.5493	0.5495	0.5496	0.5497	0.5499
93	0.5500	0.5501	0.5503	0.5504	0.5505	05506	0.5508	0.5509	0.5510	0.5512
94	0.5513	0.5514	0.5515	0.5517	0.5518	0.5519	0.5521	0.5522	0.5523	0.5524
95	0.5526	0.5527	0.5528	0.5530	0.5531	0.5532	0.5533	0.5535	0.5536	0.5537
96	0.5538	0.5540	0.5541	0.5542	0.5544	0.5545	0.5546	0.5547	0.5549	0.5550
97	0.5551	0.5552	0.5554	0.5555	0.5556	0.5557	0.5559	0.5560	0.5561	0.5562
98	0.5563	0.5565	0.5566	0.5567	0.5568	0.5570	0.5571	0.5572	0.5573	0.5575
99	0.5576	0.5577	0.5578	0.5579	0.5581	0.5582	0.5583	0.5584	0.5586	0.5587
100	0.5588	0.5589	0.5590	0.5592	0.5593	0.5594	0.5595	0.5596	0.5598	0.5599

TABLE A.11
Zero-Sequence Resistive and Inductive Factors
$R_e{}^a$, $X_e{}^a$, $\Omega/(\text{Conductor} \cdot \text{mi})$

	ρ ($\Omega \cdot$ m) All	r_e, x_e (f = 60 Hz)
r_e	All	0.2860
	1	2.050
	5	2.343
	10	2.469
x_e	50	2.762
	100[b]	2.888[b]
	500	3.181
	1,000	3.307
	5,000	3.600
	10,000	3.726

Source: Reprinted from Anderson, P.M., *Analysis of Faulted Power Systems*, The Iowa State University Press, Ames, IA, Copyright 1973. With permission.

[a] From formulas:

$$r_e = 0.004764f$$

$$x_e = 0.006985 f \log_{10} 4{,}665{,}600 \frac{r}{f}$$

where
  f is the frequency
  ρ is the resistivity (Ω m)

[b] This is an average value which may be used in the absence of definite information.

Fundamental equations:

$$z_1 = z_2 = r_a + j(x_a + x_d)$$

$$z_0 = r_a + r_e + j(x_a + x_e - 2x_d)$$

where $x_d = wk \ln d$ and d = separation (ft).

TABLE A.12

Shunt Capacitive Reactance Spacing Factor x'_d (MΩ/Conductor·mi), at 60 Hz

Ft	0.0	0.1	0.2	0.3	0.4	0.5	0.6	0.7	0.8	0.9
0		−0.0683	−0.0477	−0.0357	−0.0272	−0.0206	−0.0152	−0.0106	−0.0066	−0.0031
1	0.0000	0.0028	0.0054	0.0078	0.0100	0.0120	0.0139	0.0157	0.0174	0.0190
2	0.0206	0.0220	0.0234	0.0247	0.0260	0.0272	0.0283	0.0295	0.0305	0.0316
3	0.0326	0.0336	0.0345	0.0354	0.0363	0.0372	0.0380	0.0388	0.0396	0.0404
4	0.0411	0.0419	0.0426	0.0433	0.0440	0.0446	0.0453	0.0459	0.0465	0.0471
5	0.0477	0.0483	0.0489	0.0495	0.0500	0.0506	0.0511	0.0516	0.0521	0.0527
6	0.0532	0.0536	0.0541	0.0546	0.0551	0.0555	0.0560	0.0564	0.0569	0.0573
7	0.0577	0.0581	0.0586	0.0590	0.0594	0.0598	0.0602	0.0606	0.0609	0.0613
8	0.0617	0.0621	0.0624	0.0628	0.0631	0.0635	0.0638	0.0642	0.0645	0.0649
9	0.0652	0.0655	0.0658	0.0662	0.0665	0.0668	0.0671	0.0674	0.0677	0.0680
10	0.0683	0.0686	0.0689	0.0692	0.0695	0.0698	0.0700	0.0703	0.0706	0.0709
11	0.0711	0.0714	0.0717	0.0719	0.0722	0.0725	0.0727	0.0730	0.0732	0.0735
12	0.0737	0.0740	0.0742	0.0745	0.0747	0.0749	0.0752	0.0754	0.0756	0.0759
13	0.0761	0.0763	0.0765	0.0768	0.0770	0.0772	0.0774	0.0776	0.0779	0.0781
14	0.0783	0.0785	0.0787	0.0789	0.0791	0.0793	0.0795	0.0797	0.0799	0.0801
15	0.0803	0.0805	0.0807	0.0809	0.0811	0.0813	0.0815	0.0817	0.0819	0.0821
16	0.0823	0.0824	0.0826	0.0828	0.0830	0.0832	0.0833	0.0835	0.0837	0.0839
17	0.0841	0.0842	0.0844	0.0846	0.0847	0.0849	0.0851	0.0852	0.0854	0.0856
18	0.0857	0.0859	0.0861	0.0862	0.0864	0.0866	0.0867	0.0869	0.0870	0.0872
19	0.0874	0.0875	0.0877	0.0878	0.0880	0.0881	0.0883	0.0884	0.0886	0.0887
20	0.0889	0.0890	0.0892	0.0893	0.0895	0.0896	0.0898	0.0899	0.0900	0.0902
21	0.0903	0.0905	0.0906	0.0907	0.0909	0.0910	0.0912	0.0913	0.0914	0.0916
22	0.0917	0.0918	0.0920	0.0921	0.0922	0.0924	0.0925	0.0926	0.0928	0.0929
23	0.0930	0.0931	0.0933	0.0934	0.0935	0.0937	0.0938	0.0939	0.0940	0.0942
24	0.0943	0.0944	0.0945	0.0947	0.0948	0.0949	0.0950	0.0951	0.0953	0.0954
25	0.0955	0.0956	0.0957	0.0958	0.0960	0.0961	0.0962	0.0963	0.0964	0.0965
26	0.0967	0.0968	0.0969	0.0970	0.0971	0.0972	0.0973	0.0974	0.0976	0.0977
27	0.0978	0.0979	0.0980	0.0981	0.0982	0.0983	0.0984	0.0985	0.0986	0.0987
28	0.0989	0.0990	0.0991	0.0992	0.0993	0.0994	0.0995	0.0996	0.0997	0.0998
29	0.0999	0.1000	0.1001	0.1002	0.1003	0.1004	0.1005	0.1006	0.1007	0.1008
30	0.1009	0.1010	0.1011	0.1012	0.1013	0.1014	0.1015	0.1016	0.1017	0.1018
31	0.1019	0.1020	0.1021	0.1022	0.1023	0.1023	0.1024	0.1025	0.1026	0.1027
32	0.1028	0.1029	0.1030	0.1031	0.1032	0.1033	0.1034	0.1035	0.1035	0.1036
33	0.1037	0.1038	0.1039	0.1040	0.1041	0.1042	0.1043	0.1044	0.1044	0.1045
34	0.1046	0.1047	0.1048	0.1049	0.1050	0.1050	0.1051	0.1052	0.1053	0.1054
35	0.1055	0.1056	0.1056	0.1057	0.1058	0.1059	0.1060	0.1061	0.1061	0.1062
36	0.1063	0.1064	0.1065	0.1066	0.1066	0.1067	0.1068	0.1069	0.1070	0.1070
37	0.1071	0.1072	0.1073	0.1074	0.1074	0.1075	0.1076	0.1077	0.1078	0.1078
38	0.1079	0.1080	0.1081	0.1081	0.1082	0.1083	0.1084	0.1085	0.1085	0.1086
39	0.1087	0.1088	0.1088	0.1089	0.1090	0.1091	0.1091	0.1092	0.1093	0.1094
40	0.1094	0.1095	0.1096	0.1097	0.1097	0.1098	0.1099	0.1100	0.1100	0.1101
41	0.1102	0.1102	0.1103	0.1104	0.1105	0.1105	0.1106	0.1107	0.1107	0.1108
42	0.1109	0.1110	0.1110	0.1111	0.1112	0.1112	0.1113	0.1114	0.1114	0.1115
43	0.1116	0.1117	0.1117	0.1118	0.1119	0.1119	0.1120	0.1121	0.1121	0.1122
44	0.1123	0.1123	0.1124	0.1125	0.1125	0.1126	0.1127	0.1127	0.1128	0.1129
45	0.1129	0.1130	0.1131	0.1131	0.1132	0.1133	0.1133	0.1134	0.1135	0.1135
46	0.1136	0.1136	0.1137	0.1138	0.1138	0.1139	0.1140	0.1140	0.1141	0.1142
47	0.1142	0.1143	0.1143	0.1144	0.1145	0.1145	0.1146	0.1147	0.1147	0.1148
48	0.1148	0.1149	0.1150	0.1150	0.1151	0.1152	0.1152	0.1153	0.1153	0.1154
49	0.1155	0.1155	0.1156	0.1156	0.1157	0.1158	0.1158	0.1159	0.1159	0.1160
50	0.1161	0.1161	0.1162	0.1162	0.1163	0.1164	0.1164	0.1165	0.1165	0.1166

TABLE A.12 (continued)
Shunt Capacitive Reactance Spacing Factor x_d' (MΩ/Conductor$\cdot$mi), at 60 Hz

Ft	0.0	0.1	0.2	0.3	0.4	0.5	0.6	0.7	0.8	0.9
51	0.1166	0.1167	0.1168	0.1168	0.1169	0.1169	0.1170	0.1170	0.1171	0.1172
52	0.1172	0.1173	0.1173	0.1174	0.1174	0.1175	0.1176	0.1176	0.1177	0.1177
53	0.1178	0.1178	0.1179	0.1180	0.1180	0.1181	0.1181	0.1182	0.1182	0.1183
54	0.1183	0.1184	0.1184	0.1185	0.1186	0.1186	0.1187	0.1187	0.1188	0.1188
55	0.1189	0.1189	0.1190	0.1190	0.1191	0.1192	0.1192	0.1193	0.1193	0.1194
56	0.1194	0.1195	0.1195	0.1196	0.1196	0.1197	0.1197	0.1198	0.1198	0.1199
57	0.1199	0.1200	0.1200	0.1201	0.1202	0.1202	0.1203	0.1203	0.1204	0.1204
58	0.1205	0.1205	0.1206	0.1206	0.1207	0.1207	0.1208	0.1208	0.1209	0.1209
59	0.1210	0.1210	0.1211	0.1211	0.1212	0.1212	0.1213	0.1213	0.1214	0.1214
60	0.1215	0.1215	0.1216	0.1216	0.1217	0.1217	0.1218	0.1218	0.1219	0.1219
61	0.1220	0.1220	0.1221	0.1221	0.1221	0.1222	0.1222	0.1223	0.1223	0.1224
62	0.1224	0.1225	0.1225	0.1226	0.1226	0.1227	0.1227	0.1228	0.1228	0.1229
63	0.1229	0.1230	0.1230	0.1231	0.1231	0.1231	0.1232	0.1232	0.1233	0.1233
64	0.1234	0.1234	0.1235	0.1235	0.1236	0.1236	0.1237	0.1237	0.1237	01238
65	0.1238	0.1239	0.1239	0.1240	0.1240	0.1241	0.1241	0.1242	0.1242	0.1242
66	0.1243	0.1243	0.1244	0.1244	0.1245	0.1245	0.1246	0.1246	0.1247	0.1247
67	0.1247	0.1248	0.1248	0.1249	0.1249	0.1250	0.1250	0.1250	0.1251	0.1251
68	0.1252	0.1252	0.1253	0.1253	0.1254	0.1254	0.1254	0.1255	0.1255	0.1256
69	0.1256	0.1257	0.1257	0.1257	0.1258	0.1258	0.1259	0.1259	0.1260	0.1260
70	0.1260	0.1261	0.1261	0.1262	0.1262	0.1262	0.1263	0.1263	0.1264	0.1264
71	0.1265	0.1265	0.1265	0.1266	0.1266	0.1267	0.1267	0.1268	0.1268	0.1268
72	0.1269	0.1269	0.1270	0.1270	0.1270	0.1271	0.1271	0.1272	0.1272	0.1272
73	0.1273	0.1273	0.1274	0.1274	0.1274	0.1275	0.1275	0.1276	0.1276	0.1276
74	0.1277	0.1277	0.1278	0.1278	0.1278	0.1279	0.1279	0.1280	0.1280	.01280
75	0.1281	0.1281	0.1282	0.1282	0.1282	0.1283	0.1283	0.1284	0.1284	0.1284
76	0.1285	0.1285	0.1286	0.1286	0.1286	0.1287	0.1287	0.1288	0.1288	0.1288
77	0.1289	0.1289	0.1289	0.1290	0.1290	0.1291	0.1291	0.1291	0.1292	0.1292
78	0.1292	0.1293	0.1293	0.1294	0.1294	0.1294	0.1295	0.1295	0.1296	0.1296
79	0.1296	0.1297	0.1297	0.1297	0.1298	0.1298	0.1299	0.1299	0.1299	0.1300
80	0.1300	0.1300	0.1301	0.1301	0.1301	0.1302	0.1302	0.1303	0.1303	0.1303
81	0.1304	0.1304	0.1304	0.1305	0.1305	0.1306	0.1306	0.1306	0.1307	0.1307
82	0.1307	0.1308	0.1308	0.1308	0.1309	0.1309	0.1309	0.1310	0.1310	0.1311
83	0.1311	0.1311	0.1312	0.1312	0.1312	0.1313	0.1313	0.1313	0.1314	0.1314
84	0.1314	0.1315	0.1315	0.1316	0.1316	0.1316	0.1317	0.1317	0.1317	0.1318
85	0.1318	0.1318	0.1319	0.1319	0.1319	0.1320	0.1320	0.1320	0.1321	0.1321
86	0.1321	0.1322	0.1322	0.1322	0.1323	0.1323	0.1324	0.1324	0.1324	0.1325
87	0.1325	0.1325	0.1326	0.1326	0.1326	0.1327	0.1327	0.1327	0.1328	0.1328
88	0.1328	0.1329	0.1329	0.1329	0.1330	0.1330	0.1330	0.1331	0.1331	0.1331
89	0.1332	0.1332	0.1332	0.1333	0.1333	0.1333	0.1334	0.1334	0.1334	0.1335
90	0.1335	0.1335	0.1336	0.1336	0.1336	0.1337	0.1337	0.1337	0.1338	0.1338
91	0.1338	0.1339	0.1339	0.1339	0.1340	0.1340	0.1340	0.1340	0.1341	0.1341
92	0.1341	0.1342	0.1342	0.1342	0.1343	0.1343	0.1343	0.1344	0.1344	0.1344
93	0.1345	0.1345	0.1345	0.1346	0.1346	0.1346	0.1347	0.1347	0.1347	0.1348
94	0.1348	0.1348	0.1348	0.1349	0.1349	0.1349	0.1350	0.1350	0.1350	0.1351
95	0.1351	0.1351	0.1352	0.1352	0.1352	0.1353	0.1353	0.1353	0.1353	0.1354
96	0.1354	0.1354	0.1355	0.1355	0.1355	0.1356	0.1356	0.1356	0.1357	0.1357
97	0.1357	0.1357	0.1358	0.1358	0.1358	0.1359	0.1359	0.1359	0.1360	0.1360
98	0.1360	0.1361	0.1361	0.1361	0.1361	0.1362	0.1362	0.1362	0.1363	0.1363
99	0.1363	0.1364	0.1364	0.1364	0.1364	0.1365	0.1365	0.1365	0.1366	0.1366
100	0.1366	0.1366	0.1367	0.1367	0.1367	0.1368	0.1368	0.1368	0.1369	0.1369

TABLE A.13

Zero-Sequence Shunt Capacitive Reactance Factor x_0', MΩ/(Conductor · mi)

Conductor Height above Ground (ft)	x_0' (f = 60 Hz)
10	0.267
15	0.303
20	0.328
25	0.318
30	0.364
40	0.390
50	0.410
60	0.426
70	0.440
80	0.452
90	0.462
100	0.472

Source: Reprinted from Anderson, P.M., *Analysis of Faulted Power Systems*, The Iowa State University Press, Ames, IA, Copyright 1973. With permission.

$$x_0' = \frac{12.30}{f} \log_{10} 2h$$

where h = height above ground and f = frequency. Fundamental equations:

$$x_1' = x_2' = x_a' = x_d'$$

$$x_0' = x_a' + x_c' - 2x_d'$$

where $x_d' = (1/\omega k') \ln d$ and d = separation (ft).

TABLE A.14
Standard Impedances of Distribution Transformers

	Rating of Transformer Primary Winding																	
	2.4 kV		4.8 kV		7.2 kV		12 kV		24.9/14.4 Gnd Y		23 kV		34.5 kV		46 kV		69 kV	
kVA Rating	% R	% Z	% R	% Z	% R	% Z	% R	% Z	% R	% Z	% R	% Z	% R	% Z	% R	% Z	% R	% Z
Single phase																		
3	1.9	2.3	2.1	2.3	2.5	2.8			3.0	3.5								
10	1.7	2.1	1.8	2.1	1.9	2.3	2.1	2.6	2.2	2.9								
25	1.5	2.3	1.6	2.3	1.6	2.2	1.6	2.3	1.7	2.6	2.0	5.2	2.2	5.2				
50	1.2	2.3	1.4	2.2	1.3	2.2	1.4	2.4	1.5	2.8	1.7	5.2	1.7	5.2	1.8	5.7		
100	1.2	2.7	1.3	2.6	1.2	3.2	1.3	3.2			1.4	5.2	1.5	5.2	1.5	5.7	1.4	6.5
333	1.1	4.8	1.1	4.8	1.0	4.9	1.0	5.1			1.0	5.2	1.1	5.2	1.1	5.7	1.1	6.5
500	1.0	4.8	1.0	4.8	1.0	5.1	1.0	5.0			0.9	5.2	1.0	5.2	1.0	5.7	1.0	6.5
Three phase																		
9	2.0	2.4	2.1	2.5	2.4	2.7												
30	1.6	2.5	1.8	2.5	1.9	2.6	2.1	3.1										
75	1.5	3.2	1.6	3.1	1.6	3.2	1.6	3.3										
150	1.2	4.2	1.4	4.3	1.4	4.3	1.4	4.2			1.6	5.5						
300	1.3	4.9	1.3	4.9	1.3	4.9	1.3	5.0			1.3	5.5	1.4	5.5	1.4	6.2		
500	1.2	4.9	1.2	4.9	1.1	5.0	1.1	5.1			1.2	5.5	1.2	5.5	1.3	6.3	1.2	6.7

Source: Westinghouse Electric Corporation, *Applied Protective Relaying*, Newark, NJ, 1970. With permission.

TABLE A.15
Standard Impedances for Power Transformers 10,000 kVA and below

Highest-Voltage Winding (BIL kV)	Low-Voltage Winding, BIL kV (for Intermediate BIL, Use Value for Next Higher BIL Listed)	At kVA Base Equal to 55°C Rating of Largest Capacity Winding Self-Cooled (OA), Self-Cooled Rating of Self-Cooled/Forced-Air Cooled (OA/FA) Standard Impedance (%)	
		Ungrounded Neutral Operation	Grounded Neutral Operation
110 and below	45	5.75	
	60, 75, 95, 110	5.5	
150	45	5.75	
	60, 75, 95, 110	5.5	
200	45	6.25	
	60, 75, 95, 110	6.0	
	150	6.5	
250	45	6.75	
	60, 150	6.5	
	200	7.0	
350	200	7.0	
	250	7.5	
450	200	7.5	7.00
	250	8.0	7.50
	350	8.5	8.00
550	200	8.0	7.50
	350	9.0	8.25
	450	10.0	9.25
650	200	8.5	8.00
	350	9.5	8.50
	550	10.5	9.50
750	250	9.0	8.50
	450	10.0	9.50
	650	11.0	10.25

Source: Westinghouse Electric Corporation, *Applied Protective Relaying*, Newark, NJ, 1970. With permission.
BIL, basic impulse insulation level.

TABLE A.16
Standard Impedance Limits for Power Transformers above 10,000 kVA

		At kVA Base Equal to 55°C Rating of Largest Capacity Winding							
		Self-Cooled (OA), Self-Cooled Rating of Self-Cooled/Forced-Air Cooled (OA/FA), Self-Cooled Rating of Self-Cooled/Forced-Air, Forced-Oil Cooled (OA/FOA) Standard Impedance (%)				Forced-Oil Cooled (FOA and FOW) Standard Impedance (%)			
Highest-Voltage Winding (BIL kV)	Low-Voltage Winding, BIL kV (For Intermediate BIL, Use Value for Next Higher BIL Listed)	Ungrounded Neutral Operation		Grounded Neutral Operation		Ungrounded Neutral Operation		Grounded Neutral Operation	
		Min.	Max.	Min.	Max.	Min.	Max.	Min.	Max.
110 and below	110 and below	5.0	6.25			8.25	10.5		
150	110	5.0	6.25			8.25	10.5		
200	110	5.5	7.0			9.0	12.0		
	150	5.75	7.5			9.75	12.75		
250	150	5.75	7.5			9.5	12.75		
	200	6.25	8.5			10.5	14.25		
350	200	6.25	8.5			10.25	14.25		
	250	6.75	9.5			11.25	15.75		
450	200	6.75	9.5	6.0	8.75	11.25	15.75	10.5	14.5
	250	7.25	10.75	6.75	9.5	12.0	17.25	11.25	16.0
	350	7.75	11.75	7.0	10.25	12.75	18.0	12.0	17.25
550	200	7.25	10.75	6.5	9.75	12.0	18.0	10.75	16.5
	350	8.25	13.0	7.25	10.75	13.25	21.0	12.0	18.0
	450	8.5	13.5	7.75	11.75	14.0	22.5	12.75	19.5
650	200	7.75	11.75	7.0	10.75	12.75	19.5	11.75	18.0
	350	8.5	13.5	7.75	12.0	14.0	22.5	12.75	19.5
	450	9.25	14.0	8.5	13.5	15.25	24.5	14.0	22.5
750	250	8.0	12.75	7.5	11.5	13.5	21.25	12.5	19.25
	450	9.0	13.75	8.25	13.0	15.0	24.0	13.75	21.5
	650	10.25	15.0	9.25	14.0	16.5	25.0	15.0	24.0
825	250	8.5	13.5	7.75	12.0	14.25	22.5	13.0	20.0
	450	9.5	14.25	8.75	13.5	15.75	24.0	14.5	22.25
	650	10.75	15.75	9.75	15.0	17.25	26.25	15.75	24.0
900	250			8.25	12.5			13.75	21.0
	450			9.25	14.0			15.25	23.5
	750			10.25	15.0			16.5	25.5
1050	250			8.75	13.5			14.75	22.0
	550			10.0	15.0			16.75	25.0
	825			11.0	16.5			18.25	27.5
1175	250			9.25	14.0			15.5	23.0
	550			10.5	15.75			17.5	25.5
	900			12.0	17.5			19.5	29.0
1300	250			9.75	14.5			16.25	24.0
	550			11.25	17.0			18.75	27.0
	1050			12.5	18.25			20.75	30.5

Source: Westinghouse Electric Corporation, *Applied Protective Relaying*, Newark, NJ, 1970. With permission.
BIL, basic impulse insulation level.

TABLE A.17

60 Hz Characteristics of Three-Conductor Belted Paper-Insulated Cables

Voltage Class	Insulation Thickness (mils)		Circular Mils or AWG (B & S)	Type of Conductor[e]	Weight/1000 ft	Diameter[d] or Sector Depth (in.)	Resistance[a] (Ω/mi)	GMR of One Conductor[b] (in.)
	Conductor	Belt						
1 kV	60	35	6	SR	1,500	0.184	2.50	0.067
	60	35	4	SR	1,910	0.232	1.58	0.084
	60	35	2	SR	2,390	0.292	0.987	0.106
	60	35	1	SR	2,820	0.332	0.786	0.126
	60	35	0	SR	3,210	0.373	0.622	0.142
	60	35	00	CS	3,160	0.323	0.495	0.151
	60	35	000	CS	3,650	0.364	0.392	0.171
	60	35	0000	CS	4,390	0.417	0.310	0.191
	60	35	250,000	CS	4,900	0.455	0.263	0.210
	60	35	300,000	CS	5,660	0.497	0.220	0.230
	60	35	350,000	CS	6,310	0.539	0.190	0.249
	60	35	400,000	CS	7,080	0.572	0.166	0.265
	60	35	500,000	CS	8,310	0.642	0.134	0.297
	65	40	600,000	CS	9,800	0.700	0.113	0.327
	65	40	750,000	CS	11,800	0.780	0.091	0.366
3 kV	70	40	6	SR	1,680	0.184	2.50	0.067
	70	40	4	SR	2,030	0.232	1.58	0.084
	70	40	2	SR	2,600	0.292	0.987	0.106
	70	40	1	SR	2,930	0.332	0.786	0.126
	70	40	0	SR	3,440	0.373	0.622	0.142
	70	40	00	CS	3,300	0.323	0.495	0.151
	70	40	000	CS	3,890	0.364	0.392	0.171
	70	40	0000	CS	4,530	0.417	0.310	0.191
	70	40	250,000	CS	5,160	0.455	0.263	0.210
	70	40	300,000	CS	5,810	0.497	0.220	0.230
	70	40	350,000	CS	6,470	0.539	0.190	0.249
	70	40	400,000	CS	7,240	0.572	0.166	0.265
	70	40	500,000	CS	8,660	0.642	0.134	0.297
	75	40	600,000	CS	9,910	0.700	0.113	0.327
	75	40	750,000	CS	11,920	0.780	0.091	0.366
5 kV	105	55	6	SR	2,150	0.184	2.50	0.067
	100	55	4	SR	2,470	0.232	1.58	0.084
	95	50	2	SR	2,900	0.292	0.987	0.106
	90	45	1	SR	3,280	0.332	0.786	0.126
	90	45	0	SR	3,660	0.373	0.622	0.142
	85	45	00	CS	3,480	0.323	0.495	0.151
	85	45	000	CS	4,080	0.364	0.392	0.171
	85	45	0000	CS	4,720	0.417	0.310	0.191
	85	45	250,000	CS	5,370	0.455	0.263	0.210
	85	45	300,000	CS	6,050	0.497	0.220	0.230

Positive and Negative Sequences			Zero Sequence			Sheath	
Series Reactance (Ω/mi)	Shunt Capacitive Reactance[c] (Ω/mi)	GMR— Three Conductors[e]	Series Resistance[d] (Ω/mi)	Series Reactance[d] (Ω/mi)	Shunt Capacitive Reactance[c] (Ω/mi)	Thickness (mils)	Resistance (Ω/mi) at 50°C
0.185	6300	0.184	10.66	0.315	11,600	85	2.69
0.175	5400	0.218	8.39	0.293	10,200	90	2.27
0.165	4700	0.262	6.99	0.273	9,000	90	2.00
0.165	4300	0.295	6.07	0.256	8,400	95	1.76
0.152	4000	0.326	5.54	0.246	7,900	95	1.64
0.138	2800	0.290	5.96	0.250	5,400	95	1.82
0.134	2300	0.320	5.46	0.241	4,500	95	1.69
0.131	2000	0.355	4.72	0.237	4,000	100	1.47
0.129	1800	0.387	4.46	0.224	3,600	100	1.40
0.128	1700	0.415	3.97	0.221	3,400	105	1.25
0.126	1500	0.446	3.73	0.216	3,100	105	1.18
0.124	1500	0.467	3.41	0.214	2,900	110	1.08
0.123	1300	0.517	3.11	0.208	2,600	110	0.993
0.122	1200	0.567	2.74	0.197	2,400	115	0.877
0.121	1100	0.623	2.40	0.194	2,100	120	0.771
0.192	6700	0.192	9.67	0.322	12,500	90	2.39
0.181	5800	0.227	8.06	0.298	11,200	90	2.16
0.171	5100	0.271	6.39	0.278	9,800	95	1.80
0.181	4700	0.304	5.83	0.263	9,200	95	1.68
0.158	4400	0.335	5.06	0.256	8,600	100	1.48
0.142	3500	0.297	5.69	0.259	6,700	95	1.73
0.138	2700	0.329	5.28	0.246	5,100	95	1.63
0.135	2400	0.367	4.57	0.237	4,600	100	1.42
0.132	2100	0.396	4.07	0.231	4,200	105	1.27
0.130	1900	0.424	3.82	0.228	3,800	105	1.20
0.129	1800	0.455	3.61	0.219	3,700	105	1.14
0.128	1700	0.478	3.32	0.218	3,400	110	1.05
0.126	1500	0.527	2.89	0.214	3,000	115	0.918
0.125	1400	0.577	2.68	0.210	2,800	115	0.855
0.123	1300	0.633	2.37	0.204	2,500	120	0.758
0.215	8500	0.218	8.14	0.342	15,000	95	1.88
0.199	7600	0.250	6.86	0.317	13,600	95	1.76
0.184	6100	0.291	5.88	0.290	11,300	95	1.63
0.171	5400	0.321	5.23	0.270	10,200	100	1.48
0.165	5000	0.352	4.79	0.259	9,600	100	1.39
0.148	3600	0.312	5.42	0.263	9,300	95	1.64
0.143	3200	0.343	4.74	0.254	6,700	100	1.45
0.141	2800	0.380	4.33	0.245	8,300	100	1.34
0.138	2600	0.410	3.89	0.237	7,800	105	1.21
0.135	2400	0.438	3.67	0.231	7,400	105	1.15

(continued)

TABLE A.17 (continued)
60 Hz Characteristics of Three-Conductor Belted Paper-Insulated Cables

Voltage Class	Insulation Thickness (mils) Conductor	Belt	Circular Mils or AWG (B & S)	Type of Conductor[e]	Weight/1000 ft	Diameter[d] or Sector Depth (in.)	Resistance[a] (Ω/mi)	GMR of One Conductor[b] (in.)
5 kV	85	45	350,000	CS	6,830	0.539	0.190	0.249
	85	45	400,000	CS	7,480	0.572	0.166	0.265
	85	45	500,000	CS	8,890	0.642	0.134	0.297
	85	45	600,000	CS	10,300	0.700	0.113	0.327
	85	45	750,000	CS	12,340	0.780	0.091	0.366
8 kV	130	65	6	SR	2,450	0.184	2.50	0.067
	125	65	4	SR	2,900	0.232	1.58	0.084
	115	60	2	SR	3,280	0.292	0.987	0.106
	110	55	1	SR	3,560	0.332	0.786	0.126
	110	55	0	SR	4,090	0.373	0.622	0.142
	105	55	00	CS	3,870	0.323	0.495	0.151
	105	55	000	CS	4,390	0.364	0.392	0.171
	105	55	0000	CS	5,150	0.417	0.310	0.191
	105	55	250,000	CS	5,830	0.455	0.263	0.210
	105	55	300,000	CS	6,500	0.497	0.220	0.230
	105	55	350,000	CS	7,160	0.539	0.190	0.249
	105	55	400,000	CS	7,980	0.572	0.166	0.265
	105	55	500,000	CS	9,430	0.642	0.134	0.297
	105	55	600,000	CS	10,680	0.700	0.113	0.327
	105	55	750,000	CS	12,740	0.780	0.091	0.366
15 kV	170	85	2	SR	4,350	0.292	0.987	0.106
	165	80	1	SR	4,640	0.332	0.786	0.126
	160	75	0	SR	4,990	0.373	0.622	0.142
	155	75	00	SR	5,600	0.419	0.495	0.159
	155	75	000	SR	6,230	0.470	0.392	0.178
	155	75	0000	SR	7,180	0.528	0.310	0.200
	155	75	250,000	SR	7,840	0.575	0.263	0.218
	155	75	300,000	CS	7,480	0.497	0.220	0.230
	155	75	350,000	CS	8,340	0.539	0.190	0.249
	155	75	400,000	CS	9,030	0.572	0.166	0.265
	155	75	500,000	CS	10,550	0.642	0.134	0.297
	155	75	600,000	CS	12,030	0.700	0.113	0.327
	155	75	750,000	CS	14,190	0.780	0.091	0.366

Source: Westinghouse Electric Corporation, *Electrical Transmission and Distribution Reference Book*, East Pittsburgh, PA, 1964.

[a] AC resistance based on 100% conductivity at 65°C including 2% allowance for stranding.

[b] GMR of sector-shaped conductors is an approximate figure close enough for most practical applications.

[c] Dielectric constant = 3.7.

[d] Based on all return current in the sheath; none in ground.

[e] See Figure 7, p. 67, of Ref. [1]. The following symbols are used to designate the cable types: SR, stranded round; CS, compact sector.

Positive and Negative Sequences			Zero Sequence			Sheath	
Series Reactance (Ω/mi)	Shunt Capacitive Reactance[c] (Ω/mi)	GMR—Three Conductors[e]	Series Resistance[d] (Ω/mi)	Series Reactance[d] (Ω/mi)	Shunt Capacitive Reactance[c] (Ω/mi)	Thickness (mils)	Resistance (Ω/mi) at 50°C
0.133	2200	0.470	3.31	0.225	7,000	110	1.04
0.131	2000	0.493	3.17	0.221	6,700	110	1.00
0.129	1800	0.542	2.79	0.216	6,200	115	0.885
0.128	1600	0.587	2.51	0.210	5,800	120	0.798
0.125	1500	0.643	2.21	0.206	5,400	125	0.707
0.230	9600	0.236	7.57	0.353	16,300	95	1.69
0.212	8300	0.269	6.08	0.329	14,500	100	1.50
0.193	6800	0.307	5.25	0.302	12,500	100	1.42
0.179	6100	0.338	4.90	0.280	11,400	100	1.37
0.174	5700	0.368	4.31	0.272	10,700	105	1.23
0.156	4300	0.330	4.79	0.273	8,300	100	1.43
0.151	3800	0.362	4.41	0.263	7,400	100	1.34
0.147	3500	0.399	3.88	0.254	6,600	105	1.19
0.144	3200	0.428	3.50	0.246	6,200	110	1.08
0.141	2900	0.458	3.31	0.239	5,600	110	1.03
0.139	2700	0.489	3.12	0.233	5,200	110	0.978
0.137	2500	0.513	2.86	0.230	4,900	115	0.899
0.135	2200	0.563	2.53	0.224	4,300	120	0.800
0.132	2000	0.606	2.39	0.218	3,900	120	0.758
0.129	1800	0.663	2.11	0.211	3,500	125	0.673
0.217	8600	0.349	4.20	0.323	15,000	110	1.07
0.202	7800	0.381	3.88	0.305	13,800	110	1.03
0.193	7100	0.409	3.62	0.288	12,800	110	1.00
0.185	6500	0.439	3.25	0.280	12,000	115	0.918
0.180	6000	0.476	2.99	0.272	11,300	115	0.867
0.174	5600	0.520	2.64	0.263	10,600	120	0.778
0.168	5300	0.555	2.50	0.256	10,200	120	0.744
0.155	5400	0.507	2.79	0.254	7,900	115	0.855
0.152	5100	0.536	2.54	0.250	7,200	120	0.784
0.149	4900	0.561	2.44	0.245	6,900	120	0.758
0.145	4600	0.611	2.26	0.239	6,200	125	0.690
0.142	4300	0.656	1.97	0.231	5,700	130	0.620
0.139	4000	0.712	1.77	0.226	5,100	135	0.558

TABLE A.18
60 Hz Characteristics of Three-Conductor Shielded Paper-Insulated Cables

Voltage Class	Insulation Thickness (mils)	Circular Miles or AWG (B & S)	Type of Conductor[f]	Weight/1000 ft	Diameter or Sector Depth[b] (in.)	Resistance (Ω/mi)[a]	GMR of One Conductor[c] (in.)
15 kV	205	4	SR	3,860	0.232	1.58	0.084
	190	2	SR	4,260	0.292	0.987	0.106
	185	1	SR	4,740	0.332	0.786	0.126
	180	0	SR	5,090	0.373	0.622	0.141
	175	00	CS	4,790	0.323	0.495	0.151
	175	000	CS	5,510	0.364	0.392	0.171
	175	0000	CS	6,180	0.417	0.310	0.191
	175	250,000	CS	6,910	0.455	0.263	0.210
	175	300,000	CS	7,610	0.497	0.220	0.230
	175	350,000	CS	8,480	0.539	0.190	0.249
	175	400,000	CS	9,170	0.572	0.166	0.265
	175	500,000	CS	10,710	0.642	0.134	0.297
	175	600,000	CS	12,230	0.700	0.113	0.327
	175	750,000	CS	14,380	0.780	0.091	0.366
23 kV	265	2	SR	5,590	0.292	0.987	0.106
	250	1	SR	5,860	0.332	0.786	0.126
	250	0	SR	6,440	0.373	0.622	0.141
	240	00	CS	6,060	0.323	0.495	0.151
	240	000	CS	6,620	0.364	0.392	0.171
	240	0000	CS	7,480	0.410	0.310	0.191
	240	250,000	CS	8,070	0.447	0.263	0.210
	240	300,000	CS	8,990	0.490	0.220	0.230
	240	350,000	CS	9,720	0.532	0.190	0.249
	240	400,000	CS	10,650	0.566	0.166	0.265
	240	500,000	CS	12,280	0.635	0.134	0.297
	240	600,000	CS	13,610	0.690	0.113	0.327
	240	750,000	CS	15,830	0.767	0.091	0.366
35 kV	355	0	SR	8,520	0.288	0.622	0.141
	345	00	SR	9,180	0.323	0.495	0.159
	345	000	SR	9,900	0.364	0.392	0.178
	345	0000	CS	9,830	0.410	0.310	0.191
	345	250,000	CS	10,470	0.447	0.263	0.210
	345	300,000	CS	11,290	0.490	0.220	0.230
	345	350,000	CS	12,280	0.532	0.190	0.249
	345	400,000	CS	13,030	0.566	0.166	0.265
	345	500,000	CS	14,760	0.635	0.134	0.297
	345	600,000	CS	16,420	0.690	0.113	0.327
	345	750,000	CS	18,860	0.767	0.091	0.366

Source: Westinghouse Electric Corporation, *Electrical Transmission and Distribution Reference Book*, East Pittsburgh, PA, 1964.

[a] AC resistance based on 100% conductivity at 65°C including 2% allowance for stranding.

[b] Geometric mean radius (GMR) of sector-shaped conductors is an approximate figure close enough for most practical applications.

[c] Dielectric constant = 3.7.

[d] Based on all return current in the sheath; none in ground.

[e] See Figure 7, p. 67, of Ref. [1].

[f] The following symbols are used to designate the conductor types: SR, stranded round; CS, compact sector.

Positive and Negative Sequences			Zero Sequence			Sheath	
Series Reactance (Ω/mi)	Shunt Capacitive Reactance (Ω/mi)	GMR— Three Conductors	Series Resistance (Ω/mi)[d]	Series Reactance (Ω/mi)[d]	Shunt Capacitive Reactance (Ω/mils)[e]	Thickness (mi)	Resistance (Ω/mi) at 50°C
0.248	8200	0.328	5.15	0.325	8200	105	1.19
0.226	6700	0.365	4.44	0.298	6700	105	1.15
0.210	6000	0.398	3.91	0.285	6000	110	1.04
0.201	5400	0.425	3.65	0.275	5400	110	1.01
0.178	5200	0.397	3.95	0.268	5200	105	1.15
0.170	4800	0.432	3.48	0.256	4800	110	1.03
0.166	4400	0.468	3.24	0.249	4400	110	0.975
0.158	4100	0.498	2.95	0.243	4100	115	0.897
0.156	3800	0.530	2.80	0.237	3800	115	0.860
0.153	3600	0.561	2.53	0.233	3600	120	0.783
0.151	3400	0.585	2.45	0.228	3400	120	0.761
0.146	3100	0.636	2.19	0.222	3100	125	0.684
0.143	2900	0.681	1.98	0.215	2900	130	0.623
0.139	2600	0.737	1.78	0.211	2600	135	0.562
0.250	8300	0.418	3.60	0.317	8300	115	0.870
0.232	7500	0.450	3.26	0.298	7500	115	0.851
0.222	8800	0.477	2.99	0.290	6800	120	0.788
0.196	6600	0.446	3.16	0.285	6600	115	0.890
0.188	6000	0.480	2.95	0.285	6000	115	0.851
0.181	5600	0.515	2.64	0.268	5800	120	0.775
0.177	5200	0.545	2.50	0.261	5200	120	0.747
0.171	4900	0.579	2.29	0.252	4900	125	0.690
0.167	4600	0.610	2.10	0.249	4600	125	0.665
0.165	4400	0.633	2.03	0.240	4400	130	0.620
0.159	3900	0.687	1.82	0.237	3900	135	0.562
0.154	3700	0.730	1.73	0.230	3700	135	0.540
0.151	3400	0.787	1.56	0.225	3400	140	0.488
0.239	9900	0.523	2.40	0.330	9900	130	0.594
0.226	9100	0.548	2.17	0.322	9100	135	0.559
0.217	8500	0.585	2.01	0.312	8500	135	0.538
0.204	7200	0.594	2.00	0.290	7200	135	0.563
0.197	6800	0.628	1.90	0.280	6800	135	0.545
0.191	6400	0.663	1.80	0.273	6400	135	0.527
0.187	6000	0.693	1.66	0.270	6000	140	0.491
0.183	5700	0.721	1.61	0.265	5700	140	0.480
0.177	5200	0.773	1.46	0.257	5200	145	0.441
0.171	4900	0.819	1.35	0.248	4900	150	0.412
0.165	4500	0.879	1.22	0.243	4500	155	0.377

TABLE A.19
60 Hz Characteristics of Three-Conductor Oil-Filled Paper-Insulated Cables

Voltage Class	Insulation Thickness (mils)	Circular Mile or AWG (B & S)	Type of Conductor[f]	Weight/1000 ft	Diameter or Sector Depth[e] (in.)	Resistance (Ω/mi)[a]	GMR of One Conductor[b] (in.)
35 kV	190	00	CS	5,590	0.323	0.495	0.151
		000	CS	6,150	0.364	0.392	0.171
		0000	CS	6,860	0.417	0.310	0.191
		250,000	CS	7,680	0.455	0.263	0.210
		300,000	CS	9,090	0.497	0.220	0.230
		350,000	CS	9,180	0.539	0.190	0.249
		400,000	CS	9,900	0.572	0.166	0.265
		500,000	CS	11,550	0.642	0.134	0.297
		600,000	CS	12,900	0.700	0.113	0.327
		750,000	CS	15,660	0.780	0.091	0.366
46 kV	225	00	CS	6,360	0.323	0.495	0.151
		000	CS	6,940	0.364	0.392	0.171
		0000	CS	7,660	0.410	0.310	0.191
		250,000	CS	8,280	0.447	0.263	0.210
		300,000	CS	9,690	0.490	0.220	0.230
		350,000	CS	10,100	0.532	0.190	0.249
		400,000	CS	10,820	0.566	0.166	0.265
		500,000	CS	12,220	0.635	0.134	0.297
		600,000	CS	13,930	0.690	0.113	0.327
		750,000	CS	16,040	0.767	0.091	0.366
		1,000,000	CS				
69 kV	315	00	CR	8,240	0.370	0.495	0.147
		000	CS	8,830	0.364	0.392	0.171
		0000	CS	9,660	0.410	0.310	0.191
		250,000	CS	10,330	0.447	0.263	0.210
		300,000	CS	11,540	0.490	0.220	0.230
		350,000	CS	12,230	0.532	0.190	0.249
		400,000	CS	13,040	0.566	0.166	0.205
		500,000	CS	14,880	0.635	0.134	0.297
		600,000	CS	16,320	0.690	0.113	0.327
		750,000	CS	18,980	0.767	0.091	0.366
		1,000,000					

Source: Westinghouse Electric Corporation, *Electrical Transmission and Distribution Reference Book*, East Pittsburgh, PA, 1964.

[a] AC resistance based on 100% conductivity at 65°C, including 2% allowance for stranding.

[b] GMR of sector-shaped conductors is an approximate figure close enough for most practical applications.

[c] Dielectric constant = 3.5.

[d] Based on all return current in sheath, none in ground.

[e] See Figure 7, p. 67, of Ref. [1].

[f] The following symbols are used to designate the cable types: CR, compact round; CS, compact sector.

Positive and Negative Sequences		GMR—	Zero Sequence			Sheath	
Series Reactance (Ω/mi)	Shunt Capacitive Reactance[c] (Ω/mi)	Three Conductors	Series Resistance (Ω/mi)[d]	Series Reactance (Ω/mi)[d]	Shunt Capacitive Reactance (Ω/mi)[c]	Thickness (mils)	Resistance (Ω/mi) at 50°V
0.185	6030	0.406	3.56	0.265	6030	115	1.02
0.178	5480	0.439	3.30	0.256	5480	115	0.970
0.172	4840	0.478	3.06	0.243	4840	115	0.918
0.168	4570	0.508	2.72	0.238	4570	125	0.820
0.164	4200	0.539	2.58	0.232	4200	125	0.788
0.160	3900	0.570	2.44	0.227	3900	125	0.752
0.157	3690	0.595	2.35	0.223	3690	125	0.729
0.153	3400	0.646	2.04	0.217	3400	135	0.636
0.150	3200	0.691	1.94	0.210	3200	135	0.608
0.148	3070	0.763	1.73	0.202	3070	140	0.548
0.195	6700	0.436	3.28	0.272	6700	115	0.928
0.188	6100	0.468	2.87	0.265	6100	125	0.826
0.180	5520	0.503	2.67	0.256	5520	125	0.788
0.177	5180	0.533	2.55	0.247	5180	125	0.761
0.172	4820	0.566	2.41	0.241	4820	125	0.729
0.168	4490	0.596	2.16	0.237	4400	135	0.658
0.165	4220	0.623	2.08	0.232	4220	135	0.639
0.160	3870	0.672	1.94	0.226	3870	135	0.603
0.156	3670	0.718	1.74	0.219	3670	140	0.542
0.151	3350	0.773	1.62	0.213	3350	140	0.510
0.234	8330	0.532	2.41	0.290	8330	135	0.639
0.208	7560	0.538	2.32	0.284	7560	135	0.642
0.200	6840	0.575	2.16	0.274	6840	135	0.618
0.195	6500	0.607	2.06	0.266	6500	135	0.597
0.190	6030	0.640	1.85	0.260	6030	140	0.543
0.185	5700	0.672	1.77	0.254	5700	140	0.527
0.181	5430	0.700	1.55	0.248	5430	140	0.513
0.176	5050	0.750	1.51	0.242	5050	150	0.460
0.171	4740	0.797	1.44	0.235	4740	150	0.442
0.165	4360	0.854	1.29	0.230	4360	155	0.399

TABLE A.20

60 Hz Characteristics of Single-Conductor Concentric-Strand Paper-Insulated Cables

Voltage Class	Insulation Thickness (mils)	Circular Mils or AWG (B & S)	Weight/1000 ft	Diameter of Conductor (in.)	GMR of One Conductor[a] (in.)	x_a Reactance at 12 in. (Ω/Phase/mi)	z_a Reactance of Sheath (Ω/Phase/mi)	r_a Resistance of One Conductor (Ω/Phase/mi)[a]	r_a Resistance of Sheath (Ω/Phase/mi) at 50°C	Shunt Capacitive Reactance[c] (Ω/Phase/mi)	Lead Sheath Thickness (mils)
1 kV	60	6	560	0.184	0.067	0.628	0.489	2.50	6.20	4040	75
	60	4	670	0.232	0.084	0.602	0.475	1.58	5.56	3360	75
	60	2	880	0.292	0.106	0.573	0.458	0.987	4.55	2760	80
	60	1	990	0.332	0.126	0.552	0.450	0.786	4.25	2490	80
	60	0	1110	0.373	0.141	0.539	0.442	0.622	3.61	2250	80
	60	00	1270	0.418	0.159	0.524	0.434	0.495	3.34	2040	80
	60	000	1510	0.470	0.178	0.512	0.425	0.392	3.23	1840	85
	60	0000	1740	0.528	0.200	0.496	0.414	0.310	2.98	1650	85
	60	250,000	1930	0.575	0.221	0.484	0.408	0.263	2.81	1530	85
	60	350,000	2490	0.681	0.262	0.464	0.392	0.190	2.31	1300	90
	60	500,000	3180	0.814	0.313	0.442	0.378	0.134	2.06	1090	90
	60	750,000	4380	0.998	0.385	0.417	0.358	0.091	1.65	885	95
	60	1,000,000	5560	1.152	0.445	0.400	0.344	0.070	1.40	800	100
	60	1,500,000	8000	1.412	0.543	0.374	0.319	0.050	1.05	645	110
	60	2,000,000	10,190	1.632	0.633	0.356	0.305	0.041	0.894	555	115
3 kV	75	6	600	0.184	0.067	0.628	0.481	2.50	5.80	4810	75
	75	4	720	0.232	0.084	0.602	0.467	1.58	5.23	4020	75
	75	2	930	0.292	0.106	0.573	0.453	0.987	4.31	3300	80
	75	1	1040	0.332	0.126	0.552	0.445	0.786	4.03	2990	80
	75	0	1170	0.373	0.141	0.539	0.436	0.622	3.79	2670	80
	75	00	1320	0.418	0.159	0.524	0.428	0.495	3.52	2450	80
	75	000	1570	0.470	0.178	0.512	0.420	0.392	3.10	2210	85
	75	0000	1800	0.528	0.200	0.496	0.412	0.310	2.87	2010	85
	75	250,000	1990	0.575	0.221	0.484	0.403	0.263	2.70	1860	85
	75	350,000	2550	0.681	0.262	0.464	0.389	0.190	2.27	1610	90
	75	500,000	3340	0.814	0.313	0.442	0.375	0.134	1.89	1340	95
	75	750,000	4570	0.998	0.385	0.417	0.352	0.091	1.53	1060	100
	75	1,000,000	5640	1.152	0.445	0.400	0.341	0.070	1.37	980	100
	75	1,500,000	8090	1.412	0.543	0.374	0.316	0.050	1.02	805	110
	75	2,000,000	10,300	1,632	0,633	0.356	0.302	0.041	0.877	685	115

						x_z	z_a	r_a	r_a		
Voltage Class	Insulation Thickness (mils)	Circular Mils or AWG (B & S)	Weight/1000 ft	Diameter of Conductor (in.)	GMR of One Conductor[a] (in.)	Reactance at 12 in. (Ω/Phase/mi)	Reactance of Sheath (Ω/Phase/mi)	Resistance of One Conductor (Ω/Phase/mi)[a]	Resistance of Sheath (Ω/Phase/mi) at 50°C	Shunt Capacitive Reactance[c] (Ω/Phase/mi)[b]	Lead Sheath Thickness (mils)
15 kV	220	4	1340	0.232	0.084	0.602	0.412	1.58	2.91	8580	85
	215	2	1500	0.292	0.106	0.573	0.406	0.987	2.74	7270	85
	210	1	1610	0.332	0.126	0.552	0.400	0.786	2.64	6580	85
	200	0	1710	0.373	0.141	0.539	0.397	0.622	2.59	5880	85
	195	00	1940	0.418	0.159	0.524	0.391	0.495	2.32	5290	90
	185	000	2100	0.470	0.178	0.512	0.386	0.392	2.24	4680	90
	180	0000	2300	0.528	0.200	0.496	0.380	0.310	2.14	4200	90
	175	250,000	2500	0.575	0.221	0.484	0.377	0.263	2.06	3820	90
	175	350,000	3110	0.681	0.262	0.464	0.366	0.190	1.98	3340	95
	175	500,000	3940	0.814	0.313	0.442	0.352	0.134	1.51	2870	100
	175	750,000	5240	0.998	0.385	0.417	0.336	0.091	1.26	2420	105
	175	1,000,000	6350	1.152	0.445	0.400	0.325	0.070	1.15	2130	105
	175	1,500,000	8810	1.412	0.546	0.374	0.305	0.050	0.90	1790	115
	175	2,000,000	11,080	1.632	0.633	0.356	0.294	0.041	0.772	1570	120
23 kV	295	2	1920	0.292	0.106	0.573	0.383	0.987	2.16	8890	90
	285	1	2010	0.332	0.126	0.552	0.380	0.786	2.12	8050	90
	275	0	2120	0.373	0.141	0.539	0.377	0.622	2.08	7300	90
	265	00	2250	0.418	0.159	0.524	0.375	0.495	2.02	6580	90
	260	000	2530	0.470	0.178	0.512	0.370	0.392	1.85	6000	95
	250	0000	2740	0.528	0.200	0.496	0.366	0.310	1.78	5350	95
	245	250,000	2930	0.575	0.221	0.484	0.361	0.263	1.72	4950	95
	240	350,000	3550	0.681	0.262	0.464	0.352	0.190	1.51	4310	100
	240	500,000	4300	0.814	0.313	0.442	0.341	0.134	1.38	3720	100
	240	750,000	5630	0.998	0.385	0.417	0.325	0.091	1.15	3170	105
	240	1,000,000	6910	1.152	0.445	0.400	0.313	0.070	1.01	2800	110
	240	1,500,000	9460	1.412	0.546	0.374	0.296	0.050	0.806	2350	120
	240	2,000,000	11,790	1.632	0.633	0.356	0.285	0.041	0.697	2070	125

(continued)

TABLE A.20 (continued)

60 Hz Characteristics of Single-Conductor Concentric-Strand Paper-Insulated Cables

Voltage Class	Insulation Thickness (mils)	Circular Mils or AWG (B & S)	Weight/1000 ft	Diameter of Conductor (in.)	GMR of One Conductor[a] (in.)	x_a Reactance at 12 in. (Ω/Phase/mi)	z_a Reactance of Sheath (Ω/Phase/mi)	r_a Resistance of One Conductor (Ω/Phase/mi)[a]	r_a Resistance of Sheath (Ω/Phase/mi) at 50°C	Shunt Capacitive Reactance[c] (Ω/Phase/mi)	Lead Sheath Thickness (mils)
5 kV	120	6	740	0.184	0.067	0.628	0.456	2.50	4.47	6700	80
	115	4	890	0.232	0.084	0.573	0.447	1.58	4.17	5540	80
	110	2	1040	0.292	0.106	0.573	0.439	0.987	3.85	4520	80
	110	1	1160	0.332	0.126	0.552	0.431	0.786	3.62	4100	80
	105	0	1270	0.373	0.141	0.539	0.425	0.622	3.47	3600	80
	100	00	1520	0.418	0.159	0.524	0.420	0.495	3.09	3140	85
	100	000	1710	0.470	0.178	0.512	0.412	0.392	2.91	2860	85
	95	0000	1870	0.525	0.200	0.496	0.406	0.310	2.74	2480	85
	90	250,000	2080	0.575	0.221	0.484	0.400	0.263	2.62	2180	85
	90	350,000	2620	0.681	0.262	0.464	0.386	0.190	2.20	1890	90
	90	500,000	3410	0.814	0.313	0.442	0.396	0.134	1.85	1610	95
	90	750,000	4650	0.998	0.385	0.417	0.350	0.091	1.49	1360	100
	90	1,000,000	5850	1.152	0.445	0.400	0.339	0.070	1.27	1140	105
	90	1,500,000	8160	1.412	0.543	0.374	0.316	0.050	1.02	950	110
	90	2,000,000	10,370	1.632	0.663	0.356	0.302	0.041	0.870	820	115
8 kV	150	6	890	0.184	0.067	0.628	0.431	2.50	3.62	7780	80
	150	4	1010	0.232	0.084	0.602	0.425	1.58	3.$2	6660	85
	140	2	1150	0.292	0.106	0.573	0.417	0.987	3.06	5400	85
	140	1	1330	0.332	0.126	0.552	0.411	0.786	2.91	4920	85
	135	0	1450	0.373	0.141	0.539	0.408	0.622	2.83	4390	85
	130	00	1590	0.418	0.159	0.524	0.403	0.495	2.70	3890	85
	125	000	1760	0.470	0.178	0.512	0.397	0.392	2.59	3440	85
	120	0000	1980	0.528	0.200	0.496	0.389	0.310	2.29	3020	90
	120	250,000	2250	0.575	0.221	0.484	0.383	0.263	2.18	2790	90
	115	350,000	2730	0.681	0.262	0.464	0.375	0.190	1.90	2350	95
	115	500,000	3530	0.814	0.313	0.442	0.361	0.134	1.69	2010	95
	115	750,000	4790	0.998	0.385	0.417	0.341	0.091	1.39	1670	100
	115	1,000,000	6000	1.152	0.415	0.400	0.330	0.070	1.25	1470	105
	115	1,500,000	8250	1.412	0.543	0.374	0.310	0.050	0.975	1210	110
	115	2,000,000	10,480	1.632	0.663	0.356	0.297	0.041	0.797	1055	120

Source: Westinghouse Electric Corporation, *Electrical Transmission and Distribution Reference Book*, East Pittsburgh, PA, 1964.

[a] Conductors are standard concentric-stranded, not compact round.

[b] AC resistance based on 100% conductivity at 65°C including 2% allowance for stranding.

[c] Dielectric constant = 3.7.

Voltage Class	Insulation Thickness (mils)	Circular Mils or AWG (B & S)	Weight/1000 ft	Diameter of Conductor (in.)	GMR of One Conductor[a] (in.)	x_z Reactance at 12 in. (Ω/Phase/mi)	z_a Reactance of Sheath (Ω/Phase/mi)	r_a Resistance of One Conductor (Ω/Phase/mi)[a]	r_a Resistance of Sheath (Ω/Phase/mi) at 50°C	Shunt Capacitive Reactance[c] (Ω/Phase/mi)[b]	Lead Sheath Thickness (mils)
35 kV	395	0	2900	0.373	0.141	0.539	0.352	0.622	1.51	9150	100
	385	00	3040	0.418	0.159	0.524	0.350	0.495	1.48	8420	100
	370	000	3190	0.470	0.178	0.512	0.347	0.392	1.46	7620	100
	355	0000	3380	0.528	0.200	0.496	0.344	0.310	1.43	6870	100
	350	250,000	3590	0.575	0.221	0.484	0.342	0.263	1.39	6410	100
	345	350,000	4230	0.681	0.262	0.464	0.366	0.190	1.24	5640	105
	345	500,000	5040	0.814	0.313	0.442	0.325	0.134	1.15	4940	105
	345	750,000	5430	0.998	0.385	0.417	0.311	0.091	0.975	4250	110
	345	1,000,000	7780	1.152	0.445	0.400	0.302	0.070	0.866	3780	115
	345	1,500,000	10,420	1.412	0.546	0.374	0.285	0.050	0.700	3210	125
	345	2,000,000	12,830	1.632	0.633	0.356	0.274	0.041	0.811	2830	130
46 kV	475	000	3910	0.470	0.178	0.512	0.331	0.392	1.20	8890	105
	460	0000	4080	0.528	0.200	0.496	0.329	0.310	1.19	8100	105
	450	250,000	4290	0.575	0.221	0.484	0.326	0.263	1.16	7570	105
	445	350,000	4990	0.681	0.262	0.464	0.319	0.190	1.05	6720	110
	445	500,000	5820	0.814	0.313	0.442	0.310	0.134	0.930	5950	115
	445	750,000	7450	0.998	0.385	0.417	0.298	0.091	0.807	5130	120
	445	1,000,000	8680	1.152	0.445	0.400	0.290	0.070	0.752	4610	120
	445	1,500,000	11,420	1.412	0.546	0.374	0.275	0.050	0.615	3930	130
	445	2,000,000	13,910	1.632	0.633	0.356	0.264	0.041	0.543	3520	135
69 kV	650	350,000	6720	0.681	0.262	0.464	0.292	0.190	0.773	8590	120
	650	500,000	7810	0.814	0.313	0.442	0.284	0.134	0.695	7680	125
	650	750,000	9420	0.998	0.385	0.417	0.275	0.091	0.615	6700	130
	650	1,000,000	10,940	1.152	0.445	0.400	0.267	0.070	0.557	6060	135
	650	1,500,000	13,680	1.412	0.546	0.374	0.258	0.050	0.488	5250	140
	650	2,000,000	16,320	1.632	0.633	0.356	0.246	0.041	0.437	4710	145

TABLE A.21

60 Hz Characteristics of Single-Conductor Oil-Filled (Hollow-Core) Paper-Insulated Cables

Inside Diameter of Spring Core = 0.5 in.

Voltage Class	Insulation Thickness (mils)	Circular Mils or AWG (B & S)	Weight/1000 ft	Diameter of Conductor (in.)	Gmr of One Conductor[c] (in.)	X_a Reactance at 12 in. (Ω/Phase/mi)	X_a Reactance of Sheath (Ω/Phase/mi)	Resistance of One Conductor (Ω/Phase/mi)[a]	T_c Resistance of Sheath (Ω/Phase/mi) at 50°C	R_a Shunt Capacitive Reactance[c] (Ω/Phase/mi)[b]	Lead Sheath Thickness (mils)
69 kV	315	00	3,980	0.736	0.345	0.431	0.333	0.495	1.182	5240	110
		000	4,090	0.768	0.356	0.427	0.331	0.392	1.157	5070	110
		0000	4,320	0.807	0.373	0.421	0.328	0.310	1.130	4900	110
		250,000	4,650	0.837	0.381	0.418	0.325	0.263	1.057	4790	115
		350,000	5,180	0.918	0.408	0.410	0.320	0.188	1.009	4470	115
		500,000	6,100	1.028	0.448	0.399	0.312	0.133	0.905	4070	120
		750,000	7,310	1.180	0.505	0.384	0.302	0.089	0.838	3620	120
		1,000,000	8,630	1.310	0.550	0.374	0.294	0.068	0.752	3380	125
		1 $$ 000	11,090	1.547	0.639	0.356	0.281	0.048	0.649	2920	130
		2,000,000	13,750	1.760	0.716	0.342	0.270	0.039	0.550	2570	140
115 kV	480	0000	5720	0.807	0.373	0.421	0.305	0.310	0.805	6650	120
		250,000	5930	0.837	0.381	0.418	0.303	0.263	0.793	6500	120
		350,000	6390	0.918	0.408	0.410	0.298	0.188	0.730	6090	125
		500,000	7480	1.028	0.448	0.399	0.291	0.133	0.692	5600	125
		750,000	8950	1.180	0.505	0.381	0.283	0.089	0.625	5040	130
		1,000,000	10,350	1.310	0.550	0.374	0.276	0.068	0.568	4700	135
		1,500,000	12,960	1.547	0.639	0.356	0.265	0.048	0.500	4110	140
		2,000,000	15,530	1.760	0.716	0.342	0.255	0.039	0.447	3710	145
138 kV	560	0000	6480	0.807	0.373	0.421	0.205	0.310	0.758	7410	125
		250,000	6700	0.837	0.381	0.418	0.293	0.263	0.746	7240	125
		350,000	7460	0.918	0.408	0.410	0.288	0.188	0.690	6820	130
		500,000	8310	1.028	0.448	0.399	0.282	0.133	0.658	6260	130
		750,000	9800	1.180	0.505	0.384	0.274	0.089	0.592	5680	135
		1,000,000	11,270	1.310	0.550	0.374	0.268	0.068	0.541	5240	140
		1,500,000	13720	1.547	0.639	0.356	0.257	0.048	0.477	4670	145
		2,000,000	16080	1.760	0.716	0.342	0.248	0.039	0.427	4170	150
161 kV	650	250,000	7600	0.837	0.381	0.418	0.283	0.263	0.660	7980	130
		350,000	8390	0.918	0.408	0.410	0.279	0.188	0.611	7520	135
		500,000	9270	1.028	0.448	0.399	0.273	0.133	0.585	6980	135
		750,000	10,840	1.180	0.505	0.384	0.266	0.089	0.532	6320	140
		1,000,000	12,340	1.310	0.550	0.374	0.259	0.068	0.483	5880	145
		1,500,000	15,090	1.547	0.639	0.356	0.246	0.048	0.433	5190	150
		2,000,000	18,000	1.760	0.716	0.342	0.241	0.039	0.391	4710	155

Source: Westinghouse Electric Corporation, *Electrical Transmission and Distribution Reference Book*, East Pittsburgh, PA, 1964.

[a] AC resistance based on 100% conductivity at 65°C including 2% allowance for stranding.

[b] Dielectric constant = 3.5.

[c] Calculated for circular tube.

Voltage Class	Insulation Thickness (mils)	Circular Mils or AWG (B & S)	Weight/1000 ft	Diameter of Conductor (in.)	GMR of One Conductor[c] (in.)	X_z Reactance at 12 in. (Ω/Phase/mi)	X_a Reactance of Sheath (Ω/Phase/mi)	T_c Resistance of One Conductor (Ω/Phase/mi)[a]	R_a Resistance of Sheath (Ω/Phase/mi) at 50°C	Shunt Capacitive Reactance (Ω/Phase/mi)[2]	Lead Sheath Thickness (mils)
69 kV	315	000	4860	0.924	0.439	0.399	0.320	0.392	1.007	4450	115
		0000	5090	0.956	0.450	0.398	0.317	0.310	0.985	4350	115
		250,000	5290	0.983	0.460	0.396	0.315	0.263	0.975	4230	115
		350,000	5950	1.050	0.483	0.390	0.310	0.188	0.897	4000	120
		500,000	6700	1.145	0.516	0.382	0.304	0.132	0.850	3700	120
		750,000	8080	1.286	0.550	0.374	0.295	0.089	0.759	3410	125
		1,000,000	9440	1.416	0.612	0.360	0.288	0.067	0.688	3140	130
		1,500,000	11,970	1.635	0.692	0.346	0.276	0.047	0.601	2750	135
		2,000,000	14,450	1.835	0.763	0.334	0.266	0.038	0.533	2510	140
115 kV	480	0000	6590	0.956	0.450	0.398	0.295	0.310	0.760	5950	125
		250,000	6800	0.983	0.460	0.396	0.294	0.263	0.752	5790	125
		350,000	7340	1.050	0.483	0.390	0.290	0.188	0.729	5540	125
		500,000	8320	1.145	0.516	0.382	0.284	0.132	0.669	5150	130
		750,000	9790	1.286	0.550	0.374	0.277	0.089	0.606	4770	135
		1,000,000	11,060	1.416	0.612	0.360	0.270	0.067	0.573	4430	135
		1,500,000	13,900	1.635	0.692	0.346	0.260	0.047	0.490	3920	145
		2,000,000	16,610	1.835	0.763	0.334	0.251	0.038	0.440	3580	150
138 kV	560	0000	7390	0.956	0.450	0.398	0.786	0.310	0.678	6590	130
		250,000	7610	0.983	0.460	0.396	0.285	0.263	0.669	6480	130
		350,000	8170	1.050	0.483	0.390	0.281	0.188	0.649	6180	130
		500,000	9180	1.145	0.516	0.382	0.276	0.132	0.601	5790	135
		750,000	10,660	1.286	0.550	0.374	0.269	0.089	0.545	5320	140
		1,000,000	12,010	1.416	0.612	0.360	0.263	0.067	0.519	4940	140
		1,500,000	14,450	1.635	0.692	0.346	0.253	0.047	0.462	4460	145
		2,000,000	16,820	1.835	0.763	0.334	0.245	0.038	0.404	4060	155
161 kV	650	250,000	8560	0.983	0.460	0.396	0.275	0.263	0.596	7210	135
		350,000	9140	1.050	0.483	0.390	0.272	0.188	0.580	6860	135
		500,000	10,280	1.145	0.516	0.382	0.267	0.132	0.537	6430	140
		750,000	11,770	1.286	0.550	0.374	0.261	0.089	0.492	5980	145
		1,000,000	13,110	1.416	0.612	0.360	0.255	0.067	0.469	5540	145
		1,500,000	15,840	1.635	0.692	0.346	0.246	0.047	0.421	4980	150
		2,000,000	18,840	1.835	0.763	0.334	0.238	0.038	0.369	4600	160
230 kV	925	750,000	15,360	1.286	0.550	0.374	0.238	0.089	0.369	7610	160
		1,000,000	16,790	1.416	0.612	0.360	0.233	0.067	0.355	7140	160
		2,000,000	22,990	1.835	0.763	0.334	0.219	0.038	0.315	5960	170

TABLE A.22
Current-Carrying Capacity of Three-Conductor Belted Paper-Insulated Cables

Number of Equally Loaded Cables in Duct Bank

4,500 V

Conductor Size AWG or MCM	Conductor Type[a]	One (Amperes per Conductor[b])				Three				Six Percent Load Factor (Amperes per Conductor[a], Copper Temperature 85°C)				Nine				Twelve			
		30	50	75	100	30	50	75	100	30	50	75	100	30	50	75	100	30	50	75	100
6	S	82	80	78	75	81	78	73	68	79	74	68	63	78	72	65	58	76	69	61	54
4	SR	109	106	103	98	108	102	96	89	104	97	89	81	102	94	84	74	100	90	79	69
2	SR	143	139	134	128	139	133	124	115	136	127	115	104	133	121	108	95	130	117	101	89
1	SR	164	161	153	146	159	152	141	130	156	145	130	118	152	138	122	108	148	133	115	100
0	CS	189	184	177	168	184	175	162	149	180	166	149	134	175	159	140	122	170	152	130	114
00	CS	218	211	203	192	211	201	185	170	208	190	170	152	201	181	158	138	195	173	148	126
000	CS	250	242	232	219	242	229	211	193	237	217	193	172	229	206	179	156	223	197	167	145
0000	CS	286	276	264	249	276	260	240	218	270	246	218	194	261	234	202	176	254	223	189	163
250	CS	316	305	291	273	305	288	263	239	297	271	239	212	288	258	221	192	279	244	206	177
300	CS	354	340	324	304	340	321	292	264	332	301	264	234	321	285	245	211	310	271	227	195
350	CS	392	376	357	334	375	353	320	288	366	330	288	255	351	311	266	229	341	296	248	211
400	CS	424	406	385	359	406	380	344	309	395	355	309	272	380	334	285	244	367	317	264	224
500	CS	487	465	439	408	465	433	390	348	451	403	348	305	433	378	320	273	417	357	296	251
600	CS	544	517	487	450	517	480	430	383	501	444	383	334	480	416	350	298	462	393	323	273
750	CS	618	581	550	505	585	541	482	427	566	500	427	371	541	466	390	331	519	439	359	302
		(1.07 at 10°C,				(1.07 at 10°C,				(1.07 at 10°C,				(1.07 at 10°C,				(1.07 at 10°C,			
		0.92 at 30°C,				0.92 at 30°C,				0.92 at 30°C,				0.92 at 30°C,				0.92 at 30°C,			
		0.83 at 40°C,				0.83 at 40°C,				0.83 at 40°C,				0.83 at 40°C,				0.83 at 40°C,			
		0.73 at 50°C)[c]				0.73 at 50°C)[c]				0.73 at 50°C)[c]				0.73 at 50°C)[c]				0.73 at 50°C)[c]			

7,500 V — Copper Temperature 83°C

Size	Type																				
6	S	81	80	77	74	79	76	72	67	78	74	67	62	77	71	64	57	75	69	60	53
4	SR	107	105	101	97	104	100	94	87	103	97	87	79	100	92	82	73	98	89	77	68
2	SR	140	137	132	126	136	131	122	113	134	126	113	102	130	119	105	93	127	114	99	87
1	SR	161	156	150	143	156	149	138	128	153	143	128	115	149	136	120	105	145	130	112	98
0	CS	186	180	174	165	180	172	156	146	177	165	148	131	172	155	136	120	167	149	128	111
00	CS	214	206	198	188	206	196	181	166	202	188	166	148	196	177	155	135	191	169	145	125
000	CS	243	236	226	214	236	224	206	188	230	214	188	168	223	200	174	152	217	192	163	141
0000	CS	280	270	258	243	270	255	235	214	264	243	213	190	255	229	198	172	247	218	184	159
250	CS	311	300	287	269	300	283	259	235	293	269	235	208	282	252	217	188	273	240	202	174
300	CS	349	336	320	300	335	316	288	260	326	300	259	230	315	279	240	207	304	265	223	190
350	CS	385	369	351	328	369	346	315	283	359	328	282	249	345	305	261	224	333	289	242	206
400	CS	417	399	378	353	398	373	338	303	388	353	303	267	371	317	279	239	360	309	257	220
500	CS	476	454	429	399	454	423	381	341	440	399	340	298	422	369	312	267	406	348	288	245
600	CS	534	508	479	443	507	471	422	376	491	443	375	327	469	408	343	291	451	384	315	267
750	CS	607	576	540	497	575	532	473	413	555	497	418	363	529	455	381	323	507	428	350	295

(1.08 at 10°C,
0.92 at 30°C,
0.83 at 40°C,
0.72 at 50°C)^c

15,000 V — Copper Temperature 75°C

Size	Type																				
6	S	78	77	74	71	76	74	69	64	75	73	64	59	73	68	61	54	72	65	57	50
4	SR	102	99	96	92	98	95	89	83	97	95	83	75	95	87	78	69	93	85	73	64
2	SR	132	129	125	119	129	123	115	106	126	123	106	96	123	112	99	88	120	108	93	82
1	SR	151	147	142	135	146	140	131	120	144	140	120	109	140	128	112	99	136	122	107	92
0	CS	175	170	163	155	169	161	150	138	166	153	137	123	161	146	128	112	156	139	120	104
00	CS	200	194	187	177	194	184	170	156	189	175	156	139	183	166	145	127	178	158	135	117
000	CS	230	223	214	202	222	211	195	178	217	199	177	158	210	189	165	143	203	180	153	132
0000	CS	266	257	245	232	253	242	222	202	249	228	201	179	240	215	187	158	233	205	173	149
250	CS	295	284	271	255	281	268	245	221	276	251	220	196	266	239	204	177	257	225	189	163
300	CS	330	317	301	283	316	297	271	245	307	278	244	215	295	264	225	194	285	248	208	178

(1.08 at 10°C,
0.92 at 30°C,
0.83 at 40°C,
0.72 at 50°C)^c

(continued)

TABLE A.22 (continued)
Current-Carrying Capacity of Three-Conductor Belted Paper-Insulated Cables

Number of Equally Loaded Cables in Duct Bank

Conductor Size AWG or MCM	Conductor Type[a]	One				Three				Six Percent Load Factor				Nine				Twelve			
		30	50	75	100	30	50	75	100	30	50	75	100	30	50	75	100	30	50	75	100
		Amperes per Conductor[b]								Amperes per Conductor[a]											
										Copper Temperature 75°C											
15,000 V																					
350	CS	365	349	332	310	348	327	297	267	339	305	266	235	324	289	245	211	313	271	227	193
400	CS	394	377	357	333	375	352	319	286	365	327	285	251	349	307	262	224	336	290	241	206
500	CS	449	429	406	377	428	399	359	321	414	396	319	280	396	346	293	250	379	326	269	229
600	CS	502	479	450	417	476	443	396	352	459	409	351	306	438	380	319	273	420	358	294	249
750	CS	572	543	510	468	540	499	444	393	520	458	391	341	494	425	356	302	471	399	326	275
		(1.09 at 10°C,				(1.09 at 10°C,				(1.09 at 10°C,				(1.09 at 10°C,				(1.09 at 10°C,			
		0.90 at 30°C,				0.90 at 30°C,				0.90 at 30°C,				0.90 at 30°C,				0.90 at 30°C,			
		0.79 at 40°C,				0.79 at 40°C,				0.79 at 40°C,				0.79 at 40°C,				0.79 at 40°C,			
		0.67 at 50°C)c				0.67 at 50°C)c				0.66 at 50°C)c				0.66 at 50°C)c				0.66 at 50°C)c			

Source: Westinghouse Electric Corporation, *Electrical Transmission and Distribution Reference Book*, East Pittsburgh, PA, 1964.

[a] The following symbols are used here to designate conductor types: S, solid copper; SR, standard round concentric-stranded; CS, compact-sector stranded.
[b] Current ratings are based on the following conditions:
   1. Ambient earth temperature = 20°C.
   2. 60-cycle alternating current.
   3. Ratings include dielectric loss, and all induced AC losses.
   4. One cable per duct, all cables equally loaded and in outside ducts only.
[c] Multiply tabulated currents by these factors when earth temperature is other than 20°C.

TABLE A.23
Current-Carrying Capacity of Three-Conductor Shielded Paper-Insulated Cables

15,000 V

Conductor Size AWG or MCM	Conductor Type[a]	Number of Equally Loaded								Six Percent Load Factor				Cables in Duct Bank							
		One				Three								Nine				Twelve			
		Amperes per Conductor[b]								Amperes per Conductor[b]											
		Copper Temperature 81°C								Copper Temperature 81°C											
		30	50	75	100	30	50	75	100	30	50	75	100	30	50	75	100	30	50	75	100
6	S	94	91	88	83	91	87	81	75	89	83	74	66	87	78	69	60	84	75	64	56
4	SR	123	120	115	107	119	114	104	95	116	108	95	85	113	102	89	77	109	96	83	75
2	SR	159	154	146	137	153	144	139	121	149	136	120	107	144	129	112	97	139	123	104	90
1	SR	179	174	166	156	172	163	149	136	168	153	136	121	162	145	125	109	158	138	117	100
0	CS	203	195	182	176	196	185	169	154	190	173	154	137	183	164	141	122	178	156	131	112
00	CS	234	224	215	202	225	212	193	175	218	198	174	156	211	187	162	139	203	177	148	127
000	CS	270	258	245	230	258	242	220	198	249	225	198	174	241	212	182	157	232	202	168	144
0000	CS	308	295	281	261	295	276	250	223	285	257	224	196	275	241	205	176	265	227	189	162
250	CS	341	327	310	290	325	305	276	246	315	283	245	215	303	265	224	193	291	250	207	177
300	CS	383	365	344	320	364	339	305	272	351	313	271	236	337	293	246	211	322	276	227	194
350	CS	417	397	375	346	397	369	330	293	383	340	293	255	366	318	267	227	350	301	245	208
400	CS	453	428	403	373	429	396	354	314	413	366	313	273	394	340	285	242	376	320	262	222
500	CS	513	487	450	418	483	446	399	350	467	410	350	303	444	381	318	269	419	358	292	247
600	CS	567	537	501	460	534	491	437	385	513	450	384	330	488	416	346	293	465	390	317	269
750	CS	643	606	562	514	602	551	485	426	576	502	423	365	545	464	383	323	519	432	348	293
		(1.08 at 10°C, 0.91 at 30°C, 0.82 at 40°C, 0.71 at 50°C)[b]				(1.08 at 10°C, 0.91 at 30°C, 0.82 at 40°C, 0.71 at 50°C)[b]				(1.08 at 10°C, 0.91 at 30°C, 0.82 at 40°C, 0.71 at 50°C)[b]				(1.08 at 10°C, 0.91 at 30°C, 0.82 at 40°C, 0.71 at 50°C)[b]				(1.08 at 10°C, 0.91 at 30°C, 0.82 at 40°C, 0.71 at 50°C)[b]			

(continued)

TABLE A.23 (continued)
Current-Carrying Capacity of Three-Conductor Shielded Paper-Insulated Cables

Conductor Size AWG or MCM	Conductor Type[a]	Number of Equally Loaded								Cables in Duct Bank											
		One				Three				Six Percent Load Factor				Nine				Twelve			
		30	50	75	100	30	50	75	100	30	50	75	100	30	50	75	100	30	50	75	100
23,000 V		Amperes per Conductor[b]				Amperes per Conductor[b]				Amperes per Conductor[b]											
		Copper Temperature 77°C				Copper Temperature 77°C				Copper Temperature 77°C											
2	SR	156	150	143	134	149	141	130	117	145	132	117	105	140	125	107	84	134	119	100	86
1	SR	177	170	162	152	170	160	145	133	164	149	132	117	159	140	121	105	154	133	112	97
0	CS	200	192	183	172	192	182	166	149	186	169	147	132	178	158	136	118	173	149	126	109
00	CS	227	220	210	197	221	208	189	170	212	193	168	149	202	181	156	134	196	172	144	123
000	CS	262	251	238	223	254	238	216	193	242	220	191	169	230	206	175	150	222	195	162	139
0000	CS	301	289	271	251	291	273	246	219	278	250	215	190	264	233	197	169	255	221	182	157
250	CS	334	315	298	277	321	299	270	239	308	275	236	207	290	258	216	184	279	242	199	170
300	CS	373	349	328	306	354	329	297	263	341	302	259	227	320	283	232	202	309	266	217	186
350	CS	405	379	358	331	384	356	318	283	369	327	280	243	347	305	255	217	335	285	233	199
400	CS	434	409	386	356	412	379	340	302	396	348	298	260	374	325	273	232	359	303	247	211
500	CS	492	465	436	401	461	427	379	335	443	391	333	288	424	363	302	257	400	336	275	230
600	CS	543	516	484	440	512	470	414	366	489	428	365	313	464	396	329	279	441	367	299	248
750	CS	616	583	541	495	577	528	465	407	550	479	402	347	520	439	364	306	490	408	329	276
		(1.09 at 10°C, 0.90 at 30°C,				(1.09 at 10°C, 0.90 at 30°C,				(1.09 at 10°C, 0.90 at 30°C,				(1.09 at 10°C, 0.90 at 30°C,				(1.09 at 10°C, 0.90 at 30°C,			

34,500 V — Copper Temperature 70°C

Size	0.80 at 40°C, 0.67 at 50°C[c]				0.80 at 40°C, 0.67 at 50°C[b]				0.79 at 40°C, 0.67 at 50°C[c]				0.79 at 40°C, 0.66 at 50°C[c]				0.79 at 40°C, 0.65 at 50°C[a]			
0	193	185	176	165	184	174	158	141	178	161	140	124	171	149	129	111	164	142	119	103
00	219	209	199	187	208	197	178	160	202	182	158	140	194	170	145	126	185	161	134	115
000	250	238	225	211	238	222	202	182	229	206	179	158	220	193	165	141	209	182	152	128
0000	288	275	260	241	273	256	229	205	263	234	203	179	251	219	186	160	238	205	170	144
250	316	302	285	266	301	280	253	224	289	258	222	196	276	240	202	174	262	222	187	157
300	352	335	315	293	334	310	278	246	320	284	244	213	304	264	221	190	288	244	203	171
350	384	364	342	318	363	336	301	267	346	308	264	229	329	285	238	204	311	263	217	184
400	413	392	367	341	384	360	321	284	372	329	281	244	352	303	254	216	334	282	232	195
500	468	442	414	381	436	402	358	317	418	367	312	271	393	337	281	238	372	313	256	215
600	514	487	455	416	481	440	391	344	459	401	340	294	430	367	304	259	406	340	277	232
750	584	548	510	466	541	496	435	383	515	447	378	324	481	409	337	284	452	377	304	255

All conductors: **CS**

Correction multipliers (by column group):

- (1.10 at 10°C, 0.89 at 30°C, 0.76 at 40°C, 0.61 at 50°C)[c]
- (1.10 at 10°C, 0.89 at 30°C, 0.76 at 40°C, 0.60 at 50°C)[b]
- (1.10 at 10°C, 0.89 at 30°C, 0.76 at 40°C, 0.60 at 50°C)[1]
- (1.10 at 10°C, 0.88 at 30°C, 0.75 at 40°C, 0.58 at 50°C)[3]
- (1.10 at 10°C, 0.88 at 30°C, 0.74 at 40°C, 0.56 at 50°C)[3]

Source: Westinghouse Electric Corporation, *Electrical Transmission and Distribution Reference Book*, East Pittsburgh, PA, 1964.

[a] The following symbols are used here to designate conductor types: S, solid copper; SR, standard round concentric-stranded; CS, compact-sector-stranded.

[b] Current ratings are based on the following conditions:

1. Ambient earth temperature = 20°C.
2. 60-cycle alternating current.
3. Ratings include dielectric loss, and all induced AC losses.
4. One cable per duct, all cables equally loaded and in outside ducts only.

[c] Multiply tabulated currents by these factors when earth temperature is other than 20°C.

TABLE A.24
Current-Carrying Capacity of Single-Conductor Solid Paper-Insulated Cables

Number of Equally Loaded Cables in Duct Bank

7,500 V — Copper Temperature, 85°C — Amperes per Conductor[a]

Six Percent Load Factor

Conductor Size AWG or MCM	Three				Six				Nine				Twelve			
	30	50	75	100	30	50	75	100	30	50	75	100	30	50	75	100
6	116	113	109	103	115	110	103	96	113	107	98	90	111	104	94	85
4	164	149	142	135	152	144	134	125	149	140	128	116	147	136	122	110
2	202	196	186	175	199	189	175	162	196	183	167	151	192	178	159	142
1	234	226	214	201	230	218	201	185	226	210	190	172	222	204	181	162
0	270	262	245	232	266	251	231	212	261	242	219	196	256	234	204	184
00	311	300	283	262	309	290	270	241	303	278	250	224	295	268	236	208
000	356	344	324	300	356	333	303	275	348	319	285	255	340	308	270	236
0000	412	395	371	345	408	380	347	314	398	364	325	290	390	352	307	269
250	456	438	409	379	449	418	379	344	437	400	358	316	427	386	336	294
300	512	491	459	423	499	464	420	380	486	442	394	349	474	428	371	325
350	561	537	500	460	546	507	457	403	532	483	429	379	518	466	403	352
400	607	580	540	496	593	548	493	445	576	522	461	407	560	502	434	378
500	692	660	611	561	679	626	560	504	659	597	524	459	641	571	490	427
600	772	735	679	621	757	696	621	557	733	663	579	506	714	632	542	470
700	846	804	741	677	827	758	674	604	802	721	629	548	779	688	587	508
750	881	837	771	702	860	789	700	627	835	750	651	568	810	714	609	526
800	914	866	797	725	892	817	726	648	865	776	674	588	840	740	630	544
1000	1037	980	898	816	1012	922	815	725	980	874	758	657	950	832	705	606
1250	1176	1108	1012	914	1145	1039	914	809	1104	981	845	730	1068	941	784	673
1500	1300	1224	1110	1000	1268	1146	1000	884	1220	1078	922	794	1178	1032	855	731

Size																
1750	1420	1332	1204	1080	1382	1240	1078	949	1342	1166	992	851	1280	1103	919	783
2000	1546	1442	1300	1162	1500	1343	1162	1019	1442	1260	1068	914	1385	1190	986	839

Footnotes for rows above: (1.07 at 10°C, 0.92 at 30°C, 0.83 at 40°C, 0.73 at 50°C)b

15,000 V   **Copper Temperature, 81°C**   **Copper Temperature, 81°C**

Size																
6	113	110	105	100	112	107	100	93	110	104	96	87	108	101	92	83
4	149	145	138	131	147	140	131	117	144	136	125	114	142	132	119	107
2	195	190	180	170	193	183	170	157	189	177	161	146	186	172	154	137
1	226	218	208	195	222	211	195	179	218	204	185	167	214	197	175	157
0	256	248	234	220	252	239	220	203	247	230	209	188	242	223	198	177
00	297	287	271	254	295	278	253	232	287	265	239	214	283	257	226	202
000	344	330	312	290	341	320	293	267	333	306	274	245	327	296	260	230
0000	399	384	361	335	392	367	335	305	383	352	315	280	374	340	298	263
250	440	423	396	367	432	404	367	334	422	387	345	306	412	372	325	286
300	490	470	439	406	481	449	406	369	470	429	382	338	457	413	359	316
350	539	516	481	444	527	491	443	401	514	468	416	367	501	450	391	342
400	586	561	522	480	572	530	478	432	556	506	447	395	542	485	419	366
500	669	639	592	543	655	605	542	488	636	577	507	445	618	551	474	412
600	746	710	656	601	727	668	598	537	705	637	557	488	685	608	521	452
700	810	772	712	652	790	726	647	581	766	691	604	528	744	659	564	488
750	840	797	736	674	821	753	672	602	795	716	625	547	772	684	584	505
800	869	825	762	696	850	780	695	622	823	741	646	565	800	707	604	522
1000	991	939	864	785	968	882	782	697	933	832	724	631	903	794	675	581
1250	1130	1067	975	864	1102	1000	883	784	1063	941	816	706	1026	898	759	650
1500	1250	1176	1072	966	1220	1105	972	856	1175	1037	892	772	1133	987	828	707
1750	1368	1282	1162	1044	1330	1198	1042	919	1278	1124	958	824	1230	1063	886	755
2000	1464	1368	1233	1106	1422	1274	1105	970	1360	1192	1013	889	1308	1125	935	795

Footnotes:
(1.08 at 10°C, 0.92 at 30°C, 0.71 at 50°C)b
(1.08 at 10°C, 0.92 at 30°C, 0.82 at 40°C, 0.71 at 50°C)b
(1.07 at 10°C, 0.92 at 30°C, 0.83 at 40°C, 0.73 at 50°C)b

(continued)

TABLE A.24 (continued)

Current-Carrying Capacity of Single-Conductor Solid Paper-Insulated Cables

Number of Equally Loaded Cables in Duct Bank

23,000 V — Amperes per Conductor[a] — Copper Temperature, 77°C — Percent Load Factor

Conductor Size AWG or MCM	Three 30	Three 50	Three 75	Three 100	Six 30	Six 50	Six 75	Six 100	Nine 30	Nine 50	Nine 75	Nine 100	Twelve 30	Twelve 50	Twelve 75	Twelve 100
2	186	181	172	162	184	175	162	150	180	169	154	140	178	164	147	132
1	214	207	197	186	211	200	185	171	206	193	176	159	203	187	167	150
0	247	239	227	213	244	230	213	196	239	222	197	182	234	216	192	171
00	283	273	258	242	278	263	243	221	275	253	225	205	267	245	217	193
000	326	314	296	277	320	302	276	252	315	290	259	233	307	280	247	220
0000	376	362	340	317	367	345	315	288	360	332	297	265	351	320	281	250
250	412	396	373	346	405	380	346	316	396	365	326	290	386	351	307	272
300	463	444	416	386	450	422	382	349	438	404	360	319	428	389	340	301
350	508	488	466	422	493	461	418	380	481	442	393	347	468	424	369	326
400	548	525	491	454	536	498	451	409	521	478	423	373	507	458	398	349
500	627	600	559	514	615	570	514	464	597	546	480	423	580	521	450	392
600	695	663	616	566	684	632	568	511	663	603	529	466	645	577	496	431
700	765	729	675	620	744	689	617	554	725	656	574	503	703	627	538	467
750	797	759	702	643	779	717	641	574	754	681	596	527	732	650	558	483
800	826	786	726	665	808	743	663	595	782	706	617	540	759	674	576	500
1000	946	898	827	752	921	842	747	667	889	797	692	603	860	759	646	580
1250	1080	1020	935	848	1052	957	845	751	1014	904	781	676	980	858	725	630
1500	1192	1122	1025	925	1162	1053	926	818	1118	993	855	736	1081	940	791	682
1750	1296	1215	1106	994	1256	1130	991	875	1206	1067	911	785	1162	1007	843	720
2000	1390	1302	1180	1058	1352	1213	1053	928	1293	1137	967	831	1240	1073	893	760

Three / Six / Nine: (1.09 at 10°C, 0.90 at 30°C, 0.80 at 40°C, 0.68 at 50°C)[b]

Twelve: (1.09 at 10°C, 0.90 at 30°C, 0.80 at 40°C, 0.62 at 50°C)[b]

34,500 V

Size	Copper Temperature, 70°C								Copper Temperature, 70°C							
0	227	221	209	197	225	213	197	182	220	205	187	169	215	199	177	158
00	260	251	239	224	255	242	224	205	249	234	211	190	245	226	200	179
000	299	290	273	256	295	278	256	235	288	268	242	217	282	259	230	204
0000	341	330	312	291	336	317	291	267	328	304	274	246	321	293	259	230
250	380	367	345	322	374	352	321	294	364	337	303	270	356	324	286	253
300	422	408	382	355	416	390	356	324	405	374	334	298	395	359	315	278
350	464	446	419	389	455	426	388	353	443	408	364	324	432	392	343	302
400	502	484	451	419	491	460	417	379	478	440	390	347	466	421	368	323
500	575	551	514	476	562	524	474	429	547	500	442	392	532	479	416	364
600	644	616	573	528	629	584	526	475	610	556	491	433	593	532	459	401
700	710	675	626	577	690	639	574	517	669	608	535	470	649	580	500	435
750	736	702	651	598	718	664	595	535	696	631	554	486	675	602	518	450
800	765	730	676	620	747	690	617	555	723	654	574	503	700	624	535	465
1000	875	832	766	701	852	783	698	624	823	741	646	564	796	706	601	520
1250	994	941	864	786	967	882	782	696	930	833	722	628	898	790	670	577
1500	1098	1036	949	859	1068	972	856	760	1025	914	788	682	988	865	730	626
1750	1192	1123	1023	925	1156	1048	919	814	1109	984	845	730	1066	929	780	668
2000	1275	1197	1088	981	1234	1115	975	860	1182	1045	893	770	1135	985	824	704
2500	1418	1324	1196	1072	1367	1225	1064	936	1305	1144	973	834	1248	1075	893	760

(1.10 at 10°C, 0.89 at 30°C, 0.76 at 40°C, 0.61 at 50°C)[b]   (1.10 at 10°C, 0.89 at 30°C, 0.76 at 40°C, 0.60 at 50°C)[b]

46,000 V

Size	Copper Temperature, 65°C								Copper Temperature, 65°C							
000	279	270	256	240	274	259	230	221	268	249	226	204	262	241	214	191
0000	322	312	294	276	317	299	274	251	309	287	259	232	302	276	244	217
250	352	340	321	300	346	326	299	274	336	313	282	252	329	301	266	236
300	394	380	358	334	385	364	332	304	377	349	313	280	367	335	295	260
350	433	417	392	365	425	398	364	331	413	382	341	304	403	366	321	283
400	469	451	423	393	459	430	391	356	446	411	367	326	433	394	344	307
500	534	512	482	444	522	487	441	400	506	464	412	365	492	444	386	339

(1.10 at 10°C, 0.89 at 30°C, 0.76 at 40°C, 0.61 at 50°C)[b]   (1.10 at 10°C, 0.89 at 30°C, 0.76 at 40°C, 0.60 at 50°C)[b]

(continued)

TABLE A.24 (continued)
Current-Carrying Capacity of Single-Conductor Solid Paper-Insulated Cables

Number of Equally Loaded Cables in Duct Bank

Six Percent Load Factor

Conductor Size AWG or MCM	Three				Six				Nine				Twelve			
	30	50	75	100	30	50	75	100	30	50	75	100	30	50	75	100
46,000 V	Amperes per Conductor[a]								Amperes per Conductor[a]							
	Copper Temperature, 65°C				Copper Temperature, 65°C				Copper Temperature, 65°C				Copper Temperature, 65°C			
600	602	577	538	496	589	546	494	447	570	520	460	406	553	497	430	377
700	663	633	589	542	645	598	538	488	626	569	502	441	605	542	468	408
750	689	658	611	561	672	622	559	504	650	590	520	457	629	562	485	422
800	717	683	638	583	698	645	578	522	674	612	538	472	652	582	501	436
1000	816	776	718	657	794	731	653	585	766	691	604	528	740	657	562	487
1250	927	879	810	738	900	825	732	654	865	777	675	589	834	736	626	541
1500	1020	968	887	805	992	904	799	703	951	850	735	638	914	802	679	585
1750	1110	1047	959	867	1074	976	859	762	1028	915	788	682	987	862	726	623
2000	1184	1115	1016	918	1144	1035	909	805	1094	970	833	718	1048	913	766	656
2500	1314	1232	1115	1002	1265	1138	994	875	1205	1062	905	778	1151	996	830	708
	(1.11 at 10°C, 0.87 at 30°C, 0.73 at 40°C, 0.54 at 50°C)[b]				(1.11 at 10°C, 0.87 at 30°C, 0.72 at 40°C, 0.53 at 50°C)[b]				(1.11 at 10°C, 0.87 at 30°C, 0.72 at 40°C, 0.52 at 50°C)[b]				(1.11 at 10°C, 0.87 at 30°C, 0.70 at 40°C, 0.51 at 50°C)[b]			
69,000 V	Copper Temperature, 60°C				Copper Temperature, 60°C				Copper Temperature, 60°C				Copper Temperature, 60°C			
350	395	382	360	336	387	364	333	305	375	348	312	279	365	332	293	259
400	428	413	389	362	418	393	358	328	405	375	335	300	394	358	315	278
500	489	470	441	409	477	446	406	370	461	425	379	337	447	405	354	312
600	545	524	490	454	532	496	450	409	513	471	419	371	497	448	391	343

700	599	573	536	495	582	543	490	444	561	514	455	403	542	489	425	372
750	623	597	556	514	605	562	508	460	583	533	472	417	563	506	439	384
800	644	617	575	531	626	582	525	475	603	554	487	430	582	523	453	396
1000	736	702	652	599	713	660	592	533	685	622	547	481	660	589	508	442
1250	832	792	734	672	806	742	664	595	772	698	610	535	741	659	564	489
1500	918	872	804	733	886	814	724	647	848	763	664	580	812	718	612	529
1750	994	942	865	788	957	876	776	692	913	818	711	618	873	770	653	563
2000	1066	1008	924	840	1020	931	822	732	972	868	750	651	927	814	688	592
2500	1163	1096	1001	903	1115	1013	892	791	1060	942	811	700	1007	880	741	635

(1.13 at 10°C, 0.85 at 30°C, 0.67 at 40°C, 0.42 at 50°C)[b]

(1.13 at 10°C, 0.85 at 30°C, 0.66 at 40°C, 0.42 at 50°C)[b]

(1.13 at 10°C, 0.84 at 30°C, 0.65 at 40°C, 0.36 at 50°C)[b]

(1.14 at 10°C, 0.84 at 30°C, 0.64 at 40°C, 0.32 at 50°C)[b]

Source: Westinghouse Electric Corporation, *Electrical Transmission and Distribution Reference Book*, East Pittsburgh, PA, 1964.

[a] Current ratings are based on the following conditions:
1. Ambient earth temperature = 20°C.
2. 60-cycle alternating current.
3. Sheaths bonded and grounded at one point only (open-circuited sheaths).
4. Standard concentric stranded conductors.
5. Ratings include dielectric loss and skin effect.
6. One cable per duct, all cables equally loaded and in outside ducts only.

[b] Multiply tabulated values by these factors when earth temperature is other than 20°C.

TABLE A.25

60 Hz Characteristics of Self-Supporting Rubber-Insulated Neoprene-Jacketed Aerial Cable

Voltage Class	Conductor Size	Stranding	Insulation Thickness	Shielding	Jacket Thickness	Diameter	Messenger Used with Copper Conductors	Weight per 1000 ft; Messenger and Copper	Messenger Used with Aluminum Conductors	Weight per 1000 ft Messenger and Aluminum	Positive Sequence 60 ~ AC Ω/mi Resistance[a] Copper	Aluminum	Reactance Series Inductive	Reactance Shunt Capacitive[b]	Zero Sequence[c] 60 ~ AC Ω/mi Resistance[a] Copper	Aluminum	Reactance Series Inductive Copper	Aluminum	Shunt Capacitive[b]
3 kV ungrounded neutral 5 kV grounded neutral	6	7	10/4	No	1/4	0.59	3/4" 30% CCS	1020	3/0"30% CCS	854	2.52	4.13	0.258	—	3.592	5.082	3.712	3.712	—
	4	7	10/4	No	1/4	0.67	3/4"30% CCS	1230	3/0"30% CCS	958	1.58	2.58	0.246	—	2.632	3.572	3.662	3.662	—
	2	7	10/4	No	1/4	0.73	3/4"30% CCS	1630	3/0"30% CCS	1100	1.00	1.64	0.229	—	2.025	2.605	3.615	3.615	—
	1	19	10/4	No	1/4	0.77	3/4"30% CCS	1780	3/0"30% CCS	1250	0.791	1.29	0.211	—	1.815	2.275	3.582	3.582	—
	1/0	19	10/4	No	1/4	0.81	3/4"30% CCS	2070	3/0"30% CCS	1390	0.635	1.03	0.207	—	1.644	2.015	3.555	3.555	—
	2/0	19	10/4	No	1/4	0.85	3/4"30% CCS	2510	3/0"30% CCS	1530	0.501	0.816	0.200	—	1.622	1.803	3.162	3.526	—
	3/0	19	10/4	No	1/4	0.91	3/4"30% CCS	2890	3/0"30% CCS	1690	0.402	0.644	0.194	—	1.517	1.637	3.135	3.499	—
	4/0	19	10/4	No	1/4	0.99	3/4"30% CCS	3570	3/0"30% CCS	1900	0.318	0.518	0.191	—	1.401	1.508	2.665	3.459	—
	250	37	11/4	No	1/4	1.08	3/2"30% CCS	4080	3/0"30% CCS	2160	0.269	0.437	0.189	—	1.351	1.430	2.635	3.429	—
	300	37	11/4	No	1/4	1.13	3/2"30% CCS	4620	3/0"30% CCS	2500	0.228	0.366	0.184	—	1.308	1.465	2.612	3.042	—
	350	37	11/4	No	1/4	1.18	3/2"30% CCS	5290	3/0"30% CCS	2780	0.197	0.316	0.180	—	1.277	1.415	2.591	3.021	—
	400	37	11/4	No	1/4	1.23	3/2"30% CCS	5800	3/0"30% CCS	3040	0.172	0.276	0.176	—	1.252	1.377	2.576	3.006	—
	500	37	11/4	No	1/4	1.32	3/2"30% CCS	6860	3/0"30% CCS	3650	0.141	0.223	0.172	—	1.219	1.290	2.543	2.543	—
5 kV ungrounded neutral	6	7	11/4	Yes	1/4	0.74	3/2"30% CCS	1310	3/2"30% CCS	1140	2.52	4.13	0.292	4970	—	—	—	—	—
	4	7	11/4	Yes	1/4	0.79	3/2"30% CCS	1540	3/2"30% CCS	1270	1.58	2.58	0.272	4320	—	—	—	—	—
	2	7	11/4	Yes	1/4	0.88	3/2"30% CCS	1950	3/2"30% CCS	1520	1.00	1.64	0.257	3630	—	—	—	—	—
	1	19	11/4	Yes	1/4	0.92	3/2"30% CCS	2180	3/2"30% CCS	1640	0.791	1.29	0.241	3330	—	—	—	—	—
	1/0	19	11/4	Yes	1/4	0.96	3/2"30% CCS	2450	3/2"30% CCS	1770	0.655	1.03	0.233	3080	—	—	—	—	—
	2/0	19	11/4	Yes	1/4	1.00	3/2"30% CCS	2910	3/2"30% CCS	1930	0.501	0.816	0.223	2830	—	—	—	—	—
	3/0	19	11/4	Yes	1/4	1.06	3/2"30% CCS	3320	3/2"30% CCS	2120	0.402	0.644	0.215	2580	—	—	—	—	—
	4/0	19	11/4	Yes	1/4	1.11	3/2"30% CCS	4030	3/2"30% CCS	2350	0.318	0.518	0.207	2380	—	—	—	—	—

15 kV grounded neutral

Size																		
250	37	11/4	Yes	1.20	3/2"30% CCS	4570	3/2"30% CCS	2770	0.269	0.437	0.206	2380	—	—	—	—	—
300	37	11/4	Yes	1.29	3/2"30% CCS	5260	3/2"30% CCS	3140	0.228	0.366	0.203	2280	—	—	—	—	—
350	37	11/4	Yes	1.34	3/2"30% CCS	5840	3/2"30% CCS	3380	0.197	0.316	0.199	2090	—	—	—	—	—
400	37	11/4	Yes	1.39	3/2"30% CCS	6380	3/2"30% CCS	3610	0.172	0.276	0.194	1890	—	—	—	—	—
500	37	11/4	Yes	1.47	3/2"30% CCS	7470	3/2"30% CCS	4240	0.141	0.223	0.187	1740	—	—	—	—	—
6	19	11/4	Yes	1.05	3/2"30% CCS	2090	3/2"30% CCS	1920	2.52	4.13	0.326	7150	3.846	5.346	3.396	3.396	7150
4	19	11/4	Yes	1.10	3/2"30% CCS	2350	3/2"30% CCS	2080	1.58	2.58	0.302	6260	2.901	3.831	3.364	3.364	6260
2	19	11/4	Yes	1.16	3/2"30% CCS	2860	3/2"30% CCS	2430	1.00	1.64	0.279	5460	2.459	3.039	2.851	2.851	5460
1	19	11/4	Yes	1.20	3/2"30% CCS	3120	3/2"30% CCS	2580	0.791	1.29	0.268	5110	2.238	2.701	2.837	2.837	5110
1/0	19	11/4	Yes	1.27	3/2"30% CCS	3560	3/2"30% CCS	2880	0.655	1.03	0.260	4720	2.052	2.426	2.825	2.825	4720
2/0	19	11/4	Yes	1.32	3/2"30% CCS	4120	3/2"30% CCS	3070	0.501	0.816	0.249	4370	1.896	2.214	2.251	2.801	4370
3/0	19	11/4	Yes	1.37	3/2"30% CCS	4580	3/2"30% CCS	3510	0.402	0.644	0.241	4120	1.782	2.008	2.240	2.240	4120
4/0	19	11/4	Yes	1.43	3/2"30% CCS	5150	3/2"30% CCS	3790	0.318	0.518	0.231	3770	1.681	1.864	2.235	2.235	3770
250	37	11/4	Yes	1.47	3/2"30% CCS	5590	3/2"30% CCS	3980	0.269	0.437	0.223	3570	1.630	1.782	2.227	2.227	3570
300	37	11/4	Yes	1.53	3/2"30% CCS	6260	3/2"30% CCS	4330	0.228	0.366	0.217	3330	1.577	1.701	2.226	2.226	3330
350	37	11/4	Yes	1.59	3/2"30% CCS	6870	3/2"30% CCS	4600	0.197	0.316	0.212	3130	1.536	1.640	2.226	2.226	3130
400	37	11/4	Yes	1.63	3/2"30% CCS	7450	3/2"30% CCS	4860	0.172	0.276	0.208	2980	1.500	1.592	2.216	2.216	2980
500	37	11/4	Yes	1.75	3/2"30% CCS	8970	3/2"30% CCS	5560	0.141	0.223	0.204	2830	1.454	1.524	2.198	2.198	2830

Source: Westinghouse Electric Corporation, *Electrical Transmission and Distribution Reference Book*, East Pittsburgh, PA, 1964.

a AC resistance based on 65°C with allowance for stranding, skin effect, and proximity effect.

b Dielectric constant assumed to be 6.0.

c Zero sequence impedance based on return current both in the messenger and in 100 mΩ earth.

REFERENCES

1. Westinghouse Electric Corporation: *Electrical Transmission and Distribution Reference Book*, East Pitsburgh, PA, 1964.
2. Westinghouse Electric Corporation: *Electric Utility Engineering Reference Book—Distribution Systems*. Vol. 3, East Pitsburgh, PA, 1965.
3. Edison Electric Institute: *Transmission Line Reference Book*, New York, 1968.
4. Anderson, P. M.: *Analysis of Faulted Power Systems*, Iowa State University Press, Ames, IA, 1973.
5. Westinghouse Electric Corporation: *Applied Protective Relaying*, Newark, NJ, 1970.

Appendix B: Graphic Symbols Used in Distribution System Design

Some of the most commonly used graphic symbols for distribution systems, both in this book and in general usage, are given on the following pages.

TABLE B.1

Graphic Symbols Used in Distribution System Design

Symbol	Usage

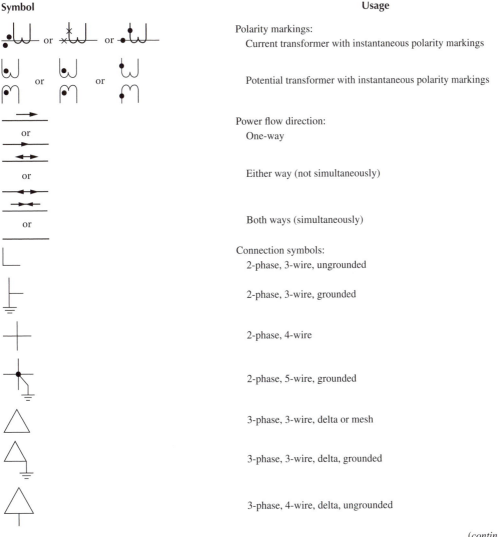

	Polarity markings:
	Current transformer with instantaneous polarity markings
	Potential transformer with instantaneous polarity markings
	Power flow direction:
	One-way
	Either way (not simultaneously)
	Both ways (simultaneously)
	Connection symbols:
	2-phase, 3-wire, ungrounded
	2-phase, 3-wire, grounded
	2-phase, 4-wire
	2-phase, 5-wire, grounded
	3-phase, 3-wire, delta or mesh
	3-phase, 3-wire, delta, grounded
	3-phase, 4-wire, delta, ungrounded

(continued)

961

TABLE B.1 (continued)
Graphic Symbols Used in Distribution System Design

Symbol	Usage
	3-phase, 4-wire, delta, grounded
	3-phase, open-delta
	3-phase, open-delta, grounded at the middle point of one winding
	3-phase, broken-delta
	3-phase, wye or star, ungrounded
	3-phase, wye, grounded neutral
	3-phase, 4-wire, ungrounded
	3-phase, zigzag, ungrounded
	3-phase, zigzag, grounded
	3-phase, Scott or T
	6-phase, double-delta
	6-phase, hexagonal (or chordal)
	6-phase, star (or diametrical)
	6-phase, star, with grounded neutral
	6-phase, double zigzag with neutral brought out and grounded
or	Resistor: Resistor (general)
or	Tapped resistor

TABLE B.1 (continued)
Graphic Symbols Used in Distribution System Design

Symbol	Usage
	Resistor with adjustable contact
	Shunt resistor
	Series resistor and path open
	Series resistor and path short-circuited
	Capacitor:
	Capacitor (general)
	Polarized capacitor
	Variable capacitor
	Series capacitor and path open
	Series capacitor and path short-circuited
	Shunt capacitor
	Capacitor bushing for circuit breaker or transformer
	Capacitor-bushing potential device
	Coupling capacitor potential device
	Battery:
	Battery (general)
	Battery with one cell
	Battery with multicell

(continued)

TABLE B.1 (continued)

Graphic Symbols Used in Distribution System Design

Symbol	Usage
	Transmission path (conductor, cable wire):
	Bus bar, with connection
	Conductor or path
	2 conductors or paths
	3 conductors or paths
	n conductors or paths
(draw individual paths)	
	Crossing of two conductors or paths not connected
	Junction
	Junction of connected paths
	Shielded single conductor
	Shielded 5-conductor cable
	Shielded 2-conductor cable with conductors separated on the diagram for convenience
	3-conductor cable
	Grouping of leads:
	General
	Interrupted
	Transmission and distribution lines:
	Telephone line
	Cable (or line) underground
	Submarine line
	Overhead line
	Loaded line

TABLE B.1 (continued)
Graphic Symbols Used in Distribution System Design

Symbol	Usage
	Ground:
	Ground (general)
	Switch:
	Single-throw switch (disconnect switch)
	Double-throw switch
	Knife switch
	Connector:
	Female contact
	Male contact
	Separable connectors
	Operating coil:
	Operating coil (general), e.g., reactor
	Transformer:
	Transformer (general)
	Adjustable mutual inductor (constant-current transformer)
	Single-phase transformer with taps
	Single-phase autotransformer
	Adjustable
	Step-voltage regulator or load-ratio control autotransformer
	Step-voltage regulator
	Load-ratio control autotransformer
	Load-ratio control transformer with taps

(*continued*)

TABLE B.1 (continued)
Graphic Symbols Used in Distribution System Design

Symbol	Usage
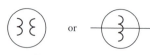 or	Single-phase induction voltage regulator
	Triplex induction voltage regulator
	3-phase induction voltage regulator
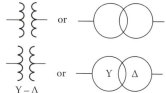	1-phase, 2-winding transformer
	3-phase bank of 1-phase, 2-winding transformers with wye–delta connections
	Polyphase transformer:
	Polyphase transformer (general)
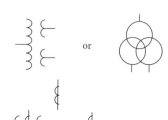	1-phase, 3-winding transformer
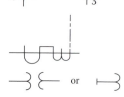	Current transformer: Current transformer (general)
	Bushing-type current transformer
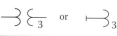	Potential transformers: Potential transformer (general)
	Outdoor metering device
	Linear coupler

TABLE B.1 (continued)
Graphic Symbols Used in Distribution System Design

Symbol	Usage
	Fuse: Fuse (general)
	Fuse (supply side indicated by a thick line)
	Isolating fuse switch (HV primary fuse cutout) dry
	HV primary fuse cutout, oil
	Isolating fuse switch for onload switching
	Current limiter (for power cable)
	Lightning arrester
	Horn gap
	Multigap (general)
	Circuit breaker: Circuit breaker, air (for dc or ac rated at 1.5 kV or less)
	Network protector
	HV circuit breaker (for ac rated at above 1.5 kV)

(*continued*)

TABLE B.1 (continued)
Graphic Symbols Used in Distribution System Design

Symbol	Usage
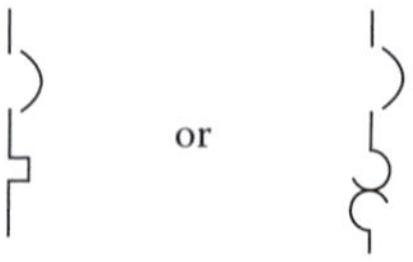	Circuit breaker with thermal-overload device
	Circuit breaker with magnetic thermal-overload device
	Circuit breaker, drawout type
G or GEN	Rotating machine: Generator (general)
$\underline{G}$	Generator, dc
G ~	Generator, ac
GS	Generator, synchronous
M or MOT	Motor (general)
$\underline{M}$	Motor, dc
M ~	Motor, ac
MS	Motor, synchronous

Appendix C: Standard Device Numbers Used in Protection Systems

Some of the frequently used device numbers are listed in the following. A complete list and definitions are given in ANSI/IEEE Standard C37.2-1079.

1. Master element: normally used for hand-operated devices
2. Time-delay starting or closing relay
3. Checking or interlocking relay
4. Master contactor
5. Stopping device
6. Starting circuit breaker
7. Anode circuit breaker
8. Control power disconnecting device
9. Reversing device
10. Unit sequence switch
12. Synchronous-speed device
14. Underspeed device
15. Speed- or frequency-matching device
17. Shunting or discharge switch
18. Accelerating or decelerating device
20. Electrically operated valve
21. Distance relay
23. Temperature control device
25. Synchronizing or synchronism-check device
26. Apparatus thermal device
27. Undervoltage relay
29. Isolating contactor
30. Annunciator relay
32. Directional power relay
37. Undercurrent or underpower relay
46. Reverse-phase or phase-balance relay
47. Phase-sequence voltage relay
48. Incomplete-sequence relay
49. Machine or transformer thermal relay
50. Instantaneous overcurrent or rate-of-rise relay
51. AC time overcurrent relay
52. AC circuit breaker: mechanism-operated contacts are as follows:
    a. 52a, 52aa: open when breaker, closed when breaker contacts closed
    b. 52b, 52bb: operates just as mechanism motion start; known as high-speed contacts
55. Power factor relay
57. Short-circuiting or grounding device
59. Overvoltage relay

60. Voltage or current balance relay
62. Time-delay stopping or opening relay
64. Ground detector relay
67. AC directional overcurrent relay
68. Blocking relay
69. Permissive control device
72. AC circuit breaker
74. Alarm relay
76. DC overcurrent relay
78. Phase-angle measuring or out-of-step protective relay
79. AC reclosing relay
80. Flow switch
81. Frequency relay
82. DC reclosing relay
83. Automatic selective control or transfer relay
84. Operating mechanism
85. Carrier or pilot-wire receiver relay
86. Lockout relay
87. Differential protective relay
89. Line switch
90. Regulating device
91. Voltage directional relay
92. Voltage and power directional relay
93. Field-changing contactor
94. Tripping or trip-free relay

Appendix D: The Per-Unit System

D.1 INTRODUCTION

Because of various advantages involved, it is customary in power system analysis calculations to use impedances, currents, voltages, and powers in per-unit values (which are scaled or normalized values) rather than in physical values of ohms, amperes, kilovolts, and megavoltamperes (or megavars, or megawatts).

A per-unit system is a means of expressing quantities for ease in comparing them. The per-unit value of any quantity is defined as the ratio of the quantity to an "*arbitrarily*" chosen base (i.e., *reference*) value having the same dimensions. Therefore, the per-unit value of any quantity can be defined as physical quantity

$$\text{Quantity in per unit} = \frac{\text{Physical quantity}}{\text{Base value of quantity}} \tag{D.1}$$

where "*physical quantity*" refers to the given value in ohms, amperes, volts, etc. The *base value* is also called unit value since in the per-unit system it has a value of 1, or unity. Therefore, a base current is also referred to as a unit current.

Since both the physical quantity and base quantity have the same dimensions, the resulting per-unit value expressed as a decimal has no dimension and, therefore, is simply indicated by a subscript pu. The base quantity is indicated by a subscript B. The symbol for per unit is pu, or 0/1. The percent system is obtained by multiplying the per-unit value by 100. Hence,

$$\text{Quantity in percent} = \frac{\text{Physical quantity}}{\text{Base value of quantity}} \times 100 \tag{D.2}$$

However, the percent system is somewhat more difficult to work with and more subject to possible error since it must always be remembered that the quantities have been multiplied by 100.

Thus, the factor 100 has to be continually inserted or removed for reasons that may not be obvious at the time. For example, 40% reactance times 100% current is equal to 4000% voltage, which, of course, must be corrected to 40% voltage. Hence, the per-unit system is preferred in power system calculations. The advantages of using the per-unit system include the following:

1. Network analysis is greatly simplified since all impedances of a given equivalent circuit can directly be added together regardless of the system voltages.
2. It eliminates the $\sqrt{3}$ multiplications and divisions that are required when balanced three-phase systems are represented by per-phase systems. Therefore, the factors $\sqrt{3}$ and 3 associated with delta and wye quantities in a balanced three-phase system are directly taken into account by the base quantities.
3. Usually, the impedance of an electrical apparatus is given in percent or per unit by its manufacturer based on its nameplate ratings (e.g., its rated voltamperes and rated voltage).
4. Differences in operating characteristics of many electrical apparatus can be estimated by a comparison of their constants expressed in per units.

5. Average machine constants can easily be obtained since the parameters of similar equipment tend to fall in a relatively narrow range and, therefore, arc comparable when expressed
as per units based on rated capacity.
6. The use of per-unit quantities is more convenient in calculations involving digital
computers.

D.2 SINGLE-PHASE SYSTEM

In the event that any two of the four base quantities (i.e., base voltage, base current, base voltamperes,
and base impedance) are "*arbitrarily*" specified, the other two can be determined immediately.

Here, the term *arbitrarily* is slightly misleading since in practice the base values are selected so
as to force the results to fall into specified ranges. For example, the base voltage is selected such that
the system voltage is normally close to unity.

Similarly, the base voltampere is usually selected as the kilovoltampere or megavoltampere rating
of one of the machines or transformers in the system, or a convenient round number such as 1, 10, 100,
or 1000 MVA, depending on system size. As aforementioned, on determining the base voltamperes
and base voltages, the other base values are fixed. For example, current base can be determined as

$$I_B = \frac{S_B}{V_B} = \frac{VA_B}{V_B} \qquad \text{(D.3)}$$

where
I_B is the current base in amperes
S_B is the selected voltampere base in voltamperes
V_B is the selected voltage base in volts

Note that

$$S_B = VA_B = P_B = Q_B = V_B I_B \qquad \text{(D.4)}$$

Similarly, the impedance base* can be determined as

$$Z_B = \frac{V_B}{I_B} \qquad \text{(D.5)}$$

where

$$Z_B = X_B = R_B \qquad \text{(D.6)}$$

Similarly,

$$Y_B = B_B = G_B = \frac{I_B}{V_B} \qquad \text{(D.7)}$$

Note that by substituting Equation D.3 into Equation D.5, the impedance base can be expressed as

$$Z_B = \frac{V_B}{VA_B/V_B} = \frac{V_B^2}{VA_B} \qquad \text{(D.8)}$$

---

* It is defined as that impedance across which there is a voltage drop that is equal to the base voltage if the current through
it is equal to the base current.

or

$$Z_B = \frac{(kV_B)^2}{MVA_B} \tag{D.9}$$

where

kV_B is the voltage base in kilovolts

MVA_B is the voltampere base in megavoltamperes

The per-unit value of any quantity can be found by the *normalization process*, that is, by dividing the physical quantity by the base quantity of the same dimension. For example, the per-unit impedance can be expressed as

$$Z_{pu} = \frac{Z_{physical}}{Z_B} \tag{D.10}$$

or

$$Z_{pu} = \frac{Z_{physical}}{V_B^2/(kVA_B \times 1000)} \tag{D.11}$$

or

$$Z_{pu} = \frac{(Z_{physical})(kVA_B)(1000)}{V_B^2} \tag{D.12}$$

or

$$Z_{pu} = \frac{(Z_{physical})(kVA_B)}{(kV_B)^2(1000)} \tag{D.13}$$

or

$$Z_{pu} = \frac{(Z_{physical})}{(kV_B)^2/MVA_B} \tag{D.14}$$

or

$$Z_{pu} = \frac{(Z_{physical})(MVA_B)}{(kV_B)^2} \tag{D.15}$$

Similarly, the others can be expressed as

$$I_{pu} = \frac{I_{physical}}{I_B} \tag{D.16}$$

or

$$V_{pu} = \frac{V_{physical}}{V_B} \tag{D.17}$$

or

$$kV_{pu} = \frac{kV_{physical}}{kV_B} \tag{D.18}$$

or

$$VA_{pu} = \frac{VA_{physical}}{VA_B} \tag{D.19}$$

or

$$kVA_{pu} = \frac{kVA_{physical}}{kVA_B} \tag{D.20}$$

or

$$MVA_{pu} = \frac{MVA_{physical}}{MVA_B} \tag{D.21}$$

Note that the base quantity is always a real number, whereas the physical quantity can be a complex number. For example, if the actual impedance quantity is given as $Z\angle\theta\ \Omega$, it can be expressed in the per-unit system as

$$\mathbf{Z}_{pu} = \frac{Z\angle\theta}{Z_B} = Z_{pu}\angle\theta \tag{D.22}$$

that is, it is the magnitude expressed in per-unit terms.

Alternatively, if the impedance has been given in rectangular form as

$$\mathbf{Z} = R + jX \tag{D.23}$$

then

$$\mathbf{Z}_{pu} = R_{pu} + jX_{pu} \tag{D.24}$$

where

$$R_{pu} = \frac{R_{physical}}{Z_B} \tag{D.25}$$

and

$$X_{pu} = \frac{X_{physical}}{Z_B} \tag{D.26}$$

Similarly, if the complex power has been given as

$$\mathbf{S} = P + jQ \tag{D.27}$$

then

$$\mathbf{S}_{pu} = P_{pu} + jQ_{pu} \tag{D.28}$$

where

$$P_{pu} = \frac{P_{physical}}{S_B} \tag{D.29}$$

and

$$Q_{pu} = \frac{Q_{physical}}{S_B} \tag{D.30}$$

If the actual voltage and current values are given as

$$\mathbf{V} = V \angle \theta_V \tag{D.31}$$

and

$$\mathbf{I} = I \angle \theta_I \tag{D.32}$$

the complex power can be expressed as

$$\mathbf{S} = \mathbf{V}\mathbf{I}^* \tag{D.33}$$

or

$$S\angle\theta = (V\angle\theta_V)(I\angle-\theta_I) \tag{D.34}$$

Therefore, dividing through by S_B,

$$\frac{S\angle\phi}{S_B} = \frac{(V\angle\theta_V)(I\angle-\theta_I)}{S_B} \tag{D.35}$$

However,

$$S_B = V_B I_B \tag{D.36}$$

Thus,

$$\frac{S\angle\theta}{S_B} = \frac{(V\angle\theta_V)(I\angle-\theta_I)}{V_B I_B} \tag{D.37}$$

or

$$S_{pu}\angle\theta = (V_{pu}\angle\theta_V)(I_{pu}\angle-\theta_I) \tag{D.38}$$

or

$$S_{pu} = V_{pu} I_{pu}^* \tag{D.39}$$

Example D.1

A 240/120 V single-phase transformer rated 5 kVA has a high-voltage winding impedance of 0.3603 Ω. Use 240 V and 5 kVA as the base quantities and determine the following:

   a. The high-voltage side base current.
   b. The high-voltage side base impedance in ohms.
   c. The transformer impedance referred to the high-voltage side in per unit.
   d. The transformer impedance referred to the high-voltage side in percent.
   e. The turns ratio of the transformer windings.
   f. The low-voltage side base current.
   g. The low-voltage side base impedance.
   h. The transformer impedance referred to the low-voltage side in per unit.

Solution:

   a. The high-voltage side base current is

$$I_{B(HV)} = \frac{S_B}{V_{B(HV)}}$$

$$= \frac{5000 \text{ VA}}{240 \text{ V}}$$

$$= 20.8333 \text{ A}$$

   b. The high-voltage side base impedance is

$$Z_{B(HV)} = \frac{V_{B(HV)}}{I_{B(HV)}}$$

$$= \frac{240 \text{ V}}{20.8333 \text{ A}}$$

$$= 11.52 \text{ }\Omega$$

   c. The transformer impedance referred to the high-voltage side is

$$Z_{pu(HV)} = \frac{Z_{HV}}{Z_{B(HV)}}$$

$$= \frac{0.3603 \text{ }\Omega}{11.51 \Omega}$$

$$= 0.0313 \text{ pu}$$

   d. The transformer impedance referred to the high-voltage side is %

$$\% Z_{HV} = Z_{pu(HV)} \times 100$$

$$= (0.0313 \text{ pu}) 100$$

$$= 3.13\%$$

e. The turns ratio of the transformer windings is

$$n = \frac{V_{HV}}{V_{LV}}$$

$$= \frac{240 \text{ V}}{120 \text{ V}}$$

$$= 2$$

f. The low-voltage side base current is

$$I_{B(LV)} = \frac{S_B}{V_{B(LV)}}$$

$$= \frac{5000 \text{ VA}}{120 \text{ V}}$$

$$= 41.6667 \text{ A}$$

or

$$I_{B(LV)} = nI_{B(LV)}$$

$$= 2(20.8333 \text{ A})$$

$$= 41.6667$$

g. The low-voltage side base impedance is

$$Z_{B(LV)} = \frac{V_{B(LV)}}{I_{B(LV)}}$$

$$= \frac{120 \text{ V}}{41.667 \text{ A}}$$

$$= 2.88 \ \Omega$$

or

$$Z_{B(HV)} = \frac{Z_{B(LV)}}{n^2}$$

$$= 2.88 \ \Omega$$

h. The transformer impedance referred to the low-voltage side is

$$Z_{LV} = \frac{Z_{HV}}{n^2}$$

$$= \frac{0.3603 \ \Omega}{2^2}$$

$$= 0.0901 \ \Omega$$

Therefore,

$$Z_{pu(LV)} = \frac{Z_{LV}}{Z_{B(LV)}}$$

$$= \frac{0.0901\,\Omega}{2.88\,\Omega}$$

$$= 0.0313\,\text{pu}$$

or

$$Z_{pu(LV)} = Z_{pu(HV)}$$

$$= 0.0313\,\text{pu}$$

Notice that in terms of per units the impedance of the transformer is the same whether it is referred to the high-voltage side or the low-voltage side.

Example D.2

Redo the Example D.1 by using MATLAB®.

    a. Write the MATLAB program script.
    b. Give the MATLAB program output.

Solution:
    a. Here is the MATLAB program script:

```
% MATLAB SCRIPT for Example D.1

clear
clc
%System Parameters
 VBhv = 240;
 VBlv = 120;
 SB = 5e3;
 Zhv = 0.3603;

%Solution for part (a)
 IBhv = SB/VBhv

%Solution for part (b)
 ZBhv = VBhv/IBhv

%Solution for part (c)
 Zpu_hv = Zhv/ZBhv

%Solution for part (d)
 percent_Zhv = Zpu_hv*100

%Solution for part (e)
 n = VBhv/VBlv
```

```
%Solution for part (f)
 IBlv = SB/VBlv
 IBlv = n*IBhv

%Solution for part (g)
 ZBlv = VBlv/IBlv
 ZBlv = ZBhv/n^2

%Solution for part (h)
 Zlv = Zhv/n^2
 Zpu_lv = Zlv/ZBlv
 Zpu_lv = Zpu_hv
```

b. Here is the MATLAB program output:

```
IBhv =

   20.8333

ZBhv =

   11.5200

Zpu_hv =

   0.0313

percent_Zhv =

   3.1276

n =

   2

IBlv =

   41.6667

IBlv =

   41.6667

ZBlv =

   2.8800

ZBlv =

   2.8800

Zlv =

   0.0901

Zpu_lv =

   0.0313

Zpu_lv =

   0.0313

>>
```

D.3   CONVERTING FROM PER-UNIT VALUES TO PHYSICAL VALUES

The physical values (or system values) and per-unit values are related by the following relationships:

$$\mathbf{I} = \mathbf{I}_{pu} \times I_B \tag{D.40}$$

$$\mathbf{V} = \mathbf{V}_{pu} \times V_B \tag{D.41}$$

$$\mathbf{Z} = \mathbf{Z}_{pu} \times Z_B \tag{D.42}$$

$$R = R_{pu} \times Z_B \tag{D.43}$$

$$X = X_{pu} \times Z_B \tag{D.44}$$

$$VA = VA_{pu} \times VA_B \tag{D.45}$$

$$P = P_{pu} \times VA_B \tag{D.46}$$

$$Q = Q_{pu} \times VA_B \tag{D.47}$$

D.4   CHANGE OF BASE

In general, the per-unit impedance of a power apparatus is given based on its own voltampere and voltage ratings and consequently based on its own impedance base. When such an apparatus is used in a system that has its own bases, it becomes necessary to refer all the given per-unit values to the system base values. Assume that the per-unit impedance of the apparatus is given based on its nameplate ratings as

$$Z_{pu(given)} = (Z_{physical}) \frac{MVA_{B(given)}}{[kV_{B(given)}]^2} \tag{D.48}$$

and that it is necessary to refer the very same physical impedance to a new set of voltage and voltampere bases such that

$$Z_{pu(new)} = (Z_{physical}) \frac{MVA_{B(new)}}{[kV_{B(new)}]^2} \tag{D.49}$$

By dividing Equation D.48 by Equation D.49 side by side,

$$Z_{pu(new)} = Z_{pu(old)} \left[\frac{MVA_{B(old)}}{MVA_{B(given)}} \right] \left[\frac{kV_{B(given)}}{kV_{B(old)}} \right]^2 \tag{D.50}$$

In certain situations, it is more convenient to use subscripts 1 and 2 instead of subscripts "*given*" and "*new*," respectively. Then Equation D.50 can be expressed as

$$Z_{pu(2)} = Z_{pu(1)} \left[\frac{MVA_{B(2)}}{MVA_{B(1)}} \right] \left[\frac{kV_{B(1)}}{kV_{B(2)}} \right]^2 \tag{D.51}$$

In the event that the kV bases are the same but the MVA bases are different, from Equation D.50,

$$Z_{pu(new)} = Z_{pu(given)} \frac{MVA_{B(new)}}{MVA_{B(given)}} \tag{D.52}$$

Similarly, if the megavoltampere bases are the same but the kilovolt bases are different, from Equation D.50,

$$Z_{pu(new)} = Z_{pu(given)} \left[\frac{kV_{B(given)}}{kV_{B(new)}} \right]^2 \tag{D.53}$$

Equations D.49 through D.52 must only be used to convert the given per-unit impedance from the base to another but not for referring the physical value of an impedance from one side of the transformer to another [3].

Example D.3

Consider Example B.1 and select 300/150 V as the base voltages for the high-voltage and the low-voltage windings, respectively. Use a new base power of 10 kVA and determine the new per-unit, base, and physical impedances of the transformer referred to the high-voltage side.

Solution:

By using Equation D.50, the new per-unit impedance can be found as

$$Z_{pu(new)} = Z_{pu(old)} \left[\frac{MVA_{B(old)}}{MVA_{B(given)}} \right] \left[\frac{kV_{B(given)}}{kV_{B(old)}} \right]^2$$

$$= (0.0313 \text{ pu}) \left(\frac{10,000 \text{ VA}}{300 \text{ V}} \right) \left(\frac{240 \text{ V}}{300 \text{ V}} \right)^2$$

$$= 33.334 \text{ A}$$

The new current base is

$$I_{B(HV)new} = \frac{S_B}{V_{B(HV)new}}$$

$$= \frac{10,000 \text{ VA}}{300 \text{ V}}$$

$$= 33,334 \text{ A}$$

Thus,

$$Z_{B(HV)new} = \frac{V_{B(HV)new}}{I_{B(HV)new}}$$

$$= \frac{300 \text{ V}}{33.334 \text{ A}}$$

$$= 9 \ \Omega$$

Therefore, the physical impedance of the transformer is still

$$Z_{HV} = Z_{pu,\,new} \times Z_{B(HV)new}$$

$$= (0.0401\,pu)(9\,\Omega)$$

$$= 0.3609\,\Omega$$

D.5 THREE-PHASE SYSTEMS

The three-phase problems involving balanced systems can be solved on a per-phase basis. In that case, the equations that are developed for single-phase systems can be used for three-phase systems as long as per-phase values are used consistently. Therefore,

$$I_B = \frac{S_{B(1\phi)}}{V_{B(L-N)}} \tag{D.54}$$

or

$$I_B = \frac{VA_{B(1\phi)}}{V_{B(L-N)}} \tag{D.55}$$

and

$$Z_B = \frac{V_{B(L-N)}}{I_B} \tag{D.56}$$

or

$$Z_B = \frac{[kV_{B(L-N)}]^2(1000)}{kVA_{B(1\phi)}} \tag{D.57}$$

or

$$Z_B = \frac{[kV_{B(L-N)}]^2}{MVA_{B(1\phi)}} \tag{D.58}$$

where the subscripts 1ϕ and $L-N$ denote per phase and line to neutral, respectively. Note that, for a balanced system,

$$V_{B(L-N)} = \frac{V_{B(L-L)}}{\sqrt{3}} \tag{D.59}$$

and

$$S_{B(1\phi)} = \frac{S_{B(3\phi)}}{3} \tag{D.60}$$

However, it has been customary in three-phase system analysis to use line-to-line voltage and three-phase voltamperes as the base values. Therefore,

$$I_B = \frac{S_{B(3\phi)}}{\sqrt{3}V_{B(L-L)}} \tag{D.61}$$

or

$$I_B = \frac{kVA_{B(3\phi)}}{\sqrt{3}kV_{B(L-L)}} \tag{D.62}$$

and

$$Z_B = \frac{V_{B(L-L)}}{\sqrt{3}I_B} \tag{D.63}$$

$$Z_B = \frac{[kV_{B(L-L)}]^2(1000)}{kVA_{B(3\phi)}} \tag{D.64}$$

or

$$Z_B = \frac{[kV_{B(L-L)}]^2}{MVA_{B(3\phi)}} \tag{D.65}$$

where the subscripts 3ϕ and $L-L$ denote per three phase and line, respectively. Furthermore, base admittance can be expressed as

$$Y_B = \frac{1}{Z_B} \tag{D.66}$$

or

$$\tag{D.67}$$

where

$$Y_B = B_B = G_B \tag{D.68}$$

The data for transmission lines are usually given in terms of the line resistance R in ohms per mile at a given temperature, the line inductive reactance X_L in ohms per mile at 60 Hz, and the line shunt capacitive reactance X_c in megohms per mile at 60 Hz. Therefore, the line impedance and shunt susceptance in per units for 1 mi of line can be expressed as

$$\mathbf{Z}_{pu} = (\mathbf{Z}, \Omega/mi)\frac{MVA_{B(3\phi)}}{[kV_{B(L-L)}]^2} pu \tag{D.69}$$

where

$$\mathbf{Z} = R + jX_L = Z\angle\theta \; \Omega/mi$$

and

$$B_{\text{pu}} = \frac{[kV_{B(L-L)}]^2 \times 10^{-6}}{[MVA_{B(3\phi)}][X_c, M\Omega/\text{mi}]} \tag{D.70}$$

In the event that the admittance for a transmission line is given in microsiemens per mile, the per-unit admittance can be expressed as

$$Y_{\text{pu}} = \frac{[kV_{B(L-L)}]^2 (Y, \mu S)}{[MVA_{B(3\phi)}] \times 10^6} \tag{D.71}$$

Similarly, if it is given as reciprocal admittance in megohms per mile, the per-unit admittance can be found as

$$Y_{\text{pu}} = \frac{[kV_{B(L-L)}]^2 \times 10^{-6}}{[MVA_{B(3\phi)}][Z, M\Omega/\text{mi}]} \tag{D.72}$$

Figure 4.29 shows conventional three-phase transformer connections and associated relationships between the high-voltage and low-voltage side voltages and currents. The given relationships are correct for a three-phase transformer as well as for a three-phase bank of single-phase transformers. Note that in the figure, n is the turns ratio, that is,

$$n = \frac{N_1}{N_2} = \frac{V_1}{V_2} = \frac{I_2}{I_1} \tag{D.73}$$

where the subscripts 1 and 2 are used for the primary and secondary sides. Therefore, an impedance Z_2 in the secondary circuit can be referred to the primary circuit provided that

$$Z_1 = n^2 Z_2 \tag{D.74}$$

Thus, it can be observed from Figure 4.29 that in an ideal transformer, voltages are transformed in the direct ratio of turns, currents in the inverse ratio, and impedances in the direct ratio squared, and power and voltamperes are, of course, unchanged. Note that a balanced delta-connected circuit of Z_Δ Ω/phase is equivalent to a balanced wye-connected circuit of Z_Y Ω/phase as long as

$$Z_Y = \frac{1}{3} Z_\Delta \tag{D.75}$$

The per-unit impedance of a transformer remains the same without taking into account whether it is converted from physical impedance values that are found by referring to the high-voltage side or low-voltage side of the transformer. This can be accomplished by choosing separate appropriate bases for each side of the transformer (whether or not the transformer is connected in wye–wye, delta–delta, delta–wye, or wye–delta since the transformation of voltages is the same as that made by wye–wye transformers as long as the same line-to-line voltage ratings are used). In other words, the designated per-unit impedance values of transformers are based on the coil ratings.

Since the ratings of coils cannot be altered by a simple change in connection (e.g., from wye–wye to delta–wye), the per-unit impedance remains the same regardless of the three-phase connection. The line-to-line voltage for the transformer will differ. Because of the method of choosing the base in various sections of the three-phase system, the per-unit impedances calculated in various sections can be put together on one impedance diagram without paying any attention to whether the transformers are connected in wye–wye or delta–wye.

Example D.4

Assume that a 19.5 kV 120 MVA three-phase generator has a synchronous reactance of 1.5/Ω and is connected to a 150 MVA 18/230 kV delta–wye connected three-phase transformer with a 0.1/Ω reactance. The transformer is connected to a transmission line at the 230 kV side. Use the new MVA base of 100 MVA and 240 kV base for the line and determine the following:

a. The new reactance value for the generator in per unit ohms.
b. The new reactance value for the transformer in per unit ohms.

Solution:

a. Using Equation B.49, the new per unit impedance of the generator is

$$Z_{pu(new)} = Z_{pu(old)} \left[\frac{MVA_{B(new)}}{MVA_{B(old)}} \right] \left[\frac{kV_{B(old)}}{kV_{B(old)}} \right]^2$$

But, first determining the new kV base for the generator,

$$kV^{gen}_{B(new)} = (240 \text{ kV}) \left(\frac{18 \text{ kV}}{230 \text{ kV}} \right) = 18.783 \text{ kV}$$

Thus, the new and adjusted synchronous reactance of the generator is

$$X^{gen}_{pu(new)} = X^{gen}_{pu(old)} \left[\frac{MVA_{B(new)}}{MVA_{B(old)}} \right] \left[\frac{kV_{B(old)}}{kV_{B(new)}} \right]^2$$

$$= (1.5 \text{ pu}) \left[\frac{100 \text{ MVA}}{120 \text{ MVA}} \right] \left[\frac{19.5 \text{ kV}}{18.783 \text{ kV}} \right]^2$$

$$= 1.347 \text{ pu}$$

b. The new reactance value for the transformer in per unit ohms, referred to high-voltage side is

$$X^{trf}_{pu(new)} = (0.1 \text{ pu}) \left[\frac{100 \text{ MVA}}{150 \text{ MVA}} \right] \left[\frac{230 \text{ kV}}{240 \text{ kV}} \right]^2$$

$$= 0.061 \text{ pu}$$

And referred to the low-voltage side is

$$X^{trf}_{pu(new)} = (0.1 \text{ pu}) \left[\frac{100 \text{ MVA}}{150 \text{ MVA}} \right] \left[\frac{18 \text{ kV}}{18.783 \text{ kV}} \right]^2$$

$$= 0.061 \text{ pu}$$

Note that the transformer reactance referred to the high-voltage side or the low-voltage side is the same, as it should be!

Example D.5

A three-phase transformer has a nameplate ratings of 20 MVA, 345Y/34.5Y kV with a leakage reactance of 12% and the transformer connection is wye–wye. Select a base of 20 MVA and 345 kV on the high-voltage side and determine the following:

a. Reactance of transformer in per units.
b. High-voltage side base impedance.
c. Low-voltage side base impedance.
d. Transformer reactance referred to high-voltage side in ohms.
e. Transformer reactance referred to low-voltage side in ohms.

Solution:

a. The reactance of the transformer in per units is 12/100, or 0.12 pu. Note that it is the same whether it is referred to the high-voltage or the low-voltage sides.
b. The high-voltage side base impedance is

$$Z_{B(HV)} = \frac{[kV_{B(HV)}]^2}{MVA_{B(3\phi)}}$$

$$= \frac{345^2}{20} = 5951.25 \ \Omega$$

c. The low-voltage side base impedance is

$$Z_{B(LV)} = \frac{[kV_{B(LV)}]^2}{MVA_{B(3\phi)}}$$

$$= \frac{34.5^2}{20} = 59.5125 \ \Omega$$

d. The reactance referred to the high-voltage side is

$$X_{(HV)} = X_{pu} \times X_{B(HV)}$$

$$= (0.12)(5951.25) = 714.15 \ \Omega$$

e. The reactance referred to the low-voltage side is

$$X_{(LV)} = X_{pu} \times X_{B(LV)}$$

$$= (0.12)(59.5125) = 7.1415 \ \Omega$$

or from

$$X_{(LV)} = \frac{X_{(HV)}}{n^2}$$

$$= \frac{714.15 \ \Omega}{\left(\frac{345/\sqrt{3}}{34.5/\sqrt{3}}\right)^2} = 7.1415 \ \Omega$$

where n is defined as the turns ratio of the windings.

Example D.6

A three-phase transformer has a nameplate ratings of 20 MVA, and the voltage ratings of 345Y/34.5Δ kV with a leakage reactance of 12% and the transformer connection is wye–delta. Select a base of 20 MVA and 345 kV on the high-voltage side and determine the following:

    a. Turns ratio of windings.
    b. Transformer reactance referred to low-voltage side in ohms.
    c. Transformer reactance referred to low-voltage side in per units.

Solution:

    a. The turns ratio of the windings is

$$n = \frac{345/\sqrt{3}}{34.5} = 5.7735$$

    b. Since the high-voltage side impedance base is

$$Z_{B(HV)} = \frac{[kV_{B(HV)}]^2}{MVA_{B(3\phi)}}$$

$$= \frac{345^2}{20} = 5951.25\ \Omega$$

and

$$X_{(HV)} = X_{pu} \times X_{B(HV)}$$

$$= (0.12)(5951.25) = 714.15\ \Omega$$

Thus, the transformer reactance referred to the delta-connected low-voltage side is

$$X_{(LV)} = \frac{X_{(HV)}}{n^2}$$

$$= \frac{714.14\ \Omega}{5.7735^2} = 21.4245\ \Omega$$

    c. The reactance of the equivalent wye connection is

$$Z_Y = \frac{Z_\Delta}{3}$$

$$= \frac{21.4245\ \Omega}{3} = 7.1415\ \Omega$$

Similarly,

$$Z_{B(LV)} = \frac{[kV_{B(LV)}]^2}{MVA_{B(3\phi)}}$$

$$= \frac{34.5^2}{20} = 59.5125\ \Omega$$

Thus,

$$X_{pu} = \frac{7.1415\,\Omega}{Z_{B(LV)}}$$

$$= \frac{7.1415\,\Omega}{59.5125\,\Omega} = 0.12\,\text{pu}$$

Alternatively, if the line-to-line voltages are used,

$$X_{(LV)} = \frac{X_{(HV)}}{n^2}$$

$$= \frac{714.14\,\Omega}{(345/34.5)^2} = 7.1415\,\Omega$$

and therefore,

$$X_{pu} = \frac{X_{(LV)}}{Z_{B(LV)}}$$

$$= \frac{7.1415\,\Omega}{59.5125\,\Omega} = 0.12\,\text{pu}$$

as before.

Example D.7

Consider a three-phase system that has a generator connected to a 2.4/24 kV, wye–wye connected, three-phase step-up transformer T_1. Suppose that the transformer is connected to three-
phase power line. The receiving end of the line is connected to a second, wye–wye connected,
three-phase 24/12 kV step-down transformer T_2. Assume that the line length between the two
transformers is negligible and the three-phase generator is rated 4160 kVA, 2.4 kV, and 1000 A
and that it supplies a purely inductive load of $I_{pu} = 2.08\angle{-90°}$ pu. The three-phase transformer T_1
is rated 6000 kVA, 2.4Y–24Y kV, with leakage reactance of 0.04 pu. Transformer T_2 is made up
of three single-phase transformers and is rated 4000 kVA, 24Y–12Y kV, with leakage reactance of
0.04 pu. Determine the following for all three circuits, 2.4, 24, and 12 kV circuits:

a. Base kilovoltampere values.
b. Base line-to-line kilovolt values.
c. Base impedance values.
d. Base current values.
e. Physical current values (neglect magnetizing currents in transformers and charging currents
   in lines).
f. Per-unit current values.
g. New transformer reactances based on their new bases.
h. Per-unit voltage values at buses 1, 2, and 4.
i. Per-unit apparent power values at buses 1, 2, and 4.
j. Summarize results in a table.

Solution:

a. The kilovoltampere base for all three circuits is arbitrarily selected as 2080 kVA
b. The base voltage for the 2.4 kV circuit is arbitrarily selected as 2.5 kV. Since the turns ratios
   for transformers T_1 and T_2 are

$$\frac{N_1}{N_2} = 10 \quad \text{or} \quad \frac{N_2}{N_1} = 0.10$$

and

$$\frac{N_1'}{N_2'} = 2$$

the base voltages for the 24 and 12 kV circuits are determined to be 25 and 12.5 kV, respectively.

c. The base impedance values can be found as

$$Z_B = \frac{[kV_{B(L-L)}]^2(1000)}{kVA_{B(3\phi)}}$$

$$= \frac{[2.5\,kV]^2 1000}{2080\,kVA} = 3.005\,\Omega$$

and

$$Z_B = \frac{[25\,kV]^2 1000}{2080\,kVA} = 300.5\,\Omega$$

and

$$Z_B = \frac{[12.5\,kV]^2 1000}{2080\,kVA} = 75.1\,\Omega$$

d. The base current values can be determined as

$$I_B = \frac{kVA_{B(3\phi)}}{\sqrt{3}kV_{B(L-L)}}$$

$$= \frac{2080\,kVA}{\sqrt{3}(2.5\,kV)} = 480\,A$$

and

$$I_B = \frac{2080\,kVA}{\sqrt{3}(25\,kV)} = 48\,A$$

and

$$I_B = \frac{2080\,kVA}{\sqrt{3}(12.5\,kV)} = 96\,A$$

e. The physical current values can be found based on the turns ratios as

$$I = 1000\,A$$

$$I = \left(\frac{N_2}{N_1}\right)(1000\,A) = 100\,A$$

$$I = \left(\frac{N\!\!\!\!c_1}{N\!\!\!\!c_2}\right)(100\,A) = 200\,A$$

TABLE D.1

Results of Example D.7

Quantity	2.4 kV Circuit	24 kV Circuit	12 kV Circuit
$kVA_{B(3\phi)}$	2080 kVA	2080 kVA	2080 kVA
$kV_{B(L-L)}$	2.5 kV	25 kV	12.5 kV
Z_B	3005 Ω	300.5 Ω	75.1 Ω
I_B	480 A	48 A	96 A
$I_{physical}$	1000 A	100 A	200 A
I_{pu}	2.08 pu	2.08 pu	2.08 pu
V_{pu}	0.96 pu	0.9334 pu	0.8935 pu
S_{pu}	2.00 pu	1.9415 pu	1.8585 pu

f. The per-unit current values are the same, 2.08 pu, for all three circuits.

g. The given transformer reactances can be converted based on their new bases using

$$Z_{pu(new)} = Z_{pu(given)} \left[\frac{kVA_{B(new)}}{kVA_{B(given)}} \right] \left[\frac{kV_{B(given)}}{kV_{B(new)}} \right]^2$$

Therefore, the new reactances of the two transformers can be found as

$$Z_{pu(T_1)} = j0.04 \left[\frac{2080\ kVA}{6000\ kVA} \right] \left[\frac{2.4\ kV}{2.5\ kV} \right]^2 = j0.0128\ pu$$

and

$$Z_{pu(T_2)} = j0.04 \left[\frac{2080\ kVA}{4000\ kVA} \right] \left[\frac{12\ kV}{12.5\ kV} \right]^2 = j0.0192\ pu$$

h. Therefore, the per-unit voltage values at buses 1, 2, and 4 can be calculated as

$$\mathbf{V}_1 = \frac{2.4\ kV\angle 0°}{2.5\ kV} = 0.96\ \angle 0°\ pu$$

$$\mathbf{V}_2 = \mathbf{V}_1 - \mathbf{I}_{pu}Z_{pu(T_1)}$$

$$= 0.96\ \angle 0° - (2.08\ \angle -90°)(0.0128\angle 90°) = 0.9334\angle 0°\ pu$$

$$\mathbf{V}_4 = \mathbf{V}_2 - \mathbf{I}_{pu}Z_{pu(T_2)}$$

$$= 0.9334\angle 0° - (2.08\angle -90°)(0.0192\ \angle 90°) = 0.8935\angle 0°\ pu$$

i. Thus, the per-unit apparent power values at buses 1, 2, and 4 are

$$S_1 = 2.00\ pu$$

$$S_2 = V_2 I_{pu} = (0.9334)(2.08) = 1.9415\ pu$$

$$S_4 = V_4 I_{pu} = (0.8935)(2.08) = 1.8585\ pu$$

j. The results are summarized in Table D.1.

PROBLEMS

D.1 Solve Example D.1 for a transformer rated 100 kVA and 2400/240 V that has a high-voltage winding impedance of 0.911.

D.2 Consider the results of Problem D.1 and use 3000/300 V as new base voltages for the high-voltage and low-voltage windings, respectively. Use a new base power of 200 kVA and determine the new per-unit, base, and physical impedances of the transformer referred to the high-voltage side.

D.3 A 240/120 V single-phase transformer rated 25 kVA has a high-voltage winding impedance of 0.65 Ω. If 240 V and 25 kVA are used as the base quantities, determine the following:

   a. The high-voltage side base current.
   b. The high-voltage side base impedance in Q.
   c. The transformer impedance referred to the high-voltage side in per unit.
   d. The transformer impedance referred to the high-voltage side in percent.
   e. The turns ratio of the transformer windings.
   f. The low-voltage side base current.
   g. The low-voltage side base impedance.
   h. The transformer impedance referred to the low-voltage side in per unit.

D.4 A 240/120 V single-phase transformer is rated 25 kVA and has a high-voltage winding impedance referred to its high-voltage side that is 0.2821 pu based on 240 V and 25 kVA. Select 230/115 V as the base voltages for the high-voltage and low-voltage windings, respectively. Use a new base power of 50 kVA and determine the new per-unit base, and physical impedances of the transformer referred to the high-voltage side.

D.5 After changing the S base from 5 to 10 MVA, redo the Example D.1 by using MATLAB.

   a. Write the MATLAB program script.
   b. Give the MATLAB program output.

Appendix E: Glossary for Distribution System Terminology

Some of the most commonly used terms, both in this book and in general usage, are defined later. Most of the definitions given in this glossary are based on Refs. [1–8].

AAAC: Abbreviation for all-aluminum-alloy conductors. Aluminum-alloy conductors have higher strength than those of the ordinary electric-conductor grade of aluminum.

AA: Abbreviation for all-aluminum conductors.

ACAR: Abbreviation for aluminum conductor alloy reinforced. It has a central core of higher-strength aluminum surrounded by layers of electric-conductor grade aluminum.

ACL cable: A cable with a lead sheath over the cable insulation that is suitable for wet locations. It is used in buildings at low voltage.

ACSR: Abbreviation for aluminum conductor steel reinforced. It consists of a central core of steel strands surrounded by layers of aluminum strands.

Active filter: Any of a number of sophisticated power electronic devices for eliminating harmonic distortion.

Admittance: The ratio of the phasor equivalent of the steady-state sine-wave current to the phasor equivalent of the corresponding voltage.

Adverse weather: Weather conditions that cause an abnormally high rate of forced outages for exposed components during the periods such conditions persist, but which do not qualify as major storm disasters. Adverse weather conditions can be defined for a particular system by selecting the proper values and combinations of conditions reported by the Weather Bureau: thunderstorms, tornadoes, wind velocities, precipitation, temperature, etc.

Aerial cable: An assembly of insulated conductors installed on a pole line or similar overhead structures; it may be self-supporting or installed on a supporting messenger cable.

Air-blast transformer: A transformer cooled by forced circulation of air through its core and coils.

Air circuit breaker: A circuit breaker in which the interruption occurs in air.

Air switch: A switch in which the interruptions of the circuit occur in air. Al: Symbol for aluminum.

Ampacity: Current rating in amperes, as of a conductor.

ANSI: Abbreviation for American National Standards Institute.

Apparent sag (at any point): The departure of the wire at the particular point in the span from the straight line between the two points of the span, at 60°F, with no wind loading.

Arcing time of fuse: The time elapsing from the severance of the fuse link to the final interruption of the circuit under specified conditions.

Arc-over of insulator: A discharge of power current in the form of an arc following a surface discharge over an insulator.

Armored cable: A cable provided with a wrapping of metal, usually steel wires, primarily for the purpose of mechanical protection.

Askarel: A generic term for a group of nonflammable synthetic chlorinated hydrocarbons used as electrical insulating media. Askarels of various compositional types are used. Under arcing conditions, the gases produced, while consisting predominantly of noncombustible hydrogen chloride, can include varying amounts of combustible gases depending upon the askarel type. Because of environmental concerns, it is not used in new installations anymore.

Automatic substations: Those in which switching operations are so controlled by relays that transformers or converting equipment are brought into or taken out of service as variations in load may require, and feeder circuit breakers are closed and reclosed after being opened by overload relays.

Autotransformer: A transformer in which at least two windings have a common section.

AWG: Abbreviation for American Wire Gauge. It is also sometimes called the Brown and Sharpe Wire Gauge.

Base load: The minimum load over a given period of time.

Benchboard: A switchboard with a horizontal section for control switches, indicating lamps, and instrument switches; may also have a vertical instrument section.

BIL: Abbreviation for basic impulse insulation levels, which are reference levels expressed in impulse-crest voltage with a standard wave not longer than 1.5×50 μs. The impulse waves are defined by a combination of two numbers. The first number is the time from the start of the wave to the instant crest value; the second number is the time from the start to the instant of half-crest value on the tail of the wave.

Billing demand: The demand used to determine the demand charges in accordance with the provisions of a rate schedule or contract.

Branch circuit: A set of conductors that extend beyond the last overcurrent device in the low-voltage system of a given building. A branch circuit usually supplies a small portion of the total load.

Breakdown: Also termed puncture, denoting a disruptive discharge through insulation.

Breaker, primary-feeder: A breaker located at the supply end of a primary feeder that opens on a primary-feeder fault if the fault current is of sufficient magnitude.

Breaker-and-a-half scheme: A scheme that provides the facilities of a double main bus at a reduction in equipment cost by using three circuit breakers for each two circuits.

Bus: A conductor or group of conductors that serves as a common connection for two or more circuits in a switchgear assembly.

Bus, transfer: A bus to which one circuit at a time can be transferred from the main bus.

Bushing: An insulating structure including a through conductor, or providing a passageway for such a conductor, with provision for mounting on a barrier, conductor or otherwise, for the purpose of insulating the conductor from the barrier and conducting from one side of the barrier to the other.

BVR: Abbreviation for bus voltage regulator or regulation.

BW: Abbreviation for bandwidth.

BX cable: A cable with galvanized interlocked steel spiral armor. It is known as ac cable and used in a damp or wet location in buildings at low voltage.

Cable: Either a standard conductor (single-conductor cable) or a combination of conductors insulated from one another (multiple-conductor cable).

Cable fault: A partial or total load failure in the insulation or continuity of the conductor.

Capability: The maximum load-carrying ability expressed in kilovoltamperes or kilowatts of generating equipment or other electric apparatus under specified conditions for a given time interval.

Capability, net: The maximum generation expressed in kilowatthours per hour that a generating unit, station, power source, or system can be expected to supply under optimum operating conditions.

Capacitor bank: An assembly at one location of capacitors and all necessary accessories (switching equipment, protective equipment, controls, etc.) required for a complete operating installation.

Capacity: The rated load-carrying ability expressed in kilovoltamperes or kilowatts of generating equipment or other electric apparatus.

Capacity factor: The ratio of the average load on a machine or equipment for the period of time considered to the capacity of the machine or equipment.

Charge: The amount paid for a service rendered or facilities used or made available for use.

Circuit, earth (ground) return: An electric circuit in which the earth serves to complete a path for current.

Circuit breaker: A device that interrupts a circuit without injury to itself so that it can be reset and reused over again.

Circuit-breaker mounting: Supporting structure for a circuit breaker.

Circular mil: A unit of area equal to 1/4 of a square mil (=0.7854 square mil). The cross-sectional area of a circle in circular mils is therefore equal to the square of its diameter in mils. A circular inch is equal to 1 million circular mils. A mil is one one-thousandth of an inch. There are 1974 circular mils in a square millimeter. Abbreviated cmil.

CL: Abbreviation for current limiting (fuse).

cmil: Abbreviation for circular mil.

Coincidence factor: The ratio of the maximum coincident total demand of a group of consumers to the sum of the maximum power demands of individual consumers comprising the group, both taken at the same point of supply at the same time.

Coincident demand: Any demand that occurs simultaneously with any other demands; also the sum of any set of coincident demands.

Component: A piece of equipment, a line, a section of a line, or a group of items that is viewed as an entity.

Condenser: Also termed capacitor; a device whose primary purpose is to introduce capacitance into an electric circuit. The term condenser is deprecated.

Conductor: A substance that has free electrons or other charge carriers that permit charge flow when an emf is applied across the substance.

Conductor tension, final unloaded: The longitudinal tension in a conductor after the conductor has been stretched by the application for an appreciable period, with subsequent release of the loadings of ice and wind, at the temperature decrease assumed for the loading district in which the conductor is strung (or equivalent loading).

Conduit: A structure containing one or more ducts; commonly formed from iron pipe or electrical metallic tubing, used in buildings at low voltage.

Connection charge: The amount paid by a customer for connecting the customer's facilities to the supplier's facilities.

Contactor: An electric power switch, not operated manually and designed for frequent operation.

Contract demand: The demand that the supplier of electric service agrees to have available for delivery.

Cress factor: A value that is displayed on many power quality monitoring instruments representing the ratio of the crest value of the measured waveform to the rms value of the waveform. For example, the cress factor of a sinusoidal wave is 1.414.

CT: Abbreviation for current transformers.

Cu: Symbol for copper.

Customer charge: The amount paid periodically by a customer without regard to demand or energy consumption.

Demand: The load at the receiving terminals averaged over a specified interval of time.

Demand charge: That portion of the charge for electric service based upon a customer's demand.

Demand factor: The ratio of the maximum coincident demand of a system, or part of a system, to the total connected load of the system, or part of the system, under consideration.

Demand, instantaneous: The load at any instant.

Demand, integrated: The demand integrated over a specified period.

Demand interval: The period of time during which the electric energy flow is integrated in determining demand.

Depreciation: The component that represents an approximation of the value of the portion of plant consumed or "used up" in a given period by a utility.

Disconnecting or isolating switch: A mechanical switching device used for changing the connections in a circuit or for isolating a circuit or equipment from the source of power.

Disconnector: A switch that is intended to open a circuit only after the load has been thrown off by other means. Manual switches designed for opening loaded circuits are usually installed in a circuit with disconnectors to provide a safe means for opening the circuit under load.

Displacement factor (DPF): The ratio of active power (watts) to apparent power (voltamperes).

Distribution center: A point of installation for automatic overload protective devices connected to buses where an electric supply is subdivided into feeders and/or branch circuits.

Distribution switchboard: A power switchboard used for the distribution of electric energy at the voltages common for such distribution within a building.

Distribution system: That portion of an electric system that delivers electric energy from transformation points in the transmission, or bulk power system, to the consumers.

Distribution transformer: A transformer for transferring electric energy from a primary distribution circuit to a secondary distribution circuit or consumer's service circuit; it is usually rated in the order of 5–500 kVA.

Diversity factor: The ratio of the sum of the individual maximum demands of the various subdivisions of a system to the maximum demand of the whole system.

Duplex cable: A cable composed of two insulated stranded conductors twisted together. They may or may not have a common insulating covering.

Effectively grounded: Grounded by means of a ground connection of sufficiently low impedance that fault grounds that may occur cannot build up voltages dangerous to connected equipment.

EHV: Abbreviation for extra high voltage.

Electric rate schedule: A statement of an electric rate and the terms and conditions governing its application.

Electric system loss: Total electric energy loss in the electric system. It consists of transmission, transformation, and distribution losses between sources of supply and points of delivery.

Electrical reserve: The capability in excess of that required to carry the system load.

Emergency rating: Capability of installed equipment for a short time interval.

EMT: Abbreviation for electrical metallic tubing. A raceway that has a thin wall that does not permit threading. Connectors and couplings are secured by either compression rings or setscrews. It is used in buildings at low voltage.

Energy: That which does work or is capable of doing work. As used by electric utilities, it is generally a reference to electric energy and is measured in kilowatthours.

Energy charge: That portion of the charge for electric service based upon the electric energy consumed or billed.

Energy loss: The difference between energy input and output as a result of transfer of energy between two points.

Express feeder: A feeder that serves the most distant networks and must traverse the systems closest to the bulk power source.

Extra high voltage: A term applied to voltage levels higher than 230 kV. Abbreviated EHV.

Facilities charge: The amount paid by the customer as a lump sum or, periodically, as reimbursement for facilities furnished. The charge may include operation and maintenance as well as fixed costs.

FCN: Abbreviation for full-capacity neutral.

Feeder: A set of conductors originating at a main distribution center and supplying one or more secondary distribution centers, one or more branch-circuit distribution centers, or any combination of these two types of load.

Feeder, multiple: Two or more feeders connected in parallel.

Feeder, tie: A feeder that connects two or more independent sources of power and has no tapped load between the terminals. The source of power may be a generating system, substation, or feeding point.

First-contingency outage: The outage of one primary feeder.

Fixed-capacitor bank: A capacitor bank with fixed, not switchable, capacitors.

Flicker: Impression of unsteadiness of visual sensation induced by a light stimulus whose luminance or spectral distribution fluctuates with time.

Flicker factor: A factor used to quantify the load impact of electric arc furnaces on the power system.

Forced interruption: An interruption caused by a forced outage.

Forced outage: An outage that results from emergency conditions directly associated with a component, requiring that it be taken out of service immediately, either automatically or as soon as switching operations can be performed; or an outage caused by improper operation of equipment or by human error.

Frequency deviation: An increase or decrease in the power frequency. Its duration varies from few cycles to several hours.

Fuel adjustment clause: A clause in a rate schedule that provides for adjustment of the amount of the bill as the cost of fuel varies from a specified base amount per unit.

Fuse: An overcurrent protective device with a circuit-opening fusible part that is heated and severed by the passage of overcurrent through it.

Fuse cutout: An assembly consisting of a fuse support and holder; it may also include a fuse link.

Ground: Also termed earth; a conductor connected between a circuit and the soil; an accidental ground occurs due to cable insulation faults, an insulator defect, etc.

Ground wire: A conductor having grounding connections at intervals that is suspended usually above but not necessarily over the line conductor to provide a degree of protection against lightning discharges.

Harmonics: Sinusoidal voltages or currents having frequencies that are an integer multiple of the fundamental frequency at which the supply system is designed to operate.

Harmonic distortion: Periodic distortion of the sign wave.

Harmonic resonance: A condition in which the power system is resonating near one of the major harmonics being produced by nonlinear elements in the system, hence increasing the harmonic distortion.

HMWPE: Abbreviation for high-molecular-weight polyethylene (cable insulation).

HV: Abbreviation for high voltage.

Impedance: The ratio of the phasor equivalent of a steady-state sine-wave voltage to the phasor equivalent of a steady-state sine-wave current.

Impulsive transient: A sudden (nonpower) frequency change in the steady-state condition of the voltage or current that is unidirectional in polarity.

Incremental energy costs: The additional cost of producing or transmitting electric energy above some base cost.

Index of reliability: A ratio of cumulative customer minutes that service was available during a year to total customer minutes demanded; can be used by the utility for feeder reliability comparisons.

Indoor transformer: A transformer that must be protected from the weather.

Installed reserve: The reserve capability installed on a system.

Interruptible load: A load that can be interrupted as defined by contract.

Interruption: The loss of service to one or more consumers. An interruption is the result of one or more component outages.

Interruption duration: The period from the initiation of an interruption to a consumer until service has been restored to that consumer.

Investment-related charges: Those certain charges incurred by a utility that are directly related to the capital investment of the utility. kcmil: Abbreviation for a thousand circular mils.

Isolated ground: It originates at an isolated ground-type receptacle or equipment input terminal block and terminates at the point where neutral and ground are bonded at the power source. Its conductor is insulated from the metallic raceway and all ground points throughout its length.

K-factor: A factor used to quantify the load impact of electric arc furnaces on the power system.

Lag: Denotes that a given sine wave passes through its peak at a later time than a reference time wave.

Lambda: The incremental operating cost at the load center, commonly expressed in mils per kilowatthour.

Lateral conductor: A wire or cable extending in a general horizontal direction or at an angle to the general direction of the line; service wires either overhead or underground are considered laterals from the street mains.

LDC: Abbreviation for line-drop compensator.

Lightning arrestor: A device that reduces the voltage of a surge applied to its terminals and restores itself to its original operating condition.

L–L: Abbreviation for line to line.

Limit switch: A switch that is operated by a moving part at the end of its travel typically to stop or reverse the motion.

Limiter: A device in which some characteristic of the output is automatically prevented from exceeding a predetermined value.

Line: A component part of a system extending between adjacent stations or from a station to an adjacent interconnection point. A line may consist of one or more circuits.

Line-drop compensator: A device that causes the voltage-regulating relay to increase the output voltage by an amount that compensates for the impedance drop in the circuit between the regulator and a predetermined location at the circuit.

Line loss: Energy loss on a transmission or distribution line.

L–N: Abbreviation for line to neutral.

Load center: A point at which the load of a given area is assumed to be concentrated.

Load diversity: The difference between the sum of the maxima of two or more individual loads and the coincident or combined maximum load, usually measured in kilowatts over a specified period of time.

Load duration curve: A curve of loads, plotted in descending order of magnitude, against time intervals for a specified period.

Load factor: The ratio of the average load over a designated period of time to the peak load occurring in that period.

Load-interrupter switch: An interrupter switch designed to interrupt currents not in excess of the continuous-current rating of the switch.

Load, interruptible: A load that can be interrupted as defined by contract.

Load losses, transformer: Those losses that are incident to the carrying of a specified load. They include I^2R loss in the winding due to load and eddy currents, stray loss due to leakage fluxes in the windings, etc., and the loss due to circulating currents in parallel windings.

Load tap changer: A selector switch device applied to power transformers to maintain a constant low-side or secondary voltage with a variable primary voltage supply, or to hold a constant voltage out along the feeders on the low-voltage side for varying load conditions on the low-voltage side. Abbreviated LTC.

Load-tap-changing transformer: A transformer used to vary the voltage, or phase angle, or both, of a regulated circuit in steps by means of a device that connects different taps of tapped winding(s) without interrupting the load.

Loop feeder: A number of tie feeders in series, forming a closed loop. There are two routes by which any point on a loop feeder can receive electric energy, so that the flow can be in either direction.

Loop service: Two services of substantially the same capacity and characteristics, supplied from adjacent sections of a loop feeder. The two sections of the loop feeder are normally tied together on the consumer's bus through switching devices.

Loss factor: The ratio of the average power loss to the peak-load power loss during a specified period of time.

Low-side surges: The current surge that appears to be injected into the transformer secondary terminals upon a lighting strike to grounded conductors in the vicinity.

LTC: Abbreviation for load tap changer.

LV: Abbreviation for low voltage.

Main distribution center: A distribution center supplied directly by mains.

Maintenance expenses: The expense required to keep the system or plant in proper operating repair.

Maximum demand: The largest of a particular type of demand occurring within a specified period.

MC: Abbreviation for metal clad (cable).

Messenger cable: A galvanized steel or Copperweld cable used in construction to support a suspended current-carrying cable.

Metal-clad switchgear, outdoor: A switchgear that can be mounted in suitable weatherproof enclosures for outdoor installations. The base units are the same for both indoor and outdoor applications. The weatherproof housing is constructed integrally with the basic structure and is not merely a steel enclosure. The basic structure, including the mounting details and withdrawal mechanisms for the circuit breakers, bus compartments, transformer compartments, etc., is the same as that of indoor metal-clad switchgear.

Metal-enclosed switchgear: Primarily indoor-type switchgear. It can, however, be furnished in weatherproof houses suitable for outdoor operation. The switchgear is suitable for 600 V maximum service.

Minimum demand: The smallest of a particular type of demand occurring within a specified period.

Momentary interruption: An interruption of duration limited to the period required to restore service by automatic or supervisory-controlled switching operations or by manual switching at locations where an operator is immediately available.

Monthly peak duration curve: A curve showing the total number of days within the month during which the net 60 min clock-hour integrated peak demand equals or exceeds the percent of monthly peak values shown.

NC: Abbreviation for normally closed.

NEC: Abbreviation for National Electric Code.

NESC: Abbreviation for National Electrical Safety Code.

Net system energy: Energy requirements of a system, including losses, defined as (1) net generation of the system, plus (2) energy received from others, less, and (3) energy delivered to other systems.

Network distribution system: A distribution system that has more than one simultaneous path of power flow to the load.

Network protector: An electrically operated low-voltage air circuit breaker with self-contained relays for controlling its operation. It provides automatic isolation of faults in the primary feeders or network transformers. Abbreviated NP.

NO: Abbreviation for normally open.

Noise: An unwanted electrical signal with a less than 200 kHz superimposed upon the power-system voltage or current in-phase conductors, or found on neutral conductors or signal lines. It is not a harmonic distortion or transient. It disturbs microcomputers and programmable controllers.

No-load current: The current demand of a transformer primary when no current demand is made on the secondary.

No-load loss: Energy losses in an electric facility when energized at rated voltage and frequency but not carrying load.

Noncoincident demand: The sum of the individual maximum demands regardless of time of occurrence within a specified period.

Nonlinear load: An electrical load that draws current discontinuously or whose impedances vary throughout the cycle of the input ac voltage waveform.

Normal rating: Capacity of installed equipment.

Normal weather: All weather not designated as adverse or major storm disaster.

Normally closed: Denotes the automatic closure of contacts in a relay when deenergized. Abbreviated NC.

Normally open: Denotes the automatic opening of contacts in a relay when deenergized. Abbreviated NO.

NP: Abbreviation for network protector.

NSW: Abbreviation for nonswitched.

Notch: A switching (or other) disturbance of the normal power voltage waveform, lasting less than a half cycle, which is initially of opposite polarity than the waveform. It includes complete loss of voltage for up to a 0.5 cycle.

Notching: A periodic disturbance caused by normal operation of a power electronic device, when its current is commutated from one phase to another.

NX: Abbreviation for nonexpulsion (fuse).

Off-peak energy: Energy supplied during designated periods of relatively low system demands.

On-peak energy: Energy supplied during designated periods of relatively high system demands.

OH: Abbreviation for overhead.

Operating expenses: The labor and material costs for operating the plant involved.

Outage: The state of a component when it is not available to perform its intended function due to some event directly associated with that component. An outage may or may not cause an interruption of service to consumers depending upon the system configuration.

Outage duration: The period from the initiation of an outage until the affected component or its replacement once again becomes available to perform its intended function.

Outage rate: For a particular classification of outage and type of component, the mean number of outages per unit exposure time per component.

Oscillatory transient: A sudden and nonpower frequency change in the steady-state condition of voltage or current that includes both positive and negative polarity values.

Overhead expenses: The costs that in addition to direct labor and material are incurred by all utilities.

Overload: Loading in excess of normal rating of equipment.

Overload protection: Interruption or reduction of current under conditions of excessive demand, provided by a protective device.

Overvoltage: A voltage that has a value at least 10% above the nominal voltage for a period of time greater than 1 min.

Pad-mounted: A general term describing equipment positioned on a surface mounted pad located outdoors. The equipment is usually enclosed with all exposed surfaces at ground potential.

Pad-mounted transformer: A transformer utilized as part of an underground distribution system, with enclosed compartment(s) for high-voltage and low-voltage cables entering from below, and mounted on a foundation pad.

Panelboard: A distribution point where an incoming set of wires branches into various other circuits.

Passive filter: A combination of inductors, capacitors, and resistors designed to eliminate one or more harmonics. The most common variety is simply an inductor in series with a shunt capacitor, which short-circuits the major distorting harmonic component from the system.

PE: Abbreviation used for polyethylene (cable insulation).

Peak current: The maximum value (crest value) of an alternating current.

Peak voltage: The maximum value (crest value) of an alternating voltage.

Peaking station: A generating station that is normally operated to provide power only during maximum load periods.

Peak-to-peak value: The value of an ac waveform from its positive peak to its negative peak. In the case of a sine wave, the peak-to-peak value is double the peak value.

Pedestal: A bottom support or base of a pillar, statue, etc.

Percent regulation: See Percent voltage drop.

Percent voltage drop: The ratio of voltage drop in a circuit to voltage delivered by the circuit, multiplied by 100 to convert to percent.

Permanent forced outage: An outage whose cause is not immediately self-clearing but must be corrected by eliminating the hazard or by repairing or replacing the component before it can be returned to service. An example of a permanent forced outage is a lightning flashover that shatters an insulator, thereby disabling the component until repair or replacement can be made.

Permanent forced outage duration: The period from the initiation of the outage until the component is replaced or repaired.

Phase: The time of occurrence of the peak value of an ac waveform with respect to the time of occurrence of the peak value of a reference waveform.

Phase angle: An angular expression of phase difference.

Phase shift: The displacement in time of one voltage waveform relative to other voltage waveform(s).

Pole: A column of wood or steel, or some other material, supporting overhead conductors, usually by means of arms or brackets.

Pole fixture: A structure installed in lieu of a single pole to increase the strength of a pole line or to provide better support for attachments than would be provided by a single pole. Examples are A fixtures, H fixtures.

Primary disconnecting devices: Self-coupling separable contacts provided to connect and disconnect the main circuits between the removable element and the housing.

Primary distribution feeder: A feeder operating at primary voltage supplying a distribution circuit.

Primary distribution mains: The conductors that feed from the center of distribution to direct primary loads or to transformers that feed secondary circuits.

Primary distribution network: A network consisting of primary distribution mains.

Primary distribution system: A system of ac distribution for supplying the primaries of distribution transformers from the generating station or substation distribution buses.

Primary distribution trunk line: A line acting as a main source of supply to a distribution system.

Primary feeder: That portion of the primary conductors between the substation or point of supply and the center of distribution.

Primary lateral: That portion of a primary distribution feeder that is supplied by a main feeder or other laterals and extends through the load area with connections to distribution transformers or primary loads.

Primary main feeder: The higher-capacity portion of a primary distribution feeder that acts as a main source of supply to primary laterals or direct connected distribution transformers and primary loads.

Primary network: A network supplying the primaries of transformers whose secondaries may be independent or connected to a secondary network.

Primary open-loop service: A service that consists of a single distribution transformer with dual primary switching, supplied from a single primary circuit that is arranged in an open-loop configuration.

Primary selective service: A service that consists of a single distribution transformer with primary throw-over switching, supplied by two independent primary circuits.

Primary transmission feeder: A feeder connected to a primary transmission circuit.

Primary unit substation: A unit substation in which the low-voltage section is rated above 1000 V.

Protective relay: A device whose function is to detect defective lines or apparatus or other power-system conditions of an abnormal or dangerous nature and to initiate appropriate control circuit action.

Power: The rate (in kilowatts) of generating, transferring, or using energy.

Power, active: The product of the rms value of the voltage and the rms value of the in-phase component of the current.

Power, apparent: The product of the rms value of the voltage and the rms value of the current.

Power, instantaneous: The product of the instantaneous voltage multiplied by the instantaneous current.

Power, reactive: The product of the rms value of the voltage and the rms value of the quadrature component of the current.

Power factor: The ratio of active power to apparent power.

Power-factor adjustment clause: A clause in a rate schedule that provides for an adjustment in the billing if the customer's power factor varies from a specified reference.

Power pool: A group of power systems operating as an interconnected system and pooling their resources.

Power transformer: A transformer that transfers electric energy in any part of the circuit between the generator and the distribution primary circuits.

PT: Abbreviation for potential transformers.

pu: Abbreviation for per unit.

Raceway: A channel for holding wires, cables, or busbars. The channel may be in the form of a conduit, electrical metallic tubing, or a square sheet-metal duct. It is used in buildings at low voltage.

Radial distribution system: A distribution system that has a single simultaneous path of power flow to the load.

Radial service: A service that consists of a single distribution transformer supplied by a single primary circuit.

Radial system, complete: A radial system that consists of a radial subtransmission circuit, a single substation, and a radial primary feeder with several distribution transformers each supplying radial secondaries; has the lowest degrees of service continuity.

Ratchet demand: The maximum past or present demands that are taken into account to establish billings for previous or subsequent periods.

Ratchet demand clause: A clause in a rate schedule that provides that maximum past or present demands be taken into account to establish billings for previous or subsequent periods.

Rate base: The net plant investment or valuation base specified by a regulatory authority upon which a utility is permitted to earn a specified rate of return.

RCN: Abbreviation for reduced-capacity neutral.

Recloser: A dual-timing device that can be set to operate quickly to prevent downline fuses from blowing.

Reclosing device: A control device that initiates the reclosing of a circuit after it has been opened by a protective relay.

Reclosing fuse: A combination of two or more fuse holders, fuse units, or fuse links mounted on a fuse support(s), mechanically or electrically interlocked, so that one fuse can be connected into the circuit at a time and the functioning of that fuse automatically connects the next fuse into the circuit, thereby permitting one or more service restorations without replacement of fuse links, refill units, or fuse units.

Reclosing relay: A programming relay whose function is to initiate the automatic reclosing of a circuit breaker.

Reclosure: The automatic closing of a circuit-interrupting device following automatic tripping. Reclosing may be programmed for any combination of instantaneous, time-delay, single-shot, multiple-shot, synchronism-check, dead-line-live-bus, or dead-bus-live-line operation.

Recovery voltage: The voltage that occurs across the terminals of a pole of a circuit-interrupting device upon interruption of the current.

Required reserve: The system planned reserve capability needed to ensure a specified standard of service.

Resistance: The real part of impedance.

Return on capital: The requirement that is necessary to pay for the cost of investment funds used by the utility.

RP: Abbreviation for regulating point.

Sag: The distance measured vertically from a conductor to the straight line joining its two points of support. Unless otherwise stated, the sag referred to is the sag at the midpoint of the span.

Sag: A decrease to between 0.1 and 0.9 pu in rms voltage and current at the power frequency for a duration of 0.5 cycles to 1 min.

Sag, final unloaded: The sag of a conductor after it has been subjected for an appreciable period to the loading prescribed for the loading district in which it is situated, or equivalent loading, and the loading removed. Final unloaded sag includes the effect of inelastic deformation.

Sag, initial unloaded: The sag of a conductor prior to the application of any external load.

Sag of a conductor (at any point in a span): The distance measured vertically from the particular point in the conductor to a straight line between its two points of support.

Sag section: The section of line between snub structures. More than one sag section may be required to properly sag the actual length of conductor that has been strung.

Sag span: A span selected within a sag section and used as a control to determine the proper sag of the conductor, thus establishing the proper conductor level and tension. A minimum of two, but normally three, sag spans are required within a sag section to sag properly. In mountainous terrain or where span lengths vary radically, more than three sag spans could be required within a sag section.

Scheduled interruption: An interruption caused by a scheduled outage.

Scheduled outage: An outage that results when a component is deliberately taken out of service at a selected time, usually for purposes of construction, preventive maintenance, or repair.

Scheduled outage duration: The period from the initiation of the outage until construction, preventive maintenance, or repair work is completed.

Scheduled maintenance (generation): Capability that has been scheduled to be out of service for maintenance.

SCV: Abbreviation for steam cured (cable insulation).

Seasonal diversity: Load diversity between two (or more) electric systems that occurs when their peak loads are in different seasons of the year.

Secondary, radial: A secondary supplied from either a conventional or completely self-protected (type CSP) distribution transformer.

Secondary current rating: The secondary current existing when the transformer is delivering rated kilovoltamperes at rated secondary voltage.

Secondary disconnecting devices: Self-coupling separable contacts provided to connect and disconnect the auxiliary and control circuits between the removable element and the housing.

Secondary distributed network: A service consisting of a number of network-transformer units at a number of locations in an urban load area connected to an extensive secondary cable grid system.

Secondary distribution feeder: A feeder operating at secondary voltage supplying a distribution circuit.

Secondary distribution mains: The conductors connected to the secondaries of distribution transformers from which consumers' services are supplied.

Secondary distribution network: A network consisting of secondary distribution mains.

Secondary distribution system: A low-voltage ac system that connects the secondaries of distribution transformers to the consumers' services.

Secondary distribution trunk line: A line acting as a main source of supply to a secondary distribution system.

Secondary fuse: A fuse used on the secondary-side circuits, restricted for use on a low-voltage secondary distribution system that connects the secondaries of distribution transformers to consumers' services.

Secondary mains: Those that operate at utilization voltage and serve as the local distribution main. In radial systems, secondary mains that supply general lighting and small power are usually separate from mains that supply three-phase power because of the dip in voltage caused by starting motors. This dip in voltage, if sufficiently large, causes an objectionable lamp flicker.

Secondary network: It consists of two or more network-transformer units connected to a common secondary system and operating continuously in parallel.

Secondary network service: A service that consists of two or more network transformer units connected to a common secondary system and operating continuously in parallel.

Secondary selective service: A service that consists of two distribution transformers, each supplied by an independent primary circuit, and with secondary main and tie breakers.

Secondary spot network: A network that consists of at least two and as many as six network-transformer units located in the same vault and connected to a common secondary service bus. Each transformer is supplied by an independent primary circuit.

Secondary system, banked: A system that consists of several transformers supplied from a single primary feeder, with the low-voltage terminals connected together through the secondary mains.

Secondary unit substation: A unit substation whose low-voltage section is rated 1000 V and below.

Secondary voltage regulation: A voltage drop caused by the secondary system, it includes the drop in the transformer and in the secondary and service cables.

Second-contingency outage: The outage of a secondary primary feeder in addition to the first one.

Sectionalizer: A device that resembles an oil circuit recloser but lacks the interrupting capability.

Service area: Territory in which a utility system is required or has the right to supply or make available electric service to ultimate consumers.

Service availability index: See Index of reliability.

Service drop: The overhead conductors, through which electric service is supplied, between the last utility company pole and the point of their connection to the service facilities located at the building or other support used for the purpose.

Service entrance: All components between the point of termination of the overhead service drop or underground service lateral and the building main disconnecting device, with the exception of the utility company's metering equipment.

Service entrance conductors: The conductors between the point of termination of the overhead service drop or underground service lateral and the main disconnecting device in the building.

Service entrance equipment: Equipment located at the service entrance of a given building that provides overcurrent protection to the feeder and service conductors, provides a means of disconnecting the feeders from energized service conductors, and provides a means of measuring the energy used by the use of metering equipment.

Service lateral: The underground conductors, through which electric service is supplied, between the utility company's distribution facilities and the first point of their connection to the building or area service facilities located at the building or other support used for the purpose.

SF$_6$: Formula for sulfur hexafluoride (gas).

St: Abbreviation for steel.

Strand: One of the wires, or groups of wires, of any stranded conductor.

Stranded conductor: A conductor composed of a group of wires, or of any combination of groups of wires. Usually, the wires are twisted together.

Submarine cable: A cable designed for service under water. It is usually a lead-covered cable with a steel armor applied between layers of jute.

Submersible transformer: A transformer so constructed as to be successfully operable when submerged in water under predetermined conditions of pressure and time.

Substation: An assemblage of equipment for purposes other than generation or utilization, through which electric energy in bulk is passed for the purpose of switching or modifying its characteristics.

Substation voltage regulation: The regulation of the substation voltage by means of the voltage regulation equipment that can be load-tap-changing (LTC) mechanisms in the substation transformer, a separate regulator between the transformer and low-voltage bus, switched capacitors at the low-voltage bus, or separate regulators located in each individual feeder in the substation.

Subtransmission: That part of the distribution system between bulk power source(s) (generating stations or power substations) and the distribution substation.

Susceptance: The imaginary part of admittance.

Swell: An increase to between 1.1 and 1.8 pu in rms voltage or current at the power frequency for durations from 0.5 cycle to 1 min.

Sustained interruption: The complete loss of voltage (<0.1 pu) on one or more phase conductors for a time greater than 1 min.

Switch: A device for opening and closing or for changing connections in a circuit.

Switch, isolating: An auxiliary switch for isolating an electric circuit from its source of power; it is operated only after the circuit has been opened by other means.

Switch, limi: A switch that is operated by some part or motion of a power-driven machine or equipment to alter the electric circuit associated with the machine or equipment.

Switchboard: A large single panel, frame, or assembly of panels on which are mounted (on the face, or back, or both) switches, fuses, buses, and usually instruments.

Switched-capacitor bank: A capacitor bank with switchable capacitors.

Switchgear: A general term covering switching or interrupting devices and their combination with associated control, instrumentation, metering, protective, and regulating devices; also assemblies of these devices with associated interconnections, accessories, and supporting structures.

Switching time: The period from the time a switching operation is required due to a forced outage until that switching operation is performed.

System: A group of components connected together in some fashion to provide flow of power from one point or points to another point or points.

System interruption duration index: The ratio of the sum of all customer interruption durations per year to the number of customers served. It gives the number of minutes out per customer per year.

Total demand distortion (TDD): The ratio of the root-mean-square (rms) of the harmonic current to the rms value of the rated or maximum demand fundamental current, expressed as a percent.

Total harmonic distortion (THD): The ratio of the root-mean-square of the harmonic content to the root-mean-square value of the fundamental quantity, expressed as a percent of the fundamental.

Triplen harmonics: A term frequently used to refer to the odd multiples of the third harmonic, which deserve special attention because of their natural tendency to be zero sequence.

True power factor (TPF): The ratio of the active power of the fundamental wave, in watts, to the apparent power of the fundamental wave, in root-mean-square voltamperes (including the harmonic components).

Underground distribution system: That portion of a primary or secondary distribution system that is constructed below the earth's surface. Transformers and equipment enclosures for such a system may be located either above or below the surface as long as the served and serving conductors are located underground.

Unit substation: A substation consisting primarily of one or more transformers that are mechanically and electrically connected to and coordinated in design with one or more switchgear or motor control assemblies or combinations thereof.

Undervoltage: A voltage that has a value at least 10% below the nominal voltage for a period of time greater than 1 min.

URD: Abbreviation for underground residential distribution.

Utilization factor: The ratio of the maximum demand of a system to the rated capacity of the system.

VD: Abbreviation for voltage drop.

VDIP: Abbreviation for voltage dip.

Voltage, base: A reference value that is a common denominator to the nominal voltage ratings of transmission and distribution lines, transmission and distribution equipment, and utilization equipment.

Voltage, maximum: The greatest 5 min average or mean voltage.

Voltage imbalance (or unbalance): The maximum deviation from the average of the three-phase voltages or currents, divided by the average of the three-phase voltages or currents, expressed in percent.

Voltage, minimum: The least 5 min average or mean voltage.

Voltage, nominal: A nominal value assigned to a circuit or system of a given voltage class for the purpose of convenient designation.

Voltage, rated: The voltage at which operating and performance characteristics of equipment are referred.

Voltage, service: Voltage measured at the terminals of the service entrance equipment.

Voltage, utilization: Voltage measured at the terminals of the machine or device.

Voltage dip: A voltage change resulting from a motor starting.

Voltage drop: The difference between the voltage at the transmitting and receiving ends of a feeder, main or service.

Voltage flicker: Voltage fluctuation caused by utilization equipment resulting in lamp flicker, that is, in a lamp illumination change.

Voltage fluctuation: A series of voltage changes or a cyclical variation of the voltage envelope.

Voltage interruption: Disappearance of the supply voltage on one or more phases. It can be momentary, temporary, or sustained.

Voltage magnification: The magnification of capacitor switching oscillatory transient voltage on the primary side by capacitors on the secondary side of a transformer.

Voltage regulation: The percent voltage drop of a line with reference to the receiving-end voltage.

$$\% \text{ regulation} = \frac{\left|\bar{E}_s\right| - \left|\bar{E}_r\right|}{\left|\bar{E}_r\right|} \times 100$$

where
$\left|\bar{E}_s\right|$ is the magnitude of the sending-end voltage
$\left|\bar{E}_r\right|$ is the magnitude of the receiving-end voltage

Voltage regulator: An induction device having one or more windings in shunt with, and excited from, the primary circuit, and having one or more windings in series between the primary circuit and the regulated circuit, all suitably adapted and arranged for the control of the voltage, or of the phase angle, or of both, of the regulated circuit.

Voltage spread: The difference between maximum and minimum voltages.

VRR: Abbreviation for voltage-regulating relay.

Waveform distortion: A steady-state deviation from an ideal sine wave of power frequency principally characterized by the special content of the deviation.

XLPE: Abbreviation for cross-linked polyethylene (cable insulation).

REFERENCES

1 IEEE Committee Report: Proposed definitions of terms for reporting and analyzing outages of electrical transmission and distribution facilities and interruptions, *IEEE Trans. Power Appar. Syst.*, PAS-87(5), May 1968, 1318–1323.

2. IEEE Committee Report: Guidelines for use in developing a specific underground distribution system design standard, *IEEE Trans. Power Appar. Syst.*, PAS-97(3), May/June 1978, 810–827.

3. *IEEE Standard Definitions in Power Operations Terminology*, IEEE Standard 346-1973, November 2, 1973.

4. Proposed standard definitions of general electrical and electronics terms, IEEE Standard 270, 1966.

5. Pender, H. and W. A. Del Mar: *Electrical Engineers' Handbook-Electrical Power*, 4th edn., Wiley, New York, 1962.

6. *National Electrical Safety Code*, 1977 edn., ANSI C2, IEEE, New York, November 1977.

7. Fink, D. G. and J. M. Carroll (eds.): *Standard Handbook for Electrical Engineers*, 10th edn., McGraw-Hill, New York, 1969.

8. *IEEE Standard Dictionary of Electrical and Electronics Terms*, IEEE, New York, 1972.

Notation

Capital English Letters

A	component availability (Chapter 11)
A	levelized annual cost, $ (Chapter 7)
A	weighted average Btu/kWh net generation (Chapter 2)
A, B, C	phase designation
A, B, C, D	general line (circuit) constants (Chapter 5)
A_{FDR}	feeder availability (Chapter 11)
A_i	availability of component i (Chapter 11)
A_n	area served by one of n substation feeders, mi^2
A_{SD}	service-drop conductor size, cmil (Chapter 6)
A_{SL}	secondary-line conductor size, cmil (Chapter 6)
A_{sys}	system availability (Chapter 11)
AEC	annual equivalent of energy cost, $
AEIC$_c$	annual equivalent of total installed capacitor bank cost, $ (Chapter 8)
AIC	annual equivalent of feeder investment cost, $
B	average fuel cost, $/MBtu (Chapter 2)
BEC	original (base) annual kWh energy consumption
BVR	bus voltage regulator
BW	bandwidth of voltage-regulating relay
C	capacitance, F
C	common winding (Chapter 3)
C_F	installed feeder cost, $/kVA (Chapter 8)
C_G	generation system cost, $/kVA
C_S	distribution substation cost, $/kVA
C_T	transmission system cost, $/kVA
C_T	transmission cost, $/kVA (Equation 8.26)
C_T	total reactive compensation ($= cn$) (Equation 8.85)
CR	corrective ratio (Chapter 9)
CT$_P$	primary-side rating of current transformer (Chapter 9)
CTR	current transformer ratio
D	distance or separation, ft
D	load density, kVA/mi^2
D	ratio of kWh losses to net system input (Chapter 2)
D_g	coincident maximum group demand, W
D_i	demand of load i, W
DF_i	demand factor of load group i
DTA(ij, k)	downtime array coefficients (Chapter 11)
E	source emf; voltage
E_i	event that component i operates successfully (Chapter 11)
EC	energy cost, $/kWh
EC$_{off}$	incremental cost of off-peak electric energy, $/kWh
EC$_{on}$	incremental cost of on-peak electric energy, $/kWh (Chapter 6)
ECL$_1$	eddy-current loss at rated fundamental current
$E(T)$	expected time during which a component will survive
F	fault point

F'_{LD}	reactive load factor (= Q/S)
F_c	coincident factor
F_D	diversity factor
F_{LD}	load factor
F_{LL}	load-location factor (Chapter 7)
F_{LS}	loss factor
F_{LSA}	loss-allowance factor (Chapter 7)
F_{PR}	peak-responsibility factor
F_R	reserve factor
F_u	utilization factor
FCAF	fuel cost adjustment factor, \$/kWh
FDR	feeder (Chapter 11)
$F(t)$	unreliability function (Chapter 11)
H	transformer higher-voltage-side winding
HF_I	current harmonic factor
HF_V	voltage harmonic factor
I	failed zone-branch array coefficient (Chapter 11)
I	rms phasor current, A
$\mathbf{I}$	current matrix
I_{AB}	current in higher-voltage-side winding between phases A and B, A
I_{ab}	current in lower-voltage-side winding between phases a and b, A
$I_{a,3\phi}$	current in phase a due to single-phase load, A
I_B	base current, A
I_C	current in common winding (Chapter 3)
I_c	core-loss component of excitation current (Chapter 3)
I_e	excitation current (Chapter 3)
I_{exc}	per unit excitation current (Chapter 6)
$I_{f,a}, I_{f,b}, I_{f,c}$	fault currents in phases a, b, and c
$I_{f,3\phi}$	three-phase fault current, A
$I_{F,3\phi}$	three-phase fault current referred to subtransmission voltage, A (Chapter 10)
$(I_{F,3\phi})_{max}$	maximum three-phase fault current, A (Chapter 10)
$I_{F,HV}$	fault current in transformer high-voltage side, A
$I_{f,L-G}$	line-to-ground fault current, A
$I_{f,LV}$	fault current in transformer low-voltage side, A
$I_{f,L-L}$	line-to-line fault current, A
$I_{\phi 1}$	current due to single-phase load, A
I_h	harmonic current
I_L	line current; load current, A
I_m	magnetizing current component of excitation current (Chapter 3)
I_m	current in feeder main at substation, A
I_N	current in primary neutral, A
I_n	current in secondary neutral, A
I_{op}	operating current, A
$I_{P,pu}$	no-load primary current at substation transformer, pu
I_{ra}	rated current, A
I_S	current in series winding (Chapter 3)
IC_c	installed cost of capacitor bank, \$/kvar (Chapter 8)
IC_{cap}	total installed cost of shunt capacitors, \$
IC_F	installed feeder cost, \$ (Chapter 7)
IC_{PH}	annual installed cost of pole and its hardware, \$ (Chapter 6)

IC_{SD}	annual installed cost of service drop, $ (Chapter 6)
IC_{SL}	annual installed cost of secondary line, $ (Chapter 6)
IC_{sys}	average investment cost of power system upstream, $/kVA
IC_T	annual installed cost of distribution transformer, $
K	percent voltage drop per kilovolt-ampere-mile
$\tilde{K}$	per unit voltage drop per 10,000 A·ft
$\hat{K}$	constant (Equation 5.63)
K_h	watt-hour meter constant (Chapter 2)
K_R	conversion factor for resistance (Chapter 7)
K_r	number of watt-hour meter disk revolutions (Chapter 2)
K_X	conversion factor for reactance (Chapter 7)
K_1	a constant to convert energy-loss savings to dollars, $/kWh (Equation 8.87)
K_2	a constant to convert power-loss savings to dollars, $/kWh (Equation 8.87)
K_3	a constant to convert total fixed capacitor size to dollars, $/kWh (Equation 8.95)
L_{sc}	system inductance, H (Equation 8.108)
LCDH	losses in capacitors due to harmonics
LD	load diversity, W
LD	load (Chapters 2 and 5)
LDC	line-drop compensator
LS	loss
LTC	load tap changer
LV	low voltage
MTTR	mean time to repair ($= \bar{r}$) (Chapter 11)
MTBF	mean time between failures ($= \bar{T}$) (Chapter 11)
MTTF	mean time to failure ($= \bar{m}$) (Chapter 11)
N	expected duration of normal weather (Chapter 11)
N	neutral primary terminal
$\mathbf{0}$	row vector of zeros (Chapter 11)
OC_{exc}	annual operating cost of transformer excitation current, $ (Chapter 6)
$OC_{SD,\,Cu}$	annual operating cost of service-drop cable due to copper losses, $ (Chapter 6)
$OC_{SL,\,Cu}$	annual operating cost of secondary line due to copper losses, $ (Chapter 6)
$OC_{T,\,Cu}$	annual operating cost of transformer due to copper losses, $ (Chapter 6)
$OC_{T,\,Fe}$	annual operating cost of transformer due to core losses, $
P	average power, W
$\mathbf{P}$	transition (or stochastic) matrix (Chapter 11)
P'_{LS}	power loss after capacitor bank addition, W (Equation 8.46)
P_{av}	average power, W
$P_{LS,\,av}$	average power loss, W
P_i	peak load i, W
P_{LD}	average power of load, W
P_{LS}	average power loss, W
$P_{LS,\,i}$	peak loss at peak load i, W
$P_{LS,\,max}$	maximum power loss, W
$P_{LS,1\phi}$	single-phase power loss, W (Chapter 7)
$P_{LS,3\phi}$	three-phase power loss, W (Chapter 7)
P_n	load at year n, W (Chapter 2)
P_0	initial load, W
P_r	receiving-end average power, VA
$P_{T,\,Cu}$	transformer copper loss, W (Chapter 6)
$P_{T,\,Fe}$	transformer core loss, W (Chapter 6)

$P_{\text{SL, Cu}}$	power loss of secondary line due to copper losses, W
PF	power factor
PTR	potential transformer ratio
PT_N	turns ratio of potential transformer (Chapter 9)
Q	average reactive power, var
Q_c	reactive power due to corrective capacitors, var (Chapter 8)
$Q_{c,3\phi}$	three-phase reactive power due to corrective capacitors, var (Equation 8.30)
Q_i	unreliability of component i (Chapter 11)
Q_r	receiving-end average reactive power, VA
Q_{sys}	system unreliability (Chapter 11)
R	resistance, Ω
R_{eff}	effective resistance, Ω (Chapter 9)
R_L	resistance of load impedance, Ω
R_{set}	R dial setting of line-drop compensator (Chapter 9)
R_{sys}	system reliability (Chapter 11)
RIA(ij, k)	recognition and isolation array coefficients (Chapter 11)
RP	regulating point
S	apparent power, VA
$\overline{S}$	$= P + jQ$, complex apparent power, VA
S	expected duration of adverse weather (Chapter 11)
S	series winding (Chapter 3)
S_B	base apparent power, VA
S_{sc}	short-circuit apparent power, VA
S_{ckt}	circuit capacity, VA (Chapter 9)
S_G	generation capacity, VA (Chapter 8)
S_L	load apparent power, VA
S_{Li}	apparent power of load i, VA
$S_{\angle-\angle}$	apparent power rating of an open-delta bank
S_{lump}	apparent power of lumped load, VA
$S_{L,3\phi}$	three-phase apparent power of load, VA (Chapter 8)
S_m	total kVA load served by one feeder main
S_n	kVA load served by one of n substation feeders
S_{PK}	feeder apparent power at peak load, VA
S_{reg}	regulator capacity, VA (Chapter 9)
S_S	substation capacity, VA (Equation 8.27)
S_T	transformer apparent power, VA
S_T	transmission capacity, VA (Equation 8.24)
$S_{T,\, ab}$	apparent power rating of single-phase transformer connected between phases a and b, VA
S_{Ti}	apparent power rating of transformer i, VA
$S_{T,3\phi}$	three-phase transformer apparent power, VA
$S_{1\phi}$	single-phase VA rating
$S_{3\phi}$	three-phase VA rating
$S_{\Delta-\Delta}$	apparent power rating of a delta–delta bank
SD	service drop (Chapter 6)
SW	switchable capacitors (Chapter 8)
T	a random variable representing failure time (Chapter 11)
T	time
T	transformer
TA_n	total area served by all n feeders, mi²
TAC	total annual cost, $
TAEL_{Cu}	total annual energy loss due to copper losses, W

TCD_i	total connected group demand i, W
TD	time delay
TECL	total eddy-current loss (Chapter 8)
TS_n	total kVA load served by a substation with n feeders
U	component unavailability (Chapter 11)
U_{FDR}	feeder unavailability (Chapter 11)
UG	underground
URD	underground residential distribution
V	volt, unit symbol abbreviation for voltage
V	voltage matrix
$V_{ab,pu}$	voltage between phases a and b, pu
$V_{B,\phi}$	single-phase base voltage, V
$V_{B,3\phi}$	three-phase base voltage, V
V_C	voltage across common winding (Chapter 3)
V_H	higher-voltage-side voltage, V (Chapter 3)
V_h	rms voltage of hth harmonic
$V_{L\text{-}L}$	line-to-line distribution voltage, V (Chapter 10)
$V_{L\text{-}L}$	line-to-line voltage, V
$V_{L\text{-}N}$	line-to-neutral voltage, V
$Y_{l,pu}$	per unit voltage at feeder end (Chapter 9)
V_P	primary distribution voltage, V (Chapter 9)
$V_{P,max}$	maximum primary distribution voltage, V
V_r	receiving-end voltage
V_{reg}	output voltage of regulator, V
V_{RP}	voltage at regulating point, V
V_S	voltage across series winding (Chapter 3)
V_s	sending-end voltage
V_{ST}	subtransmission voltage, V (Chapter 9)
$V_{ST,L\text{-}L}$	line-to-line subtransmission voltage, V (Chapter 10)
V_X	lower-voltage-side voltage, V (Chapter 3)
VD	voltage drop, V
VD_{pu}	per unit voltage drop
$VD_{pu,1\phi}$	single-phase voltage drop, pu
$VD_{pu,3\phi}$	three-phase voltage drop, pu
VD_{SD}	voltage drop in service-drop cable, V
VD_{SL}	voltage drop in secondary line, V
VD_T	voltage drop in transformer, V
$\% VD_{ab}$	percent voltage drop between a and b
$\% VD_m$	percent voltage drop in feeder main
VDIP	voltage dip, V
$VDIP_{SD}$	voltage dip in service-drop cable, V
$VDIP_{SL}$	voltage dip in secondary line, V
$VDIP_T$	voltage dip in transformer, V
$VR_{l,pu}$	per unit voltage rise at distance l (Chapter 9)
VR_{pu}	per unit voltage regulation
$\% VR$	percent voltage regulation
$\% VR$	percent voltage rise (Chapter 8)
$\%VR_{NSW}$	percent voltage rise due to nonswitchable capacitors (Chapter 8)
$\% VR_{SW}$	percent voltage rise due to switchable capacitors (Chapter 8)
VRR	voltage-regulating relay
VRR_{pu}	per unit setting of voltage-regulating relay

W	wire (in transformer connections) (Chapter 3)
X	reactance, Ω; transformer lower-voltage-side winding
X_c	capacitive reactance
X_L	reactance of load impedance, Ω
X_{sc}	system reactance, Ω (Equation 8.107)
X_{set}	X dial setting of line-drop compensator (Chapter 9)
$X(t_n)$	sequence of discrete-valued random variables (Chapter 11)
Y	admittance, Ω; wye connection
Y	admittance matrix
Z	impedance, Ω
Z	secondary-winding impedance, Ω (Equation 10.71)
$\mathbf{Z}$	impedance matrix
Z_{eq}	equivalent (total) impedance to fault, Ω (Chapter 10)
Z_f	fault impedance, Ω
Z_G	impedance to ground, Ω
$Z_{G, ckt}$	impedance to ground of circuit, Ω
Z_{LD}	load impedance, Ω
Z_M	impedance of secondary main, Ω
ZT	transformer impedance, Ω
Z_T	equivalent impedance of distribution transformer, Ω
$Z_{T, pu}$	per unit transformer impedance
Z_Δ	equivalent delta impedance, Ω (Equation 10.54)
Z_0	zero-sequence impedance, Ω
$Z_{0, ckt}$	zero-sequence impedance of circuit, Ω
Z_1	positive-sequence impedance, Ω
$Z_{1, ckt}$	positive-sequence impedance of circuit, Ω
$Z_{1, SL}$.	positive-sequence impedance of secondary line, Ω (Chapter 10)
$Z_{1, ST}$	positive-sequence impedance of subtransmission line, Ω
$Z_{1, sys}$	positive-sequence impedance of system, Ω
$Z_{1, T}$	positive-sequence impedance of transformer, Ω
Z_2	negative-sequence impedance, Ω

Lowercase English Letters

a, b, c	phase designation
c	capacitor compensation ratio (Chapter 8)
c_i	contribution factor of load i
dn	mutual geometric mean distance of phase and neutral wires, ft
dp	mutual geometric mean distance between phase wires, ft
$\bar{f}$	mean failure frequency (Chapter 11)
f_p	parallel resonant frequency, Hz
f_1	fundamental frequency, Hz
f_{sys}	average failure frequency of a system (Chapter 11)
$f(t)$	probability density function
h	harmonic order
$h(t)$	hazard rate (Chapter 11)
i	investment fixed charge rate (Chapter 6)
i_c	annual fixed charge rate for capacitors
i_F	annual fixed rate for feeder
i_G	annual fixed charge rate for generation system
i_s	annual fixed charge rate for distribution substation

i_T	annual fixed charge rate for transmission system
k	constant used in computing loss factor (Chapter 2)
l	inductance per unit length; leakage inductance
l	feeder length, mi
l_n	linear dimension of primary-feeder service area, mi
m_i	observed time to failure for cycle i (Chapter 11)
$\bar{m}_s$	mean time to failure of series system (Chapter 11)
n	total number of cycles (Chapter 11)
n	transfer ratio (inverse of turns ratio) (Chapter 10)
n	$= n_1/n_2$, turns ratio; neutral secondary terminal; number of feeders emanating from a substation
n_1	number of turns in primary winding
n_2	number of turns in secondary winding
$\mathbf{P}^{(n)}$	vector of state probabilities at time t_n (Chapter 11)
p_{ij}	transition probabilities (Chapter 11)
p_{ij}	probability of proper operation of isolating equipment in zone branch ij (Chapter 11)
q	probability of component failure (Chapter 11)
q_{ij}	probability of failure of isolating equipment in zone branch ij (Chapter 11)
	receiving end
r	radius; internal (source) resistance; resistance per unit length
r_a	resistance of phase wires, Ω/1000 ft
r_e	earth resistance, Ω/1000 ft
r_{eq}	transformer equivalent resistance, Ω
r_i	observed time to repair for cycle i (Chapter 11)
r_l	lateral resistance per unit length
r_m	resistance of feeder main, Ω/mi
$\bar{r}_s$	mean time to repair of series system (Chapter 11)
s	sending end; effective feeder (main) length, mi (Chapter 4)
s	series system (Chapter 11)
t	time
x	line reactance per unit length; internal (source) reactance
x_a	self-inductive reactance of a phase conductor, Ω/mi
x_{ap}	reactance of phase wire with 1 ft spacing, Ω/1000 ft
x_{an}	reactance of neutral wire with 1 ft spacing, Ω/1000 ft
x_d	inductive reactance spacing factor, Ω/mi
x_{dn}	mutual reactance between phase and neutral wires, Ω/1000 ft
x_{dp}	mutual reactance of phase wires, Ω/1000 ft
x_e	earth reactance, Ω/1000 ft
x_{eq}	transformer equivalent reactance, Ω
$x_{i,\,opt}$	optimum location of capacitor bank i in per unit length
x_L	inductive line reactance (Chapters 5 and 8)
x_l	lateral reactance per unit length
x_m	reactance of feeder main, Ω/mi
x_{RP}	regulating point distance from substation, mi (Chapter 9)
x_T	transformer reactance, % Ω (Chapter 8)
z	impedance per unit length
z_l	lateral impedance per unit length
z_m	impedance of feeder main, Ω/mi
$z_{0,\,a}$	zero-sequence self-impedance of phase circuit, Ω/1000 ft
$z_{0,\,ag}$	zero-sequence mutual impedance between phase and ground wires, Ω/1000 ft
$z_{0,\,g}$	zero-sequence self-impedance of ground wire, Ω/1000 ft

Capital Greek Letters

Δ	delta connection; determinant
Δ	difference; increment; savings; benefits
ΔACE	annual conserved energy, Wh (Chapter 8)
ΔBEC	additional energy consumption increase
ΔEL	energy-loss reduction
ΔP_{LS}	additional decrease in power loss, W (Chapter 8)
$\Delta P_{LS,\ opt}$	optimum loss reduction, W (Chapter 8)
ΔQ_c	required additional capacitor size, var (Chapter 8)
ΔS_F	released feeder capacity, VA (Chapter 8)
ΔS_G	released generation capacity, VA (Chapter 8)
ΔS_S	released substation capacity, VA (Equation 8.29)
ΔS_T	released transmission capacity, VA (Equation 8.24)
ΔS_{sys}	released system capacity, W (Chapter 8)
$\Delta \$_{ACE}$	annual benefits due to conserved energy, \$ (Chapter 8)
$\Delta \$_F$	annual benefits due to released feeder capacity, \$ (Equation 8.36)
$\Delta \$_G$	annual benefits due to released generation capacity, \$ (Chapter 8)
$\Delta \$_S$	annual benefits due to released substation capacity, \$ (Equation 8.29)
$\Delta \$_T$	annual benefits due to released transmission capacity, \$ (Equation 8.26)
Λ	transition rate matrix (Chapter 11)
Π	unconditional steady-state probability matrix (Chapter 11)
Σ	total savings due to capacitor installation, \$ (Equation 8.86)

Lowercase Greek Letters

α	a constant $[=(1 + \lambda + \lambda^2)]^{-1}$ (Chapter 11) 6
δ	power angle
θ	power-factor angle
θ_{max}	power-factor angle at maximum voltage drop
λ	ratio of reactive current at line end to reactive current at line beginning
λ	failure rate (Chapter 11)
$\overline{\lambda}$	complex flux linkages, $(Wb \cdot T)/m$
λ_{CT}	annual fault rate of cable terminations (Chapter 11)
λ_{FDR}	annual feeder fault rate (Chapter 11)
λ_{ij}	total failure rate of zone branch ij (Chapter 11)
λ_{ijB}	breaker failure rate in zone i branch j (Chapter 11)
λ_{ijM}	zone branch ij failure rate due to preventive maintenance (Chapter 11)
λ_{ijW}	zone branch ij failure rate due to adverse weather (Chapter 11)
λ_{OH}	annual fault rate of overhead feeder section (Chapter 11)
λ_s	failure rate of supply (substation) (Chapter 11)
λ_{UG}	annual fault rate of underground feeder section (Chapter 11)
μ	mean repair rate (Chapter 11)
μ_{ij}	zone branch ij repair rate (Chapter 11)
μ_{sijc}	reclosing rate of reclosing equipment in zone branch ij (Chapter 11)
μ_{sijo}	isolation rate of isolating equipment in zone branch ij (Chapter 11)
ϕ	$= \tan^{-1}(X/R)$, impedance angle
ϕ	magnetic flux; phase angle
ω	radian frequency

Subscripts

A	phase *a*
a	phase *a*
B	phase *b*
b	phase *b*
B	base quantity
C	phase *C*; common winding (Chapter 3)
c	phase *c*
c	capacity; capacitive; coincident (Chapter 3)
cap	shunt capacitor
ckt	circuit
CT	cable termination (Chapter 11)
Cu	copper
eff	effective
eq	equivalent circuit quantity
exc	excitation
D	diversity
F	feeder; fault point; referring to fault
f	referring to fault
FDR	feeder
Fe	iron
H	high-voltage side (HV)
L	inductive (reactance); load (Chapter 3)
L	line; load
l	lateral; inductive (reactance); length
LD	load
L-G	line-to-ground
L-L	line-to-line
LL	load location (Chapter 7)
L-N	line-to-neutral
LS	loss (Chapters 2 and 5)
LSA	loss allowance
M	secondary main
m	feeder main
max	maximum
min	minimum
N	turns ratio
N	primary neutral
n	number of feeders emanating from a substation
n	neutral
NSW	nonswitchable (fixed) capacitors (Chapter 9)
off	off-peak
OH	overhead (Chapter 11)
op	operating
opt	optimum
on	on-peak
•	primary
PK	peak

PR	peak responsibility (Chapter 7)
pu	per unit
r	receiving end
ra	rated
reg	regulator
S	substation; series winding (Chapter 3)
s	sending end
sc	short circuit
SD	service drop
set	dial setting (line-drop compensator)
SL	secondary line (Chapter 6)
ST	subtransmission
SW	switchable (capacitors)
sys	power system
T	transformer
T_i	transformer *i*
X	low-voltage side (LV)
Y	wye connection
$1\phi, 3\phi$	single-phase, three-phase
0, 1, 2	zero-, positive-, negative-sequence quantity
Δ	delta connection
$\angle$	open-delta connection

Answers to Selected Problems

Chapter 2

2.1 (a) 1112.5 kW; (b) 10.08 kW; (c) 88,300.8 kWh
2.2 0.62
2.3 (a) 1.0; (b) 0.50; (c) 0.60; (d) 0.44
2.5 0.64
2.6 (a) 1.0; (b) 0.55; (c) 0.65; (d) 0.48
2.7 0.46
2.8 0.75 and 0.33
2.9 (a) 131,400 kWh; (b) $3285
2.11 (a) 0.40; (b) 0.50
2.13 (a) $370.40; (b) 0.11; (c) 0.34; (d) justifiable
2.15 (a) 0.41 and 0.29; (b) 30 and 60 kVA; (c) $196 and $262; (d) 4.937 kvar; (e) not justifiable

Chapter 3

3.1 (a) $7.62\angle 60°$ kV; (b) $13.2\angle 30°$ kV
3.2 (a) 27,745.67 VA; (b) 13,872.83 VA; (c) 41,618.5 VA; (d) 24 kVA; (e) 24 kVA; (f) 48 kVA; (g) 1.078
3.4 (a) $100\angle -30°$ A, $100\angle 30°$ A, and $100\angle 90°$ A
(b) $\bar{I}_A = 11.54\angle 0°$ A, $\bar{I}_B = 11.53\angle -120°$ A, and $\bar{I}_C = 11.53\angle 120°$ A
3.5 0 var and 83,040 W
  a. 24, 13.8 and 27.6 kVA
  b. 1.075
3.9 $113.93 and $117.65
3.10 a. 24.305 A, 0.7 pu A, and 69.444 A
  b. 7.2576 and 0.8891 Ω
  c. 218.2526 and 623.5792 A
  d. 0.042 pu V and 302.4 V

Chapter 4

4.4 (a) 3.5935×10^{-5}% VD/(kVA·mi); (b) 3.5×10^{-5}% VD/(kVA·mi)
4.5 4.7524×10^{-5}% VD/(kVA·mi)
4.6 5.5656×10^{-5}% VD/(kVA·mi)
4.7 7.7686×10^{-5}% VD/(kVA·mi)
4.8 0.02% VD/(kVA·mi) and 10% VD
4.9 5%
4.10 6.667%
4.11 (a) 2.0; (b) 1.5; (c) 1.33
4.14 2.7 mi, 5.4 mi, and 14,670 kVA

Chapter 5

5.2 1.39%
5.3 (a) 2.98%; (b) it meets the VD criterion; (c) $\Delta X_d = 0.0436$ Ω/mi
5.4 (a) 2.31%; (b) yes; (c) the same
5.5 (a) The same; (b) 1.49%; (c) 4.66%

Chapter 6

6.1 (*a*) 1/0 AWG or 105.5 kcmil; (*e*) $935.47/block/year;
(*b*) Not applicable; (*f*) $950.25/block/year;
(*c*) 25 kVA; (*g*) $2.32/customer/month
(*d*) $920.55/block/year (*h*) $0.93/customer/month

6.4 (*a*) 1/0 AWG; (*b*) not applicable; (*c*) $935.47/block/year
6.5 a. 113.3−*j*5.09 A and −78.6 + *j*1.5 A
b. 35.03∠−6.45° A
c. 113.4∠−2.6° and 117.9∠−1.1° V
d. 231.3∠−1.8° V
6.7 a. 112.1∠−35.4° and 116.3∠175.9° A
b. 24.6 + *j*56.6 A
c. 112.1∠−1.5° and 116.3∠−4.1° V
d. 228.1∠−1.3° V

Chapter 7

7.1 (*a*) 8.94 pu V; (*b*) 24.24 kW; (*c*) 12.342 kvar; (*d*) 81.6 kVA and 0.891 lagging
7.4 (*a*) 0.0106 pu V; (b) 0.0135 pu V; (*c*) not applicable
7.5 (*a*) 0.72 pu A; (*b*) 0.050 pu V

Chapter 8

8.1 (*a*) 0.4863 leading; (*b*) 0.95 leading
8.2 a. 1200 kvar and 0.975 lagging
b. 84.7 A and 609.2 kVA
c. 224.5 V or 9.37% V
d. 0.97
8.4 (*a*) 228.5 kvar; (*b*) 0.86; (*c*) 4644.1 kvar; (*d*) 7493.6 kVA
8.5 (*a*) 8 kW; (*g*) $11,237.40/year;
(*b*) 12 k W; (*h*) $1340.30/year;
(*c*) 20 kW; (*i*) $6390.70/year;
(*d*) 2081 kVA; (*j*) $7127/year;
(*e*) 6727 kvar; (*k*) No
(*f*) $940/year
8.6 (*a*) 283 A; (*b*) 235/11.4° A; (*c*) 164.5 kvar/phase

Chapter 9

9.3 a. 124.2 V
b. 4 and 12 steps
c. At peak load, 1.0667 and 1.0083 pu V
At no load, 1.035 and 1.0083 pu V
9.4 (*a*) 1.74 mi; (*b*) 2.55 mi; (*c*) it takes into account the future growth
9.5 167 kVA
9.9 12.5 A
9.10 (*a*) Bus *a*; (*b*) 76.2 kVA; (*c*) bus *c*; (*d*) 0.0455 pu V; (*e*) 0.0314 and 0.0404 pu V
9.11 (*a*) 0.0304 pu V; (*b*) 0.98″/8 pu V; (*c*) 1.0011 pu V
9.13 No
9.14 (*a*) 3 V; (*b*) objectionable

Chapter 10

10.3   It lacks 30%
10.4   (*a*) 7-A tap;   (*b*) 8.93
10.5   a. 1.3422 + *j*3.5281 and 0.7698 + *j*1.3147 Ω
      b. 0.9696 + *j*2.0525 Ω
      c. 1.4897 + *j*4.7766 Ω
      d. 1.5797 + *j*5.3922 Ω
      e. 479.66 A
      f. 415.4 A
      g. 427.14 A
10.7   (*a*) 2020, 1750, and 1450 A;   (*b*) 870 A
10.8   a. 0.2083+ *j*1.5485 0
      b. 0.0409 + *j*0.1168 and 0.1336 + *j*0.3753 Ω
      c. 0.0718 + *j*0.2030 Ω
      d. 0.2801 + *j*1.7515 Ω
      e. 0.0003112 + *j*0.0019461 Ω
      f. 0.00768 + *j*0.0133025 Ω
      g. 0.009215 + *j*0.001365 Ω
      h. 0.0172062 + *j*0.0166136 Ω
      i. 10,068 A

Chapter 11

11.2   0.0045
11.3   (*a*) 0.0588;   (*b*) 0.0625
11.5   0.8347
11.6   200 days
11.8   0.0000779264
11.12   (*a*) 0.8465;   (*b*) 0.9925;   (*c*) 0.9988;   (*d*) 0.996
11.13   (*a*) 0.9703;   (*b*) 0.49
11.14   0.912
11.15   (*a*) 0.14715;   (*b*) 0.00635
11.16   (*a*) 0.15355;   (*b*) 0.00640
11.17   (*a*) 1.206 faults/year;   (*b*) 15.44 h;   (*c*) 0.18%;   (d) 99.93
11.19   (*a*) 0.6, 0.7, and 0.8 faults/year;   (*b*) 0.916, 0.8071, and 0.725 h
11.23   0.345
11.24   0.345

Index